Global Environment Outlook

GEO₄

environment for development

UNEP   United Nations Environment Programme

First published by the United Nations Environment Programme in 2007

Copyright © 2007, United Nations Environment Programme

ISBN: 978-92-807-2836-1 (UNEP paperback)
       DEW/0962/NA
       978-92-807-2872-9 (UNEP hardback)
       DEW/1001/NA

Printed and bound in Malta by Progress Press Ltd, Malta

PROGRESS PRESS LTD
P.O. BOX 328 341 ST. PAUL STREET
CMR 01
VALLETTA, MALTA

UNEP promotes environmentally sound practices globally and in its own activities. This publication is printed on chlorine free, acid free paper made of wood pulp from sustainably managed forests. Our distribution policy aims to reduce UNEP's carbon footprint.

University of Kassel, Germany

University of South Pacific, Fiji Islands

University of the West Indies, Centre for Environment and Development (UWICED), Jamaica

University of the West Indies, St. Augustine Campus, Trinidad and Tobago

World Resources Institute, United States

FUNDING

The Governments of Belgium, The Netherlands, Norway and Sweden together with the UNEP Environment Fund funded the GEO-4 assessment and outreach activities.

High-Level Consultative Group

Jacqueline McGlade (Co-Chair), Agnes Kalibbala (Co-Chair), Ahmed Abdel-Rehim (Alternate), Svend Auken, Philippe Bourdeau, Preety Bhandari, Nadia Makram Ebeid, Idunn Eidheim, Exequiel Ezcurra, Peter Holmgren, Jorge Illueca, Fred Langeweg, John Matuszak, Jaco Tavenier, Dan Tunstall, Vedis Vik, Judi Wakhungu, Toral Patel-Weynand (Alternate)

Coordinating Lead Authors

John Agard, Joseph Alcamo, Neville Ash, Russell Arthurton, Sabrina Barker, Jane Barr, Ivar Baste, W. Bradnee Chambers, David Dent, Asghar Fazel, Habiba Gitay, Michael Huber, Jill Jäger, Johan C. I. Kuylenstierna, Peter N. King, Marcel T. J. Kok, Marc A. Levy, Clever Mafuta, Diego Martino, Trilok S. Panwar, Walter Rast, Dale S. Rothman, George C. Varughese, and Zinta Zommers

Outreach Group

Richard Black, Quamrul Chowdhury, Nancy Colleton, Heather Creech, Felix Dodds, Randa Fouad, Katrin Hallman, Alex Kirby, Nicholas Lucas, Nancy MacPherson, Patricia Made, Lucy O'Shea, Bruce Potter, Eric Quincieu, Nick Rance, Lakshmi M. N. Rao, Solitaire Townsend, Valentin Yemelin

GLOBAL ENVIRONMENT OUTLOOK 4

GEO Coordination section

Sylvia Adams, Ivar Baste, Munyaradzi Chenje, Harsha Dave, Volodymyr Demkine, Thierry De Oliveira, Carolyne Dodo-Obiero, Tessa Goverse, Elizabeth Migongo-Bake, Neeyati Patel, Josephine Wambua

GEO Regional Coordinating team

Adel Abdelkader, Salvador Sánchez Colón, Joan Eamer, Charles Sebukeera, Ashbindu Singh, Kakuko Nagatani Yoshida, Ron Witt, and Jinhua Zhang

UNEP Extended Team

Johannes Akiwumi, Joana Akrofi, Christopher Ambala, Benedicte Boudol, Christophe Bouvier, Matthew Broughton, Edgar Arredondo Casillas, Juanita Castano, Marion Cheatle, Twinkle Chopra, Gerard Cunningham, Arie de Jong, Salif Diop, Linda Duquesnoy, Habib N. El-Habr, Norberto Fernandez, Silvia Giada, Peter Gilruth, Gregory Giuliani, Maxwell Gomera, Teresa Hurtado, Priscilla Josiah, Charuwan Kalyangkura, Nonglak Kasemsant, Amreeta Kent, Nipa Laithong, Christian Lambrechts, Marcus Lee, Achira Leophairatana, Arkadiy Levintanus, Monika Wehrle MacDevette, Esther Mendoza, Danapakorn Mirahong, Patrick M'mayi, Purity Muguku, John Mugwe, Josephine Nyokabi Mwangi, Bruce Pengra, Daniel Puig, Valarie Rabesahala, Anisur Rahman, Priscilla Rosana, Hiba Sadaka, Frits Schlingemann, Meg Seki, Nalini Sharma, Gemma Shepherd, Surendra Shrestha, James Sniffen, Ricardo Sánchez Sosa, Anna Stabrawa, Gulmira Tolibaeva, Sekou Toure, Brennan Van Dyke, Hendricus Verbeek, Anne-Marie Verbeken, Janet Waiyaki, Mick Wilson, Kaveh Zahedi

Production Coordination: Neeyati Patel

GEO-4 e-peer-review coordination: Herb Caudill, Shane Caudill, Sylvia Adams, Harsha Dave

Data Support: Jaap van Woerden, Stefan Schwarzer, Andrea DeBono and Diawoye Konte

Maps: Bounford.com and UNEP/GRID-Arendal

Editors: Mirjam Schomaker, Michael Keating, and Munyaradzi Chenje

Design and layout: Bounford.com

Cover Design: Audrey Ringler

Outreach and Communications: Jacquie Chenje, Eric Falt, Elisabeth Guilbaud-Cox, Beth Ingraham, Steve Jackson, Mani Kabede, Fanina Kodre, Angele Sy Luh, Danielle Murray, Francis Njoroge, Nick Nuttall, Naomi Poulton, David Simpson, Jennifer Smith

Contents

LIST OF ILLUSTRATIONS

Chapter 7 Vulnerability of People and the Environment: Challenges and Opportunities

LIST OF BOXES

Chapter 8 Interlinkages: Governance for Sustainability

Chapter 9 The Future Today

Chapter 10 From the Periphery to the Core of Decision Making – Options for Action

LIST OF TABLES

Chapter 1 Environment for Development

Chapter 2 Atmosphere

Chapter 3 Land

Foreword

Few global issues are more important than the environment and climate change. Since taking office, I have consistently emphasized the dangers of global warming, environmental degradation, the loss of biodiversity and the potential for conflict growing out of competition over dwindling natural resources such as water – the topics which are analysed in the GEO-4 report. Dealing with these issues is the great moral, economic and social imperative of our time.

ENVIRONMENT

Rapid environmental change is all around us. The most obvious example is climate change, which will be one of my top priorities as Secretary-General. But that is not the only threat. Many other clouds are on the horizon, including water shortages, degraded land and the loss of biodiversity. This assault on the global environment risks undermining the many advances human society has made in recent decades. It is undercutting our fight against poverty. It could even come to jeopardize international peace and security.

These issues transcend borders. Protecting the global environment is largely beyond the capacity of individual countries. Only concerted and coordinated international action will be sufficient. The world needs a more coherent system of international environmental governance. And we need to focus in particular on the needs of the poor, who already suffer disproportionately from pollution and disasters. Natural resources and ecosystems underpin all our hopes for a better world.

ENERGY AND CLIMATE CHANGE

Issues of energy and climate change can have implications for peace and security. This is especially true in vulnerable regions that face multiple stresses at the same time – pre-existing conflict, poverty and unequal access to resources, weak institutions, food insecurity, and incidence of diseases such as HIV/AIDS.

We must do more to use and develop renewable energy sources. Greater energy efficiency is also vital. So are cleaner energy technologies, including advanced fossil fuel and renewable energy technologies, which can create jobs, boost industrial development, reduce air pollution and help to mitigate greenhouse gas emissions. This is a matter of urgency that requires sustained, concerted and high-level attention. It has a broad impact not just on the environment but also on economic and social development, and needs to be considered in the context of sustainable development. It should be a concern to all countries, rich or poor.

Energy, climate change, industrial development and air pollution are critical items on the international agenda. Addressing them in unison creates many win-win opportunities and is crucial for sustainable development. We need to take joint action on a global scale to address climate change. There are many policy and technological options available to address the impending crisis, but we need the political will to seize them. I ask you to join the fight against climate change. If we do not act, the true cost of our failure will be borne by succeeding generations, starting with yours. That would be an unconscionable legacy; one which we must all join hands to avert.

BIODIVERSITY

Biodiversity is the foundation of life on earth and one of the pillars of sustainable development. Without the conservation and sustainable use of biodiversity, we will not achieve the Millennium Development Goals. The conservation and sustainable use of biodiversity is an essential element of any strategy to adapt to climate change. Through the Convention on Biological Diversity and the United Nations Framework Convention on Climate Change, the international community is committed to conserving biodiversity and combating climate change. The global response to these challenges needs to move much more rapidly, and with more determination at all levels – global, national and local. For the sake of current and future generations, we must achieve the goals of these landmark instruments.

WATER

The state of the world's waters remains fragile and the need for an integrated and sustainable approach to water resource management is as pressing as ever. Available supplies are under great duress as a result of high population growth, unsustainable consumption patterns, poor management practices, pollution, inadequate investment in infrastructure and low efficiency in water-use. The water-supply-demand gap is likely to grow wider still, threatening economic and social development and environmental sustainability. Integrated water resource management will be of crucial importance in overcoming water scarcity. The Millennium Development Goals have helped to highlight the importance of access to safe drinking water supplies and adequate sanitation, which undeniably separates people living healthy and productive lives from those living in poverty and who are most vulnerable to various life-threatening diseases. Making good on the global water and sanitation agenda is crucial to eradicating poverty and achieving the other development goals.

INDUSTRY

Increasingly, companies are embracing the Global Compact not because it makes for good public relations, or because they have paid a price for making mistakes. They are doing so because in our interdependent world, business leadership cannot be sustained without showing leadership on environmental, social and governance issues.

Ban Ki-moon
Secretary-General of the United Nations
United Nations Headquarters, New York,
October 2007

Preface

The *Global Environment Outlook: environment for development (GEO-4)* report is published in what may prove to be a remarkable year – a year when humanity faced up to the scale and pace of environmental degradation with a new sense of realism and honesty matched by firm, decisive and above all, imaginative action.

It highlights the unprecedented environmental changes we face today and which we have to address together. These changes include climate change, land degradation, collapse of fisheries, biodiversity loss, and emergence of diseases and pests, among others. As society, we have the responsibility to tackle these and the development challenges we face. The trigger propelling countries and communities towards a rediscovery of collective responsibility is the most overarching challenge of this generation: climate change.

Humanity's capacity to order its affairs in a stable and sustainable way is likely to prove impossible if greenhouse gases are allowed to rise unchecked.

Attempts to meet the Millennium Development Goals relating to poverty, water and other fundamental issues may also fail without swift and sustained action towards de-carbonizing economies.

The difference between this GEO and the third report, which was released in 2002, is that claims and counter claims over climate change are in many ways over. The Intergovernmental Panel on Climate Change (IPCC) has put a full stop behind the science of whether human actions are impacting the atmosphere and clarified the likely impacts – impacts not in a far away future but within the lifetime of our generation.

The challenge now is not whether climate change is happening or whether it should be addressed. The challenge now is to bring over 190 nations together in common cause. The prize is not just a reduction in emissions of greenhouse gases, it is a comprehensive re-engagement with core objectives and principles of sustainable development.

For climate change, by its very nature cannot be compartmentalized into one ministerial portfolio, a single-line entry in corporate business plans or a sole area of NGO activism. Climate change, while firmly an environmental issue is also an environmental threat that impacts on every facet of government and public life – from finance and planning to agriculture, health, employment and transport.

If both sides of the climate coin can be addressed – emission reductions and adaptation – then perhaps many of the other sustainability challenges can also be addressed comprehensively, cohesively and with a long-term lens rather than in the segmented, piecemeal and short-sighted ways of the past.

GEO-4 underlines the choices available to policy-makers across the range of environmental, social and economic challenges – both known and emerging. It underlines not only the enormous, trillion-dollar value of the Earth's ecosystems and the goods-and-services they provide, but also underscores the central role the environment has for development and human well-being.

The year 2007 is also momentous because it is the 20th anniversary of the report by the World Commission on Environment and Development, *Our Common Future*. It augurs well that the report's principal architect and a person credited with popularizing the term sustainable development as the chair of the Commission – former Norwegian Prime Minister Gro Harlem Brundtland – is one of three special climate envoys appointed this year by UN Secretary-General Ban Ki-moon.

The *GEO-4* report is a living example of international cooperation at its best. About 400 individual scientists and policy-makers, and more than 50 GEO Collaborating Centres and other partner institutions around the world participated in the assessment with many of them volunteering their time and expertise. I would like to thank them for their immense contribution.

I would also like to thank the governments of Belgium, Norway, The Netherlands, and Sweden for their financial support to the *GEO-4* assessment that was invaluable in, for example, funding global and regional meetings and the comprehensive peer review process of 1 000 invited experts. My thanks are also extended to the *GEO-4* High-level Consultative Group whose members offered their invaluable policy and scientific expertise.

Achim Steiner
United Nations Under-Secretary General and Executive
Director, United Nations Environment Programme

Reader's guide

The fourth *Global Environment Outlook – environment for development* (*GEO-4*) places sustainable development at the core of the assessment, particularly on issues dealing with intra- and intergenerational equity. The analyses include the need and usefulness of valuation of environmental goods and services, and the role of such services in enhancing development and human well-being, and minimizing human vulnerability to environmental change. The *GEO-4* temporal baseline is 1987, the year in which the World Commission on Environment and Development (WCED) published its seminal report, *Our Common Future*. The Brundtland Commission was established in 1983, under UN General Assembly resolution 38/161 to look at critical environment and development challenges. It was established at a time of an unprecedented rise in pressures on the global environment, and when grave predictions about the human future were becoming commonplace.

The year 2007 is a major milestone in marking what has been achieved in the area of sustainable development and recording efforts – from local to global – to address various environmental challenges. It will be:

■ Twenty years since the launch of *Our Common Future*, which defined sustainable development as a blueprint to address our interlinked environment and development challenges.
■ Twenty years since the UNEP Governing Council adopted the "Environmental Perspective to the Year 2000 and Beyond", to implement the major findings of the WCED and set the world on a sustainable development path.
■ Fifteen years since the World Summit on Environment and Development (the Rio Earth Summit), adopted Agenda 21, providing the foundation on which to build intra- and intergenerational equity.
■ Five years since the World Summit on Sustainable Development (WSSD) in 2002, which adopted the Johannesburg Plan of Implementation.

The year 2007 is also the halfway point to the implementation of some of the internationally recognized development targets, including the Millennium Development Goals (MDGs). These and other issues are analysed in the report.

The *GEO-4* assessment report is the result of a structured and elaborate consultative process, which is outlined at the end of this report. *GEO-4* has 10 chapters, which provide an overview of global social and economic trends, and the state-and-trends of the global and regional environments over the past two decades, as well as the human dimensions of these changes. It highlights the interlinkages as well as the challenges of environmental change and opportunities that the environment provides for human well-being. It provides an outlook for the future, and policy options to address present and emerging environmental issues. The following are the highlights of each chapter:

Chapter 1: Environment for Development – examines the evolution of issues since *Our Common Future* popularized "sustainable development," highlighting institutional developments and conceptual changes in thought since then, as well as the major environmental, social and economic trends, and their influence on human well-being.

Chapter 2: Atmosphere – highlights how atmospheric issues affect human well-being and the environment. Climate change has become the greatest challenge facing humanity today. Other atmospheric issues, such as air quality and ozone layer depletion are also highlighted.

Chapter 3: Land – addresses the land issues identified by UNEP regional groups, and highlights the pressures of human demands on the land resource as the cause of land degradation. The most dynamic elements of land-use change are the far-reaching changes in forest cover and composition, cropland expansion and intensification, and urban development.

Chapter 4: Water – reviews the pressures that are causing changes in the state of the Earth's water environment in the context of global and regional drivers. It describes the state-and-trends in changes in the water environment, including its ecosystems and their fish stocks, emphasizing the last 20 years, and the impacts of changes on the environment and human well-being at local to global scales.

Chapter 5: Biodiversity – highlights biodiversity as a key pillar of ecologically sustainable development, providing a synthesis of the latest information on the state-and-trends of global biodiversity. It also links trends in biodiversity to the consequences for sustainable development in a number of key areas.

Chapter 6: Sustaining a Common Future – identifies and analyses priority environmental issues between 1987–2007 for each of the seven GEO regions: Africa, Asia and the Pacific, Europe, Latin America and the Caribbean, North America, West Asia and the Polar Regions. The chapter points out that for the first time since the GEO report series was first published in 1997, all seven regions recognize climate change as a major issue.

Chapter 7: Vulnerability of People and the Environment: Challenges and Opportunities – identifies challenges to and opportunities for improving human well-being through analyses of the vulnerability of some environmental systems and groups in society to environmental and socio-economic changes. The export and import of human vulnerability have grown as a result of the phenomenal global consumption, increased poverty and environmental change.

Chapter 8: Interlinkages: Governance for Sustainability – presents an assessment of the interlinkages within and between the biophysical components of the Earth system, environmental change, the development challenges facing human society, and the governance regimes developed to address such challenges. These elements are interlinked through significant systemic interactions and feedbacks, drivers, policy and technology synergies and trade-offs. Governance approaches that are flexible, collaborative and learning-based may be more responsive and adaptive to change, and therefore, better able to cope with the challenges of linking environment to development.

Chapter 9: The Future Today – builds on previous chapters by presenting four scenarios to the year 2050 – *Markets First, Policy First, Security First* and *Sustainability First* – which explore how current social, economic and environmental trends may unfold, and what this means for the environment and human well-being. The scenarios examine different policy approaches and societal choices. They are presented using narrative storylines and quantitative data at both global and regional levels. The degree of many environmental changes differs over the next half-century across the scenarios as a result of differences in policy approaches and societal choices.

Chapter 10 From the Periphery to the Core of Decision Making – Options for Action – discusses the main environmental problems highlighted in earlier chapters, and categorizes them along a continuum from *problems with proven solutions* to *problems for which solutions are emerging*. It also describes the adequacy of current policy responses, and possible barriers to more effective policy formulation and implementation. It then outlines the future policy challenges, pointing to the need for a two-track approach: extending policies that have been demonstrated to work for conventional environmental problems into regions lagging behind, and beginning to tackle the emerging environmental problems through structural reforms to social and economic systems.

FOURTH GLOBAL ENVIRONMENT OUTLOOK (*GEO-4*) ASSESSMENT
GEO-4 conceptual framework

The *GEO-4* assessment uses the drivers-pressures-state-impacts-responses (DPSIR) framework in analysing the interaction between environmental change over the past two decades as well as in presenting the four scenarios in Chapter 9.

The concepts of human well-being and ecosystem services are core in the analysis. However, the report broadens its assessment from focusing exclusively on ecosystems to cover the entire environment and the interaction with society. The framework attempts to reflect the key components of the complex and multidimensional, spatial and temporal chain of cause-and-effect that characterizes the interactions between society and the environment. The *GEO-4* framework is generic and flexible, and recognizes that a specific thematic and geographic focus may require a specific and customized framework.

The *GEO-4* conceptual framework (Figure 1), therefore, contributes to society's enhanced understanding of the links between the environment and development, human well-being and vulnerability to environmental change. The framework places, together with the environment, the social issues and economic sectors in the 'impacts' category rather than just exclusively in the 'drivers' or 'pressures' categories (Figure 1). The characteristics of the components of the *GEO-4* analytical framework are explained below.

Drivers

Drivers are sometimes referred to as indirect or underlying drivers or driving forces. They refer to fundamental processes in society, which drive activities with a direct impact on the environment. Key drivers include: demographics; consumption and production patterns; scientific and technological innovation; economic demand, markets and trade; distribution patterns; institutional and social-political frameworks and value systems. The characteristics and importance of each driver differ substantially from one region to another, within regions and within and between nations. For example, in the area of population dynamics, most developing countries are still facing population growth while developed countries are faced with a stagnant and

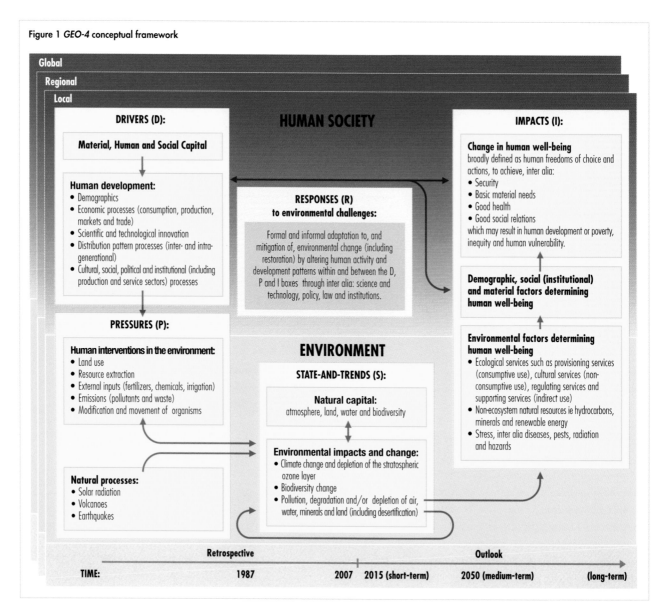

Figure 1 *GEO-4 conceptual framework*

ageing population. The resource demand of people influence environmental change.

Pressures

Key pressures include: emissions of substances which may take the form of pollutants or waste; external inputs such as fertilizers, chemicals and irrigation; land use; resource extraction; and modification and movement of organisms. Human interventions may be directed towards causing a desired environmental change such as land use, or they may be intentional or unintentional by-products of other human activities, for example, pollution. The characteristics and importance of each pressure may vary from one region to another, but is often a combination of pressures that lead to environmental change. For example, climate change is a result of emissions of different greenhouse gases, deforestation and land-use practices. Furthermore, the ability to create and transfer environmental pressures onto the environment of other societies varies from one region to another. Affluent societies with high levels of production, consumption and trade tend to contribute more towards global and transboundary environmental pressures than the less affluent societies which interact in more direct fashion with the environment in which they live.

State-and-trends

Environmental state also includes trends, which often refers to environmental change. This change may be natural, human-induced or both. Examples of natural processes include solar radiation, extreme natural events, pollination, and erosion. Key forms of human induced environmental change include climate change, desertification and land degradation, biodiversity loss, and air and water pollution, for example.

Different forms of natural or human-induced changes interact. One form of change, for example, climate change, will inevitably lead to ecosystem change, which may result in desertification and/or biodiversity loss. Different forms of environmental change can reinforce or neutralize each other. For example, a temperature increase due to climate change can, in Europe, partly be offset by changes in ocean currents triggered by climate change. The complexity of the physical, chemical and biological systems constituting the environment makes it hard to predict environmental change, especially when

it is subject to multiple pressures. The state of the environment and its resilience to change varies greatly within and among regions due to different climatic and ecological conditions.

Impacts

The environment is directly or indirectly affected by the social and economic sectors, contributing to change (either negative or positive) in human well-being and in the capacity/ability to cope with environmental changes. Impacts, be they on human well-being, the social and economic sectors or environmental services, are highly dependent on the characteristics of the drivers and, therefore, vary markedly between developing and developed regions.

Responses

Responses address issues of vulnerability of both people and the environment, and provide opportunities for reducing human vulnerability and enhancing human well-being. Responses take place at various levels: for example, environmental laws and institutions at the national level, and multilateral environmental agreements and institutions at the regional and global levels. The capacity to mitigate and/or adapt to environmental change differs among and within regions, and capacity building is, therefore, a major and overarching component of the response components.

The *GEO-4* framework has been used in the analyses of issues in all the 10 chapters, both explicitly and implicitly. Its utility is in integrating the analyses to better reflect the cause-and-effect, and ultimately society's response in addressing the environmental challenges it faces.

A variation of Figure 1 is presented in Chapter 8 as Figure 8.2 to better highlight the dual roles of economic sectors such as agriculture, forestry, fisheries and tourism – in contributing to development and human well-being, and also in exerting pressure on the environment and influencing environmental change, and in some cases, to human vulnerability to such change.

ARCTIC OCEAN

ARCTIC OCEAN

Laptev Sea

East Siberian Sea

Barents Sea

Kara Sea

Arctic Circle

80°

Eastern Europe

North Sea

Baltic Sea

Sea of Okhots

Bering Sea

Western Europe

Central Asia

Central Europe

Aral Sea

Black Sea

Caspian Sea

Northwest Pacific and East Asia

Sea of Japan

40°

Mediterranean Sea

Mashriq

Yellow Sea

East China Sea

P A C I F I C

Northern Africa

Red Sea

South Asia

O C E A N

Tropic of Cancer

Arabian Peninsula

Gulf of Aden

Western Africa

Bay of Bengal

South China Sea

Philippine Sea

Central Africa

Eastern Africa

Celebes Sea

Gulf of Guinea

■ UNEP Headquarters

0°

I N D I A N

O C E A N

South East Asia

South Pacific

Western Indian Ocean

Mozambique Channel

Australia

Coral Sea

Southern Africa

Tropic of Capricorn

New Zealand

40°

Tasman Sea

● UNEP Regional Offices

○ Collaborating centres

Polar (Antarctic)

Antarctic Circle

0°   45°   90°   135°

GEO-4 REGIONS

Name	Region	Sub-region
AFRICA		
Cameroon	Africa	Central Africa
Central African Republic	Africa	Central Africa
Chad	Africa	Central Africa
Congo	Africa	Central Africa
Democratic Republic of the Congo	Africa	Central Africa
Equatorial Guinea	Africa	Central Africa
Gabon	Africa	Central Africa
São Tomé and Príncipe	Africa	Central Africa
Burundi	Africa	Eastern Africa
Djibouti	Africa	Eastern Africa
Eritrea	Africa	Eastern Africa
Ethiopia	Africa	Eastern Africa
Kenya	Africa	Eastern Africa
Rwanda	Africa	Eastern Africa
Somalia	Africa	Eastern Africa
Uganda	Africa	Eastern Africa
Algeria	Africa	Northern Africa
Egypt	Africa	Northern Africa
Libyan Arab Jamahiriya	Africa	Northern Africa
Morocco	Africa	Northern Africa
Sudan	Africa	Northern Africa
Tunisia	Africa	Northern Africa
Western Sahara	Africa	Northern Africa
Angola	Africa	Southern Africa
Botswana	Africa	Southern Africa
Lesotho	Africa	Southern Africa
Malawi	Africa	Southern Africa
Mozambique	Africa	Southern Africa
Namibia	Africa	Southern Africa
Saint Helena (United Kingdom)	Africa	Southern Africa
South Africa	Africa	Southern Africa
Swaziland	Africa	Southern Africa
United Republic of Tanzania	Africa	Southern Africa
Zambia	Africa	Southern Africa
Zimbabwe	Africa	Southern Africa
Benin	Africa	Western Africa
Burkina Faso	Africa	Western Africa
Cape Verde	Africa	Western Africa
Cote d'Ivoire	Africa	Western Africa

Name	Region	Sub-region
Gambia	Africa	Western Africa
Ghana	Africa	Western Africa
Guinea	Africa	Western Africa
Guinea-Bissau	Africa	Western Africa
Liberia	Africa	Western Africa
Mali	Africa	Western Africa
Mauritania	Africa	Western Africa
Niger	Africa	Western Africa
Nigeria	Africa	Western Africa
Senegal	Africa	Western Africa
Sierra Leone	Africa	Western Africa
Togo	Africa	Western Africa
Comoros	Africa	Western Indian Ocean
Madagascar	Africa	Western Indian Ocean
Mauritius	Africa	Western Indian Ocean
Mayotte (France)	Africa	Western Indian Ocean
Réunion (France)	Africa	Western Indian Ocean
Seychelles	Africa	Western Indian Ocean

ASIA AND THE PACIFIC

Name	Region	Sub-region
Australia	Asia and the Pacific	Australia and New Zealand
New Zealand	Asia and the Pacific	Australia and New Zealand
Kazakhstan	Asia and the Pacific	Central Asia
Kyrgyzstan	Asia and the Pacific	Central Asia
Tajikistan	Asia and the Pacific	Central Asia
Turkmenistan	Asia and the Pacific	Central Asia
Uzbekistan	Asia and the Pacific	Central Asia
China	Asia and the Pacific	NW Pacific and East Asia
Democratic People's Republic of Korea	Asia and the Pacific	NW Pacific and East Asia
Japan	Asia and the Pacific	NW Pacific and East Asia
Mongolia	Asia and the Pacific	NW Pacific and East Asia
Republic of Korea	Asia and the Pacific	NW Pacific and East Asia
Afghanistan	Asia and the Pacific	South Asia
Bangladesh	Asia and the Pacific	South Asia
Bhutan	Asia and the Pacific	South Asia
India	Asia and the Pacific	South Asia
Iran (Islamic Republic of)	Asia and the Pacific	South Asia
Maldives	Asia and the Pacific	South Asia
Nepal	Asia and the Pacific	South Asia
Pakistan	Asia and the Pacific	South Asia
Sri Lanka	Asia and the Pacific	South Asia

Name	Region	Sub-region
Brunei Darussalam	Asia and the Pacific	South East Asia
Cambodia	Asia and the Pacific	South East Asia
Christmas Island (Australia)	Asia and the Pacific	South East Asia
Indonesia	Asia and the Pacific	South East Asia
Lao People's Democratic Republic	Asia and the Pacific	South East Asia
Malaysia	Asia and the Pacific	South East Asia
Myanmar	Asia and the Pacific	South East Asia
Philippines	Asia and the Pacific	South East Asia
Singapore	Asia and the Pacific	South East Asia
Thailand	Asia and the Pacific	South East Asia
Timor-Leste	Asia and the Pacific	South East Asia
Viet Nam	Asia and the Pacific	South East Asia
American Samoa (United States)	Asia and the Pacific	South Pacific
Cocos (Keeling) Islands (Australia)	Asia and the Pacific	South Pacific
Cook Islands	Asia and the Pacific	South Pacific
Fiji	Asia and the Pacific	South Pacific
French Polynesia (France)	Asia and the Pacific	South Pacific
Guam (United States)	Asia and the Pacific	South Pacific
Johnston Atoll (United States)	Asia and the Pacific	South Pacific
Kiribati	Asia and the Pacific	South Pacific
Marshall Islands	Asia and the Pacific	South Pacific
Micronesia (Federated States of)	Asia and the Pacific	South Pacific
Midway Islands (United States)	Asia and the Pacific	South Pacific
Nauru	Asia and the Pacific	South Pacific
New Caledonia (France)	Asia and the Pacific	South Pacific
Niue	Asia and the Pacific	South Pacific
Norfolk Island (Australia)	Asia and the Pacific	South Pacific
Northern Mariana Islands (United States)	Asia and the Pacific	South Pacific
Palau (Republic of)	Asia and the Pacific	South Pacific
Papua New Guinea	Asia and the Pacific	South Pacific
Pitcairn Island (United Kingdom)	Asia and the Pacific	South Pacific
Samoa	Asia and the Pacific	South Pacific
Solomon Islands	Asia and the Pacific	South Pacific
Tokelau (New Zealand)	Asia and the Pacific	South Pacific
Tonga	Asia and the Pacific	South Pacific
Tuvalu	Asia and the Pacific	South Pacific
Vanuatu	Asia and the Pacific	South Pacific
Wake Island (United States)	Asia and the Pacific	South Pacific
Wallis and Futuna (France)	Asia and the Pacific	South Pacific

Name	Region	Sub-region
EUROPE		
Albania	Europe	Central Europe
Bosnia and Herzegovina	Europe	Central Europe
Bulgaria	Europe	Central Europe
Croatia	Europe	Central Europe
Cyprus	Europe	Central Europe
Czech Republic	Europe	Central Europe
Estonia	Europe	Central Europe
Hungary	Europe	Central Europe
Latvia	Europe	Central Europe
Lithuania	Europe	Central Europe
Montenegro	Europe	Central Europe
Poland	Europe	Central Europe
Romania	Europe	Central Europe
Serbia	Europe	Central Europe
Slovakia	Europe	Central Europe
Slovenia	Europe	Central Europe
The former Yugoslav Republic of Macedonia	Europe	Central Europe
Turkey	Europe	Central Europe
Armenia	Europe	Eastern Europe
Azerbaijan	Europe	Eastern Europe
Belarus	Europe	Eastern Europe
Georgia	Europe	Eastern Europe
Moldova, Republic of	Europe	Eastern Europe
Russian Federation	Europe	Eastern Europe
Ukraine	Europe	Eastern Europe
Andorra	Europe	Western Europe
Austria	Europe	Western Europe
Belgium	Europe	Western Europe
Denmark	Europe	Western Europe
Faroe Islands (Denmark)	Europe	Western Europe
Finland	Europe	Western Europe
France	Europe	Western Europe
Germany	Europe	Western Europe
Gibraltar (United Kingdom)	Europe	Western Europe
Greece	Europe	Western Europe
Guernsey (United Kingdom)	Europe	Western Europe
Holy See	Europe	Western Europe
Iceland	Europe	Western Europe
Ireland	Europe	Western Europe
Isle of Man (United Kingdom)	Europe	Western Europe
Israel	Europe	Western Europe
Italy	Europe	Western Europe
Jersey (United Kingdom)	Europe	Western Europe

Name	Region	Sub-region
Liechtenstein	Europe	Western Europe
Luxembourg	Europe	Western Europe
Malta	Europe	Western Europe
Monaco	Europe	Western Europe
Netherlands	Europe	Western Europe
Norway	Europe	Western Europe
Portugal	Europe	Western Europe
San Marino	Europe	Western Europe
Spain	Europe	Western Europe
Svalbard and Jan Mayen Islands (Norway)	Europe	Western Europe
Sweden	Europe	Western Europe
Switzerland	Europe	Western Europe
United Kingdom of Great Britain and Northern Ireland	Europe	Western Europe

LATIN AMERICA AND THE CARIBBEAN

Name	Region	Sub-region
Anguilla (United Kingdom)	Latin America and the Caribbean	Caribbean
Antigua and Barbuda	Latin America and the Caribbean	Caribbean
Aruba (Netherlands)	Latin America and the Caribbean	Caribbean
Bahamas	Latin America and the Caribbean	Caribbean
Barbados	Latin America and the Caribbean	Caribbean
British Virgin Islands (United Kingdom)	Latin America and the Caribbean	Caribbean
Cayman Islands (United Kingdom)	Latin America and the Caribbean	Caribbean
Cuba	Latin America and the Caribbean	Caribbean
Dominica	Latin America and the Caribbean	Caribbean
Dominican Republic	Latin America and the Caribbean	Caribbean
Grenada	Latin America and the Caribbean	Caribbean
Guadeloupe (France)	Latin America and the Caribbean	Caribbean
Haiti	Latin America and the Caribbean	Caribbean
Jamaica	Latin America and the Caribbean	Caribbean
Martinique (France)	Latin America and the Caribbean	Caribbean
Montserrat (United Kingdom)	Latin America and the Caribbean	Caribbean
Netherlands Antilles (Netherlands)	Latin America and the Caribbean	Caribbean
Puerto Rico (United States)	Latin America and the Caribbean	Caribbean
Saint Kitts and Nevis	Latin America and the Caribbean	Caribbean
Saint Lucia	Latin America and the Caribbean	Caribbean
Saint Vincent and the Grenadines	Latin America and the Caribbean	Caribbean
Trinidad and Tobago	Latin America and the Caribbean	Caribbean
Turks and Caicos Islands (United Kingdom)	Latin America and the Caribbean	Caribbean
United States Virgin Islands (United States)	Latin America and the Caribbean	Caribbean
Belize	Latin America and the Caribbean	Meso-America
Costa Rica	Latin America and the Caribbean	Meso-America
El Salvador	Latin America and the Caribbean	Meso-America
Guatemala	Latin America and the Caribbean	Meso-America
Honduras	Latin America and the Caribbean	Meso-America

Name	Region	Sub-region
Mexico	Latin America and the Caribbean	Meso-America
Nicaragua	Latin America and the Caribbean	Meso-America
Panama	Latin America and the Caribbean	Meso-America
Argentina	Latin America and the Caribbean	South America
Bolivia	Latin America and the Caribbean	South America
Brazil	Latin America and the Caribbean	South America
Chile	Latin America and the Caribbean	South America
Colombia	Latin America and the Caribbean	South America
Ecuador	Latin America and the Caribbean	South America
French Guiana (France)	Latin America and the Caribbean	South America
Guyana	Latin America and the Caribbean	South America
Paraguay	Latin America and the Caribbean	South America
Peru	Latin America and the Caribbean	South America
Suriname	Latin America and the Caribbean	South America
Uruguay	Latin America and the Caribbean	South America
Venezuela	Latin America and the Caribbean	South America

NORTH AMERICA

Canada	North America	North America
United States of America	North America	North America

POLAR

Antarctic	Polar	Antarctic
Arctic (The eight Arctic countries are: Alaska (United States), Canada, Finland, Greenland (Denmark), Iceland, Norway, Russia, Sweden)	Polar	Arctic

WEST ASIA

Bahrain	West Asia	Arabian Peninsula
Kuwait	West Asia	Arabian Peninsula
Oman	West Asia	Arabian Peninsula
Qatar	West Asia	Arabian Peninsula
Saudi Arabia	West Asia	Arabian Peninsula
United Arab Emirates	West Asia	Arabian Peninsula
Yemen	West Asia	Arabian Peninsula
Iraq	West Asia	Mashriq
Jordan	West Asia	Mashriq
Lebanon	West Asia	Mashriq
Occupied Palestinian Territory	West Asia	Mashriq
Syrian Arab Republic	West Asia	Mashriq

Section **A**

Overview

Chapter 1 **Environment for Development**

"The 'environment' is where we live; and development is what we all do in attempting to improve our lot within that abode. The two are inseparable."

Our Common Future

Environment for Development

Coordinating lead authors: Diego Martino and Zinta Zommers

Lead authors: Kerry Bowman, Don Brown, Flavio Comim, Peter Kouwenhoven, Ton Manders, Patrick Milimo, Jennifer Mohamed-Katerere, and Thierry De Oliveira

Contributing authors: Dan Claasen, Simon Dalby, Irene Dankelman, Shawn Donaldson, Nancy Doubleday, Robert Fincham, Wame Hambira, Sylvia I. Karlsson, David MacDonald, Lars Mortensen, Renata Rubian, Guido Schmidt-Traub, Mahendra Shah, Ben Sonneveld, Indra de Soysa, Rami Zurayk, M.A. Keyzer, and W.C.M. Van Veen

Chapter review editor: Tony Prato

Chapter coordinators: Thierry De Oliveira, Tessa Goverse, and Ashbindu Singh

Main messages

It is 20 years since the report of the World Commission on Environment and Development (WCED), *Our Common Future*, emphasized the need for a sustainable way of life which not only addresses current environmental challenges but also ensures a secure society well into the future. This chapter analyses the evolution of such ideas as well as global trends in relation to environment and socio-economic development. The following are its main messages:

The world has changed radically since 1987 – socially, economically and environmentally. Global population has grown by more than 1.7 billion, from about 5 billion people. The global economy has expanded and is now characterized by increasing globalization. Worldwide, GDP per capita (purchasing power parity) has increased from US$5 927 in 1987 to US$8 162 in 2004. However, growth has been distributed unequally between regions. Global trade has increased during the past 20 years, fuelled by globalization, better communication, and low transportation costs. Technology has also changed. Communications have been revolutionized with the growth of telecommunications and the Internet. Worldwide, mobile phone subscribers increased from 2 people per 1 000 in 1990 to 220 per 1 000 in 2003. Internet use increased from 1 person per 1 000 in 1990 to 114 per 1 000 in 2003. Finally, political changes have also been extensive. Human population and economic growth has increased demand on resources.

The World Commission on Environment and Development (WCED) recognized 20 years ago that the environment, economic and social issues are interlinked. It recommended that the three be integrated into development decision making. In defining sustainable development, the Commission acknowledged the need for both intra- and intergenerational equity – development that meets not only today's human needs but also those of more people in the future.

Changing drivers, such as population growth, economic activities and consumption patterns, have placed increasing pressure on the environment. Serious and persistent barriers to sustainable development remain. In the past 20 years, there has been limited integration of environment into development decision making.

Environmental degradation is therefore undermining development and threatens future development progress. Development is a process that enables people to better their well-being. Long-term development can only be achieved through sustainable management of various assets: financial, material, human, social and natural. Natural assets, including water, soils, plants and animals, underpin people's livelihoods.

Environmental degradation also threatens all aspects of human well-being. Environmental degradation has been demonstrably linked to human health problems, including some types of cancers, vector-borne diseases, emerging animal to human disease transfer, nutritional deficits and respiratory illnesses. The environment provides essential material assets and an economic base for human endeavour. Almost half the jobs worldwide depend on fisheries, forests or agriculture. Non-sustainable use of natural resources, including land, water, forests and fisheries, can threaten individual livelihoods as well as local, national and international economies. The environment can play a significant role in contributing to development and human well-being, but can also increase human vulnerability, causing human migration and insecurity, such as in the case of storms, droughts or environmental mismanagement. Environmental scarcity can foster cooperation, but also contribute to tensions or conflicts.

Environmental sustainability, Millennium Development Goal 7, is critical to the attainment of the other MDG goals. Natural resources are the basis of subsistence in many poor communities. In fact, natural capital accounts for 26 per cent of the wealth of low-income countries. Up to 20 per cent of the total burden of disease in developing countries is associated with environmental risks. Poor women are particularly vulnerable to respiratory infections related to exposure to indoor air pollution. Acute respiratory infections are the leading cause of death in children, with pneumonia killing more children under the age of five than any other illness. A combination of unsafe water and poor sanitation is the world's second biggest killer of children. About 1.8 million children die annually and about 443 million school days are missed due to diarrhoea. Clean water and air are powerful preventative medicines. Sustainable management of natural resources contributes to poverty alleviation, helps reduce diseases and child mortality, improves maternal health, and can contribute to gender equity and universal education.

Some progress towards sustainable development has been made since 1987 when the WCED report, *Our Common Future*, was launched. The number of meetings and summits related to the environment and development has increased (for example, the 1992 Rio Earth Summit and the 2002 World Summit on Sustainable Development), and there has been a rapid growth in multilateral environmental agreements (for example, the Kyoto Protocol and the Stockholm Convention on Persistent Organic Pollutants). Sustainable development strategies have been implemented at local, national, regional and international levels. An increasing number of scientific assessments (for example, the Intergovernmental Panel on Climate Change) have contributed to a greater understanding of environmental challenges. In addition, proven and workable solutions have been identified for environmental problems that are limited in scale, highly visible and acute, (for example, industrial air and water pollution, local soil erosion and vehicle exhaust emission).

However, some international negotiations have stalled over questions of equity and responsibility sharing. Interlinkages between drivers and pressures on the global environment make solutions complex. As a result, action has been limited on some issues, for example, climate change, persistent organic pollutants, fisheries management, invasive alien species and species extinction.

Effective policy responses are needed at all levels of governance. While proven solutions continue to be used, action should also be taken to address both the drivers of change and environmental problems themselves. A variety of tools that have emerged over the past 20 years may be strategic. Economic instruments, such as property rights, market creation, bonds and deposits, can help correct market failures and internalize costs of protecting the environment. Valuation techniques can be used to understand the value of ecosystem services. Scenarios can provide insights on the future impacts of policy decisions. Capacity building and education are critical to generate knowledge and inform the decision making process.

Society has the capacity to make a difference in the way the environment is used to underpin development and human well-being. The following chapters highlight many of the challenges society faces today and provides signposts towards sustainable development.

INTRODUCTION

Imagine a world in which environmental change threatens people's health, physical security, material needs and social cohesion. This is a world beset by increasingly intense and frequent storms, and by rising sea levels. Some people experience extensive flooding, while others endure intense droughts. Species extinction occurs at rates never before witnessed. Safe water is increasingly limited, hindering economic activity. Land degradation endangers the lives of millions of people.

This is the world today. Yet, as the World Commission on Environment and Development (Brundtland Commission) concluded 20 years ago "humanity has the ability to make development sustainable." The fourth *Global Environment Outlook* highlights imperative steps needed to achieve this vision.

The fourth GEO assesses the current state of the world's atmosphere, land, water and biodiversity, providing a description of the state of environment, and demonstrating that the environment is essential for improving and sustaining human well-being. It also shows that environmental degradation is diminishing the potential for sustainable development. Policies for action are highlighted to facilitate alternative development paths.

This chapter examines developments since the landmark 1987 Brundtland Commission report – *Our Common Future* – placed sustainable development much higher on the international policy agenda. It examines institutional developments and changes in thought since the mid-1980s, and explores the relationships involving environment, development and human well-being, reviews major environmental, social and economic trends, and their impacts on environment and human well-being, and provides options to help achieve sustainable development.

Subsequent chapters will analyse of environmental changes in the atmosphere, land, water and biodiversity, both at global and regional levels, and will highlight human vulnerability and strategic policy interlinkages for effective responses. Positive developments since 1987 are described. These include progress towards meeting the goals of the Montreal Protocol, and the reduction in emissions of chemicals that deplete the stratospheric ozone layer. Yet, the chapters also highlight current environmental trends that threaten human well-being:

- In some cases, climate change is having severe effects on human health, food production, security and resource availability.
- Extreme weather conditions are having an increasingly large impact on vulnerable human communities, particularly the world's poor.
- Both indoor and outdoor pollution is still causing many premature deaths.
- Land degradation is decreasing agricultural productivity, resulting in lower incomes and reduced food security.
- Decreasing supplies of safe water are jeopardizing human health and economic activity.
- Drastic reductions of fish stocks are creating both economic losses and a loss of food supply.
- Accelerating species extinction rates are threatening the loss of unique genetic pools, possible sources for future medical and agricultural advances.

Choices made today will determine how these threats will unfold in the future. Reversing such adverse environmental trends will be an immense challenge. Ecosystem services collapse is a distinct possibility if action is not taken. Finding solutions to these problems today is therefore urgent.

This chapter provides a message for action today: The Earth is our only home. Its well-being, and our own, is imperilled. To ensure long-term well-being, we must take an alternative approach to development, one that acknowledges the importance of environment.

OUR COMMON FUTURE: EVOLUTION OF IDEAS AND ACTIONS

Two decades ago the Brundtland Commission report – *Our Common Future* – addressed the links between development and environment, and challenged policy-makers to consider the interrelationships among environment, economic and social issues when it comes to solving global problems. The report examined emerging global challenges in:

- population and human resources;
- food security;
- species and ecosystems;
- energy;
- industry; and
- urbanization.

The commission recommended institutional and legal changes in six broad areas to address these challenges:

- getting at the sources;
- dealing with the effects;
- assessing global risks;
- making informed choices;
- providing the legal means; and
- investing in our future.

Recommendations emphasized the expansion of international institutions for cooperation, and the creation of legal mechanisms for environmental protection and sustainable development, and also stressed the links between poverty and environmental degradation. They also called for increased capacity to assess and report on risks of irreversible damage to natural systems, as well as threats to human survival, security and well-being.

The work of the commission was built on the foundation of, among others, the 1972 UN Conference on the Human Environment in Stockholm and the 1980 World Conservation Strategy, which emphasized conservation as including both protection and the rational use of natural resources (IUCN and others 1991). The Brundtland Commission is widely attributed with popularizing sustainable development internationally (Langhelle 1999). It defined sustainable development as "development that meets the needs of the present generation without compromising the ability of future generations to meet their own needs." The commission further explained that, "the concept of sustainable development implies limits – not absolute limits but limitations imposed by the present state of technology and social organization on environmental resources and by the ability of the biosphere to absorb the effects of human activities." It was argued that, "technology and social organization can be both managed and improved to make way for a new era of economic growth" (WCED 1987).

The most immediate and perhaps one of the most significant results of *Our Common Future* was the organization of the UN Conference on Environment and Development (UNCED), also known as the Earth Summit, which gathered many heads of state in Rio de Janeiro in 1992. Not only did this meeting bring together 108 government leaders,

more than 2 400 representatives from non-governmental organizations (NGOs) attended, and 17 000 people participated in a parallel NGOs event. The Earth Summit strengthened interaction among governments, NGOs and scientists, and fundamentally changed attitudes towards governance and the environment. Governments were encouraged to rethink the concept of economic development, and to find ways to halt the destruction of natural resources and reduce pollution of the planet.

The summit resulted in several important steps towards sustainable development. Through the adoption of the Rio Declaration and Agenda 21, it helped formalize an international institutional framework to implement the ideas highlighted in *Our Common Future*. The Rio Declaration contains 27 principles that nations agreed to follow to achieve the goals articulated by the

Gro Harlem Brundtland introduces to the General Assembly, the report of the World Commission on Environment and Development in 1987, which she chaired. The work of the Brundtland Commission challenged policy-makers to consider the interrelationships among environment, economic and social issues in efforts to solve global problems.

Credit: UN Photo/Milton Grant

Brundtland Commission. Key commitments in the Rio Declaration included integration of environment and development in decision making, provision for polluters to pay for costs of pollution, recognition of common but differentiated responsibilities, and application of the precautionary approach to decision making.

Agenda 21 articulated a comprehensive plan of action towards sustainable development. It contains 40 chapters, which can be divided into four main areas:

- social and economic issues, such as poverty, human health and population;
- conservation and management of natural resources including the atmosphere, forests, biological diversity, wastes and toxic chemicals;
- the role of nine major groups in implementing the sustainable development agenda (local authorities, women, farmers, children and youth, indigenous peoples, workers and trade unions, NGOs, the scientific and technological community, and business and industry); and
- means of implementation, including technology transfer, financing, science, education and public information.

Embedded in these four main areas of Agenda 21 are the environmental challenges, as well as the broad governance issues highlighted in the Brundtland Commission report. As the blueprint for sustainable development, Agenda 21 remains the most significant non-binding instrument in the environmental field (UNEP 2002).

Funding for the implementation of Agenda 21 was to be obtained from the Global Environment Facility (GEF). As a partnership involving the UNEP, UNDP and World Bank, GEF was established the year before the Earth Summit to mobilize resources for projects that seek to protect the environment. Since 1991, the GEF has provided US$6.8 billion in grants, and generated more than US$24 billion in co-financing from other sources to support about 2 000 projects that produce global environmental benefits in more than 160 developing countries and countries with economies in transition. GEF funds are contributed by donor countries, and in 2006, 32 countries pledged a total of US$3.13 billion to fund various environment-related initiatives over four years (GEF 2006).

The turn of the century brought a sense of urgency to attempts to address environment and development challenges. World leaders sought to ensure a world free from want. In the Millennium Declaration, adopted in 2000, world leaders committed to free their people from the "threat of living on a planet irredeemably spoilt by human activities, and whose resources would no longer be sufficient for their needs" (UN 2000). The Millennium Summit, adopted the declaration and created time-bound goals and targets – the Millennium Development Goals (MDGs) – to better human well-being.

Two years after the Millennium Declaration and a decade after the Rio Earth Summit, world leaders reaffirmed sustainable development as a central goal on the international agenda at the 2002 Johannesburg World Summit on Sustainable Development (WSSD). More than 21 000 participants attended the summit, along with representatives of more than 191 governments. The UN Secretary-General designated five priority areas for discussion: water, including sanitation, energy, health, agriculture and biodiversity. These became to be known by the acronym WEHAB. These issues can also be traced back to initiatives such as the Brundtland Commission. The WSSD outcomes include the Johannesburg Declaration on Sustainable Development, and a 54-page plan of implementation. World leaders committed themselves, "to expedite the achievement of the time-bound, socio-economic and environmental targets" contained within the Plan of Implementation (Johannesburg Declaration on Sustainable Development). This historic summit also achieved new commitments on water and sanitation, poverty eradication, energy, sustainable production and consumption, chemicals, and management of natural resources (UN 2002).

The last 20 years has also seen a growth in the number of scientific assessments, such as the Intergovernmental Panel on Climate Change, the Millennium Ecosystem Assessment and the Global Environment Outlook. The Intergovernmental Panel on Climate Change was established in 1988 to assess on an objective, open and transparent basis the scientific, technical and socio-economic information relevant to climate change. In 2007, the IPCC released its Fourth Assessment Report. The Millennium Ecosystem Assessment was called for by the then UN Secretary-General Kofi Annan, to assess the consequences of ecosystem change for human well-being. These scientific assessments reflect the work of thousands of experts worldwide, and have led to greater understanding of environmental problems.

As a result of the conferences and assessments highlighted above, a diversity of multilateral environmental agreements (MEAs) have been adopted (see Figure 1.1), and these and several others are analysed in relevant chapters throughout this report. The Convention on Biological Diversity (CBD) was signed by 150 government leaders at the Rio Earth Summit. The CBD sets out commitments for conserving biodiversity, the sustainable use of its components, and fair and equitable sharing of its benefits. The Cartagena Protocol on Biosafety is based on the precautionary approach from the Rio Declaration. Principle 15 of the Rio Declaration states that, "where there are threats of serious and irreversible damage lack of full scientific certainty shall not be used as a reason for postponing cost effective measures to prevent biological degradation" (UNGA 1992). The Protocol promotes biosafety in the handling, transfer and use of living modified organisms.

Two agreements that have drawn significant attention during the last 20 years are the Montreal Protocol to the Vienna Convention on Substances that Deplete the Ozone Layer and the Kyoto Protocol to the

UN Framework Convention on Climate Change. The Montreal Protocol, which became effective in 1989 and had 191 parties at the beginning of 2007, has helped decrease or stabilize atmospheric concentrations of many of the ozone-depleting substances, including chlorofluorocarbons. The protocol is regarded as one of the most successful international agreements to date. By contrast, despite the urgency of climate change, it has been much more difficult to get some countries responsible for significant emissions of greenhouse gases to ratify the Kyoto Protocol.

Environmental governance has changed since the Brundtland Commission. Today, a broader scope of issues related to environment and development are discussed. The issues of trade, economic development, good governance, transfer of technology, science and education policies, and globalization, which links them together, have become even more central to sustainable development.

Different levels of government participate in environmental policy. The post-WCED period saw a strong increase in sub-national and local government action, for example, through local Agenda 21 processes. The Johannesburg Plan of Implementation stressed that the role of national policies and development strategies "cannot be overemphasized." It also strengthened the role of the regional level,

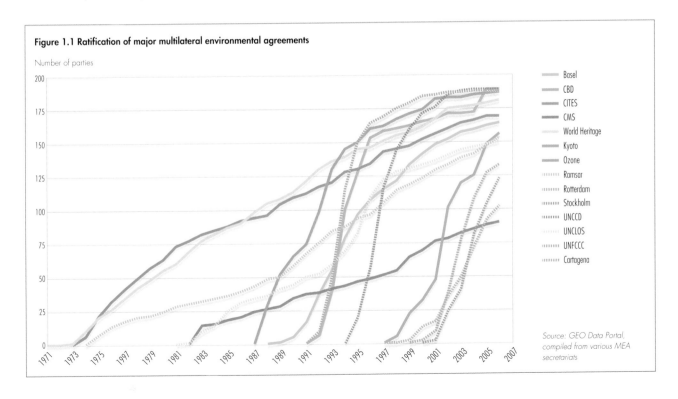

Figure 1.1 Ratification of major multilateral environmental agreements

Number of parties

Legend:
- Basel
- CBD
- CITES
- CMS
- World Heritage
- Kyoto
- Ozone
- Ramsar
- Rotterdam
- Stockholm
- UNCCD
- UNCLOS
- UNFCCC
- Cartagena

Source: GEO Data Portal, compiled from various MEA secretariats

for example by giving the regional UN economic commissions new tasks, and establishing a regional preparation process for the Commission on Sustainable Development (CSD) (UN 2002).

The number of non-governmental stakeholders involved in environmental governance has grown considerably, with organizations playing key roles from local to global levels. NGOs and advocacy groups devoted to public interest and environmental causes have multiplied exponentially, particularly in countries undertaking democratic transitions (Carothers and Barndt 2000).

The private sector should also take action to help protect the environment. Even though business was "given little attention by the WCED ..., more boards and executive committees are trying to consider all dimensions of their impacts at once, on the same agenda, in the same room" (WBCSD 2007). As consumer demand for "green" products arose, some businesses developed voluntary environmental codes, or followed codes developed by non-governmental organizations and governments (Prakash 2000). Other companies began to monitor and report on their sustainability impacts. A study by eight corporate leaders on what business success would look like in the future concluded that it would be tied to helping society cope with challenges such as poverty, globalization, environmental decline and demographic change (WBCSD 2007).

Finally, decision making is increasingly participatory. Stakeholder groups interact with each other and with governments through networks, dialogues and partnerships. Interaction among groups at local, national and global levels was institutionalized in the action plans of UNCED and WSSD. Chapter 37 of Agenda 21 urged countries to involve all possible interest groups in building national consensus on Agenda 21 implementation, and Chapter 28 encouraged local authorities to engage in dialogue with their citizens.

Environment as the foundation for development

Before the Brundtland Commission, "development progress" was associated with industrialization, and measured solely by economic activity and increases in wealth. Environmental protection was perceived by many as an obstacle to development. However, *Our Common Future* recognized "environment or development" as a false dichotomy. Focus shifted to "environment and development," and then to "environment for development (see Box 1.1)." Principle 1 of Agenda 21 states: "Human beings are at the centre of concerns for sustainable development. They are entitled to a healthy and productive life in harmony with nature."

The normative framework for human development is reflected by the MDGs (UNDP 2006). In signing on to the MDGs, nations explicitly recognized that achieving Goal 7 on environmental sustainability is key to achieving poverty eradication. However, environmental issues are not highly integrated into other MDGs (UNDP 2005a). A healthy environment is essential for achieving all the goals (see Table 1.1). To achieve real progress, the interlinkages between MDG 7 and the other MDGs need to be acknowledged and integrated into all forms of planning.

While a healthy environment can support development, the relationship is not always reciprocal. Many alternative views exist on the benefits and disadvantages of modern development (Rahnema 1997). It has been argued that development is destructive, even violent, to nature (Shiva 1991). As *GEO-4* illustrates, past development practices have often not been beneficial to the environment. However, opportunities exist to make development sustainable.

Environmental degradation due to development raises deep ethical questions that go beyond economic

---

Box 1.1 Environment as the foundation for development

Development is the process of furthering people's well-being. Good development entails:
- increasing the asset base and its productivity;
- empowering poor people and marginalized communities;
- reducing and managing risks; and
- taking a long-term perspective with regard to intra- and intergenerational equity.

The environment is central to all four of these requirements. Long-term development can only be achieved through sustainable management of various assets: financial, material, human, social and natural. Natural assets, including water, soils, plants and animals, underpin the livelihoods of all people. At the national level, natural assets account for 26 per cent of the wealth of low-income countries. Sectors such as agriculture, fishery, forestry, tourism and minerals provide important economic and social benefits to people. The challenge lies in the proper management of these resources. Sustainable development provides a framework for managing human and economic development, while ensuring a proper and optimal functioning over time of the natural environment.

Sources: Bass 2006, World Bank 2006a

Table 1.1 Links between the environment and the Millennium Development Goals

Millennium Development Goal	Selected environmental links
1. Eradicate extreme poverty and hunger	Livelihood strategies and food security of the poor often depend directly on healthy ecosystems, and the diversity of goods and ecological services they provide. Natural capital accounts for 26 per cent of the wealth of low-income countries. Climate change affects agricultural productivity. Ground-level ozone damages crops.
2. Achieve universal primary education	Cleaner air will decrease the illnesses of children due to exposure to harmful air pollutants. As a result, they will miss fewer days of school. Water-related diseases such as diarrhoeal infections cost about 443 million school days each year, and diminish learning potential.
3. Promote gender equality, and empower women	Indoor and outdoor air pollution is responsible for more than 2 million premature deaths annually. Poor women are particularly vulnerable to respiratory infections, as they have high levels of exposure to indoor air pollution. Women and girls bear the brunt of collecting water and fuelwood, tasks made harder by environmental degradation, such as water contamination and deforestation.
4. Reduce child mortality	Acute respiratory infections are the leading cause of death in children. Pneumonia kills more children under the age of 5 than any other illness. Environmental factors such as indoor air pollution may increase children's susceptibility to pneumonia. Water-related diseases, such as diarrhoea and cholera, kill an estimated 3 million people/year in developing countries, the majority of whom are children under the age of five. Diarrhoea has become the second biggest killer of children, with 1.8 million children dying every year (almost 5 000/day).
5. Improve maternal health	Indoor air pollution and carrying heavy loads of water and fuelwood adversely affect women's health, and can make women less fit for childbirth and at greater risk of complications during pregnancy. Provision of clean water reduces the incidence of diseases that undermine maternal health and contribute to maternal mortality.
6. Combat major diseases	Up to 20 per cent of the total burden of disease in developing countries may be associated with environmental risk factors. Preventative environmental health measures are as important and at times more cost-effective than health treatment. New biodiversity-derived medicines hold promises for fighting major diseases.
7. Ensure environmental sustainability	Current trends in environmental degradation must be reversed in order to sustain the health and productivity of the world's ecosystems.
8. Develop a global partnership for development	Poor countries and regions are forced to exploit their natural resources to generate revenue and make huge debt repayments. Unfair globalization practices export their harmful side-effects to countries that often do not have effective governance regimes.

Source: Adapted from DFID and others 2002, UNDP 2006, UNICEF 2006

cost-benefit ratios. The question of justice is perhaps the greatest moral question emerging in relation to environmental change and sustainable development. Growing evidence indicates that the burden of environmental change is falling far from the greatest consumers of environmental resources, who experience the benefits of development. Often, people living in poverty in the developing world, suffer the negative effects of environmental degradation. Furthermore, costs of environmental degradation will be experienced by humankind in future generations. Profound ethical questions are raised when benefits are extracted from the environment by those who do not bear the burden.

Barriers to sustainable development

Despite changes in environmental governance, and greater understanding of the links between environment and development, real progress towards sustainable development has been slow. Many governments continue to create policies concerned with environmental, economic and social matters as single issues. There is a continued failure to link environment and development in decision making (Dernbach 2002). As a result, development strategies often ignore the need to maintain the very ecosystem services on which long-term development goals depend. A notable example, made apparent in the aftermath of the

Women and girls bear the brunt of collecting fuelwood, tasks made harder by environmental degradation.

Credit: Christian Lambrechts

make public participation design daunting. If participation is treated superficially, and embodied merely as a quota of specified groups in decision making processes, it could easily be no more than "lip service." The task of designing modern, cross-cutting, transparent, evidence-based interdisciplinary decision making is not only conceptually challenging, but also necessitates a huge increase in local capacity for democracy and decision making (MacDonald and Service 2007).

Many social, economic and technological changes described later in this chapter have made implementation of the recommendations in *Our Common Future* difficult. As also illustrated in other chapters, changes such as a growing population and increased consumption of energy have had a huge impact on the environment, challenging society's ability to achieve sustainable development.

Finally, the nature of the environmental problems has influenced the effectiveness of past responses. Environmental problems can be mapped along a continuum from "problems with proven solutions" to "less known emerging (or persistent) problems" (Speth 2004). With problems with proven solutions, the cause-and-effect relationships are well known. The scale tends to be local or national. Impacts are highly visible and acute, and victims are easily identified. During the past 20 years, workable solutions have been identified for several such problems, for example industrial air and water pollution, local soil erosion, mangrove clearance for aquaculture, and vehicle exhaust emissions.

the 2005 Hurricane Katrina, is the failure of some government agencies to see the link between destruction of coastal wetlands and the increased vulnerability of coastal communities to storms (Travis 2005, Fischetti 2005). For many, acknowledging that environmental change could endanger future human well-being is inconvenient, as it requires an uncomfortable level of change to individual and working lives (Gore 2006).

International negotiations on solutions to global environmental problems have frequently stalled over questions of equity (Brown 1999). For instance, in the case of climate change, international negotiations have slowed down over the question on how to share responsibilities and burden among nations, given different historic and current levels of national emissions.

Providing widespread participation in sustainable development decision making called for by Agenda 21 has also raised significant challenges. The enormous diversity of issues that need to be considered in sustainable development policy making, together with aspirations for transparency,

However, progress has been limited on harder to manage environmental issues, which can also be referred to as "persistent" problems (Jänicke and Volkery 2001). These are deeply rooted structural problems, related to the ways production and consumption are conducted at the household, national, regional and global levels. Harder to manage problems tend to have multiple dimensions and be global in scale. Some of the basic science of cause-and-effect relationships is known, but often not enough to predict when a tipping point or a point of no return will be reached. There is often a need to implement measures on a very large-scale. Examples of such problems include global climate change, persistent organic pollutants and heavy metals, ground level ozone, acid rain, large-scale deterioration of fisheries, extinction of species, or introductions of alien species.

Awareness of the nature of an environmental problem provides a basis for creating strategies, targeting efforts, and finding and implementing a sustainable solution. Possible solutions to different types of environmental problems are introduced in the last section of this chapter, highlighted in the rest of the report, and discussed further in Chapter 10.

HUMAN WELL-BEING AND THE ENVIRONMENT

For sustainable development to be achieved, links between the environment and development must be examined. It is also important to consider the end point of development: human well-being. The evolution of ideas on development has made the concept of human well-being central to the policy debate. Human well-being is the outcome of development. Human well-being and the state of the environment are strongly interlinked. Establishing how environmental changes have impacts on human well-being, and showing the importance of environment for human well-being, are among the core objectives of this report.

Defining human well-being

Defining human well-being (see Box 1.2) is not easy, due to alternative views on what it means. Simply put, human well-being can be classified according to three views, each of which has different implications for the environment:

- The resources people have, such as money and other assets. Wealth is seen as conducive to well-being. This view is closely linked to the concept of weak sustainability, which argues that environmental losses can be compensated for by increases in physical capital (machines) (Solow 1991). The environment can only contribute to development as a means to promote economic growth.

- How people feel about their lives (their subjective views). Individuals' assessments of their own living conditions take into account the intrinsic importance that environment has for life satisfaction. According to this view, people value the environment for its traditional or cultural aspects (Diener 2000, Frey and Stutzer 2005).

- What people are able to be and to do. This view focuses on what the environment allows individuals to be and to do (Sen 1985, Sen 1992, Sen 1999). It points out that the environment provides the basis for many benefits, such as proper nourishment, avoiding unnecessary morbidity and premature mortality, enjoying security and self-respect, and taking part in the life of the community. The environment is appreciated beyond its role as income generator, and its impacts on human well-being are seen as multidimensional.

Box 1.2 Human well-being

Human well-being is the extent to which individuals have the ability and the opportunity to live the kinds of lives they have reason to value.

People's ability to pursue the lives that they value is shaped by a wide range of instrumental freedoms. Human well-being encompasses personal and environmental security, access to materials for a good life, good health and good social relations, all of which are closely related to each other, and underlie the freedom to make choices and take action:

- Health is a state of complete physical, mental and social well-being, and not merely the absence of disease or illness. Good health not only includes being strong and feeling well, but also freedom from avoidable disease, a healthy physical environment, access to energy, safe water and clean air. What one can be and do include among others, the ability to keep fit, minimize health-related stress, and ensure access to medical care.

- Material needs relate to access to ecosystem goods-and-services. The material basis for a good life includes secure and adequate livelihoods, income and assets, enough food and clean water at all times, shelter, clothing, access to energy to keep warm and cool, and access to goods.

- Security relates to personal and environmental security. It includes access to natural and other resources, and freedom from violence, crime and wars (motivated by environmental drivers), as well as security from natural and human-caused disasters.

- Social relations refer to positive characteristics that define interactions among individuals, such as social cohesion, reciprocity, mutual respect, good gender and family relations, and the ability to help others and provide for children.

Increasing the real opportunities that people have to improve their lives requires addressing all these components. This is closely linked to environmental quality and the sustainability of ecosystem services. Therefore, an assessment of the impact of the environment on individuals' well-being can be done by mapping the impact of the environment on these different components of well-being.

Sources: MA 2003, Sen 1999

The evolution of these ideas has progressed from the first to the third, with increasing importance being given to the real opportunities that people have to achieve what they wish to be and to do. This new understanding of human well-being has several important aspects. First, multidimensionality is viewed as an important feature of human well-being. Consequently, the impact of the environment on human well-being is seen according to many different dimensions.

Second, autonomy is considered a defining feature of people, and of well-being. Autonomy can be defined broadly as allowing people to make individual or collective choices. In other words, to know whether an individual is well requires considering his or her resources, subjective views, and the ability to choose and act. This concept of human well-being highlights the importance of understanding whether individuals are simply passive spectators of policy interventions, or, in fact, active agents of their own destiny.

Context of human well-being

The potential for individuals, communities and nations to make their own choices, and maximize opportunities to achieve security and good health, meet material needs and maintain social relations is affected by many interlinked factors, such as poverty, inequality and gender. It is important to note how these factors relate to each other, and to the environment.

Poverty and inequality

Poverty is understood as a deprivation of basic freedoms. It implies a low level of well-being, with such outcomes as poor health, premature mortality and morbidity, and illiteracy. It is usually driven by inadequate control over resources, discrimination (including by race or gender), and lack of access to material assets, health care and education (UN 2004).

Inequality refers to the skewed distribution of an object of value, such as income, medical care or clean water, among individuals or groups. Unequal access to environmental resources remains an important source of inequality among individuals. Equity is the idea that a social arrangement addresses equality in terms of something of value. Distributive analysis is used to assess features of human well-being that are unequally distributed among individuals according to arbitrary factors, such as gender, age, religion and ethnicity. When an analysis of this distribution focuses on its lower end, it refers to poverty.

Mobility

When seen in a dynamic perspective, inequality and poverty are better understood through the concepts of social mobility and vulnerability. Mobility relates to the ability of people to move from one social group, class or level to another. Environmental degradation may be responsible for locking individuals within low-mobility paths, limiting opportunities to improve their own well-being.

Vulnerability

Vulnerability involves a combination of exposure and sensitivity to risk, and the inability to cope or adapt to environmental change. Most often, the poor are more vulnerable to environmental change. Broad patterns of vulnerability to environmental and socio-economic changes can be identified so that policy-makers can respond, providing opportunities for reducing

Individuals' assessments of their own living conditions take into account the intrinsic importance that the environment has for life satisfaction.

Credit: Mark Edwards/Still Pictures

vulnerability, while protecting the environment. Chapter 7 assesses the vulnerability of the human-environment system to multiple stresses (drivers and pressures).

Gender inequality

An analysis of distributive impacts of the environment on human well-being cannot ignore features such as gender. Gender inequality is one of the most persistent inequalities in both developed and developing countries, with the majority of people living in poverty being women (UNDP 2005b). Women and girls often carry a disproportionate burden from environmental degradation compared to men. Understanding the position of women in society, and their relationship with the environment is essential for promoting development. In many cases, women and girls assume greater responsibilities for environmental management, but have subordinate positions in decision making (Braidotti and others 1994). Women need to be at the centre of policy responses (Agarwal 2000). At the same time, it is important to avoid stereotyping these roles, and to base responses on the complexities of local realities (Cleaver 2000).

Environmental change and human well-being

One of the main findings of the Millennium Ecosystem Assessment is that the relationship between human well-being and the natural environment is mediated by services provided by ecosystems (see Box 1.3). Changes to these services, as a result of changes in the environment, affect human well-being through

impacts on security, basic material for a good life, health, and social and cultural relations (MA 2003). All people – rich and poor, urban and rural, and in all regions – rely on natural capital.

The world's poorest people depend primarily on environmental goods-and-services for their livelihoods, which makes them particularly sensitive and vulnerable to environmental changes (WRI 2005). Furthermore, many communities in both developing and developed countries derive their income from environmental resources, which include fisheries, non-timber forest products and wildlife.

Health

Shortly before the publication of *Our Common Future*, the nuclear accident at Chernobyl illustrated the catastrophic impact pollution can have on health. Twenty years later, as victims of Chernobyl still struggle with disease, the health of countless other people around the world continues to be affected by human-induced changes to the environment. Changes

The relationship between human well-being and the natural environment is influenced by services provided by ecosystems.

Credit: Joerg Boethling/Still Pictures

affecting provisioning services, including water, can influence human health. Changes affecting regulating services influence health via the distribution of disease transmitting insects or pollutants in water and air (MA 2003). Almost one-quarter of all diseases are caused by environmental exposure (WHO 2006).

As described in Chapter 2, urban air pollution is one of the most widespread environmental problems, affecting health in almost all regions of the world. While air pollution has decreased in many industrialized countries, it has increased in other regions, particularly in Asia. Here, rapid population growth, economic development and urbanization have been associated with increasing use of fossil fuels, and a deterioration of air quality. WHO estimates that more than 1 billion people in Asian countries are exposed to air pollutant levels exceeding their guidelines (WHO 2000). In 2002, WHO estimated that more than 800 000 people died prematurely due to PM_{10} (particulate

---

Box 1.4 Wild meat trade

The bushmeat trade in Central Africa, and wildlife markets in Asia are examples of activities that both have impacts on the environment, and carry risk of disease emergence. In Viet Nam, the illegal trade in wildlife currently generates US$20 million/year. Wild meat is a critical source of protein and income for forest dwellers and rural poor. However, commercial demand for wild meat has been growing as a result of urban consumption, from wildlife restaurants and medicine shops, but also from markets in neighbouring countries. Rates of wildlife harvesting are unsustainable, and threaten species such as the small-toothed palm civet with extinction.

In wildlife markets, mammals, birds and reptiles come in contact with dozens of other species and with countless numbers of people, increasing opportunity for disease transmission. Not surprisingly, during the 2003 Sudden Acute Respiratory Syndrome (SARS) epidemic, several of the early patients in Guangdong Province, China, worked in the sale or preparation of wildlife for food. The disease may have first spread to humans from civet cats or bats in local wildlife markets. Through human air travel, SARS quickly spread to 25 countries across five continents. With more than 700 million people travelling by air annually, disease outbreaks can easily grow into worldwide epidemics.

It is estimated that every year between 1.1 and 3.4 million tonnes of undressed wild animal biomass, or bushmeat, are consumed by people living the Congo Basin. The wild meat trade, commercial hunting of wild animals for meat, has decimated endangered populations of long-lived species such as chimpanzees. Trade is global in nature, and primate meat has even been found in markets in Paris, London, Brussels, New York, Chicago, Los Angeles, Montreal and Toronto. Contact with primate blood and bodily fluids during hunting and butchering has exposed people to novel viruses. Between 2000 and 2003, 13 of 16 Ebola outbreaks in Gabon and the Republic of Congo resulted from the handling of gorilla or chimpanzee carcasses. A recent study documents simian foamy virus (SFV) and human T-lymphotic (HTLV) viruses in individuals engaged in bushmeat hunting in rural Cameroon.

Sources: Bell and others 2004, Brown 2006, Goodall 2005, Fa and others 2007, Karesh and others 2005, Leroy and others 2004, Li and others 2005, Peiris and others 2004, Peterson 2003, Wolfe and others 2004, Wolfe and others 2005

Commercial demand for wild meat has been growing and rates of wildlife harvesting are unsustainable.

Credit: Lise Albrechtsen

matter with a diameter less than 10 micrometers) outdoor pollution and 1.6 million due to PM_{10} indoor air pollution (WHO 2002) (see Chapter 2).

Chapter 4 highlights how the overexploitation and pollution of freshwater ecosystems – rivers, lakes, wetlands and groundwater – has direct impacts on human well-being. Although access to clean water and sanitation has improved, in 2002 more than 1.1 billion people lacked access to clean water, and 2.6 billion lacked access to improved sanitation (WHO and UNICEF 2004). Annually, 1.8 million children die from diarrhoea, making the disease the world's second biggest killer of children (UNDP 2006).

Many heavy metals, such as mercury and lead, are found in water and sediments, and are a major concern as they can accumulate in the tissues of humans and other organisms (UNESCO 2006). Numerous activities contribute to heavy metal contamination. Burning coal, incineration, urban and agricultural run-off, industrial discharges, small-scale industrial activities, mining, and landfill leakages are among the main ones described in Chapters 2, 3 and 4.

Changes in the environment have also resulted in the emergence of diseases. Since 1980, more than 35 infectious diseases have emerged or taken on new importance. These include previously unknown, emerging diseases, such as HIV, SARS and avian influenza (H5N1), as well as diseases once thought controllable, such as dengue fever, malaria and bubonic plague (Karesh and others 2005, UNEP 2005a). Human-induced changes to the environment, such as climate change, land use change and interaction with wildlife (see Box 1.4), have driven this recent epidemiological transition (McMichael 2001, McMichael 2004). Growing human contact with wildlife, caused by population pressure on remaining relatively undisturbed environmental resources, increases the opportunity for pathogen exchange (Wolfe and others 1998). Globalization, in turn, has an effect on disease emergence as disease agents have the opportunity to move into new niches, and meet new, vulnerable populations. A recent UNEP report on Avian Influenza and the Environment states: "If the transfer of Asian lineage H5N1 between domestic flocks and wild birds is to be reduced, it will become essential to take measures to minimize their contact. Restoring wetland health will reduce the need for migrating wild birds to share habitat with domestic poultry" (UNEP 2006).

Material needs

People depend on natural resources for their basic needs, such as food, energy, water and housing. In many communities, particularly in developing countries, environmental resources, including fisheries, timber, non-timber forest products and wildlife, directly contribute to income and other material assets required to achieve a life that one values. The ability to meet material needs is strongly linked to the provisioning, regulating and supporting services of ecosystems (MA 2003).

More than 1.3 billion people depend on fisheries, forests and agriculture for employment – close to half of all jobs worldwide (see Box 1.5) (FAO 2004a). In Asia and the Pacific, small-scale fisheries contributed 25 per cent to the total fisheries production of Malaysia, the Philippines, and Thailand for the decade ending in 1997 (Kura and others 2004). In Africa, more than 7 in 10 people live in rural areas, with most engaged in resource-dependent activities (IFAD 2001). The corresponding small-scale production accounts for a significant percentage of the GDP in many African countries (IFPRI 2004). Moreover, small-scale agriculture accounts for more than 90 per cent of Africa's agricultural production (Spencer 2001). A study of households in the Masvingo province in southeast Zimbabwe indicates that 51 per cent of incomes are from agriculture, and that the total income from the environment averages 66 per cent (Campbell and others 2002). Where resources are degraded, livelihoods are placed at risk. Forest loss may reduce

Box 1.5 Material well-being from fisheries

The fisheries sector plays an important role in material well-being, providing income generation, poverty alleviation and food security in many parts of the world. Fish is an important protein source, especially in the developing world, providing more than 2.6 billion people with at least 20 per cent of their average per capita animal protein intake. The world's population growth outpaced that of total fish supply and FAO projections indicate that a global shortage is expected (see Chapter 4).

While fish consumption increased in some regions, such as South East Asia, Europe and North America, it declined in others, including sub-Saharan Africa and Eastern Europe. The collapse of the Canadian east coast cod fishery in the late 1980s had devastating impacts on local fishing communities, and illustrates that developed countries are not immune to the economic implications of mismanaging natural resources. It resulted in unemployment for 25 000 fishers and 10 000 other workers (see Box 5.2 and Figure 7.17 in Chapters 5 and 7 respectively).

Sources: Delgado and others 2003, FAO 2004b, Matthews 1995

the availability of food, energy resources and other forest products, which, in many communities, support trade and income earning opportunities.

Increasing evidence shows that investment in ecosystem conservation, such as watershed management, results in increased income for the rural poor. In the Adgaon watershed in India, the annual days of employment (wage labour) per worker increased from 75 days before watershed rehabilitation to 200 days after restoration was completed (Kerr and others 2002). In Fiji, strengthening the traditional "no-take" management system to promote recovery of marine life has resulted in a 35–43 per cent increase in income over a period of three years (see Box 7.13) (WRI 2005). In a pioneering people-led watershed management project in India, the implementation of a participatory restoration scheme led to halving the distance to the water table, a doubling of land under irrigation, and an increase in the total agricultural income of the village from about US$55 000 in 1996, before watershed regeneration, to about US$235 000 in 2001 (D'Souza and Lobo 2004, WRI 2005).

Security

Security incorporates economic, political, cultural, social and environmental aspects (Dabelko and others 2000). It includes freedom from threats of bodily harm, and from violence, crime and war. It means having stable and reliable access to resources, the ability to be secure from natural and human disasters, and the ability to mitigate and respond to shocks and stresses. Environmental resources are a critical part of the livelihoods of millions of people, and when these resources are threatened through environmental change, people's security is also threatened. "At the centre of sustainable development is the delicate balance between human security and the environment" (CHS 2006).

The Earth has shown clear signs of warming over the past century. Eleven of the last 12 years (1995–2006) rank among the 12 warmest years in the instrumental record of global surface temperature (since 1850) (IPCC 2007). As Chapter 2 describes, climate change is very likely to affect ecological regulating services, resulting in increased frequency and intensity of extreme weather hazards in many regions around the globe (IPCC 2007), and greater insecurity for much of the world's population (Conca and Dabelko 2002). The impacts of extreme weather events will fall disproportionately upon developing countries, such as Small Island Developing States (SIDS) (see Figure 1.2), as well as on the poor in all countries (IPCC 2007). During Hurricane Katrina in the United States in 2005, impoverished people without access to private transportation were unable

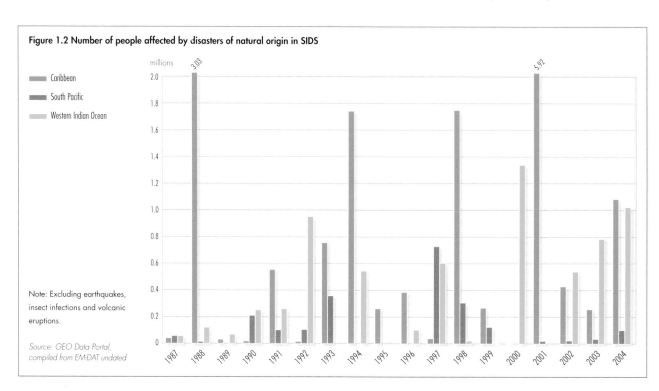

Figure 1.2 Number of people affected by disasters of natural origin in SIDS

- Caribbean
- South Pacific
- Western Indian Ocean

Note: Excluding earthquakes, insect infections and volcanic eruptions.

Source: GEO Data Portal, compiled from EM-DAT undated

to leave the city. People in poor health or lacking bodily strength were less likely to survive the Indian Ocean tsunami in 2004. For example, in villages in North Aceh, Indonesia, women constituted up to 80 per cent of deaths (Oxfam 2005). In Sri Lanka, a high mortality rate was also observed among other vulnerable groups: children and the elderly (Nishikiori and others 2006).

Environmental change can also affect security through changes in provisioning services, which supply food and other goods. Scarcity of shared resources has been a source of conflict and social instability (deSombre and Barkin 2002). Disputes over water quantity and quality are ongoing in many parts of the world. The apparent degradation of Easter Island's natural resources by its Polynesian inhabitants, and the ensuing struggle between clans and chiefs, provides a graphic illustration of a society that destroyed itself by overexploiting scarce resources (Diamond 2005). Natural resources can play an important role in armed conflicts. They have often been a means of funding war (see Box 1.6). Armed conflicts have also been used as a means to gain access to resources (Le Billion 2001), and they can destroy environmental resources.

Box 1.6 Conflict in Sierra Leone and Liberia, and refugee settlement in Guinea

Natural resources, including diamonds and timber, helped fuel civil war in Liberia and Sierra Leone during the 1990s. Diamonds were smuggled from Sierra Leone into Liberia and onto the world market. In the mid-1990s, Liberia's official diamond exports ranged between US$300 and US$450 million annually. These diamonds have been referred to as "blood diamonds," as their trade helped finance rebel groups and the continued hostilities. By the end of the war in 2002, more than 50 000 people had died, 20 000 were left mutilated and three-quarters of the population had been displaced in Sierra Leone alone.

As civil wars raged in Sierra Leone and Liberia, hundreds of thousands of refugees fled to safety in Guinea. In 2003, about 180 000 refugees resided in Guinea. Between Sierra Leone and Liberia, there is a small strip of land belonging to Guinea known as the "Parrot's Beak," because of the parrot shape contour of the international border between the countries (depicted as a black line on both images). This strip is where refugees constituted up to 80 per cent of the local population.

The 1974 image shows small, evenly spread, scattered flecks of light green in the dark green forest cover of the Parrot's Beak and surrounding forests of Liberia and Sierra Leone. These flecks are village compounds, with surrounding agricultural plots. The dark areas in the upper left of the image are most likely burn scars.

In the 2002 image Parrot's Beak is clearly visible as a more evenly spread light grey and green area surrounded by darker green forest of Liberia and Sierra Leone. The light colours show deforestation in the "safe area" where refugees had set up camp. Many of the refugees integrated into local villages, creating their own family plots by cutting more trees. As a result the isolated flecks merged into one larger area of degraded forest. The forest devastation is especially obvious in the upper left part, where areas that were green in 1974 now appear grey and brown, also due to expanded logging.

Sources: Meredith 2005, UNEP 2005b, UNHCR 2006a

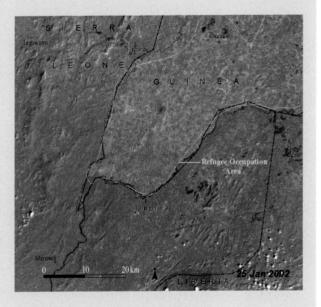

Credit: UNEP 2005b

Insecurity caused by bad governance or war can contribute to environmental degradation. Security requires the current and future availability of environmental goods-and-services, through good governance, mechanisms for conflict avoidance and resolution, and for disaster prevention, preparedness and mitigation (Dabelko and others 2000, Huggins and

others 2006, Maltais and others 2003). Inequitable governance and institutions may prevent people from having secure livelihoods, as illustrated by land tenure conflicts in Southern Africa (Katerere and Hill 2002), and by poor management in Indonesia's peat swamps (Hecker 2005). In both examples, the resource is closely linked to local livelihoods, and insecurity is a result not so much of scarcity but of unequal access to and distribution of these vital resources. In other cases, as illustrated in Box 1.6, degradation may result from changes in settlement patterns as people are forced to flee an area due to hostilities or war.

It has become clear in recent years that joint management on environmental matters is needed to facilitate cooperation across societal and international boundaries to avoid conflict (Matthew and others 2002; UNEP 2005b). The case of cooperative endeavours to deal with fisheries decline in Lake Victoria is an excellent example. Cooperation on water management and transnational ecosystems can also foster diplomatic habits of consultation and dialogue with positive political results, suggesting that human and environmental security are very closely linked (Dodds and Pippard 2005).

Social relations

The environment also affects social relations by providing cultural services, such as the opportunity to express aesthetic, cultural or spiritual values associated with ecosystems (MA 2005a). The natural world provides opportunities for observation and education, recreation and aesthetic enjoyment, all of which are of value to a given society. In some communities, the environment underpins the very structure of social relations. As described in Chapter 5, many cultures, particularly indigenous ones, are deeply interwoven with the local environment.

Climate change is a major concern for SIDS and their high cultural diversity; SIDS are imperilled by sea-level rise and increases in the intensity and number of storms (Watson and others 1997) (see Chapter 7). Tuvalu is an example of an island vulnerable to environmental change. Even though its culture is strongly related to the local environment, the islanders may have to consider relocating to other countries to escape rising sea level as a result of climate change. Coping mechanisms embedded in such cultures might be lost, making society less resilient to future natural disasters (Pelling and Uitto 2001).

Box 1.7 Chemicals affect Arctic peoples

As described in Chapters 5 and 6, the relationships that indigenous peoples have with the environment play an important role in their identity and overall well-being. Scientific assessments have detected persistent organic pollutants (POPs) and heavy metals in all components of the Arctic ecosystem, including in people. The majority of these substances are present in the ecosystems and diets of Arctic peoples as a result of choices (such as using the insecticide toxaphene on cotton fields) by industrial societies elsewhere. Contaminants reach the Arctic from all over the world through wind, air and water currents (see Figure 1.3), entering the food chain.

Inuit populations in the eastern Canadian Arctic and Greenland have among the highest exposures to POPs and mercury from a traditional diet of populations anywhere. A sustainable lifestyle, with ancient roots in the harvesting, distribution and consumption of local renewable resources, is endangered as a result.

Sources: Doubleday 1996, Van Oostdam 2005

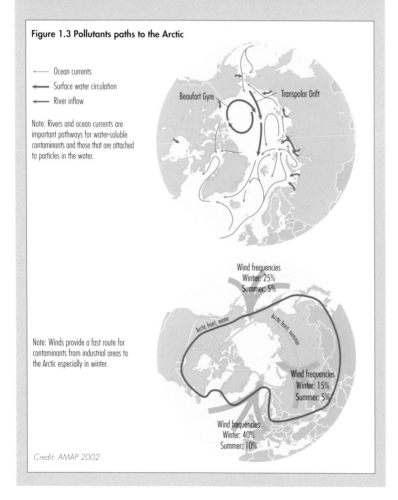

Figure 1.3 Pollutants paths to the Arctic

⟵ Ocean currents
⟵ Surface water circulation
⟵ River inflow

Note: Rivers and ocean currents are important pathways for water-soluble contaminants and those that are attached to particles in the water.

Beaufort Gyre   Transpolar Drift

Note: Winds provide a fast route for contaminants from industrial areas to the Arctic especially in winter.

Wind frequencies
Winter: 25%
Summer: 5%

Arctic front, winter   Arctic front, summer

Wind frequencies
Winter: 15%
Summer: 5%

Wind frequencies
Winter: 40%
Summer: 10%

Credit: AMAP 2002

A diet of traditional foods plays a particularly important role in the social, cultural, nutritional and economic health of indigenous peoples living in the Arctic (Donaldson 2002). Hunting, fishing, and the gathering of plants and berries are associated with important traditional values and practices that are central to their identity as indigenous peoples. Their traditional food is compromised by environmental contaminants (see Box 1.7 and Figure 1.3) and climate change (see Chapter 6), and this affects all dimensions of indigenous well-being. The issue becomes magnified in light of the lack of accessible, culturally acceptable and affordable alternatives. Store food is expensive, and lacks cultural significance and meaning. Long-term solutions require that Arctic lifestyles be considered when development choices are made in industrial and agricultural regions around the world (Doubleday 2005).

DRIVERS OF CHANGE AND PRESSURES

Environmental changes and the effects on human well-being are induced by various drivers and pressures. Drivers such as demographic changes, economic demand and trade, science and technology, as well as institutional and socio-political frameworks induce pressures which, in turn, influence the state of the environment with impacts on the environment itself, and on society and economic activity. Most pressures on ecosystems result from, for example, changes in emissions, land use and resource extraction. Analyses of the linkages shown by the drivers-pressures-state-impacts-responses (DSPIR) framework (described in the Reader's Guide to the report) form the foundation on which the GEO-4 assessment is constructed. In the two decades since the Brundtland Commission, these drivers and pressures have changed, often at an increasing rate. The result is that the environment has changed dramatically. No region has been spared the reality of a changing environment, and its immediate, short- and long-term impacts on human well-being.

Population

Population is an important driver behind environmental change, leading to increased demand for food, water and energy, and placing pressure on natural resources. Today's population is three times larger than it was at the beginning of the 20th century. During the past 20 years global population has continued to rise, increasing from 5 billion in 1987 to 6.7 billion in 2007 (see Figure 1.4), with an average annual growth rate of 1.4 per cent. However, large differences in growth are evident across regions, with Africa

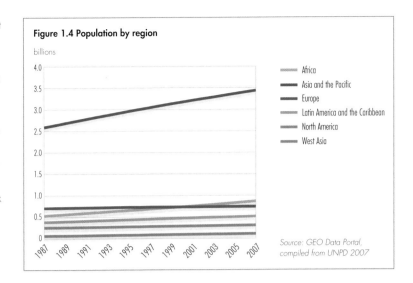

Figure 1.4 Population by region

billions

Africa
Asia and the Pacific
Europe
Latin America and the Caribbean
North America
West Asia

Source: GEO Data Portal, compiled from UNPD 2007

and West Asia recording high growth rates, and the European population stabilizing (see Chapter 6 for more detail). Although the world population is increasing, the rate of increase is slowing (see Box 1.8).

Forced and economic migrations influence demographic changes and settlement patterns, particularly at the regional level. There were 190 million international migrants in 2005, compared to 111 million in 1985. About one-third of migrants in the world have moved from one developing country to another, while another third have moved from a developing country to a developed country (UN 2006). Many migrants are refugees, internally displaced or stateless persons. At the end of 2005, more than 20.8 million people were classified as "of concern" to the UN High Commission for Refugees (UNHCR 2006b). These included refugees, internally displaced and stateless persons. Worldwide refugee numbers have decreased since 2000, but there has been an upward trend in numbers of other displaced groups (UNHCR 2006b).

The term *ecomigrant* has been used to describe anyone whose need to migrate is influenced by environmental factors (Wood 2001). It has been claimed that during the mid-1990s up to 25 million people were forced to flee as a result of environmental change, and as many as 200 million people could eventually be at risk of displacement (Myers 1997). Other analyses indicated that while the environment may play a role in forced migration, migration is usually also linked to political divisions, economic interests and ethnic rivalries (Castles 2002). A clear separation between factors is often difficult.

Box 1.8 Demographic transition

The annual global population growth rate declined from 1.7 per cent in 1987 to 1.1 per cent in 2007. Significant regional variations are analysed in chapter 6. Demographic transition, the change from high birth and death rates to low birth and death rates, can explain these changes in population. As a result of economic development, fertility rates are falling in all regions. In the period between 2000 and 2005, the world recorded a fertility rate of 2.7 children per woman, compared to a fertility rate of 5.1 children per woman 50 years before. Ultimately, fertility may even drop below 2, the replacement rate, leading to a global population decline. Some European countries are at this stage, and have ageing populations.

Improved health has led to lower mortality rates and higher life expectancies in most regions (see Figure 1.5). However, life expectancy in many parts of Africa has decreased during the last 20 years, partly as a result of the AIDS pandemic. Around the world, more than 20 million people have died since the first cases of AIDS were identified in 1981. It is estimated that 39.5 million adults and children where living with HIV in 2005, of which 24.7 million were in sub-Saharan Africa. In hardest-hit countries, the pandemic has reduced life expectancy, lessening the number of healthy agricultural workers and deepening poverty.

Sources: GEO Data Portal, from UNPD 2007, UNAIDS 2006

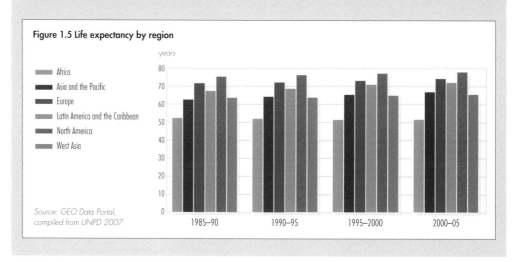

Figure 1.5 Life expectancy by region

- Africa
- Asia and the Pacific
- Europe
- Latin America and the Caribbean
- North America
- West Asia

Source: GEO Data Portal, compiled from UNPD 2007

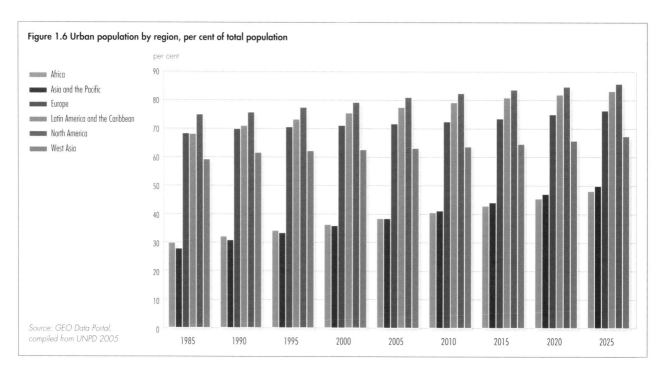

Figure 1.6 Urban population by region, per cent of total population

- Africa
- Asia and the Pacific
- Europe
- Latin America and the Caribbean
- North America
- West Asia

Source: GEO Data Portal, compiled from UNPD 2005

Urbanization continues around the world, particularly in developing countries, where rural migration continues to fuel urban growth (see Figure 1.6). By the end of 2007, more people will be living in cities than in rural areas for the first time in history (UN-HABITAT 2006). In North East Asia and South East Asia, the population living in urban areas increased from 28–29 per cent in 1985 to 44 per cent in 2005, and is projected to reach 59 per cent by 2025 (GEO Data Portal, from UNPD 2005). In some places, the urban area is increasing faster than the urban population, a process known as urban sprawl. For example, between 1970 and 1990, the total area of the 100 largest urban areas in the United States increased by 82 per cent. Only half of this increase was caused by population growth (Kolankiewicz and Beck 2001) (see Box 1.9). A growing number of people living in urban areas are living in slums – inadequate housing with no or few basic services (UN-HABITAT 2006). In many sub-Saharan African cities, children living in slums are more likely to die from water-borne and respiratory illnesses than rural children. For 2005, the number of slum dwellers was estimated at almost 1 billion (UN-HABITAT 2006).

Migration and urbanization have complex relationships with environmental change. Natural disasters, and degradation of land and local ecosystems are among the causes of migration (Matutinovic 2006). Changing demographic patterns, caused by migration or urbanization, alter land use and demand for ecosystem services (see Box 1.9).

Box 1.9 Urban sprawl, Las Vegas

Las Vegas, the fastest growing metropolitan area in the United States, exemplifies the problems of rampant urban sprawl. As the gaming and tourism industry blossomed, so has the city's population. In 1985, Las Vegas was home to 557 000 people, and was the 66th largest metropolitan area in the United States. In 2004, the Las Vegas-Paradise area was ranked 32nd in size, with a permanent population nearing 1.7 million. According to one estimate, it may double by 2015. Population growth has put a strain on water supplies.

Satellite imagery of Las Vegas provides a dramatic illustration of the spatial patterns and rates of change resulting from the city's urban sprawl. The city covers the mainly green and grey areas in the centre of these images recorded in 1973 and 2000. Note the proliferation of roads and other infrastructure (the rectangular pattern of black lines) and the dramatic increase in irrigated areas.

Source: UNEP 2005b

Credit: UNEP 2005b

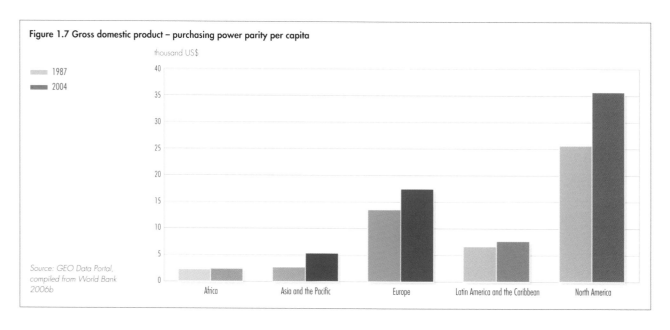

Figure 1.7 Gross domestic product – purchasing power parity per capita

thousand US$

- 1987
- 2004

Source: GEO Data Portal, compiled from World Bank 2006b

Africa | Asia and the Pacific | Europe | Latin America and the Caribbean | North America

---

Urbanization in particular can exert significant pressure on the environment (see Chapter 6). Coastal urban areas often cause offshore water pollution. Coastal populations alone are expected to reach 6 billion by 2025 (Kennish 2002). In these areas, large-scale development results in excessive nutrient inputs from municipal and industrial waste. As described in Chapter 4, eutrophication contributes to the creation of dead zones, areas of water with low or no dissolved oxygen. Fish cannot survive, and aquatic ecosystems are destroyed. Dead zones are an emerging problem in Asia, Africa and South America, but are present around the world. With

population growth, and increasing industrialization and urbanization, dead zones can only continue to expand. Properly managed, cities can also become a solution for some of the environmental pressures. They provide economies of scale, opportunities for sustainable transport and efficient energy options.

Economic growth

Global economic growth has been spectacular during the last two decades. Gross domestic product per capita (at purchasing power parity) increased by almost 1.7 per cent annually, but this growth was unevenly spread (see Figure 1.7). People in Africa, Eastern Europe and Central Asia, and certain areas of Latin America and the Caribbean are worse off than those in North America and Central and Western Europe. Many countries in these regions experienced no growth and some even a clear economic decline between 1987 and 2004. Especially in Africa there are large differences within the region, and even where there is growth, countries are faced with a heavy debt burden (see Box 1.10). Income in Asia and the Pacific is still well below the global average, but its growth rate was twice the global average. These sub-regional differences are highlighted in Chapter 6.

Economic growth and unsustainable consumption patterns represent a growing pressure on the environment, though this pressure is often distributed unequally. Dasgupta (2002) argues that economic growth is unsustainable in poor countries, partly

---

Box 1.10 Debt repayments continue to be a major impediment to growth

Even though Africa has only 5 per cent of the developing world's income, it carries about two-thirds of the Global South's debt burden – over US$300 billion. Despite extreme poverty, sub-Saharan Africa transfers US$14.5 billion a year to rich nations in external debt repayments. The average sub-Saharan African country, therefore, spends three times more on repaying debt than it does on providing basic services to its people. By the end of 2004, Africa spent about 70 per cent of its export earnings on external debt servicing. At the 2005 Gleneagles Summit, G8 countries cancelled 100 per cent of debts of a number of eligible Heavily Indebted Poor Countries to three multilateral institutions – the International Monetary Fund (IMF), International Development Association (IDA) and African Development Fund. This was a step towards relieving the burden that debt repayment places on growth and social services. As a result of debt cancellation and targeted aid increases between 2000 and 2004, 20 million more children in Africa are in school. While G8 countries reaffirmed Gleneagles commitments at the 2007 Heiligendamm Summit, their ability to fulfill these promises has been questioned.

Sources: Christian Reformed Church 2005, DATA 2007, Katerere and Mohamed-Katerere 2005

because it is sustainable in wealthy countries. Countries that export resources are subsidizing the consumption of importing countries (Dasgupta 2002). However, consumption patterns among regions are changing with the emergence of new economies and powers such as China, India, Brazil, South Africa and Mexico. China, for example, is expected to become the world's largest economy between 2025 and 2035. Its rapid economic development is influencing global patterns of resource production and consumption, with both environmental and geopolitical consequences (Grumbine 2007). Vehicle ownership patterns illustrate the impact of changing consumption patterns (see Chapter 2). China had some 27.5 million passenger vehicles and 79 million motorcycles in use by 2004 (CSB 1987–2004). The growing trend in vehicle ownership affects urban air quality, which has clear consequences for human health.

Globalization

The world's economy has been characterized by growing globalization, which is spurring the increasing integration of the global economy through trade and financial flows, and in the integration of knowledge through the transfer of information, culture and technology (Najam and others 2007). Governance has also become globalized, with increasingly complex interstate interactions, and with a growing role for non-state actors. International companies have become influential economic actors in a global governance context traditionally dominated by nations. While states "rule the world," corporations have publicly sought the global political stage at gatherings such as the World Economic Forum and at multilateral negotiations, such as the Multilateral Agreement on Investment (De Grauwe and Camerman 2003, Graham 2000). Advances in technology and communications, such as the Internet, have also boosted the role of individuals and organizations as key players in a globalized world (Friedman 2005).

Globalization raises both fears and expectations. Some suggest that increasing interdependence is good for cooperation, peace and solving common problems (Bhagwati 2004, Birdsall and Lawrence 1999, Russett and Oneal 2001). Economic integration may offer dynamic benefits, such as higher productivity. The exchange of goods-and-services also helps the exchange of ideas and knowledge. A relatively open economy is better able to learn and adopt foreign, state-of-the-art technologies than is a relatively closed economy (Coe and Helpman 1995, Keller 2002). Others, however, view growing economic interdependence as destabilizing. They say that rapid flows of investment into and out of countries cause job losses, increase inequality, lower wages (Haass and Litan 1998) and result in harm to the environment. It is argued that globalization is exploitative, and is creating a murkier future for global cooperation and justice (Falk 2000, Korten 2001, Mittelman 2000).

The environment and globalization are intrinsically linked. The globalization of trade has facilitated the spread of exotic species, including the five most important freshwater suspension feeding invaders (*Dressena polymorpha, D. bugensis, Corbicula fluminea, C. fluminalis* and *Limoperna fortunei*). The zebra mussel (*Dressena polymorpha*) has spread through North America during the last 20 years, resulting in significant ecological and economic impacts. Its introduction corresponds with dramatic increase in wheat shipments between the US, Canada and the former Soviet Union (Karatayev and others 2007). In a globalized world, important decisions related to environmental protection may have more to do with corporate management and market outcomes than with state-level, political factors. Countries may be reluctant to enforce strict environmental laws, fearing that companies would relocate elsewhere. However, it is often forgotten that the environment itself can have an impact on globalization. Resources fuel global economic growth and trade. Solutions to environmental crises, such as climate change, require coordinated global action and greater globalization of governance (Najam and others 2007).

Trade

World trade has continued to grow over the past 20 years, as a result of lower transport and communication costs, trade liberalization and multilateral trade agreements, such as the North American Free Trade Agreement. Between 1990 and 2003, trade in goods increased from 32.5 to 41.5 per cent of world GDP. Differences exist between regions. In North East Asia, trade in goods increased from 47 to 70.5 per cent of GDP, and high technology exports increased from 16 to 33 per cent of manufactured exports. By contrast, trade in goods in West Asia and Northern Africa only increased from 46.6 per cent to 50.4 per cent of GDP. High technology exports only accounted for 2 per cent of manufactured exports in 2002 (World Bank 2005). Since 1990, least developed countries (LDC) have increased their share of world merchandise trade, but still accounted for only 0.6 per cent of world exports and 0.8 per cent of world imports in 2004 (WTO 2006).

As with globalization, a two-way relationship exists between the environment and trade. Transport has increased as a result of increasing flows of goods and global production networks. Transport is now one of the most dynamic sectors in a modern economy, and has strong environmental impacts (Button and Nijkamp 2004) (see Chapters 2 and 6). Trade itself can exert pressures on the environment. Increases in international

grain prices may increase the profitability of agriculture, and result in the expansion of farming into forested areas in Latin America and the Caribbean, for example (see Box 1.11). The wildlife trade in Mongolia, valued at US$100 million annually, is contributing to the rapid decline of species such as saiga antelope (World Bank and WCS 2006). In the presence of market or intervention failures, international trade may also exacerbate environmental problems indirectly. For example, production subsidies in the fishing sector can promote overfishing (OECD 1994). Natural disasters, in turn, can have an impact on trade at the national level, when exports fall as a result of physical damage. One example of this linkage is the hurricane damage to oil refineries in the Gulf of Mexico in 2005. Oil production in the Gulf of Mexico, which supplies 2 per cent of the world's crude

Box 1.11 Trade, growth and the environment

In recent years, Chile has been considered one of the most economically competitive countries in Latin America and the Caribbean. Rapid growth in Chile's production and export of forest products is based on the expansion and management of exotic species in newly planted forests over the past 30 years. To do so, the traditional land-use practices in small-scale logging of native forests, livestock raising and agricultural cultivation have been

replaced by large-scale timber production. Many endangered tree and shrub species have been affected by this growth of planted forest, which has also led to a dramatic reduction of landscape diversity as well as goods-and-services from forests. The two images, taken in 1975 (left) and 2001 (right), show clear reductions in forested land on the one hand (red arrows), and new forest areas on the other (yellow arrows).

Source: UNEP 2005b

Credit: UNEP 2005b

oil, slowed following Hurricane Katrina, and crude oil prices jumped to over US$70 a barrel (WTO 2006).

Trade may also be positive for the environment. Debate rages over whether or not free trade will raise incomes to a point where environmental protection becomes a priority (Gallagher 2004). At the 2002 WSSD in Johannesburg, commitments were made to expand markets for environmental goods-and-services. Liberalization of trade in goods that protect the environment may help spur the creation of industry dedicated to environmental improvements (OECD 2005). Consumer preferences can influence production standards, which can be used to improve environmental conditions. In 2006, a large grain distributor imposed a moratorium on the purchase of soy produced on deforested areas of the Amazon, as a result of a Greenpeace campaign in Europe (Cargill 2006, Greenpeace 2006).

Energy

The world is facing twin threats: inadequate and insecure supplies of energy at affordable prices, and environmental damage due to overconsumption of energy (IEA 2006a). Global demand for energy keeps growing, placing an ever-increasing burden on natural resources and the environment. For about three decades, world primary energy demand grew by 2.1 per cent annually, rising from 5 566 million tonnes oil equivalent (Mtoe) in 1971 to 11 204 Mtoe in 2004 (IEA 2006b). Over two-thirds of this increase came from developing countries, but OECD countries still account for almost 50 per cent of world energy demand. In 2004, primary energy use per capita in OECD countries was still 10 times higher than in sub-Saharan Africa. Figure 1.8 highlights primary energy supply per capita.

Global increases in carbon dioxide emissions are primarily due to fossil fuel use (IPCC 2007), the fuels that met 82 per cent of the world's energy demand in 2004. Traditional biomass (firewood and dung) remains an important energy source in developing countries, where 2.1 billion people rely on it for heating and cooking (IEA 2002). Use of cleaner energy sources, such as solar and wind power, remains minimal overall (see Figure 5.5, Chapter 5 for energy supply by source). The need to curb growth in energy demand, increase fuel

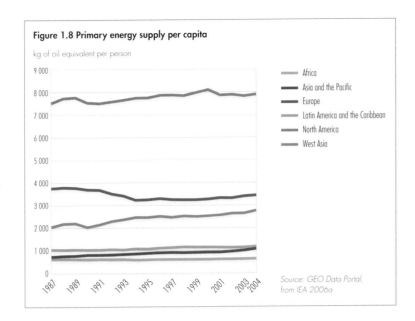

Figure 1.8 Primary energy supply per capita

kg of oil equivalent per person

- Africa
- Asia and the Pacific
- Europe
- Latin America and the Caribbean
- North America
- West Asia

Source: GEO Data Portal, from IEA 2006a

supply diversity and mitigate climate destabilizing emissions is more urgent than ever (IEA 2006a). However, expansion of alternative energy sources, such as biofuels, must also be carefully planned. Brazil expects to double the production of ethanol, a "modern" biofuel, in the next two decades (Government of Brazil 2005). In order to produce enough crops to reach production targets, the cultivated area is increasing rapidly. The growth of farming jeopardizes entire ecoregions, like the Cerrado, one of the world's biodiversity hot spots (Klink and Machado 2005).

Global demand for energy keeps growing, placing an ever-increasing burden on natural resources and the environment.

Credit: Ngoma Photos

Technological innovation

Advances in agriculture, energy, medicine and manufacturing have offered hope for continued human development and a cleaner environment. New farming technologies and practices related to water use, fertilizer and plant breeding have transformed agriculture, increasing food production and addressing undernutrition and chronic famine in some regions. Since 1970, food consumption is increasing in all regions, and is expected to continue to increase as a result of economic development and population growth. Concerns have been raised over the ability to meet future demand: 11 per cent of the world's land is already used for agriculture, and in many places little room exists for agricultural expansion due to land or water shortages. Biotechnology, including genetic modification, as well as nanotechnology, has the potential to increase production in agriculture and contribute to advances human health (UNDP 2004), but remains subject to much controversy over effects on health and the environment. Earlier lessons from new technologies show the importance of applying the precautionary approach (CIEL 1991), because unintended effects of technological advances can lead to the degradation of ecosystem services. For example, eutrophication of freshwater systems and hypoxia in coastal marine ecosystems result from excess application of inorganic fertilizers. Advances in fishing technologies have contributed significantly to the depletion of marine fish stocks.

Communications and cultural patterns have also been revolutionized in the last 20 years, with the exponential growth of the Internet and telecommunications (see Figure 1.9). Worldwide, mobile phone subscribers increased from 2 per 1 000 people in 1990 to 220 per 1 000 in 2003 and worldwide Internet use increased from 1 in 1 000 in 1990 to 114 per 1 000 in 2003 (GEO Data Portal, from ITU 2005). Many developed countries lead the way in the number of Internet users, hosts and secure servers, prompting some to claim that there is a digital divide between different regions of the world. In Australia and New Zealand, for example, only 4 per cent of the population used Internet in 1996, but by 2003, that had risen to 56 per cent of the population By contrast in 2003, in poor countries such as Bangladesh, Burundi, Ethiopia, Myanmar and Tajikistan only 1 or 2 people per 1 000 used the Internet (GEO Data Portal, from ITU 2005).

Governance

The global and regional political context has changed considerably since the Brundtland Commission, with the end of the Cold War triggering renewed optimism in multilateral and

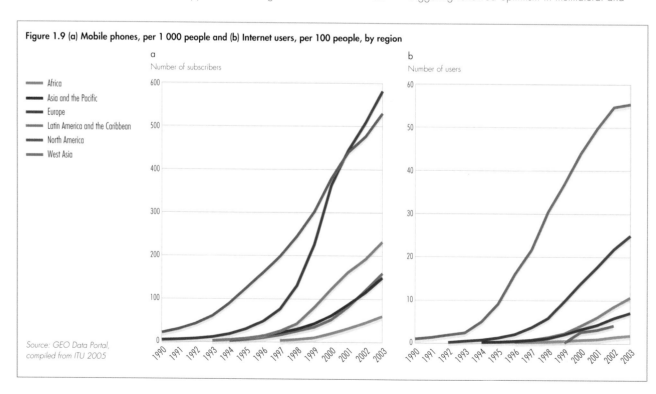

Figure 1.9 (a) Mobile phones, per 1 000 people and (b) Internet users, per 100 people, by region

Africa
Asia and the Pacific
Europe
Latin America and the Caribbean
North America
West Asia

a
Number of subscribers

b
Number of users

Source: GEO Data Portal, compiled from ITU 2005

global governance. The 1990s was a decade of global summits on a diversity of issues, including children (1990), sustainable development (1992), human rights (1994), population (1994), social development (1995), gender equality (1995) and human settlements (1996). The new millennium has been equally active and agenda-setting, starting with the Millennium Summit in 2000, and its follow-up in 2005. Normative declarations and ambitious action plans from all these summits illustrate an emerging unity in how governments and the international community understand complex and global problems and formulate appropriate responses. The establishment of the World Trade Organization in 1994 strengthened global governance through its considerable authority in the areas of trade, while the establishment of the International Criminal Court of Justice in 2002 attempted to do the same for crimes against humanity. Some important reforms have happened within the UN system, including an approach that increasingly uses partnerships (such as the Global Water Partnership) and institutionalized processes to strengthen the participation of civil society (such as the UNEP's Global Civil Society Forum and Global Women's Assembly on Environment).

At the regional level, countries have expanded or established institutions to enhance cooperation, including the European Union (EU), the North American Free Trade Agreement (NAFTA), the Southern Common Market (MERCOSUR), the Association of Southeast Asian Nations (ASEAN) and the African Union (AU). Regions became more visible in global deliberations, through, for example, the emphasis on regional preparation meetings for the World Summit on Sustainable Development.

The national level remains central in governance, despite discussions in the context of globalization and regionalization. Some countries are adopting innovative governance systems and there has been a trend towards both political and fiscal decentralization of governance to sub-national levels. This does not necessarily mean that local authorities have been empowered. It has been argued that decentralization without devolution of power can be a way to strengthen the presence of the central authority (Stohr 2001). Local governments have also engaged much more widely in international cooperation in various arenas,

and their role has been strengthened at the global level through the establishment in 2000 of the UN Advisory Committee of Local Authorities (UNACLA) and the World Urban Forum in 2002, as well as the founding of the United Cities and Local Governments Organization in 2004.

RESPONSES

Interactions between drivers and pressures, and their consequent impacts on ecosystem services and human well-being present challenges that could not be foreseen in 1987. There is an urgent need for effective policy responses at all levels – international, regional, national and local. As highlighted in the other chapters of this report, the range and scope of response options available to policy-makers has progressively evolved over the past 20 years (see Box 1.12), with a diversity of multilateral environmental agreements and institutions now involved in trying to address the challenges. The increase in governance regimes has brought about its own challenges, including competition and overlap. An interlinkages approach is essential to managing the environment, not in its individual parts but more holistically. This approach recognizes that the environment itself is interlinked; land, water and atmosphere are connected in many ways, particularly through the carbon, nitrogen and water cycles. Chapter 8 highlights both the biophysical and governance regimes interlinkages.

Chapter 10 highlights the evolution of policy response measures – from a focus on command-and-

---

Box 1.12 Types of responses

Command-and-control regulation includes standards, bans, permits and quotas, zoning, liability systems, legal redress, and flexible regulation.

Direct provisions by government deal with environmental infrastructure, eco-industrial zones or parks, protected areas and recreation facilities, and ecosystem rehabilitation.

Public and private sector engagement relates to public participation, decentralization, information disclosure, eco-labelling, voluntary agreements and public-private partnerships.

Market use includes environmental taxes and charges, user charges, deposit-refund systems, targeted subsidies, and the removal of perverse subsidies.

Market creation addresses issues of property rights, tradeable permits and rights, offset programmes, green programmes, environmental investment funds, seed funds and incentives.

control policies to creating markets and incentives, particularly for industry to implement voluntary measures aimed at minimizing environmental damage. For conventional, well-known environmental problems with proven solutions, it is necessary to continue to apply, and to further improve upon previously successful approaches. Countries that have yet to address such problems should apply these proven, workable solutions to current problems. Previously successful approaches have generally addressed changes to pressures, for example by regulating emission levels, land use or resource extraction. In order to address less-known persistent (or emerging) problems, transformative policies are needed. These policies address the drivers of environmental problems, such as demographic change and consumption patterns. Adaptive management is essential, to enable policy-makers to learn from previous experience as well as to make use of a variety of new tools that may be needed.

Economic instruments

Today, greater emphasis is being placed on the potential use of economic instruments to help correct market failures. These instruments were promoted by Principle 16 of the Rio Declaration: "National authorities should endeavour to promote the internalization of environmental costs and the use of economic instruments."

Natural resources can be seen as a capital asset belonging to a general portfolio, which is comprised of other assets and capitals, including material, financial, human and social. Managing this portfolio in a good and sustainable manner to maximize its returns and benefits over time is good investment. It is also central to sustainable development.

A variety of economic instruments exist, including property rights, market creation, fiscal instruments, charge systems, financial instruments, liability systems, and bonds and deposits. There is a mix of so-called market-based instruments (MBIs) and command-and-control instruments to enable policy-makers to better manage and get more accurate information regarding the portfolio of capital assets. Table 1.2 summarizes different economic instruments, and how they can be applied to different environmental sectors. One of the tools is

valuation, which can be used to help better assess the value of ecosystem services, and the costs of human-induced changes to the environment.

Valuation

Environmental ministries and agencies are often the last to benefit from investments, because economics and growth generation take precedence in government spending decisions. This is often due to lack of information on the value and carrying limits of the Earth's ecosystems. Measurement of economic development and progress has often been linked to measures of economic output such as Gross National Product (GNP). Such aggregate measurements do not consider the depletion of natural capital caused by the consumption and production of goods-and-services. National accounting systems need revision to better include the value of the changes in the environmental resource base due to human activities (Mäler 1974, Dasgupta and Mäler 1999).

Valuing different goods-and-services involves comparisons across different sets of things. How these things are accounted for, and how the services provided by the ecosystems, for example, improve well-being is called the accounting price. Table 1.3 illustrates different approaches to valuation, and how these approaches might be used to help assess the impact of policies on environmental change and human well-being.

A "set of institutions capable of managing the natural resources, legal frameworks, collecting resource rents, redirecting these rents into profitable investments" is key to effective use of valuation (World Bank 2006a). Valuing natural resources and evaluating policies where institutions such as markets do not exist, and where there is a lack of individual property rights, pose challenges. Under such uncertainties, and where divergent sets of values exist, the economic value of common resources can be measured by the maximum amount of other goods-and-services that individuals are willing to give up to obtain a given good or service. Therefore, it is possible to weigh the benefits from an activity such as the construction of a dam against its negative impacts on fishing, livelihoods of nearby communities, and changes to scenic and aesthetic values. Box 1.13 provides an example of non-market valuation using the contingent valuation method (CVM).

Table 1.2 Economic instruments and applications

	Property rights	Market creation	Fiscal instruments	Charge systems	Financial instruments	Liability systems	Bonds and deposits
Forests	Communal rights	Concession building	Taxes and royalties		Reforestation incentives	Natural resource liability	Reforestation bonds, forest management bonds
Water resources	Water rights	Water shares	Capital gains tax	Water pricing Water protection charges			
Oceans and seas		Fishing rights, Individual transferable quotas Licensing					Oil spill bonds
Minerals	Mining rights		Taxes and royalties				Land reclamation bonds
Wildlife		Access fees				Natural resource liability	
Biodiversity	Patents Prospecting rights	Transferable development rights		Charges for scientific tourism		Natural resource liability	
Water pollution		Tradeable effluent permits	Effluent charges	Water treatment fees	Low-interest loans		
Land and soils	Land rights, use rights		Property taxes, land-use taxes		Soil conservation incentives (such as loans)		Land reclamation bonds
Air pollution		Tradeable emission permits	Emission charges	Technology subsidies, low-interest loans			
Hazardous waste				Collection charges			Deposit refund systems
Solid waste			Property taxes	Technology subsidies, low-interest loans			
Toxic chemicals			Differential taxation			Legal liability, liability insurance	Deposit refund
Climate	Tradeable emission entitlements Tradeable forest protection obligations	Tradeable CO_2 permits Tradeable CFC quotas CFC quota auction Carbon offsets	Carbon taxes BTU tax		CFC replacement incentives Forest compacts		
Human settlements	Land rights	Access fees Tradeable development quotas Transferable development rights	Property taxes, land-use taxes	Betterment charges Development charges Land-use charges Road tolls Import fees			Development completion bonds

Source: Adapted from Panayotou 1994

Table 1.3 Purpose and application of different valuation approaches

Approach	Why it is done	How it is done
Determining the total value of the current flow of benefits from an ecosystem.	To understand the contribution that ecosystems make to society and to human well-being.	Identify all mutually compatible services provided. Measure the quantity of each service provided, and multiply by the value of each service.
Determining the net benefits of an intervention that alters ecosystem conditions.	To assess whether the intervention is worthwhile.	Measure how the quantity of each service would change as a result of the intervention, as compared to their quantity without the intervention. Multiply by the marginal value of each service.
Examining how the costs and benefits of an ecosystem (or an intervention) are distributed.	To identify winners and losers, for ethical and practical reasons.	Identify relevant stakeholder groups. Determine which specific services they use, and the value of those services to that group (or changes in values resulting from an intervention).
Identifying potential financing sources for conservation.	To help make ecosystem conservation financially self-sustaining.	Identify groups that receive large benefit flows from which funds could be extracted, using various mechanisms.

Source: Adapted from Stephano 2004

Valuation presents a set of challenges beyond conflicting value systems or lack of existing market institutions. It uses notional and proxy measures to estimate the economic values of tangible and intangible services provided by the environment. An increasing body of valuation work has been undertaken on provisioning services of ecosystems. It has produced estimates of the value of non-timber forest products, forestry, and the health impacts of air pollution and water-borne diseases. However, studies on less tangible but yet important services, such as water purification and the prevention of natural disasters, as well as recreational, aesthetic and cultural services, have been hard to get. To get objective monetary estimates of these services remains a challenge. Market data is limited to a small number of services provided by ecosystems. Furthermore, methodologies such as cost-benefit analysis and CVM may raise problems of bias.

The use of market and non-market-based instruments has also shown gaps in addressing distributional and intergenerational equity issues (MA 2005b), notably with regard to poverty-related issues. Finally, many valuation studies estimating the impact of policies or projects on human well-being fail due to the lack of sufficiently precise estimates of the consequences of these policies or projects now and in the future. Despite these flaws, valuation may be a useful tool with which to examine the complex relationships and feedback involving the environment, economic growth and human well-being.

Box 1.13 Valuing the removal of the Elwha and Glines Dams

An environmental impact analysis using CVM was conducted in the 1990s to explore the removal of the Elwha and Glines dams in Washington State in the United States. These two 30- and 60-metre-high dams, respectively, are old, and block the migration of fish to 110 km of pristine water located in the Olympic National Park. The dams also harm the Lower Elwha Klallam Tribe which relies on the salmon and river for their physical, spiritual and cultural well-being. Dam removal could bring substantial fishing benefits, more than tripling the salmon populations. The cost of removing the dams, and especially the sediment build-up is estimated at about US$100–$125 million. Recreational and commercial fishing benefits resulting from dam removal would not be sufficient to cover these costs.

A CVM survey was conducted and yielded a 68 per cent response in Washington State, and 55 per cent response for the rest of the United States. Willingness to pay for dam removal ranged from US$73 per household for Washington to US$68 for the rest of the United States. If every household in Washington State were to pay US$73, the cost of dam removal and river restoration could be covered. If the return stemming from Washington residents' willingness to pay was added to the rest of the US willingness to pay (the 86 million households and their willingness to pay an average of US$68 per head) in excess of US$1 billion dollars would result.

After years of negotiations it has been decided that the dams will be removed, and the Elwha Restoration Project will go forward. This is the biggest dam-removal project in history, and an event of national significance in the United States. It is expected that the two dams will be removed in stages over the course of three years, between 2009 and 2011.

Source: American Rivers 2006, Loomis 1997, USGS 2006

Non-economic instruments

In addition to economic instruments, a variety of non-economic instruments have been employed to address both well-known proven and less clear emerging (or persistent) environmental problems. Today, the emerging understanding of human well-being increasingly influences our choice of instruments.

Public participation

Human well-being depends on the unconstrained ability of people to participate in decisions, so that they can organize society in a way that is consistent with their highest values and aspirations. In other words, public participation is not only a matter of procedural justice, but also a precondition for achieving well-being. While this is challenging, managers should involve civil society in policy interventions. The Convention on Biological Diversity offers several examples of possible stakeholder engagement in decision making. These include CBD VII/12, The Addis Ababa Guidelines on the sustainable use of the components of biodiversity; CBD VII/14 guidelines on sustainable tourism development; and the CBD VII/16 Akwe, on voluntary guidelines for the conduct of cultural, environmental and social impact assessments for development proposals on sacred sites, lands and waters traditionally occupied or used by indigenous and local communities. The development of similar agreements and protocols that enhance effective engagement of all sectors of society should be encouraged.

Education

Access to information and education is a basic human right, and an important aspect of human well-being. It is also an important tool for generating knowledge that links ecological analyses to societal challenges, and is critical to the decision making process. Women and marginalized communities must be ensured access to education. The United Nations launched its Decade of Education for Sustainable Development (DESD) in 2005 and designated UNESCO as lead agency for the promotion of the Decade (see Box 1.14).

Justice and ethics

Since the environment affects the very basis of human well-being, it is a matter of justice to consider the impacts of environmental degradation on others, and attempt to minimize harm for both current and future

generations. It has been argued that a "global ethic" is required to address the problems of the 21st century (Singer 2002). The intrinsic value of species has also been recognized (IUCN and others 1991). The pursuit of some people's opportunities and freedoms may harm or limit those of others. It is important that policy-makers consider the adverse effects their decisions have on people and the environment in other areas or regions, since such communities do not participate in local decision making.

Scenario development

The use of scenarios to inform policy processes is growing, providing policy-makers with opportunities to explore the likely impacts and outcomes of various policy decisions. The goal of developing scenarios "is often to support more informed and rational decision making that takes both the known and unknown into account" (MA 2005c). Their purpose is to widen perspectives and illuminate key issues that might

---

Box 1.14 The UN Decade of Education for Sustainable Development

The overall goal of the DESD is "to integrate the principles, values, and practices of sustainable development into all aspects of education and learning."

This educational effort will encourage changes in behaviour that will create a more sustainable future in terms of environmental integrity, economic viability, and a just society for present and future generations.

In the long-term, education must contribute to government capacity building, so that scientific expertise can inform policy.

Source: UNESCO 2007

otherwise be missed or dismissed. Chapter 9 uses four plausible scenarios to explore the impact of different policy decisions on environmental change and future human well-being.

CONCLUSION

Two decades after *Our Common Future* emphasized the urgency of sustainable development, environmental degradation continues to threaten human well-being, endangering health, physical security, social cohesion and the ability to meet material needs. Analyses throughout *GEO-4* also highlight rapidly disappearing forests, deteriorating landscapes, polluted waters and urban sprawl. The objective is not to present a dark and gloomy scenario, but an urgent call for action.

While progress towards sustainable development has been made through meetings, agreements and changes in environmental governance, real change has been slow. Since 1987, changes to drivers, such as population growth, consumption patterns and energy use, have placed increasing pressure on the state of the environment. To effectively address environmental problems, policy-makers should design policies that tackle both pressures and the drivers behind them. Economic instruments such as market creation and charge systems may be used to help spur environmentally sustainable behaviour. Valuation can help policy-makers make informed decisions about the value of changes to ecosystem services. Non-economic instruments should be used to address both well-known problems with proven solutions and less

clear emerging problems. This chapter has provided an overview of the challenges of the 21st century, highlighted conceptual ideas that have emerged to analyse and understand these environmental problems, and indicated options on the way forward.

The following chapters highlight areas where society has contributed to environmental degradation and human vulnerability. Everyone depends on the environment. It is the foundation of all development, and provides opportunities for people and society as a whole to achieve their hopes and aspirations. Current environmental degradation undermines natural assets, and negatively affects human well-being. It is clear that a deteriorating environment is an injustice to both current and future generations.

The chapters also emphasize that alternative development paths that protect the environment are available. Human ingenuity, resilience and capacity to adapt are powerful forces from which to draw to effect change.

Imagine a world in which human well-being for all is secure. Every individual has access to clean air and water, ensuring improvements in global health. Global warming has been addressed, through reductions in energy use, and investment in clean technology. Assistance is offered to vulnerable communities. Species flourish as ecosystem integrity is assured. Transforming these images into reality is possible, and it is this generation's responsibility to start doing so.

References

Agarwal, B. (2000). Conceptualizing Environmental Collective Action: Why Gender Matters. In *Cambridge Journal of Economics* 24(3):283-310

AMAP (2002). *Persistent Organic Polluatants, Heavy Metals, Radioactivity, Human Health, Changing Pathways.* Arctic monitoring and Assessment Programme, Oslo

American Rivers (2006). *Elwha River Restoration.* http://www.americanrivers.org/site/PageServer?pagename=AMR_elwharestoration (last accessed 12 June 2007)

Bass, S. (2006). *Making poverty reduction irreversible: development implications of the Millennium Ecosystem Assessment.* IIED Environment for the MDGs' Briefing Paper. International Institute on Environment and Development, London

Bell, D., Robertson, S. and Hunter, P. (2004). Animal origins of SARS coronavirus: possible links with the international trade of small carnivores. In *Philosophical Transactions of the Royal Society London* 359:1107-1114

Bhagwati, J. (2004). *In Defense of Globalization.* Oxford University Press, Oxford

Birdsall, N. and Lawrence, R. (1999). Deep Integration and Trade Agreements: Good for Developing Countries? In Grunberg, K. and Stern, M (eds.). In *Global Public Goods: International Cooperation in the 21st Century.* Oxford University Press, New York, NY

Braidotti, R., Charkiewicz, E., Hausler, S. and Wieringa, S. (1994). *Women, the Environment and Sustainable Development.* Zed, London

Brown, D. (1999). Making CSD Work. In Earth Negotiations Bulletin 3(2):2-6

Brown, S. (2006). The west develops a taste for bushmeat. In *New Scientist* 2559:8

Button, K. and Nijkamp, P. (2004). Introduction: Challenges in conducting transatlantic work on sustainable transport and the STELLA/STAR Initiative. In *Transport Reviews* 24 (6):635-643

Campbell, B., Jeffrey, S., Kozanayi, W., Luckert, M., Mutamba, M. and Zindi, C. (2002). *Household livelihoods in semi-arid regions: options and constraints.* Center for International Forestry Research, Bogor

Castles, S. (2002). *Environmental change and forced migration: making sense of the debate.* New Issues in Refugee Research, Working Paper No. 70. United Nations High Commission for Refugees, Geneva

Cargill (2006). *Brazilian Soy Industry Announces Initiative Designed To Curb Soy-Related Deforestation in the Amazon.* http://www.cargill.com/news/issues/issues_soyannouncement_en.htm (last accessed 11 June 2007)

Carothers, T. and Barndt, W. (2000). Civil Society. In *Foreign Policy* (11):18-29

China Statistical Bureau (1987-2004). *China Statistical Yearbook (1987-2004).* China 28 Statistics Press (in Chinese), Beijing

CIEL (1991). *The Precautionary Principle: A Policy for Action in the Face of Uncertainty.* King's College, London

Cleaver, F. (2000). Analysing Gender Roles in Community Natural Resource Management: Negotiation, Life Courses, and Social Inclusion. In *IDS Bulletin* 31(2):60-67

Coe, D. T. and Helpman, E. (1995). *International R&D Spillovers.* NBER Working Papers 4444. National Bureau of Economic Research, Inc, Cambridge, MA

Christian Reformed Church (2005). *Global Debt. An OSJHA Fact Sheet.* Office of Social Justice and Hunger Action http://www.crcna.org/site_uploads/uploads/factsheet_globaldebt.doc (last accessed 21 April 2007)

CHS (2006). Outline of the Report of the Commission on Human Security. Commission on Human Security http://www.humansecurity-chs.org/finalreport/Outlines/outline.pdf (last accessed 1 May 2007)

Conca, K. and Dabelko, G. (2002). *Environmental Peacemaking.* Woodrow Wilson Center Press, Washington, DC

Dabelko, D., Lonergan, S. and Matthew, R. (2000). *State of the Art Review of Environmental Security and Co-operation.* Organisation for Economic Cooperation and Development, Paris

Dasgupta, P. (2002). Is contemporary economic development sustainable? In *Ambio* 31(4):269-271

Dasgupta, P. and Mäler, K.G. (1999). Net National Product, Wealth, and Social Well-Being. In *Environment and Development Economics* 5:69-93

DATA (2007). *The DATA Report 2007: Keep the G8 Promise to Africa.* Debt AIDS Trade Africa, London

De Grauwe, P. and Camerman, F. (2003). Are Multinationals Really Bigger Than Nations? In *World Economics* 4 (2):23-37

Delgado, C., Wada, N., Rosegrant, M., Meijer, S. and Ahmed, M. (2003). Outlook for fish to 2020. In *Meeting Global Demand. A 2020 Vision for Food, Agriculture, and the Environment Initiative,* International Food Policy Research Institute, Washington, DC

Dernbach, J. (2002). *Stumbling Toward Sustainability.* Environmental Law Institute, Washington, DC

DeSombre, E.R. and Barkin, S. (2002). Turbot and Tempers in the North Atlantic. In Matthew, R. Halle, M. and Switzer, J (eds.). In *Conserving the Peace: Resources, Livelihoods, and Security* 325-360. International Institute for Sustainable Development and The World Conservation Union, Winnipeg, MB

Diamond, J. (2005). *Collapse: How Societies Choose to Fail or Survive.* Penguin Books, London

Diener, E. (2000). Subjective well-being. The science of happiness and a proposal for a national index. In *The American Psychologist* 55:34-43

Dodds, F. and Pippard, T. (eds.) (2005). *Human and Environmental Security: An agenda for change.* Earthscan, London

Donaldson, S. (2002). Re-thinking the mercury contamination issue in Arctic Canada. M.A. Thesis (Unpublished). Carleton University, Ottawa, ON

Doubleday, N. (1996). "Commons" concerns in search of uncommon solutions: Arctic contaminants, catalyst of change? In *The Science of the Total Environment* 186:169-179

Doubleday, N. (2005). Sustaining Arctic visions, values and ecosystems: writing Inuit identity, reading Inuit Art. In Williams, M. and Humphrys, G. (eds.). *Cape Dorset, Nunavut' in Presenting and Representing Environments: Cross-Cultural and Cross-Disciplinary Perspectives.* Springer, Dordrecht

EM-DAT (undated). *Emergency Events Database: The OFDA/CRED International Disaster Database* (in GEO Data Portal). Université Catholique de Louvain, Brussels

Fa, J., Albrechtsen, L. and Brown. D. (2007). Bushmeat: the challenge of balancing human and wildlife needs in African moist tropical forests. In Macdonald, D. and Service, K. (eds.) *Key Topics in Conservation Biology* 206-221. Blackwell Publishing, Oxford

Falk, R. (2000). *Human rights horizons: the pursuit of justice in a globalizing world.* Routledge, New York, NY

FAO (2004a). *The State of Food and Agriculture 2003-2004: Agriculture Biotechnology-Meeting the Needs of the Poor?* Food and Agriculture Organization of the United Nations, Rome http://www.fao.org/WAICENT/FAOINFO/ECONOMIC/ESA/en/pubs_sofa.htm (last accessed 11 June 2007)

FAO (2004b). *The State of the World's Fisheries and Aquaculture 2004.* Food and Agriculture Organization of the United Nations, Rome

Fischetti, M. (2005). Protecting against the next Katrina: Wetlands mitigate flooding, but are they too damaged in the gulf? In *Scientific American* October 24

Frey, B and Stutzer, A. (2005). Beyond Outcomes: Measuring Procedual Utility. In *Oxford Economic Papers* 57(1):90-111

Friedman, T. (2005). *The World is Flat: A Brief History of the Twenty-First Century.* Farrar, Straus, and Giroux, New York, NY

Gallagher, K. (2004). *Free Trade and the Environment: Mexico, NAFTA and Beyond.* Stanford University Press, Stanford

GEF (2006). *What is the GEF?* The Global Environment Facility, Washington, DC http://www.gefweb.org/What_is_the_GEF/what_is_the_gef.html (last accessed 1 May 2007)

GEO Data Portal. *UNEP's online core database with national, sub-regional, regional and global statistics and maps, covering environmental and socio-economic data and indicators.* United Nations Environment Programme, Geneva http://www.unep.org/geo/data or http://geodata.grid.unep.ch (last accessed 12 June 2007)

Goodall, J. (2005). Introduction. In Reynolds, V. (ed.). *The Chimpanzees of the Budongo Forest.* Oxford University Press, Oxford

Gore, A. (2006). *An Inconvenient Truth: the planetary emergency of global warming and what we can do about it.* Bloomsbury, London

Graham, E. (2000). *Fighting the Wrong Enemy: Antiglobal Activists and Multinational Enterprises.* Institute of International Economics, Washington, DC

Greenpeace (2006). *The future of the Amazon hangs in the balance.* http://www.greenpeace.org/usa/news/mcvictory (last accessed 11 June 2007)

Grumbine, R. (2007). China's emergence and the prospects for global sustainability. In *BioScience* 57 (3):249-255

Haass, R., and Litan, R. (1998). Globalization and Its Discontents: Navigating the Dangers of a Tangled World. In *Foreign Affairs* 77(3):2-6

Hecker, J.H. (2005). *Promoting Environmental Security and Poverty Alleviation in the Peat Swamps of Central Kalimantan, Indonesia.* Institute of Environmental Security, The Hague

IEA (2002). *World Energy Outlook 2003.* International Energy Agency, Paris

IEA (2006a). *World Energy Outlook 2006.* International Energy Agency, Paris

IEA (2006b). *Key Energy Statistics.* International Energy Agency, Paris

IFAD (2001). *Rural Poverty Report 2001. The Challenge of Ending Rural Poverty.* International Fund for Agricultural Development, Rome http://www.ifad.org/poverty/index.htm (last accessed 1 May 2007)

IFPRI (2004). *Ending Hunger in Africa: Prospects for the Small Farmer.* International Food Policy Research Institute, Washington, DC http://www.ifpri.org/pubs/ib/ib16.pdf (last accessed 1 May 2007)

IPCC (2001). Technical Summary, *Climate Change 2001: Impacts, Adaptation and Vulnerability.* Contribution of Working Group II to the Third Assessment Report of the Intergovernmental Panel on Climate Change. Intergovernmental Panel on Climate Change. Cambridge University Press, New York, NY

IPCC (2007). *Climate change 2007: The Physical Science Basis. Summary for Policymakers.* Contribution of Working Group 1 to the Fourth Assessment Report of the Intergovernmental Panel on Climate Change, Geneva

ITU (2005). *ITU Yearbook of Statistics.* International Telecommunication Union (in GEO Data Portal).

IUCN, UNEP and WWF (1991). *Caring for the Earth: A Strategy for Sustainable Living.* The World Conservation Union, United Nations Environment Programme and World Wide Fund for Nature, Gland

Jänicke, M. and Volkery, A. (2001). Persistente Probleme des Umweltschutzes. In *Natur und Kultur* 2(2001):45-59

Karatayev, A., Padilla, D., Minchin, D., Boltovskoy, D. and Burlakova, L. (2007). Changes in global economies and trade: the potential spread of exotic freshwater bivalves. In *Bio Invasions* 9:161-180

Karesh, W., Cook, R., Bennett, E. and Newcomb, J. (2005). Wildlife Trade and Global Disease Emergence. In *Emerging Infectious Diseases* 11 (7):1000-1002

Katerere, Y. and Hill, R. (2002). Colonialism and inequality in Zimbabwe. In Matthew, R., Halle, M. and Switzer, J. (eds.). *Conserving the Peace: Resources, Livelihoods, and Security* 247-71 International Institute for Sustainable Development and The World Conservation Union, Winnipeg and Gland

Katerere, Y. and Mohamed-Katerere, J. (2005). From Poverty to Prosperity: Harnessing the Wealth of Africa's Forests. In Mery, G., Alfaro, R., Kanninen, M. and Lobovikov, M. (eds.). *Forests in the Global Balance – Changing Paradigms.* IUFRO World Series Vol. 17. International Union of Forest Research Organizations, Helsinki

Keller, W. (2002). Trade and the Transmission of Technology. In *Journal of Economic Growth* 7:5-24

Kennish, M. (2002). Environmental Threats and Environmental Future of Estuaries. In *Environmental Conservation* 29 (1):78 – 107

Kerr J., Pangare G., and Pangare V. (2002). Watershed development projects in India: An evaluation. In *Research Report of the International Food Policy Research Institute* 127:1-90

Klink, C. and Machado, R. (2005). Conservation of the Brazilian Cerrado. In *Conservation Biology* 19 (3):707-713

Kolankiewicz, L. and Beck, R. (2001). Weighing Sprawl Factors in Large U.S. Cities, Analysis of U.S. Bureau of the Census Data on the 100 Largest Urbanized Areas of the United States. http://www.sprawlcity.org (last accessed 1 May 2007)

Korten, D. (2001). *When Corporations Rule the World, 2nd edition.* Kumarian Press, Bloomfield

Kura,Y., Revenga, C., Hoshino, E. and Mock, G. (2004). *Fishing for Answers: Making Sense of the Global Fish Crisis.* World Resources Institute, Washington, DC

Langhelle, O. (1999). Sustainable development: exploring the ethics of Our Common Future. In *International Political Science Review* 20 (2):129-149

Le Billion, P. (2001). The political Ecology of war: natural resources and armed conflict. In *Political Geography* 20:561-584

LeRoy, E., Rouquet, P., Formenty, P., Souquière, S., Kilbourne, A., Froment, J., Bermejo, M., Smit, S., Karesh, W., Swanepoel, R., Zaki, S. and Rollin, P. (2004). Multiple Ebola virus transmission events and rapid decline of central African wildlife. In *Science* 303:387-390

Li, W., Shi, Z., Yu, M., Ren, W., Smith, C., Epstein, J. Wang, H. Crameri, G., Hu., Z., Zhang, H., Zhang, J., McEachern, J., Field, H., Daszak, P., Eaton, B., Zhang, S. and Wang, L. (2005). Bats are natural reservoirs of SARS-like coronavirsues. In *Science* 310:676-679

Loomis, J. (1997). Use of Non-Market Valuations Studies. Water Resources Management Assessments. In *Water Resources Update* 109:5-9

MA (2003). *Ecosystems and Human Well-being; a framework for assessment.* Millennium Ecosystem Assessment. Island Press, Washington, DC

MA (2005a). *Ecosystems and Human well-being: Biodiversity Synthesis.* Millennium Ecosystem Assessment. World Resources Institute. Island Press, Washington, DC

MA (2005b). *Ecosystems and Human Well-Being. Synthesis Report.* Millennium Ecosystem Assessment. Island Press, Washington, DC

MA (2005c). *Ecosystems and Human Well-being: Volume 2 – Scenarios.* Millennium Ecosystem Assessment. Island Press, Washington, DC

MacDonald, D. and Service, K (2007). *Key Topics in Conservation Biology.* Blackwell Publications, Oxford

Mäler, K-G. (1974). *Environmental Economics: A Theoretical Enquiry.* John Hopkins University Press, Baltimore, MB

Maltais, A., Dow, K. and Persson, A. (2003). *Integrating Perspectives on Environmental Security.* SEI Risk and Vulnerability Programme, Report 2003-1. Stockholm Environment Institute, Stockholm

Matthews, D. (1995). Common versus open access. The collapse of Canada's east coast fishery. In *The Ecologist* 25:86-96

Matthew, R., Halle, M. and Switzer, J. (eds.) (2002). *Conserving the Peace: Resources, Livelihoods and Security.* International Institute for Sustainable Development, Winnipeg, MB

Matutinovic, I. (2006). Mass migrations, income inequality and ecosystem health in the second wave of globalization. In *Ecological Economics* 59:199 – 203

McMichael, A. (2001). Human culture, ecological change and infectious disease: are we experiencing history's fourth great transition? In *Ecosystem Health* (7):107-115

McMichael, A. (2004). Environmental and social influences on emerging infectious disease: past, present and future. *Philosophical Transactions of the Royal Society of London Biology* 10:1-10

Meredith, M. (2005).*The State of Africa: A history of fifty years of independence.* Free Press, London

Government of Brazil (2005). *Diretrizes de Política de Agroenergia 2006–2011.* Ministério da Agricultura, Pecuária e Abastecimento, Ministério da Ciência e Tecnologia, Ministério de Minas e Energia, Ministério do Desenvolvimento, Indústria e Comércio Exterior, Brasilia

Mittelman, J. (2000). *Capturing Globalization.* Carfax, Abingdon

Myers, N. (1997). Environmental Refugees. In *Population and Environment* 19(2):167-82

Najam, A., Runnalls, D. and Halle, M. (2007). *Environment and Globalization: Five Propositions.* International Institute for Sustainable Development, Winnipeg

Nishikiori, N., Abe, T., Costa, D., Dharmaratne, S., Kunii, O. and Moji, K. (2006). Who died as a result of the tsunami? Risk factors of mortality among internally displaced persons in Sri Lanka: a retrospective cohort analysis. In *BMC Public Health* 6:73

OECD (2005). *Trade that Benefits the Environment and Development: Opening Markets for Environmental Goods and Services.* Organisation for Economic Co-operation and Development, Paris

OECD (1994). *The Environmental Effects of Trade.* Organisation for Economic Co-operation and Development, Paris

Oxfam (2005). *The Tsunami's Impact on Women.* Oxfam Briefing Note. http://www.oxfam.org.uk/what_we_do/issues/conflict_disasters/bn_tsunami_women.htm (last accessed 11 June 2007)

Panayotou, T. (1994). *Economic Instruments for environmental Management and Sustainable Development.* Environmental Economics series Paper No.1, United Nations Environment Programme, Nairobi

Peiris, J., Guan, Y. and Yuen, K. (2004). Severe acute respiratory syndrome. In *Nature Medicine* 10 (12):S88- S97

Pelling, M. and Uitto, J. (2001). Small island developing states: natural disaster vulnerability and global change. In *Environmental Hazards* 3:49-62

Peterson, D. (2003). *Eating Apes.* University of California Press, London

Prakash, A. (2000). Responsible Care: An Assessment. In *Business and Society* 39(2):183-209

Rahnema, M. (Ed.) (1997). *The Post-Development Reader.* Zed Books, London

Russett, B. and Oneal, J. (2001). *Triangulating Peace: Democracy, Interdependence, and International Organizations,* The Norton Series in World Politics. W. W. Norton and Company, London

Sen, A. (1985). *Commodities and Capabilities,* Oxford University Press, Oxford

Sen, A. (1992). *Inequality Re-examined.* Clarendon Press, Oxford

Sen, A. (1999). *Development as Freedom.* Oxford University Press, Oxford

Shiva, V. (1991). *The Violence of the Green Revolution: Third World Agriculture, Ecology and Politics.* Zed Books, London

Singer, P. (2002). *One World.* Yale University Press, London

Smith, K. (2006). Oil from bombed plant left to spill. In *Nature* 442:609

Solow, R. M. (1991), Sustainability: An Economist's Perspective. The Eighteen J. Seward Johnson Lecture to the Marine Policy Center, Woods Hole Oceanographic Insitution. In *Economics of the Environment: Selected Readings* (ed. R. Dorfman and N.s Dorfman) 179-187. Norton, New York, NY

D'Souza, M. and Lobo, C. (2004). Watershed Development, Water Management and the Millennium Development Goals. *Paper presented at the Watershed Summit, Chandigarh, November 25-27, 2004.* Watershed Organization Trust, Ahmednagar

Spencer, D. (2001). Will They Survive? Prospects for Small farmers in sub-Saharan Africa. *Paper Presented in Vision 2020: Sustainable food Security for All by 2020.* International Conference Organized by the International Food Policy Research Institute (IFPRI), September 4-6, 2001, Bonn

Speth, J. (2004). *Red Sky at Morning: America and the Crisis of the Global Environment.* Yale University Press, New Haven and London

Stefano, P., Von Ritter, K. and Bishop, J. (2004). *Assessing the Economic Value of Ecosystem Conservation.* Environment Development Paper No.101. The World Bank, Washington, DC

Stohr, W. (2001). Introduction. In *New Regional Development Paradigms: Decentralization, Governance and the New Planning for Local-Level Development.* (ed. Stohr, W., Edralin, J. and Mani, D). Contributions in Economic History Series (225). Published in cooperation with the United Nations and the United Nations Centre for Regional Development. Greenwood Press, Westport, CT

UN (2000). *United Nations Millennium Declaration.* United Nations, New York, NY http://www.un.org/millennium/declaration/ares552e.htm (last accessed 1 May 2007)

UN (2002). *Report of the World Summit on Sustainable Development.* Johannesburg, South Africa, 26 August - 4 September. A/CONF.199/20. United Nations, New York, NY

UN(2004). *Human Rights and Poverty Reduction. A conceptual framework.* United Nations Office of the High Commissioner for Human Rights. United Nations, New York and Geneva

UN (2006). *Trends in Total Migrant Stock: The 2005 Revision.* Population Division of the Department of Economic and Social Affairs of the United Nations Secretariat, New York, NY http://www.un.org/esa/population/publications/migration/UN_Migrant_Stock_Documentation_2005.pdf (last accessed 1 May 2007)

UNAIDS (2006). *2006 Report on Global AIDS Epidemic.* United Nations Programme on HIV/AIDS, Geneva

UNDP (2004). *Human Development Report 2001: Making New Technologies Work for Human Development.* United Nations Development Programme, New York, NY

UNDP (2005a). *Environmental Sustainability in 100 Millennium Development Goal Country Report.* United Nations Development Programme, New York, NY

UNDP (2005b) *Human Development Report 2005: International Cooperation at a Crossroads.* United Nations Development Programme, New York, NY

UNDP (2006) *Human Development Report 2006. Beyond Scarcity: power, poverty and the global water crisis.* United Nations Development Programme, New York, NY

UNEP (2002). *Global Environment Outlook (GEO-3).* United Nations Environment Programme, Nairobi

UNEP (2004b). *GEO Year Book 2003.* United Nations Environment Programme, Nairobi

UNEP (2005a). *GEO yearbook 2004/2005.* United Nations Environment Programme, Nairobi

UNEP (2005b). *One Planet Many People: Atlas of our Changing Environment.* United Nations Environment Programme, Nairobi

UNEP (2006). Avian Influenza and the Environment: An Ecohealth Perspective. Paper prepared by David J. Rapport on behalf of UNEP, United Nations Environment Programme and EcoHealth Consulting, Nairobi

UNESCO (2007). *United Nations Decade of Education for Sustainable Development (2005-2014)* http://portal.unesco.org/education/en/ev.php-URL_ID=27234&URL_DO=DO_TOPIC&URL_SECTION=201.html/ (last accessed June 25)

UNESCO-WWAP (2006). *Water for People. Water for Life, The United Nations World Water Development Report.* United Nations Educational, Scientific and Cultural Organization, Paris and Berghahn Books, Oxford and New York, NY

UN-Habitat (2006). *State of the World's Cities 2006/7.* United Nations-Habitat, Nairobi

UNHCR (2006a). *Statistical Yearbook 2004 Country Data Sheets: Guinea.* United Nations High Commission for Refugees, Geneva

UNHCR (2006b). *2005 Global refugee trends statistical overview of populations of refugees, asylum-seekers, internally displaced persons, stateless persons, and other persons of concern to UNHCR.* United Nations High Commission for Refugees, Geneva

UNICEF (2006). *Pneumonia: The forgotten killer of children.* United Nations Childrens Fund and World Health Organization, New York, NY

UNPD (2005). *World Urbanisation Prospects: The 2005 Revision* (in GEO Data Portal). UN Population Division, New York, NY http://www.un.org/esa/population/unpop.htm (last accessed 4 June 2007)

UNPD (2007). *World Population Prospects: The 2006 Revision* (in GEO Data Portal). UN Population Division, New York, NY http://www.un.org/esa/population/unpop.htm (last accessed 4 June 2007)

USGS (2006). Studying the Elwha River, Washington, in Preparation for Dam Removal. In *Sound Waves Monthly Newsletter.* US Geological Survey, Washington, DC http://soundwaves.usgs.gov/2006/11/fieldwork3.html (last accessed 12 June 2007)

Van Oostdam, J., Donaldson, S., Feeley, M., Arnold, D., Ayotte, P., Bondy, G., Chan, L., Dewaily, E., Furgal, C.M., Kuhnlein, H., Loring, E., Muckle, G., Myles, E., Receveur, O., Tracy, B., Gill, U., Kalhok, S. (2005). Human health implications of environmental contaminants in Arctic Canada: A review. In *Science of the Total Environment* 351–352:165–246

Watson, R., Zinyower, M. and Dokken, D. (eds.) (1997). *The regional impacts of climate change: an assessment of vulnerability. Summary for Decision Makers.* Special Report of IPCC Working Group II. Intergovernmental Panel on Climate Change

WBCSD (2007). *Then & Now: Celebrating the 20th Anniversary of the "Brundtland Report" – 2006 WBCSD Annual Review.* World Business Council for Sustainable Development, Geneva

WCED (1987). *Our Common Future.* Oxford University Press, Oxford

WHO (2000). *Guidelines for Air Quality.* WHO/SDE/OEH/00.02, World Health Organization, Geneva

WHO (2002) *The World Health Report. Reducing risks, promoting healthy life.* World Health Organization, Geneva

WHO (2006). *Preventing disease through healthy environments: Towards an estimate of the environmental burden of disease.* World Health Organization, Geneva

WHO and UNICEF (2004). *Meeting the MDG drinking-water and sanitation target: A mid-term assessment of progress.* World Health Organization and United Nations Children's Fund, Geneva and New York, NY

Wolfe, N., Escalante, A., Karesh, W., Kilbourn, A., Spielman, A. and Lal, A. (1998). Wild Primate Populations in Emerging Infectious Disease Research: The Missing Link? In *Emerging Infectious Diseases* 4 (2):148-159

Wolfe, N., Heneine, W., Carr, J., Garcia, A., Shanmugam, V., Tamoufe, U., Torimiro, J., Prosser, T., LeBreton, M., Mpoudi-Ngole, E., McCutchan, F., Birx, D., Folks, T., Burke, D. and Switzer, W. (2005). Emergence of unique primate T-lymphotropic viruses among central African bushmeat hunters. In *Proceedings of the National Academy of Sciences* 102 (22):7994 – 7999

Wolfe, N., Switzer. W., Carr, J., Bhullar, V., Shanmugam, V., Tamoufe, U., Prosser, A., Torimiro, J., Wright, A., Mpoudi-Ngole, E., McCutchan, F., Birx. D., Folks, T., Burke, D. and Heneine, W. (2004). Naturally acquired simian retrovirus infections in central African hunters. In *The Lancet* 363:932- 937

Wood, W.B. (2001). Ecomigration: Linkages between environmental change and migration. In Zolberg, A.R. and Benda, P. M.(eds.) *Global Migrants, Global Refugees.* Berghahn, Oxford

World Bank (2005). *The Little Data Book 2005.* The World Bank, Washington, DC

World Bank (2006a). *Where is the Wealth of Nations? Measuring Capital for the 21st Century.* The World Bank, Washington, DC

World Bank (2006b). *World Development Indicators 2006* (in GEO Data Portal). The World Bank, Washington, DC

World Bank and Wildlife Conservation Society (2006). *The Silent Steppe: the Illegal Wildlife Trade Crisis.* The World Bank, Washington DC

WTO (2006). *World Trade Report 2006: Exploring the Links Between Subsidies, Trade and the WTO.* World Trade Organization, Geneva

WRI (2005). *World Resources 2005: The Wealth of the Poor – Managing Ecosystems to Fight Poverty.* World Resources Institute in collaboration with the United Nations Development Programme, the United Nations Environment Programme and The World Bank. World Resources Institute, Washington, DC

State-and-Trends of the Environment: 1987–2007

Climate change affects the warming and acidification of the global ocean, it influences the Earth's surface temperature, the amount, timing and intensity of precipitation, including storms and droughts. On land, these changes affect freshwater availability and quality, surface water run-off and groundwater recharge, and the spread of water-borne disease vectors and it is likely to play an increasing role in driving changes in biodiversity and species' distribution and relative abundance.

Chapter 2

Atmosphere

Coordinating lead authors: Johan C.I. Kuylenstierna and Trilok S. Panwar

Lead authors: Mike Ashmore, Duncan Brack, Hans Eerens, Sara Feresu, Kejun Jiang, Héctor Jorquera, Sivan Kartha, Yousef Meslmani, Luisa T. Molina, Frank Murray, Linn Persson, Dieter Schwela, Hans Martin Seip, Ancha Srinivasan, and Bingyan Wang

Chapter review editors: Michael J. Chadwick and Mahmoud A.I. Hewehy

Chapter coordinator: Volodymyr Demkine

Main messages

A series of major atmospheric environment issues face the world, with both short- and long-term challenges, that are already affecting human health and well-being. Impacts are changing in their nature, scope and regional distribution, and there is a mixture of both worrying developments and substantial progress.

Climate change is a major global challenge. Impacts are already evident, and changes in water availability, food security and sea-level rise are projected to dramatically affect many millions of people. Anthropogenic greenhouse gas (GHG) emissions (principally CO_2) are the main drivers of change. There is now visible and unequivocal evidence of climate change impacts. There is confirmation that the Earth's average temperature has increased by approximately 0.74°C over the past century. The impacts of this warming include sea-level rise and increasing frequency and intensity of heat waves, storms, floods and droughts. The best estimate for warming over this century is projected by the Intergovernmental Panel on Climate Change (IPCC) to be between a further 1.8 and 4°C. This will intensify the impacts, leading to potentially massive consequences, especially for the most vulnerable, poor and disadvantaged people on the planet. There is increasing concern about the likelihood of changes in rainfall patterns and water availability, thereby affecting food security. Major changes are projected for regions, such as Africa, that are least able to cope. Sea-level rise threatens millions of people and major economic centres in coastal areas and the very existence of small island states. Adaptation to anticipated climate change is now a global priority.

To prevent future severe impacts from climate change, drastic steps are necessary to reduce emissions from energy, transport, forest and agricultural sectors. There has been a remarkable lack of urgency in tackling GHG emissions during most of the past two decades. Since the 1987 report of the World Commission on Environment and Development (Brundtland Commission), there has been a sharp and continuing rise in the emissions. There is an agreement in force, the Kyoto Protocol, but the global response is far from adequate. Recent studies show that the total cost of measures to mitigate climate change would be a small fraction of the global economy. Mainstreaming climate concerns in development planning is urgent, especially in sectors such as energy, transport, agriculture, forests and infrastructure development, at both policy and implementation levels. Likewise, policies facilitating adaptation to climate change in vulnerable sectors, such as agriculture, are crucial to minimize adverse impacts. Transformations in social and economic structures, with broad stakeholder participation toward low carbon societies, are critical.

More than 2 million people globally are estimated to die prematurely each year due to indoor and outdoor air pollution. Although air quality has improved dramatically in some cities, many areas still suffer from excessive air pollution. The situation on air pollution is mixed, with some successes in both developed and developing countries, but major problems remain. Air pollution has decreased in some cities in different parts of the world through a combination of technology improvement and policy measures. However, increasing human activity is offsetting some of the gains. Transport demand increases every year, and is responsible for a substantial part of both anthropogenic GHG emissions and health effects due to air pollution. Many people, especially in Asia where the most polluted cities are now found, still suffer from very high levels of pollutants in the air they breathe, particularly from very fine particulate matter, the main air pollutant affecting human health. This is also related to the massive industrial expansion in many Asian cities that are producing goods for the global economy. This pollution also reduces visibility by creating urban and regional haze. Many poor communities are still dependent on traditional biomass and coal for cooking. The health of women and children in particular suffers as a result of indoor air pollution, and a total of 1.6 million people are estimated to die prematurely each year. Many air pollutants, including sulphur and nitrogen oxides, accelerate damage to materials, including historic buildings. Long-range transport of a variety of air pollutants remains an issue of concern for human and ecosystem health, and for the provision of ecosystem services. Tropospheric (ground-level) ozone is increasing throughout the northern hemisphere, and is a regional pollutant affecting human health and crop yields. Persistent organic pollutants from industrial economies accumulate in the Arctic, affecting people not responsible for the emissions.

The "hole" over the Antarctic in the stratospheric ozone layer that gives protection from harmful ultraviolet radiation is now the largest ever. Emissions of ozone-depleting substances (ODS) have decreased over the last 20 years, yet the concern about the state of stratospheric ozone still persists. On the positive side, precautionary action on stratospheric ozone depletion was taken by some industrialized countries before the impacts were evident. Their leadership was key to making the reductions in the manufacture and consumption of ODS a global success story. Although emissions of ODS have decreased over the last 20 years, it is estimated that the ozone layer over the Antarctic will not fully recover until between 2060 and 2075, assuming full Montreal Protocol compliance.

Rapid growth in energy demand, transport and other forms of consumption continue to result in air pollution, and are responsible for unprecedented absolute growth in anthropogenic GHG emissions. Since the Brundtland Commission emphasized the urgent need for addressing these problems, the situation has changed, in some cases for the better, but in others for the worse. A number of pressures are still building, driving up the emissions. The population is increasing, and people use more and more fossil fuel-based energy, consume more goods and travel further, increasingly using cars as their favourite transport mode. Aviation is growing rapidly and increased trade, as part of the globalized economy, leads to growth in the transport of goods by sea, where fuel quality and emissions are currently not strictly regulated. These pressures are being somewhat offset by increases in efficiency and/or from implementation of new or improved technology.

Measures to address harmful emissions are available and cost-effective, but require leadership and collaboration. Existing mechanisms to tackle ODS are adequate, while air quality management in many parts of the world requires the strengthening of institutional, human and financial resources for implementation. Where air pollution has been reduced, the economic benefits associated with reduced impacts have far outweighed the costs of action. For climate change, more innovative and equitable approaches for mitigation and adaptation are crucial, and will require systemic changes in consumption and production patterns. Many policies and technologies required to address emissions of GHGs and air pollutants are currently available and are cost-effective. Some nations have started to implement changes. While additional research and assessment efforts should continue, dynamic leadership and international collaboration, including technological transfer and effective financial mechanisms, are required to accelerate policy implementation around the world. The long-term risks from emissions of substances with long residence times, especially those that are also GHGs, should strongly encourage the use of a precautionary approach now.

INTRODUCTION

In 1987 the World Commission on Environment and Development (WCED), also known as the Brundtland Commission, recognized problems of regional air pollution, with its impacts on environmental and cultural values (see Chapter 1). It stated that burning fossil fuels gives rise to carbon dioxide (CO_2) emissions, and that the consequent greenhouse effect "may by early next century have increased average global temperatures enough to shift agricultural production areas, raise sea levels to flood coastal cities and disrupt national economies." It also said that "Other industrial gases threaten the planet's protective ozone shield," and "Industry and agriculture put toxic substances into the human food chain," highlighting the lack of an approach to effective chemicals management.

Key conclusions of *Our Common Future*, the Brundtland Commission report, were that while economic activity, industrial production and consumption have profound environmental impacts, "poverty is a major cause-and-effect of global environmental problems." Human well-being, especially poverty and equity, are affected by all of the atmospheric environment issues addressed in this chapter. It is clear that air pollution from human activities constitutes one of the most important environmental issues affecting development across the world. Climate change threatens coastal areas, as well as the food security and livelihoods of people in the most vulnerable regions. Indoor air pollution, from burning biomass or coal for cooking, particularly affects women and young children. Outdoor air pollution in cities or near major industries disproportionately kills or harms the health of poorer people. Tackling emissions will contribute to the attainment of the Millennium Development Goals (UN 2007), especially the goals of eradicating hunger, ensuring good health for all and ensuring environmental sustainability.

Atmospheric environment issues are complex. Different primary pollutants emitted, and secondary pollutants formed in the atmosphere, have very different residence times, and are transported to varying distances, and this affects the scale at which their impact is felt (see Figure 2.1). Those substances that have very short residence times affect indoor and local air quality. Substances with residence times of days to weeks give rise to local and regional problems, those with residence

Although there have been some important pollution control success stories, the atmospheric problems highlighted by the Brundtland Commission still exist (such as here in Santiago de Chile).

Credit: Luis A. Cifuentes

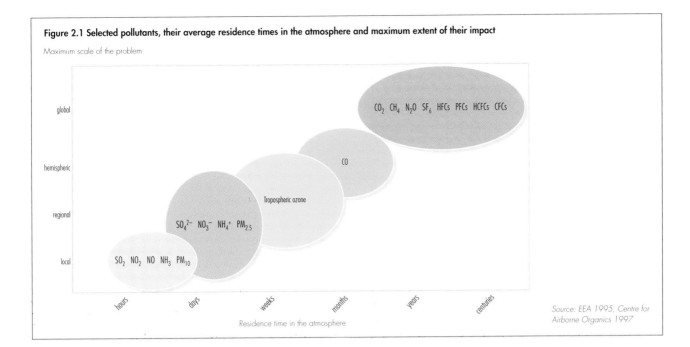

Figure 2.1 Selected pollutants, their average residence times in the atmosphere and maximum extent of their impact

Maximum scale of the problem

global — CO_2 CH_4 N_2O SF_6 HFCs PFCs HCFCs CFCs

hemispheric — CO

regional — Tropospheric ozone

SO_4^{2-} NO_3^- NH_4^+ $PM_{2.5}$

local — SO_2 NO_2 NO NH_3 PM_{10}

hours   days   weeks   months   years   centuries

Residence time in the atmosphere

Source: EEA 1995, Centre for Airborne Organics 1997

times from weeks to months give rise to continental and hemispheric problems, and those with residence times of years give rise to global problems. Some greenhouse gases may last up to 50 000 years in the atmosphere.

There is now a consensus amongst the vast majority of scientists that anthropogenic emissions of greenhouse gases, of which CO_2 and methane are the most significant, are already causing climate change. The global emissions are still increasing and the impact will be felt by all regions of the world, with changing weather patterns and sea-level rise affecting coastal human settlements, disease patterns, food production and ecosystem services.

Air pollution is still leading to the premature death of a large number of people. Although the air quality of some cities has improved dramatically over the last 20 years, mainly in the richer nations, the air quality of many cities in developing nations has deteriorated to extremely poor levels. Even in richer countries, in recent years, improvements in levels of particulate matter and tropospheric ozone have stagnated, and further measures are needed. Regional air pollution problems of acidification have been reduced in Europe and North America, but are now a growing policy focus in parts of Asia, where acidic deposition has increased. Tropospheric (ground-level) ozone pollution causes significant reductions in crop yield and quality. The transfer of pollutants across the northern hemisphere, especially tropospheric ozone, is becoming an

increasingly important issue. Despite efforts to tackle air pollution since 1987, emissions of various air pollutants to the atmosphere are still having dramatic impacts on human health, economies and livelihoods, as well as on ecosystem integrity and productivity.

Emissions of ozone-depleting substances (ODS), such as chlorofluorocarbons, lead to thinning of the stratospheric ozone layer, resulting in increased ultraviolet (UV-B) radiation reaching the Earth's surface. The ozone hole, or seasonal ozone depletion over the Antarctic, still occurs. Increasing UV-B radiation affects skin cancer rates, eyes and immune systems, thus having important public health implications (WHO 2006b). There are concerns about the UV-B effect on ecosystems, for example through impacts on phytoplankton and marine food webs (UNEP 2003).

Since 1987, it has become clear that there are high levels of persistent organic pollutants (POPs) and mercury in food chains, with the potential to affect the health of humans and wildlife, especially species higher in food chains. POPs represent a global problem. Some have low residence times in the atmosphere, but are re-volatilized, and can migrate over long distances and persist in the environment. Many POPs are transported through the atmosphere, but their impacts are mediated by aquatic and land-based food chains (see Chapters 3 and 4) and accumulated in Polar Regions (see Chapter 6).

DRIVERS OF CHANGE AND PRESSURES

Atmospheric composition is affected by virtually all human activities. Population increases, income growth and the global liberalization of trade in goods-and-services all stimulate an increase in energy and transport demand. These are drivers of emissions of substances into the atmosphere and, as many cost-benefit studies have shown (Stern 2006), the costs to our collective well-being often outweigh the individual benefits of the high-consumption lifestyles people have or aspire to (see Chapter 1). In many cases, emissions result from satisfying the wants of a rising affluent class rather than from fulfilling basic needs (see Box 2.1). Significant downward pressure on emissions has come from increases in efficiency and/or from implementation of new or improved technology.

The developed world is still the main per capita user of fossil fuel, and often exports long-lived, outdated and polluting technology to developing countries. The wealthier nations also "transfer" pollution by purchasing goods that are produced in a less environmentally friendly manner in lower-income countries. As a consequence, vulnerable communities in developing countries are most exposed to the adverse health effects caused by air pollution (see Chapters 6, 7 and 10).

Due to inertia in economic, social, cultural and institutional systems, transitions to more sustainable modes of production and consumption are slow and cumbersome. Typically, it takes 30–50 years or more before such changes are fully implemented, although the first improvements can be seen at a much earlier stage (see Box 2.2). Understanding how policy decisions will affect economic activities, and their associated emissions and impacts can facilitate early-warning signals and timely actions. Table 2.1 presents the main drivers affecting the atmosphere.

Production, consumption and population growth

Ultimately, the drivers for atmospheric environment impacts are the increasing scale and changing form of human activity. The increasing population on the planet contributes to the scale of activity but, of even greater importance, the continuing expansion of the global economy has led to massive increases in production and consumption (see Chapter 1), indirectly or directly causing emissions to the atmosphere.

Since the Brundtland Commission report, the Earth's population has risen by almost 30 per cent (see Chapter 1), with regional increases ranging from 5.1 per cent in Europe to 57.2 per cent in Africa

---

Box 2.1 Energy use in the context of Millennium Development Goals (MDGs)

Currently, access to energy for heating, cooking, transport and electricity is considered a basic human right. Various studies have investigated the consequences of meeting the minimum standards set out in the MDGs, and found that the total amount of primary energy required to meet the minimum standards is negligible on the global scale. Electricity for lighting (in homes, schools and rural health facilities), liquefied petroleum gas (LPG) for cooking fuel (for 1.7 billion urban and rural dwellers), and diesel used in cars and buses for transport (for 1.5 million rural communities) would require less than 1 per cent of total annual global energy demand, and would generate less than 1 per cent of current annual global CO_2 emissions. This shows that energy services could be provided to meet the MDGs without significantly increasing the global energy sector's environmental impacts.

Sources: Porcaro and Takada 2005, Rockström and others 2005

Box 2.2 Examples of inertia in drivers

Energy supply

The energy sector requires massive investments in infrastructure to meet projected demand. The International Energy Agency (IEA) estimates that the investments will total around US$20 trillion from 2005 to 2030, or US$800 billion/year, with the electricity sector absorbing the majority of this investment. Developing countries, where energy demand is expected to increase quickly, will require about half of such investments. Often, these investments are long term. Nuclear plants, for example, are designed for a lifetime of 50 years or more. Decisions made today will have effects well into our future.

Transport

Production of road vehicles, aircraft and ships are all examples of steadily growing mature markets. It will take time for new concepts, such as hybrid or hydrogen fuel cell cars, or high-speed magnetic trains, to massively penetrate markets. Technology barriers and standards, cost reductions, new production plants and, finally, market penetration are all challenging obstacles. Old production facilities often remain operational until they are economically outdated, and the lifetime of a new car is well over a decade. The penetration time of a new technology, such as the hydrogen fuel cell car will, even under the most optimistic projections, take at least 40 years.

Source: IEA 2006

(GEO Data Portal, from UNPD 2007). Global economic output (measured in purchasing power parity or PPP) has increased by 76 per cent, almost doubling the average per capita gross national income from about US$3 300 to US$6 400. This average increase in per capita income masks large regional variations, ranging from virtual stagnation in Africa to a doubling in some countries in Asia and the Pacific. Over the same period, urban populations have risen to include half of humanity. Although the rate of population growth is expected to continue to slow, the world population is still expected to be 27 per cent above current levels by 2030 (GEO Data Portal, from UNPD 2007 medium variant). Nearly all population growth expected for the world in that period will be concentrated in urban areas (see Chapter 1).

In line with population and GDP growth, there is an increase in production and consumption. Energy use has been partly decoupled from the growth of GDP (see Figure 2.2), due to increased efficiency in energy and electricity production, improved production processes and a reduction in material

Table 2.1 Trends and relevance of drivers for atmospheric issues						
Driver	Stratospheric ozone depletion		Climate change		Air pollution	
	Situation in 1987	Relevance/trend 2007	Situation in 1987	Relevance/trend 2007	Situation in 1987	Relevance/trend 2007
Population	Important	Emission per capita reduced dramatically	Important	Increases in demand lead to increased emissions	Important, with urban areas affected most	Increasing urbanization has put more people at risk
Agricultural production	Negligible source	Methyl bromide is now a more significant proportion of remaining ODS emissions	Important due to methane and N$_2$O emissions, and land-use change	Increases in production cause increased emissions	Ammonia and pesticide emissions	Emissions have grown with increasing production
Deforestation (including forest fires)	Negligible source	Negligible source	Important contributor of GHG emissions	Continuing deforestation contributes significantly to GHG emissions	CO, PM and NO$_X$ emissions	Increasing frequency of forest fires
Industrial production	Largest emission source	Strong decline in ODS production	Important	Important, but share of emission decreasing	Important emissions source	Production decreases in some regions, increases in others
Electricity production	Negligible source	Negligible source	Important	Increasingly important driver	Important emissions source	Share of emission decreases in some regions, increases in others
Transport	Relevant	Decline in relevance, but still a source	Important	Strong increase in transport and its emissions	Lead, CO, PM, NO$_X$ emissions	Varies by region and pollutant
Consumption of basic goods	Relevant	Decline in relevance	Small share in emissions	Constant	Large emissions from traditional biomass	Continued high share in rural communities
Consumption of luxury goods	Important	Strong decline in relevance	Important	Share of emissions increasing	Moderate share of emissions	Increasing share of emissions
Scientific and technological innovation	Innovation starting	Very important for solutions	Important for energy efficiency improvements	High relevance for efficiency and energy generation	Important for all emissions	Crucial for improvements in all sectors
Institutional and socio-political frameworks	Frameworks initiating	Highly advanced	Non-existent	Considerable improvement	Established in developed countries	Increasing number of regions tackling problems

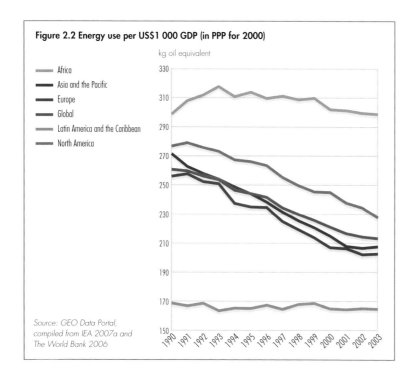

Figure 2.2 Energy use per US$1 000 GDP (in PPP for 2000)

kg oil equivalent

Legend:
- Africa
- Asia and the Pacific
- Europe
- Global
- Latin America and the Caribbean
- North America

Source: GEO Data Portal, compiled from IEA 2007a and The World Bank 2006

2.7 per cent in 2004 (GEO Data Portal, from IEA 2007a) (see Chapter 5).

The energy intensity of our society (defined as energy use per unit of GDP in PPP units) has decreased since Brundtland by an average of 1.3 per cent per year (see Figure 2.2). However, the impact of total GDP growth on energy use has outweighed these mitigating efficiency improvements.

Manufacturing processes can also cause direct emissions, such as CO_2 from steel and cement production, SO_2 from copper, lead, nickel and zinc production, NO_x from nitric acid production, CFCs from refrigeration and air conditioning, SF_6 from electricity equipment use, and perfluorocarbons (PFCs) from the electronic industry and aluminium production.

Humanity's footprint on the planet has grown correspondingly larger. Natural resource demands have expanded, the burden on the environment has grown heavier, and this trend looks set to continue although there have been shifts in the sources of the pressures. The share of total GDP of the agriculture and industry sectors has decreased from 5.3 and 34.2 per cent in 1987 to 4 and 28 per cent of GDP in 2004 (GEO Data Portal, from World Bank 2006). The transport sector has shown a consistently high growth rate over the same period, with a 46.5 per cent increase in energy used globally by road transport between 1987 and 2004 (GEO Data Portal, from IEA 2007a). Reducing the impacts

intensity. Nevertheless, the major proportion of pollutant emissions result from energy-related activities, especially from the use of fossil fuel. The global primary energy supply has increased by 4 per cent/ year between 1987 and 2004 (GEO Data Portal, from IEA 2007a) since Brundtland, and fossil fuels still supply over 80 per cent of our energy needs (see Figure 2.3). The contribution of non-biomass renewable energy sources (solar, wind, tidal, hydro and geothermal) to the total global energy supply has increased very slowly, from 2.4 per cent in 1987 to

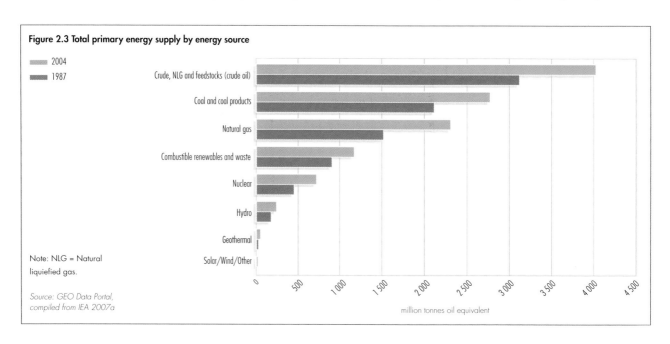

Figure 2.3 Total primary energy supply by energy source

Legend:
- 2004
- 1987

Categories:
- Crude, NLG and feedstocks (crude oil)
- Coal and coal products
- Natural gas
- Combustible renewables and waste
- Nuclear
- Hydro
- Geothermal
- Solar/Wind/Other

Note: NLG = Natural liquefied gas.

Source: GEO Data Portal, compiled from IEA 2007a

million tonnes oil equivalent

of these major drivers of atmospheric pollution will involve multiple transitions in sectors such as energy, transport, agricultural land use and urban infrastructure. The right mix of appropriate government regulation, greater use of energy saving technologies and behavioural change can substantially reduce CO_2 emissions from the building sector, which accounts for 30–40 per cent of global energy use. An aggressive energy efficiency policy in this sector might deliver billions of tonnes of emission reductions annually (UNEP 2007a).

Increasing demand for such products and services as refrigeration, air-conditioning, foams, aerosol sprays, industrial solvents and fire suppressants led to increasing production of a variety of chemicals. Some of them, after being released into the atmosphere, can rise into the stratosphere, where they break apart, releasing chlorine or bromine atoms, which can destroy ozone molecules. Though the physical volume of emissions of ozone-depleting substances has never been very large in comparison to other anthropogenic emissions to the atmosphere, the risks associated with potential impacts are enormous. Fortunately, the response to this problem has been a success story.

Sectors and technology
Transport

The relatively high growth in passenger car sales reveals that people put a high preference on car ownership as they become more affluent (see Figure 2.4). Moreover, there has been a shift to heavier cars, equipped with an increasing number of energy demanding features (for example air conditioning and power windows), which add to a greater than expected growth in energy use by the transport sector.

Atmospheric emissions from the transport sector depend upon several factors, such as vehicle fleet size, age, technology, fuel quality, vehicle kilometres travelled and driving modes. The low fleet turnover rate, especially for diesel-powered vehicles, and the export of older vehicles from rich to poor countries, slows progress in curbing emissions in developing countries. In some parts of Asia, a majority of road vehicles consist of two- and three-wheelers powered by small engines. They provide mobility for millions of families. Although inexpensive, and with lower fuel consumption than cars or light trucks on a per vehicle basis, they contribute disproportionately to particulate,

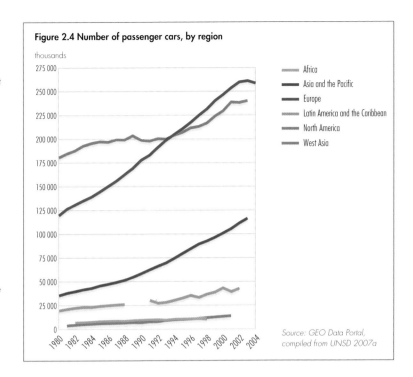

Figure 2.4 Number of passenger cars, by region

thousands

- Africa
- Asia and the Pacific
- Europe
- Latin America and the Caribbean
- North America
- West Asia

Source: GEO Data Portal, compiled from UNSD 2007a

hydrocarbon and carbon monoxide emissions (World Bank 2000, Faiz and Gautam 2004).

Shifting from public transport systems to private car use increases congestion and atmospheric emissions. Poor urban land-use planning, which leads to high levels of urban sprawl (spreading the urban population over a larger area), results in more car travel (see Figure 2.5) and higher energy consumption. The lack of adequate infrastructure

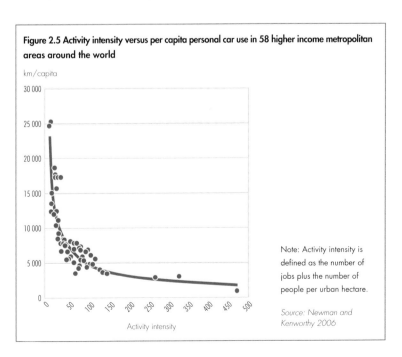

Figure 2.5 Activity intensity versus per capita personal car use in 58 higher income metropolitan areas around the world

km/capita

Activity intensity

Note: Activity intensity is defined as the number of jobs plus the number of people per urban hectare.

Source: Newman and Kenworthy 2006

Credit: Press-Office City of Münster, Germany

for walking and cycling, which are the most environmentally-friendly transport modes, also contributes to increased vehicle use. Figure 2.6 shows the relative space required to accommodate people driving cars, using buses or cycling, with clear implications for transport strategy and planning.

Air transport is one the fastest rising transport modes, with an 80 per cent increase in kilometres flown between 1990 and 2003 (GEO Data Portal, from UNSD 2007b). This dramatic increase was driven by growing affluence, more airports, the rise in low-cost airlines and the promotion of overseas tourism. Economic efficiency is driving improvements in energy efficiency, and new commercial aircraft are claimed to use up to 20 per cent less fuel than those sold 10 years ago (IATA 2007). Shipping has also grown remarkably since Brundtland, mirroring the increase in global trade. It has risen from 4 billion tonnes in 1990 to 7.1 billion tonnes total goods loaded in 2005 (UNCTAD 2006). Improvements in the environmental performance of the shipping industry have been less pronounced than for air transport.

Industry

The shift in the regional character of industrial production, which has decreased in developed countries and increased in the developing world, can be illustrated by the changes in secondary energy use by the industrial sector. In the United States, increased energy use in the transport and service sectors has been partially counterbalanced

by the decrease (0.48 tonne oil equivalent/capita) in the industrial sector. In contrast, in Asia and the Pacific, and Latin America and the Caribbean, there has been an increase in per capita energy use in all sectors (GEO Data Portal 2006).

Atmospheric emissions from large stationary sources in developed countries have been reduced by using cleaner fuels, end-of-pipe controls, relocating or shutting down high-emitting sources and promoting more efficient energy use. In many developing countries such measures have not been fully implemented, but have the potential to rapidly reduce emissions. If 20 per cent of energy was saved in existing energy generation and industrial facilities in developing countries through use of currently available technologies, the increase in CO_2 emissions from developing countries from 2000 to 2020 would only be about half of what it otherwise be (METI 2004). Industrial sources that use obsolete technology, lack emission controls and are not subject to effective enforcement measures, contribute significantly to the emission load. In general, the implementation of governmental regulations has stimulated the use of technologies that often reduce costs, and result in greater benefits than originally foreseen.

Emissions from small factories and commercial sources are much more difficult to control. Enforcement of compliance with emission standards is politically difficult and expensive. Technology solutions are more challenging, and there is no simple way to check that best management practices are being used.

Energy

In the industrialized world, large power plants are confronted with increasingly tight environmental standards. A wide range of options for the production of clean energy exists, and has started to penetrate the market, often stimulated by government subsidies. High growth rates in clean energy options since 1987 have been observed, especially for solar and wind energy. Energy supply from wind power increased 15 times by 2004, with an average growth of approximately 30 per cent per year, although its share in global electricity supply is still very small at about 0.5 per cent in 2004 (IEA 2007b).

Energy efficiency improvements and energy conservation are given high priority in the energy development strategies of many countries, including developing countries. High efficiency and clean technology will be crucial to achieve a low-emission development path, combined with security of supply. Among the factors that define the level of emissions are fuel quality, technology, emission control measures, and operation and maintenance practices. Energy security considerations and fuel costs often determine the choice of fuels, such as coal and nuclear (see Chapter 7). Thermal power plants burning coal are major air pollution sources, and emit higher levels of many pollutants than gas-fired power plants to produce the same amount of energy. Clean energy sources, such as geothermal, wind energy and solar power, are still underutilized. With the recent high oil prices, more efficient power plants have become more cost-effective, but still require substantial investment in infrastructure. Many countries in, for example, sub-Saharan Africa, cannot cope with the rising energy demand, and continue to rely on obsolete, low-efficiency power plants that emit high levels of pollutants.

Land-use practices

In rural areas, customary land-use practices also drive atmospheric emissions. The clearance of forested land, and its subsequent use for cattle and crop production, releases carbon stored in the trees and soils, and depletes its potential as a CO_2 sink (see Chapter 3). It may also increase methane, ammonia and nitrogen oxide emissions. Deforestation is known to contribute as much as 20–25 per cent to annual atmospheric emissions of CO_2 (IPCC 2001a). Normal agricultural land-use practices, such as burning crop residues and other intentional fires, increase emissions of CO_2, particulate matter and other pollutants (Galanter and

others 2000). Wildfires and forest fires used for land clearance also release very high levels of particulates. The Southeast Asian haze of 1997, produced by land clearance, cost the people of that region an estimated US$1.4 billion, mostly in short-term health costs (ADB 2001). Since 1987, there has been little progress in mitigating these unwanted effects. Fine dust particles from the ground are also a major concern in arid or semi-arid areas subject to seasonal or periodic high winds.

The clearance of forested land and its subsequent use for cattle and crop production, releases carbon stored in the trees and soils, and depletes its potential as a CO_2 sink.

Credit: Ngoma Photos

Urban settlements

Emissions in densely populated areas tend to be higher due to the total level of emission-related activity, even though the per capita emissions are reduced by higher efficiency and shorter travel distances using personal transport (see Figure 2.5). In combination with low dispersion conditions, this results in exposure of large populations to poor air quality. Urbanization, seen in such forms as urban population growth in Latin America, Asia and Africa, and urban sprawl in North America and Europe, is continuing as a result of a combination of social and economic drivers. Urban areas concentrate energy demands for transport, heating, cooking, air conditioning, lighting and housing. Despite the obvious opportunities that cities offer, such as their economic and cultural benefits, they are often associated with problems that are aggravated by large increases in population and limited financial means, which force city authorities to accept unsustainable short-term solutions. For example, there is pressure to use land reserved as green areas and for future public transport systems for houses, offices, industrial complexes or other uses with a high economic value. Moreover, cities create heat islands that alter regional meteorological conditions and affect atmospheric chemistry and climate. Reversing the trend of unsustainable development is a challenge for many city authorities.

Technological innovation

Technological innovation, coupled with technology transfer and deployment, is essential for reducing emissions. A broad portfolio of technologies is necessary, as no single technology will be adequate to achieve the desired level of emissions. Desulphurization technologies, low nitrogen combustors and end-of-pipe particulate capture devices are examples of technologies that have contributed considerably to SO_2, NO_x and PM emission reduction. A number of technologies may play key roles in reducing GHG emissions. They include those for improved energy efficiency, renewable energy, integrated gasification combined cycle (IGCC), clean coal, nuclear and carbon sequestration (Goulder and Nadreau 2002).

A "technology push" approach, based on large-scale research and technology deployment programmes and new breakthrough technologies, is needed to achieve deeper GHG emission cuts in the long run (2050 and beyond).

In addition to government and private sector investment in technology research and development, regulations for energy, environment and health are key drivers for stimulating the deployment of cleaner technologies in developing countries. It is also important to lower the risk of locking in more CO_2-intensive energy technologies in developing countries.

ENVIRONMENTAL TRENDS AND RESPONSES

In this chapter three major atmosphere-related environmental issues are analysed in detail: air pollution, climate change and stratospheric ozone depletion. For each issue the changes in the environmental state are related to the impacts on both the environment and on human well-being for the period since 1987. This is followed by descriptions of what has been done to curb emissions. Table 2.2 below summarizes the interconnections between changes in the atmosphere and human well-being, including changes in state of the atmosphere, the mechanisms through which impacts occur and changes in well-being over time.

Table 2.2 Linkages between state changes in the atmospheric environment and environmental and human impacts

State changes	Mediated environmental/ ecosystem impacts	Impacts on human well-being				
		Human health	Food security	Physical security and safety	Socio-economic	Other impacts
Outdoor air pollution related issues						
Concentration/ deposition of criteria pollutants (not tropospheric ozone) ⇓ Developed countries ⇕ Developing countries	Exposure to poor air quality: ⇑ developing countries ⇓ developed countries	⇕ Respiratory and cardiac diseases ⇕ Premature deaths and morbidity ⇑ Childhood asthma	⇕ Crop yields	⇕ Conflict over transboundary movement	⇕ Health costs ⇕ DALYs ⇑ Cost for control of pollution	⇓ Tourism potential ⇓ Visibility ⇑ Haze
	⇕ Acidification		⇑ Decline of forests and natural ecosystems	⇑ Corrosion of materials	⇑ Maintenance costs for physical infrastructure	⇓ Tourism potential
	⇑ Eutrophication		⇓ Fish supply when nutrients enter surface waters	⇑ Loss of biodiversity	⇑ Odour nuisance	

Table 2.2 Linkages between state changes in the atmospheric environment and environmental and human impacts *continued*

State changes	Mediated environmental/ ecosystem impacts	Impacts on human well-being				
		Human health	Food security	Physical security and safety	Socio-economic	Other impacts
Outdoor air pollution related issues						
Tropospheric ozone formation and concentrations ⇧ Northern Hemisphere	⇧ Exposure of crops, natural ecosystems and humans	⇧ Respiratory inflammation ⇧ Mortality and morbidity	⇩ Crop yields	⇧ Loss of biodiversity	⇩ Income generation (particularly for the poor) ⇧ Restricted activity days	
⇕ Concentrations of air toxics (heavy metals, PAHs, VOC)	⇕ Air quality	⇕ Incidence of carcinogenic diseases	⇧ Food chain contamination		⇧ Health costs	
⇧ POPs emissions	⇧ Deposition on natural ecosystems ⇧ Bioaccumulation in food chain	⇩ Food safety ⇩ Human health	⇩ Sustainability of fish resources		⇩ Commercial fish value ⇧ Vulnerability of Polar communities	
Indoor air pollution related issues						
Criteria pollutants and air toxics ⇧ Developing countries	⇧ Exposed population	⇧ Mortality and respiratory diseases			⇧ Vulnerability of poor communities	⇧ Impact on women and children
Climate change related issues						
⇕ GHG concentrations	⇧ Temperature ⇧ Extreme weather events	⇧ Deaths due to heat stress ⇧ Diseases (diarrhoea and vector-borne diseases)	⇧ Risk of hunger ⇕ Crop production (see Chapters 3 and 6)	⇧ Human vulnerability (see Chapters 6 and 8)	⇧ Energy requirement for cooling ⇧ Loss of economic properties	⇧ Threatened livelihood of communities ⇧ Vulnerability of poor communities
	⇧ Sea surface temperature ⇕ Precipitation ⇧ Land and sea ice melting ⇧ Ocean acidification	See Table 4.2				
Stratospheric ozone related issues						
⇩ ODS emissions ⇕ ODS concentrations in stratosphere	⇕ UV-B radiation ⇧ Stratospheric ozone depletion at the poles	⇧ Skin cancer ⇧ Damage to eyes and immune systems	⇩ Fish stocks (impact on phytoplankton and other organisms) (see Chapter 4) ⇩ Food production (altered disease intensity)		⇩ Time spent outdoors (lifestyle change) ⇧ Expenditure on preventing exposure to UV-B radiation	⇕ Global warming (due to long residence times)

⇧ increasing

⇩ decreasing

⇕ variable depending on location

AIR POLLUTION

Human and environmental exposure to air pollution is a major challenge, and an issue of global concern for public health. The World Health Organization (WHO) estimated that about 2.4 million people die prematurely every year due to fine particles (WHO 2002, WHO 2006c). This includes about 800 000 deaths due to outdoor urban PM_{10} (see Box 2.3 for an explanation), and 1.6 million due to indoor PM_{10}, even though the study did not include all mortality causes likely to be related to air pollution. Figure 2.7 shows the annual mortality that is attributable to outdoor PM_{10} for different world regions. The highest number of estimated annual premature deaths occurs in developing countries of Asia and the Pacific (Cohen and others 2004).

Beside effects on human health, air pollution has adverse impacts on crop yields, forest growth, ecosystem structure and function, materials and visibility. Once released into the atmosphere, air pollutants can be carried by winds, mix with other pollutants, undergo chemical transformations and eventually be deposited on various surfaces (see Box 2.3).

Atmospheric emissions and air pollution trends

Emissions in the various regions show different trends for SO_2 and NO_X (see Figure 2.8). There have been decreases in the national emissions in the more affluent countries of Europe and North America since 1987. More recently Europe is as concerned with unregulated sulphur emissions from international shipping as it is

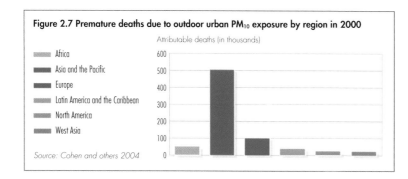

Figure 2.7 Premature deaths due to outdoor urban PM_{10} exposure by region in 2000

Attributable deaths (in thousands)

- Africa
- Asia and the Pacific
- Europe
- Latin America and the Caribbean
- North America
- West Asia

Source: Cohen and others 2004

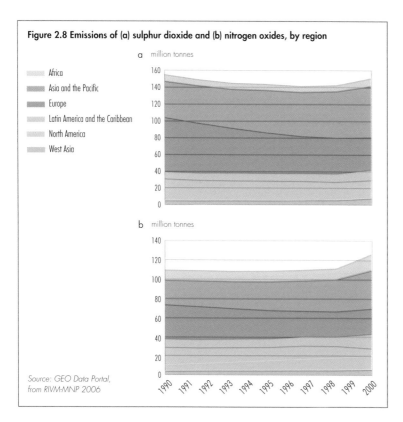

Figure 2.8 Emissions of (a) sulphur dioxide and (b) nitrogen oxides, by region

a   million tonnes

b   million tonnes

- Africa
- Asia and the Pacific
- Europe
- Latin America and the Caribbean
- North America
- West Asia

Source: GEO Data Portal, from RIVM-MNP 2006

Box 2.3 Features of different air pollutants

Six common pollutants – suspended particulate matter (SPM), sulphur dioxide (SO_2), nitrogen dioxide (NO_2), carbon monoxide (CO), tropospheric ozone (O_3) and lead (Pb) – harm human health, and are used as indicators of air quality by regulatory agencies. They are known as criteria pollutants, for which health-based ambient air quality guidelines have been recommended by WHO. PM is distinguished as different inhalable fractions that are classified as coarse and fine particulates with aerodynamic diameters below 10 µm (PM_{10}) and 2.5 µm ($PM_{2.5}$) respectively.

Air pollutants may be considered primary – emitted directly into the air – or secondary pollutants that are formed in the air by chemical and/or photochemical reactions on primary pollutants. The formation of secondary pollutants, such as tropospheric ozone and secondary aerosols, from primary pollutants such as SO_2, NO_X, NH_3 and volatile organic compounds (VOCs) is strongly dependent on climate and atmospheric composition. Due to atmospheric transport, their impacts can occur far from their sources.

The major chemical components of PM are sulphate, nitrate, ammonium, organic carbon, elemental carbon and soil dust (consisting of several mineral elements). Other important primary pollutants include heavy metals, such as mercury, cadmium and arsenic; VOCs, such as benzene, toluene, ethylbenzene and xylenes; polycyclic aromatic hydrocarbons (PAHs); and some persistent organic pollutants (POPs), such as dioxins and furans. These air pollutants result from the burning of fossil fuels, biomass and solid waste. Ammonia (NH_3) is emitted primarily from agricultural sources.

Source: Molina and Molina 2004, WHO 2006a

with the regulated land-based sources (EEA 2005). For the industrializing nations of Asia, emissions have increased, sometimes dramatically, over the last two decades. There are no aggregate data for regions after 2000, and therefore recent changes in emissions of developing countries are not displayed, especially in Asia. For instance, from 2000 to 2005 the Chinese SO_2 emissions increased by approximately 28 per cent (SEPA 2006), and satellite data suggest that NO_X emissions in China have grown by 50 per cent between 1996 and 2003 (Akimoto and others 2006). The main result is that global emissions of SO_2 and NO_X are increasing with respect to 1990 levels. In Africa, and in Latin America and the Caribbean, small increases have been reported.

In many large cities in developing countries, current air pollution concentrations are very high, especially for PM_{10} (see Figures 2.9 and 2.10). However, pollutant levels are decreasing, usually because of controls on emission sources, changing fuel use patterns, and closures of obsolete industrial facilities. For lead, the trends are decreasing, and ambient levels in most cities are currently below the WHO guideline (WHO 2006a). In general, PM_{10} and SO_2 levels have been decreasing, although levels of PM_{10} are still many times higher than the WHO guideline in many developing countries, and SO_2 levels are above the WHO guideline in a number of cities and differences are considerable in different regions. Most large cities exceed the WHO guideline for NO_2, and the levels are not showing any significant decreases.

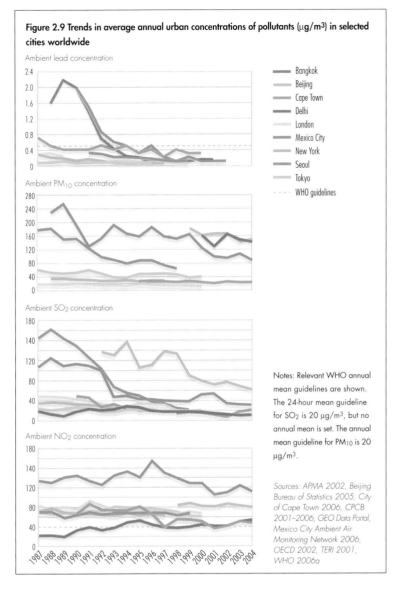

Figure 2.9 Trends in average annual urban concentrations of pollutants ($\mu g/m^3$) in selected cities worldwide

Ambient lead concentration

Ambient PM_{10} concentration

Ambient SO_2 concentration

Ambient NO_2 concentration

Bangkok
Beijing
Cape Town
Delhi
London
Mexico City
New York
Seoul
Tokyo
WHO guidelines

Notes: Relevant WHO annual mean guidelines are shown. The 24-hour mean guideline for SO_2 is 20 $\mu g/m^3$, but no annual mean is set. The annual mean guideline for PM_{10} is 20 $\mu g/m^3$.

Sources: APMA 2002, Beijing Bureau of Statistics 2005, City of Cape Town 2006, CPCB 2001–2006, GEO Data Portal, Mexico City Ambient Air Monitoring Network 2006, OECD 2002, TERI 2001, WHO 2006a

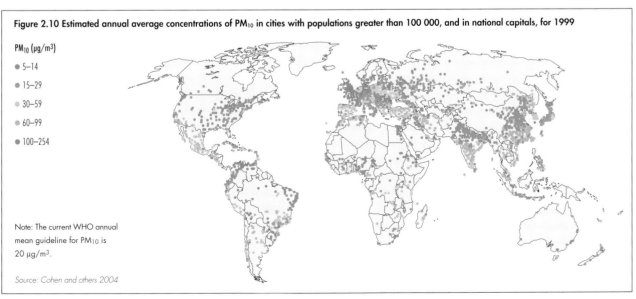

Figure 2.10 Estimated annual average concentrations of PM_{10} in cities with populations greater than 100 000, and in national capitals, for 1999

PM_{10} ($\mu g/m^3$)

5–14
15–29
30–59
60–99
100–254

Note: The current WHO annual mean guideline for PM_{10} is 20 $\mu g/m^3$.

Source: Cohen and others 2004

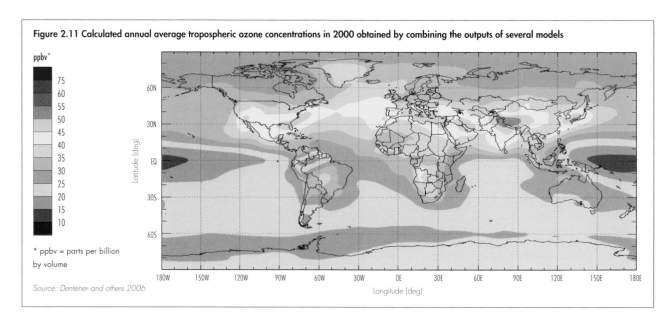

Figure 2.11 Calculated annual average tropospheric ozone concentrations in 2000 obtained by combining the outputs of several models

ppbv*

75
60
55
50
45
40
35
30
25
20
15
10

* ppbv = parts per billion by volume

Source: Dentener and others 2006

Modelling indicates the highest levels of tropospheric ozone – a major component of photochemical smog – are in a subtropical belt that includes southeastern parts of North America, southern Europe, northern Africa, the Arabian Peninsula, and the southern and northeastern parts of Asia (see Figure 2.11). However, there is currently a lack of rural measurements in Asia, Africa and Latin America that could validate these results. There is a trend of rising annual mean tropospheric ozone concentrations across the northern hemisphere (Vingarzan 2004) that implies that several regions may need to cooperate to address the problem.

In addition, clouds of tiny aerosol particles from emissions hang over a number of regions (known as Atmospheric Brown Clouds). These seasonal layers of haze reduce the amount of sunlight that can reach the Earth's surface, which has potential direct and indirect impacts on the water cycle, agriculture and human health (Ramanathan and others 2002). The aerosols and other particulate air pollutants in the atmosphere absorb solar energy and reflect sunlight back into space (Liepert 2002).

Effects of air pollution

Air pollution is one of the major environmental factors causing adverse impacts on human health, crops, ecosystems and materials, with priorities varying among regions (see Box 2.4). Both indoor and outdoor air pollution are associated with a broad range of acute and chronic impacts on health, with the specific type of the impact depending on the characteristics of the pollutant. The developing nations of northeast, southeast and southern Asia are estimated to suffer about two-thirds of the world's premature deaths due to indoor and outdoor air pollution (Cohen and others 2005).

Box 2.4 Key air pollution issues differ around the world

(See graphs presented throughout this chapter and Chapter 6 for details)

Africa, Asia and the Pacific, Latin America and the Caribbean and West Asia

■ The highest priority issue for these regions is the effect of indoor and outdoor particulates on human health, especially for women and young children exposed to indoor smoke when cooking.

■ The widespread use of poor quality fuels for industrial processes and transport represent a critical outdoor urban air pollution issue for the regions' policy-makers, especially in Asia and the Pacific.

■ Food security issues caused by growing levels of tropospheric ozone represent future challenges for parts of the regions.

■ The risks of acidic deposition are not yet well understood, but acidification already is a policy focus in parts of Asia and the Pacific.

Europe and North America

■ The priority issues for these regions are impacts of fine particulates and tropospheric ozone on human health and agricultural productivity, and the effects of nitrogen deposition on natural ecosystems.

■ The effects of SO_2 and coarse particles emissions, and acidic deposition are well understood in these regions. They have been generally successfully addressed and are of decreasing importance (see Chapter 3).

The most important air pollutant from a disease perspective is fine particulate matter. WHO estimated that particulates (see Box 2.5) in urban areas worldwide cause about 2 per cent of mortality from cardiopulmonary disease in adults, 5 per cent of mortality from cancers of the trachea, bronchus and lung, and about 1 per cent of mortality from acute respiratory infections in children, amounting to about 1 per cent of premature deaths in the world each year (WHO 2002). In addition, the WHO estimated that indoor smoke from solid fuel causes about one-third of lower respiratory infections, about one-fifth of chronic obstructive pulmonary disease, and approximately 1 per cent of cancers of the trachea, bronchus and lung (WHO 2002). Figure 2.12 presents global estimates of the burden of disease attributable to indoor and urban PM_{10} pollution.

The health impacts of air pollution are closely linked with poverty and gender issues. Women in poor families bear a disproportionate burden of the impacts of air pollution due to their greater exposure to smoke from poor quality fuel for cooking. In general, the poor are more exposed to air pollution due to the location of their residences and workplaces, and their increased susceptibility due to such factors as poor nutrition and medical care (Martins and others 2004).

Air pollution also adversely affects agriculture. Measurable, regional-scale impacts on crop yields caused by tropospheric ozone have been estimated to cause economic losses for 23 arable crops in Europe in the range US$5.72–12 billion/year (Holland and others 2006). There is evidence of significant adverse effects on staple crops in some

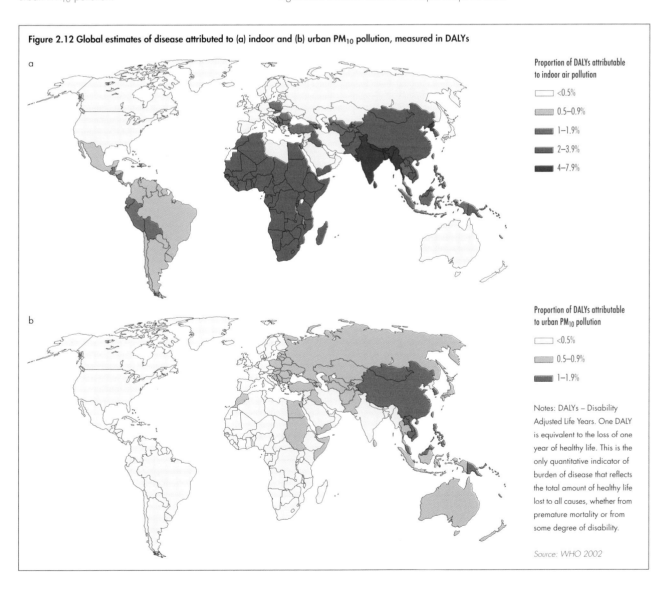

Figure 2.12 Global estimates of disease attributed to (a) indoor and (b) urban PM$_{10}$ pollution, measured in DALYs

a

Proportion of DALYs attributable to indoor air pollution

☐ <0.5%
▨ 0.5–0.9%
▨ 1–1.9%
▨ 2–3.9%
■ 4–7.9%

b

Proportion of DALYs attributable to urban PM$_{10}$ pollution

☐ <0.5%
▨ 0.5–0.9%
■ 1–1.9%

Notes: DALYs – Disability Adjusted Life Years. One DALY is equivalent to the loss of one year of healthy life. This is the only quantitative indicator of burden of disease that reflects the total amount of healthy life lost to all causes, whether from premature mortality or from some degree of disability.

Source: WHO 2002

Box 2.5 The health impacts of fine particles

The health impacts of particles depend considerably on their physical and chemical characteristics. Particle size is important, as this influences how easily and deeply the particles get into the lungs. The ability of the body to protect itself against inhaled particles, and the susceptibility of individuals to particles are closely linked with particle size and chemical composition. Particles larger than 10 μm in diameter generally do not penetrate into the lungs, and have a short residence time in the atmosphere. As a consequence, epidemiological evidence generally links PM_{10} and $PM_{2.5}$ particles with adverse effects on health.

There has been more interest recently in ultrafine particles (those having a diameter less than 0.1 μm), because poorly soluble ultrafine particles can move from the lung to the blood, and then to other parts of the body. Scientists know that chemical composition and size of particles are often linked to health effects, and that particle number and surface area are also important factors in assessing the health effects of particles. However, there is little detailed understanding yet of the specific chemical components of particles responsible for adverse health outcomes.

Source: Lippmann 2003, Pope and Dockery 2006

developing countries, such as India, Pakistan and China, which are now starting to deal with this issue (Emberson and others 2003) (see also the example in Figure 2.13).

In 1987 the regional impacts of acid rain caused by sulphur and nitrogen deposition were of major importance in Europe and North America, causing lake acidification and forest decline, mainly due to soil acidification. More recently, such declines have also been documented in Mexico and China, and are probably occurring in many other countries (Emberson and others 2003). There is recent evidence that emission controls led to a reversal of freshwater acidification (Skjelkvåle and others 2005), and the dire warnings related to widespread forest decline across Europe and North America at the time of the Brundtland Commission have not materialized. There is now a risk of acidification in other areas of the world, particularly Asia (Ye and others 2002, Kuylenstierna and others 2001, Larssen and others 2006) (see Chapter 3 and 6).

Over recent decades the eutrophying effect of nitrogen deposition has also caused significant loss of biodiversity in some sensitive, nutrient limited ecosystems, such as heaths, bogs and mires in northern Europe and North America (Stevens and others 2004). Nitrogen deposition has been recognized within the Convention on Biological Diversity as a significant driver of species loss. Several major global biodiversity hot spots have been identified as being at significant risk because of nitrogen deposition (Phoenix and others 2006) (see Chapters 4, 5 and 6).

Figure 2.13 The impact of local air pollution on the growth of wheat in suburban Lahore, Pakistan

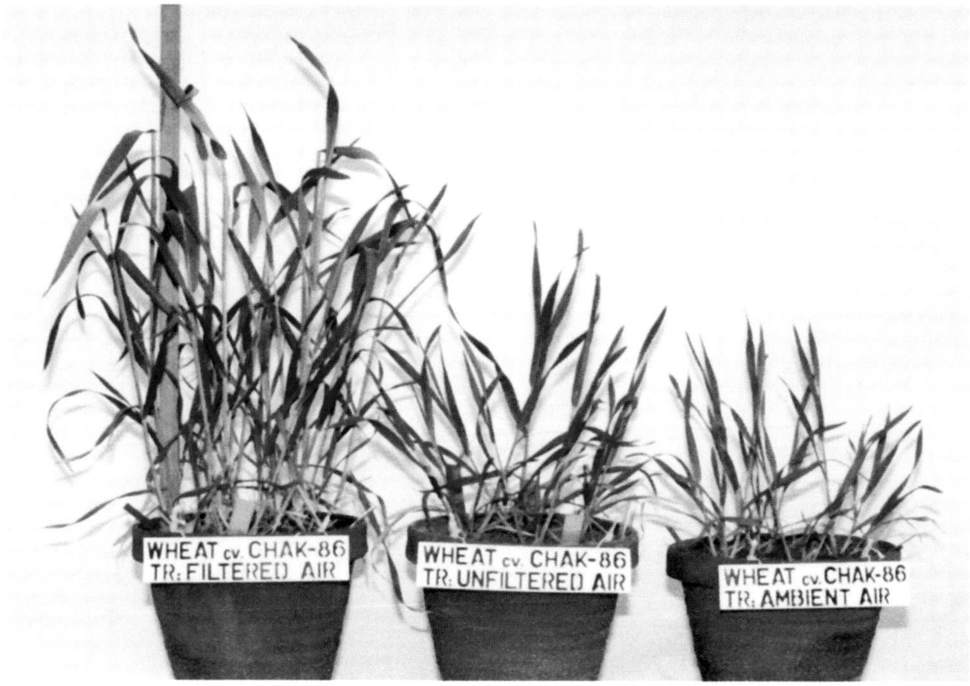

Note: The plants in the centre and on the right were both grown in local air, while the plant on the left was grown in filtered air. The effect of filtering the polluted air increased grain yield by about 40 per cent.

Credit: A. Wahid

The built environment is affected by air pollution in several ways. Soot particles and dust from transport are deposited on monuments and buildings, SO_2 and acid deposition induces corrosion of stone and metal structures and ozone attacks many synthetic materials, decreasing their useful life, and degrading their appearance. All these effects impose significant costs for maintenance and replacement. In addition, fine particles in urban environments typically reduce visibility by one order of magnitude (Jacob 1999).

Persistent organic pollutants (POPs) and mercury have emerged as important issues since 1987. These toxic substances become volatile when emitted to the environment, and can then be transported over long distances. When pollutants are persistent, concentrations will build up in the environment, causing a risk of bioaccumulation in food chains. Many POPs are now found around the globe, even far from their sources. In the Arctic environment, harmful health effects have been observed in northern wildlife, and the pollution threatens the integrity of traditional food systems and the health of indigenous peoples (see Chapter 6).

Managing air pollution

Progress in managing air pollution presents a mixed picture. Urban air pollution remains a critical issue, affecting people's health in many developing countries, although progress is evident in high-income countries. Some regional air pollution issues, such as acid rain, have been successfully addressed in Europe, but they pose a threat in parts of Asia. Tropospheric ozone has emerged as a particularly intractable problem, mainly in the northern hemisphere, where it affects crops and health. Burning biomass fuels indoors in developing regions imposes an enormous health burden on poor families, especially women and young children. Action in developing countries has been inadequate to date, but there remains an opportunity to improve health and reduce premature mortality.

The considerable progress that has been made in preventing and controlling air pollution in many parts of the world has been achieved largely through command-and-control measures, both at the national and regional levels. At the national level, many countries have clean air legislation that set emission and ambient air quality standards to protect public health and the environment. At the regional level, examples include the Convention on Long-Range Transboundary Air

Pollution (UNECE 1979–2005), the Canada-U.S. Air Quality Agreement (Environment Canada 2006) and European Union legislation (EU 1996, EU 1999, EU 2002). Other emerging regional intergovernmental agreements include the ASEAN Haze Agreement (ASEAN 2003), the Malé Declaration on the Control and Prevention of Air Pollution in South Asia (UNEP/RRC-AP 2006), and the Air Pollution Information Network for Africa (APINA), a regional science-policy network. At the global level, the Stockholm Convention on Persistent Organic Pollutants (Stockholm Convention 2000) regulates the use and emission of certain pollutants (POPs). Although the Brundtland Commission highlighted the issue of mercury in the environment, no global agreement to limit mercury contamination has been reached. There has been a global mercury programme operational since 2001, and changes in technology and the use of alternative compounds seem to have reduced emissions (UNEP/Chemicals 2006).

Transport emissions

Fuel and vehicle technologies have improved substantially during the last two decades, driven both by technological and legislative developments. Vehicle emissions have been partially controlled by the removal of lead from gasoline, requirements for catalytic converters, improved evaporative emission controls, fuel improvements, on-board diagnostic systems and other measures. Diesel vehicle emissions have been reduced by improved engine design and, for some vehicles, particle traps. Widespread use of particle traps will await reductions of sulphur in diesel fuel to below 15 ppm. Current diesel fuel sulphur levels differ considerably among regions (see Figure 2.14). Reducing sulphur in gasoline to low levels enables use of more effective catalytic converters, thus leading to improved emission control. Hybrid gasoline-electric vehicles, which tend to be more fuel efficient in urban traffic than gasoline-only vehicles, have been introduced in many developed countries, but their use is still very limited.

Most developed countries have made substantial progress in reducing per vehicle emissions, and many middle-income countries have implemented significant measures to control vehicle emissions. In addition to improved vehicle technologies, effective vehicle inspection and maintenance programmes have helped to control vehicle emissions and enforce emission standards (Gwilliam and others 2004). However, progress in some low-income countries has been slow. Developing countries will not achieve benefits of advanced emission control technologies unless they implement cleaner fuel options.

In some Asian countries motorized two- and three-wheeled vehicles contribute disproportionately to emissions. However, regulations in some nations are reducing emissions from these vehicles. The shift from two-cycle to four-cycle engines, and the introduction of emission standards that effectively ban the sale of new vehicles powered by two-cycle engines will, in time, lead to a significant improvement in vehicle emissions (WBCSD 2005, Faiz and Gautam 2004).

Mass transport is an important alternative to private vehicles, and has been successfully implemented in many cities by using light rail, underground and rapid bus transit systems (Wright and Fjellstrom 2005). Fuel switching from diesel to compressed natural gas has been implemented for public transport vehicles in cities such as Delhi and Cairo, leading to reductions in emissions of particulate matter and SO_2. In many countries, widespread use of mass transport continues to be hampered, however, by inefficiency and negative perceptions.

Industrial and energy sector emissions
In many developed countries emissions from large industrial sources have been controlled by fuel changes and emission control laws. The reduction of SO_2 emissions in Europe and North America has been one of the success stories of recent decades. Agreements such as the 1979 UNECE Convention on Long-Range Transboundary Air Pollution played an important role in this success. The ECE convention adopted the concept of critical loads (thresholds in the environment) in 1988 and, in 1999, the Gothenburg Protocol set targets for national emissions of SO_2, NO_X, NH_X and VOCs. In Europe, SO_2 has been reduced considerably, partly due to these agreements. It is also the result of policies calling for cleaner fuels, flue gas desulphurization and new industrial processes. Emissions also fell as the result because of the demise of many heavy industries, particularly in Eastern Europe and the former Soviet Union. However, SO_2 emissions have increased in many developing country regions.

Stricter environmental regulation and economic instruments, such as emissions trading, have triggered the introduction of cleaner technologies, and promoted further technological innovation.

Economic policies send important signals to producers and consumers. For example, Europe is shifting from taxing labour to taxing energy use to better reflect the impacts of emissions (Brown 2006). Other successful examples include cap-and-trade policies in the United States to reduce SO_2 emissions from power plants (UNEP 2006). International use of such economic instruments is growing (Wheeler 1999). Many cleaner technologies and cleaner production options are mature and commercially available, but there is great need for global cooperation regarding technology transfer to make them more widely available.

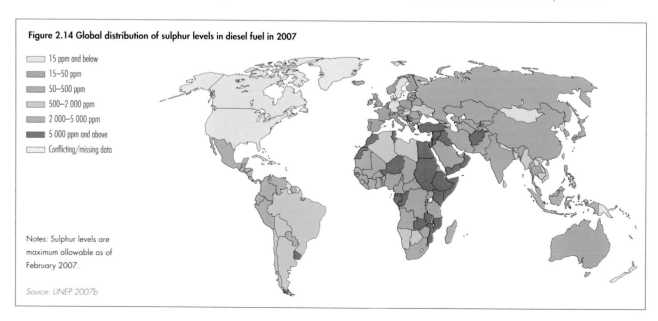

Figure 2.14 Global distribution of sulphur levels in diesel fuel in 2007

- 15 ppm and below
- 15–50 ppm
- 50–500 ppm
- 500–2 000 ppm
- 2 000–5 000 ppm
- 5 000 ppm and above
- Conflicting/missing data

Notes: Sulphur levels are maximum allowable as of February 2007.

Source: UNEP 2007b

Indoor air quality

With some 1.6 million people dying prematurely each year from exposure to polluted indoor air (WHO 2006c), many developing countries in Africa, Asia and Latin America have attempted to address the emissions from the burning of biomass fuels and coal indoors. Responses include providing households with improved stoves, cleaner fuels, such as electricity, gas and kerosene, and information and education to make people aware of the impacts of smoke on the health of those exposed, especially women and young children. A modest shift from solid biomass fuels, such as wood, dung and agricultural residues, to cleaner fuels has been achieved, and governments have supported such measures, but further progress along such lines is urgently necessary if any major advances are to be realized (WHO 2006c).

CLIMATE CHANGE

The trend of global warming is virtually certain, with 11 of the last 12 years (1995–2006) ranking among the 12 warmest years since 1850, from which time there has been systematic temperature keeping (IPCC 2007). The evidence of this warming includes a number of shrinking mountain glaciers (Oerlemans 2005), thawing permafrost (ACIA 2005), earlier breakup of river and lake ice, lengthening of mid- to high-latitude growing seasons, shifts of plant, insect and animal ranges, earlier tree flowering, insect emergence and egg laying in birds (Menzel and others 2006), changes in precipitation patterns and ocean currents (Bryden and others 2005), and, possibly, increasing intensity and lifetimes of tropical storms in some regions (IPCC 2007, Webster and others 2005, Emanuel 2005).

Poor communities are most directly dependent for their livelihoods on a stable and hospitable climate. In developing countries the poor, often relying on rain-fed subsistence agriculture and gathered natural resources, are deeply dependent on climate patterns, such as the monsoons, and are most vulnerable to the devastation of extreme weather events, such as hurricanes. Vulnerable communities already suffer from climate variability, for example due to increasing frequency of droughts in Africa (AMCEN and UNEP 2002) and, as was demonstrated by the effects of Hurricane Katrina in 2005, and by the European heat wave of 2003, it is the poor or vulnerable who suffer most from weather extremes, even within relatively affluent societies.

While the Earth's climate has varied throughout the prehistoric ages, the last few decades have witnessed a global climate disruption that is unprecedented over the recent millennia, a period of relative climatic stability during which civilization emerged (Moberg and others 2005, IPCC 2007). Some regions, particularly the Arctic, will be more affected by climate change than others closer to the equator (see Polar Regions section of Chapter 6). In many regions, the agricultural sector will be particularly affected. The combination of high temperatures and decreased soil moisture projected for parts of Africa will be particularly hard to adapt to. With the majority of the world's population struggling to meet basic development needs, such as those identified in the Millennium Development Goals, humanity can ill afford this additional burden of climate change impacts (Reid and Alam 2005).

Many developing countries have attempted to address health concerns from the burning of biomass fuels and coal indoors, through responses such as providing households with improved, fuel saving stoves.

Credit: Charlotte Thege/Das Fotoarchiv/Still Pictures

Figure 2.15 Atmospheric concentrations of CO₂ over the last 10 000 years

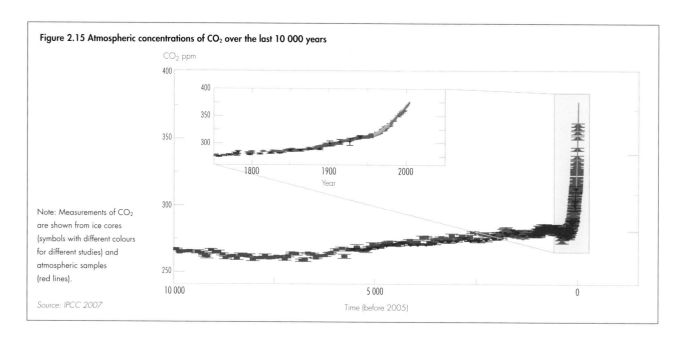

CO₂ ppm

Note: Measurements of CO₂ are shown from ice cores (symbols with different colours for different studies) and atmospheric samples (red lines).

Source: IPCC 2007

Time (before 2005)

Figure 2.16 CO₂ emissions from fossil fuels by region

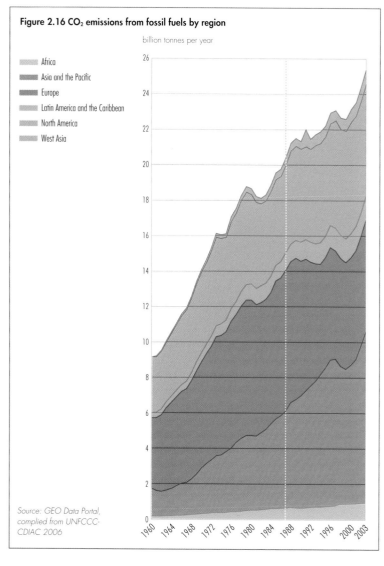

billion tonnes per year

- Africa
- Asia and the Pacific
- Europe
- Latin America and the Caribbean
- North America
- West Asia

Source: GEO Data Portal, complied from UNFCCC-CDIAC 2006

Greenhouse gas concentrations and anthropogenic warming

The greatest direct human pressure on the climate system arises from the emission of greenhouse gases, chief of which is CO_2, mainly originating from fossil fuel consumption. Since the dawn of the industrial age, the concentrations of these gases have been steadily increasing in the atmosphere. Figure 2.15 shows the atmospheric concentration of CO_2 over the past 10 000 years. The unprecedented recent rise has resulted in a current level of 380 parts per million, much higher than the pre-industrial (18th century) level of 280 ppm. Since 1987, annual global emissions of CO_2 from fossil fuel combustion have risen by about one-third (see Figure 2.16), and the present per capita emissions clearly illustrate large differences among regions (see Figure 2.17).

There has also been a sharp rise in the amount of methane, another major greenhouse gas, with an atmospheric level 150 per cent above that of the 19th century (Siegenthaler and others 2005, Spahni and others 2005). Examination of ice cores has revealed that levels of CO_2 and methane are now far outside their ranges of natural variability over the preceding 500 000 years (Siegenthaler and others 2005).

There are other atmospheric pollutants that affect the planet's heat balance. They include industrial gases, such as sulphur hexafluoride, hydrofluorocarbons and perfluorocarbons; several ozone-depleting gases that are regulated under the Montreal Protocol; tropospheric ozone; nitrous oxide; particulates; and sulphur- and carbon-based aerosols from burning fossil fuels and biomass. Elemental carbon aerosols (soot or "black carbon") contribute to global warming by absorbing short-wave radiation, while also contributing to local air pollution. Removing such pollutants will be beneficial both with respect to climate change and health effects. Sulphur-based aerosol pollutants, on the other hand, cool the planet through their influence on the formation of clouds, and by scattering incoming sunlight, and are thus currently "shielding" the planet from the full warming effect of greenhouse gas emissions (IPCC 2007). In the future, the policy measures needed to reduce public health problems and local environmental impacts associated with sulphur-based pollutants will weaken this unintended but fortunate shielding.

The Earth's surface temperature has increased by approximately 0.74°C since 1906, and there is very high confidence among scientists that the globally averaged net effect of human activities since 1750 has been one of warming (IPCC 2007). The warming of the last few decades is exceptionally rapid in comparison to the changes in climate during the past two millennia. It is very likely that the present temperature has not been exceeded during this period.

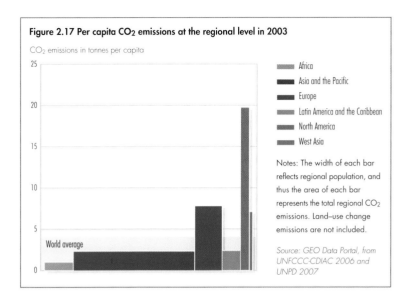

Figure 2.17 Per capita CO$_2$ emissions at the regional level in 2003

CO$_2$ emissions in tonnes per capita

- Africa
- Asia and the Pacific
- Europe
- Latin America and the Caribbean
- North America
- West Asia

Notes: The width of each bar reflects regional population, and thus the area of each bar represents the total regional CO$_2$ emissions. Land–use change emissions are not included.

Source: GEO Data Portal, from UNFCCC-CDIAC 2006 and UNPD 2007

Earlier discrepancies between surface temperature measurements and satellite measurements have been largely resolved (Mears and Wentz 2005). Model calculations including both natural and anthropogenic drivers give quite good agreement with the observed changes since the beginning of the industrial age (see Figure 2.18). Most of the warming over the last century has occurred in recent decades, and this more rapid warming cannot be accounted for by changes in solar radiation or any other effects related to the sun that have been examined (IPCC 2007).

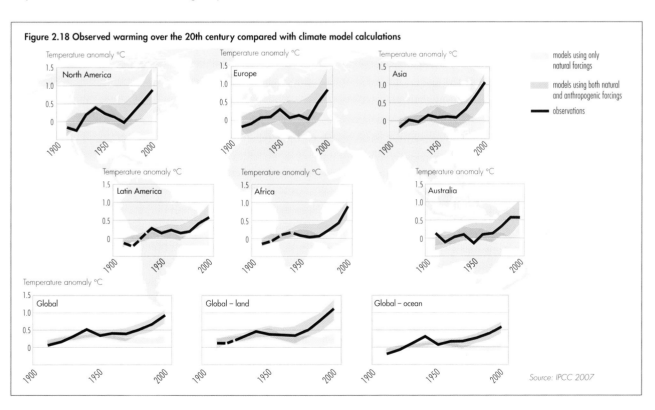

Figure 2.18 Observed warming over the 20th century compared with climate model calculations

Source: IPCC 2007

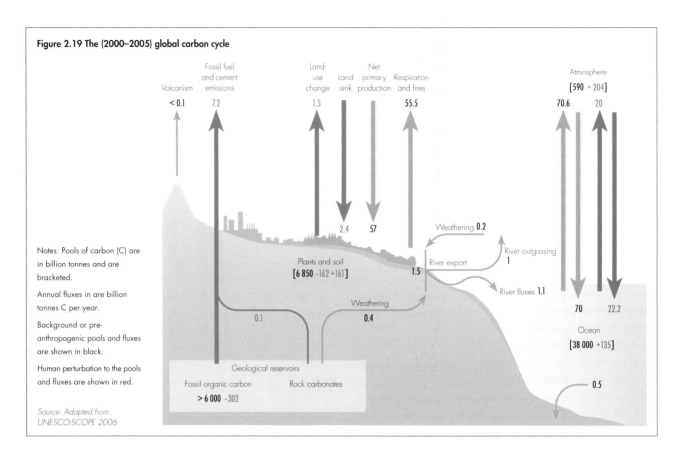

Figure 2.19 The (2000–2005) global carbon cycle

Notes: Pools of carbon (C) are in billion tonnes and are bracketed.

Annual fluxes in are billion tonnes C per year.

Background or pre-anthropogenic pools and fluxes are shown in black.

Human perturbation to the pools and fluxes are shown in red.

Source: Adapted from UNESCO-SCOPE 2006

The climate system possesses intrinsic positive and negative feedback mechanisms that are generally beyond society's control. The net effect of warming is a strong positive feedback (IPCC 2001b), with several processes within the Earth's complex climate system (see Figure 2.19 for the stocks and flows of carbon on a global scale) acting to accelerate warming once it starts (see Box 2.6 below). The magnitude of such feedbacks is the subject of intense study.

What is known is that the Earth's climate has entered a state that has no parallel in the recent prehistory. The cumulative result of these feedbacks will be far greater than the "direct" warming caused by the increase in greenhouse gas emissions alone.

Effects of climate change

Spells of very high temperatures appear to be increasing as global temperatures increase. A notable

---

Box 2.6 Positive feedbacks in the Earth system

A first important positive feedback is the increase in the amount of water vapour in the atmosphere that will result from higher air and ocean temperatures. The ability of air to hold moisture increases exponentially with temperature, so a warming atmosphere will contain more water vapour, which in turn will enhance the greenhouse effect. Recent observations confirm that the atmosphere water vapour concentration increases with a warming planet.

Another important feedback is the loss of snow and sea ice that results from rising temperatures, exposing land and sea areas that are less reflective, and hence more effective at absorbing the sun's heat. Over the last few decades, there is a documented decline in alpine glaciers, Himalayan glaciers and Arctic sea ice, (see Chapters

3 and 6). A third feedback is the melting of permafrost in boreal regions, resulting in the release of methane, a potent greenhouse gas, and CO_2 from soil organic matter. Recent studies in Siberia, North America and elsewhere have documented the melting of permafrost. A fourth important feedback is the release of carbon from ecosystems due to changing climatic conditions. The dieback of high-carbon ecosystems, such as the Amazon, due to changes in regional precipitation patterns, has been predicted from some models, but it has not yet been observed. Laboratory studies have indicated accelerated decomposition of soil organic matter in temperate forests and grasslands due to temperature and precipitation changes, or the CO_2-induced enhancement of decomposition by mycorrhizae.

Sources: ACIA 2005, Cox and others 2004, Heath and others 2005, Soden and others 2005, Walter and others 2006, Zimov and others 2006

recent case is the exceptional heat wave experienced in much of Europe in the summer of 2003, with over 30 000 estimated premature deaths from heat stress and associated air pollution (UNEP 2004). In the Arctic, average temperatures are rising almost twice as rapidly as in the rest of the world. Widespread melting of glaciers and sea ice, and rising permafrost temperatures present further evidence of strong Arctic warming. Since 1979, satellite observation has allowed scientists to carefully track the extent of seasonal melting of the surface of the Greenland Ice Sheet (see Figure 2.20). There is now also evidence of widespread melting of permafrost, both in Alaska and Siberia, which is expected to increase the release of methane from frozen hydrates, giving rise to a significant positive feedback (see Box 2. 6 above and the Polar Regions section in Chapter 6). This phenomenon has a precedent, as a vast amount of methane was emitted some 55 million years ago, and was associated with a temperature increase of 5–7°C (Dickens 1999, Svensen and others 2004). It took approximately 140 000 years from the start of the emission period to return to a "normal" situation.

Trends in global patterns (Dore 2005) reveal increased variance in precipitation everywhere: wet areas are becoming wetter, and dry and arid areas are becoming dryer. It is notable that the regions with the lowest contribution to anthropogenic GHG emissions, such as Africa, are those projected to be most vulnerable to their negative consequences, especially in the form of water stress (IPCC 2001b) (see Chapters 4 and 6).

There is observational evidence for an increase of intense tropical cyclone activity in the North Atlantic since about 1970, correlated with increases in tropical sea surface temperatures. There are also suggestions of more intense tropical cyclone activity in some other regions, where concerns over data quality are greater (IPCC 2007). The number of the most intense tropical storms (Class 4 and 5) has nearly doubled over the past 35 years, increasing in every ocean basin. This is consistent with model results that suggest this trend will continue in a warming world (Emanuel 2005, Trenberth 2005, Webster and others 2005). If correct, this would suggest an increasing frequency in the future

Figure 2.20 Seasonal melting of the Greenland Ice Sheet

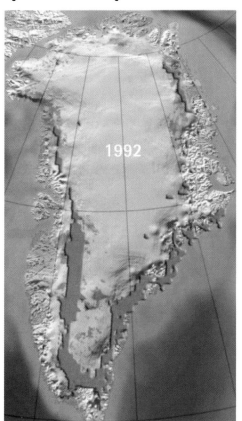

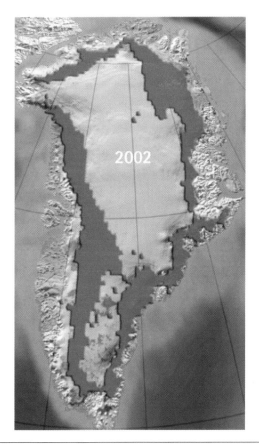

Note: The areas in orange/red are the areas where there is seasonal melting at the surface of the ice sheet.

Source: Steffen and Huff 2005

of devastatingly intense hurricanes, such as Katrina (in 2005) and Mitch (in 1998), and cyclones such as the super cyclone of Orissa in India in 1999. However, there has been recent controversy over these conclusions (Landsea and others 2006), and the IPCC and WMO suggest that more research is necessary (IPCC 2007, WMO 2006a).

It is believed that the 20th century's anthropogenic greenhouse gas emissions, which are blamed for most of the warming up to now, have also committed the Earth to an additional 0.1°C of warming per decade that is "in the pipeline," owing to the climate system's inertia. Some warming would have occurred even if the concentrations of all greenhouse gases and aerosols in the atmosphere had been kept constant at year 2000 levels, in which case the estimated increase would be 0.3–0.9°C by the end of this century. The actual temperature change will depend critically on choices that society makes regarding the reduction in greenhouse gas emissions. The potential future scenarios span a wide range. The increase in the global mean temperature by 2090–99 is estimated to be 1.8–4.0°C, relative to 1980–99 (IPCC 2007). This is the best estimate, drawing on six emissions marker scenarios, while the likely range is 1.1–6.4°C. If CO_2 concentrations in the atmosphere double, the global average surface warming would likely be in the range 2–4.5°C, with the best estimate of about 3°C above pre-industrial levels, although values substantially higher than 4.5°C cannot be excluded (IPCC 2007). These figures are for global averages, while the predicted temperature increases will be greater in some regions.

Sea-level rise is caused by thermal expansion of water, and melting of glaciers and ice sheets. Projections by IPCC (IPCC 2007) for a rise by the end of this century, corresponding to those for temperature changes described above, range from 0.18–0.59 m. It is important to note that possible future rapid dynamic changes in ice flow are not included in these estimates. (The majority of the impact will, however, be post-2100 (see Figure 2.21). It is estimated that the Greenland Ice Sheet will become unstable if the global average temperature increases above 3°C, which may well occur in this century (Gregory and others 2004, Gregory and Huybrechts 2006). The melting would raise sea levels by about 7 metres over the next 1 000 years. However, the mechanisms involved in melting of ice sheets are not well understood, and some scientists argue that melting may be much quicker due to dynamic process not yet incorporated in model predictions (such as Hansen 2005). Research is continuing to evaluate the further potential impacts on sea levels from the West Antarctic Ice Sheet (Zwally and others 2005). There are a number of small island states whose very existence is already being threatened by sea-level rise associated with climate change (IPCC 2001c).

The future temperatures in northern Europe are dependent on the fate of the North Atlantic Current (Gulf Stream) that transports warm water to the Norwegian Sea, and further northwards. Model predictions vary, but in general forecast a weakening, but no total shutdown in this century (Curry and Maurtizen 2005, Hansen and others 2004). A significant shift could greatly affect

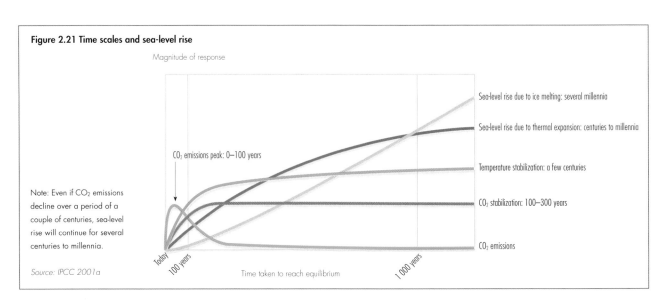

Figure 2.21 Time scales and sea-level rise

Magnitude of response

Note: Even if CO_2 emissions decline over a period of a couple of centuries, sea-level rise will continue for several centuries to millennia.

Source: IPCC 2001a

CO_2 emissions peak: 0–100 years

Sea-level rise due to ice melting: several millennia

Sea-level rise due to thermal expansion: centuries to millennia

Temperature stabilization: a few centuries

CO_2 stabilization: 100–300 years

CO_2 emissions

Today

100 years

1 000 years

Time taken to reach equilibrium

regional weather patterns, with major ramifications for ecosystems and human activities (see Chapter 4 and Chapter 6, Polar Regions).

Over the past 200 years the oceans have absorbed nearly half the CO_2 produced by human activities. One effect has been to produce carbonic acid, thus increasing acidity and lowering the pH of surface seawater by 0.1 pH unit. Projections based on different emission scenarios give additional reductions in average global surface ocean pH of between 0.14 and 0.35 units by the year 2100 (IPCC 2007). This seawater acidity is probably higher than has been experienced for hundreds of millennia, and there is convincing evidence that such acidification will impair the process of calcification by which animals, such as corals and molluscs, make their shells from calcium carbonate (Royal Society 2005b, Orr and others 2005).

Initially a slight warming, together with the fertilizing effects of more atmospheric CO_2, may increase crop yields in some areas, but the negative effects are expected to dominate as warming increases (IPCC 2001c). Some sub-regions in Africa (see Chapter 6) are especially vulnerable, and studies warn that there may be an alarming increase in the risk of hunger (Royal Society 2005a, Royal Society 2005b, Huntingford and Gash 2005).

Using projections of species distributions for future climate scenarios, Thomas and others (Thomas and others 2004a, Thomas and others 2004b) assessed extinction risks for 20 per cent of the Earth's terrestrial surface. They estimated that a climate warming of 2°C by 2050 would cause 15–37 per cent of species and taxa in these regions to be "committed to extinction." Certain extinctions have already been attributed to climate change, such as the loss of numerous species of Harlequin frog in mountainous parts of South America (Pounds and others 2006) (see Chapter 5).

Although higher CO_2 levels promote photosynthesis, and may help to maintain rain forests in the next few decades, continued warming and drying could eventually lead to abrupt reductions in forest cover (Gash and others 2004). Some models predict a dramatic dieback of Amazonian rain forests, which will release CO_2, and cause a positive feedback to climate change. In addition to adding considerably to global CO_2 emissions, the loss of large tracts of the Amazon would radically transform the habitat,

and threaten the livelihoods of local indigenous communities. Similarly, the melting of the permafrost will dramatically change the ecosystems and livelihoods in northern latitudes (see Chapter 6).

In 2000, climate change was estimated to be responsible for approximately 2.4 per cent of worldwide diarrhoea, and 6 per cent of malaria in some middle-income countries (WHO 2002). Diarrhoea and malaria are already devastating forces in developing countries, and the likelihood that they will be exacerbated by climate change is of significant concern. Continued warming is expected to cause shifts in the geographic range (latitude and altitude) and seasonality of certain infectious diseases, including vector-borne infections, such as malaria and dengue fever, and food-borne infections, such as salmonellosis, which peak in the warmer months. Some health impacts will be beneficial. For example, milder winters will reduce the winter peak in deaths that occurs in temperate countries. However, overall it is likely that negative health impacts of climate change will by far outweigh the positive ones. WHO and Patz and others give estimates of changes in morbidity and mortality due to changes in climate by the year 2000, compared with the baseline climate of 1961–1990 (Patz and others 2005, WHO 2003). They estimated there were 166 000 more deaths worldwide, mostly in Africa and some in Asian countries, and mainly from malnutrition, diarrhoea and malaria. The largest increase in the risks by 2025 will be from flooding, with more modest increases in diseases such as diarrhoea and malaria. The regions facing the greatest burden from climate-sensitive diseases are also the regions with the lowest capacity to adapt to such new hazards.

Managing climate change

Climate change is a major challenge to society's existing policy making apparatus, as it presents a threat whose precise magnitude is unknown, but is potentially massive. The conventional cost-benefit framework is difficult to apply to climate policy. Not only are both the costs and impacts highly uncertain, but the cost-benefit analyses are critically sensitive to parameters, such as the choice of discount rate, which reflect the relative importance placed on climate damages suffered by future generations, and the temperature increase expected. There is no consensus on the best approach(es) to use in such cases, and they are inherently value laden (Groom and others 2005, Stern 2006).

The impacts of decisions made today will continue to emerge for decades or centuries. Faced with such a challenge, a precautionary approach seems inevitable. A minimal response would involve setting a threshold for intolerable impacts. Various scientists, analysts and policy making bodies have identified a 2°C increase in the global mean temperature above pre-industrial levels as a threshold beyond which climate impacts become significantly more severe, and the threat of major, irreversible damage more plausible. Some argue for an even lower threshold (Hansen 2005). Hare and Meinshausen have concluded that staying under the 2°C threshold will require a very stringent GHG concentration goal, and the longer the delay in implementation, the steeper the reduction trajectory required (see path 2 in Figure 2.22) (Hare and Meinshausen 2004).

Governments worldwide, in cooperation with the private sector and the public, have been implementing various policies and measures to mitigate climate change (see Table 2.3). These actions comprise a crucial first wave of efforts to limit GHG emissions, and to ultimately achieve a transition away from carbon intensive economies. While there are many important actions to address climate change, such as carbon taxes and carbon trading in Europe, and the coming into force of the Kyoto Protocol, the net effect of current actions is woefully inadequate (see Chapter 6).

A comprehensive system of actions and measures, including public-private partnerships is required (see Chapter 10). Achieving the required global emission reductions will clearly require a concerted global effort by both industrialized and developing countries. Even though per capita emissions in some rapidly industrializing developing countries are far lower than in industrialized countries, their emissions are rising as their economies grow, and their living standards rise.

Several technologically feasible options are available to address climate change in all countries, and many of them are economically competitive, especially when the co-benefits of increased energy security, reduced energy costs and lower impacts of air pollution on health are considered (Vennemo and others 2006, Aunan and others 2006). These include improvements in energy efficiency and a shift to low-carbon and renewable resources, such as solar, wind, biofuels and geothermal energy. Social changes that make less consumptive, less material-intensive lifestyles possible may also be necessary. Carbon capture and storage, for example by storing CO_2 deep underground, and other technological options, such as nuclear energy, may play significant roles in the future, although some questions remain regarding widespread application of such options, such as public concerns and political debate over nuclear energy related to the future of used nuclear fuel, the risk of accidents, high costs and proliferation of nuclear weapons.

Recent studies show that measures to mitigate climate change do not necessarily imply exorbitant costs, and that total cost would remain a very small fraction of the global economy (Stern 2006, Edenhofer and others 2006). Azar and Schneider reported that the increase in the global economy expected over the coming century would not be compromised, even by the most stringent stabilization targets (350–550 ppm), and the point at which the global economy would reach its 2100 level of wealth, according to business-as-usual projections, would be delayed by only a few years (Azar and Schneider 2002). DeCanio attributes the common perception of high mitigation costs to the fact that current modelling frameworks tend to be biased strongly towards overestimating costs (DeCanio 2003).

Some impacts of climate change are inevitable in the coming decades due to the inertia of the climate system. Adaptation is necessary, even if major mitigation measures are rapidly

---

Figure 2.22 Paths to reach a 400 ppm CO₂-equivalent greenhouse gas concentration target (Kyoto gas emissions plus land use CO₂)

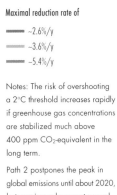

Maximal reduction rate of
- ~2.6%/y
- ~3.6%/y
- ~5.4%/y

Notes: The risk of overshooting a 2°C threshold increases rapidly if greenhouse gas concentrations are stabilized much above 400 ppm CO₂-equivalent in the long term.

Path 2 postpones the peak in global emissions until about 2020, but requires subsequent annual emissions reductions at an exceptionally challenging pace of more than 5 per cent/year.

Source: Den Elzen and Meinshausen 2005

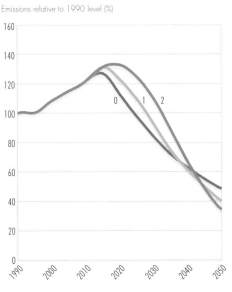

implemented. Adaptation to climate change is defined as "adjustment in natural or human systems in response to actual or expected climatic stimuli or their effects, which moderates harm or exploits beneficial opportunities" (IPCC 2001b). Developing new varieties of crops that resist droughts and floods, and climate proofing infrastructure to cope with future impacts of climate change are a few examples. Adaptation is often site-specific, and must be designed on the basis of local circumstances. National and international policies and financial mechanisms are crucial to facilitate such efforts. However, weak institutional mechanisms, inadequate financial resources, insufficient research on adaptation and the failure to mainstream adaptation concerns in development planning have so far hampered progress on adaptation. Adaptation responses call for additional financial resources, and

the polluter-pays-principle would in general imply that countries should provide resources in proportion to their contribution to climate change.

An extensive multilateral infrastructure exists to address climate change at the international level. The United Nations Framework Convention on Climate Change (UNFCCC) was signed in 1992 at the UN Earth Summit, and has been ratified by 191 counties. It encourages countries to work together to stabilize GHG emissions "at a level that would prevent dangerous anthropogenic interference with the climate system." Recognizing that binding obligations are necessary to achieve the objective, countries adopted the Kyoto Protocol in 1997, and more than 160 have ratified it. The protocol acknowledges that the industrialized countries must lead efforts to address climate

Table 2.3 Selected policies and measures to mitigate climate change

Nature	Policies	Measures
Target-oriented GHG emissions reduction measures	International	36 countries and the European Community accepted targets under the Kyoto Protocol
	State or province	14 states in the United States, and many provinces in other countries adopted targets (Pew Centre on Global Climate Change 2007)
	City or local government	>650 local governments worldwide, and 212 US cities in 38 states adopted targets (Cities for Climate Protection – CCP)
	Private sector	For example, Climate Leaders Programme of USEPA – 48 companies (USEPA 2006)
Regulatory measures	Energy process and efficiency improvements	Energy efficiency portfolio standards, appliance efficiency standards, building codes, interconnection standards
	Renewable energy improvement	Renewable energy portfolio standards (RPS) Biofuels standard (for example, US Energy Policy Act of 2005 mandates 28.4 billion litres of biofuel/year in 2012) (DOE 2005)
	Raw material improvements	Industrial standards, research development and demonstration (RD&D)
	Fuel switching	Mandatory standards, RD&D
	Recycling and reuse	Mandatory standards, awareness creation, pollution tax
Economic measures	Taxation polices	Carbon taxes, pollution tax, fuel taxes, public benefit funds
	Subsidy policies	Equipment subsidies for promotion of renewable energy sources
Technological measures	Technology commitments	Initiatives on strategic technologies, such as Generation IV Nuclear Partnership, Carbon Sequestration Leadership Forum, International Partnership for the Hydrogen Economy, Asia Pacific Partnership on Clean Development and Climate (USEIA 1999)
	New technology penetration	Technology standards Technology transfer, RD&D
	Carbon sequestration	Technology transfer, emission taxes
	Nuclear	Emission taxes, socio-political consensus
Others	Awareness raising	"Cool Biz" or "Warm Biz" campaigns

change, and commits those included in Annex B to the protocol to emissions targets. The United States and Australia (both included in Annex B) have chosen not to ratify, so far. The 36 countries with binding commitments comprise roughly 60 per cent of total industrialized country baseline emissions.

Besides the actions and measures to be taken by parties at the national level, the Kyoto Protocol allows for three flexible implementation mechanisms: emissions trading, Joint Implementation and the Clean Development Mechanism (CDM). International emissions trading is an approach under which Annex B countries can supplement domestic reductions. Under the latter two mechanisms, Annex I parties may invest in mitigation activities in other countries, and thereby generate emission reduction credits that can be used toward compliance with their own obligations. Many but not all countries appear to be on track to meet their targets during the 2008–2012 compliance period (UNFCCC 2007).

The CDM had been advanced as a unique opportunity for promoting sustainable development in developing countries in return for undertaking emission reductions, with financial and technological assistance from developed countries. However, progress to date suggests that the emphasis has been more on reducing the cost of mitigation rather than on facilitating sustainable development. There are growing calls to strengthen the CDM beyond 2012 to secure more sustainable development benefits (Srinivasan 2005).

Kyoto commitments end in 2012 and early clarification of the post-2012 regime is required. At the second meeting of parties in Nairobi in 2006, countries agreed in principle that there should be no gap between the 2012 commitments and the next period of commitments. To that end, they set a target of completing a review of the Kyoto Protocol by 2008, in preparation for establishing the next set of commitments. With regard to adaptation, the parties agreed on principles for governing the Adaptation Fund – the Kyoto instrument for distributing resources to developing countries to support adaptation – with hopes that funds might be disbursed within the next few years.

The ultimate success of global efforts in mitigation and adaptation can be realized only if climate concerns are mainstreamed in development planning at national and local levels. Since most GHG emissions are from energy, transport and agricultural land use, it is crucial to integrate climate concerns in these sectors, both at policy and operational levels, to achieve maximum co-benefits, such as improvements in air quality, generation of employment and economic gains. Setting mandatory targets for renewable energy and energy efficiency in these sectors may be an example of policy-level mainstreaming. The replacement of fossil fuels with biofuels to reduce air pollution and GHG emissions is an example of mainstreaming at an operational level. Integrating climate concerns in planning for sectors such as agriculture and water resources is crucial to facilitate adaptation of communities and ecosystems.

Although political actions to cut greenhouse gases were slow in starting, a major change in the political climate began in late 2006 and early 2007. At least two events played a role in sensitizing public and political opinion: parts of Europe and North America had a very mild winter, and the IPCC released its 2007 assessment report, saying that climate change was real and evident. Many influential speakers were carrying the message, using photographs and images of melting glaciers and thinning ice in the Arctic to present visible evidence of climate warming unprecedented in the Earth's recent history. In late 2006, the US state of California passed legislation mandating a 25 per cent cut below its current emissions of greenhouse gases by 2020.

STRATOSPHERIC OZONE DEPLETION
The ozone layer
Stratospheric ozone depletion (see Box 2.7) is present everywhere to some degree, except over the tropics. Seasonal stratospheric ozone depletion is at its worst over the poles, particularly the Antarctic, and the inhabited areas most affected by the resulting increase in ultraviolet (UV-B) radiation include parts of Chile, Argentina, Australia and New Zealand.

Antarctic ozone depletion in the southern hemisphere spring has been large and increasing in extent since the Brundtland Commission report. The average area covered by the ozone hole (an area of almost total ozone depletion) has increased, though not as rapidly as it did during the 1980s, before the Montreal Protocol entered into force.

The area under the ozone hole varies from year to year (see Figure 2.23), and it is not yet possible to say whether it has hit its peak. The largest "holes" occurred in 2000, 2003 and 2006. On 25 September 2006, it extended over 29 million square kilometres and the total ozone loss was the largest on record (WMO 2006b). Chemistry-climate models predict that recovery to pre-1980 Antarctic ozone levels can be expected around 2060–2075 (WMO and UNEP 2006).

The atmosphere above the Arctic is not as cold as that above the Antarctic, so ozone depletion there is not as severe. Ozone depletion during the Arctic winter and spring is highly variable, due to changes in stratospheric meteorological conditions from one winter to another, as can be seen from the unexpected ozone losses over central Europe in the summer of 2005. A future Arctic ozone hole as severe as that of the Antarctic appears unlikely, but the population at risk from stratospheric ozone depletion in the Arctic is much higher than in the Antarctic (WMO and UNEP 2006).

Effects of stratospheric ozone depletion

UV-B radiation (medium wavelength ultraviolet radiation) causes adverse effects on human eyes, skin and immune systems, and the understanding of the mechanism through which UV-B affects health has improved in recent years (UNEP 2003). Specific mechanisms for the development of skin cancer have been identified. Quantifying the increased incidence of skin cancer cases due to

stratospheric ozone depletion is difficult, as other factors, such as lifestyle changes (for example, spending more time outdoors), also have an impact. However, in the case of Australia, where skin-reddening radiation is estimated to have increased by 20 per cent from 1980 to 1996, it is deemed probable that some of the increase in cancer incidence is due to stratospheric ozone depletion (ASEC 2001).

Managing stratospheric ozone depletion

The international community reacted to the threat of ozone depletion with the Montreal Protocol on Substances that Deplete the Ozone Layer. This led to a phase-out of production and consumption of CFCs and other ODS. The protocol was signed by governments in 1987, and entered into force two years later. Initially, it called for a 50 per cent reduction in the manufacture of CFCs by the end of the century. This was strengthened through the London (1990), Copenhagen (1992), Montreal

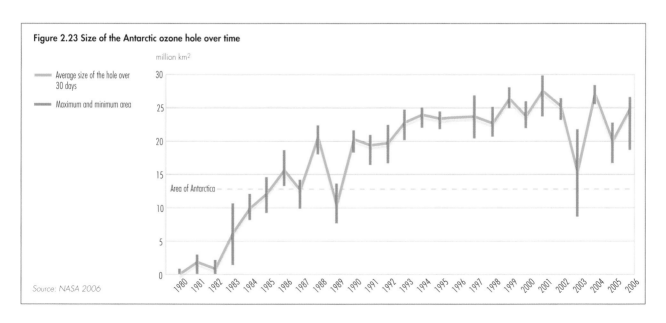

Figure 2.23 Size of the Antarctic ozone hole over time

million km²

— Average size of the hole over 30 days

— Maximum and minimum area

Area of Antarctica

Source: NASA 2006

(1997) and Beijing (1999) amendments. It is now widely regarded as one of the most effective multilateral environmental agreements in existence. In addition to CFCs, the protocol covers substances such as halons, carbon tetrachloride, methyl chloroform, hydrochlorofluorocarbons (HCFCs), methyl bromide and bromochloromethane. The latter was added to the protocol's control schedules in 1999, through the Beijing Amendment. Such amendments require a lengthy process of ratification, and other ODS with no commercial significance have not been added, though five such substances have been identified in recent years (Andersen and Sarma 2002).

The phase-out schedules under the Montreal Protocol have reduced the consumption of many ODS (see Figure 2.24). The main exceptions are HCFCs (transitional replacements for CFCs, with much lower ozone-depleting potentials) and methyl bromide. Observations in the troposphere confirm a fall in ODS levels over recent years. Changes in the stratosphere lag by a few years, but chlorine levels there are declining. The bromine concentrations in the stratosphere have still not decreased (WMO and UNEP 2006).

Other than for a few essential uses, consumption of CFCs in the industrialized world was phased out completely by 1996, except in some countries with economies in transition. By 2005, consumption of all categories of ODS, other than HCFCs and methyl bromide for approved critical uses, ended in industrialized countries. Although the protocol allows developing countries a buffer period for phasing out CFCs and halons, by 2005 they were already significantly ahead of schedule. Among the success factors behind the progress made under the Montreal Protocol (see Figure 2.25) is the principle of common but differentiated responsibility, and the financial mechanism of the protocol (Brack 2003).

Furthermore it is clear that continued decreases in ODS production and use, following the Montreal Protocol provisions, are important for ozone layer recovery, and such measures will also reduce the ODS contribution to climate change. However, detailed knowledge concerning such interlinkages is still lacking (see Box 2.9 on interlinkages between climate change and ozone depletion below).

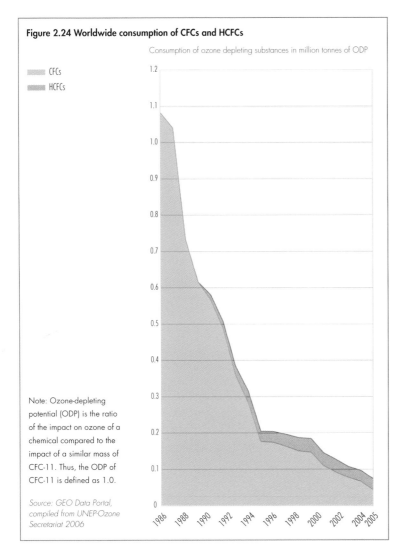

Figure 2.24 Worldwide consumption of CFCs and HCFCs

Consumption of ozone depleting substances in million tonnes of ODP

CFCs
HCFCs

Note: Ozone-depleting potential (ODP) is the ratio of the impact on ozone of a chemical compared to the impact of a similar mass of CFC-11. Thus, the ODP of CFC-11 is defined as 1.0.

Source: GEO Data Portal, compiled from UNEP-Ozone Secretariat 2006

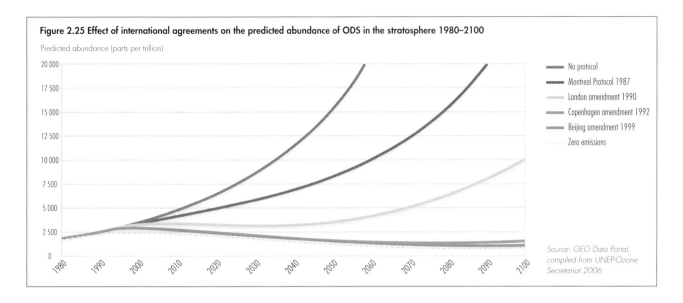

Figure 2.25 Effect of international agreements on the predicted abundance of ODS in the stratosphere 1980–2100

Predicted abundance (parts per trillion)

- No protocol
- Montreal Protocol 1987
- London amendment 1990
- Copenhagen amendment 1992
- Beijing amendment 1999
- Zero emissions

Source: GEO Data Portal, compiled from UNEP-Ozone Secretariat 2006

Despite the success of the protocol, the struggle against stratospheric ozone depletion is not yet over, and the ozone regime still faces a number of key challenges. Phasing out production and use of methyl bromide, a gaseous pesticide used mainly in agriculture, crop storage, buildings and transport, is one challenge. The development of alternatives to methyl bromide has been more complex than for most other ODS. Although alternatives exist, replacement has been slow. The protocol has a "critical use" exemption process where alternatives are not technologically and economically feasible, and there have been a large number of nominations for such critical uses by industrialized countries for the period after phase-out (2005 onwards).

Another challenge is the problem of illegal trade in ODS, mostly for servicing air conditioning and refrigeration. As the phase-out of CFCs neared completion in industrialized countries, a thriving black market in these chemicals started in the mid-1990s. It was reduced when the demand from end users for CFCs steadily dropped, and law enforcement improved. Illegal trade is, however, widespread in the developing world, as it proceeds through its own phase-out schedules (UNEP 2002). The main response at the global level, an amendment of the protocol in 1997 to introduce a system of export and import licenses, has had some effect. The Multilateral Fund and the Global Environment Facility (GEF) have also provided assistance with the establishment of licensing systems and training for customs officers.

Box 2.9 Climate change and stratospheric ozone – interlinked systems

Stratospheric ozone depletion and global warming share many common physical and chemical processes. Many categories of ODS, and several of their substitutes are, like CFCs, greenhouse gases that contribute to climate change. The efforts undertaken under the Montreal Protocol have reduced the atmospheric abundances of CFCs, but global observations confirm increasing atmospheric concentrations of some of the common CFC alternatives, such as HCFCs.

Overall, the understanding of the impact of stratospheric ozone depletion on climate change has been strengthened, although there are still many aspects of these complex systems where knowledge is lacking. The same is true for the effects of climate change on stratospheric ozone recovery. Different processes are simultaneously acting in different directions. Climate change is projected to lead to stratospheric cooling, which, in turn, is predicted both to enhance ozone concentrations in the upper stratosphere, but at the same time delay ozone recovery in the lower stratosphere. It is not yet possible to predict the net effect of these two processes.

Sources: IPCC/TEAP 2005, WMO and UNEP 2006

UNEP's Green Customs Initiative has established cooperation among the secretariats of the Montreal Protocol and those of other multilateral environmental agreements, such as the Basel, Stockholm and Rotterdam conventions, and CITES. This also involves Interpol and the World Customs Organization (Green Customs 2007).

CHALLENGES AND OPPORTUNITIES

Our Common Future, the 1987 Brundtland Commission report, encouraged policy efforts to avoid adverse effects from climate change and air pollution, and it called on the international community to develop follow-up activities. The report was followed by renewed commitments to solving these issues at the summits in Rio de Janeiro in 1992 and in Johannesburg in

2002. *Agenda 21* and the *Johannesburg Plan of Implementation* were created to guide the international community. Several global conventions have been developed to deal with the atmospheric environment issues, and all have set targets for the reduction of the causes and impacts of the emissions. In Table 2.4 some of the major targets are summarized. In addition to global and regional policy initiatives, there have been numerous national initiatives.

Two decades of mixed progress

Despite the many efforts initiated the atmospheric environment issues identified in 1987 still pose problems today. Responses to the challenges of air pollution and climate change have been patchy. The reduction in the emission of the stratospheric ozone-depleting substances has been impressive. Without this rapid and precautionary action, the health and environmental consequences would have been dire. In contrast, there is a remarkable lack of urgency in tackling the anthropogenic emissions of greenhouse gases. Every year of delayed effort will entail the need for more drastic annual reductions in the future, if the climate is to be stabilized at a "relatively safe" level. Since the impacts of climate change are already evident on vulnerable communities and ecosystems, more effort on adaptation to climate change is urgent. The means to make rapid progress exist, but if this is to be achieved, political will and leadership will be crucial. The following discussion assesses national and international policy development and other responses to air pollution, climate change and stratospheric ozone depletion.

Table 2.4 The most recent targets set by international conventions for substances emitted to the atmosphere

Convention/Year of signature	Protocol	Controlled substances	Geographical coverage	Target year	Reduction target/Main component
Long-Range Transboundary Air Pollution (LRTAP), 1979	1998 Aarhus Protocol	Heavy metals (cadmium, lead and mercury)	UNECE region (targets not applied to North America)	2005–2011	Each party to reduce its emissions below the level in 1990 (or an alternative year between 1985 and 1995), by taking effective measures, appropriate to its particular circumstances.
	1998 Aarhus Protocol	POPs	UNECE region (targets not applied to North America)	2004–2005	Eliminate any discharges, emissions and losses of POPs. Parties to reduce their emissions of dioxins, furans, polycyclic aromatic hydrocarbons (PAHs) and hexachlorobenzene below their levels in 1990 (or alternative year between 1985 and 1995).
	1999 Gothenburg Protocol	SO_x, NO_x, VOCs and ammonia	UNECE region (targets not applied to North America)	2010	Cut sulphur compound emissions by at least 63 per cent, NO_x emissions by 41 per cent, VOC emissions by 40 per cent, and ammonia emissions by 17 per cent, compared to 1990 levels.
Vienna Convention, 1985	1987 Montreal Protocol and amendments	ODS	Global	2005–2010	Developing countries to reduce the consumption of CFCs by 50 per cent by 1 January 2005, and to fully eliminate CFCs by 1 January 2010. Earlier phase-out for developed countries. Other control measures apply to other ODS, such as methyl bromide and HCFCs.
United Nations Framework Convention on Climate Change (UNFCCC), 1992	1997 Kyoto Protocol	GHG emissions (CO_2, CH_4, N_2O, HFCs, PFCs, SF_6)	36 countries accepted emissions targets	2008–2012	Kyoto Protocol. The individual commitments add up to a total cut in greenhouse gas emissions of at least 5 per cent from 1990 levels from Annex 1 countries in the commitment period 2008–2012.
Stockholm Convention, 2000		POPs	Global		Reduce or eliminate the most dangerous POPs (Dirty Dozen)*.

* Dirty Dozen: PCBs (polychlorinated biphenyls), dioxins, furans, aldrin, dieldrin, DDT, endrin, chlordane, hexachlorobenzene (HCB), mirex, toxaphene, heptachlor.

Sources: UNECE 1979–2005, Vienna Convention 1987, UNFCCC 1997, Stockholm Convention 2000

Comparing the responses to different atmospheric environment issues

Substantial reductions in emissions to the atmosphere are feasible if all stakeholders act to remove barriers and promote sustainable solutions. The removal of lead from gasoline by almost all nations during the past 20 years is an outstanding example of a successful measure to reduce air pollution, with considerable benefits for human health and the environment (see Box 2.10).

The successful reductions in emissions of SO_2, mainly in Europe and North America, is also noteworthy. It was achieved through a range of different pollution prevention and control strategies, including changes in fuel type (from coal to natural gas), desulphurization of emissions, coal washing, use of fuels with a lower sulphur content and improved energy efficiency (UNECE 1979–2005). Despite enormous economic growth in China, India and elsewhere, Figure 2.8a shows that global sulphur emissions have changed little since 1990. NO_X has proved to be more difficult to address than sulphur, and Figure 2.8b shows an overall global increase in emissions. Even though vehicle technology has improved, with lower NO_X emissions per vehicle, the number of passenger kilometres has increased. As a result, total NO_X emissions in different countries have increased, stabilized or, at best, fallen slightly. Shipping and aviation emissions of NO_X are increasing globally, while power station emissions have stabilized or reduced.

The Montreal Protocol is a good demonstration of the precautionary approach in action, as governments agreed to respond to stratospheric ozone depletion

Box 2.10 Worldwide ban on leaded gasoline within reach, with progress in sub-Saharan African countries

Emission of lead from gasoline has adverse impacts on human health, especially the intellectual development of children. Countries in North America, Europe and Latin America have phased out leaded gasoline, and the global phase-out of lead in gasoline has accelerated dramatically over the last decade. However, some countries in Asia, West Asia and Africa still use lead additives to boost octane levels in gasoline. Representatives from 28 sub-Saharan countries adopted the Dakar declaration in June 2001, to commit to national programmes to phase out leaded gasoline by 2005 (see Figure 2.26). The refinery conversion costs were often cheaper to implement than first assumed. For example, the Kenya refinery in Mombasa is expected to produce unleaded gasoline for an investment of US$20 million, down from an original estimate of US$160 million.

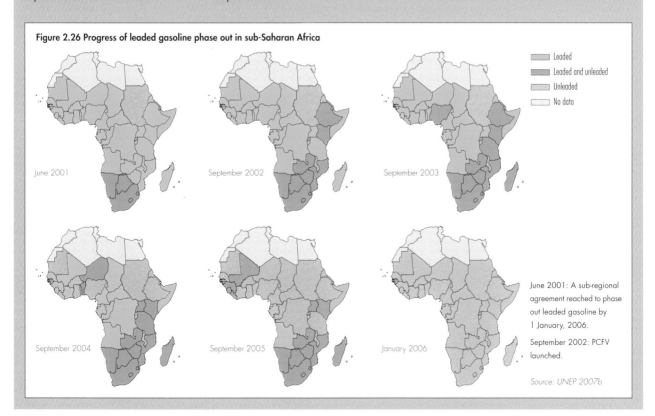

Figure 2.26 Progress of leaded gasoline phase out in sub-Saharan Africa

Leaded
Leaded and unleaded
Unleaded
No data

June 2001
September 2002
September 2003
September 2004
September 2005
January 2006

June 2001: A sub-regional agreement reached to phase out leaded gasoline by 1 January, 2006.

September 2002: PCFV launched.

Source: UNEP 2007b

before its effects were fully clear. Even though CO_2 and CFCs are both long-lived gases, and their potential consequences are severe, the precautionary approach has not been sufficiently implemented in the response to climate change. The reasons for this and the factors that affect the successful responses are summarized in Table 2.5.

The timing of negotiations on the Montreal Protocol was fortunate. The 1980s saw growing public concern over the state of the natural environment, and the dramatic illustrations of the ozone hole above the Antarctic demonstrated the impact of human activities. The number of key actors involved in the negotiations was small, which made agreement easier, and there was a clear leadership role exercised, first by the United States, and, subsequently, by the European Union. The success of the protocol was largely attributable to the flexibility designed into it to allow for its further development with evolving scientific knowledge and technologies. Since entering into force in 1989, the protocol has been adjusted on five occasions, allowing parties to accelerate phase-outs, without the need for repeated amendments to the treaty.

The recognition of the special needs of developing countries through slower phase-out schedules was important in encouraging low-income countries to adhere. In addition, the development of an effective financial mechanism, the Multilateral Fund, which has disbursed almost US$2 billion to developing countries to meet the incremental costs of phase-outs, also enabled institutional strengthening to carry out the phase-out process, and was an important contribution to its success (Bankobeza 2005). Alongside the financial mechanism, the trade measures of the protocol required signatories not to trade in ODS with non-parties, providing an incentive for countries to join. In addition, the non-compliance mechanism has proved to be flexible and highly effective. In contrast, the climate convention and Kyoto Protocol express intentions for technology transfer and assistance, but, to date, there has been limited implementation or provision of financial and technological resources to enable reductions in developing countries.

An important element underlying the success of the Montreal Protocol lies in the extent to which industry responded to the control schedules. Despite initial resistance, companies raced to compete in the markets for non-ozone-depleting substances and technologies, developing cheaper and more effective alternatives more rapidly and at lower costs than expected. In the case of climate change the same market conditions do not exist. In contrast, after the UNFCCC in 1992, the timing of the Kyoto Protocol was less fortunate, as it coincided with declining public and political interest in global environmental issues in the mid-1990s. The number of key stakeholders was large, and, with powerful opposition in some sectors, it proved difficult to reach an agreement.

Despite the fact that the design of the climate protection regime was broadly similar to that developed for ozone, the level of support from developed to developing countries, in relation to the scale of the task, was less generous. Although alternative, complementary approaches, such as the Asia-Pacific Partnership for Clean Development and Climate, and the G8 Gleneagles Programme of Action, which focus on technological development and deployment, have been put forward, the progress is far from satisfactory.

Only a limited amount of international cooperation has occurred through the CDM, although this could potentially be considerably higher in subsequent periods, if developed country targets are substantially more stringent. A second major weakness was the ease with which countries could opt out of the protocol with no adverse responses. This encouraged "free rider" behaviour, in which nations that chose not to ratify the protocol benefit doubly. They share the climate benefits of mitigation occurring in other countries, and have a competitive advantage that arises from avoiding the sometimes costly implementation measures that some Kyoto signatories are adopting. Thus, some industrial sectors that were unfavourable to the Kyoto Protocol managed successfully to undermine the political will to ratify. Even for signatories, incentives are weak, as the protocol does not yet have a substantial compliance regime.

Finally, the issue of the future evolution of the climate regime has been the focus of considerable discussion, and numerous approaches have been proposed (Bodansky 2003) (see Chapter 10). The parties to the UNFCCC have agreed that they should act to protect the climate system "on

Table 2.5 Progress from 1987 to 2007 in key factors for successful management of stratospheric ozone, climate change and air pollution

	Success factors	Stratospheric ozone		Climate change		Air pollution	
		1987	2007	1987	2007	1987	2007
Problem identification	Confidence in science	Broadly accepted Public identification "ozone hole = CFCs in spray cans"	Problem persists, but under control	First signals, potential threat	Broadly accepted	Wide range of air pollution problems, publicly understood	Reduced to fewer, but harder to solve pollution problems
Economic evaluations	Social benefits should greatly outweigh the costs	Costly measures but worthwhile	Cost more modest than foreseen	Little information	Numerous studies, with varying costs of both mitigation and impacts	Technology options are available, with modest increases in the costs of products	Further reductions available at higher costs; benefits far outweigh the costs
Negotiation	Leadership, small number of key actors	Strong leadership (first US, then EU)	n.a.	n.a.	Complex process, many stakeholders, strong vested interests	Variable at national level	Increasing regional level, start global level
Solution	Convention, then increasingly tight protocols	Protocol in place, but with insufficient measures	One protocol and four amendments; sufficient action	None in sight	First steps: UNFCCC, 1992 Kyoto Protocol, 1997	Few at national or regional level	Increasing number of standards, mature technologies available, some regional level agreements
Implementation and control	Financial support fund for measures and institutions, "sticks" and "carrots"	Scheme in place	Improved global implementation, 191 countries ratified	n.a.	Legally binding emissions commitments for 2008–12 for industrialized countries; 166 countries have ratified	Mainly national level	Variable Some regional/global harmonization (for example, lead free gasoline)
Treaties realised	Diplomatic negotiations	Vienna Convention, 1985 Montreal Protocol, 1987	Four amendments added to the protocol; stabilization reached		UNFCCC, 1992 Kyoto Protocol, 1997	UNECE CLRTAP Convention, 1979 UNECE Protocol for SO_2, NO_X	LRTAP strengthened Other regional agreements emerging
Outlook	Political leadership, efficient control mechanisms	Phasing out methyl bromide Development of economically feasible alternative uses, prevention of illegal trade		Risk of irreversible effects growing Urgent to successfully define post-Kyoto commitment Equity and burden issues remains to be solved		Challenge to disseminate solutions (acceptable levels, institutions and mechanism, technologies), at the global level; minimum global standards	

the basis of equality and in accordance with their common but differentiated responsibilities and respective capabilities," (UNFCCC 1997) but are still struggling to put this into practice. It remains the case that those who are primarily responsible for causing climate change are energy users and their customers, while those who will primarily bear the brunt of a changing climate are vulnerable communities with relatively little responsibility. As Agarwal and Narain (1991) expressed it, people have an equal right to the atmospheric commons, and a climate regime must recognize the vast differences between those who gain from overexploiting the atmospheric commons, and those who bear the costs.

The foregoing analysis suggests that existing mechanisms of the Montreal Protocol and its implementation are largely adequate to tackle the remaining emissions of ODS, while air quality management in many parts of the world requires the strengthening of institutional, human and financial resources for implementing policies. For climate change, however, current global approaches are not effective. More innovative and equitable approaches for mitigation and adaptation at all levels of society, including fundamental changes to social and economic structures, will be crucial to adequately address the climate change issue.

Reducing emissions of chemicals with long residence times in the atmosphere

The production and release of these substances constitutes a special challenge. The impacts often manifest themselves long after emissions commenced, as was the case with mercury and POPs. Some GHGs, such as perfluorocarbons and sulphur hexafluoride, have estimated lifetimes of many thousands of years in the atmosphere. The amount of fluorinated gases used is small relative to the emissions of other GHGs. However, their very long lifetimes in the atmosphere together with their high global warming potentials add to their contribution to climate change. The costs of remediation and damage repair, if possible, are often higher than the costs of preventing the release of hazardous substances (see Chapters 3, 4 and 6).

The global emissions of mercury represent an important issue, with inadequate international and national responses. The most significant releases

of mercury are emissions to air, and once added to the global environment, mercury is continuously mobilized, deposited and re-mobilized. Burning coal and waste incineration account for about 70 per cent of total quantified emissions. As combustion of fossil fuels is increasing, mercury emissions can be expected to increase, in the absence of control technologies or prevention (UNEP 2003). Current concentrations in the environment are already high, and have reached levels in some foods that can cause health impacts (see Chapter 6).

Opportunities to deal with atmospheric environment challenges

The major instrument used to address atmospheric issues has been government regulation. This instrument of policy has achieved considerable successes in some areas, such as the removal of lead from gasoline, reductions in sulphur in diesel fuel, the widespread adoption of tighter emission standards (such as the Euro standards) for vehicles around the world, and, most importantly, the virtual elimination of production of CFCs. However, the use of regulation has many limitations, and there is a growing additional use of other instruments as part of a tool box of policy approaches around the world.

In some circumstances economic instruments have been useful in applying the principle of polluter pays, addressing market failures and harnessing the power of markets to find the cheapest way to achieve policy targets. Examples include the cap-and-trade approach used in the United States as one way to achieve major reductions in emissions of SO_2. Other approaches include load-based emissions charges that provide a direct economic incentive to reduce emissions, and the removal of subsidies that encourage use of high-emitting fuels in some countries.

Self-regulation and co-regulation are increasingly being used by large corporations as tools to improve the environmental performance of their operations, wherever they are located. Environmental management systems, such as the ISO 14000 series, and industry codes, such as Responsible Care, are being used as voluntary tools, often going beyond simple compliance with government regulations to reduce impacts of

operations on the environment, and at the same time protect corporate brands.

In some circumstances information and education can also be powerful tools to mobilize public opinion, communities, civil society and the private sector to achieve environmental goals. They can be effective where government regulations are weak or not implemented. They are usually most successful when used in combination with other approaches, including regulations and economic instruments, to make selected high-emission activities both expensive, and their negative impacts well-known to the national and international community.

The success of policy development and implementation to control atmospheric emissions is largely determined by effective multistakeholder participation at different scales, and mobilization of public-private partnerships. Many countries have extensive regulations, but too often they are not applied effectively because of a lack of proper institutions, legal systems, political will and competent governance. Strong political leadership is essential to develop institutional capacity and effective outreach to the public, to ensure adequate funding, and to increase local, national and international coordination.

Most economic studies following government actions to address air pollution, even using conservative methodologies and cost estimates, generally show that the costs associated with impacts far outweigh the costs of these action, often by an order of magnitude (Watkiss and others 2004, USEPA 1999, Evans and others 2002). Furthermore, in most cases the costs of action are considerably lower than anticipated (Watkiss and others 2004). In addition, the social distribution of the burden of pollution falls on poorer people, children, older people and those with pre-existing health conditions. Emissions can be reduced in a manner that will protect the climate without major disruptions to the socio-economic structures (Azar and Schneider 2002, Edenhofer and others 2006, Stern 2006).

The future success of efforts to control atmospheric emissions will ultimately depend on strong involvement of stakeholders at all levels, coupled with suitable mechanisms for facilitating

technological and financial flows, and the strengthening of human and institutional capacities. Besides the development of innovative clean technologies, efforts to rapidly deploy currently available technologies in developing countries would go a long way to addressing these issues. Fundamental changes in social and economic structures, including lifestyle changes, are crucial if rapid progress is to be achieved.

The future success of efforts to control atmospheric emissions will heavily depend on the involvement of stakeholders at all levels.

Credit (top): Ngoma Photos

Credit (bottom): Mark Edwards/ Still Pictures

References

ACIA (2005). *Arctic Climate Impact Assessment*. Arctic Council and the International Arctic Science Committee, Cambridge University Press, Cambridge http://www.acia.uaf.edu/pages/scientific.html (last accessed 14 April 2007)

ADB (2001). *Asian Environmental Outlook 2001*. Asian Development Bank, Manilla

Agarwal, A. and Narain, S. (1991). *Global Warming in an Unequal World: A Case of Environmental Colonialism*. Centre for Science and Environment, New Delhi

Akimoto, H., Oharaa, T., Kurokawa, J. and Horii, N. (2006). Verification of energy consumption in China during 1996–2003 by using satellite observational data. In *Atmospheric Environment* 40:7663-7667

AMCEN and UNEP (2002). *Africa environment outlook: Past, Present and future perspectives*. Earthprint Limited, Stevenage, Hertfordshire

Andersen, S. O. and Sarma, M. (2002). *Protecting the Ozone Layer: The United Nations History*. Earthscan Publications, London

APMA (2002). *Benchmarking Urban Air Quality Management and Practice in Major and Mega Cities of Asia – Stage 1*. Air Pollution in the Megacities of Asia Project. Stockholm Environment Institute, York http://www.york.ac.uk/inst/sei/rapidc2/benchmarking.html (last accessed 11 April 2007)

ASEAN (2003). *ASEAN Haze Agreement*. The Association of South East Asian Nations, Jakarta http://www.aseansec.org/10202.htm (last accessed 11 April 2007)

ASEC (2001). *Australia: State of Environment 2001*. Australian State of the Environment Committee, Department of the Environment and Heritage. CSIRO Publishing, Canberra http://www.environment.gov.au/soe/2001/index.html (last accessed 15 April 2007)

Aunan, K., Fang, J. H., Hu, T., Seip, H. M. and Vennemo, H. (2006). Climate change and air quality - Measures with co-benefits in China. In *Environmental Science & Technology* 40(16):4822-4829

Azar, C. and Schneider, S. H. (2002). Are the economic costs of stabilising the atmosphere prohibitive? In *Ecological Economics* 42(1-2):73-80

Bankobeza, G. M. (2005). *Ozone protection – the international legal regime*. Eleven International Publishing, Utrecht

Beijing Bureau of Statistics (2005). *Beijing Statistical Yearbook 2005*. Beijing Bureau of Statistics, Beijing

Bodansky, D. (2003). Climate Commitments: Assessing the Options. In *Beyond Kyoto: Advancing the International Effort Against Climate Change*. Pew Center on Global Climate Change

Brack, D. (2003). Monitoring the Montreal Protocol. In *Verification Yearbook 2003*. VERTIC, London

Brown, L. R. (2006). Building a New Economy. In *Plan B 2.0: Rescuing a Planet Under Stress and a Civilization in Trouble*. W. W. Norton, Exp Upd edition

Bryden, H., Longworth, H. and Cunningham, S. (2005). Slowing of the Atlantic meridional overturning circulation at 25° N. In *Nature* 438:455-457

CAI (2003). *Phase-Out of Leaded Gasoline in Oil Importing Countries of Sub-Saharan Africa – The case of Tanzania – Action plan*. Clean Air Initiative, The World Bank, Washington, DC http://wbln0018.worldbank.org/esmap/site.nsf/files/tanzania+final.pdf/$FILE/tanzania+final.pdf (last accessed 14 April 2007)

Centre on Airborne Organics (1997). *Fine particles in the Atmosphere. 1997 Summer Symposium Report*. MIT, Boston http://web.mit.edu/airquality/www/rep1997.html (last accessed 1 May 2007)

Cities for Climate Protection (2007). http://www.iclei.org/index.php?id=809 (last accessed 18 July 2007)

City of Cape Town (2006). *Air Quality Monitoring Network*. City of Cape Town, Cape Town http://www.capetown.gov.za/airqual/ (last accessed 11 April 2007)

Cohen, A. J., Anderson, H. R., Ostro, B., Pandey, K., Krzyzanowski, M., Künzli, N., Gutschmidt, K., Pope, C. A., Romieu, I., Samet, J. M. and Smith, K. R. (2004). Mortality impacts of urban air pollution. In *Comparative quantification of health risks: global and regional burden of disease attributable to selected major risk factor* Vol. 2, Chapter 17. World Health Organization, Geneva

Cohen, A.J., Anderson, H.R., Ostro, B., Pandey, K.D., Krzyzanowski, M., Kunzli, N., Gutschmidt, K., Pope, A., Romieu, I., Samet, J.M. and Smith, K. (2005). The global burden of disease due to outdoor air pollution. In *Journal of toxicology and environmental health* Part A (68):1–7

Cox, P. M., Betts, R. A., Collins, M., Harris, P. P., Huntingford, C. and Jones, C. D. (2004). Amazonian forest dieback under climate-carbon cycle projections for the 21st century. In *Theoretical and Applied Climatology* 78(1-3):137-156

CPCB (2001-2006). *National Ambient Air Quality- Status & Statistics 1999-2004*. Central Pollution Control Board, New Delhi

Curry, R. and Mauritzen, C. (2005). Dilution of the northern North Atlantic Ocean in recent decades. In *Science* 308(5729):1772-1774

DeCanio, S. J. (2003). *Economic Models of Climate Change: A Critique*. Palgrave Macmillan, Hampshire

Den Elzen, M. G. J. and Meinshausen, M. (2005). *Meeting the EU 2·C climate target: global and regional emission implications*. RIVM report 728007031/2005. The Netherlands Environmental Assessment Agency, Bilthoven

Dentener, F., Stevenson, D., Ellingsen, K., van Noije, T., Schultz, M., Amann, M., Atherton, C., Bell, N., Bergmann, D., Bey, I., Bouwman, L., Butler, T., Cofala, J., Collins, B., Drevet, J., Doherty, R., Eickhout, B., Eskes, H., Fiore, A., Gauss, M., Hauglustaine, D., Horowitz, L., Isaksen, I. S. A. , Josse, B., Lawrence, M., Krol, M., Lamarque, J. F., Montanaro, V., Muller, J. F., Peuch, V. H., Pitari, G., Pyle, J., Rast, S., Rodriguez, J., Sanderson, M., Savage, N. H., Shindell, D., Strahan, S., Szopa, S., Sudo, K., Van Dingenen, R., Wild, O. and Zeng, G. (2006). The global atmospheric environment for the next generation. In *Environmental Science & Technology* 40(11):3586-3594

Dickens, G. R. (1999). Carbon cycle - The blast in the past. In *Nature* 401(6755):752

DOE (2005). *US Energy Policy Act of 2005*. http://genomicsgtl.energy.gov/biofuels/legislation.shtml (last accessed 11 April 2007)

Dore, M. H. I. (2005). Climate change and changes in global precipitation patterns: What do we know? In *Environment International* 31(8):1167-1181

Edenhofer, O., Kemfert, C., Lessmann, K., Grubb, M. and Koehler, J. (2006). Induced Technological Change: Exploring its implications for the Economics of Atmospheric Stabilization: Synthesis Report from the innovation Modeling Comparison Project. In *The Energy Journal Special Issue, Endogenous Technological Change and the Economics of Atmospheric Stabilization* 57:107

EDGAR (2005). *Emission Database for Global Atmospheric Research (EDGAR) information system*. Joint project of RIVM-MNP, TNO-MEP, JRC-IES and MPIC-AC, Bilthoven http://www.mnp.nl/edgar/ (last accessed 14 April 2007)

EEA (1995). *Europe's Environment - The Dobris Assessment*. European Environment Agency, Copenhagen

EEA (2005). *The European Environment. State and Outlook 2005*. European Environment Agency, Copenhagen

Emanuel, K. (2005). Increasing destructiveness of tropical cyclones over the past 30 years. In *Nature* 436(7051):686-688

Emberson, L., Ashmore, M. and Murray, F. eds. (2003). *Air Pollution Impacts on Crops and Forests - a Global Assessment*. Imperial College Press, London

Environment Canada (2006). *Canada-U.S.* Air Quality Agreement. Environment Canada, Gatineau, QC http://www.ec.gc.ca/pdb/can_us/canus_links_e.cfm (last accessed 11 April 2007)

EU (1996). Council Directive 96/62/EC of 27 September 1996 on ambient air quality assessment and management. Environment Council, European Commission, Brussels. In *Official Journal of the European Union* L 296, 21/11/1996:55- 63

EU (1999). Council Directive 1999/30/EC of 22 April 1999 relating to limit values for sulphur dioxide, nitrogen dioxide and oxides of nitrogen, particulate matter and lead in ambient air. Environment Council, European Commission, Brussels. In *Official Journal of the European Union* L 163, 29/06/1999:41-60

EU (2002). Council Directive 2002/3/EC of the European parliament and of the council of 12 February 2002 relating to ozone in ambient air. Environment Council, European Commission, Brussels. In *Official Journal of the European Union* L 67, 09/03/2002:14-30

Evans, J., Levy, J., Hammitt, J., Santos-Burgoa, C., Castillejos, M., Caballero-Ramirez, M., Hernandez-Avila, M., Riojas-Rodriguez, H., Rojas-Bracho, L., Serrano-Trespalacios, P., Spengler, J.D. and Suh, H. (2002). Health benefits of air pollution control. In Molina, L.T. and Molina, M.J. (eds.) *Air Quality in the Mexico Megacity: An Integrated Assessment*. Kluwer Academic Publishers, Dordrecht

Faiz, A. and Gautam , S. (2004). Technical and policy options for reducing emissions from 2-stroke engine vehicles in Asia. In *International Journal of Vehicle Design* 34(1):1-11

Galanter, M. H., Levy, I. I. and Carmichael, G. R. (2000). Impacts of Biomass Burning on Tropospheric CO, NO_X, and O_3. In *J. Geophysical Research* 105:6633-6653

Gash, J.H.C., Huntingford, C., Marengo, J.A., Betts, R.A., Cox, P.M., Fisch, G., Fu, R., Gandu, A.W., Harris, P.P., Machado, L.A.T., von Randow, C. and Dias, M.A.S. (2004). Amazonian climate: results and future research. In *Theoretical and Applied Climatology* 78(1-3):187-193

GEO Data Portal. *UNEP's online core database with national, sub-regional, regional and global statistics and maps, covering environmental and socio-economic data and indicators*. United Nations Environment Programme, Geneva http://www.unep.org/geo/data or http://geodata.grid.unep.ch (last accessed 7 June 2007)

Goulder, L. H. and Nadreau, B. M. (2002). International Approaches to Reducing Greenhouse Gas Emissions. In Schneider, S. H., Rosencranz, A, and Niles, J.O. (eds.) In *Climate Change Policy: A Survey*. Island Press, Washington, DC

Green Customs (2007). http://www.greencustoms.org/ (last accessed 7 June 2007)

Gregory, J. M., Huybrechts, P. and Raper, S. C. B. (2004). Threatened loss of the Greenland ice-sheet. In *Nature* 428:616

Gregory, J.M. and Huybrechts P. (2006). Ice-sheet contributions to future sea-level change. In *Phil.Trans. R. Soc.* A 364:1709-1731

Groom, B., Hepburn, C., Koundour, P. and Pearce, D. (2005). Declining Discount Rates: The Long and the Short of it. In *Environmental & Resource Economics* 32:445-493

Gwilliam, K., Kojima, M. and Johnson, T. (2004). *Reducing air pollution from urban transport*. The World Bank, Washington, DC

Hansen, B., Østerhus, S., Quadfasel, D. and Turrell, W. (2004). Already the day after tomorrow? In *Science* 305:953-954

Hansen, J. E. (2005). A slippery slope: How much global warming constitutes "dangerous anthropogenic interference?" In *Climatic Change* 68(3):269-279

Hansson, L. (2000). Induced pigmentation in zooplankton: a trade-off between threats from predation and ultraviolet radiation. In *Proc. Biol. Science 267(1459):2327–2331*

Hare, B. and Meinshausen, M. (2004). *How much warming are we committed to and how much can be avoided?* Report 93. Potsdam Institute for Climate Impact Research, Potsdam

Heath, J., Ayres, E., Possell, M., Bardgett, R.D., Black, H.I.J., Grant, H., Ineson, P. and Kerstiens, G. (2005). Rising Atmospheric CO2 Reduces Sequestration of Root-Derived Soil Carbon. In *Science* 309: 1711-1713

Holland, M., Kinghorn, S., Emberson, L., Cinderby, S., Ashmore, M., Mills, G. and Harmens, H. (2006). *Development of a framework for probabilistic assessment of the economic losses caused by ozone damage to crops in Europe*. CEH project No. C02309NEW. Centre for Ecology and Hydrology, Natural Environment Research Council, Bangor, Wales

Huntingford, C. and Gash, J. (2005). Climate equity for all. In *Science* 309(5742):1789

IEA (2006). *World Energy Outlook*. International Energy Agency, Paris

IEA (2007a). *Energy Balances of OECD Countries and Non-OECD Countries 2006 edition*. International Energy Agency, Paris (in GEO Data Portal)

IEA (2007b). *International Energy Agency Online Energy Statistics*. International Energy Agency, Paris http://www.iea.org/Textbase/stats/electricitydata.asp?COUNTRY_CODE=29 (last accessed June 21 2007)

IATA (2007). *Fuel Efficiency*. International Air Transport Association, Montreal and Geneva http://www.iata.org/whatwedo/environment/fuel_efficiency.htm (last accessed 7 June 2007)

IPCC (2001a). *Climate Change 2001: Synthesis Report*. Intergovernmental Panel on Climate Change. Cambridge University Press, Cambridge

IPCC (2001b). *Climate Change 2001: The Scientific Basis*. Intergovernmental Panel on Climate Change. Cambridge University Press, Cambridge

IPCC (2001c). *Climate Change 2001: Impacts, Adaptation and Vulnerability*. Intergovernmental Panel on Climate Change. Cambridge University Press, Cambridge

IPCC (2007). *Climate Change 2007: The Physical Science Basis*. Contribution of Working Group I to the Fourth Assessment Report of the Intergovernmental Panel on Climate Change, Geneva http://ipcc-wg1.ucar.edu/wg1/docs/WG1AR4_SPM_Approved_05Feb.pdf (last accessed 11 April 2007)

IPCC/TEAP (2005). *IPCC Special Report on Safeguarding the Ozone Layer and the Global Climate System. Issues related to Hydrofluorocarbons and Perfluorocarbons*. Approved and accepted in April 2005. Intergovernmental Panel on Climate Change, Geneva

Jacob, D. (1999). *Introduction to Atmospheric Chemistry*. Princeton University Press, New York, NY

Kuylenstierna, J.C.I., Rodhe, H., Cinderby, S. and Hicks, K. (2001). Acidification in developing countries: ecosystem sensitivity and the critical load approach on a global scale. In *Ambio* 30:20-28

Landsea, C.W., Harper, B.A., Hoarau, K. and Knaff, J.A. (2006). Can we detect trends in extreme tropical cyclones? In *Science* 313:452-454

Larssen, T., Lydersen, E., Tang, D.G., He, Y., Gao, J.X., Liu, H.Y., Duan, L., Seip, H.M., Vogt, R.D., Mulder, J., Shao, M., Wang, Y.H., Shang, H., Zhang, X.S., Solberg, S., Aas, W., Okland, T., Eilertsen, O., Angell, V., Liu, Q.R., Zhao, D.W., Xiang, R.J., Xiao, J.S. and Luo, J.H. (2006). Acid rain in China. In *Environmental Science & Technology* 40(2):418-425

Liepert, B. G. (2002). Observed refuctions of surface solar radiation at sites in the United States and worldwide from 1961 to 1990. In *Geophysical Research Letters* 29(10):1421

Lippmann, M. (2003). Air pollution and health – studies in the Americas and Europe. In *Air pollution and health in rapidly developing countries*, G. McGranahan and F. Murray (eds.). Earthscan, London

Martins, M. C. H., Fatigati, F. L., Vespoli, T. C., Martins, L. C., Pereira, L. A. A., Martins, M. A., Saldiva, P. H. N. and Braga, A.L.F. (2004). The influence of socio-economic conditions on air pollution adverse health effects in elderly people: an analysis of six regions in Sao Paulo, Brazil. In *Journal of epidemiology and community health* 58:41-46

Mears, C. A. and Wentz, F. J. (2005). The effect of diurnal correction on satellite-derived lower tropospheric temperature. In *Science* 309(5740):1548-1551

Menzel, A., Sparks, T.H., Estrella, N., Koch, E., Aasa, A., Aha, R., Alm-Kubler, K., Bissolli, P., Braslavska, O., Briede, A., Chmielewski, F.M., Crepinsek, Z., Curnel, Y., Dahl, A., Defila, C., Donnelly, A., Filella, Y., Jatcza, K., Mage, F., Mestre, A., Nordli, O., Penuelas, J., Pirinen, P., Remisova, V., Scheifinger, H., Striz, M., Susnik, A., Van Vliet. A.J.H., Wielgolaski, F.E., Zach, S. and Zust, A. (2006). European phenological response to climate change matches warming pattern. In *Global Change Biology* 12:1969-1976

METI (2004). *Sustainable future framework on climate change. Interim report by special committee on a future framework for addressing climate change.* Global Environmental Sub-Committee, Industrial Structure Council, Japan. Ministry of Economy, Trade and Industry, Tokyo

Mexico City Ambient Air Monitoring Network (2006). Federal District Government, Mexico DF http://www.sma.df.gob.mx/simat/ (last accessed 11 April 2007)

Moberg, A., Sonechkin, D.M., Holmgren, K., Datsenko, N.M. and Karlen, W. (2005). Highly variable Northern Hemisphere temperatures reconstructed from low- and high-resolution proxy data. In *Nature* 433(7026):613-617

Molina, M. J. and Molina, L. T. (2004). Megacities and atmospheric pollution. In *Journal of the Air & Waste Management Association* 54(6):644-680

NASA (2006). *Ozone Hole Monitoring. Total Ozone Mapping Spectrometer.* NASA Website http://toms.gsfc.nasa.gov/eptoms/dataqual/oz_hole_avg_area_v8.jpg (last accessed 1 May 2007)

Newman, P. and Kenworthy, J. (2006). Urban Design to Reduce Automobile Dependence. In *Opolis: An International Journal of Suburban and Metropolitan Studies* 2(1): Article 3 http://repositories.cdlib.org/cssd/opolis/vol2/iss1/art3 (last accessed 1 May 2007)

OECD (2002). *OECD Environmental Data Compendium 2002.* Organisation for Economic Cooperation and Development, Paris

Oerlemans, J. (2005). Extracting a climate signal from 169 glacier records. In *Science* 308(5722):675-77

Orr, J.C., Fabry, V.J., Aumont, O., Bopp, L., Doney, S.C., Feely, R.A., Gnanadesikan, A., Gruber, N., Ishida, A., Joos, F., Key, R.M., Lindsay, K., Maier-Reimer, E., Matear, R., Monfray, P., Mouchet, A., Najjar, R.G., Plattner, G.K., Rodgers, K.B., Sabine, C.L., Sarmiento, J.L., Schlitzer, R., Slater, R.D., Totterdell, I.J., Weirig, M.F., Yamanaka, Y. and Yool, A. (2005). Anthropogenic ocean acidification over the twenty-first century and its impact on calcifying organisms. In *Nature* 437:681-686

Patz, J. A., Lendrum, D. C., Holloway, T. and Foley, J. A. (2005). Impact of regional climate change on human health. In *Nature* 438:310-317

Perin, S. and Lean, D.R.S. (2004). The effects of ultraviolet-B radiation on freshwater ecosystems of the Arctic: Influence from stratospheric ozone depletion and climate change. In *Environment Reviews* 12:1-70

Pew Centre on Global Climate Change (2007). *Emission Targets* http://www.pewclimate.org/what_s_being_done/targets/index.cfm (last accessed 7 June 2007)

Phoenix, G.K., Hicks, W.K., Cinderby, S., Kuylenstierna, J.C.I., Stock, W.D., Dentener, F.J., Giller, K.E., Austin, A.T., Lefroy, R.D.B., Gimeno, B.S., Ashmore, M.R. and Ineson, P. (2006). Atmospheric nitrogen deposition in world biodiversity hotspots: the need for a greater global perspective in assessing N deposition impacts. In *Global Change Biology* 12:470-476

Pope, A.C., III and Dockery, D.W. (2006). Critical Review: Health Effects of Fine Particulate Air Pollution: Lines that Connect. In *J. Air & Waste Manage. Assoc.* 56:709-742

Porcaro, J. and Takada, M. (eds.) (2005). *Achieving the Millennium Development Goals: Case Studies from Brazil, Mali, and the Philippines.* United Nations Development Programme, New York, NY

Pounds, J.A., Bustamante, M.R., Coloma, L.A., Consuegra, J.A., Fogden, M.P.L., Foster, P.N., La Marca, E., Masters, K.L., Merino-Viteri, A., Puschendorf, R., Ron, S.R., Sanchez-Azofeifa, G.A., Still, C.J. and Young, B.E. (2006). Widespread amphibian extinctions from epidemic disease driven by global warming. In *Nature* 439:161-167

Ramanathan, V., Crutzen, P. J. , Mitra A. P. and Sikka, D. (2002). The Indian Ocean Experiment and the Asian Brown Cloud. In *Current Science* 83(8):947-955

Reid, H. and Alam, M. (2005). Millennium Development Goals. Stockholm Environment Institute, York. In *Tiempo* 54

RIVM-MNP (2006). *Emission Database for Global Atmospheric Research - EDGAR 3.2 and EDGAR 32FT2000.* The Netherlands Environmental Assessment Agency, Bilthoven (in GEO Data Portal)

Rockström, J. Axberg, G.N., Falkenmark, M. Lannerstad, M., Rosemarin, A., Caldwell, I., Arvidson, A. and Nordström, M. (2005). *Sustainable Pathways to Attain the Millennium Development Goals: Assessing the Key Role of Water, Energy and Sanitation.* Stockholm Environment Institute, Stockholm

Royal Society (2005a). *Food crops in a changing climate: Report of a Royal Society Discussion Meeting. 26-27 April 2005.* Policy report from the meeting, launched 20 June 2005. The Royal Society, London

Royal Society (2005b). *Full text of open letter to Margaret Beckett and other G8 energy and environment ministers from Robert May, President.* The Royal Society, London http://www.royalsoc.ac.uk/page.asp?id=3834 (last accessed 11 April 2007)

SEPA (2006). 2005 Report on the State of the Environment in China. State Environmental Protection Agency http://english.sepa.gov.cn/ghjh/hjzkgb/200701/P020070118528407141643.pdf (last accessed 1 May 2007)

Siegenthaler, U., Stocker, T.F., Monnin, E., Luthi, D., Schwander, J., Stauffer, B., Raynaud, D., Barnola, J.M., Fischer, H., Masson-Delmotte, V. and Jouzel, J. (2005). Stable carbon cycle-climate relationship during the late Pleistocene. In *Science* 310(5752):1313-1317

Skjelkvåle, B.L., Stoddard, J.L., Jeffries, D.S., Torseth, K., Hogasen, T., Bowman, J., Mannio, J., Monteith, D.T., Mosello, R., Rogora, M., Rzychon, D., Vesely, J,. Wieting, J., Wilander, A. and Worsztynowicz, A. (2005). Regional scale evidence for improvements in surface water chemistry 1990-2001. In *Environmental Pollution* 137(1):165-176

Soden, B.J., Jackson, D.L., Ramaswamy, V., Schwarzkopf, M.D. and Huang, X.L. (2005). The radiative signature of upper tropospheric moistening. In *Science* 310:841-844

Spahni, R., Chappellaz, J., Stocker, T.F., Loulergue, L., Hausammann, G., Kawamura, K., Fluckiger, J., Schwander, J., Raynaud, D., Masson-Delmotte, V. and Jouzel, J. (2005). Atmospheric methane and nitrous oxide of the late Pleistocene from Antarctic ice cores. In *Science* 310(5752):1317-1321

Srinivasan, A. (2005). Mainstreaming climate change concerns in development: Issues and challenges for Asia. In *Sustainable Asia 2005 and beyond: In the pursuit of innovative policies.* IGES White Paper, Institute for Global Environmental Strategies, Tokyo

Steffen, K. and Huff, R. (2005). *Greenland Melt Extent, 2005* http://cires.colorado.edu/science/groups/steffen/greenland/melt2005 (last accessed 11 April 2007)

Stern, N. (2006). *The Economics of Climate Change – The Stern Review.* Cambridge University Press, Cambridge

Stevens, C.J., Dise, N.B., Mountford, J.O. and Gowing, D.J. (2004). Impact of nitrogen deposition on the species richness of grasslands. In *Science* 303(5665):1876-1879

Stockholm Convention (2000). *Stockholm Convention on Persistent Organic Pollutants* http://www.pops.int/ (last accessed 11 April 2007)

Svensen, H., Planke, S., Malthe-Sorenssen, A., Jamtveit, B., Myklebust, R., Eidem, T.R. and Rey, S.S. (2004). Release of methane from a volcanic basin as a mechanism for initial Eocene global warming. In *Nature* 429(6991):542-545

Tarnocai, C. (2006). The effect of climate change on carbon in Canadian peatlands. In *Global and Planetary Change* 53(4):222-232

TERI (2001). *State of Environment Report for Delhi 2001.* Supported by the Department of Environment, Government of National Capital Territory. Tata Energy Research Institute, New Delhi

Thomas, C.D., Cameron, A., Green, R.E., Bakkenes, M., Beaumont, L.J., Collingham, Y.C., Erasmus, B.F.N., de Siqueira, M.F., Grainger, A., Hannah, L., Hughes, L., Huntley, B., van Jaarsveld, A.S., Midgley, G.F., Miles, L., Ortega-Huerta, M.A., Peterson, A.T., Phillips, O.L. and Williams, S.E. (2004a). Extinction risk from climate change. In *Nature* 427:145-148

Thomas, C. D., Williams, S. E., Cameron, A., Green, R. E., Bakkenes, M., Beaumont, L. J., Collingham, Y. C., Erasmus, B. F. N., De Siqueira, M. F., Grainger, A., Hannah, L., Hughes, L., Huntley, B., van Jaarsveld, A. S., Midgley, G. F., Miles, L., Ortega-Huerta, M. A., Peterson, A. T. and Phillips, O. L. (2004b). Biodiversity conservation - Uncertainty in predictions of extinction risk -Effects of changes in climate and land use - Climate change and extinction risk – Reply. In *Nature* 430: Brief Communications

Trenberth, K. (2005). Uncertainty in hurricanes and global warming. In *Science* 308(5729):1753-1754

UN (2007). *UN Millennium Development Goals.* United Nations Department of Public Information http://www.un.org/millenniumgoals (last accessed 7 June 2007)

UNCTAD (2006). *Review of Maritime Transport 2006.* United Nations Conference on Trade and Development, New York and Geneva http://www.unctad.org/en/docs/rmt2006_en.pdf (last accessed 14 April 2007)

UNECE (1979-2005). *The* Convention on Long-range Transboundary Air pollution website. United Nations Economic Commission for Europe, Geneva http://unece.org/env/lrtap/lrtap_h1.htm (last accessed 14 April 2007)

UNEP (2002). *Study on the monitoring of international trade and prevention of illegal trade in ozone-depleting substances.* Study for the Meeting of the Parties. UNEP/OzL.Pro/WG.1/22/4. United Nations Environment Programme, Nairobi http://ozone.unep.org/Meeting_Documents/oewg/22oewg/22oewg-2.e.pdf (last accessed 17 April 2007)

UNEP (2003). Environmental effects of ozone depletion and its interactions with climate change: 2002 assessment. In *Photochemical & Photobiological Science* 2:1-4

UNEP (2004). Impacts of summer 2003 heat wave in Europe. In: *Environment Alert Bulletin 2* UNEP Division of Early Warning and Assessment/GRID Europe, Geneva http://www.grid.unep.ch/product/publication/download/ew_heat_wave.en.pdf (last accessed 14 April 2007)

UNEP (2006). *GEO Year Book 2006.* United Nations Environment Programme, Nairobi

UNEP (2007a). *Buildings and Climate Change: Status, Challenges and Opportunities.* United Nations Environment Programme, Nairobi http://www.unep.fr/pc/sbc/documents/Buildings_and_climate_change.pdf (last accessed 14 April 2007)

UNEP (2007b). *Partnership for clean fuels and vehicles.* United Nations Environment Programme, Nairobi http://www.unep.org/pcfv (last accessed 7 June 2007)

UNEP/Chemicals (2006). *The Mercury Programme.* http:///www.chem.unep.ch/mercury/ (last accessed 14 April 2007)

UNEP/RRC-AP (2006). *Malé Declaration on the Control and Prevention of Air Pollution in South Asia and its Likely Transboundary Effects.* UNEP Regional Resource Centre for Asia and the Pacific, Bangkok http://www.rrcap.unep.org/issues/air/maledec/baseline/indexpak.html (last accessed 7 June 2007)

UNESCO-SCOPE (2006). *The Global Carbon Cycle.* UNESCO-SCOPE Policy Briefs October 2006 – No. 2. United Nations Educational, Scientific and Cultural Organization, Scientific Committee on Problems of the Environment, Paris http://unesdoc.unesco.org/images/0015/001500/150010e.pdf (last accessed 14 April 2007)

UNFCCC (1997). *United Nations Framework Convention on Climate Change.* United Nations Conference on Environment and Development, 1997 http://unfccc.int/kyoto_protocol/items/2830.php (last accessed 17 April 2007)

UNFCCC (2006). Peatland degradation fuels climate change. Wetlands International and Delft Hydraulics. Presented at *The UN Climate Conference, 7 November 2006, Nairobi* http://www.wetlands.org/ckpp/publication.aspx?id=1f64f9b5-debc-43f5-8c79-b1280f0d4b9a) (last accessed 10 April 2007)

UNFCCC-CDIAC (2006). *Greenhouse Gases Database.* United Nations Framework Convention on Climate Change, Carbon Dioxide Information Analysis Centre (in GEO Data Portal)

UNFCCC (2007). *The Kyoto Protocol website* http://unfccc.int/kyoto_protocol/items/2830.php (last accessed 10 April 2007)

UNPD (2005). *World Urbanization Prospects: The 2005 Revision.* UN Population Division, New York, NY (in GEO Data Portal)

UNPD (2007). *World Population Prospects: the 2006 Revision Highlights.* United Nations Department of Social and Economic Affairs, Population Division, New York, NY (in GEO Data Portal)

UNSD (2007a). *Transport Statistical Database* (in GEO Data Portal)

UNSD (2007b). *International Civil Aviation Yearbook: Civil Aviation Statistics of the World* (in GEO Data Portal)

USEIA (1999). *Analysis of the Climate Change Technology Initiative.* Report SR/OIAF/99-01. US Energy Information Administration, US Department of Energy, Washington, DC http://www.eia.doe.gov/oiaf/archive/climate99/climaterpt.html (last accessed 7 June 2007)

USEPA (1999). *The benefits and costs of the Clean Air Act 1990 to 2010.* US Environmental Protection Agency, Washington, DC http://www.epa.gov/air/sect812/prospective1.html (last accessed 14 April 2007)

USEPA (2006). *Climate Leaders Partners.* US Environmental Protection Agency, Washington, DC http://www.epa.gov/climateleaders/partners/index.html (last accessed 14 April 2007)

Vennemo, H., Aunan, K., Fang, J., Holtedahl, P., Hu, T. and Seip, H. M. (2006). Domestic environmental benefits of China's energy-related CDM potential. In *Climate Change* 75:215-239

Vienna Convention (2007). *The Vienna Convention website* http://ozone.unep.org/Treaties_and_Ratification/2A_vienna_convention.asp (last accessed 7 June 2007)

Vingarzan, R. (2004). A review of surface ozone background levels and trends. In *Atmospheric Environment* 38:3431-3442

Wahid, A., Maggs, R., Shamsi, S. R. A., Bell, J. N. B. and Ashmore, M. R. (1995). Air pollution and its impacts on wheat yield in the Pakistan Punjab. In *Environmental Pollution* 88(2):47-154

Walter, K.M., Zimov, S.A., Chanton, J.P., Verbyla, D. and Chapin, F.S. (2006). Methane bubbling from Siberian thaw lakes as a positive feedback to climate warming. In *Nature* 443:71-75

Watkiss, P., Baggot, S., Bush, T., Cross, S., Goodwin, J., Holland, M., Hurley, F., Hunt, A., Jones, G., Kollamthodi, S., Murrells, T., Stedman, J. and Vincent, K. (2004). *An evaluation of air quality strategy.* Department for Environment, Food and Rural Affairs, London http://www.defra.gov.uk/environment/airquality/publications/stratevaluation/index.htm (last accessed 17 April 2007)

WBCSD (2005). *Mobility 2030: Meeting the Challenges to Sustainability.* World Business Council for Sustainable Development, Geneva

Webster, P.J., Holland, G.J., Curry, J.A. and Chang, H.R. (2005). Changes in Tropical Cyclone Number, Duration, and Intensity in a Warming Environment. In *Science* 309:1844-1846

Wheeler, D. (1999). *Greening industry: New roles for communities, markets and governments.* The World Bank, Washington, DC and Oxford University Press, New York, NY

WHO (2002). *The World Health Report 2002. Reducing risks, promoting healthy life.* World Health Organization, Geneva http://www.who.int/whr/previous/en/index.html (last accessed 14 April 2007)

WHO (2003). *Climate Change and Human Health – Risks and Responses.* McMichael, A.J., Campbell-Lendrum, D.H., Corvalan, C.F., Ebi, K.L., Githeko, A.K., Scheraga, J.D. and Woodward, A. (eds.). World Health Organization, Geneva

WHO (2006a). *WHO Air quality guidelines for particulate matter, ozone, nitrogen dioxide and sulfur dioxide, Global update 2005: Summary of risk assessment.* World Health Organization, Geneva

WHO (2006b). *Solar ultraviolet radiation: global burden of disease from solar ultraviolet radiation.* Environmental Burden of Disease Series No 13. World Health Organization, Geneva

WHO (2006c). *Fuel of life: household energy and health.* World Health Organization, Geneva

WMO (2006a). *Commission for Atmospheric Sciences, Fourteenth session, 2006. Abridged final report with resolutions and recommendations.* WMO No.-1002. World Meteorological Organization, Geneva

WMO (2006b). *WMO Antarctic Ozone Bulletin #4/2006.* World Meteorological Organization, Geneva http://www.wmo.ch/web/arep/06/ant-bulletin-4-2006.pdf (last accessed 17 April, 2007)

WMO and UNEP (2003). *Twenty questions and answers about the ozone layer. Scientific assessment of ozone depletion: 2002.* http://www.wmo.int/web/arep/reports/ozone_2006/twenty-questions.pdf (last accessed 18 April 2007)

WMO and UNEP (2006). *Executive Summary of the Scientific Assessment of Ozone Depletion: 2006.* Scientific Assessment Panel of the Montreal Protocol on Substances that Deplete the Ozone Layer, Geneva and Nairobi http://ozone.unep.org/Publications/Assessment_Reports/2006/Scientific_Assessment_2006_Exec_Summary.pdf (last accessed 14 April 2007)

World Bank (2000). *Improving Urban Air Quality in South Asia by Reducing Emissions from Two-Stroke Engine Vehicles.* The World Bank, Washington, DC http://www.worldbank.org/transport/urbtrans/e&ei/2str1201.pdf (last accessed April 14, 2007)

World Bank (2006). *World Development Indicators 2006* (in GEO Data Portal)

Wright, L. and Fjellstrom, K. (2005). *Sustainable Transport: A Sourcebook for Policymakers in developing countries, Module 3a: Mass Transit Options.* German Technical Cooperation (GTZ), Bangkok http://eprints.ucl.ac.uk/archive/00000113/01/Mass_Rapid_Transit_guide,_GTZ_Sourcebook._Final,_Feb_2003,_Printable_version.pdf (last accessed 17April 2007)

Ye, X.M., Hao, J.M., Duan, L. and Zhou, Z.P. (2002). Acidification sensitivity and critical loads of acid deposition for surface waters in China. In *Science of the Total Environment* 289(1-3):189-203

Zellmer, I. D. (1998). The effect of solar UVA and UVB on subarctic Daphnia pulicaria in its natural habitat. In *Hydrobiologia* 379:55-62

Zimov, S.A., Schuur, E.A.G. and Chapin, F.S. (2006). Permafrost and the global carbon budget. In *Science* 312:1612-1613

Zwally, H.J., Giovinetto, M.B., Li, J., Cornejo, H.G., Beckley, M.A., Brenner, A.C., Saba, J.L. and Yi, D.H. (2005). Mass changes of the Greenland and Antarctic ice sheets and shelves and contributions to sea-level rise: 1992-2002.In *J Glaciol* 51(175):509-527

Chapter 3

Land

Coordinating lead author: David Dent

Lead authors: Ahmad Fares Asfary, Chandra Giri, Kailash Govil, Alfred Hartemink, Peter Holmgren, Fatoumata Keita-Ouane, Stella Navone, Lennart Olsson, Raul Ponce-Hernandez, Johan Rockström, and Gemma Shepherd

Contributing authors: Gilani Abdelgawad, Niels Batjes, Julian Martinez Beltran, Andreas Brink, Nikolai Dronin, Wafa Essahli, Göram Ewald, Jorge Illueca, Shashi Kant, Thelma Krug, Wolfgang Kueper, Li Wenlong, David MacDevette, Freddy Nachtergaele, Ndegwa Ndiang'ui, Jan Poulisse, Christiane Schmullius, Ashbindu Singh, Ben Sonneveld, Harald Sverdrup, Jo van Brusselen, Godert van Lynden, Andrew Warren, Wu Bingfang, and Wu Zhongze

Chapter review editor: Mohamed Kassas

Chapter coordinators: Timo Maukonen and Marcus Lee

Credit: Christian Lambrechts

Main messages

The demands of a burgeoning population, economic development and global markets have been met by unprecedented land-use change. The following are the main messages of this chapter:

During the last 20 years, the exponential expansion of cropland has slackened, but land is now used much more intensively: globally in the 1980s, on average a hectare of cropland produced 1.8 tonnes, but now it produces 2.5 tonnes. For the first time in history, more than half of the world's population lives in cities, which are growing rapidly, especially in developing countries. Cities draw upon extensive rural hinterlands for water and disposal of waste, while their demands for food, fuel and raw materials have a global reach.

Unsustainable land use is driving land degradation. Land degradation ranks with climate change and loss of biodiversity as a threat to habitat, economy and society, but society has different perspectives on various aspects of land degradation, according to political visibility. Inaction means a cumulative addition to a long historical legacy of degradation, from which recovery is difficult or impossible.

Harmful and persistent pollutants, such as heavy metals and organic chemicals, are still being released to the land, air and water from mining, manufacturing, sewage, energy and transport emissions; from the use of agrochemicals and from leaking stockpiles of obsolete chemicals. This issue is politically visible, effects on human health are direct and increasingly well understood, and better procedures and legislation to address chemical contamination are being developed. There has been progress in dealing with pollution in the industrialized countries, where the problem first emerged, but the shift of industry to newly-industrialized countries is yet to be followed by implementation of adequate measures to protect the environment and human health. Achievement of an acceptable level of safety, worldwide, requires strengthening of institutional and technical capacity in all countries, and the integration and effective implementation of existing controls at all levels. There remains an unacceptable lack of data, even for proxies, such as total production and application of chemicals.

Forest ecosystem services are threatened by increasing human demands. Exploitation of forests has been at the expense of biodiversity and natural regulation of water and climate, and has undermined subsistence support and cultural values for some peoples. These issues are increasingly acknowledged, prompting a range of technical responses, legislation and non-binding agreements (such as the United Nations Forum on Forests) to conserve forests, and financial mechanisms to support them. The historical decline in the area of temperate forest has been reversed, with an annual increase of 30 000 km² between 1990 and 2005. Deforestation in the tropics, having begun later, continued at an annual rate of 130 000 km² over the same period. The decline in forest area may be countered by investment in planted forest and more efficient use of wood. More forest is being designated for ecosystem services, but innovative management is required to maintain and restore ecosystems. There is an urgent need to build institutional capacity, in particular community-based management; the effectiveness of this response depends on good governance.

Land degradation in the form of soil erosion, nutrient depletion, water scarcity, salinity and disruption of biological cycles is a fundamental and persistent problem. Land degradation diminishes productivity, biodiversity and other ecosystem services, and contributes to climate change. It is a global development issue – degradation and poverty are mutually reinforcing – but is politically invisible and largely ignored. The damage can be arrested, even reversed, but this requires concerted, long-term investment across sectors, by all levels of government and by individual land users, research to provide reliable data, and adaptation of technologies appropriate to local circumstances. Such a package of measures has rarely been attempted.

Depletion of nutrients by continued cropping with few or no inputs limits productivity over vast tropical and subtropical upland areas. Research has shown the benefits of biological nutrient cycling by integration of legumes into the cropping system, improved fallows and agroforestry. However, widespread adoption is yet to be achieved, and for severely nutrient-deficient soils, there is no remedy except external nutrient inputs. The simple addition of manure or fertilizer may raise crop yields from as little as 0.5 to between 6 and 8 tonnes of grain/ha. In contrast to intensive farming systems that pollute streams and groundwater by excessive fertilizer application, many smallholders in poor countries do not have the means to purchase fertilizer, despite favourable benefit-cost ratios.

Increasing water scarcity is undermining development, food security, public health and ecosystem services. Globally, 70 per cent of available freshwater is held in the soil and accessible to plants, whereas only 11 per cent is accessible as stream flow and groundwater. Better soil and water management can greatly increase the resilience of farming systems and the availability of water downstream, but nearly all investment goes into the withdrawal of water, of which 70–80 per cent is used for irrigation. Meeting the Millennium Development Goal on hunger will require doubling of water use by crops by 2050. Even with much-needed improvements in efficiency, irrigation cannot do it alone. A policy shift is needed towards greater water-use efficiency in rain-fed farming, which will also replenish water supplies at source.

Desertification occurs when land degradation processes, acting locally, combine to affect large areas in drylands. Some 2 billion people depend on drylands, 90 per cent of them in developing countries. Six million km^2 of drylands bear a legacy of land degradation. It is hard to deal with the problem, because of cyclical swings in rainfall, land tenure that is no longer well adjusted to the environment, and because local management is driven by regional and global forces. These forces have to be addressed by national, regional and global policies. Local responses need to be guided by consistent measurement of indicators of long-term ecosystem change.

Demands on land resources and the risks to sustainability are likely to intensify. There are opportunities to meet this challenge, and to avoid potentially unmanageable threats. Population growth, economic development and urbanization will drive demands for food, water, energy and raw materials; the continued shift from cereal to animal products and the recent move towards biofuels will add to the demand for farm production. At the same time, climate change will increase water demands, and increasing variability of rainfall may increase water scarcity in drylands. Opportunities to meet these challenges include application of existing knowledge, diversification of land use, in particular to farming systems that mimic natural ecosystems and closely match local conditions instead of ignoring them, technological advances, harnessing markets to the delivery of ecosystem services, and independent initiatives by civil society and the private sector. Potentially unmanageable threats include runaway biological cycles, climate-related tipping points, conflict and breakdown of governance.

INTRODUCTION

Twenty years ago, *Our Common Future*, the report of the World Commission on Environment and Development, stated: "If human needs are to be met, the Earth's natural resources must be conserved and enhanced. Land use in agriculture and forestry must be based on a scientific assessment of land capacity and the annual depletion of topsoil." Such a scientific assessment has yet to be undertaken and significant data uncertainties remain; the fundamental principles of sustainable land management, established at the 1992 UN Conference on Environment and Development (UNCED), notably in the Agenda 21 Programme of Action for Sustainable Development, are yet to be translated into globally effective policies and tools. Sustainable development remains one of the greatest challenges, although there have been some successes: at regional scales there are the rehabilitation of much of the Loess Plateau in China and the Great Plains of the United States, as a result of long-term, concerted action.

Over the last 20 years, increasing human population, economic development and emerging global markets have driven unprecedented land-use change. Anticipated human population increases and continued economic growth are likely to further increase exploitation of land resources over the next 50 years (see Chapter 9). The most dynamic changes have been in forest cover and composition, expansion and intensification of cropland, and the growth of urban areas. Unsustainable land use drives land degradation through contamination and pollution, soil erosion and nutrient depletion. In some areas there is an excess of nutrients causing eutrophication, and there can be water scarcity and salinity. Beneath land degradation lies disturbance of the biological cycles on which life depends, as well as social and development issues. The term desertification was coined to convey this drama of pressing and interconnected issues in drylands, but human-induced land degradation extends beyond drylands or forests.

Many issues interact with the atmosphere or water, or both. This chapter covers those aspects of water resources that are intimately linked to land management, ranging from rainfall to run-off, infiltration, storage of water in the soil and its use by plants (*green* water), as well as the uptake of salt, agrochemicals and suspended sediment. Aspects related to the recharge of groundwater and stream flow (*blue* water) are covered in Chapter 4, while carbon storage and emissions are dealt with mainly in Chapter 2. The *green-blue* water flows are highlighted in Figure 3.1, below.

DRIVERS OF CHANGE AND PRESSURES

Drivers of land-use change include great increases in the human population and density, increased productivity, higher incomes and consumption patterns, and technological, political and climate change. Individual land-use decisions are also motivated by collective memory and personal histories, values, beliefs and perceptions. Table 3.1 summarizes pressures and drivers of land-use change, distinguishing between slow drivers that result in gradual impacts over decades, and fast drivers that may have impacts in one year (see the section on desertification).

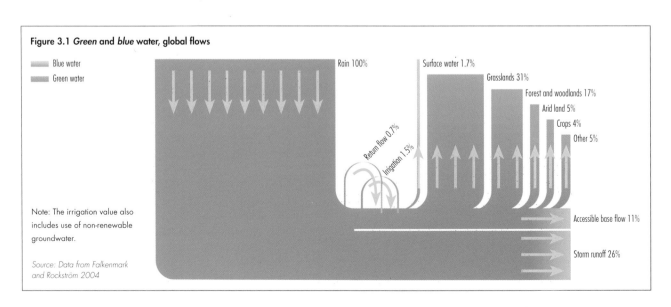

Figure 3.1 *Green* and *blue* water, global flows

Blue water
Green water

Rain 100%
Surface water 1.7%
Grasslands 31%
Forest and woodlands 17%
Arid land 5%
Crops 4%
Other 5%
Return flow 0.7%
Irrigation 1.5%
Accessible base flow 11%
Storm runoff 26%

Note: The irrigation value also includes use of non-renewable groundwater.

Source: Data from Falkenmark and Rockström 2004

Drivers of land-use change themselves change over time. For instance, the Brazilian Amazon was exploited from the late-19th to mid-20th century to supply rubber to the world market. In the second half of the 20th century, the region was drawn into the national economy, with large areas cleared for cattle ranching. Currently, it is responding to national and international markets, resulting in more intensive land use and continued forest conversion, mainly to farmland, including grassland for beef production.

Land-use change is influenced by local needs, as well as by nearby urban demands and remote economic forces (see Box 3.1 under Forests). At the global level, reliable historical data are scarce, but the available information indicates that the greatest changes over the last 20 years have been in forests, especially by conversion to cropland, woodland or grassland and also by new planted forests. Estimates of global land-use changes since 1987 are shown in Table 3.2 in terms of area change by category (the table does not show change of composition within these categories).

Since 1987, the largest forest conversions have occurred in the Amazon Basin, South East Asia, and Central and West Africa. Forest area increased in the Eurasian boreal forest, and in parts of Asia, North America, and Latin America and the Caribbean, mainly due to new planted forests (FAO 2006a). Forest degradation, from both human and natural causes, is widespread. For instance, 30 000 km^2 of forest in the Russian far east have been degraded over the past 15 years by illegal logging and fires (WWF 2005).

Cropland has expanded significantly in South East Asia, and in parts of West and Central Asia, the

Table 3.1 Pressures and drivers of land-use change					
	Changes in human population and management	Changing opportunities created by markets	Policy and political changes	Problems of adaptive capacity and increased vulnerability	Changes in social organization, resource access and attitudes
Slow	Natural population growth; subdivision of land parcels Domestic life cycles that lead to changes in labour availability Excessive or inappropriate use of land	Commercialization and agro-industrialization Improvement in accessibility through road construction Changes in market prices for inputs or outputs, such as erosion of prices of primary products, unfavourable global or urban-rural terms of trade Off-farm wages and employment opportunities	Economic development programmes Perverse subsidies, policy-induced price distortions and fiscal incentives Frontier development (for example, for geopolitical reasons, or to promote interest groups) Poor governance and corruption Insecurity in land tenure	Financial problems, such as creeping household debts, no access to credit, lack of alternative income sources Breakdown of informal social networks Dependence on external resources, or on assistance Social discrimination against ethnic minorities, women, members of lower classes or castes	Changes in institutions governing access to resources by different land managers, such as shifts from communal to private rights, tenure, holdings and titles Growth of urban aspirations Breakdown of extended families Growth of individualism and materialism Lack of public education, and poor flow of information about the environment
Fast	Spontaneous migration, forced population displacement Decrease in land availability due to encroachment of other uses, such as natural reserves	Capital investments Changes in national or global macro-economic and trade conditions that lead to changes in prices, such as a surge in energy prices, or global financial crisis New technologies for intensification of resource use	Rapid policy changes, such as devaluation Government instability War	Internal conflicts Diseases, such as malaria, and illnesses, such as HIV/AIDS Natural hazards	Loss of entitlements to environmental resources through, for example, expropriation for large-scale agriculture, large dams, forest projects, tourism and wildlife conservation

Source: Adapted from Lambin and others 2003

Table 3.2 Global land use – areas unchanged (thousands km²) and conversions 1987–2006 (thousands km²/yr)							
From \ To	Forest	Woodland/ Grassland	Farmland	Urban areas	Losses	Gains	Net change
Forest	39 699	30	98	2	−130	57	−73
Woodland/Grassland	14	34 355	10	2	−26	50	24
Farmland	43	20	15 138	16	−79	108	29
Urban areas	n.s.	n.s.	n.s.	380	0	20	20
Total					−235	235	

n.s. = not significant; farmland includes cropland and intensive pasture

Source: Holmgren 2006

Great Lakes region of Eastern Africa, the southern Amazon Basin, and the Great Plains of the United States. In contrast, some croplands have been converted to other land uses: to forests in the southeastern United States, eastern China and southern Brazil, and to urban development around most major cities. Viewed in a wider historical context, more land was converted to cropland in the 30 years after 1950, than in the 150 years between 1700 and 1850 (MA 2005a).

Even more significant than the change in cropland area, is that land-use intensity has increased dramatically since 1987, resulting in more production per hectare. Cereal yields have increased by 17 per cent in North America, 25 per cent in Asia, 37 per cent in West Asia, and by 40 per cent in Latin America and the Caribbean. Only in Africa have yields remained static and low. Globally, adding together production of cereals, fruit, vegetables and meat, output per farmer and unit of land has increased. In the 1980s, one farmer produced one tonne of food, and one hectare of arable land produced 1.8 tonnes, annually on average. Today, one farmer produces 1.4 tonnes, and one hectare of land produces 2.5 tonnes. The average amount of land cultivated per farmer remained the same, at about 0.55 ha (FAOSTAT 2006). However, world cereal production per person peaked in the 1980s, and has since slowly decreased despite the increase in average yields.

Towns and cities are expanding rapidly. They occupy only a few per cent of the land surface, but their demand for food, water, raw materials and sites for waste disposal dominate the land around them. Urban expansion occurred at the expense of farmland rather than forest, and is currently highest in developing countries.

ENVIRONMENTAL TRENDS AND RESPONSES
Land-use changes have had both positive and negative effects on human well-being, and on the provision of ecosystem services. The enormous increase in the production of farm and forest products has brought greater wealth and more secure livelihoods for billions, but often at the cost of land degradation, biodiversity loss and disruption of biophysical cycles, such as the water and nutrient cycles. These impacts create many challenges and opportunities. Table 3.3 summarizes positive and negative links between changes in land and human well-being.

Table 3.3 Links between land changes and human well-being					
Change in land	Environmental impact	Material needs	Human health	Safety	Socio-economic
Cropland expansion and intensification	Loss of habitat and biodiversity; soil water retention and regulation; disturbance of biological cycle; increase of soil erosion, nutrient depletion, salinity, and eutrophication	Increased food and fibre production – such as doubling world grain harvest in last 40 years Competing demands for water	Spread of disease vectors related to vegetation and water (such as irrigation associated with schistosomiasis) Exposure to agrochemicals in air, soil and water	Increased hazards from flood, dust and landslides during extreme weather	More secure livelihoods and growth in agricultural output Changes in social and power structures

Table 3.3 Links between land changes and human well-being, *continued*

Change in land		Environmental impact	Material needs	Human health	Safety	Socio-economic
Loss of forest, grassland and wetlands		Loss of habitat, biodiversity, stored carbon, soil water retention and regulation Disturbance of biological cycles and food webs	Diminished variety of resources Diminished water resources and water quality	Loss of forest ecosystem services, including potential new medicinal products	Increased hazard of flooding and landslides during extreme weather and tsunamis	Loss of forest products, grazing, fisheries and drought reserves Loss of livelihood, cultural values and support for traditional lifestyles of indigenous and local communities Loss of recreation opportunities and tourism
Urban expansion		Disruption of hydrological and biological cycles; loss of habitat and biodiversity; concentration of pollutants, solid and organic wastes; urban heat islands	Increased access to food, water and shelter; increased choice, but satisfaction of material needs highly dependent on income	Respiratory and digestive-tract diseases due to air pollution, poor water supply and sanitation Higher incidence of stress- and industry-related diseases Higher incidence of heat stroke	Increased exposure to crime Traffic and transport hazards Increased hazard of flooding caused by soil sealing and occupation of hazardous sites	Increased opportunity for social and economic interaction and access to services Increased competition for financial resources Diminished sense of community; increased sense of isolation
Land Degradation	Chemical contamination	Polluted soils and water	Water scarcity and non-potable water	Poisoning, accumulation of persistent pollutants in human tissue with potential genetic and reproductive consequences	Increased risk of exposure and of contamination of food chains; in severe cases, areas become uninhabitable	Loss of productivity due to ill health Diminished productivity of contaminated systems
	Soil erosion	Loss of soil, nutrients, habitat and, property; siltation of reservoirs	Loss of food and water security	Hunger, malnutrition, exposure to diseases due to weakened immune system Turbidity and contaminated water	Risk of floods and landslides Accidents due to damage to infrastructure, particularly in coastal and riverine areas	Loss of property and infrastructure Decreasing hydro-power generation due to siltation of reservoirs Diminished development in farm and forest sectors
	Nutrient depletion	Impoverished soils	Diminished farm and forest production	Malnutrition and hunger		Lack of development in farm sector, poverty
	Water scarcity	Diminished stream flow and groundwater recharge	Loss of food and water security	Dehydration Inadequate hygiene, water-related diseases	Conflict over water resources	Lack of development, poverty
	Salinity	Unproductive soils, unusable water resources, loss of freshwater habitat	Diminished farm production	Non-potable water		Loss of farm production Increased industrial costs of corrosion and water treatment Damage to infrastructure

Table 3.3 Links between land changes and human well-being, *continued*

Change in land	Environmental impact	Material needs	Human health	Safety	Socio-economic
Desertification	Loss of habitat and biodiversity Reduced groundwater recharge, water quality and soil fertility Increased soil erosion, dust storms, and sand encroachment	Diminished farm and rangeland production Loss of biodiversity Water scarcity	Malnutrition and hunger Water-borne diseases, respiratory problems	Conflict over land and water resources Increasing flash floods, dust hazard	Poverty, marginalization, decreased social and economic resilience, population movements
Carbon cycle	Climate change, acidification of ocean surface waters (see mainly Chapter 2)	Shift from fossil fuels to biofuels conflicts with food production Shift in growing seasons and risk of crop failure	Respiratory diseases related to air pollution	Risk of flood-related damage to property, particularly in coastal and riverine areas	Up to 80 per cent of energy supply is derived through manipulation of the carbon cycle
Nutrient cycles	Eutrophication of inland and coastal waters, contaminated groundwater Depletion of phosphate resources		Health effects from bioaccumulation of N or P in food chains Non-potable water		Benefits of food security and biofuel production
Acidifying cycles	Acid depositions and drainage damaging land and water ecosystems Acidification of ocean and freshwaters	Freshwater fish resources declining; risk of further collapse of marine fisheries	Poisoning from increased plant and animal uptake of toxic metals		Economic damage to forests, fisheries and tourism Corrosion of infrastructure and industrial facilities

FORESTS

Forests are not just trees, but part of ecosystems that underpin life, economies and societies. Where forests are privately owned, they are often managed mainly for production. Yet, in addition to directly supporting such industries as timber, pulp and biotechnology, all forests provide a wide range of ecosystem services. These services include prevention of soil erosion, maintenance of soil fertility, and fixing carbon from the atmosphere as biomass and soil organic carbon. Forests host a large proportion of terrestrial biodiversity, protect water catchments and moderate climate change. Forests also support local livelihoods, provide fuel, traditional medicines and foods to local communities, and underpin many cultures. The harvesting of forest products is putting severe stress on the world's forests. Box 3.1 describes some of the main pressures that drive changes in forest ecosystems.

Changes in forest ecosystems

Between 1990 and 2005, the global forest area shrank at an annual rate of about 0.2 per cent. Losses were greatest in Africa, and Latin America and the Caribbean. However, forest area expanded in Europe and North America. In Asia and the Pacific, forest area expanded after 2000 (see the FAO data in Figure 3.2 and in Figure 6.31 on annual forest change in the biodiversity and ecosystems section of Latin America and the Caribbean in Chapter 6).

In addition to the changes in global forest area, significant changes also occurred in forest composition, particularly in the conversion of primary forest to other types of forests (especially in Asia and the Pacific). It is estimated that over the past 15 years there has been an annual loss of 50 000 km^2 of primary forest, while there has been an average annual increase of 30 000 km^2

Changes in forest ecosystems, particularly the conversion from forest to other land uses and vice-versa, are driven by the harvesting of forest products and associated management activities, as well as by natural forest dynamics such as changes in age class and structure, and natural disturbance. Other drivers include climate change, diseases, invasive species, pests, air pollution and pressures from economic activities, such as agriculture and mining.

There are a number of drivers and pressures causing changes in forests.
■ Demographic trends include changes in human population density, movement, growth rates, and urban-rural distribution. These trends exert pressures on forests through demands for goods such as

timber and firewood, and for services such as regulation of water resources and recreation. The demand for services is increasing faster than the supply.
■ Economic growth is reflected in the prices of forest products and international trade. The relative contribution of the forestry sector to global GDP declined in the last decade, from 1.6 per cent in 1990 to 1.2 per cent in 2000.
■ Cultural preferences are shifting demands towards cultural services provided by forests.
■ Science has helped to improve forest management, while both science and technology have improved the productivity and the efficiency of production and utilization of forests.

Sources: Bengeston and Kant 2005, FAO 2004, FAO 2006a

of planted and semi-natural forests. Primary forests now comprise about one-third of global forest area (see Figure 3.3).

Forests are managed for various functions (see Figure 3.4): in 2005, one-third of global forests were managed primarily for production, one-fifth for conservation and protection, and the remaining forests for social and multiple services. The proportion allocated primarily for production is largest in Europe (73 per cent) and least in North America (7 per cent) and West Asia (3 per cent). Of the total wood production, 60 per cent was industrial wood and 40 per cent was fuel; 70 per cent of industrial wood is produced in North America and Europe, while 82 per cent of fuelwood is produced in the developing world (FAO 2006a). Non-wood forest products, including food, fodder, medicine, rubber and handicrafts, are increasingly acknowledged in forest assessments and, in some countries, are more valuable than wood products.

More and more forest areas are being designated for conservation and protection, partly in recognition of their valuable ecosystem services such as soil and water protection, absorption of pollution, and climate regulation through carbon fixation. However, these services have been reduced by the decline in total forest area and by continued forest degradation, especially in production and multipurpose forests. For example, the rate of decline in fixed carbon has been greater than the rate of decline in forest area (see Figure 3.5).

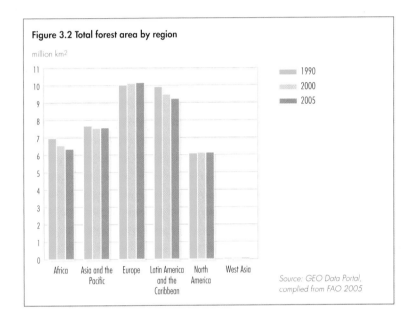

Figure 3.2 Total forest area by region

million km²

Source: GEO Data Portal, complied from FAO 2005

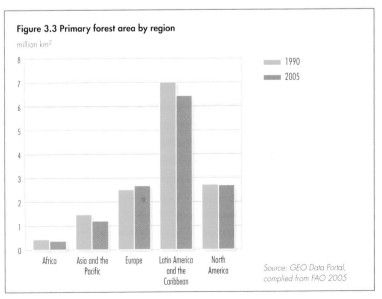

Figure 3.3 Primary forest area by region

million km²

Source: GEO Data Portal, complied from FAO 2005

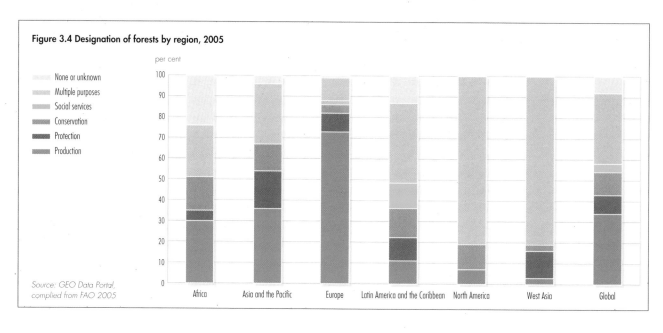

Figure 3.4 Designation of forests by region, 2005

per cent

Legend:
- None or unknown
- Multiple purposes
- Social services
- Conservation
- Protection
- Production

Source: GEO Data Portal, complied from FAO 2005

Regions: Africa, Asia and the Pacific, Europe, Latin America and the Caribbean, North America, West Asia, Global

Ensuring a continued flow of goods-and-services from forests is essential for human well-being and national economies. Greater emphasis on conservation of biodiversity may lead to increased benefits in terms of resilience, social relations, health, and freedom of choice and action (MA 2005a, FAO 2006a). Many of the world's poor are directly and intensely affected by changes in forest use. A recent synthesis of data from 17 countries found that 22 per cent of rural household income in forested regions comes from harvesting wild food, firewood, fodder and medicinal plants, generating a much higher proportion of income for the poor than for wealthy families. For the poor, this is crucial when other sources of income are scarce (Vedeld and others 2004).

Managing forests

Despite the extensive impacts of changes in forest cover and use, forest issues continue to be addressed piecemeal in multilateral conventions and other legally and non-legally binding instruments and agreements. However, some regional initiatives in forest law enforcement and governance break new ground in addressing illegal activities. Regional ministerial conferences on forests have taken place in East Asia (2001), Africa (2003), and Europe and North America (2005), jointly organized by the governments of producing and consuming countries (World Bank 2006).

The concept of sustainable forest management has evolved over the last two decades, but remains hard to define. The Forest Principles developed for UNCED state: "Forest resources and forest lands should be sustainably managed to meet the social, economic, ecological, cultural and spiritual needs of present and future generations." Alternative frameworks to assess and monitor the status and trends of different elements of sustainable forest management include criteria and indicators, forest certification and environmental accounting. At the methodological level, it is difficult to integrate information on forest state and trends, and the contribution of non-marketed, non-consumptive and intangible forest goods-and-services. A further difficulty lies in defining thresholds beyond which changes in values can be regarded as being significant. At the practical level, spatial and temporal data for assessing sustainability are often incompatible, inconsistent and insufficient. Policies to promote the fixing of atmospheric carbon by agricultural, pastoral and forest systems have been more seriously considered, because fixing carbon by forest plantations is

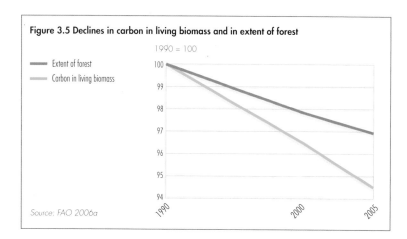

Figure 3.5 Declines in carbon in living biomass and in extent of forest

1990 = 100

Legend:
- Extent of forest
- Carbon in living biomass

Source: FAO 2006a

Table 3.4 Progress towards sustainable forest management

Thematic element	Trends in FRA 2005 variables or derivatives	Data availability	1990–2005 Annual change rate (per cent)	1990–2005 Annual change	Unit
Extent of forest resources	■ Area of forest	H	−0.21	−8 351	1 000 ha
	■ Area of other wooded land	M	−0.35	−3 299	1 000 ha
	■ Growing stock of forests	H	−0.15	−570	million m³
	■ Carbon stock per hectare in forest biomass	H	−0.02	−0.15	tonnes/ha
Biological diversity	■ Area of primary forest	H	−0.52	−5 848	1 000 ha
	■ Area of forest designated primarily for conservation of biological diversity	H	1.87	6 391	1 000 ha
	■ Total forest area excluding area of productive forest plantations	H	−0.26	−9 397	1 000 ha
Forest health and vitality	■ Area of forest affected by fire	M	−0.49	−125	1 000 ha
	■ Area of forest affected by insects, diseases and other disturbances	M	1.84	1 101	1 000 ha
Productive functions of forest resources	■ Area of forest designated primarily for production	H	−0.35	−4 552	1 000 ha
	■ Area of productive forest plantations	H	2.38	2 165	1 000 ha
	■ Commercial growing stock	H	−0.19	−321	million m³
	■ Total wood removals	H	−0.11	−3 199	1 000 m³
	■ Total NWFP removals	M	2.47	143 460	tonnes
Protective functions of forest resources	■ Area of forest designated primarily for protection	H	1.06	3 375	1 000 ha
	■ Area of protective forest plantations	H	1.14	380	1 000 ha
Socio-economic functions	■ Value of total wood removals	L	0.67	377	million US$
	■ Value of total NWFP removals	M	0.80	33	million US$
	■ Total employment	M	−0.97	−102	1 000 pers. yrs
	■ Area of forest under private ownership	M	0.76	2 737	1 000 ha
	■ Area of forest designated primarily for social services	H	8.63	6 646	1 000 ha

FRA = FAO Global Forest Resources Assessment   NWFP = non-wood forest products

■ = Positive change (greater than 0.5 per cent)   ■ = No major change (between −0.5 and 0.5 per cent)   ■ = Negative change (less than −0.5 per cent)

Source: FAO 2006a

eligible for trading under the Kyoto Protocol. Table 3.4 summarizes progress towards sustainable forest management against measures of forest extent, biodiversity, forest health, and productive, protective and socio-economic functions.

At the local level, there are many examples of innovative management, especially community-based approaches that are arresting trends in forest degradation and loss of forest ecosystem services (see Box 3.2).

Box 3.2 Sustainable forest management by smallholders in the Brazilian Amazon

Since 1998, Brazilian farmers have had to maintain 80 per cent of their land as forest (50 per cent in some special areas) as a legal forest reserve. Small-scale forest management enables smallholders to make economic use of their forest reserves.

Since 1995, a group of smallholders in the state of Acre, supported by Embrapa (the Brazilian Agricultural Research Corporation), has developed sustainable forest management systems based on traditional forest practices as a new source of income. Forest structure and biodiversity are maintained by low-impact disturbance at short intervals, combined with silvicultural practices, matching the circumstances of the smallholders (small management area, limited labour availability and investment) with appropriate management techniques (short cutting cycles, low intensity harvesting and animal traction).

The system described here is practised in forest holdings averaging 40 ha each. Cooperative agreements among neighbours facilitate the acquisition of oxen, small tractors and solo-operated sawmills, yielding higher prices in local markets and reducing transportation costs. As a result, farmers' incomes have risen 30 per cent. In 2001, the smallholders created the Association of Rural Producers in Forest Management and Agriculture to market their products nationwide and, in 2003, they won Forest Stewardship Council certification from SmartWood. Surveys have been conducted to monitor biodiversity. IBAMA (the Brazilian Environment and Renewable Natural Resource Institute) and BASA (the Bank of the Amazon) use the sustainable forest management system as a benchmark for development and financial policies for similar natural resource management schemes.

Sources: D'Oliveira and others 2005, Embrapa Acre 2006

LAND DEGRADATION

Land degradation is a long-term loss of ecosystem function and services, caused by disturbances from which the system cannot recover unaided. It blights a significant proportion of the land surface, and as much as one-third of the world's population – poor people and poor countries suffer disproportionately from its effects. Established evidence links land degradation with loss of biodiversity and climate change, both as cause-and-effect (Gisladottir and Stocking 2005). Direct effects include losses of soil organic carbon, nutrients, soil water storage and regulation, and below-ground biodiversity. Indirectly, it means a loss of productive capacity and wildlife habitat. For instance, in rangelands it disrupts wildlife migration, brings changes in forage, introduces pests and diseases, and increases competition for food and water. Water resources are diminished by disruption of the water cycle, off-site pollution and sedimentation. The threat to sustainable development posed by land degradation has been recognized for decades, including by the 1992 Earth Summit and the 2002 World Summit on Sustainable Development, but responses have been hamstrung by weaknesses in available data, particularly in relation to the distribution, extent and severity of the various facets of degradation.

The only comprehensive source of information has been the Global Assessment of Human-induced Soil Degradation (GLASOD), which assessed the severity and kind of land degradation for broadly-defined landscape units at a scale of 1:10 million

(Oldeman and others 1991). It was compiled from expert judgements and, while invaluable as a first global assessment, it has since proven to be not reproducible and inconsistent. In addition, the relationships between land degradation and policy-relevant criteria, such as crop production and poverty, were unverified (Sonneveld and Dent 2007).

A new, quantitative global assessment under the GEF/UNEP/FAO project Land Degradation Assessment in Drylands (LADA) identifies black spots of land degradation by trends analysis of the last 25 years' net primary productivity (NPP or biomass production). NPP is derived from satellite measurements of the normalized difference vegetation index (NDVI or greenness index). A negative trend in NPP does not necessarily indicate land degradation, since it depends on several other factors, especially rainfall. Figure 3.6 combines the recent trend of NPP with rain-use efficiency (NPP per unit of rainfall). Critical areas are identified as areas with a declining trend of NPP and declining rain-use efficiency over the past 25 years, excluding the simple effects of drought. For irrigated areas, only the biomass is considered and urban areas are excluded. The case study on Kenya highlights some of the results of the study (see Box 3.3).

By contrast with previous assessments, such as GLASOD, this new measure does not compound the legacy of historical land degradation with what is happening now. It shows that between 1981

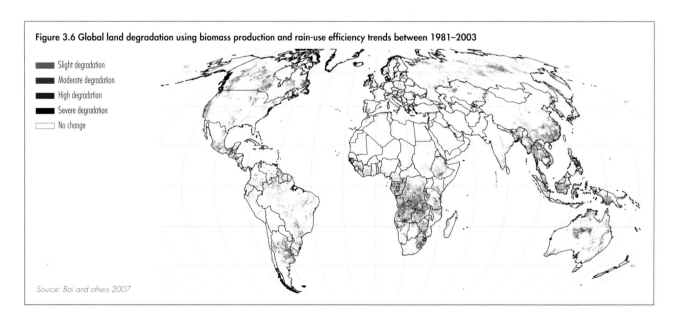

Figure 3.6 Global land degradation using biomass production and rain-use efficiency trends between 1981–2003

- Slight degradation
- Moderate degradation
- High degradation
- Severe degradation
- No change

Source: Bai and others 2007

and 2003 there was an absolute decline in NPP across 12 cent of the global land area, with a strong negative change in a further 1 per cent of the land area. In respect of rain-use efficiency, there was an absolute decrease on 29 per cent of the land area and strong negative change on 2 per cent. The areas affected are home to about 1 billion people, some 15 per cent of the global population. Apart from the loss of farm and forest production, the degraded areas represent a loss of NPP of about 800 million tonnes of carbon over the period, meaning this amount was not fixed·from the atmosphere. In addition, there were emissions to the atmosphere of one or two orders of magnitude more than this from the loss of soil organic carbon and standing biomass (Bai and others 2007).

Areas of concern include tropical Africa south of the equator and southeast Africa, southeast Asia (especially steeplands), south China, north-central Australia, Central America and the Caribbean (especially steeplands and drylands), southeast Brazil and the Pampas, and boreal forests in Alaska, Canada and eastern Siberia. In areas of historical land degradation around the Mediterranean and West Asia, only relatively small areas of change are visible, such as in southern Spain, the Maghreb and the Iraqi marshlands. Comparison of black spots with land cover reveals that 18 per cent of land degradation by area is associated with cropland, 25 per cent is in broad-leaved forests and 17 per cent in boreal forests. This is consistent with trends in forest degradation, even as the area of boreal forests has increased (see section on Drivers and pressures). This preliminary analysis will need to be validated on the ground by the country-level case studies being undertaken by LADA, which will also determine the different types of degradation.

Changes in land
Chemical contamination and pollution
Chemicals are used in every aspect of life, including industrial processes, energy, transport, agriculture, pharmaceuticals, cleaning and refrigeration. More than 50 000 compounds are used commercially, hundreds are added every year, and global chemical production is projected to increase by 85 per cent over the next 20 years (OECD 2001). The production and use of chemicals have not always been accompanied by adequate safety measures. Releases, by-products and degradation of chemicals, pharmaceuticals and other

Box 3.3 Land degradation in Kenya

About 80 per cent of Kenya is dryland. The 25-year trends of biomass and rain-use efficiency highlight two *black spots* of land degradation: the drylands around Lake Turkana, and a swath of cropland in Eastern Province, corresponding to the recent extension of cropping into marginal areas (see the red areas in the bottom map).

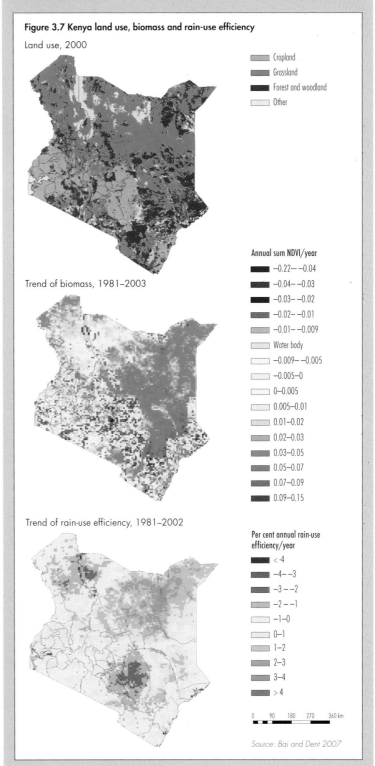

Figure 3.7 Kenya land use, biomass and rain-use efficiency

Land use, 2000

- Cropland
- Grassland
- Forest and woodland
- Other

Trend of biomass, 1981–2003

Annual sum NDVI/year
- −0.22−−0.04
- −0.04−−0.03
- −0.03−−0.02
- −0.02−−0.01
- −0.01−−0.009
- Water body
- −0.009−−0.005
- −0.005−0
- 0−0.005
- 0.005−0.01
- 0.01−0.02
- 0.02−0.03
- 0.03−0.05
- 0.05−0.07
- 0.07−0.09
- 0.09−0.15

Trend of rain-use efficiency, 1981–2002

Per cent annual rain-use efficiency/year
- < -4
- −4−−3
- −3−−2
- −2−−1
- −1−0
- 0−1
- 1−2
- 2−3
- 3−4
- > 4

0   90   180   270   360 km

Source: Bai and Dent 2007

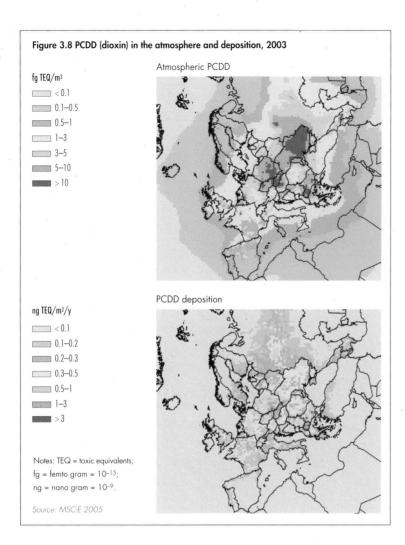

Figure 3.8 PCDD (dioxin) in the atmosphere and deposition, 2003

Atmospheric PCDD

fg TEQ/m³
- < 0.1
- 0.1–0.5
- 0.5–1
- 1–3
- 3–5
- 5–10
- > 10

PCDD deposition

ng TEQ/m²/y
- < 0.1
- 0.1–0.2
- 0.2–0.3
- 0.3–0.5
- 0.5–1
- 1–3
- > 3

Notes: TEQ = toxic equivalents;
fg = femto gram = 10^{-15};
ng = nano gram = 10^{-9}.

Source: MSC-E 2005

agriculture. There are persistent organic pollutants (POPs) such as DDT, brominated flame retardants and polyaromatic hydrocarbons heavy metals, such as lead, cadmium and mercury, and oxides of nitrogen and sulphur. In mining, for instance, toxic substances such as cyanide, mercury and sulphuric acid are used to separate metal from ores, leaving residues in the tailings. Toxic chemicals may be emitted from identifiable point sources, such as stockpiles of hazardous waste, power generation, incineration and industrial processes. They also come from diffuse sources, such as vehicle emissions, the agricultural application of pesticides and fertilizers, as well as in sewage sludge containing residues of process chemicals, consumer products and pharmaceuticals.

Many chemicals persist in the environment, circulating between air, water, sediments, soil and biota. Some pollutants travel long distances to supposedly pristine areas (De Vries and others 2003). For example, POPs and mercury are now found in high concentrations in both people and wildlife in the Arctic (Hansen 2000) (see Figure 6.57 in the Polar section of Chapter 6). Chemical emissions to the atmosphere often become fallout on land or water. Figure 3.8 shows modelling results of the distribution of polychlorodibenzodioxins (PCDD) emissions and deposition in Europe for 2003.

Chemical wastes from industry and agriculture are a big source of contamination, particularly in developing countries and countries with economies in transition. The concentrations of persistent toxic substances observed in many parts of sub-Saharan Africa indicate that this contamination is widespread across the region. Stockpiles containing at least 30 000 tonnes of obsolete pesticides were recorded in Africa (FAO 1994). These stockpiles, often leaking, are up to 40 years old, and contain some pesticides banned long ago in industrialized countries. Environmental levels of toxic chemicals will increase in countries still using them in large quantities (such as Nigeria, South Africa and Zimbabwe), and in countries without effective regulation of their use (GEF and UNEP 2003). In addition, toxic wastes are still being exported to and dumped in developing countries. The dumping of hazardous wastes, such as the 2006 dumping of poisonous oil refinery waste containing hydrogen sulphide and organochloride in Abidjan, Ivory Coast, is still a major problem. This is despite

commodities contaminate the environment, and there is growing evidence of their persistence and their detrimental effects on ecosystems and on human and animal health.

Currently, there is insufficient information on the amounts released, their toxic properties, effects on human health and safe limits for exposure to fully evaluate their environmental and human health impacts. The magnitude of chemical contamination can be measured or estimated by the residue levels and spatial concentration of substances, but data are incomplete globally and for many regions. Proxies that provide some indication include total production of chemicals, total use of pesticides and fertilizers, generation of municipal, industrial and agricultural wastes, and the status of implementation of multilateral environmental agreements relating to chemicals.

Land is subjected to a wide range of chemicals from many sources, including municipalities, industries and

such efforts as the 1991 Bamako Convention on the Ban of the Import into Africa and the Control of Transboundary Movement and Management of Hazardous Wastes within Africa.

A legacy of contaminated industrial and urban sites is common to all old industrial heartlands, particularly in the United States, Europe and the former Soviet Union. Across Europe, it is estimated that there may be more than 2 million such sites, containing hazardous substances such as heavy metals, cyanide, mineral oil and chlorinated hydrocarbons. Of these, some 100 000 require remediation (EEA 2005). See Chapter 7 for additional information on exposing people and the environment to contaminants.

Increasingly, some of the chemical waste stream comes from everyday products; increasing consumption remains coupled to increased generation of wastes, including chemical wastes. Most domestic waste still goes into landfills, although in Europe there is a shift to incineration (EEA 2005).

There are growing differences in pollution trends between industrialized and developing countries. Between 1980 and 2000, control measures resulted in lower emissions of pollutants into the atmosphere and reduced deposition over most of Europe. Now, pollution as a result of consumer activities is outpacing pollution from primary industrial sources. While OECD countries are still the largest producers and consumers of chemicals, there has been a shift of chemical production to newly industrializing countries that, 30 years ago, had little or no chemicals industry. This shift in production has not always been accompanied by control measures, increasing the risks of release of hazardous chemicals into the environment.

The last 25 years have seen accumulating evidence of the serious consequences of chemicals for the environment and human well-being. In addition to directly harming human health, atmospheric pollutants have been implicated in increasing soil acidity and forest decline, and acidification of streams and lakes (see section on acidifying cycles), and have been linked to the burden of chronic diseases such as asthma. WHO estimates that each year, 3 million people suffer from severe pesticide

poisoning, with as many as 20 000 unintentional deaths (Worldwatch Institute 2002). (See Chapter 2 under effects of air pollution).

Soil erosion

Erosion is the natural process of removal of soil by water or wind. Soil erosion becomes a problem when the natural process is accelerated by inappropriate land management, such as clearance of forest and grasslands followed by cropping which results in inadequate ground cover, inappropriate tillage and overgrazing. It is also caused by activities such as mining, infrastructural and urban developments without well-designed and well-maintained conservation measures.

Loss of topsoil means loss of soil organic matter, nutrients, water holding capacity (see section on water scarcity) and biodiversity, leading to reduced production on-site. Eroded soil is often deposited where it is not wanted, with the result that the off-site costs, such as damage to infrastructure, sedimentation of reservoirs, streams and estuaries, and loss of hydropower generation, may be much higher than the losses in farm production.

Although there is consensus that soil erosion is often a severe problem, there are few systematic measurements of its extent and severity. Indicators include barren ground, removal of topsoil as sheet erosion over a wide area or concentrated as rills and gullies, or through landslides. Wind erosion is the major problem in West Asia, with as much as 1.45 million km² – one-third of the region – affected. In extreme cases, mobile dunes encroach upon farmland and settlements (Al-Dabi and others 1997, Abdelgawad 1997). Regional or even global estimates have, quite wrongly, scaled up measurements made on small plots, arriving at huge masses of eroded soil that would reshape whole landscapes within a few decades. Erosion rates reported from Africa range from 5–100 tonnes/ ha/yr, depending on the country and assessment method (Bojö 1996). Authors including den Biggelaar and others (2004) estimate that globally, 20 000–50 000 km² is lost annually through land degradation, chiefly soil erosion, with losses 2–6 times higher in Africa, Latin America and Asia than in North America and Europe. Other global and regional spatial data present vulnerability to erosion, modelled from topographic, soil, land cover and climatic variables, but vulnerability is not the same thing as actual erosion: the most important factor determining actual erosion is the level of land management (see Box 3.4).

Nutrient depletion

Nutrient depletion is a decline in the levels of plant nutrients, such as nitrogen, phosphorous and potassium, and in soil organic matter, resulting in declining soil fertility. It is commonly accompanied by soil acidification, which increases the solubility of toxic elements, such as aluminium. The causes and consequences of nutrient depletion are well-established: in a wet climate, soluble nutrients are leached from the soil, and everywhere crops take up nutrients. The removal of the harvest and crop residues

depletes the soil, unless the nutrients are replenished by manure or inorganic fertilizers (Buresh and others 1997). Nutrient mining refers to high levels of nutrient removal and no inputs.

Deficiency of plant nutrients in the soil is the most significant biophysical factor limiting crop production across very large areas in the tropics, where soils are inherently poor. Several studies in the 1990s indicated serious nutrient depletion in many tropical countries, particularly in sub-Saharan Africa. Most calculations drew up nutrient budgets in which fluxes and pools were estimated from published data at country or sub-regional level. For example, the influential 1990 study by Stoorvogel and Smaling calculated budgets for nitrogen, phosphorus and

Box 3.4 Soil erosion in the Pampas

Soil erosion by water is the main form of land degradation in Latin America. The more extensive the area under cultivation, the more serious the erosion, even in the fertile Pampas. It has been an intractable problem, leading to the abandonment of farmland, for example, in northwest Argentina.

The most promising development has been the large-scale adoption of conservation tillage, which

increases infiltration of rain into the soil compared to conventional ploughing. The area under conservation tillage in Latin America increased from almost zero in the 1980s to 250 000 km2 in 2000, with an adoption rate of 70–80 per cent among large, mechanized farms in Argentina and Brazil, although the adoption rate by small farms is lower.

Sources: FAO 2001, KASSA 2006, Navone and Maggi 2005

In the Pampas, rills form during rainstorms when ground cover is sparse, and gradually turn into large gullies.

Credit: J.L. Panigatti

potassium for the cropland of 38 countries in sub-Saharan Africa for the years since 1983, and projected the data to 2000. In nearly every case, the nutrient inputs were less than the outputs. Some 950 000 km² of land in the region is threatened with irreversible degradation if nutrient depletion continues (Henao and Baanante 2006).

There has been criticism of the basis for such calculations, and debate on the extent and impact of nutrient depletion (Hartemink and van Keulen 2005), but broad agreement on the phenomenon. In some areas, nutrients have been depleted because of reduced fallow periods in shifting cultivation systems, and little or no inorganic fertilizer inputs. In other areas, soil fertility of cropland may be maintained or improved through biomass transfer at the expense of land elsewhere. Where such differences are explored in more detail, there are complex explanations including non-agronomic factors, such as infrastructure, access to markets, political stability, security of land tenure and investments.

Across most of the tropics, the use of inorganic fertilizers is limited by availability and cost, although inorganic fertilizers often have favourable value-to-cost ratios (van Lauwe and Giller 2006). In parts of sub-Saharan Africa, as little as 1 kilogramme of nutrients is applied per hectare. This compares with nutrient additions around 10–20 times higher in industrialized

countries – and also much higher rates in most other developing countries (Borlaug 2003), where there is established evidence that leaching of nitrates into surface and groundwater, and wash-off of phosphates into streams and estuaries, can cause eutrophication (see Chapter 4).

Water scarcity

By 2025, about 1.8 billion people will be living in countries or regions with absolute water scarcity, and two-thirds of the world population could be under conditions of water stress – the threshold for meeting the water requirements for agriculture, industry, domestic purposes, energy and the environment (UN Water 2007). This will have major impacts on activities such as farming (see Chapter 4).

The source of all freshwater is rainfall, most of which is held in the soil, and returns to the atmosphere by evapotranspiration (*green water*). Globally, only 11 per cent of the freshwater flow is available as usable stream flow and groundwater that can be tapped for irrigation, urban and industrial use, potable and stock water (see Figure 3.1). Yet, nearly all investment goes into the management of the water withdrawn from streams and groundwater. While irrigated agriculture is overwhelmingly the biggest user of freshwater, and already draws substantially on groundwater that is not being replenished, it faces increasing competition from other claims (see Figure 4.4). To meet the Millennium

Poor crop performance due to nutrient deficiency compared with enhanced fertility around a farmstead, Zimbabwe.

Credit: Ken Giller

Development Goal (MDG) of halving the proportion of people suffering from hunger by 2015, it will be necessary to manage freshwater resources from the moment that rainwater hits the land surface. This is where soil management determines whether rain runs off the surface, carrying topsoil with it, or infiltrates the soil to be used by plants or to replenish groundwater and stream flows.

Ecosystems and farming systems have adapted to water scarcity in various ways (see Table 3.5). Outside arid and semi-arid areas, absolute lack of water is not the issue; there is enough water to produce a crop in most years. For example in Eastern Africa, meteorological drought (a period when there is not enough water to grow crops because of much below average rainfall) happens every decade. Dry spells of 2–5 weeks in the growing season happen every 2–3 years (Barron and others 2003). Agricultural drought (drought in the root zone) is much more frequent, while political drought, where various failings are attributed to drought, is commonplace. Agricultural drought is more common

than meteorological drought because, on cultivated land, most rainfall runs off the surface, and soil water storage is diminished by soil erosion, resulting in poor soil structure, loss of organic matter, unfavourable texture and impeded rooting. Farmers' field water balances show that only 15–20 per cent of rainfall actually contributes to crop growth, falling to as little as 5 per cent on degraded land (Rockström 2003).

Rainfall may not be the main factor limiting crop production. Tracts of land also suffer from nutrient deficiency (see section on nutrient depletion). While commercial farmers maintain nutrient status by applying fertilizer, risk-averse subsistence farmers do not invest in overcoming other constraints unless the risk of drought is under control.

Irrigation is arguably the most successful insurance against drought. Irrigated land produces 30–40 per cent of global farm output, and a far higher proportion of high-value crops, from less than 10 per cent of the farmed area. Water withdrawals for irrigation have increased dramatically, to about 70 per cent of global

Table 3.5 Ecosystem and farming system responses to water scarcity							
Zone	Extent (per cent of global land surface)	Rainfall (mm) (Aridity index) (Rainfall/Potential evaporation)	Growing season (days)	Water-related risks	Ecosystem type	Rain-fed farming system	Risk management strategies
Hyper-arid	7	<200 (<0.05)	0	Aridity	Desert	None	None
Arid	12	<200 (0.05–0.2)	1–59	Aridity	Desert- desert scrub	Pastoral, nomadic or transhumance	Nomadic society, water harvesting
Semi-arid	18	200–800 (0.2–0.5)	60–119	Drought 1 year in 2, dry spells every year, intense rainstorms	Grassland	Pastoral and agro-pastoral: rangeland, barley, millet, cow-pea	Transhumance, water harvesting, soil and water conservation, irrigation
Dry sub-humid	10	800–1 500 (0.5–0.65)	120–179	Drought, dry spells, intense rainstorms, floods	Grassland and woodland	Mixed farming: maize, beans, groundnut, or wheat, barley and peas	Water harvesting, soil and water conservation, supplementary irrigation
Moist sub-humid	20	1 500–2 000 (0.65–1)	180–269	Floods, waterlogging	Woodland and forest	Multiple cropping, mostly annuals	Soil conservation, supplementary irrigation
Humid	33	>2 000 (>1)	>270	Floods, waterlogging	Forest	Multiple cropping, perennials and annuals	Soil conservation, drainage

Note: Drought-susceptible drylands are highlighted (see Figure 3.9)

Source: Adapted from Rockström and others 2006

water withdrawals (see Figure 4.4). One-tenth of the world's major rivers no longer reach the sea during some part of the year, because water is extracted upstream for irrigation (Schiklomanov 2000). However, limits to the growth of irrigation are in sight, and much of further development is likely to be marginal in terms of returns on investment (Fan and Haque 2000), and in terms of trade-offs against salinity (see section on *Salinity*) and ecosystem services.

Salinity

Soils, streams and groundwater in drylands contain significant amounts of naturally-occurring salt, which inhibits the absorption of water by plants and animals, breaks up roads and buildings, and corrodes metal. Soils containing more than 1 per cent soluble salt cover 4 million km2, or about 3 per cent of the land (FAO and UNESCO 1974–8). Salinity is defined by the desired use of land and water; it is salt in the wrong place when found in farmland, drinking and irrigation water, and in freshwater habitats. It is caused by inappropriate forms of land use and management. Irrigation applies much more water than rainfall and natural flooding and, nearly always, more than can be used by crops. The added water itself contains salt, and it mobilizes more salt that is already in the soil. In practice, leakage from irrigation canals, ponding because of poor land levelling and inadequate drainage raise the water table. Once the water table rises close to the soil surface, water is drawn to the surface by evaporation, further concentrating the salt, which may eventually create a salt crust on the soil surface.

Increasing water withdrawals for irrigation increase the likelihood of salinity (see Box 3.5) when there is inadequate drainage to carry the salt out of the

Salinity induced by irrigation in the Euphrates basin in Syria.

Credit: Mussaddak Janat, Atomic Energy Commission of Syria

soil. This is a threat to livelihoods and food security in dry areas, where most farm production is from irrigation and farmers use whatever water is available, however marginal, even on land with a high, saline water table. In the long run, this renders the land unproductive. Salinity will increase unless the efficiency of irrigation networks, in particular, is greatly improved.

Dryland salinity, as distinct from irrigation-induced salinity, is caused by the replacement of natural vegetation with crops and pastures that use less water, so that more water infiltrates to the groundwater than before. The rising, saline groundwater drives more salt into streams, and, where the water table comes close to the surface, evaporation pulls salt to the surface.

Worldwide, some 20 per cent of irrigated land (450 000 km2) is salt-affected, with 2 500–5 000 km2 lost from production every year as a result of salinity (FAO 2002, FAO 2006b). In Australia, for example, the National Land and Water Resources Audit (NLWRA 2001) estimated 57 000 km^2 of land to be at risk of dryland salinity, and projected three times as much in 50 years. There is underlying concern about the inexorable increase in river water salinity driven by rising water tables; it is predicted that up to 20 000 kilometres of streams may be significantly salt-affected by 2050 (Webb 2002).

Disturbances in biological cycles

Water, carbon and nutrient cycles are the basis of life. The integrity of these cycles determines the health and resilience of ecosystems, and their capacity to provide goods-and-services. Agriculture depends on manipulating parts of these cycles, often at the expense of other parts of the same cycle. Links between the carbon cycle and climate change are now well established (see Box 3.6). While the burning of fossil fuels has greatly disturbed the carbon cycle, land-use change has been responsible for about one-third of the increase in atmospheric carbon dioxide over the last 150 years, mainly through loss of soil organic carbon. Also well established are the links between soil erosion and sediment deposition, between fertilizers and eutrophication, and between emissions of sulphur and nitrogen oxides to the atmosphere and acid contamination of land and water.

Nutrient cycles: Soil fertility and chemistry are closely interwoven. Many elements in the soil participate in cycles of plant nutrition and growth, decomposition of organic matter, leaching to surface water and groundwater, and transport to the oceans. Nitrogen and phosphorus are the nutrients required in largest amounts, and there is concern about both the prospects of continued availability of chemical supplements and the resulting disturbance of these cycles.

The tiny fraction of atmospheric nitrogen made available to biological cycles through natural fixation restricted plant production until the industrial production of nitrogen fertilizers in the early 20th century. Today, the food security of two-thirds of the world's population depends on fertilizers, particularly nitrogen fertilizer. In Europe, 70–75 per cent of nitrogen comes from synthetic fertilizers; at the global scale, the proportion is about half. Some nitrogen is also fixed by legumes, with the balance of nitrogen coming mostly from crop residues and manure. However, crops take up only about half of the applied nitrogen. The rest is leached into streams and groundwater, or lost to the atmosphere. Losses of nitrogen from animal wastes account for 30–40 per cent, half of this escaping into the atmosphere as ammonia. Very high emissions are recorded from the Netherlands, Belgium, Denmark and the province of Sichuan in China. Annual emissions of reactive nitrogen from combustion of fossil fuels amount to about 25 million tonnes (Fowler and others 2004, Li 2000, Smil 1997, Smil 2001).

Enhanced levels of reactive nitrogen are now found from deep aquifers to cumulonimbus clouds, and even in the stratosphere, where N_2O attacks the ozone layer. There are concerns that elevated levels of nitrates in drinking water are a health hazard, particularly to very young children. Established evidence links enhanced concentrations of nitrates and phosphates to algal blooms in shallow lakes and coastal waters. Two of the largest blooms are in the Baltic Sea (Conley and others 2002) and in the Gulf of Mexico, off the mouth of the Mississippi River (Kaiser 2005). By-products of the algae are toxic to animals, while the decomposition of these huge masses of organic matter depletes the oxygen dissolved in water, causing fish kills (see Chapter 4).

Acidifying cycles: Oxides of carbon (CO_2), nitrogen (NO_X) and sulphur (SO_X) are released to the

---

Box 3.6 Disturbances in the carbon cycle due to losses of soil organic matter

Land-use change over the past two centuries has caused significant increases in the emissions of CO_2 and methane into the atmosphere. There are large uncertainties in the estimates, though, especially for soils. Clearance of forests causes a significant initial loss of biomass, and, where native soil organic content is high, soil organic carbon declines in response to conversion to pasture and cropland. Under cultivation, soil organic matter declines to a new, lower equilibrium, due to oxidation of organic matter.

Significant emissions also result from drainage of wet, highly organic soils and peat, as well as from peat fires. Higher temperatures, for example associated with forest fires and climate change, increase the rate of breakdown of soil organic matter and peat. Half of the organic carbon in Canadian peatlands will be severely affected, and permafrost carbon is likely to be more actively cycled. Warming will also release significant stores of methane presently trapped in permafrost.

While there has been a decline in the emissions from Europe and North America since the mid-20th century, emissions from tropical developing countries have been increasing, resulting in continued increases in overall global emissions due to land use change. The region of Asia and the Pacific accounts for roughly half of global emissions.

Sources: Houghton and Hackler 2002, Prentice and others 2001, Tarnocai 2006, UNFCCC 2006, Zimov and others 2006

atmosphere by decomposing organic matter and burning fossil fuels (see Chapter 2). SO_X are also produced by the smelting of sulphidic ores. Total emissions of SO_X from human activities are about equal to natural production, but they are concentrated in northern mid-latitudes. Large areas of eastern North America, western and central Europe, and eastern China experience SO_X deposition in the range of 10–100 kg S/ha/yr. In addition, NO_X deposition now exceeds 50 kg/ha/yr in central Europe and parts of North America.

As a result of such emissions, the pH of rainfall in polluted areas can be as low as 3.0–4.5. Where soils are weakly buffered, this translates to more acid streams and lakes, associated with increased solubility of toxic aluminium and heavy metals. Since 1800, soil pH values have fallen by 0.5–1.5 pH units over large parts of Europe and eastern North America. They are expected to fall by a further pH unit by 2100 (Sverdrup and others 2005). Canada and Scandinavia have been most severely affected by acidic precipitation in recent decades, suffering loss of phytoplankton, fish, crustaceans, molluscs and amphibians. Emission controls and rehabilitation efforts have slowed or even reversed freshwater acidification in some areas (Skjelkvåle and others 2005). The jury is still out on the forest decline predicted in the mid-1980s for Europe and North America, but acidification may be contributing to the biomass losses in boreal forests indicated in Figure 3.3. However, the risks of acidification from coal-powered industry are rising elsewhere, particularly in China and India.

Acidification is not just a problem arising from air pollution. Extreme cases develop when soils and sediments rich in sulphides are drained and excavated, for example, through the conversion of mangroves to aquaculture ponds or urban developments. In these acid sulphate soils, sulphuric acid produces pH values as low as 2.5, mobilizing aluminium, heavy metals and arsenic, which leak into the adjacent aquatic environment, causing severe loss of biodiversity (van Mensvoort and Dent 1997).

Managing land resources
Chemical contamination and pollution
Increasing awareness of the negative effects of chemical contamination and pollution is leading to stringent regulations in many industrialized countries. Since the 1992 UN Conference on Environment and

Box 3.7 Soil protection from chemicals in the European Union

In the European Union, evaluation of the effects of chemical pollutants on soil communities and terrestrial ecosystems provides a basis for soil protection policy. The Soil Framework Directive will require member states to take appropriate measures to limit the introduction of dangerous chemicals to the soil, and to identify and remediate contaminated sites.

The new REACH legislation (Registration, Evaluation, Authorisation and Restriction of Chemicals), that entered into force in June 2007, requires manufacturers and importers of chemicals to prove that substances in widely-used products, such as cars, clothes or paint, are safe, while the properties of chemicals produced or imported into the European Union have to be registered with a central agency.

Source: European Commission 2007

Development, the risks associated with chemicals and the transboundary movements of pollutants have been widely recognized. Chemicals management is now addressed by 17 multilateral agreements and 21 intergovernmental organizations and coordination mechanisms. The Basel Convention on the International Movement of Hazardous Wastes, the Rotterdam Convention on Certain Hazardous Chemicals in International Trade, and the Stockholm Convention on Persistent Organic Pollutants aim to control international traffic of hazardous chemicals and wastes that cannot be managed safely. Regional agreements include the Bamako Convention, which was adopted by African governments in 1991 and the European Union's REACH (see Box 3.7).

There has been a significant reduction in the use of some toxic chemicals, and safer alternatives are being identified. Voluntary initiatives, such as the chemical industry's Responsible Care programme encourage companies to work towards continuous improvement of their health, safety and environmental performance. A number of major chemical industries have made significant reductions in their emissions.

A Strategic Approach to International Chemicals Management (SAICM) was agreed to by more than 100 environment and health ministers in Dubai in 2006, following the ninth Special Session of the UNEP Governing Council/Global Ministerial Environment Forum. It provides a non-binding policy framework for achieving the goal of the Johannesburg Plan of Implementation: that, by 2020, chemicals are produced and used in ways that minimize adverse effects on the environment and human health. This requires responsibility for and reductions in pollution.

Chemicals and materials are to be selected for use on the basis of their non-toxicity, waste should be minimized, and products at the end of their useful life should re-enter production as raw materials for the manufacture of new products.

All these instruments depend on institutional capacity and political will. They are undermined by limited political commitment, legislative gaps, weak inter-sectoral coordination, inadequate enforcement, poor training and communication, lack of information, and failure to adopt a precautionary approach. (Until the 1990s, chemicals were considered "innocent" until proven "guilty"). While regulations to control environmental loadings have established maximum allowable limits for releases of certain chemicals, observed concentrations are often still much higher than the set limits. In addition, there are areas of uncertainty that argue for a precautionary approach. These areas of uncertainty include trigger mechanisms that may suddenly cause potentially toxic contaminants to become more harmful; triggers include a change of location, for instance through the rupture of a retaining dam, or change of chemical state, such as through oxidation of excavated materials.

Existing multilateral and regional agreements offer an opportunity to arrest and eventually reverse the increasing releases of hazardous chemicals. Prerequisites for success include:

- full integration of a precautionary approach in the marketing of chemicals, shifting the burden of proof from regulators to industry;
- development of adequate chemicals management infrastructure in all countries, including laws and regulations, mechanisms for effective enforcement and customs control, and capacity to test and monitor;
- substitution with less-hazardous materials, adoption of best available technologies and environmental practices, and easy access to these approaches for developing countries and countries with economies in transition;
- encouragement of innovation in manufacturing, non-chemical alternatives in agriculture, and waste avoidance and minimization; and
- inclusion of environmental issues related to chemicals in regular educational curricula, and in partnership processes between academia and industry.

Soil erosion

Widespread attempts to mitigate soil erosion have met with mixed success. National responses have been directed towards legislation, information, credits and subsidies, or specific conservation programmes. Local responses have been generated by land users themselves (Mutunga and Critchley 2002), or introduced by projects. At the technical level, there is a wealth of proven approaches and technologies, from improved vegetation cover and minimum tillage to terracing (see photos on facing page). These useful experiences (both positive and negative) are not well documented. The World Overview of Conservation Approaches and Technologies network (WOCAT 2007) aims to fill this gap through collection and analysis of case studies from different agro-ecological and socio-economic conditions. But the usual focus on technical aspects misses the more complex, underlying political and economic issues that must also be addressed, an issue already advocated since the early 1980s (Blaikie 1985).

Substantial investment in soil conservation over past decades has yielded some local successes, but, except for conservation tillage (see Box 3.4), adoption of recommended practices has been slow and seldom spontaneous. A historic success story is the programme undertaken in the United States following the Dust Bowl in the 1930s, when drought triggered massive soil erosion in the US Midwest, and millions of people lost their livelihoods and were forced to migrate (see Box 3.8). The way the issue was handled provides an object lesson and inspiration for today. The clear message is that effective prevention and control of soil erosion needs knowledge, forceful social and economic policy, well-founded institutions maintaining supporting services, involvement of all parties, and tangible benefits to the land users. Nothing less than the whole package, continuing over generations, will be effective (see Box 3.10 and the section on responses to desertification).

Nutrient depletion

There is no remedy for soils that are deficient in nutrients other than adding the necessary inputs. Efforts to improve soil fertility have focused on the replenishment of nutrients by the judicious use of inorganic fertilizers and organic manure. This has been very successful in many parts of the world, and is responsible for a very large increase in agricultural production. Yields may be doubled or tripled on a sustained basis by even modest

Soil and water management measures against erosion and water scarcity.
Left: Micro-basins;
Centre: Mulch;
Right: Conservation tillage.

Credit: WOCAT

application of fertilizer (Greenland 1994). In Niger, for instance, sorghum yields without fertilizer (about 600 kg/ha) were doubled by application of 40 kg/ha of nitrogen fertilizer (Christianson and Vlek 1991). However, the use of inorganic fertilizers requires cash, which can be an insurmountable barrier for most smallholders in developing countries, where inputs are rarely subsidized.

There are myriad indigenous practices to mitigate nutrient constraints, such as bush fallow, biomass transfer to home fields, and adding compost and manure on favoured plots. However, these are failing to keep up with production needs in the face of

increasing population pressure, and lack of adequate funds for labour or mechanization. In recent years, significant research efforts have focused on biological processes to optimize nutrient cycling, minimize external inputs and maximize nutrient use efficiency. Several techniques have been developed, including the integration of multipurpose legumes, agroforestry and improved fallows, but scientific breakthroughs and large-scale adoption by smallholder farmers are yet to materialize.

Nutrient depletion is not the same everywhere, because it depends on a series of interacting causes, and depletion processes are different for different nutrients. There is a need for much better spatial information at regional and local scales, and for better soil management technologies to improve responses. Techniques to reduce nutrient depletion and enhance soil fertility vary, depending on the soils and farming systems. Improved soil management, including rotation of annual with perennial crops, and the integration of trees into farming systems, can improve the efficiency of nutrient cycling by maintaining the continuity of uptake, and reducing leaching losses. Nitrogen stocks can be maintained through biological nitrogen fixation (by integrating legumes into cropping systems), but nitrogen fixing is limited by available phosphorus, which is very low in many tropical soils. For severely nutrient-deficient soils, there is no remedy other than additions from outside sources.

Water scarcity

Achieving the MDG on reducing hunger will require an increase of 50 per cent in water use by agriculture by 2015, and a doubling by 2050, whether by farming more land or by withdrawing more water for irrigation (SEI 2005). For developing countries, FAO (2003) projects an increase of 6.3 per cent in rain-fed cropland area between 2000 and 2015, and of 14.3 per cent by 2030. It also projects an increase in irrigated area of almost 20 per cent from 2000 to 2015, and to just over 30 per cent by 2030. Large dams continue to be built, because they promise certainty of supply of water and power to downstream interests, but the same investment has not gone into the catchments that supply the water. On the contrary, the last 20 years have seen continued squandering of the *green water* resource through soil erosion, and higher rates of run-off, which has increased floods at the expense of base flow. This has also resulted in siltation of reservoirs, such as those behind the Victoria Dam on the Mahaweli River in Sri Lanka (Owen and others 1987) and the Akasombo Dam on the Volta River in Ghana (Wardell 2003).

While irrigated yields will always be higher than rain-fed yields, there is much scope for improving rain-fed farming on vast areas. In Africa, average cereal yields range from 0.91 tonnes/ha in Western Africa to 1.73 tonnes/ha in Northern Africa (GEO Data Portal, from FAOSTAT 2004), while commercial farmers operating in the same soil and climatic conditions achieve 5 tonnes/ha

or more. Established, though incomplete evidence, suggests that two-thirds of the necessary increase in production needed from rain-fed farming can be achieved through better rain-use efficiency (SEI 2005). Analysis of more than 100 agricultural development projects (Pretty and Hine 2001), found a doubling of yields in rain-fed projects, compared with a 10 per cent increase for irrigation (see Box 3.9).

More crop production means more water use by crops, whether through irrigation or increase in the cropped area. However, established evidence also shows that investment in water productivity

---

Box 3.9 Gains can be made through better water use efficiency

Low yields with sparse ground cover result in a large, unproductive loss of water by run-off and evaporation from bare soil. In semi-arid areas, doubling of yield from 1–2 tonnes/ha may increase water productivity from 3 500 m^3/tonne of grain to 2 000 m^3/tonne. Improvements in water use efficiency may be achieved in various ways, some of which are illustrated, under responses to soil erosion.

- Short-duration, drought-resistant crops can be matched to a short growing season.
- Water can be funnelled to crops from micro-catchments in the field, which can increase crop water use by 40–60 per cent without any loss in groundwater recharge, simply by reducing evaporation and allowing micro-basins to hold run-off until it can infiltrate.
- Mulch can be used to absorb raindrop impact, and provide organic matter and insulation against high surface temperatures, enabling soil animals to create a permeable soil structure.
- Conventional ploughing can be replaced by deep ripping with minimum disturbance of the topsoil; such conservation tillage improves infiltration while greatly reducing draught-power requirements.
- Dramatic improvements in yield and water-use efficiency may be achieved by supplementary irrigation, not to provide the crop's full water requirements but to bridge dry spells. At Aleppo, Syria, application of 180, 125 and 75 mm of water in dry, median and wet years, respectively, increased wheat yield by 400, 150 and 30 per cent. Such volumes of water can be harvested in micro-catchments outside the cropped area, using many local systems that can be affordable as household or small community ventures.

Sources: Oweis and Hachum 2003, Rockström and others 2006

– achieving more crop per drop – can help maintain water supply downstream (Rockström and others 2006), and appropriate land use and soil management can increase groundwater recharge and stream base flow (Kauffman and others 2007).

Responses to water scarcity have focused on run-off management, water abstraction, and demand management. New policies need to focus on rainwater management, and address the competing claims on water resources. In practice, a package of mutually supporting measures and concerted action from interested parties should include:

■ capacity building for land and water management institutions;

■ investment in education, and training of land and water managers; and

■ a mechanism to reward land users for managing water supply at the source, involving payments for environmental services (Greig-Gran and others 2006).

Salinity

FAO and regional organizations have established collaborative programmes to reduce water losses from canals, match field application with the needs of the crop and drain surplus water to arrest rising water tables (FAO 2006b). However, investment in management and improvement of irrigation networks, especially in drainage and on-farm water use, has rarely been commensurate with the capital investment in water distribution.

Dryland salinity is caused by changes in the hydrological balance of the landscape, which are driven as much by fluctuations in rainfall as by land-use change. Piecemeal tree planting and crop management to reduce groundwater recharge has no chance of arresting groundwater flow systems that are orders of magnitude bigger. In both cases, successful intervention depends on information on the architecture and dynamics of groundwater flow systems (Dent 2007), and the technical capacity to act on this information. As with soil erosion, a focus on technical issues has diverted attention from wider issues of water rights and payments, the need for capacity building in managing institutions, and implementation of national and transboundary agreements. Salinity is sometimes only a symptom of underlying failures in management of common resources.

Disturbances in biological cycles

The excess of nutrients in regions such as Europe and North America has prompted the setting of legal limits on the application of manure and fertilizers. For instance, under the EU Nitrate Directive (Council Directive 91/676/EEC), the application of nitrate fertilizers has been restricted in some areas susceptible to groundwater pollution by nitrates. An evaluation 10 years after the directive went into force concluded that some farm practices have positive effects on water quality, but emphasized that there is a considerable time lag between improvements at farm level and measurable improvements in water quality (European Commission 2002).

The investment in the management and improvement of irrigation networks has rarely been commensurate with the capital investment in water distribution.

Credit: Joerg Boethling/ Still Pictures

Helicopter spreading lime over
an acidified lake in Sweden.

*Credit: Andre Maslennikov/
Still Pictures*

The reductions in acid gas emissions in Europe and
North America has been one of the success stories
of recent decades. It involved domestic regulations,
innovations by some industries and international
coordination (see mainly Chapter 2). This included
agreements such as the 1979 UN/ECE Convention
on Large Transboundary Air Pollution and the Canada-
US Air Quality Agreement. The ECE convention
adopted the concept of critical loads in 1988, and
the Gothenburg Protocol in 1999 regulated emissions
of SO_X and NO_X, defining critical loads according to
best current evidence.

Global emissions of SO_2 were reduced by about
2.5 per cent between 1990 and 2000 (GEO
Data Portal, from RIVM-MNP 2005), as a result of
clean air acts promoting a switch to cleaner fuels
and flue gas desulphurization, and the demise of
heavy industries, particularly in Eastern Europe and
the former Soviet Union. However, many areas still
receive acid deposition well in excess of critical loads
(for example, Nepal, China, Korea and Japan) and
total global emissions are rising again, driven by
newly industrialized countries (see Figure 2.8). China
alone accounts for about one-quarter of global SO_2
emissions (GEO Data Portal, from RIVM-MNP 2005),
and its coal-fired industrial development is likely to
significantly increase acid emissions (Kuylenstierna
and others 2001). Long-term liming programmes are

in place in several countries to mitigate enhanced
acidic inputs into inland waters.

When it comes to controlling acid soil drainage, only
Australia has enacted specific planning regulations
to prevent the formation of acid sulphate soils. Any
response to acid sulphate drainage from mines
and soil has usually been restricted to liming of the
acidified soil or spoil heaps but Trinity Inlet in North
Queensland, Australia, provides a recent example
of remediation through controlled restoration of tidal
flooding, whereby existing acidity is neutralized by
tidewater, and re-establishment of a tidal regime stops
further acid generation (Smith and others 2003).

DESERTIFICATION

Extent and impacts

Desertification occurs when individual land degradation
processes, acting locally, combine to affect large
areas of drylands. As defined by the UN Convention
to Combat Desertification (UNCCD), desertification
is land degradation in arid, semi-arid and dry sub-
humid areas resulting from various factors, including
climatic variations and human activities (UNGA 1994).
It is most sharply expressed in poor countries where
intertwined socio-economic and biophysical processes
adversely affect both land resources and human well-
being. Drylands cover about 40 per cent of the Earth's
land surface (see Figure 3.9) and support 2 billion
people, 90 per cent of them in developing countries
(MA 2005b). But desertification is not confined to
developing countries; one-third of Mediterranean
Europe is susceptible (DISMED 2005) as well as 85
per cent of rangelands in the United States (Lal and
others 2004). (See Chapter 7 for more information on
issues related to drylands).

Desertification endangers the livelihoods of
rural people in drylands, particularly the poor,
who depend on livestock, crops and fuelwood.
Conversion of rangelands to croplands without
significant new inputs brings about a significant,
persistent loss of productivity and biodiversity,
accompanied by erosion, nutrient depletion, salinity
and water scarcity. In 2000, the average availability
of freshwater for each person in drylands was
1 300 m3/year, far below the estimated minimum
of 2 000 m3/year needed for human well-being,
and it is likely to be further reduced (MA 2005b).
Measured by indicators of human well-being and
development, dryland developing countries lag

far behind the rest of the world. For instance, the average infant mortality rate (54 per thousand) is 23 per cent higher than in non-dryland developing countries and 10 times that of industrialized countries.

The seriousness of the issue is recognized by the UNCCD, the Convention on Biological Diversity (CBD) and the UN Framework Convention on Climate Change (UNFCCC). The New Partnership for Africa's Development also stresses the need to combat desertification as an essential component of poverty-reduction strategies. However, investment and action to combat desertification have been held back by the isolation of drylands from mainstream development, and even by controversy over the use of the term. Debate about desertification has been fuelled by alarming articles in the popular media about "encroaching deserts," reinforced by a series of droughts from the 1960s through the 1980s (Reynolds and Stafford Smith 2002).

Desertification is determined by various social, economic and biophysical factors, operating at local, national and regional scales (Geist and Lambin 2004). A recurring combination embraces national agricultural policies, such as land redistribution and market liberalization, systems of land tenure that are no longer suited to management imperatives, and the introduction of inappropriate technologies. Usually, the direct cause has been the expansion of cropping, grazing or wood exploitation. National and local policies to promote sustainable practices must take account of a hierarchy of drivers, from

the household to the international level. This can be difficult where the indirect drivers, such as global trade imbalances, seem remote from these marginal lands, and when mechanisms for bottom-up decision-making are poorly developed.

Desertification is a continuum of degradation, crossing thresholds beyond which the underpinning ecosystem cannot restore itself, but requires ever-greater external resources for recovery. Resilience is lost when a disturbance, which a system used to be able to absorb, tips the system to a less desirable state from which it cannot easily recover (Holling and others 2002). Loss of ecosystem resilience is often accompanied by a breakdown in social resilience and adaptive capacity, when vulnerable people are forced to draw on limited resources with diminished coping strategies (Vogel and Smith 2002). For example, loss of resilience of parklands (integrated tree-crop-livestock systems) may result when the trees are cleared, exposing the land to erosion. Adaptive management aims to prevent ecosystems from crossing these thresholds by maintaining ecosystem resilience as opposed to seeking only narrow, production or profit objectives (Gunderson and Pritchard 2002).

Although indicators of desertification have been proposed ever since the term was introduced (Reining 1978), lack of consistent measurement over large areas and over time has prevented reliable assessment. Over the long term, ecosystems are governed by slowly-changing biophysical and socio-economic factors. Measurable indicators for these

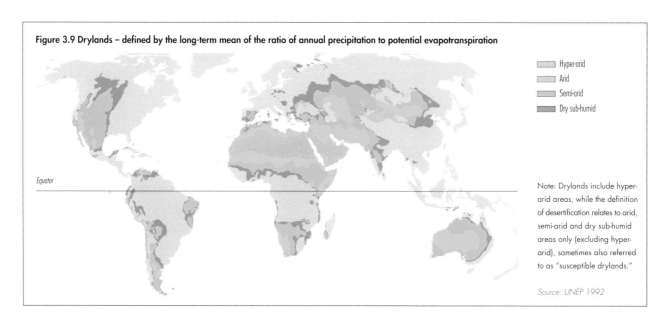

Figure 3.9 Drylands – defined by the long-term mean of the ratio of annual precipitation to potential evapotranspiration

Hyper-arid
Arid
Semi-arid
Dry sub-humid

Equator

Note: Drylands include hyper-arid areas, while the definition of desertification relates to arid, semi-arid and dry sub-humid areas only (excluding hyper-arid), sometimes also referred to as "susceptible drylands."

Source: UNEP 1992

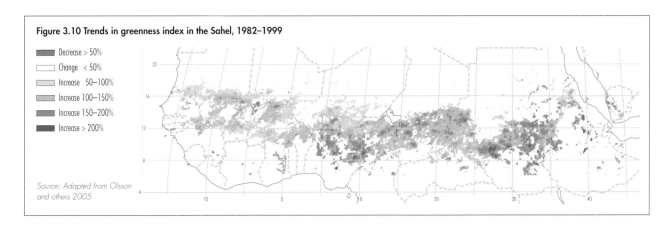

Figure 3.10 Trends in greenness index in the Sahel, 1982–1999

■ Decrease > 50%
□ Change < 50%
■ Increase 50–100%
■ Increase 100–150%
■ Increase 150–200%
■ Increase > 200%

Source: Adapted from Olsson and others 2005

slow variables (such as changes in woody vegetation cover and soil organic matter) better characterize the state of ecosystems than fast variables (such as crop or pasture yields), which are sensitive to short-term events. No systematic national or global assessment of desertification has been made using measurement of slow variables. Some areas thought to have been permanently degraded during droughts have subsequently recovered, at least in terms of the amount of green vegetation, although species composition may have changed. For instance in the Sahel, coarse-resolution satellite data show significant greening during the 1990s, following the droughts in the early 1980s (see Figure 3.10). This can be explained by increased rainfall in some areas but not in others; land-use changes as a result of urban migration and improved land management may have played a part (Olsson and others 2005). Systematic, interdisciplinary approaches are needed to provide more clarity and empirical evidence, which should enable more focused and effective interventions.

The argument that regional climate is affected by desertification through reduction in vegetation and soil water retention, and by the generation of dust (Nicholson 2002, Xue and Fennessey 2002), remains speculative. Desert dust has long-range impacts, both good and bad. It is a global fertilizer, as a source of iron and possibly phosphorus, contributing to the farmlands and forests of Western Africa (Okin and others 2004), the forests of the northeast Amazon Basin and Hawaii (Kurtz and others 2001), and the oceans (Dutkiewicz and others 2006). However, it has also been linked to toxic algal blooms, negative impacts on coral reefs, and respiratory problems (MA 2005b). Generally, dust from degraded farmland probably contributes less than 10 per cent to the global dust load (Tegen and others 2004). Natural processes create about 90 per cent of dust in areas like northern Chad and western China (Giles 2005, Zhang and others 2003).

Combating desertification

The international response to desertification has been led since 1994 by UNCCD, which has been ratified by 191 countries. It has evolved as a process seeking to integrate good governance, involvement of non-governmental organizations (NGOs), policy improvement, and the integration of science and technology with traditional knowledge. National action programmes have been drawn up by 79 countries, there are nine sub-regional programmes

Sand encroachment and land reclamation in China.

Left, 2000; right, 2004 planted with Xinjiang poplar (*Populus alba*).

Credit: Yao Jianming

Responses to desertification have focused on drought, shortfalls of food and the death of livestock, aspects that reflect inherently variable climatic cycles. Experience shows that policy and action must address long-term issues by combining a number of elements.

1. Direct action by governments
- Effective early warning, assessment and monitoring – combine remote sensing with field surveys of key indicators. Measure indicators consistently, at different scales, over the long-term.
- Integrate environmental issues into the mainstream of decision making at all levels – aim to increase system resilience and adaptive capacity, intervene before a system has crossed key biophysical or socio-economic thresholds (prevention is better and more cost-effective than cure). Include valuation of all ecosystem services in policy development.

2. Engagement of the public and private sectors
- Science and communication – integrate science, technology and local knowledge for better monitoring, assessment and adaptive learning, especially where uncertainty is impeding action. Communicate the knowledge effectively to all stakeholders, including youth, women and NGOs.
- Strengthen institutional capacity for ecosystem management – support institutions that can operate at the various scales at which ecosystems function (local catchments to river basins), and promote institutional learning, capacity building and the participation of all stakeholders. Create synergies among UNCCD, CBD, the Ramsar Convention, the Convention on International Trade in Endangered Species, Convention on Migratory Species and UNFCCC. Identify the overlaps, enhance capacity building and use demand-driven research.

3. Develop economic opportunities and markets
- Promote alternative livelihoods – grasp economic opportunities that do not depend directly on crops and livestock, but take advantage of the abundant sunlight and space in drylands, with approaches such as solar energy, aquaculture and tourism.

Sources: Reynolds and Stafford Smith 2002, UNEP 2006

targeting transboundary issues and three regional thematic networks (UNCCD 2005). Activities are now moving beyond awareness-raising and programme formulation to providing the financial resources for and implementing land reclamation projects (see sections on Africa and West Asia in Chapter 6). Starting much earlier, a national effort in reclamation of the severely degraded Loess Plateau in China now shows up in the Global Assessment of Land Degradation and Improvement as a 20-year trend of increasing biomass, in spite of a decrease in rainfall across the region during the same period (Bai and others 2005). In China in the 1990s, about 3 440 km^2 of land was affected annually by sand encroachment. Since 1999, 1 200 km^2 has been reclaimed annually (Zhu 2006).

Desertification is a global development issue, driving an exodus from the regions affected, yet policy and action are becalmed by uncertainty about the nature and extent of the problem, and about what policies and management strategies will be effective in different settings. Rigorous, systematic studies of the processes of desertification and the effects of intervention at different scales and different settings are urgently needed to guide future efforts. There is a great need to build local technical and management capacity (see Box 3.10) and applied science needs to focus on resolving the uncertainties that are impeding action, and on integrating science and technology with local knowledge to improve rigour in assessment, monitoring and adaptive learning.

CHALLENGES AND OPPORTUNITIES

Since the publication of *Our Common Future* (the Brundtland Commission report), economic growth has led to improvement of the environment in many ways, for instance by enabling investment in better technologies and some conspicuous improvements in developed countries. But many global trends are still strongly negative.

In the face of mounting evidence that much present development is unsustainable, global attention has focused on national strategies to promote sustainable development, foreshadowed at UNCED. The UN General Assembly Special Session Review Meeting in 1997 set a target date of 2002 for the introduction of such strategies. However, effective responses are still held back by limited access to information, inadequate institutional capacity faced with complex land-use issues, and the absence of broad participation or ownership of the responses. Future costs to others can be offset only by political cost to decision-makers now. Sustainability strategies need to be backed up by research to provide reliable data on biophysical, economic and social indicators of long-term change, and they require development or adaptation of technologies appropriate to local circumstances.

Strategies that are environment-driven, rather than focusing on sustainable development, rarely command the support needed to put them into action (Dalal-Clayton and Dent 2001). A successful approach deals not only with the environment, but also with the connections between environment and the economic and social issues to which people relate. For example, watershed development plans are being implemented in many places to secure water supplies and to protect hydropower facilities, and many multistakeholder projects on sustainable use of biosphere and forest reserves, some of which take account of the rights and needs of indigenous peoples.

The outlook to 2050 sees the emergence of two major sets of land-related challenges: dominating trends that are largely unavoidable, and caveats of risks that are very unpredictable, but which have serious implications for society that warrant precautions.

Challenges: Dominant land-use trends
Competing claims on the land

Given projections that the world's population will increase to over 9 billion by 2050, and to meet the MDG on hunger, a doubling of global food production will be required. In addition, a continuation of the shift from cereal to meat consumption, combined with overconsumption and waste, will increase food demand to between 2.5 and 3.5 times the present figure (Penning de Vries and others 1997). Yet, the production of cereals per person peaked in the 1980s and has since slowly decreased despite the increase in average yields. Reasons may include agricultural policies in regions of surplus, such as the European Union, ceilings to current technology, loss of farmland through land degradation and the growth of cities and infrastructure, and market competition from other land uses (Figure 3.11).

Our capacity to meet these future agricultural demands is contested. The main biophysical constraints are related to water, nutrients and land itself. Water

scarcity is already acute in many regions, and farming already takes the lion's share of water withdrawn from streams and groundwater. Other claims on water resources are growing, particularly for urban water supplies (see section on Water scarcity).

Increases in production over recent decades have come mainly from intensification rather than from an increase in the area under cultivation. Intensification has involved improved technologies, such as plant breeding, fertilizers, pest and weed control, irrigation, and mechanization; global food security now depends to a large extent on fertilizers and fossil fuels. Limits to current technologies may have been reached in mature farming systems, where they have been applied for several decades, and yields may have peaked. While there is land in poor countries that could respond to such technologies, most smallholders cannot afford fertilizers now, and the prices are being driven up by rising energy costs and the depletion of easily exploited stocks of phosphate. Food production is also constrained by the competing claims of other land uses, not least for maintenance of ecosystem services, and large areas may be reserved for conservation.

There is consensus that climate change over the next 20 years will affect farm production, with many regional differences in impacts. Changes may increase water requirements of crops, and increasing rainfall variability may exacerbate water scarcity in drylands (Burke and others 2006). Quantifying the current biological production for human consumption requires better estimates of global productivity of agricultural, grazed and human-occupied lands (Rojstaczer and others 2001). In the face of current uncertainty, it would be prudent to conserve good farmland, counteract tendencies of overconsumption, and undertake further needed research.

Bioenergy production

In most global energy scenarios that meet stringent carbon emissions constraints, biofuel is assumed to be a significant new source of energy. The World Energy Outlook 2006 (IEA 2006) forecasts an increase in the area devoted to biofuels from the current 1 per cent of cropland to 2–3.5 per cent by 2030 (when using current technologies). A major shift in agricultural production from food to biofuel presents an obvious conflict, which is already reflected in the futures market for food grains (Avery 2006). Forest products and the non-food cellulose component of food crops have a

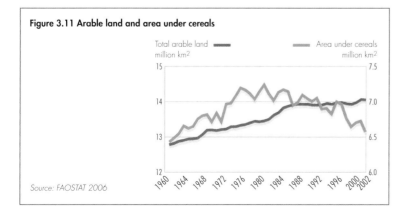

Figure 3.11 Arable land and area under cereals

Total arable land
million km² ▬▬▬   Area under cereals
million km² ▬▬▬

Source: FAOSTAT 2006

huge potential as an energy source, but technologies are still too costly to compete with fossil fuels at current prices, and the non-food component of crops also has a vital role in maintaining soil organic matter status.

Urbanization and infrastructure development

Half the world population now lives in urban areas, with positive and negative implications for the environment and human well-being. Densely populated cities use less land than do sprawling suburbs, they are easier to serve with public transportation, and can be more efficient in energy use, such as for transport and heating, and for waste reduction and recycling. The construction of housing and infrastructure in rural areas is often in conflict with other land uses, such as agriculture, recreation and other ecosystem services, particularly in rapidly industrializing countries (IIASA 2005).

However, cities are often built on prime farmland, and nutrients are being transferred from farms to cities with little or no return flow. The concentration of excrement and waste from food is often a source of pollution as well as a waste of resources. Urban areas become the source of sewage flows, run-off and other forms of waste that become environmental problems, often affecting the surrounding rural areas, as well as degrading water quality.

Challenges: Unpredictable risks to land

Tipping points

Tipping points occur when the cumulative effects of steady environmental changes reach thresholds that result in dramatic and often rapid changes. There is concern that a number of environmental systems may be heading toward such tipping points. One example is the bi-stability of the Amazon Basin, implying the possibility of a flip from a current wet phase to a dry phase, with profound implications beyond the basin (Schellnhuber and others 2006, Haines-Young and others 2006). Another very different tipping point with global implications might be simultaneous crop failures in different regions.

Runaway carbon cycle

The global carbon cycle is by no means fully understood. The missing sink for forty per cent of known carbon dioxide emissions is generally thought to be terrestrial ecosystems (Watson and others 2000, Houghton 2003). Vast areas of peat and tundra are reservoirs of stored organic carbon

(one-third of all terrestrial organic carbon is peat) and methane, and they continue to fix carbon. With global warming, there is a risk of unexpected sudden increases in the atmospheric levels of carbon dioxide, if these sinks become saturated. The peat and tundra areas might transform from being a sink of carbon to become sources of greenhouse gases (Walter and others 2006).

Eutrophication

Rivers, lakes and coastal waters receive large amounts of nutrients from the land, and overloading of nutrients often results in algal blooms. If this increases in intensity and frequency, whole ecosystems may be subject to hypoxia (dead zones due to lack of oxygen) as seen already in the Gulf of Mexico (Kaiser 2005) and the Baltic Sea (Conley and others 2002).

Breakdown of governance, conflict and war

Land-use changes are usually associated with gains in livelihoods, income opportunities, food security or infrastructure. Illegal operations do not yield these long-term benefits, so good governance is vital to protect long-term values from short-term exploitation. Areas of exceptional environmental value, such as tropical rain forest and wetlands, as well as boreal forests, are in special need of strong structures of governance. War and civil conflict are always associated with rapid and far-reaching destruction of environmental values.

Opportunities to tackle these challenges

While the dominant trends are driven by demography, the global state of the environment and decisions already taken, there are several opportunities to steer or oppose them, not least by harnessing existing knowledge. Chapter 7 analyses successful strategies that offer opportunities for reducing human vulnerability, Chapter 8 goes further into biophysical and societal interlinkages that offer opportunities for more effective policy responses and Chapter 10 summarizes a range of innovative approaches to help improve responses. Some land-specific opportunities are described below.

Precision farming

Precision farming refers to optimizing production through site-specific choices of crop varieties, fertilizer placement, planting and water management, taking advantage of the variability of soil and terrain in a field rather than ignoring it. It is also describes the automation of techniques employed to do this, such as recording crop yields with a continuously recording monitor. However, the principle can be applied equally to low capital-input farming, where crops are intensively managed, manually: water harvesting is an example. Precise monitoring of crop performance will enable farmers to economize on their inputs in terms of labour, water, nutrients and pest management. The advent of reliable and inexpensive electronic devices offers the opportunity to extend advanced, information-based agriculture to new areas. Barriers to the wider application of precision farming include the scarcity and high cost of subtle management skills compared with using chemical inputs, and, among poor farmers, insecure tenure, lack of credit and low farm-gate prices.

Multifunctional landscapes

Agroforestry is one of several promising developments that can simultaneously generate livelihoods and preserve environmental quality. Successful examples include palm oil production in semi-natural rain forests, and gum arabic production in drylands. Carbon fixation through land management is another opportunity. Since fixing carbon by planted forests is eligible for trading under the Kyoto Protocol, most attention has been directed towards capturing carbon by forests and storing it as the standing crop. But carbon can also be stored in the longer term as soil organic matter, which is a much larger and more stable pool of carbon. At the same time, it would contribute to more sustainable agriculture by increasing resistance to erosion, add to water and nutrient reserves in the soil, and increase infiltration capacity. Low capital input farming systems may have a higher potential for net carbon accumulation than intensive forms of agriculture, where the inputs (such as fertilizer and energy) are associated with high carbon costs (Schlesinger 1999). Putting organic carbon back into soils, where it will be useful, is a challenge to soil science and management.

Uptake of agroforestry has slowly increased in recent years, and further development may be expected if soil carbon is recognized as an eligible sink by climate change legislation. Other market mechanisms, such as Green Water Credits for water management services in farmed landscapes, would be required to promote such multifunctional landscapes.

Ecosystem mimicry

Multiple cropping in the same field is well established in smallholder farming systems. However, very complex multi-layered perennial cropping systems, such as the Kandyan home gardens in Sri Lanka, demand rare skills and knowledge (Jacow and Alles 1987). Such biologically diverse systems provide both high productivity and better insurance against the risks of erosion, weather, pests and disease. Aquaculture is an important contribution to the world protein supply, but is often associated with high environmental costs and risks. One option to reduce the negative impact on aquatic ecosystems is to transfer such schemes to land, where tanks or reservoirs might be better suited for cultivation of protein (Soule and Piper 1992). There is also rich experience of fish and shrimp production in rice paddies (Rothuis and others 1998).

Crop breeding

One area with significant potential, but which is contested in several aspects, is the development and use of genetically modified (GM) crops (Clark and Lehman 2001). In contrast to the development of Green Revolution crops, the development of GM crops is almost exclusively privately funded, and focuses on crops with commercial potential. There are several sources of uncertainty, including unwanted environmental impacts, social acceptance of the technologies and their agronomic potential. Currently, there is polarization between proponents of the technology, mainly from the fields of genetics and plant physiology, and sceptics, mainly from the fields of ecology and environmental sciences. Outcomes to date mainly concern crop traits related to herbicide tolerance and resistance to pests. These may be significant, because losses due to insect pests have been estimated at about 14 per cent of total global farm production (Sharma and others 2004). Negatives include higher costs to farmers, dependency on big companies and specific agrochemicals, and the fact that, over time, cross-fertilization will mean that there will be no non-GM crops.

As an alternative to introducing new genes into crop species, the new technology of marker-assisted selection assists the location of desirable traits in other varieties, or in wild relatives of existing crops, which can then be cross-bred in the conventional way to improve the crop, halving the time required to develop new plant varieties (Patterson 2006) and avoiding the possible harm associated with GM crops. However achieved, salt- and drought-tolerance would be valuable for increasing food security in drylands, but we are far from understanding the mechanisms of such adaptations, let alone operational seed technologies (Bartels and Sunkar 2005).

References

Abdelgawad, G. (1997). Degradation of soil and desertification in the Arab countries. In *J. Agriculture and Water* 17:28-55

ACSAD, CAMRE and UNEP (2004). *State of Desertification in the Arab World (Updated Study)*. Arab Center for the Studies in Arid Zones and Drylands, Damascus

Al-Dabi, H., Koch, M., Al-Sarawi, M. and El-Baz, F. (1997). Evolution of sand dune patterns in space and time in north-western Kuwait using Landsat images. In *J. Arid Environments* 36:15–24

Al-Mooji, Y. and Sadek, T. (2005). *State of Water Resources in the ESCWA Region*. UN Economic and Social Commission for West Asia, Beirut

Avery, D. (ed.) (2006). *Biofuels, Food or Wildlife? The Massive Land Costs of U.S. Ethanol*. Issue Analysis 2006:5. Competitive Enterprise Institute, Washington, DC

Bai, Z.G., Dent, D.L. and Schaepman, M.E. (2005). *Quantitative Global Assessment of Land Degradation and Improvement: Pilot Study in North China*. Report 2005/6, World Soil Information (ISRIC), Wageningen

Bai, Z.G. and Dent, D.L. (2007). *Global Assessment of Land Degradation and Improvement: Pilot Study in Kenya*. ISRIC Report 2007/03, World Soil Information (ISRIC), Wageningen

Bai, Z.G., Dent, D.L., Olsson, L. and Schaepman, M.E. (2007). *Global Assessment of Land Degradation and Improvement*. FAO LADA working paper. Food and Agriculture Organization of the United Nations, Rome

Barron, J., Rockström, J., Gichuki, F. and Hatibu, N. (2003). Dry spell analysis and maize yields for two semi-arid locations in East Africa. In *Agricultural and Forest Meteorology* 117 (1-2):23-37

Bartels, D. and Sunkar, R. (2005). Drought and salt tolerance in plants. In *Critical Reviews in Plant Sciences* 24:23–58

Bengeston, D. and Kant, S. (2005). Recent trends and issues concerning multiple values and forest management in North America. In Mery, G., Alfaro, R., Kanninen, M., and Lobovikov, M. (eds.) *Forests in the Global Balance: Changing Paradigms* . International Union of Forest Research Organizations, Vienna

Blaikie, P. (1985). *The Political Economy of Soil Erosion in Developing Countries*. Longman, London

Bojö, J. (1996). Analysis – the cost of land degradation in Sub-Saharan Africa. In *Ecological Economics* 16 (2):161-173

Borlaug, N.E. (2003). *Feeding a world of 10 billion people – the TVA/IFDC Legacy*. IFDC, Muscle Shoals, AL

Buresh, R.J., Sanchez, P.A. and Calhoun, F. eds. (1997). *Replenishing Soil Fertility in Africa*. SSSA Special Publication 51, Madison, WI

Burke, E.J., Brown, S.J. and Christidis, N. (2006). Modelling the recent evolution of global drought and projections for the twenty-first century with the Hadley Centre climate model. In *Journal of Hydrometeorology* 7:1113-1125

Christianson, C.B. and Vlek, P.L.G. (1991). Alleviating soil fertility constraints to food production in West Africa: Efficiency of nitrogen fertilizers applied to food crops. In *Fertilizer Research* 29:21-33

Clark, E. A. and Lehman, H. (2001). Assessment of GM crops in commercial agriculture. In *Journal of Agricultural and Environmental Ethics* 14:3-2

Conley, D.J., Humborg, C., Rahm, L., Savchuk, O.P. and Wulff, F. (2002). Hypoxia in the Baltic Sea and basin-scale changes in phosphorus biochemistry. In *Environmental Science and Technology* 36 (24):5315-5320

Dalal-Clayton, B.D. and Dent, D.L. (2001). *Knowledge of the Land: Land Resources Information and its Use in Rural Development*. Oxford University Press, Oxford

Den Biggelaar, C., Lal, R., Weibe, K., Eswaran, H., Breneman, V. and Reich, P. (2004). The global impact of soil erosion on productivity I: Absolute and relative erosion-induced yield losses. II: Effects on crop yields and production over time. In *Adv. Agronomy* 81:1-48, 49-95

Dent, D.L. (2007). Environmental geophysics mapping salinity and fresh water resources. In *Int. J. App. Earth Obs. and Geoinform* 9:130-136

De Vries, W., Schütze, G., Lofts, S., Meili, M., Römkens, P.F.A.M., Farret, R., De Temmerman, L. and Jakubowski, M. (2003). Critical limits for cadmium, lead and mercury related to ecotoxicological effects on soil organisms, aquatic organisms, plants, animals and humans. In Schütze, G., Lorenz, U. and Spranger, T. (eds.) *Expert meeting on critical limits for heavy metals and methods for their application, 2–4 December 2002 in Berlin, Workshop Proceedings*. UBA Texte 47/2003. Federal Environmental Agency (Umweltbundesamt), Berlin

DISMED (2005). *Desertification Information System for the Mediterranean*. European Environment Agency, Copenhagen

Dutkiewicz, S., Follows, M.J., Heimbach, P., Marshall, J. (2006). Controls on ocean productivity and air-sea carbon flux: An adjoint model sensitivity study. In *Geophysical Research Letters* 33 (2) Art. No. L02603

EEA (2005). *The European Environment – State and Outlook 2005*. European Environment Agency, Copenhagen

EMBRAPA Acre (2006). *Manejo Florestal Sustentavel*. Empresa Brasileira de Pesquisa Agropecuária, Acre

European Commission (2002). *The Implementation of Council Directive 91/676/EEC concerning the Protection of Waters against Pollution caused by Nitrates from Agricultural Sources. Synthesis from year 2000 member States reports*. Report COM(2002)407. Brussels http://ec.europa.eu/environment/water/water-nitrates/report.html (last accessed 29 June 2007)

European Commission (2007). The New EU Chemicals Legislation – REACH http://ec.europa.eu/enterprise/reach/index_en.htm (last accessed 29 June 2007)

Falkenmark, M. and Rockström, J. (2004). *Balancing water for humans and nature*. Earthscan, London

Fan, P.H. and Haque, T. (2000). Targeting public investments by agro-ecological zone to achieve growth and poverty alleviation goals in rural India. In *Food Policy* 25:411-428

FAO (1994). Prevention and disposal of obsolete and unwanted pesticide stocks in Africa and the Near East. http://www.fao.org/docrep/w8419e/w8419e00.htm (last accessed 29 June 2007)

FAO (2001). *Conservation Agriculture Case Studies in Latin America and Africa*. Soils Bulletin 78. Food and Agriculture Organization of the United Nations,Rome

FAO (2002). *Crops and drops: making the best use of water for agriculture*. Food and Agriculture Organization of the United Nations, Rome

FAO (2003). *World Agriculture: Towards 2015/2030 – An FAO Perspective*. Earthscan, London

FAO (2004). *Trends and Current Status of the Contribution of the Forestry Sector to National Economies*. Forest Products and Economics Division Working Paper, FSFM/ACC/007. Food and Agriculture Organization of the United Nations, Rome

FAO (2005). *Global Forest Resources Assessment 2005 (FRA 2005) database*. Food and Agriculture Organization of the United Nations, Rome (in GEO Data Portal)

FAO (2006a). *Global Forest Resources Assessment 2005 – Progress Towards Sustainable Forest Management*. Forestry Paper 147. Food and Agriculture Organization of the United Nations, Rome

FAO (2006b). FAO-AGL Global Network on Integrated Soil Management for Sustainable Use of Salt-Affected Soils In Participating Countries (SPUSH) http://www.fao.org/AG/AGL/agll/spush/intro.htm (last accessed 29 June 2007)

FAOSTAT (2006). FAO Statistics Database http://faostat.org (last accessed 29 June 2007)

FAO and UNESCO (1974-8). *Soil Map of the World*. Food and Agriculture Organization of the United Nations and United Nations Educational, Scientific and Cultural Organization, Paris

Fowler, D., Muller, J.B.A. and Sheppard, L.J. (2004). Water, air, and soil pollution. In *Focus* 4:3-8

GEF and UNEP (2003). *Regionally Based Assessment of Persistent Toxic Substances – Global Report 2003*. UNEP Chemicals, Geneva

Geist, H.J. and Lambin, E.F. (2004). Dynamic causal patterns of desertification. In *Bioscience* 54 (9):817-829

GEO Data Portal. *UNEP's online core database with national, sub-regional, regional and global statistics and maps, covering environmental and socio-economic data and indicators*. United Nations Environment Programme, Geneva http://www.unep.org/geo/data or http://geodata.grid.unep.ch (last accessed 1 June 2007)

Giles, J. (2005). The dustiest place on Earth. In *Nature* 434 (7035):816-819

Gisladottir, G. and Stocking, M.A. (2005). Land degradation control and its global environmental benefits. In *Land Degradation and Development* 16:99-112

Greenland, D.J. (1994). Long-term cropping experiments in developing countries: the need, the history and the future. In Leigh, R.A. and Johnston, A.E. (eds.) *Long-term Experiments in Agriculture and Ecological Sciences*. CAB, Wallingford

Greig-Gran, M., Noel, S. and Porras, I. (2006). *Lessons Learned from Payments for Environmental Services*. Green Water Credits Report 2. World Soil Information (ISRIC), Wageningen

Gunderson, L.H. and Pritchard, L.P. eds. (2002). *Resilience and the Behaviour of Large-Scale Systems*. SCOPE 60. Island Press, Washington, DC and London

Haines-Young, R., Potschin, M. and Cheshire, D. (2006). *Defining and identifying environmental limits for sustainable development – a scoping study*. Final report to UK Department for Environment, Food and Rural Affairs, Project code: NR0102

Hansen, J.C. (2000). Environmental contaminants and human health in the Arctic. In *Toxicol. Lett.* 112:119-125

Hansen, Z.K. and Libecap, G. D. (2004). Small farms, externalities and the Dust Bowl of the 1930s. In *J. Political Economy* 112 (3):665-694

Hartemink, A. and van Keulen, H. (2005). Soil degradation in Sub-Saharan Africa. In *Land Use Policy* 22 (1)

Henao, J. and Baanante, C. (2006). *Agricultural Production and Soil Nutrient Mining in Africa – Implications for Resource Conservation and Policy Development*. IFDC, Muscle Shoals, AL

Holling, C.S., Gunderson, L.H. and Ludwig, D. (2002). In quest of a theory of adaptive change. In Gunderson, L.H. and Holling, C.S. (eds.) *Panarchy: Understanding Transformations in Human and Natural Systems*. Island Press, Washington, DC

Holmgren, P. (2006). *Global Land Use Area Change Matrix: Input to GEO-4*. FAO Forest Resources Assessment Working Paper 134. Food and Agriculture Organization of the United Nations, Rome

Houghton, R.A. (2003). Why are estimates of the terrestrial carbon balance so different? In *Global Change Biology* 9:500-509

Houghton, R.A. and Hackler, J.L. (2002). Carbon flux to the atmosphere from land-use changes. In *Trends: A Compendium of Data on Global Change*. Carbon Dioxide Information Analysis Center, Oak Ridge National Laboratory, US Department of Energy, Oak Ridge, TN

IEA (2006). *World Energy Outlook 2006*. International Energy Agency. International Press, London

IIASA 2005 (2005). Feeding China in 2030. In *Options* Autumn 2005:12-15

Jacow, V.J. and Alles, W.S. (1987). Kandyan gardens of Sri Lanka. In *Agroforestry Systems* 5:123-137

Kauffman, J.H., Droogers P., and Immerzeel, W.W. (2007). *Green and blue water services in the Tana Basin, Kenya: assessment of soil and water management scenarios*. Green Water Credits Report 3. World Soil Information (ISRIC), Wageningen

Kaiser, J. (2005). Gulf's dead zone worse in recent decades. In *Science*, 308:195

KASSA (2006). *The Latin American Platform*. CIRAD, Brussels

Kurtz, A.C., Derry, L.A. and Chadwick, O.A. (2001). Accretion of Asian dust to Hawaiian soils: isotopic, elemental and mineral mass balances. In *Geochimica et Cosmochimica Act* 65 (12):1971-1983

Kuylenstierna, J.C.I., Rodhe, H., Cinderby S. and Hicks, K. (2001). Acidification in developing countries: Ecosystem sensitivity and the critical load approach on a global scale. In *Ambio* 30 (1):20-28

Lal, R., Sobecki, T.M., Iivari, T. and Kimble, J.M. (2004). Desertification. In *Soil Degradation in the United States: Extent Severity and Trends*. Lewis Publishers, CRC Press, Boca Raton

Lambin, E.F., Geist, H. and Lepers, E. (2003). Dynamics of land use and cover change in tropical regions. In *Annual Review of Environment and Resources* 28:205-241

Li, Y. (2000). Improving the Estimates of GHG Emissions from Animal Manure Management Systems in China. Proceedings of the IGES/NIES Workshop on GHG Inventories for Asia-Pacific Region, Hayama, Japan, 9-10 March

MA (2005a). *Ecosystems and Human Well-being: Synthesis*. Millennium Ecosystem Assessment. World Resources Institute, Island Press, Washington, DC

MA (2005b). *Ecosystems and Human Well-being: Desertification Synthesis*. Millennium Ecosystem Assessment World Resources Institute, Island Press, Washington, DC

NLWRA (2001). *Australian Dryland Salinity Assessment 2000 National Land and Water Resources Audit*. National Land and Water Resources Audit, Land & Water Australia, Canberra

MSC-E (2005). *Persistent Organic Pollutants in the Environment*. EMEP Status Report 3/2005. Meteorological Synthesising Centre-East, Moscow and Chemical Coordinating Centre, Kjeller

Mutunga, K. and Critchley, W.R.S. (eds.) (2002). *Farmers' Initiatives in Land Husbandry: Promising Technologies for the Drier Areas of East Africa*. Regional Land Management Unit, Nairobi

Navone, S. and Maggi, A.J. (2005). *La Inundación del Año 2001 en el Oeste de la Prov. De Buenos Aires: Potencial Productivo de las Tierras Afectadas y las Consecuencias Sobre la Produción de Granos Para el Periodo 1993-2002*. Fundación Hernandarias, Buenos Aires

Nicholson, S.E. (2002). What are the key components of climate as a driver of desertification? In Reynolds, J.F. and Stafford Smith, D.M. (eds.) *Global Desertification: Do Humans Cause Deserts?* Dahlem University Press, Berlin

OECD (2001). *OECD Environmental Outlook for the Chemicals Industry*. Organisation for Economic Co-operation and Development, Paris

Okin, G.S., Mahowald, N. Okin G., Mahowald, S.N., Chadwick, O.A. and Artaxo, P. (2004). Impact of desert dust on the biogeochemistry of phosphorus in terrestrial ecosystems. In *Global Biogeochemical Cycles* 18:2

Oldeman, L.R., Hakkeling, R.T.A. and Sombroek, W.G. (1991). *World Map of the Status of Human-Induced Soil Degradation: A Brief Explanatory Note*. World Soil Information (ISRIC), Wageningen

D'Oliveira, M.V.N., Swaine, M.D., Burslem, D.F.R.P., Braz, E.M. and Araujo, H.J.B. (2005). Sustainable forest management for smallholder farmers in the Brazilian Amazon. In Palm, C.A., Vosti, S.A., Sanchez, P.A. and Ericksen, P.J. (eds.). *Slash and Burn: the Search for Alternatives*. Columbia University Press, New York, NY

Olsson, L., Eklundh, L. and Ardö, J. (2005). A recent greening of the Sahel: trends, patterns and potential causes. In *Journal of Arid Environments* 63:556

Oweis, T.Y. and Hachum, A.Y. (2003). Improving water productivity in the dry areas of West Asia and North Africa. In Kijne, J.W., Barker, R. and Molden, D. (eds.). *Water Productivity in Agriculture.* CABI, Wallingford

Owen, P.L., Muir, T.C., Rew, A.W. and Driver, P.A. (1987). *Evaluation of the Victoria Dam Project in Sri Lanka, 1978-1985. Vol. 3, Social and Environmental Impact.* Evaluation Rept 392, Overseas Development Administration, London

Patterson, A.H. (2006). Leafing through the genome of our major crop plants: strategies for capturing unique information. In *Nature Reviews: Genetics* 7:174-184

Penning de Vries, F.W.T., Rabbinge, R. and Groot J.J.R. (1997). Potential and attainable food production and food security in different regions. In *Philosophical Translations: Biological Sciences* 352 (1356):917-928

Prentice, I.C., Farquhar, G.D., Fasham, M.J.R., Goulden, M.L., Heimann, M., Jaramillo, V.J., Kheshgi, H.S., Le Quéré, C., Scholes, R.J. and Wallace, D.W.R. (2001). The carbon cycle and atmospheric carbon dioxide. In Houghton, J.T., Ding, Y., Griggs, D.J., Noguer, M., van der Linden, P.J., Dai, X., Maskell, K., and Johnson, C.A (eds.). *Climate Change: IPCC Third assessment report* pp.881. Cambridge University Press, Cambridge and New York, NY

Pretty, J. and Hine, R. (2001). *Reducing Food Poverty with Sustainable Agriculture: A Summary of New Evidence.* Final report, Safe World Research Project. University of Essex, Colchester

Reining, P. (1978). *Handbook on Desertification Indicators.* American Association for the Advancement of Science, Washingtgon, DC

Reynolds, J.F. and Stafford Smith, M.D. eds. (2002): *Global Desertification — Do Humans Cause Deserts?* Dahlem Workshop Report 88. Dahlem University Press, Berlin

RIVM-MNP (2005). Emission Database for Global Atmospheric Research (EDGAR 3.2 and EDGAR 32FT2000). Netherlands Environmental Assessment Agency, Bilthoven

Rockström, J. (2003). Water for food and nature in the tropics: vapour shift in rain-fed agriculture. In *Transactions of the Royal Society B, special issue: Water Cycle as Life Support Provider.* The Royal Society, London

Rockström, J., Hatibu, N., Oweis, T. and Wani, Suhas (2006). Chapter 4: Managing water in rain-fed agriculture. In *Comprehensive Assessment of Water Management in Agriculture,* International Water Management Institute, Colombo

Rojstaczer, S., Sterling, S.M. and Moore, N.J. (2001). Human appropriation of photosynthesis products. In *Science* 294:2549-2552

Rothuis, A., Nhan, D.K., Richter, C.J.J. and Ollevier, F. (1998). Rice with fish culture in semi-deep waters of the Mekong delta, Vietnam. In *Aquaculture Research* 29 (1):59-66

Schellnhuber, H.J., Cramer, W., Nakicenovic, N., Wigley, T. and Yohe, G.W. (eds.) (2006). *Avoiding Dangerous Climate Change.* Cambridge University Press, Cambridge

Schiklomanov, I. (2000). World water resources and water use: present assessment and outlook for 2025. In Rijsberman, F. (ed.) *World Water Scenarios: Analysis.* Earthscan, London

Schlesinger, W. H. (1999). Carbon Sequestration in Soils. In *Science* 284:2095

SEI (2005). *Sustainable Pathways to Attain the Millennium Development Goals — Assessing the Key Role of Water, Energy and Sanitation.* Stockholm Environmental Institute, Stockholm

Sharma, H.C., Sharma, K.K. and Crouch, J.H. (2004). Genetic transformation of crops for insect resistance: potential and limitations. In *Critical Reviews in Plant Sciences* 23 (1):47-72

Skjelkvåle, B.L., Stoddard, J.L., Jeffries, D.S., Torseth, K., Hogasen, T., Bowman, J., Mannio, J., Monteith, D.T., Mosello, R., Rogora, M., Rzychon, D., Vesely, J,. Wieting, J., Wilander, A. and Worsztynowicz, A. (2005). Regional scale evidence for improvements in surface water chemistry 1990-2001. In *Environmental Pollution* 137 (1):165-176

Smil, V. (1997). Cycles of life: Civilization and the Biosphere. In *Scientific American Library Series* 63

Smil, V. (2001). *Enriching the Earth.* MIT Press, Cambridge, MA

Smith, C., Martens, M., Ahern, C., Eldershaw, V., Powell, B., Barry, E., Hopgood, G. and Watling, K. (eds.) (2003). *Demonstration of management and rehabilitation of acid sulphate soils at East Trinity.* Australian Department of Natural Resources and Mines, Indooroopilly

Sonneveld, B.G.J.S. and Dent, D.L. (2007). How good is GLASOD? In *Journal of Environmental Management,* in press

Soule J.D. and Piper, J.K. (1992). *Farming in Nature's Image: an Ecological Approach to Agriculture.* Island Press, Washington, DC

Stoorvogel, J.J. and Smaling, E.M.A. (1990). *Assessment of Soil Nutrient Decline in Sub-Saharan Africa, 1983-2000.* Rept 28. Winand Staring Centre-DLO, Wageningen

Sverdrup, H. Martinson, L., Alveteg, M., Moldan, F., Kronnäs, V. and Munthe, J. (2005). Modelling the recovery of Swedish ecosystems from acidification. In *Ambio* 34 (1):25-31

Tarnocai. C. (2006). The effect of climate change on carbon in Canadian peatlands. In *Global and Planetary Change* 53 (4):222-232

Tegen, I., Werner, M., Harrison, S.P. and Kohfeld, K.E. (2004). Relative importance of climate and land use in determining the future of global dust emission. In *Geophysical Research Letters* 31 (5) art. L05105

UNCCD (2005). *Economic Opportunities in the Drylands Under the United Nations Convention to Combat Desertification.* Background Paper 1 for the Special Segment of the 7th Session of the Conference of Parties, Nairobi, 24-25 Oct 2005

UNEP (1992). *Atlas of desertification.* United Nations Environment Programme and Edward Arnold, Sevenoaks

UNEP (2006). *Global Deserts Outlook.* United Nations Environment Programme, Nairobi

UNGA (1994). United Nations General Assembly Document A/AC.241/27

UN Water (2007). *Coping with water scarcity: challenge of the twenty-first century.* Prepared for World Water Day 2007. http://www.unwater.org/wwd07/downloads/ documents/escarcity.pdf (last accessed 29 June 2007)

Van Lauwe, B.and Giller, K.E. (2006). Popular myths around soil fertility management in sub-Saharan Africa. In *Agriculture Ecosystems and Environment* 116 (1-2):34-46

Van Mensvoort, M.E.F. and Dent, D.L. (1997). Assessment of the acid sulphate hazard. In *Advances in Soil Science* 22:301-335

Vedeld, P., Angelsen, A., Sjastad, E. and Berg, G.K. (2004). *Counting on the Environment. Forest Incomes and the Rural Poor.* Environmental Economics Series, Environment Department Paper No. 98. World Bank, Washington, DC

Vogel, C.H. and Smith, J. (2002). Building social resilience in arid ecosystems. In Reynolds, J.F. and Stafford Smith, M.D. (eds.) *Global Desertification — Do Humans Cause Deserts?* Dahlem Workshop Report 88, Dahlem University Press, Berlin

Walter, K.M., Zimov, S.A., Chanton, J.P., Verbyla, D. and Chapin III, F.S. (2006). Methane bubbling from Siberian thaw lakes as a positive feedback to climate warming. In *Nature* 443:71-74

Wardell, D.A. (2003). Estimating watershed service values of savannah woodlands in West Africa using the effect on production of hydro-electricity. Sahel-Sudan Environmental Research Initiative. http://www.geogr.ku.dk/research/serein/docs/ WP_42 (last accessed 29 June 2007)

Watson, R.T., Noble, I.R., Bolin, B., Ravindranath, N.H., Verardo, D.J. and Dokken, J. (2000). *Land Use, Land Use Change and Forestry (A Special report of IPCC).* Cambridge University Press, Cambridge

Webb, A. (2002). *Dryland Salinity Risk Assessment in Queensland.* Consortium for Integrated Resource Management. Occ. Papers ISSN 1445-9280, Consortium for Integrated Resource Management, Indooroopilly

WOCAT (2007). *Where the land is greener — case studies and analysis of soil and water conservation initiatives worldwide* Liniger, H. and Critchley, W. (eds.). CTA, FAO, UNEP and CDE, Wageningen

World Bank (2005). *Water Sector Assessment Report on the Countries of the Cooperation Council of the Arab States of the Gulf.* Rept No32539-MNA, Water, Environment, Social and Rural Development Department, Middle East and North Africa Region. The World Bank, Washington, DC

World Bank (2006). *Strengthening Forest Law Enforcement and Governance — Addressing a Systemic Constraint to Sustainable Development.* The World Bank, Washington, DC

Worldwatch Institute (2002). *State of the World 2002.* W.W. Norton, New York, NY

WWF (2005). *Failing the Forests — Europe's Illegal Timber Trade.* World Wildlife Fund, Godalming, Surrey

Xue, Y. and Fennessy, M.J. (2002). Under what conditions does land cover change impact regional climate? In *Global Desertification — Do Humans Cause Deserts?* (eds. Reynolds, J.F. and Stafford Smith, M.D.) pp.59-74. Dahlem Workshop Report 88, Dahlem University Press, Berlin

Zhang, X.Y., Gong, S.L., Zhao, T.L., Arimoto, R., Wang, Y.Q. and Zhou, Z.J. (2003). Sources of Asian dust and the role of climate change versus desertification in Asian dust emission. In *Geophysics Research Letters* 20(23), Art. No 2272

Zhu, L.K. (2006). *Dynamics of Desertification and Sandification in China.* China Agricultural Publishing, Beijing

Zimov, S.A., Schuur, E.A.G. and Chapin, F.S. (2006). Permafrost and the global carbon budget. In *Science* 312:1612-1613

Chapter **4**

Water

Coordinating lead authors: Russell Arthurton, Sabrina Barker, Walter Rast, and Michael Huber

Lead authors: Jacqueline Alder, John Chilton, Erica Gaddis, Kevin Pietersen, and Christoph Zöckler

Contributing authors: Abdullah Al-Droubi, Mogens Dyhr-Nielsen, Max Finlayson, Matthew Fortnam (GEO fellow), Elizabeth Kirk, Sherry Heileman, Alistair Rieu-Clark, Martin Schäfer (GEO fellow), Maria Snoussi, Lingzis Danling Tang, Rebecca Tharme, Rolando Vadas, and Greg Wagner

Chapter review editor: Peter Ashton

Chapter coordinators: Salif Diop, Patrick M'mayi, Joana Akrofi, and Winnie Gaitho

Main messages

Human well-being and ecosystem health in many places are being seriously affected by changes in the global water cycle, caused largely by human pressures. The following are the main messages of this chapter:

Climate change, human use of water resources and aquatic ecosystems, and overexploitation of fish stocks influence the state of the water environment. This affects human well-being and the implementation of internationally agreed development goals, such as those in the Millennium Declaration. Evidence shows that implementing policy responses to environmental problems enhances human health, socio-economic growth and aquatic environmental sustainability.

The world's oceans are the primary regulator of global climate, and an important sink for greenhouse gases. At continental, regional and ocean basin scales, the water cycle is being affected by long-term changes in climate, threatening human security. These changes are affecting Arctic temperatures, sea- and land ice, including mountain glaciers. They also affect ocean salinity and acidification, sea levels, precipitation patterns, extreme weather events and possibly the ocean's circulatory regime. The trend to increasing urbanization and tourism development has considerable impacts on coastal ecosystems. The socio-economic consequences of all these changes are potentially immense. Concerted global actions are needed to address the root causes, while local efforts can reduce human vulnerability.

Freshwater availability and use, as well as the conservation of aquatic resources, are key to human well-being. The quantity and quality of surface- and groundwater resources, and life-supporting ecosystem services are being jeopardized by the impacts of population growth, rural to urban migration, and rising wealth and resource consumption, as well as by climate change. If present trends continue, 1.8 billion people will be living in countries or regions with absolute water scarcity by 2025, and two-thirds of the world population could be subject to water stress.

Practical implementation of Integrated Water Resource Management (IWRM) at the basin scale, including consideration of conjunctive groundwater aquifers and downstream coastal areas, is a key response to freshwater scarcity. Because agriculture accounts for more than 70 per cent of global water use, it is a logical target for water savings and demand management efforts. Stakeholders who pay attention to increasing the productivity of rain-fed agriculture and aquaculture, which can contribute to improved food security, are proving to be successful.

Water quality degradation from human activities continues to harm human and ecosystem health. Three million people die from water-borne diseases each year in developing countries, the majority of whom are children under the age of five. Pollutants of primary concern include microbial pathogens and excessive nutrient loads. Water contaminated by microbes remains the greatest single cause of human illness and death on a global scale. High nutrient loads lead to eutrophication of downstream and coastal waters, and loss of beneficial human uses. Pollution from diffuse land sources, particularly agriculture and urban run-off, needs urgent action by governments and the agricultural sector. Pesticide pollution, endocrine-disrupting substances and suspended sediments are also hard to control. There is evidence that IWRM at the basin scale, improved effluent treatment and wetland restoration, accompanied by improved education and public awareness, are effective responses.

Aquatic ecosystems continue to be heavily degraded, putting many ecosystem services at risk, including the sustainability of food supplies and biodiversity. Global marine and freshwater fisheries show large-scale declines, caused mostly by persistent overfishing. Freshwater stocks also suffer from habitat degradation and altered thermal regimes related to climate change and water impoundment. Total marine catches are being sustained only by fishing ever further offshore and deeper in the oceans, and progressively lower on the food chain. The trend of fish stock degradation can be reversed when governments, industry and fishing communities work together to reduce excess fishing effort, subsidies and illegal fishing.

A continuing challenge for the management of water resources and aquatic ecosystems is to balance environmental and developmental needs. It requires a sustained combination of technology, legal and institutional frameworks, and, where feasible, market-based approaches. This is particularly true where efforts are designed to share the benefits of water-related ecosystem services rather than merely sharing the water resource alone. In addition to capacity building, the challenge is not only to develop new approaches, but also to facilitate the practical, timely and cost-effective implementation of existing international and other agreements, policies and targets, which can provide a basis for cooperation on many levels. Although many coastal environments are benefiting from existing Regional Seas agreements, there is a paucity of international agreements addressing transboundary freshwater systems, a significant source of potential conflict in the future. A range of perverse subsidies also hampers the development and implementation of effective management measures at many levels. The benefits of tackling well-understood problems, especially those at the basin scale, are likely to be greatest when efforts are coordinated effectively among different levels of society.

INTRODUCTION

In 1987 the World Commission on Environment and Development (Brundtland Commission) warned in the final report, *Our Common Future*, that water was being polluted and water supplies were overused in many parts of the world. This chapter assesses the state of the water environment since the mid-1980s, and its impacts on human well-being with respect to human health, food security, human security and safety, and livelihoods and socio-economic development.

The ocean is the source of most of the world's precipitation (rainfall and snowfall), but people's freshwater needs are met almost entirely by precipitation on land (see Figure 4.1), with a small though increasing amount by desalination. Due to changes in the state of the ocean, precipitation patterns are altering, affecting human well-being. Ocean changes are also affecting marine living resources and other socio-economic benefits on which many communities depend. The availability, use and management of freshwater, and of aquatic ecosystems in general, are key to development and human well-being.

Solar energy absorbed by the Earth's surface, particularly the ocean, drives the circulation of the globe's water. Most water transfer occurs between ocean and atmosphere by evaporation and precipitation. Ocean circulation – the global ocean conveyor (see Figure 4.2) – is driven by differences in seawater density, determined by temperature and salt content. Heat moves via warm surface water flows towards the poles, and returns in cooler, deep water towards the equator. The cooler returning water is saltier and denser through evaporation, and, as it sinks, it is replaced by warmer water flowing poleward. This circulation is of enormous significance to the world, carrying carbon dioxide (CO_2) to the deep ocean (see Chapter 2), distributing heat and dissolved matter, and strongly influencing climate regimes and the availability of nutrients to marine life. The 1982–1983 intense El Niño provided the evidence that large-scale fluctuations in ocean and atmosphere circulation are coupled, having profound global climatic impacts (Philander 1990). There are concerns that climate change might alter global ocean circulation patterns, possibly reducing the amount of heat that is carried north in the Gulf Stream, warming western Europe and the Arctic (see Chapters 2 and 6).

The water environment and development are strongly interdependent. The state of the hydrological regime, its water quality and ecosystems are major factors contributing to human well-being. These linkages are shown in Tables 4.1 and 4.4, demonstrating the implications of the state of water in meeting the Millennium Development Goals (MDGs). The world's inland and marine fisheries are a crucial part of aquatic living resources that are vital to human well-being. The chapter assesses how these have responded,

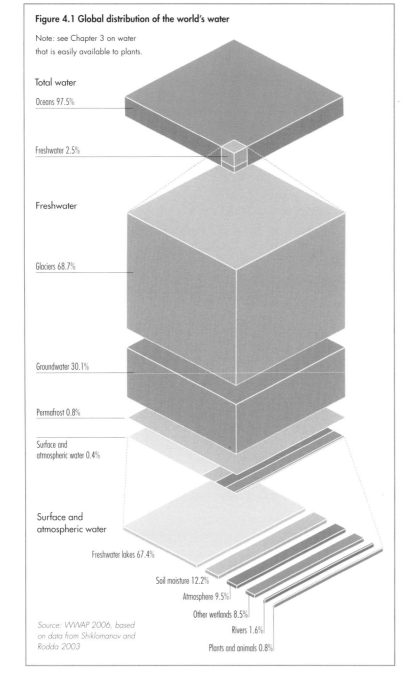

Figure 4.1 Global distribution of the world's water

Note: see Chapter 3 on water that is easily available to plants.

Total water
Oceans 97.5%

Freshwater 2.5%

Freshwater

Glaciers 68.7%

Groundwater 30.1%

Permafrost 0.8%

Surface and atmospheric water 0.4%

Surface and atmospheric water

Freshwater lakes 67.4%

Soil moisture 12.2%

Atmosphere 9.5%

Other wetlands 8.5%

Rivers 1.6%

Plants and animals 0.8%

Source: WWAP 2006, based on data from Shiklomanov and Rodda 2003

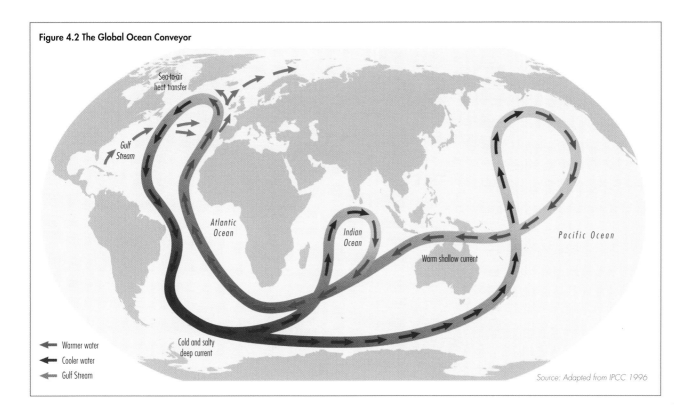

Figure 4.2 The Global Ocean Conveyor

Sea-to-air heat transfer

Gulf Stream

Atlantic Ocean

Indian Ocean

Warm shallow current

Pacific Ocean

Cold and salty deep current

← Warmer water
← Cooler water
← Gulf Stream

Source: Adapted from IPCC 1996

and are responding, to the impacts of environmental change. The range of international, regional and national policies and management responses, and indications of their success are summarized in Table 4.5 at the end of the Chapter.

International water policy is increasingly emphasizing the need to improve governance as it relates to water resources management. A global consensus has emerged on the need to implement ecosystem-based management approaches to address sustainable water resource needs. Through responses such as Integrated Water Resources Management (IWRM), social and economic development goals can be achieved in a manner that gives the world sustainable aquatic ecosystems to meet the water resource needs of future generations. An increasing realization of the limits of traditional regulation has led to the introduction of more participatory regulatory approaches, such as demand management and voluntary agreements. These necessitate education and public involvement.

DRIVERS OF CHANGE AND PRESSURES

The Earth system is modified by natural factors, but human activities have increasingly driven change over the last few decades. The drivers of change in the water environment are largely the same as those influencing change in the atmosphere and on land (see Chapters 2 and 3). The world's population, consumption and poverty have continued to grow, along with technological advances. Increased human activities are putting pressures on the environment, causing global warming, altering and intensifying freshwater use, destroying and polluting aquatic habitats, and overexploiting aquatic living resources, particularly fish. The modification of the Earth system is taking place both at the global scale, notably through increasing greenhouse gas emissions, leading to climate change, and at the scale of discrete river basins and their associated coastal areas (Crossland and others 2005).

Human pressures at global to basin scales are substantially modifying the global water cycle, with some major adverse impacts on its interconnected aquatic ecosystems – freshwater and marine – and therefore on the well-being of people who depend on the services that they provide.

Overexploitation and pollution of water, and degradation of aquatic ecosystems directly affect human well-being. Although the situation has improved (see Figure 4.3), an estimated 2.6 billion people are without improved sanitation facilities. And if the 1990–2002 trend holds, the world will miss the sanitation target of the Millennium Development Goals by half a billion people (WHO and UNICEF 2004).

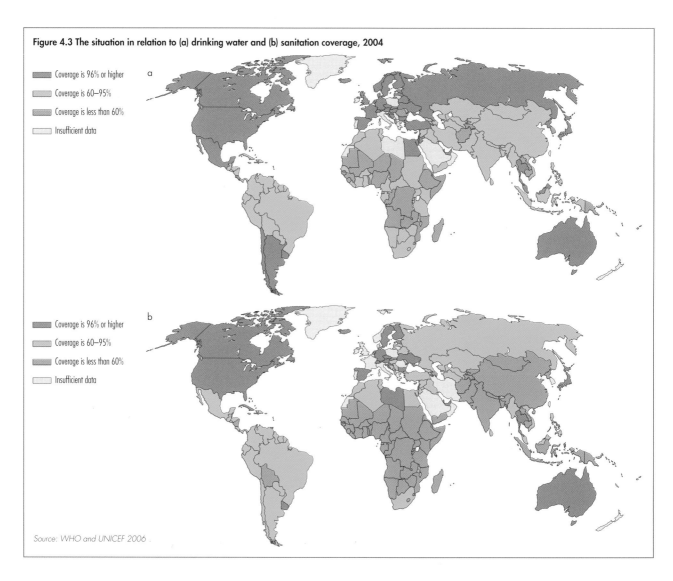

Figure 4.3 The situation in relation to (a) drinking water and (b) sanitation coverage, 2004

Coverage is 96% or higher
Coverage is 60–95%
Coverage is less than 60%
Insufficient data

a

Coverage is 96% or higher
Coverage is 60–95%
Coverage is less than 60%
Insufficient data

b

Source: WHO and UNICEF 2006 .

Climate change

Warming of the climate system is unequivocal (IPCC 2007). Climate change affects the warming and acidification of the global ocean (see Chapters 2 and 6). It influences the Earth's surface temperature, as well as the amount, timing and intensity of precipitation, including storms and droughts. On land, these changes affect freshwater availability and quality, surface water run-off and groundwater recharge, and the spread of water-borne disease vectors (see Chapters 2 and 3). Some of the most profound climate-driven changes are affecting the cryosphere, where water is in the form of ice. In the Arctic, the increase in temperature is 2.5 times the global average, causing extensive melting of sea- and land ice as well as thawing of permafrost (ACIA 2004) (see Chapters 2 and 6). Climate change is expected to exacerbate pressure, directly or indirectly, on all aquatic ecosystems.

Water use

The past 20 years have seen increasing water use for food and energy production to meet the demands of a growing population and to enhance human well-being, a continuing global trend (WWAP 2006). However, the changes in the way water is used have significant adverse impacts, which require urgent attention to ensure sustainability. Unlike the pressures of climate change, those of water use are exerted mostly within basins. Some of their drivers are global, but their remedies may be local, though enabled by transboundary conventions.

Current freshwater withdrawals for domestic, industrial and agricultural use, as well as the water evaporated from reservoirs, are shown in Figure 4.4. Agriculture is by far the biggest user. The expansion of hydropower generation and irrigated agriculture, now happening mostly in developing countries, is vital for economic development and food production. But, the consequent changes in land- and water use by agriculture, as well as for urban and industrial growth, have major adverse impacts on freshwater and coastal ecosystems.

In addition to agricultural demands, pressures on water resources are compounded by the physical alteration and destruction of habitats by urban and industrial development, and, especially in coastal areas, tourism. Invasive species, introduced to waterbodies intentionally (fish stocking) or inadvertently (ships' ballast discharges), are also a factor. Modifications of the water cycle through irrigation works and water supply schemes have benefited society for centuries. However, the global impacts of human interventions in the water cycle, including land cover change, urbanization, industrialization and water resources development, are likely to surpass those of recent or anticipated climate change, at least over decades (Meybeck and Vörösmarty 2004).

Human activities at basin scales cause increased water-borne pollution from point and diffuse sources, affecting inland and coastal aquatic ecosystems. The diffuse sources are more difficult to identify, quantify and manage. Agricultural run-off containing nutrients and agrochemicals is the main source of water pollutants in many countries (US EPA 2006). Domestic and industrial effluents also are major sources, with inadequately treated wastewater discharged directly into waterways. Virtually all industrial activities generate water pollutants, as do unsustainable forestry (land clearing, forest fires and increased erosion), mining (mine and leachate drainage), waste disposal (landfill leachate, land and sea litter disposal), aquaculture and mariculture (microbes, eutrophication and antibiotics), and hydrocarbon (oil) production and use.

Water withdrawals are predicted to increase by 50 per cent by 2025 in developing countries, and 18 per cent in developed countries (WWAP 2006). Since nearly all industrial and manufacturing activities require adequate water supplies, this situation is likely to impede socio-economic development, and increase pressures on freshwater ecosystems. At the global scale, the integrity of aquatic ecosystems – the state of their physical elements, their biodiversity and their processes – continues to decline (MA 2005), reducing their capacity to provide clean freshwater, food and other services such as contaminant attenuation, and to buffer against extreme climatic

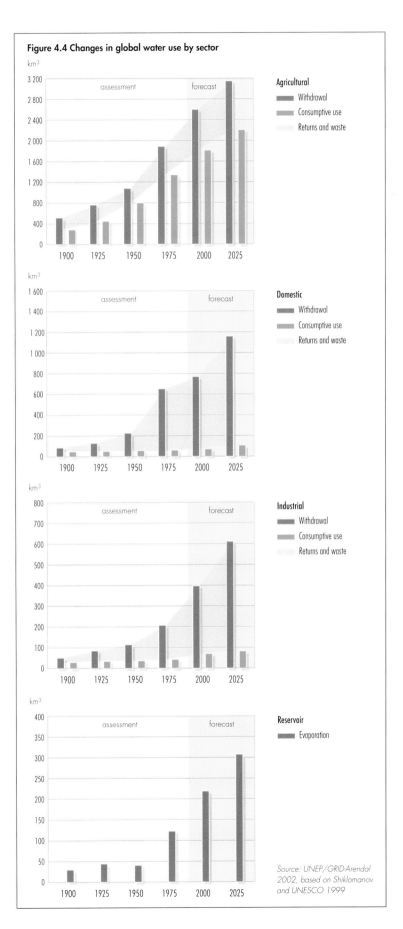

Figure 4.4 Changes in global water use by sector

Agricultural
— Withdrawal
— Consumptive use
— Returns and waste

Domestic
— Withdrawal
— Consumptive use
— Returns and waste

Industrial
— Withdrawal
— Consumptive use
— Returns and waste

Reservoir
— Evaporation

Source: UNEP/GRID-Arendal 2002, based on Shiklomanov and UNESCO 1999

Suspended mud trails made by shrimp trawlers (the small black dots) as they churn along in the ocean off the mouth of the Yangtze River.

Credit: DigitalGlobe and MAPS geosystems

events. Therefore, changes in the hydrosphere bear heavily on achieving the clean water, health and food security targets of the MDGs.

Fisheries

Several direct pressures contribute to overexploitation of fish stocks, and to the decline of marine mammals and turtles around the world. Population growth and rising wealth have resulted in an almost 50 per cent increase in fish production from 95 million tonnes in 1987 to 141 million tonnes in 2005 (FAO 2006c). The demand, especially for high-value seafood and to meet population growth, is expected to increase by about 1.5 per cent annually in coming decades. Meeting this demand will be a challenge. For instance, rapid income growth and urbanization in China from the early 1980s to the late 1990s were accompanied by a 12 per cent/year rise in consumption (Huang and others 2002). Another factor is changing food preferences as a result of the marketing of fish in developed countries as part of a healthy diet. Aquaculture continues to grow and, with it, the demand for fish meal and fish oil for use as feed, both of which are derived and primarily available only from wild fish stocks (Malherbe 2005). Fish represent the fastest-growing food commodity traded internationally, causing increasingly serious ecological and management problems (Delgado and others 2003).

Subsidies, estimated at 20 per cent of the value of the fisheries sector (WWF 2006), have created excess fishing capacity, which is outstripping available fisheries resources. Global fishing fleets are estimated to have a capacity 250 per cent greater than needed to catch what the ocean can sustainably produce (Schorr 2004). Furthermore, technological advances have allowed industrial and artisanal fleets to fish with greater precision and efficiency, and further offshore and in deeper water. This affects the spawning and

nursery grounds of many species, and decreases the economic possibilities of fishers in developing countries, who are unable to afford such technology (Pauly and others 2003). Destructive fishing gear and practices, such as bottom trawlers, dynamite and poison, also compromise the productivity of global fisheries. Trawlers in particular produce by-catch, often consisting of large quantities of non-target species, with an estimated 7.3 million tonnes/year discarded globally (FAO 2006a).

Inland fish stocks are subject to a combination of direct pressures, including habitat alteration, and loss, altered flows and habitat fragmentation due to dams and other infrastructure. They also face pollution, exotic species and overfishing. With much of inland fisheries catches destined for subsistence consumption or local markets, food demand for growing populations is a major factor driving exploitation levels in inland waters.

Superimposed on unsustainable fishing practices and other pressures is global climate change. This may affect aquatic ecosystems in many ways, although the capacity of fish species to adapt to such change is not fully understood. Changes in water temperatures and especially in wind patterns, however, suggest climate change can disturb fisheries, an important emerging issue with potentially serious impacts on global fishery resources.

ENVIRONMENTAL TRENDS AND RESPONSES

Human well-being and environmental sustainability are intrinsically interconnected. The state of the global water environment is related to climate change, changes in water use and the exploitation of aquatic living resources, notably fisheries. The consequences of environmental change for human well-being are analysed in relation to these three issues. Table 4.1 highlights major links between water and human well-being.

Table 4.1 Linkages between state changes in the water environment and environmental and human impacts

STATE CHANGES	Mediating environmental/ ecosystem impacts	HUMAN WELL-BEING IMPACTS			
		Human health	Food security	Physical security and safety	Socio-economic
Climate change related issues – disturbances to the hydrological regime mainly at the global scale					
⇧ Sea surface temperature	⇔ Trophic structure and food web	⇩ Food safety[1]	⇔ Fishery species distribution[2] ⇩ Aquaculture production[2]		⇩ Profits (loss of product sales)[2]
	⇧ Coral bleaching		⇔ Artisanal fishers[2]	⇩ Coast protection[3]	⇩ Tourism attraction[2]
	⇧ Sea-level rise		⇔ Aquaculture facilities[2]	⇧ Coastal/inland flooding[1]	⇧ Damage to property, infrastructure and agriculture[1]
	⇧ Tropical storm and hurricane frequency and intensity	⇧ Disruption of utility services[1]	⇧ Crop damage[1] ⇧ Aquaculture damage[1]	⇧ Drowning and flood damage[1] ⇩ Coast protection[1]	⇩ Energy production[1] ⇩ Law and order[1] ⇧ Damage to property and infrastructure[1]
⇧ ⇩ Precipitation	⇧ Flood damage	⇧ Water-related diseases[1]	⇧ Crop destruction[1]	⇧ Drowning and flood damage[1]	⇧ Property damage[1]
	⇧ Drought	⇧ Malnutrition[1]	⇧ Crop reduction[1]		
⇧ Land- and sea ice wasting	⇔ Ocean circulation change ⇧ Mountain glacier wasting ⇧ Sea-level		⇔ Traditional food sources[1] ⇩ Available irrigation water[2]	⇧ Coastal erosion and inundation[2]	⇧ Improved shipping access[1] ⇩ Downstream livelihoods[1]
⇧ Permafrost thaw	⇧ Tundra ecosystem changes		⇧ Agricultural development possibilities[2]	⇩ Ground stability[1]	⇩ Land transportation[1] ⇧ Buildings and infrastructure damage[1]
⇧ Ocean acidification	⇩ Biocalcifying organisms including reef coral		⇩ Coastal fisheries[3]	⇩ Coastal protection[3]	⇩ Reef tourism[3] ⇩ Fisheries as livelihoods[3]
Human water-use related issues – disturbance to the hydrological regime at basin and coastal scale					
⇔ Stream flow modification		⇩ Downstream drinking water[1] ⇧ Water-borne diseases[1]	⇧ Irrigated agriculture[1] ⇩ Inland fish stocks[1] ⇧ Salinization[1] ⇩ Floodplain cultivation[1]	⇧ Flood control[1] ⇧ Community displacement[1]	⇩ Freshwater fisheries[1] ⇩ Transportation by water[1] ⇧ Hydropower[1] ⇧ Irrigated agriculture[1] ⇧ Allocation conflicts[1]
	⇧ Ecosystem fragmentation, wetland infilling and drainage		⇩ Coastal wetland food resources[2] ⇩ Prawn fishery[1]		
	⇩ Sediment transport to coasts		⇩ Reduces floodplain sediment[1]	⇧ Coastal erosion[1]	⇩ Reservoir lifecycle[1]

STATE CHANGES	Mediating environmental/ ecosystem impacts	HUMAN WELL-BEING IMPACTS			
		Human health	Food security	Physical security and safety	Socio-economic
Human water-use related issues – disturbance to the hydrological regime at basin and coastal scale					
⇩ Groundwater levels	⇧ Drying of shallow wells[1] ⇧ Salinity and pollution		⇩ Available irrigation water[1] ⇩ Water quality[1]	⇧ Competition for groundwater[1]	⇧ Access costs[1] ⇧ Premature well abandonment[1] ⇧ Inequity[1]
	⇩ Discharge to surface water	⇩ Available surface water[1]	⇩ Freshwater for irrigation[1]		
	⇧ Land subsidence				⇧ Buildings and infrastructure damage[1]
	⇧ Saline water intrusion	⇩ Available drinking water[1]	⇩ Available irrigation water[1] ⇧ Salinization[1] ⇩ Water quality[1]		⇧ Water treatment costs[1]
	Reverse groundwater flow ⇧ Downward movement	⇧ Pollution from land surface and canals[1]	⇩ Water quality[1]		⇧ Treatment costs for public supply[1]
Human water-use related issues – water quality changes at the basin and coastal scale					
⇧ Microbial contamination		⇧ Water-borne diseases[1] ⇧ Fish, shellfish contamination[1]			⇩ Working days[2] ⇩ Recreation and tourism[1]
⇧ Nutrients	⇧ Eutrophication	⇧ Nitrate contamination of drinking water[1]	⇧ Production of macrophytes for animal fodder[1]		⇧ Cost of water treatment[1]
	⇧ Harmful algal blooms	⇧ Fish and shellfish contamination[1] ⇧ Neurological and gastrointestinal illnesses[1]	⇩ Livestock health[1] ⇩ Food available for humans[1]		⇩ Recreation and tourism[1] ⇩ Livelihood income[1]
⇧ Oxygen-demanding materials	⇩ Dissolved oxygen in waterbodies		⇩ High oxygen-demanding species[1]		⇩ Recreation and tourism[3]
⇧ Suspended sediment	⇩ Ecosystem integrity		⇩ Fish and livestock health[1]		⇧ Cost of water treatment[1]
Persistent organic pollutants (POPs)		⇧ Fish and livestock contamination[1] ⇧ Chronic disease[2]			⇩ Commercial fish value[1]
Heavy metal pollution		⇧ Seafood contamination[1] ⇧ Chronic disease[1]	⇧ Flood contamination of agricultural lands[1]		⇧ Cost of water treatment[1]
⇧ Solid waste	⇧ Ecosystem and wildlife damage	⇧ Threat to human health (infections and injuries)[1]			⇩ Recreation and tourism[2] ⇩ Fisheries[2]

Arrows show trends of state and impact changes

⇧ increase   ⇩ decrease   ⇔ no statistically proven change

[1] well established   [2] established but incomplete   [3] speculative

MDG Goal 1, Target 1: Halve, between 1990 and 2015, the proportion of people whose income is less than US$1 a day.
Target 2: Halve, between 1990 and 2015, the proportion of people who suffer from hunger.
MDG Goal 6, Target 8: Halt by 2015 and begin to reverse the incidence of malaria and other major diseases.
MDG Goal 7, Target 9: Integrate the principles of sustainable development into country policies and programmes, and reverse the loss of environmental resources.
MDG Goal 7, Target 10: Halve, by 2015, the proportion of people without sustainable access to safe drinking water and basic sanitation.

Various management responses have been adopted to address the water environment challenges. Although actions that should be taken by individuals and agencies at different levels have been identified, the primary focus is on decision-makers facing water-related challenges. In providing management guidance, the linkages and interactions between the water environment and other components of the global environment (atmosphere, land and biodiversity) must also be considered. For example, the quantity and quality of water resources can determine the types of fisheries that occur. The management options include actions and strategies for prevention, and for mitigation and adaptation (the former seeks to solve the problems and the latter focuses on adjustment to the problems).

CLIMATE CHANGE INFLUENCE

Ocean temperature and sea level

At the global scale, ocean temperatures and sea level continue their rising trends. Observations since 1961 show that the average temperature of the global ocean has increased at depths of at least 3 000 metres, and that the ocean has been absorbing more than 80 per cent of the heat added to the climate system. Such warming causes seawater to expand, contributing to sea-level rise (IPCC 2007). The global sea level rose at an average of 1.8 mm/year from 1961 to 2003, and the rate of increase was faster (about 3.1 mm/year) from 1993 to 2003 (see Table 4.2). Whether the faster rate reflects decadal variability or an increase in the longer-term trend is unclear.

There is *high confidence* that the rate of observed sea-level rise increased from the 19th to the 20th century. The total 20th century rise is estimated to be 0.17 m (IPCC 2007).

Sea surface temperatures and surface currents influence wind patterns in the lower atmosphere, and so determine regional climates. Warming ocean waters and changes in surface currents directly affect marine plant and animal communities, altering fish species distribution and stock abundance. In the tropics, unusually high sea surface water temperatures are becoming increasingly frequent, causing widespread coral bleaching and mortality (Wilkinson 2004). There is observational evidence for an increase of intense tropical cyclone activity in the North Atlantic since about 1970, correlated with increases of tropical sea surface temperatures, but there is no clear trend in the annual numbers of tropical cyclones (IPCC 2007) (see Chapter 2).

The warming of the ocean, in particular its surface waters, and the feedback of heat to the atmosphere are changing rainfall patterns, affecting the availability of freshwater and food security, and health. Due to the ocean's great heat storage capacity and slow circulation, the consequences of its warming for human well-being will be widespread. Both past and future anthropogenic greenhouse gas emissions will continue to contribute to warming and sea-level rise for more than a millennium, due to the timescales required for removal of this gas from the atmosphere (IPCC 2007).

Table 4.2 Observed sea-level rise, and estimated contributions from different sources		
	Average annual sea-level rise (mm/year)	
Source of sea-level rise	1961–2003	1993–2003
Thermal expansion	0.42 ± 0.12	1.6 ± 0.5
Glaciers and ice caps	0.50 ± 0.18	0.77 ± 0.22
Greenland ice sheet	0.05 ± 0.12	0.21 ± 0.07
Antarctic ice sheet	0.14 ± 0.41	0.21 ± 0.35
Sum of individual climate contributions to sea-level rise	1.1 ± 0.5	2.8 ± 0.7
Observed total sea-level rise	1.8 ± 0.5	3.1 ± 0.7
Difference (Observed minus sum of estimated climate contributions)	0.7 ± 0.7	0.3 ± 1.0

Note: Data prior to 1993 are from tide gauges; those from 1993 onwards are from satellite altimetry.

Source: IPCC 2007

Precipitation

Since at least the 1980s, the average atmospheric water vapour content has increased over land and ocean, and in the upper troposphere. The increase is broadly consistent with the extra water vapour that warmer air can hold (IPCC 2007). There is increasing evidence that precipitation patterns have changed worldwide as a result of atmospheric responses to climatic change (see Figure 4.5) (see Chapter 2). Significantly increased precipitation has been observed in the eastern parts of North and South America, northern Europe and northern and central Asia (IPCC 2007). Although precipitation patterns are believed to be increasingly influenced by large-scale warming of ocean and land surfaces, the exact nature of the change is uncertain, though knowledge is improving. Global land precipitation has increased by about 2 per cent since the beginning of the 20th century. While this is statistically significant, it is neither spatially nor temporally uniform. Such spatial and temporal variability is well illustrated in the Sahel region of Africa, which has experienced a succession of comparatively rainy periods alternating with droughts. Following droughts in the 1980s, changes in monsoon dynamics resulted in increased rainfall over the African Sahel and the Indian subcontinent in the 1990s, leading to increased vegetation cover in those areas (Enfield and Mestas-Nuñez 1999) (see Figure 3.10 – greenness index Sahel).

More intense and longer droughts have been observed over wider areas since the 1970s, particularly in the tropics and subtropics, and drying has been observed in the Sahel, the Mediterranean, southern Africa and parts of southern Asia (IPCC 2007). The decreasing rainfall and devastating droughts in the Sahel since the 1970s are among the least disputed and largest recent climate changes recognized by the global climate research community (Dai and others 2004, IPCC 2007) (see Figure 4.5). The reduced rainfall has been attributed to ocean surface temperature changes, particularly to warming of the southern hemisphere oceans and the Indian Ocean, leading to changes in atmospheric circulation (Brooks 2004). In 2005, the Amazon region suffered one of its worst droughts in 40 years.

For many mid- and high-latitude regions, there has been a 2–4 per cent increase in the frequency of heavy precipitation events over the latter half of the 20th century. An increased frequency and intensity of drought in parts of Asia and Africa was observed over the same period (Dore 2005). Increasing variance of continental precipitation is likely, with wet areas becoming wetter and dry areas drier. Recent trends are likely to continue. Increases in the amount of precipitation are very likely in high latitudes, while decreases are likely in most subtropical land regions. It is very likely that heat waves and heavy precipitation events will continue to become more frequent. The frequency of heavy precipitation events has increased over most land areas, consistent with warming and observed increases of atmospheric water vapour (IPCC 2007).

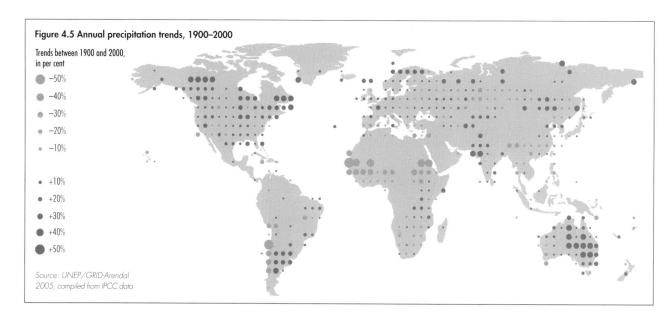

Figure 4.5 Annual precipitation trends, 1900–2000

Trends between 1900 and 2000, in per cent

- −50%
- −40%
- −30%
- −20%
- −10%

- +10%
- +20%
- +30%
- +40%
- +50%

Source: UNEP/GRID-Arendal 2005, compiled from IPCC data

The roles of soil moisture and terrestrial biomes, such as forests, in regulating global water quality and quantity are described in Chapter 3. Depending on local conditions, the effects of irrigation on water vapour flows may be as important as those of deforestation when accounting for the climatic effects of human modification of the land surface that lead to major regional transformations of vapour flow patterns (Gordon and others 2005).

An increasing frequency and severity of droughts and floods is leading to malnutrition and water-borne diseases, threatening human health and destroying livelihoods. In developing countries, an increase in droughts may lead, by 2080, to a decrease of 11 per cent in land suitable for rain-fed agriculture (FAO 2005). The likely increase of torrential rains and local flooding will affect the safety and livelihoods of mostly poor people in developing countries, as their homes and crops will be exposed to these events (WRI 2005).

Cryosphere

Continental ice sheets and mountain glaciers have continued to melt and retreat over the last 20 years (see Figure 4.6) (see Chapters 2 and 6). Losses from the ice sheets of Greenland and Antarctica have very likely contributed to global sea-level rise between 1993 and 2003 (see Table 4.2). Flow speed has increased for some Greenland and Antarctic outlet glaciers that drain ice from the interior of the ice sheets (IPCC 2007). Arctic average temperatures are rising about twice as

rapidly as temperatures in the rest of the world, attributed mainly to feedback related to shrinking ice and snow cover (ACIA 2005) (see Chapter 6). The total Arctic land ice volume, an estimated 3.1 million cubic kilometres, has declined since the 1960s, with increasing quantities of meltwater discharged to the ocean (Curry and Mauritzen 2005). The Greenland ice sheet has been melting for several decades at a rate greater than that at which new ice is being formed (see Chapter 2). The extent of ice sheet melting was a record high in 2005 (Hanna and others 2005). Sea ice cover and thickness have also declined significantly (NSIDC 2005) (see Chapter 6).

Permafrost also is thawing at an accelerating rate, with an increase in temperature of 2°C over the last few decades. The maximum area covered by seasonally frozen ground has decreased by about 7 per cent in the northern Hemisphere since 1900, with a decrease in spring of up to 15 per cent (IPCC 2007). The thawing is causing the drainage of many tundra lakes and wetlands in parts of the Arctic, and is releasing greenhouse gases – especially methane and CO_2 – to the atmosphere. The winter freezing period for Arctic rivers is becoming shorter (ACIA 2005) (see Chapters 2 and 6).

The effects of global warming on the state of the cryosphere – increasing permafrost thaw-depth, reducing sea ice cover and accelerating land ice (including mountain glacier) melting – are

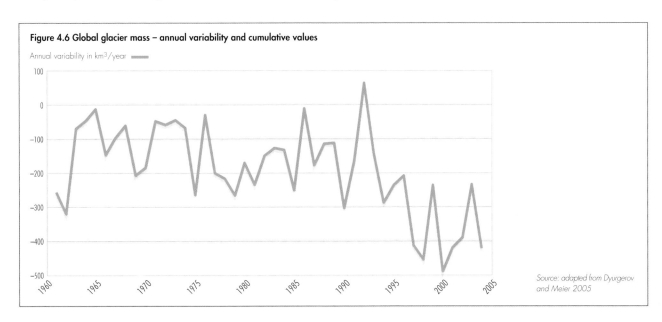

Figure 4.6 Global glacier mass – annual variability and cumulative values

Annual variability in km³/year

Source: adapted from Dyurgerov and Meier 2005

already having major impacts on human well-being (see Chapter 6). The predicted rise in sea level due to melting land ice will have huge global economic consequences. Over 60 per cent of the global population lives within 100 kilometres of the coastline (WRI 2005), and sea-level rise is already threatening the security and socio-economic development of communities and cities inhabiting low-lying coastal areas. It affects whole nations comprising small islands, including Small Island Developing States (SIDS). There is likely to be a need for major adaptation, with the relocation of millions of people in coming decades (IPCC 2001) (see Chapter 7).

While the progressive thawing of permafrost is increasing opportunities for agriculture and the commercial capture of methane gas, it is restricting road transportation, and creating instability in the built environment (ACIA 2004). It is very likely that the circulation of the North Atlantic will slow down during the 21st Century (Bryden and others 2005, IPCC 2007), with possible significant impacts on human well-being in northwestern Europe (see Chapter 6).

Rainwater and ocean acidification
Acidity in rainwater is caused by the dissolution of atmospheric CO_2, as well as by atmospheric transport and deposition of nitrogen and sulphur compounds (see Chapters 2 and 3). This is important because biological productivity is closely linked to acidity (see Chapter 3). The box on acidifying cycles in Chapter 3 describes some of the impacts of acid deposition on the world's forests and lakes.

The oceans have absorbed about half of the global CO_2 emissions to the atmosphere over the past 200 years (see Chapter 2), resulting in the increasing acidification of ocean waters (The Royal Society 2005). Acidification will continue, regardless of any immediate reduction in emissions. Additional acidification would take place if proposals to release industrially produced and compressed CO_2 at or above the deep sea floor are put into practice (IPCC 2005). To date, injection of CO_2 into seawater has been investigated only in small-scale laboratory experiments and models. Although the effects of increasing CO_2 concentration on marine

organisms would have ecosystem consequences, no controlled ecosystem experiments have been performed in the deep ocean nor any environmental thresholds identified.

The impacts of ocean acidification are speculative, but could be profound, constraining or even preventing the growth of marine animals such as corals and plankton. They could affect global food security via changes in ocean food webs, and, at the local scale, negatively affect the potential of coral reefs for dive tourism and for protecting coastlines against extreme wave events. It is presently unclear how species and ecosystems will adapt to sustained, elevated CO_2 levels (IPCC 2005). Projections give reductions in average global surface ocean pH (acidity) values of between 0.14 and 0.35 units over the 21st century, adding to the present decrease of 0.1 units since pre-industrial times (IPCC 2007).

Managing water issues related to climate change
Global-scale changes to the water environment associated with climate change include higher sea surface temperatures, disruption of global ocean currents, changes in regional and local precipitation patterns, and ocean acidification. These issues are typically addressed through global efforts, such as the UN Framework Convention on Climate Change and its Kyoto Protocol (see Chapter 2). Management at the global level involves numerous actions at regional, national and local scales. Many global conventions and treaties are implemented on this basis, with their effectiveness depending on the willingness of individual countries to contribute to their achievement. Because these changes are linked to other environmental issues (for example, land use and biodiversity), they must also be addressed by other binding or non-binding treaties and instruments (see Chapter 8).

Major responses to the drivers of climate change – primarily the increased burning of fossil fuels for energy – are analysed in Chapter 2. These responses are generally at the international level, and require concerted action by governments over the long-term, involving legal and market-driven approaches. Focus is on responses to climate change-related impacts affecting the water environment that involve regulation, adaptation and restoration (see Table 4.5 at the end of this Chapter). These actions are implemented mostly at

national or even local levels, although usually in accord with regional or international conventions. All such responses should be considered in the context of continuing climate change and its consequences, particularly the longer-term impacts of global sea-level rise on human safety, security and socio-economic development.

At the global level, measures to adapt to climate change are being addressed by the Intergovernmental Panel on Climate Change (IPCC). At regional and local levels, measures include wetland and mangrove restoration and other ecohydrological approaches, as well as carbon sequestration, flood control and coastal engineering works (see Table 4.5). Some responses, such as the restoration of coastal wetlands by the managed retreat of sea defences can serve several purposes. These include reducing the impacts of storm surges, recreating coastal and inland ecosystems, and enhancing or restoring ecosystem services, such as the provision of fish nurseries, water purification and recreational and tourism qualities, particularly for the benefit of local communities.

WATER RESOURCES AND USE

Freshwater availability and use

Available water resources continue to decline as a result of excessive withdrawal of both surface- and groundwater, as well as decreased water run-off due to reduced precipitation and increased evaporation attributed to global warming. Already, in many parts of the world, such as West Asia, the Indo-Gangetic Plain in South Asia, the North China Plain and the High Plains in North America, human water use exceeds annual average water replenishment. Use of freshwater for agriculture, industry and energy has increased markedly over the last 50 years (see Figure 4.4).

Freshwater shortage has been assessed as moderate or severe in more than half the regions studied in the Global International Waters Assessment (GIWA) assessment (UNEP-GIWA 2006a). By 2025, 1.8 billion people will be living in countries or regions with absolute water scarcity, and two-thirds of the world population could be under conditions of water stress, the threshold for meeting the water requirements for agriculture, industry, domestic purposes, energy and the environment (UN Water 2007).

Left of a breached sea-wall near Tollesbury, UK, a managed retreat site with recurring wetland; to the right a natural marsh.

Credit: Alastair Grant

An average of 110 000 km³ of rain falls on the land annually (SIWI and others 2005). About one-third of this reaches rivers, lakes and aquifers (blue water), of which only about 12 000 km³ is considered readily available for human use. The remaining two-thirds (green water) forms soil moisture or returns to the atmosphere as evaporation from wet soil and transpiration by plants (Falkenmark 2005) (see Chapter 3). Changes in land and water use are altering the balance between, and availability of, "blue" and "green" water. They are also exacerbating fragmentation of riverine ecosystems, reducing river flows and lowering groundwater levels. Increasing water loss through evaporation from reservoirs contributes to downstream flow reductions (see Figure 4.4).

Alteration of river systems, especially flow regulation by impoundment, is a global phenomenon of staggering proportions (Postel and Richter 2003). Sixty per cent of the world's 227 largest rivers are moderately to greatly fragmented by dams, diversions and canals, with a high rate of dam construction threatening the integrity of the remaining free-flowing rivers in the developing world (Nilsson and others 2005). Major changes in drainage systems will result from the engineered transfer of water between basins currently being advocated or undertaken in parts of South America, southern Africa, China and India. In southern Africa, water transfers have altered water quality, and introduced new species into the recipient basins. Excessive upstream water use or pollution can have adverse consequences for downstream water demand. In transboundary systems, such as the Nile basin, downstream water uses can threaten the stability of upstream states by constraining their development options. Some large rivers, such as the Colorado (see Box 6.32), Ganges and Nile, are so heavily

used that none of their natural run-off reaches the sea (Vörösmarty and Sahagian 2000). The boundaries of major aquifer systems often do not reflect national borders. The political changes in the former Soviet Union and the Balkans, for example, have greatly increased the number of such transboundary situations (UNESCO 2006), and emphasize the need to jointly manage water resources.

There are more than 45 000 large dams in 140 countries, about two-thirds of these in the developing world (WCD 2000), with half in China. These dams, with an estimated potential storage volume of 8 400 km³, impound about 14 per cent of global run-off (Vörösmarty and others 1997). New dam construction is limited largely to developing regions, particularly Asia. In the Yangtze River basin in China, for example, 105 large dams are planned or under construction (WWF 2007). In some developed countries, such as the United States, construction of new large dams has declined in the past 20 years. A few dams have even been decommissioned successfully to benefit humans and nature. In many reservoirs, siltation is a growing problem. Changes in land use, notably deforestation, have led to increased sediment transport through soil erosion and increased run-off. More than 100 billion tonnes of sediment are estimated to have been retained in reservoirs constructed in the past 50 years, shortening the dams' lifespans, and significantly reducing the flux of sediment to the world's coasts (Syvitski and others 2005) (see Table 4.1).

Reductions in freshwater discharge and seasonal peak flows caused by damming and withdrawal are lowering downstream agricultural yields and fish productivity, and causing the salinization of estuarine land. In Bangladesh, the livelihoods and nutrition of up to 30 million people have declined because of stream-flow modifications (UNEP-GIWA 2006a). Over the last two decades, reservoir development in tropical areas, particularly in Africa, has exacerbated water-related diseases, including malaria, yellow fever, guinea worm and schistosomiasis, for example in the Senegal River basin (Hamerlynck and others 2000). Reduced sediment discharge to coastal areas is contributing to the vulnerability of low-lying coastal communities to inundation, for example, in Bangladesh. Where reservoir lifespan is being reduced by sediment trapping (see Box 4.1), irrigation schemes and hydropower production

Box 4.1 Sediment trapping is shortening the useful lifespan of dams

In the Moulouya basin of Morocco, annual rainfall is scarce and concentrated over a few days. Construction of dams has many socio-economic benefits, boosting the economy through agricultural development, improving living standards through hydropower and controlling floodwaters. Because of high rates of natural and human-induced soil erosion, however, the reservoirs are quickly becoming silted. It is estimated the Mohammed V reservoir will be completely filled with sediment by 2030, causing an estimated loss of 70 000 ha of irrigated land and 300 megawatts of electricity. The dams have also modified the hydrological function of the Moulouya coastal wetlands, and caused biodiversity losses, salinization of surface- and groundwater, and beach erosion at the river's delta, affecting tourism.

Source: Snoussi 2004

Table 4.3 Impacts of excessive groundwater withdrawal		
Consequences of excessive withdrawal		**Factors affecting susceptibility**
Reversible interference	Pumping lifts and costs increase Borehole yield reduction Spring flow and river base flow reduction	Aquifer response characteristics Drawdown below productive horizon Aquifer storage characteristics
Reversible/irreversible	Phreatophytic vegetation stress (both natural and agricultural) Ingress of polluted water (from perched aquifer or river)	Depth to groundwater table Proximity of polluted water
Irreversible deterioration	Saline water intrusion Aquifer compaction and transmissivity reduction Land subsidence and related impacts	Proximity of saline water Aquifer compressibility Vertical compressibility of overlying and/or interbedded aquitards

Source: Foster and Chilton 2003

will be constrained over the coming decades. Decommissioning silted-up dams may restore sediment fluxes, but is likely to be difficult and costly, and alternative reservoir sites may be difficult to find.

Severe groundwater depletion, often linked with fuel subsidies, is apparent at aquifer or basin scales in all regions. Excessive groundwater withdrawal, and associated declining water levels and discharges, can have serious human and ecosystem impacts that must be weighed against anticipated socio-economic benefits. Increasing competition for groundwater also can worsen social inequity where deeper, larger-capacity boreholes lower regional water levels, increasing water costs, and eliminating access by individuals with shallower wells. This may provoke an expensive and inefficient cycle of well deepening, with the premature loss of financial investment as existing, shallower wells are abandoned. Severe, essentially irreversible effects, such as land subsidence and saline water intrusion, can also occur (see Table 4.3). In the Azraq basin in Jordan, for example, average groundwater withdrawal has risen gradually to 58 million cubic metres/year, with 35 million m³ used for agriculture and 23 million m³ for drinking water supply. This has decreased the level of the water table by up to 16 m between 1987 and 2005. By 1993, springs and pools in the Azraq Oasis had dried up completely. The reduced groundwater discharge also resulted in increased water salinity (Al Hadidi 2005).

Water quality

Changes in water quality are primarily the result of human activities on land that generate water pollutants, or that alter water availability. Increasing evidence that global climate change can change precipitation patterns, affecting human activities on land and the associated water run-off, suggests global warming also can cause or contribute to degraded water quality. The highest water quality is typically found upstream and in the open oceans, while the most degraded is found downstream and in estuarine and coastal areas. As well as absorbing vast quantities of atmospheric gases as the global climate regulator (see Chapter 2), the ocean's huge volume provides a buffer against degradation from most water pollutants. This is in contrast to inland freshwater systems and downstream estuarine and coastal systems. Point and non-point sources of pollution in drainage basins ensure a steady pollutant load into these water systems, highlighting river basin–coastal area linkages.

Human health is the most important issue related to water quality (see Table 4.1). Pollutants of primary concern include microbial contaminants and excessive nutrient loads. Groundwater in parts of Bangladesh and adjacent parts of India has a high natural arsenic content (World Bank 2005), and in many areas fluoride of geological origin produces problematic groundwater concentrations; both have major health impacts. Important point-source pollutants are microbial pathogens, nutrients, oxygen-consuming materials, heavy metals and persistent organic pollutants (POPs). Major non-point-source pollutants are suspended sediments, nutrients, pesticides and oxygen-consuming materials. Although not global-scale problems, highly saline water and radioactive materials may be pollutants in some locations.

Microbial pollution, primarily from inadequate sanitation facilities, improper wastewater disposal and animal waste, is a major cause of human illness and death. The health impacts of wastewater pollution on coastal waters have an economic cost of US$12 billion/year (Shuval 2003). In at least eight of UNEP's Regional Seas Programme regions, over 50 per cent of the wastewater discharged into freshwater and coastal areas is untreated, rising to over 80 per cent in five of the regions (UNEP-GPA 2006a). This untreated waste has major impacts on aquatic ecosystems and their biodiversity. In some developing countries, only about 10 per cent of domestic wastewater is collected for treatment and recycling, and only about 10 per cent of wastewater treatment plants operate efficiently. The number of people without, or served by inefficient, domestic wastewater treatment systems is likely to grow if investment in wastewater management is not significantly increased (WHO and UNICEF 2004). This would make it harder to achieve the MDG target on sanitation (see Figure 4.3).

An estimated 64.4 million Disability Adjusted Life Years (DALYs) are attributed to water-related pathogens (WHO 2004). The prevalence of hepatitis A (1.5 million cases), intestinal worms (133 million cases), and schistosomiasis (160 million cases) has been linked to inadequate sanitation. Swimming in wastewater-contaminated coastal waters causes more than 120 million cases of gastrointestinal disease, and 50 million cases of respiratory diseases annually. A strong increase in cholera cases, caused by ingestion of food or water containing the bacterium *Vibrio cholerae*, was reported between 1987 and 1998 (see Figure 4.7) (WHO 2000). It is estimated that in developing countries some 3 million people die of water-related diseases every year, the majority of whom are children under the age of five (DFID and others 2002). The predictions that global warming may change habitats, leading to the spread of water-related disease vectors, poses risks for human health, something that warrants increased concern.

The pH of an aquatic ecosystem, a measure of the acidity or alkalinity of water, is important because it is closely linked to biological productivity. Although the tolerance of individual species varies, water of good quality typically has a pH value between 6.5 and 8.5 in most major drainage basins. Significant improvements in pH have been made in parts of the world, likely as a result of global and

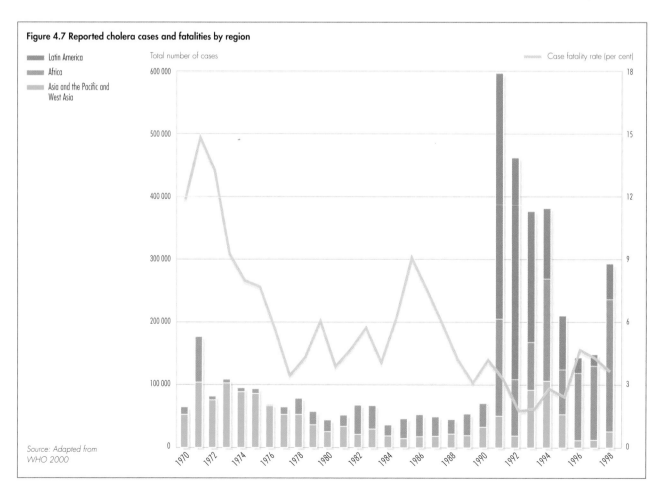

Figure 4.7 Reported cholera cases and fatalities by region

Latin America
Africa
Asia and the Pacific and West Asia

Total number of cases

Case fatality rate (per cent)

Source: Adapted from WHO 2000

regional efforts to reduce sulphur emissions (UNEP-GEMS/Water 2007).

The most ubiquitous freshwater quality problem is high concentrations of nutrients (mainly phosphorus and nitrogen) resulting in eutrophication, and significantly affecting human water use. Increasing phosphorus and nitrogen loads to surface- and groundwater come from agricultural run-off, domestic sewage, industrial effluents and atmospheric inputs (fossil fuel burning, bush fires and wind-driven dust). They affect inland and downstream (including estuarine) water systems around the world (see Chapters 3 and 5). Direct wet and dry atmospheric nutrient inputs are similarly problematic in some waterbodies, such as Lake Victoria (Lake Basin Management Initiative 2006). Projected increases in fertilizer use for food production and in wastewater effluents over the next three decades suggest there will be a 10–20 per cent global increase in river nitrogen flows to coastal ecosystems, continuing the trend of an increase of 29 per cent between 1970 and 1995

(MA 2005). Nitrogen concentrations exceeding 5 mg/l indicate pollution from such sources as human and/or animal wastes, and fertilizer run-off due to poor agricultural practices. This results in aquatic ecosystem degradation, with adverse effects on ecosystem services and human well-being (see Figure 4.8 and Table 4.4).

Nutrient pollution from municipal wastewater treatment plants, and from agricultural and urban non-point source run-off remains a major global problem, with many health implications. Harmful algal blooms, attributed partly to nutrient loads, have increased in freshwater and coastal systems over the last 20 years (see Figure 4.9 in Box 4.2). The algal toxins are concentrated by filter-feeding bivalves, fish and other marine organisms, and they can cause fish and shellfish poisoning or paralysis. Cyanobacterial toxins can also cause acute poisoning, skin irritation and gastrointestinal illnesses in humans. Global warming may be exacerbating this situation, in view of the competitive advantage of cyanobacteria over green algae at higher temperatures.

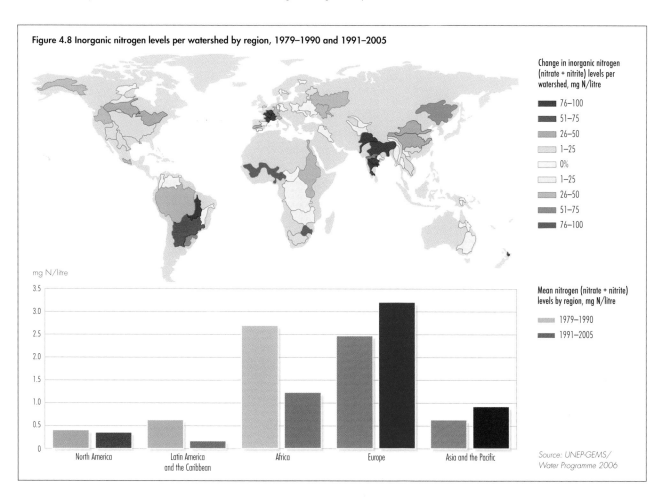

Figure 4.8 Inorganic nitrogen levels per watershed by region, 1979–1990 and 1991–2005

Change in inorganic nitrogen (nitrate + nitrite) levels per watershed, mg N/litre
- 76–100
- 51–75
- 26–50
- 1–25
- 0%
- 1–25
- 26–50
- 51–75
- 76–100

Mean nitrogen (nitrate + nitrite) levels by region, mg N/litre
- 1979–1990
- 1991–2005

mg N/litre

Source: UNEP-GEMS/Water Programme 2006

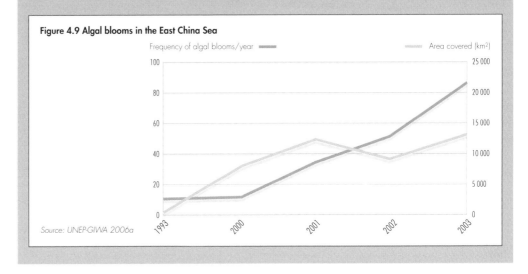
A harmful algal bloom of the dinoflagellates *Noctiluca scintillans*, known as a red tide (note the scale in relation to the boat).

Credit: J.S.P. Franks

Organic materials, from such sources as algal blooms and discharges from domestic wastewater treatment plants and food-processing operations, are decomposed by oxygen-consuming microbes in waterbodies. This pollution is typically measured as the biochemical oxygen demand (BOD). High BOD levels can cause oxygen depletion, jeopardizing fish and other aquatic species. Lake Erie's oxygen-depleted bottom zone, for example, has expanded since 1998, with negative environmental impacts. Some coastal areas also undergo oxygen depletion, including the eastern and southern coasts of North America, southern coasts of China and Japan, and large areas around Europe (WWAP 2006). Oxygen depletion in the Gulf of Mexico has created a huge "dead zone," with major negative impacts on biodiversity and fisheries (MA 2005) (see Chapter 6).

Persistent organic pollutants (POPs) are synthetic organic chemicals that have wide-ranging human and environmental impacts (see Chapters 2, 3 and 6). In the late 1970s, studies of the North American Great Lakes highlighted the existence of older, obsolete chlorinated pesticides (so-called legacy chemicals) in sediments and fish (PLUARG 1978). As regulations curtailing their use were implemented, chemical levels have declined in some water systems since the early 1980s (see Chapter 6) (see Box 6.28). Similar declines have since been observed in China and the Russian Federation (see Figure 4.10). The estimated production of hazardous organic chemical-based pollutants in the United States by industry alone is more than 36 billion kilogrammes/year, with about 90 per cent of these chemicals not being disposed of in an environmentally responsible manner (WWDR 2006).

The chemicals in pesticides can also contaminate drinking water through agricultural run-off. There is growing concern about the potential impacts on aquatic ecosystems of personal-care products and pharmaceuticals such as birth-control residues, painkillers and antibiotics. Little is known about their long-term impacts on human or ecosystem health, although some may be endocrine disruptors.

Some heavy metals in water and sediments accumulate in the tissues of humans and other organisms. Arsenic, mercury and lead in drinking water, fish and some crops consumed by humans have caused increased rates of chronic diseases. Marine monitoring conducted since the early 1990s in Europe indicates decreasing cadmium, mercury and lead concentrations in mussels and fish from both the northeast Atlantic Ocean and Mediterranean Sea. Most North Sea states achieved the 70 per cent reduction target for these metals, except for copper, and tributyltin (EEA 2003).

Although occurring in some inland locations, such as the Upper Amazon, oil pollution remains primarily a marine problem, with major impacts on seabirds and other marine life, and on aesthetic quality. With reduced oil inputs from marine transportation, and with vessel operation and design improvements, estimated oil inputs into the marine environment are declining (UNEP-GPA 2006a) (see Figure 4.11), although in the ROPME Sea Area about 270 000 tonnes of oil are still spilled annually in ballast water.

The total oil load to the ocean includes 3 per cent from accidental spills from oil platforms, and 13 per cent from oil transportation spills (National Academy of Sciences 2003).

Despite international efforts, solid waste and litter problems continue to worsen in both freshwater and marine systems, as a result of inappropriate disposal of non- or slowly degradable materials from land-based and marine sources (UNEP 2005a).

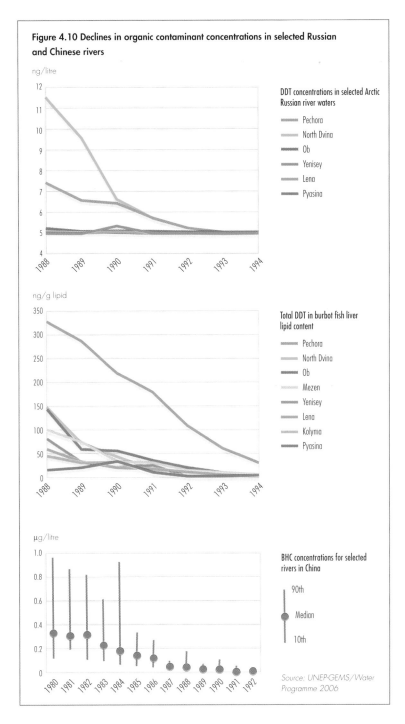

Figure 4.10 Declines in organic contaminant concentrations in selected Russian and Chinese rivers

ng/litre

DDT concentrations in selected Arctic Russian river waters
- Pechora
- North Dvina
- Ob
- Yenisey
- Lena
- Pyasina

ng/g lipid

Total DDT in burbot fish liver lipid content
- Pechora
- North Dvina
- Ob
- Mezen
- Yenisey
- Lena
- Kolyma
- Pyasina

µg/litre

BHC concentrations for selected rivers in China
- 90th
- Median
- 10th

Source: UNEP-GEMS/Water Programme 2006

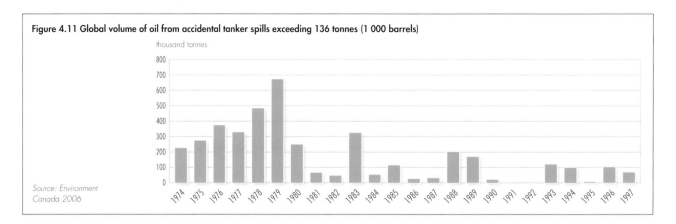

Figure 4.11 Global volume of oil from accidental tanker spills exceeding 136 tonnes (1 000 barrels)

thousand tonnes

Source: Environment
Canada 2006

Ecosystem integrity

Since 1987, many coastal and marine ecosystems and most freshwater ecosystems have continued to be heavily degraded, with many completely lost, some irreversibly (Finlayson and D'Cruz 2005, Argady and Alder 2005) (see Box 4.3). It has been projected that many coral reefs will disappear by 2040 because of rising seawater temperatures (Argady and Alder 2005). Freshwater and marine species are declining more rapidly than those of other ecosystems (see Figure 5.2d). Wetlands, as defined by the Ramsar Convention, cover 9–13 million km2 globally, but more than 50 per cent of inland waters (excluding lakes and rivers) have been lost in parts of North America, Europe, and Australia (Finlayson and D'Cruz 2005). Although data limitations preclude an accurate assessment of global wetland losses, there are many well-documented examples of dramatic degradation or loss of individual wetlands. The surface area of the Mesopotamian marshes, for example, decreased from 15 000–20 000 km2 in the 1950s to less than 400 km2 around the year 2000 because of excessive water withdrawals, damming and industrial development (UNEP 2001) but is now recovering (see Figure 4.12). In Bangladesh, more than 50 per cent of mangroves and coastal mudflats outside the protected Sunderbans have been converted or degraded.

Reclamation of inland and coastal water systems has caused the loss of many coastal and floodplain ecosystems and their services. Wetland losses have changed flow regimes, increased flooding in some places, and reduced wildlife habitat. For centuries, coastal reclamation practice has been to reclaim as much land from the sea as possible. However, a major shift in management

practice has seen the introduction of managed retreat for the marshy coastlines of Western Europe and the United States.

Although limited in area compared to marine and terrestrial ecosystems, many freshwater wetlands are relatively species-rich, supporting a disproportionately large number of species of certain faunal groups. However, populations of freshwater vertebrate species suffered an average decline of almost 50 per cent between 1987 and 2003, remarkably more dramatic than for terrestrial or marine species over the same time scale (Loh and Wackernagel 2004). Although freshwater invertebrates are less well assessed, the few available data suggest an even more dramatic decline, with possibly more than 50 per cent being threatened (Finlayson and D'Cruz 2005). The continuing loss and degradation of freshwater and coastal habitats is likely to affect aquatic biodiversity more strongly, as these habitats, compared to many terrestrial ecosystems, are disproportionately species-rich and productive, and also disproportionately imperilled.

The introduction of invasive alien species, via ship ballast water, aquaculture or other sources, has disrupted biological communities in many coastal and marine aquatic ecosystems. Many inland ecosystems have also suffered from invasive plants and animals. Some lakes, reservoirs and waterways are covered by invasive weeds, while invasive fish and invertebrates have severely affected many inland fisheries.

Declines in global marine and freshwater fisheries are dramatic examples of large-scale ecosystem degradation related to persistent overfishing, pollution, and habitat disturbance and losses. Although there are limited data, marine fish stock losses and declines in marine trophic levels suggest large areas of marine shelf areas have been degraded by trawling over the last few decades. While most deep-sea communities are likely to remain relatively pristine, seamount and cold-water coral communities in the deep sea are being severely disrupted by trawling, and urgently require protection (see Chapter 5) (see Box 5.4).

Aquatic ecosystems provide many services contributing to human well-being (see Table 4.4). Maintenance of the integrity and the restoration of these ecosystems are vital for services such as water replenishment and purification, flood and drought

Box 4.3 Physical destruction of coastal aquatic ecosystems in Meso-America

Coastal development represents one of the main threats to the Meso-American coral reefs and mangroves. Construction and the conversion of coastal habitat has destroyed sensitive wetlands (mangroves) and coastal forests, and led to an increase in sedimentation. The effects of coastal development are compounded by insufficient measures for the treatment of wastewater.

Tourism

Tourism, particularly when it is coastal- and marine-based, is the fastest growing industry in the region. The state of Quintana Roo in Mexico is experiencing significant growth in the tourism infrastructure all along the Caribbean coast to Belize. The conversion of mangrove forest into beachfront tourist resorts along the Mayan Riviera, south of Cancun, has left coastlines vulnerable. Playa del Carmen, at 14 per cent, has the fastest growth in tourism infrastructure in Mexico. Threats to the aquifers come from increasing water use, of which 99 per cent is withdrawn from groundwater,

and wastewater disposal. Much of the attraction of the Quintana Roo coast is provided by its cavern systems, and their preservation is a major challenge. This trend is echoed in Belize, where ecotourism appears to be giving way to large-scale tourism development, involving the transformation of entire cays, lagoons and mangrove forests to accommodate cruise ships, recreational facilities and other tourism demands.

Aquaculture

The rapid growth of shrimp aquaculture in Honduras has had serious impacts on the environment and local communities. The farms deprive fishers and farmers of access to the mangroves, estuaries and seasonal lagoons; they destroy the mangrove ecosystems and the habitats of fauna and flora, thus reducing the biodiversity; they alter the hydrology of the region and contribute to degraded water quality; and they contribute to the decline of fish stocks through the indiscriminate capture of fish for feed.

Sources: CNA 2005, INEGI 2006, UNEP 2005b, World Bank 2006

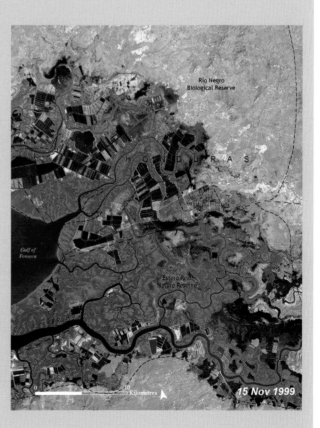

Credit: UNEP 2005b

Table 4.4 Linkages between state changes in aquatic ecosystems and environmental and human impacts						
			HUMAN WELL-BEING IMPACTS			
Aquatic ecosystems	**Pressures**	**SELECTED STATE CHANGES**	**Human health**	**Food security**	**Physical security and safety**	**Socio-economic**
Inland ecosystems						
Rivers, streams and floodplains	Flow regulation by damming and withdrawal Water loss by evaporation Eutrophication Pollution	↓ Water residence time ↑ Ecosystem fragmentation ↑ Disruption of dynamic between river and floodplain ↑ Disruption to fish migration ↑ Blue-green algal blooms	↓ Freshwater quantity[1] ↓ Water purification and quality[1] ↑ Incidences of some water-borne diseases[1]	↓ Inland and coastal fish stocks[1]	↑ Flood protection[1]	↓ Tourism[3] ↓ Small-scale fisheries[1] ↑ Poverty[1] ↓ Livelihoods[1]
Lakes and reservoirs	Infilling and drainage Eutrophication Pollution Overfishing Invasive species Global warming induced changes in physical and ecological properties	↓ Habitat ↑ Algal blooms ↑ Anaerobic conditions ↑ Alien fish species ↑ Water hyacinth	↓ Water purification and quality[1]	↓ Inland fish stocks[1]		↓ Small-scale fisheries[2] ↑ Displacement of human communities[1] ↓ Tourism[2] ↓ Livelihoods[1]
Seasonal lakes, marshes and swamps, fens and mires	Conversion through infilling and drainage Change in flow regimes Change in fire regimes Overgrazing Eutrophication Invasive species	↓ Habitat and species ↓ Flow and water quality ↑ Algal blooms ↑ Anaerobic conditions ↑ Threat to indigenous species	↓ Water replenishment[1] ↓ Water purification and quality[1]		↑ Flash flood frequency and magnitude[1] ↓ Mitigation of floodwaters[1] ↓ Mitigation of droughts[1]	↓ Flood, drought and flow-related buffering effects[1] ↓ Livelihoods[1]
Forested marshes and swamps	Conversion through tree felling, drainage and burning	Partly irreversible ecosystem loss Direct contact between wild birds and domesticated fowl	↓ Water replenishment[1] ↓ Water purification and quality[1]		↑ Flash flood frequency and magnitude[2]	↓ Flood, drought, and flow-related buffering effects[2] ↓ Livelihoods[2]
Alpine and tundra wetlands	Climate change Habitat fragmentation	Expansion of scrubland and forest Shrinking of surface waters in tundra lakes	↓ Water purification and quality[1]	↓ Reindeer herding[2] ↓ Inland fish stocks[2]	↑ Flash flood frequency and magnitude[2]	↓ Livelihoods[2]
Peatlands	Drainage Withdrawal	↓ Habitat and species ↑ Soil erosion ↑ Loss of carbon storage	↓ Water replenishment[1] ↓ Water purification and quality[1]		↑ Flash flood frequency and magnitude[2]	

Aquatic ecosystems	Pressures	SELECTED STATE CHANGES	HUMAN WELL-BEING IMPACTS			
			Human health	Food security	Physical security and safety	Socio-economic
Inland ecosystems						
Oases	Water withdrawal Pollution Eutrophication	⇧ Degradation of water resources	⇩ Water availability and quality[1]		⇧ Conflicts and instability[1]	⇧ Drought events[1] ⇩ Livelihoods[1]
Aquifers	Water withdrawal Pollution		⇩ Water availability and quality[1]	⇩ Reduced agriculture[1]	⇧ Conflicts and instability[1]	⇩ Livelihoods[1]
Coastal and marine ecosystems						
Mangrove forests and salt marshes	Conversion to other uses Freshwater scarcity Overexploitation of timber Storm surges and tsunamis Reclamation	⇩ Mangroves ⇩ Tree density, biomass, productivity and species diversity	⇧ Risk of malaria due to standing water[1]	⇩ Coastal fish and shellfish stocks[1]	⇩ Buffer capacity along coast[2]	⇩ Timber products[1] ⇩ Small-scale fisheries[1] ⇧ Displacement of human communities[2] ⇩ Tourism[3] ⇩ Livelihoods[2]
Coral reefs	Eutrophication Sedimentation Overfishing Destructive fishing High sea surface temperature Ocean acidification Storm surges	⇧ Reef coral bleaching and mortality ⇧ Associated fisheries loss		⇩ Coastal fish and shellfish stocks[1]	⇩ Buffer capacity along coast[2]	⇩ Tourism[1] ⇩ Small-scale fisheries[1] ⇧ Poverty[1] ⇩ Livelihoods[1]
Estuaries and intertidal mudflats	Reclamation Eutrophication Pollution Overharvesting Dredging	⇔ Intertidal sediments and nutrient exchange ⇧ Oxygen depletion ⇩ Shellfish	⇩ Coastal water quality and purification[1] ⇧ Sedimentation[1]	⇩ Coastal fish and shellfish stocks[1]	⇩ Buffering capacity along coasts[2]	⇩ Tourism[3] ⇩ Small-scale fisheries[1] ⇧ Poverty[1] ⇩ Livelihoods[1]
Seagrass and algal beds	Coastal development Pollution Eutrophication Siltation Destructive fishing practices Dredging Conversion for algal and other mariculture	⇩ Habitat		⇩ Coastal fish stocks[1]	⇩ Buffer capacity along coast[2]	⇩ Livelihoods[1]
Soft-bottom communities	Trawling Pollution Persistent organics and heavy metals Mineral extraction	⇩ Habitat	⇩ Coastal water quality[2]	⇩ Fish stocks and other livelihoods[1]		⇩ Shellfish production[1]
Subtidal hard-bottom communities	Trawling Pollution (as for soft-bottom communities) Mineral extraction	Seamount and cold-water coral communities severely disrupted		⇩ Fish stocks[1]		

			HUMAN WELL-BEING IMPACTS			
Aquatic ecosystems	Pressures	SELECTED STATE CHANGES	Human health	Food security	Physical security and safety	Socio-economic
Coastal and marine ecosystems						
Pelagic ecosystems	Overfishing Pollution Sea surface temperature change Ocean acidification Invasive species	Disturbance of trophic level balance, changes in plankton communities	⇩ Coastal water quality[1]	⇩ Fish stocks[1]		⇩ Livelihoods[1]

Table 4.4 Linkages between state changes in aquatic ecosystems and environmental and human impacts *continued*

Arrows show trends of state and impact changes

⇧ increase   ⇩ decrease   ⇔ no statistically proven change

[1] well established   [2] established but incomplete   [3] speculative

MDG Goal 1, Target 1: Halve, between 1990 and 2015, the proportion of people whose income is less than US$1 a day.

Target 2: Halve, between 1990 and 2015, the proportion of people who suffer from hunger.

MDG Goal 6, Target 8: Halt by 2015 and begin to reverse the incidence of malaria and other major diseases.

MDG Goal 7, Target 9: Integrate the principles of sustainable development into country policies and programmes, and reverse the loss of environmental resources.

MDG Goal 7, Target 10: Halve, by 2015, the proportion of people without sustainable access to safe drinking water and basic sanitation.

mitigation, and food production. Fish production is among the most prominent of the services from inland and marine aquatic ecosystems, with an estimated 250 million people dependent upon small-scale fisheries for food and income (WRI 2005). Change in the flow regime of the Lower Mekong basin, due to such factors as the construction of dams for hydropower, the diversion of river water for irrigation, industrial development and human settlements, affects the well-being of 40 million people who depend on seasonal flooding for fish breeding (UNEP-GIWA 2006b). Loss and degradation of mangroves, coral reefs and intertidal mudflats reduces their value for human well-being, mainly affecting the poor, who are reliant on their ecosystem services. Coastal wetlands on the Yellow Sea have suffered losses of more than 50 per cent over the last 20 years (Barter 2002).

The primary functions of aquatic ecosystems are commonly compromised by the development of one single service, as for example the protective function of mangrove forest that is lost due to aquaculture development. The protection of coastal communities from marine flooding has become less effective with wetland loss, mangrove clearance and the destruction of coral reefs. Reefs are losing their value for human well-being in terms of

diminished food security and employment, coastal protection, and reduced potential for tourism and pharmaceutical research and production (see Chapter 5) (see Box 5.5). The bleaching of corals due to climate change may result in global economic losses of up to US$104.8 billion over the next 50 years (IUCN 2006).

In cases such as the impacts of dam building on fish migration and breeding, conflicting water interests are often evident, even if not transparent. Many become apparent only after catastrophic events, when the wider functions and values of these ecosystems become more obvious. Prominent examples include the devastating hurricane-induced flooding of New Orleans in August 2005 (see Box 4.4), and the tsunami-induced inundation in southern Asia in December 2004. In both cases, the impacts were worsened because human alterations had reduced coastal wetland functions. Numerous other examples, from Asia to Europe, demonstrate increased risks of flash floods caused by land-use changes, including the infilling and loss of wetlands. Changes in water flows from increased urban drainage can also increase the severity of such floods. An increase in flooding events in London has been linked to the paving of front gardens for car parking.

Managing water resources and ecosystems

Human water use issues relate to the quantity and quality of the available water resources, as well as to the aquatic ecosystems that provide life-supporting ecosystem services to humanity. Good governance for addressing these issues in a context of matching water demands to the supply of water resources and related ecosystem services, requires attention to three major groups of approaches:

- suitable laws and policies and effective institutional structures;
- effective market mechanisms and technologies; and
- adaptation and restoration (see Table 4.5 at the end of this Chapter).

A variety of regional level treaties strengthen cooperation among states on such water resource issues. Examples are the 1992 OSPAR Convention, the 1992 Helsinki Convention for the Baltic Sea and its additional protocols, the 1986 Cartagena Convention for the Wider Caribbean Region and its additional protocols, and the 1995 African Eurasian Waterbird Agreement (AEWA). The

European Union has made water protection a priority of its member states (see Box 4.5). These examples highlight the importance of regional framework agreements in strengthening national and local laws and policies (the enabling environment) and institutional structures, such as cooperation among states. Another example is the UN Watercourses Convention, signed by 16 parties to date. A recent action plan by the UN Secretary-General's advisory board calls upon national governments to ratify the 1997 UN Watercourses Convention as a means of applying IWRM principles to international basins (UN Secretary-General's Advisory Board on Water and Sanitation 2006). However, there still are many regions that urgently require binding agreements and institutions, and need to strengthen existing frameworks, including those relating to transboundary aquifers and regional seas.

Collaboration among institutions with complementary environmental and economic development functions is equally important. Institutional integration for managing extreme hydrological events, for example, is found with the EU (2006) and UN ECE (2000) approaches to flood management, and with the 1998 Rhine and 2004 Danube basin action plans. All emphasize cooperation among various organizations, institutions, users and uses of the river basin, including (APFM 2006):

- clearly-established roles and responsibilities;
- availability and accessibility of basic data and information for informed decision making; and
- an enabling environment for all stakeholders to participate in collective decision making.

In addition, public-private partnerships can be employed in water supply and demand management. This could be done by increasing supply (through dams, for example), by reducing demand (through technological improvements and increased efficiency in the delivery of water services), or by appropriate pricing of water resources and metering of water use as a means of recovering the costs of providing water supplies. Other market-based instruments may include (tradeable) quotas, fees, permits, subsidies and taxation.

Market-based instruments can operate by valuing public demand for a good or service, then paying suppliers directly for changes in management practices or land use. These instruments may have positive or negative impacts. "Watershed markets" is a positive example involving payments from downstream users to upstream landowners to maintain water quality or quantity (see Box 4.6). But agricultural subsidies, for instance to increase food production, may lead to inefficient water uses, and pollution and habitat degradation.

Since the Brundtland Commission report, tradeable quota systems and permits have emerged as effective tools for encouraging users to develop and use more efficient technologies and techniques to reduce water demand and pollutant emissions, and achieving the sustainable use of common resources and ecosystems. Some examples are:

- the Total Maximum Daily Load (TMDL) programme in the United States;
- reducing fishing pressure on inland and marine fisheries (Aranson 2002);
- managing groundwater salinity (Murray-Darling river basin in Australia); and
- optimizing groundwater withdrawal.

To be effective, such approaches require monitoring use of the resource. If monitoring results show negative trends, quotas or permits may have to be revoked. The Dutch government, for example, put a complete ban on cockle fishing in 2005, after it was demonstrated that cockle dredging caused degradation of mudflats and other adverse effects on the coastal ecosystems and their species in the Dutch Wadden Sea (Piersma and others 2001).

Quota systems may be particularly useful in managing water demand in arid and semi-arid areas with limited supplies, but they can be problematic where resources are undervalued, leading to overuse and degradation. Quota mechanisms are best suited to countries with high levels of institutional development. They can prove problematic for economically stressed states and communities that lack the financial base to invest in compliance and enforcement.

Technological responses to water scarcity (see Table 4.5) include reducing water consumption with such approaches as more efficient irrigation and water distribution techniques, wastewater recycling and reuse. Water availability can be increased through artificial groundwater recharge, damming, rainwater harvesting and desalination. Rainwater harvesting (see Chapter 3) has been used successfully in China (20 per cent of the land relies on it), as well as in Chile and India (to recharge underground aquifers) (WWAP 2006). Japan and Korea have systems for harvested rainwater use in disaster situations. Managed aquifer recovery (MAR) and artificial storage and recovery (ASR) have also been used with some success. Another low-tech solution for reducing water demand is the use of reclaimed water instead of potable water for irrigation, environmental restoration, toilet flushing and industry. This approach has gained significant public acceptance, having been used successfully in Israel, Australia and Tunisia (WWAP 2006). Environmental problems arising from large-scale damming are being addressed by a number of approaches. They include the increasing use of smaller dams, fish ladders and managed environmental flows that keep freshwater, estuarine and coastal ecosystems healthy and productive, maintaining ecosystem services (IWMI 2005).

Technology has long been an important tool in preventing and remediating water quality

Box 4.6 Watershed markets

Watershed markets are a mechanism, typically involving payments for ecosystem services, such as water quality. This mechanism can take the form of upstream conservation and restoration actions. As an example, farmers' associations in the Valle del Cauca in Colombia pay upstream landowners to implement conservation practices, revegetate land and protect critical source areas, all of which reduce the downstream sediment loads. About 97 000 families participate in this effort, the funds being collected through user charges based on water use. Similar water user associations have been formed across Colombia. Sixty one examples of watershed protection markets in 22 countries were identified, many focusing on water quality improvement.

Source: Landell-Mills and Porras 2002

degradation (see Table 4.5), particularly to facilitate industrial and agricultural development. Its use has been recognized in international agreements, which, over the last 20 years, have often evolved from reactive responses to proactive approaches. There is also increasing use of standards such as Best Available Technology, Best Environmental Practice and Best Environmental Management Practice. These approaches are intended to stimulate improved technology and practices, rather than to set inflexible standards. Technological responses are best known in water and wastewater treatment and re-use applications (mainly point-source controls). They range from source control of contaminants (composting toilets, clean technology, recycling municipal and industrial wastes) to centralized, high-tech wastewater treatment plants, utilizing energy and chemicals to clean water prior to its discharge to natural watercourses (Gujer 2002). Access to wastewater treatment and disinfection technology (using low- and high-tech methods) is largely responsible for the reduction in water-borne diseases since 1987. Other treatment technologies remove hazardous materials before discharge. Non-point source pollution is less readily addressed by high-tech approaches, and its effective control requires improved education and public awareness.

Justifying technology-based interventions in the decision making process should include consideration of the long-term values of the aquatic resources being managed. Technological approaches to pollution reduction may be ineffective over the long-term unless the underlying root causes of problems are addressed.

The economic valuation of ecosystem services provided by the water environment (such as water filtration, nutrient cycling, flood control and habitat for biodiversity) can provide a powerful tool for mainstreaming aquatic ecosystem integrity into development planning and decision making.

Ecological restoration efforts have also become important management responses since the Brundtland Commission report, especially for disturbances to the hydrological regime, water

Fields under plastic with drip irrigation in Israel's drylands.

Credit: Fred Bruemmer/
Still Pictures

Box 4.7 Restoration of ecosystems

Mauritania and Senegal

The Diawling Delta has been virtually destroyed by a combination of continuing low rainfall, and construction of a dam in 1985, leading to loss of wetland-dependent livelihoods and the mass migration of its inhabitants. Beginning in 1991, IUCN and local communities worked together in restoration efforts covering 50 000 hectares, with the primary goal being to bring back flooding and saltwater inflows, restoring a diverse delta ecosystem. The positive results of this effort include increasing fish catches from less than 1 000 kg in 1992 to 113 000 kg in 1998. Bird counts also rose from a mere 2 000 in 1992 to more than 35 000 in 1998. The total value added to the region's economy from this restoration effort is approximately US$1 million/year.

North America

More than half the major North American rivers have been dammed, diverted or otherwise controlled. While the structures provide hydropower, control floods, supply irrigation and increase navigation,

they have changed the hydrological regime, damaging aquatic life, recreational opportunities and livelihoods of some indigenous peoples. The ecological and economic costs of dams are being increasingly evaluated in comparison to their anticipated benefits, and some have been removed. At least 465 dams have been decommissioned in the United States, with about 100 more planned for removal. There has also been a trend towards river restoration in the United States since 1990, with most projects directed to enhancing water quality, managing riparian zones, improving in-stream habitat, allowing fish passage and stabilizing stream banks. However, of over 37 000 restoration projects, only 10 per cent indicated that any assessment or monitoring took place as part of the projects, and many of these activities were not designed to assess the outcome of the restoration efforts. Although large-scale dam building still takes place in Canada, there has been a recent trend towards small-scale hydro projects, with more than 300 plants with a capacity of 15 megawatts or less in operation, and many others under consideration.

Source: Bernhardt and others 2005, Hamerlynck and Duvail 2003, Hydropower Reform Coalition n.d., Prowse and others 2004

quality and ecosystem integrity. Efforts are usually directed to restoring degraded ecosystems to enhance the services they provide. Examples include ecological engineering, controlling invasive species, reintroducing desired species, restoring hydrological flow patterns, canalization, damming and reversing the impacts of drainage (see Table 4.5). Restoration of riverine ecosystem integrity has also been achieved in Europe and in the United States by the removal of existing dams that are no longer economically or ecologically justifiable (see Box 4.7).

Although global statistics on riparian, wetland and lake restoration are difficult to obtain, the US National River Restoration Science Synthesis database identifies over 37 000 river and stream restoration projects. It shows the number of projects increased exponentially between 1995 and 2005, and most were local initiatives not recorded in national databases. The primary listed river and stream restoration goals are: improved water quality, management of riparian zones, improved in-stream habitats, fish passage and bank stabilization (Bernhardt and others 2005). Estimated costs of these projects between 1990 and

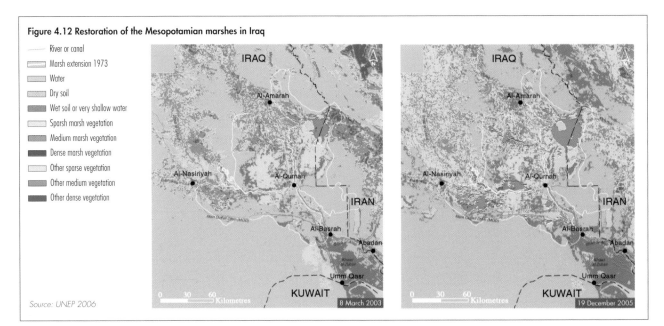

Figure 4.12 Restoration of the Mesopotamian marshes in Iraq

- River or canal
- Marsh extension 1973
- Water
- Dry soil
- Wet soil or very shallow water
- Sparsh marsh vegetation
- Medium marsh vegetation
- Dense marsh vegetation
- Other sparse vegetation
- Other medium vegetation
- Other dense vegetation

Source: UNEP 2006

2003 were at least US$14 billion. Although global estimates of restoration efforts are not readily available, several large projects have been undertaken since 1987 in Europe, Africa and Asia. These involve the Danube River delta in Romania, Aral Sea in Central Asia and, most recently, the Mesopotamian marshes in Iraq (Richardson and others 2005) (see Figure 4.12). In the last case, more than 20 per cent of the original marshland area was re-flooded between May 2003 and March 2004, with the marshlands exhibiting a 49 per cent extension of wetland vegetation and water surface area in 2006, compared to that observed in the mid-1970s. Another example is the Waza Lagone floodplain in Cameroon, where restoration measures have produced an annual benefit of approximately US$3.1 million in fish catches and productivity, availability of surface freshwater, flood farming, wildlife and a range of plant resources (IUCN 2004). But restoration is more costly than prevention, and should be a response of last resort (see Chapter 5).

FISH STOCKS

Marine and inland fish stocks show evidence of declines from a combination of unsustainable fishing pressures, habitat degradation and global climate change. Such declines are major factors in terms of biodiversity loss. They also have serious implications for human well-being. Fish provide more than 2.6 billion people with at least 20 per cent of their average per capita animal protein intake. Fish account for 20 per cent of animal-derived protein in Low-Income Food Deficit (LIFD) countries, compared to 13 per cent in industrialized countries, with many

countries where overfishing is a concern also being LIFD countries (FAO 2006b). While fish consumption increased in some regions, such as southeast Asia and Western Europe, and in the United States, it declined in other regions, including sub-Saharan Africa and Eastern Europe (Delgado and others 2003). According to FAO projections, a global shortage of fish supply is expected. Although its severity will differ among countries, the forecast is for an average increase in fish prices, in real terms, of 3 per cent by 2010 and 3.2 per cent by 2015 (FAO 2006a).

Marine fisheries

The mid-20th century saw the rapid expansion of fishing fleets throughout the world, and an increase in the volume of fish landed. These trends continued until the 1980s, when global marine landings reached slightly over 80 million tonnes/year, following which they either stagnated (FAO 2002) or began to slowly decline (Watson and Pauly 2001). Aquaculture accounts for the further increase in seafood production. Output (excluding aquatic plants) grew at a rate of 9.1 per cent/year between 1987 and 2004, reaching 45 million tonnes in 2004 (FAO 2006a). However, this growth has not improved food security in places where aquaculture products are primarily for export (Africa, Latin America).

Data on fish stocks (in terms of volume) exploited for at least 50 years within a single FAO area highlight an increase in the number of stocks either overexploited or that have crashed over the last few years (see Figure 4.13). Based on refined definitions,

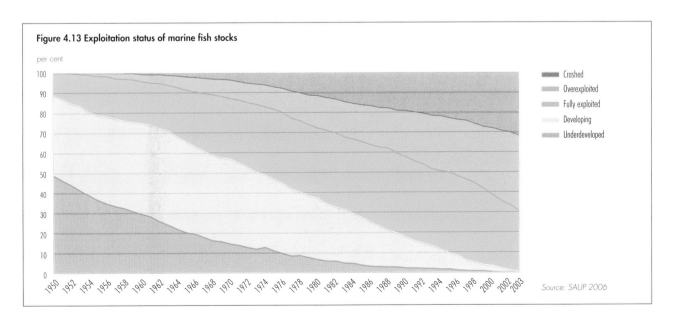

Figure 4.13 Exploitation status of marine fish stocks

per cent

Crashed
Overexploited
Fully exploited
Developing
Underdeveloped

Source: SAUP 2006

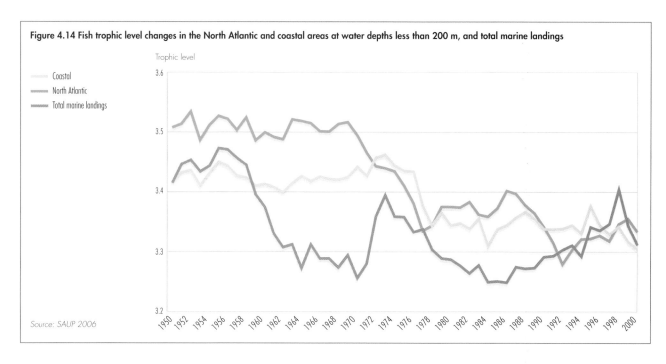

Figure 4.14 Fish trophic level changes in the North Atlantic and coastal areas at water depths less than 200 m, and total marine landings

Trophic level

Coastal
North Atlantic
Total marine landings

Source: SAUP 2006

more than 1 400 stocks were fished. Of the 70 stocks fished in 1955, at most only one per cent had crashed, compared to nearly 20 per cent of at least 1 400 stocks fished in 2000 (240 stocks crashed). Many areas have passed their peak fish production, and are not returning to the maximum catch levels seen in the 1970s and 1980s. Another important trend is the declining trophic levels of fish captured for human consumption (see Figure 4.14), indicating a decline in top predator fish catches (marlin, tuna) and groupers (Myers and Worm 2003). These stocks are being replaced by generally less desirable, less valuable fish (mackerel and hake), higher-valued invertebrates (shrimp and squid) or higher-valued aquaculture products (salmon, tuna and invertebrates).

More recently, some deep-sea fish stocks, such as the Patagonian toothfish, deepwater sharks, roundnose grenadier and orange roughy, have been severely overfished. Orange roughy stocks off New Zealand, for example, were fished to 17 per cent of their original spawning biomass within eight years (Clarke 2001), with recovery taking much longer. Deep-sea species possess biological characteristics (long lifespan, late maturity and slow growth) that make them highly vulnerable to intensive fishing pressure (see Chapter 5) (see Box 5.1).

Exploitation of West Africa's fish resources by EU, Russian and Asian fleets has increased sixfold between the 1960s and 1990s. Much of the catch is exported

or shipped directly to Europe, and compensation for access is often low compared to the value of the landed fish. Such agreements adversely affect fish stocks, reducing artisanal catches, affecting food security and the well-being of coastal West African communities (Alder and Sumaila 2004). The overexploitation of fish is forcing artisanal fishers from coastal West Africa to migrate to some of the regions that are exploiting their resources. Senegalese fishers emigrating to Spain claim the reason for leaving their homes is the lack of their traditional fisheries livelihood. Based on FAO profiles, countries in Africa with high per capita fish consumption, including Ghana, Nigeria, Angola and Benin, are now importing large quantities of fish to meet domestic demands.

A major issue is lost opportunities in jobs and hard currency revenues (Kaczynski and Fluharty 2002). After processing in Europe, the end value of seafood products from these resources is estimated at about US$110.5 million, illustrating a huge disparity in value of the resources taken by EU companies, and the licence fee paid to the countries, which is only 7.5 per cent of the value of the processed products (Kaczynski and Fluharty 2002). Fisheries sector employment has also decreased. In Mauritania, the number of people employed in traditional octopus fishing decreased from nearly 5 000 in 1996 to about 1 800 in 2001 because of the operation of foreign vessels (CNROP 2002). In 2002, fisheries provided direct employment to about 38 million people, especially in developing

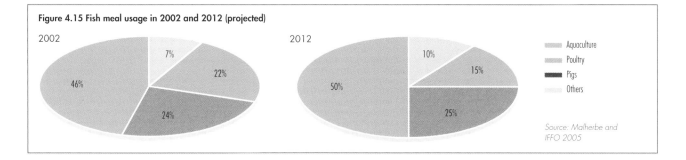

Figure 4.15 Fish meal usage in 2002 and 2012 (projected)

2002

46%
7%
22%
24%

2012

50%
10%
15%
25%

Aquaculture
Poultry
Pigs
Others

Source: Malherbe and IFFO 2005

regions such as Asia (87 per cent of world total) and Africa (7 per cent of world total) (FAO 2006a). In developing countries, however, fisheries employment has decreased. In many industrialized countries, notably Japan and European countries, employment in fishing and associated land-based sectors has been declining for several years, in part because of lower catches (Turner and others 2004).

Aquaculture and fish meal

While output from capture fisheries grew at an annual average rate of 0.76 per cent (total fish captures during 1987–2004, including freshwater), output from aquaculture (excluding aquatic plants) grew at a rate of 9.1 per cent, reaching 45 million tonnes in 2004 (FAO 2006c). Aquaculture produced 71 per cent of the total growth in food fish production by weight during 1985–1997. Although the catch is stable, the use and/or demand for wild-caught fish as feed in aquaculture is changing, being more than 46 per cent of fish meal in 2002 (Malherbe 2005), and over 70 per cent of fish oil used in aquaculture. About two-thirds of the world's fish meal is derived from fisheries devoted entirely to its production (New and Wijkstrom 2002).

Growth in aquaculture will help compensate for some shortfall in wild-caught fish, although much of the aquaculture increase has been in high-value species that meets the demands of affluent societies, and the use of fish meal from wild-caught fish for aquaculture is predicted to increase at the expense of fish meal for poultry feed (see Figure 4.15). Aquaculture growth in Africa and Latin America (for example, Chile) (Kurien 2005) is primarily for export, doing little to improve food security in these regions. The trophic level of species used for fish meal also is increasing (see Figure 4.16), implying that some fish species previously destined for human consumption are being diverted to fish meal. Therefore, food production and food security in other countries could be affected.

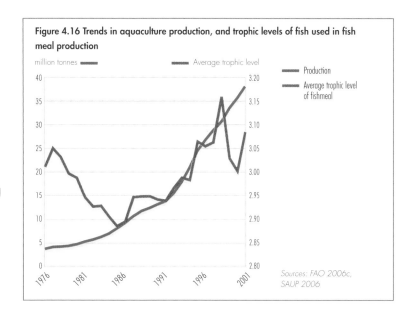

Figure 4.16 Trends in aquaculture production, and trophic levels of fish used in fish meal production

million tonnes — Average trophic level — Production — Average trophic level of fishmeal

Sources: FAO 2006c, SAUP 2006

Inland water fisheries

In 2003, the estimated total catch from inland waters (excluding aquaculture) was 9 million tonnes (FAO 2006a). Most inland capture fisheries based on wild stocks are overfished or are being fished at their biological limits (Allan and others 2005). For instance, in Lake Victoria, the Nile perch fishery decreased from a record catch of 371 526 tonnes in 1990 to 241 130 tonnes in 2002. Sturgeon catches in the countries surrounding the Caspian Sea have also decreased from about 20 000 tonnes in 1988 to less than 1 400 tonnes in 2002. In the Mekong River, there is evidence that stocks are being overfished and threatened by damming, navigation projects and habitat destruction. Several species are now endangered, with at least one, the Mekong giant catfish, close to extinction (FAO 2006a).

Inland fishes have been characterized as the most threatened group of vertebrates used by humans (Bruton 1995). Allan and others (2005) suggest that the collapse of particular inland fish stocks, even as overall fish production rises, is a biodiversity crisis more than a fisheries crisis (see Chapter 5). Increasing catches have been accompanied by changes in species composition, as catches of large and late-maturing species have declined (FAO 2006a). According to the IUCN Red List, most of the world's largest freshwater fish are at risk, and in a number of cases, overfishing has been a contributing factor. Recovery of fish stocks from historical overfishing is hindered or even impossible because of a host of current pressures. Now living in altered conditions, these native stocks are more vulnerable to disturbances, such as species invasions and diseases. Some inland fisheries have been enhanced through stocking programmes, the introduction of alien species, habitat engineering and habitat improvement.

At the global scale, inland fisheries represent an important source of nutrition. In the Lower Mekong River basin, for example, 40 million fishers and fish farmers are dependent on such fisheries for their livelihood (see Box 4.8).

Managing the world's fish stocks

Fisheries management involves ecosystem maintenance and efforts to reduce overfishing. Since the Brundtland Commission report, efforts in improved fisheries management have focused on three main themes: governance, economic incentives and property rights. Global responses include reducing fishing efforts, implementing ecosystem-based management (ESBM) approaches, property rights, economic and market incentives, marine protected areas (MPAs), and enforcement of fishing regulations (see Table 4.5). International governance initiatives, including establishment of conventions and associated regional fish management bodies (RFMOs), have facilitated

negotiations among countries exerting pressure on fish stocks. Their effectiveness in addressing declining stocks has been highly variable, depending on the stock and location. In northern Europe, where members of the Northeast Atlantic Fisheries Council reached consensus on reducing fishing efforts for species such as herring, the rebuilding of sustainable stocks has been effective. Where no agreement has been reached (such as for blue whiting), stocks are at risk of collapse.

The FAO's 1988 International Plan of Action to tackle seabird by-catch has been effective in reducing seabird mortality associated with long fishing lines used to capture tuna. Other international governance initiatives (such as managing tuna in the Atlantic) have been less successful, with many stocks in danger of crashing. Well-financed RFMOs, mostly in developed countries, are generally more effective than those that are less well financed, mostly in developing countries.

Further action is needed to induce governments to increase their political commitment to reduce fishing efforts globally, and to provide funds for RFMOs to develop and implement new approaches, such as ESBM and benefit-sharing models. RFMOs providing services to developing nations must receive increasing levels of catalytic funding assistance. Funding to the fisheries sector has declined since the 1990s, with far less support for improving fisheries management, compared to capital, infrastructure and technical assistance transfers.

At the national level, many countries have revised or rewritten their fisheries legislation and policies to reflect current trends, including multi-species fish management, ESBM, greater stakeholder participation in decision making and property rights. The FAO Code of Conduct for Responsible Fisheries Management provides ample guidance for incorporating these measures into legislation and policy. The Faroe Islands, for example, highly dependant on their marine fishery resources, have embraced ESBM (UNEP-GIWA 2006a). However, many developing and developed countries are still struggling with methods for implementing ESBM for both marine and freshwater fisheries. Further development and testing of models for implementing ESBM are needed.

National and state fisheries management agencies also are implementing programmes to rebuild declining or crashed stocks through fishing effort reductions, including closures of fishing grounds and effective enforcement of regulations (such as with the Namibian hake fisheries), as well as habitat protection by MPAs. Habitat restoration, such as mangrove rehabilitation in tsunami-affected areas, and enhancement, using, for example, fish aggregation devices (FADs), is also underway in some countries. While habitat restoration can be effective in providing fish habitat, it requires significant financial and human resources. Thailand, for example, has major efforts underway with public and industry finance and support. Habitat enhancement, using structures such as artificial reefs and FADs, must be undertaken with caution. In the tropical Pacific (as in the Philippines and Indonesia) FADs used to improve pelagic catches also capture large numbers of juvenile tuna, highlighting the need to carefully consider the impacts of proposed responses (Bromhead and others 2003).

Over the last 20 years, the number and sizes of MPAs have been increasing, contributing to effective fisheries management by protecting existing stocks or rebuilding depleted stocks. In the Philippines, many MPAs have been effective in rebuilding stocks, but further research is needed to assess their overall contribution to fisheries management. Despite calls under the Convention on Biological Diversity and the World Summit on Sustainable Development for more and larger MPAs, none of the targets will be met within the deadlines, given current trends. Other management responses include increased enforcement of fishing regulations through the use of technology, especially vessel monitoring systems using satellite technology. Despite the training and costs involved, this approach is effective in covering large areas of the ocean under all weather conditions, and helps in the effective and efficient deployment of enforcement officers.

The removal of market-distorting subsidies, as discussed at World Trade Organization negotiations, is being promoted to address concerns of overfishing. The EU Common Fisheries Policy has provided subsidies, resulting in increased fishing effort and distorted competition. Progress in removing subsidies has been slow, with many developing countries requesting subsidies to better manage their fisheries. There also is considerable debate among governments regarding what constitutes "good" versus "bad" subsidies. Certification schemes, such as the one used by the Marine Stewardship Council (MSC), are influencing wholesale and consumer purchases. Certification of farmed fish is an emerging issue but, since fish used for feed for many farmed species are not certified, it will be difficult for these fisheries to meet MSC criteria.

Some countries have been successful in reducing fishing efforts through a range of schemes, including buying out licenses, transferring property rights, and using alternative income-generating options to compensate fishers leaving the industry. But, buyouts are expensive, and must be carefully crafted to keep fishing effort from regrowing, or shifting to other sectors within the industry. Another response, considered effective in New Zealand, but less so in Chile, where small-scale fishers have been marginalized, is the transfer of property rights to fishers in various forms, such as individual transferable quotas (as discussed under Managing water resources and ecosystems).

CHALLENGES AND OPPORTUNITIES

As the Earth's primary integrating medium, water has a wide potential to reduce poverty, increase food security, improve human health, contribute to sustainable energy sources, and strengthen ecosystem integrity and sustainability. These water-related goods-and-services represent significant opportunities for society and governments to jointly achieve the goals of sustainable development, as recognized in the Millennium Declaration and at the World Summit on Sustainable Development, in the context of the MDGs. Table 4.5, at the end of this chapter, summarizes the relative effectiveness of existing responses.

Water for poverty and hunger eradication

There is compelling evidence that a substantial increase in global food production is needed to feed growing populations, and to reduce or eliminate situations where people have insufficient food for their daily needs. This increase in production will require more water (see Figure 4.4). On a global scale, the agricultural sector uses the vast majority of freshwater resources, and so is a logical target for economizing water use and developing methodologies for growing more food with less water (more crop per drop). Because agriculture and healthy ecosystems can be compatible goals, the major challenge is to improve irrigation for food production by increasing water and land productivity, supporting ecosystem services and building resilience, while mitigating environmental damage, especially within the context of ecosystem-based IWRM approaches (see Box 4.9).

Since groundwater levels are falling, and aquifer water stores are shrinking in many highly-populated countries, much of the additional water required for agricultural production must come from dammed

As promulgated by the Global Water Partnership (GWP) in 2000, IWRM is based on three pillars: the enabling environment, institutional roles and management instruments. In 2002, the Johannesburg Plan of Implementation (adopted at the World Summit on Sustainable Development) recommended that all countries "develop integrated water resource management and water efficiency plans by 2005." This was to include identifying actions needed to reform policies, legislation and financing networks, institutional roles and functions, and enhancing relevant management instruments to address water resource issues. The GWP (2006) subsequently surveyed 95, primarily developing, countries regarding the status of IWRM policies, laws, plans and strategies within their water resource management efforts in response to the WSSD mandate. Although the concept of an ecosystem-based approach for addressing water resources management and use issues is, like IWRM, a recent introduction to the international water arena, the survey revealed that 21 per cent of the surveyed countries had plans or strategies in place or well underway, and a further 53 per cent had initiated a process for formulating IWRM strategies. For example, South Africa has developed legislation translating IWRM into law, including provisions for its implementation. Burkina Faso defines IWRM within its national water policy. It is supporting enhanced IWRM awareness among its population, and the creation of local water committees including the private sector.

The IWRM approach embraces variants such as Integrated River Basin Management (IRBM), Integrated Lake Basin Management (ILBM) and Integrated Coastal Management (ICM), all of which represent a fundamental change from single issue, command-and-control regulatory approaches for managing the water environment. A global-scale, GEF-funded ILBM project highlighting this integrated approach to lake and reservoir basin management was conducted by The World Bank and the International Lake Environment Committee. These integrated, adaptive management approaches share common principles, while also being tailored to the unique characteristics, problems and management possibilities of specific aquatic ecosystems. IWRM incorporates social dimensions, such as gender equity and empowerment of women, cultural factors and the ability to make choices. Integrated Coastal Area and River Basin Management (ICARM) is an even more comprehensive approach which links the management needs of inland freshwater basins and their downstream coastal ecosystems, while Large Marine Ecosystem initiatives represent another important step, moving from single stock to ecosystem-based fisheries management. However, it has been difficult to transform these principles and recommendations into practical actions at the international, national and local levels, due partly to a lack of experience in their application, and the challenges in overcoming institutional, scientific and other significant barriers to integration.

Source: ILEC 2005, GWP 2006, WWAP 2006, UNEP-GPA 2006b

rivers. While acknowledging the environmental damage and socio-economic dislocation associated with construction of some dams, the building of more dams cannot be dismissed, since they can provide significant sources of water. But, more attention must be directed to understanding and balancing the environmental and socio-economic impacts associated with dam construction and operation against the benefits to be derived from them. Augmenting the resources of water-scarce regions by interbasin transfer is another established option, although proposed schemes must demonstrate the social, environmental and economic benefits to both the donating and the receiving basins.

While the impacts of increasing water demand for agriculture may be acceptable in countries with ample water resources, the escalating burden of water demand will become intolerable in water-scarce countries. Such situations can be alleviated to some degree by water-scarce countries shifting their food production to "water-rich" ones, deploying their own limited water resources into more productive economic sectors. This would address the need for energy- and technology intensive transport of water to distant areas of demand. Although globalization in the agriculture and related food production sectors already facilitates such changes, these approaches require close cooperation between producing and receiving countries.

Better management of marine, coastal and inland waters and their associated living resources improves the integrity and productivity of these ecosystems. Although there is little scope to expand or develop new fisheries, there is considerable opportunity to improve the management of existing fisheries and food production. Governments, industry and fishing communities can cooperate in reducing fish stock losses by making much needed changes to reduce excess fishing effort, subsidies and illegal fishing. Aquaculture currently helps to address the issue of food security, and has the potential to contribute further both by increasing fish supplies cost effectively, and by generating foreign income by exporting increased fish production, which can improve local livelihoods. But, aquaculture development to meet food security needs must include species that are not dependent on fish meal and fish oil, and that are palatable to a wide range of consumers.

Combating water-borne diseases

Although safeguarding human health ranks first among the priorities of water resources management, direct

human consumption and sanitation are among the smaller uses of freshwater in terms of volume. Even though the percentage of the world's population with access to improved water supply rose from 78 to 82 per cent between 1990 and 2000, and the percentage with access to improved sanitation rose from 51 to 61 per cent during this same period, contaminated water remains the greatest single cause of human sickness and death on a global scale. In 2002, then UN Secretary-General, Kofi Annan, pointed out that "no single measure would do more to reduce disease and save lives in the developing world than bringing safe water and sanitation to all" (UN 2004). Improved sanitation alone could reduce related deaths by up to 60 per cent, and diarrhoeal episodes by up to 40 per cent. The UN has designated 2008 as the International Year of Sanitation, in recognition of its key role in human well-being.

Controlling many diseases that are either water-borne or closely linked to water supplies depends on the use of specific technological measures, the maintenance or restoration of aquatic ecosystems, and public education and awareness. Technological approaches, such as the construction and operation of cost-effective water treatment plants and sanitation facilities for treating human wastes, provide effective measures against water-borne diseases. Many industrial water pollutants with human health implications also are amenable to treatment with technologies that capture materials from water. These technologies can sometimes recover

useful products (such as sulphur) from waste streams. Ecosystem restoration may reduce the incidence of some water-borne diseases, but it can also lead to an increase in the incidence of others. This negative aspect may be countered by improved understanding of the ecological requirements of disease vectors, and incorporating this knowledge into restoration projects. Traditional approaches, such as rainwater harvesting, can provide sources of safe drinking water, particularly in water-scarce areas or locations that experience natural disasters and other emergencies.

Properly managed fish farms have much potential to address food security and improve local livelihoods.

Credit: UNEP/Still Pictures

Safe drinking water saves lives.

Credit: I. Uwanaka – UNEP/ Still Pictures

Global water responses and partnerships

The world's oceans remain a huge, almost entirely untapped reserve of energy. Governments and the private sector can cooperate in exploring the energy production possibilities of the oceans, including the development of more efficient technologies for harnessing tidal and wave power as renewable sources of hydropower. The use of the oceans for large-scale carbon sequestration is another area of active investigation, though the potential impacts on the chemical composition of the oceans and its living resources remain unknown.

International water policy is increasingly emphasizing the need to improve governance as it relates to water resources management. The 2000 Ministerial Declaration of The Hague on Water Security in the 21st Century identified

Table 4.5 Selected responses to water issues addressed in this chapter					
Issue	**Key Institutions**	**Law, policy and management**	**Market-based instruments**	**Technology and adaptation**	**Restoration**
Climate change related issues					
Rising ocean temperature Ocean acidification Precipitation change	Intergovernmental Panel on Climate Change International Research frameworks International advocacy NGOs (such as WWF) Local authorities	■ International agreements, (such as Kyoto) ■ National CO_2 reduction and adaptation law and policy	■ International emissions capping and trading	■ Carbon sequestration (see Chapter 2) ■ Rainwater harvesting ■ Factoring climate change in planning future water development projects	■ Coral reef restoration
Increasing storminess rising sea level		■ Land-use zoning and regulation	■ Insurance instruments	■ Flood and coastal protection	■ Coastal managed retreat ■ Wetland restoration
Freshwater acidification				■ Industrial nitrogen and sulphur scrubbing	
Human water use and related ecosystem impact issues					
Clean water supply	Water and sanitation service delivery authorities River basin organizations	■ National policy and law ■ IWRM Catchment management ■ Improved water distribution ■ Stakeholder participation Empowering women	■ Private sector involvement ■ Private-public partnership ■ Tariffs and taxes ■ Agricultural and other subsidies as incentives	■ Water re-use ■ Low cost water and sanitation ■ Desalination	■ Catchment rehabilitation
Stream-flow modification Excessive surface water withdrawal Excessive groundwater withdrawal – ecosystem fragmentation – physical alteration and destruction of habitats	International and Regional organizations (such as UN-Water, MRC) International research frameworks, (such as CGIAR) International advocacy NGOs (GWF, WWC, IUCN, WWF) National water apex bodies River basin organizations	■ IWRM, ILBM, IRBM, ICARM, ICAM ■ International agreements ■ National policy and law ■ Strategic planning ■ Ecosystem approaches ■ Protected areas	■ Licensing supply sources and withdrawals ■ Realistic water pricing ■ Reduce or eliminate energy and agricultural subsidies and subsidized credit facilities ■ Valuing ecosystem services	■ Construction of large dams ■ Artificial recharge ■ More efficient irrigation techniques ■ Less water-demanding crops (see Chapter 3) ■ Improved rain-fed agriculture (see Chapter 3) ■ Environmental flows ■ Fish ladders	■ Dam removal ■ Wetland restoration ■ Basin reforestation ■ Upland habitat restoration ■ Coastal restoration ■ Coastal managed retreat

Table 4.5 Selected responses to water issues addressed in this chapter

Issue	Key Institutions	Law, policy and management	Market-based instruments	Technology and adaptation	Restoration
Human water use and related ecosystem impact issues					
Water-borne diseases	Health care extension organizations	■ IWRM, ILBM, IRBM, ICARM, ICAM ■ International agreements ■ National policy and law (such as Regional Seas, Helsinki Convention) ■ Enforceable water quality standards, land-use controls and best practices ■ Ecosystem approaches ■ Adherence to published guidelines International agreements, (such as Ramsar, AEWA)	■ Agricultural and other subsidies as incentives for clean water ■ Tradeable emission permits Organic farming certification	■ Wastewater treatment and re-use	■ Wetland restoration
Nutrient pollution – ecosystem pollution	Municipalities, wastewater treatment River basin organizations Farming, forestry and other stakeholder organizations			■ Wastewater treatment and re-use ■ Source reduction ■ Fertilizer application methods	■ Wetland restoration and creation ■ Ecohydrology
Pesticide pollution				Integrated pest management ■ Development of safer pesticides	
Suspended sediments – ecosystem pollution				■ Soil conservation (see Chapter 3) and other sediment control efforts	■ Reforestation ■ Dam removal
Hazardous chemicals	Disaster preparedness organizations	■ International agreements (such as Basel Convention) International agreements, (such as MARPOL) ■ National law	■ Regulation and penalties	■ Clean production technology ■ Treatment technology ■ Accident and disaster preparedness	
Fish stocks issues					
Pollution and habitat degradation	UNESCO/IOC, UNEP-GPA, Local stakeholders, (such as LMMA) (see Chapter 6)	International agreements, (such as MARPOL) ■ OSPAR	■ Private-public partnerships for MPAs, (such as Komodo, Chimbe)	■ Source reduction ■ Double-hulled vessels ■ Restocking programmes	■ Coastal habitat restoration ■ Fish ladders
Overexploitation	Regional, national and local fisheries management bodies Traditional communities	■ Licensing, gear restrictions ■ Ecosystem-based management ■ Marine Protected Areas (MPAs) ■ International agreements, (such as, UNCLOS, EC, CITES)	■ Individually tradeable quotas (ITQ) ■ Adequate pricing ■ Elimination of subsidies ■ Certification	■ Breeding and releasing young fish ■ By-catch reduction devices and other gear modifications (such as circle hooks for tuna)	■ Ecosystem rebuilding
Illegal Unreported Unregulated Fisheries (IUU)	Judiciary (such as fishery courts in South Africa) Fishery commissions (such as in European Union)	■ FAO International Plan of Action ■ Improved surveillance and enforcement including harsher penalties	■ Supply chain documentation (such as Patagonian toothfish)	■ Vessel monitoring systems (satellite technology)	

■ Particularly successful responses

■ Responses partially successful, successful in some places, or with a potential for success

■ Less successful responses

■ Responses with insufficient information, or not yet adequately tested

inadequate water governance as a main obstacle to water security for all. The 2001 International Conference on Freshwater in Bonn stressed that the essential issue was a need for stronger, better performing governance arrangements, noting that the primary responsibility for ensuring sustainable and equitable management of water resources rests with governments. Governance and water policy reforms were a core element of the Johannesburg plan for sustainable development in 2002. The Global Water Partnership (GWP) defined water governance as the exercise of economic, political and administrative authority to manage a nation's water affairs at all levels. It consists of the mechanisms, processes and institutions through which citizens and groups define their water interests, exercise their legal rights and obligations, promote transparency and mediate their differences. The need to strengthen existing legal and institutional frameworks for water management at both the national and international levels is central to all these efforts. The acknowledgement of the centrality of integrated approaches, full implementation, and compliance and enforcement mechanisms is also key to success.

Decision-makers are increasingly adopting integrated, adaptive management approaches, such as IWRM (see Box 4.9), rather than single issue, command-and-control regulatory approaches that previously dominated water resources management efforts. An integrated approach is fundamental in achieving social and economic development goals, while working for the sustainability of aquatic ecosystems to meet the water resource needs of future generations. To be effective, such approaches must consider the linkages and interactions between hydrological entities that cross multiple "boundaries," be they geographic, political or administrative. Ecosystem-based management approaches also provide a basis for cooperation in addressing common water resources management issues, rather than allowing such issues to become potential sources of conflict between countries or regions.

There are a number of key components for achieving cooperation among water stakeholders. They include international agreements, such as the 1997 UN Watercourses Convention, the Ramsar Convention, the Convention on Biological Diversity, and the Global Programme of Action for the Protection of the Marine Environment from Land-

based Activities. There is also a need to apply an ecosystem-based approach, as promulgated in the principles of IWRM, as well the Good Governance approach, developed at 1992 Rio Earth Summit and the 2002 World Summit on Sustainable Development. These approaches facilitate the sustainable and equitable management of common or shared water resources and, help achieve the goal of sustainable development in protecting freshwater and coastal sites to secure their vital ecosystem services.

More participatory regulatory approaches, such as demand management and voluntary agreements, have been introduced, due to an increasing realization of the limits of traditional regulation. These necessitate education and public involvement. Accordingly, public education curricula at all levels should vigorously address the issues of the water environment.

To enhance international cooperation in addressing the exploitation and degradation of water resources, the United Nations proclaimed 2005–2015 as the International Decade for Action, "Water for Life." A major challenge is focusing attention on action-oriented activities and policies directed to sustainable management of the quantity and quality of water resources. In 2004, the United Nations established UN-Water as its system-wide mechanism for coordinating its agencies and programmes involved in water-related issues. A complementary mechanism will facilitate integrative cross-linkages between activities coordinated under UN-Water with UN-Oceans, strengthening coordination and cooperation of UN activities related to oceans, coastal areas and Small Island Developing States.

In developing responses to the impacts of change in the water environment, national governments and the international community face a major challenge. They need to not only develop new approaches, but also to facilitate the practical, timely and cost-effective implementation of existing international and other agreements, policies and targets (see Table 4.5). Continuous monitoring and evaluation of the responses – with adjustments as necessary – are required to secure the sustainable development of the water environment for the benefit of humans, and for the maintenance of life-supporting ecosystems over the long-term.

References

ACIA (2004). *Impacts of a warming Arctic*. Arctic Monitoring and Assessment Programme. Cambridge University Press, Cambridge

ACIA (2005). *Arctic Climate Impact Assessment, Scientific Report*. Cambridge University Press, Cambridge

Alder, J. and Sumaila, R.U. (2004). Western Africa: a fish basket of Europe past and present. In *Journal of Environment and Development* 13:156-178

Al Hadidi, K. (2005). Groundwater Management in the Azraq Basin. In *ACSAD-BGR Workshop on Groundwater and Soil Protection, 27-30 June 2005, Amman*

Allan, J.D., Abell, R., Hogan, Z., Revenga, C., Taylor, B.W., Welcomme, R. L. and Winemiller, K. (2005). Overfishing of inland waters. In *Bioscience* 55(12):1041-1051

America's Wetlands (2005). *Wetland issues exposed in wake of Hurricane Katrina*. http://www.americaswetland.com/article.cfm?id=292&cateid=2&pageid=3&cid=16 (last accessed 31 March 2007)

APFM (2006). Legal and Institutional Aspects of Integrated Flood Management. WMO/GWP Associated Programme on Flood Management. Flood Management Policy Series, WMO No. 997. World Meteorological Organization, Geneva http://www.apfm.info/pdf/ifm_legal_aspects.pdf (last accessed 31 March 2007)

Aranson, R. (2002). A review of international experiences with ITQs: an annex to *Future options for UK fish quota management*. CERMARE Report 58. University of Portsmouth, Portsmouth http://statistics.defra.gov.uk/esg/reports/fishquota/revitqs.pdf (last accessed 10 April 2007)

Argady, T. and Alder, J. (2005). Coastal Systems. In *Ecosystems and human well-being, Volume 1: Current status and trends: findings of the Condition and Trends Working Group* (eds. R. Hassan, Scholes, R.and Ash, N.) Chapter 19. Island Press, Washington, DC

Barter, M.A. (2002). *Shorebirds of the Yellow Sea: Importance, threats and conservation status*. Wetlands International Global Series 9, International Wader Studies 12, Canberra

Bernhardt, E.S., Palmer, M.A., Allan, J.D., Alexander, G., Barnas, K., Brooks, S., Carr, J., Clayton, S., Dahm, C., Follstad-Shah, J., Galat, D., Gloss, S., Goodwin, P., Hart, D., Hassett, B., Jenkinson, R., Katz, S., Kondolf, G.M., Lake, P.S., Lave, R., Meyer, J.L., O'Donnell, T.K., Pagano, L., Powell, B. and Sudduth, E. (2005). Synthesizing U.S. River restoration efforts. In *Science* 308:636-637

Bromhead, D., Foster, J., Attard, R., Findlay, J., and Kalish, J. (2003). *A review of the impact of fish aggregating devices (FADs) on tuna fisheries*. Final Report to Fisheries Resources Research Fund. Commonwealth of Australia, Department of Agriculture, Fisheries and Forestry, Bureau of Rural Sciences, Canberra http://affashop.gov.au/PdfFiles/PC12777.pdf (last accessed 31 March 2007)

Brooks N. (2004). *Drought in the African Sahel: long term perspectives and future prospects*. Working Paper No. 61, Tyndall Centre for Climate Change Research, University of East Anglia, Norwich http://test.earthscape.org/r1/ES15602/wp61.pdf (last accessed 31 March 2007)

Bruton, M.N. (1995). Have fishes had their chips? The dilemma of threatened fishes. In *Environmental Biology of Fishes* 43:1-27

Bryden, H.L., Longworth, H.R. and Cunningham, S.A. (2005). Slowing of the Atlantic meridional overturning circulation at 25° N. In *Nature* 438(7068):655

Choowaew, S. (2006). What wetlands can do for poverty alleviation and economic development. Presented at *Regional IWRM 2005 Meeting, 11-13 September 2006, Rayong*

Clark, M. (2001). Are deepwater fisheries sustainable? – the example of orange roughy (*Hoplostethus atlanticus*) in New Zealand. In *Fisheries Research* 51:123-135

CNA (2005). Information provided by the Comisión Nacional del Agua (National Water Commission, Mexican government) from their Estadísticas del Agua en México

CNROP (2002). *Environmental impact of trade liberalization and trade-linked measures in the fisheries sector*. National Oceanographic and Fisheries Research Centre, Nouadhibou

Crossland, C.J., Kremer, H.H., Lindeboom, H.J., Marshall Crossland, J.I. and Le Tissier, M.D.A., eds (2005). *Coastal Fluxes in the Anthropocene. The Land-Ocean Interactions in the Coastal Zone Project of the International Geosphere-Biosphere Programme*. Global Change, IGBP Series. Springer-Verlag, Berlin

Curry, R. and Mauritzen, C. (2005). Dilution of the northern North Atlantic Ocean in recent decades. In *Science* 308(5729):1772-1774

Dai, A., Lamb, P.J., Trenberth, K.E., Hulme, M., Jones, P.D. and Xie, P. (2004). The recent Sahel drought is real. In *International Journal of Climatology* 24:1323-1331

Delgado, C.L., Wada, N., Rosegrant, M.W., Meijer, S. and Ahmed M. (2003). *Fish to 2020: Supply and demand in changing global markets*. Technical Report 62. International Food Policy Research Institute, Washington, DC and WorldFish Centre, Penang http://www.ifpri.org/pubs/books/fish2020/oc44front.pdf (last accessed 31 March 2007)

DIFD, EC, UNDP and World Bank (2002). *Linking Poverty Reduction and Environmental Management: Policy Challenges and Opportunities*. The World Bank, Washington, DC

Dore, M.H.I. (2005). Climate change and changes in global precipitation patterns: What do we know? In *Environment International* 31(8):1167-1181

Dyurgerov, M.B. and Meier, M.F. (2005). *Glaciers and the Changing Earth System: A 2004 Snapshot*. INSTAAR, University of Colorado at Boulder, Boulder, CO

EEA (2003). *Europe's environment: the third assessment*. Environmental assessment report 10, European Environment Agency, Copenhagen

Enfield D.B. and Mestas-Nuñez, A.M. (1999). Interannual-to-multidecadal climate variability and its relationship to global sea surface temperatures. In *Present and Past Inter-Hemispheric Climate Linkages in the Americas and their Societal Effects* (ed. V. Markgraf), Cambridge University Press, Cambridge http://www.aoml.noaa.gov/phod/docs/enfield/full_ms.pdf (last accessed 31 March 2007)

Environment Canada (2006). *TAG tanker spill database*. Environmental Technology Centre, Environment Canada http://www.etc-cte.ec.gc.ca/databases/TankerSpills/Default.aspx (last accessed 31 March 2007)

EU (2006). *Proposal for a Directive of the European Parliament and of the Council on the Assessment and Management of Floods*. COM(2006) 15 final. European Commission, Brussels http://ec.europa.eu/environment/water/flood_risk/pdf/com_2006_15_en.pdf (last accessed 31 March 2007)

Falkenmark, M. (2005). Green water—conceptualising water consumed by terrestrial ecosystems. *Global Water News* 2: 1-3 http://www.gwsp.org/downloads/GWSP_Newsletter_no2Internet.pdf (last accessed 31 March 2007)

FAO (2002). *The State of World Fisheries and Aquaculture 2002*. Fisheries Department, Food and Agriculture Organization of the United Nations, Rome

FAO (2005). Special Event on Impact of Climate Change, Pests and Diseases on Security and Poverty Reduction. Paper presented to the *31st Session of the Committee on World Food Security, 23-26 May 2005 in Rome*. Food and Agriculture Organization of the United Nations, Rome

FAO (2006a). *State of the World Fisheries and Aquaculture (SOFIA) 2004*. Food and Agriculture Organization of the United Nations, Rome http://www.fao.org/documents/show_cdr.asp?url_file=/DOCREP/007/y5600e/y5600e00.htm (last accessed 2 April 2007)

FAO (2006b). *Low-Income Food-Deficit Countries (LIFDC)*. Food and Agriculture Organization of the United Nations, Rome http://www.fao.org/countryprofiles/lifdc.asp?lang=en (last accessed 16 November 2006)

FAO (2006c). FISHSTAT. Food and Agriculture Organization of the United Nations, Rome http://www.fao.org/fi/statist/FISOFT/FISHPLUS.asp (last accessed 2 April 2007)

Finlayson, C.M. and D'Cruz, R. (2005). Inland Water Systems. In Hassan, R., Scholes, R. and Ash, N. (eds.) *Ecosystems and human well-being, Volume 1: current status and trends: findings of the Condition and Trends Working Group*. Chapter 20, Island Press, Washington, DC

Foster S.S.D. and Chilton P.J. (2003). Groundwater: the processes and global significance of aquifer degradation. In *Philosophical Transactions of the Royal Society* 358:1957-1972

Gordon, L.J., Steffen, W., Jönsson, B.F., Folke, C., Falkenmark, M. and Johannessen, A. (2005). Human modification of global water vapor flows from the land surface. *PNAS* Vol. 102, No. 21: 7612–7617 http://www.gwsp.org/downloads/7612.pdf (last accessed 3 May 2007)

Gujer, W. (2002). *Siedlungswasserwirtschaft*. 2nd Edition. Springer-Verlag, Heidelberg

GWP (2006). *Setting the Change for Change*. Technical Report, Global Water Partnership, Stockholm

Hamerlynck, O. and Duvail, S. (2003). *The rehabilitation of the delta of the Senegal River in Mauritania*. Fielding an ecosystem approach. The World Conservation Union, Gland http://www.iucn.org/themes/wetlands/pdf/diawling/Diawling_GB.pdf (last accessed 5 April 2007)

Hamerlynck, O., Duvail, S. and Baba, M.L. Ould (2000). *Reducing the environmental impacts of the Manantali and Diama dams on the ecosystems of the Senegal river and estuary: alternatives to the present and planned water management schemes*. Submission to World Commission on Dams, Serial No: ins131 http://www.dams.org/kbase/submissions/showsub.php?rec=ins131 (last accessed 5 April 2007)

Hanna, E., Huybrechts, P., Cappelen, J., Steffen, K. and Stephens, A. (2005). Runoff and mass balance of the Greenland ice sheet. In *Journal of Geophysical Research* 110(D13108):1958-2003

Huang, J., Liu, H. and Li, L. (2002). Urbanization, income growth, food market development, and demand for fish in China. Presented at *Biennial Meeting, International Institute of Fisheries Economics and Trade, Wellington, New Zealand, 19-23 August 2002*

Hydropower Reform Coalition (n.d.). Hydropower Reform. Washington, DC http://www.hydroreform.org/about (last accessed 5 April 2007)

ILEC (2005). *Managing lakes and their basins for sustainable use: A report for lake basin managers and stakeholders*. International Lake Environment Committee Foundation, Kusatsu, Shiga

INEGI (2006). *II Conteo de población y vivienda. Resultados definitivos, 2005*. Instituto Nacional de Geografía Estadística e Informática, Mexico, DF http://www.inegi.gob.mx/est/default.aspx?c=6789 (last accessed 2 April 2007)

IPCC (1996). *Climate change 1995: Impacts, adaptations and mitigation of climate change: scientific-technical analyses*. Contribution of Working Group II to the Second Assessment Report. Intergovernmental Panel on Climate Change. Cambridge University Press, Cambridge

IPCC (2001). *Climate Change 2001: The Scientific Basis*. Contribution of Working Group I to the Third Assessment Report of the Intergovernmental Panel on Climate Change. Cambridge University Press, Cambridge

IPCC (2005). *IPCC Special Report on Carbon Dioxide Capture and Storage*. Prepared by Working Group III of the Intergovernmental Panel on Climate Change, Geneva. Metz, B., Davidson, O., de Coninck, H. C., Loos, M. and Meyer L. A. (eds.). Cambridge University Press, Cambridge http://www.ipcc.ch/activity/srccs/index.htm (last accessed 6 June 2007)

IPCC (2007). *Climate Change 2007: The Physical Science Basis. Summary for Policymakers*. Contribution of Working Group I to the Fourth Assessment Report of the Intergovernmental Panel on Climate Change, Geneva http://ipcc-wg1.ucar.edu/wg1/docs/WG1AR4_SPM_Approved_05Feb.pdf (last accessed 5 April 2007)

IUCN (2004). Fact Sheet - Ecosystem Management. http://www.iucn.org/congress//2004/documents/fact_ecosystem.pdf (accessed 30 March 2007)

IUCN (2006). News release - Coral bleaching will hit the world's poor. http://www.iucn.org/en/news/archive/2006/11/17_coral_bleach.htm (last accessed 2 April 2007)

IWMI (2005). *Environmental flows: Planning for environmental water allocation*. Water Policy Briefing 15. International Water Management Institute, Battaramulla http://www.iwmi.cgiar.org/waterpolicybriefing/files/wpb15.pdf (last accessed 2 April 2007)

Kaczynski, V.M. and Fluharty, D.L. (2002). European policies in West Africa: who benefits from fisheries agreements? In *Marine Policy* 26:75-93

Kurien, J. (2005). Comercio Internacional en la Pesca y Seguridad Alimentaria (Fish trade and food security). Presented at *ICSF-CeDePesca Seminar, Santa Clara del Mar, Argentina, March 1-4, 2005* http://oldsite.icsf.net/cedepesca/presentaciones/kurien_comercio/kurien_comercio.htm (last accessed 13 April 2007)

Lake Basin Management Initiative (2006). *Managing Lakes and their Basins for Sustainable Use*. A Report for Lake Basin Managers and Stakeholders. Lake Basin Management Initiative. International Lake Environment Committee Foundation, Kusatsu, Shiga, and The World Bank, Washington, DC

Landell-Mills, N. and Porras, I.T. (2002). *Silver bullet or fools' gold? A global review of markets for forest environmental services and their impact on the poor*. International Institute for Environment and Development, London

Loh, J., and Wackernagel, M. (eds.) (2004). *Living Planet Report 2004*. World Wide Fund for Nature, Gland

MA (2005). *Ecosystem Services and Human Well-being: Wetlands and Water Synthesis*. Millennium Ecosystem Assessment. World Resources Institute, Washington, DC

Malherbe, S. and IFFO (2005). The world market for fishmeal. In *Proceedings of World Pelagic Conference, Cape Town, South Africa, 24-25 October 2005*. Agra Informa Ltd., Tunbridge Wells

Meybeck, M. and Vörösmarty, C. (2004). The Integrity of River and Drainage Basin Systems: Challenges from Environmental Changes. In Kabat, P., Claussen, M., Dirmeyer, P.A., Gash, J.H.C., Bravo de Guenni, L., Meybeck, M., Pielke, R.S., Vörösmarty, C.J., Hutjes, R.W.A. and Lütkemeier, S. (eds.). *Vegetation, Water, Humans and the Climate: a New Perspective on an Interactive System* IGBP Global Change Series. International Geosphere-Biosphere Programme and Springer-Verlag, Berlin

Myers, R.A. and Worm, B. (2003). Rapid worldwide depletion of predatory fish communities. In *Nature* 423(6937):280-283

National Academy of Sciences (2003). *Oil in the SEA III: Inputs, Fates and Effects*. National Research Council. National Academies Press, Washington, DC

New, M.B. and Wijkstrom, U.S. (2002). *Use of fishmeal and fish oil in aquafeeds: Further thoughts on the fishmeal trap*. Fisheries Circular 975. Food and Agriculture Organization of the United Nations, Rome

Nilsson, C., Reidy, C.A., Dynesius, M., and Revenga, C. (2005). Fragmentation and flow regulation of the world's large river systems. In *Science* 308:305-308

NOAA (2004). Louisiana Scientists Issue "Dead Zone" Forecast http://www.noaanews.noaa.gov/stories2004/s2267.htm (last accessed 2 April 2007)

NSIDC (2005). Sea Ice Decline Intensifies. National Snow and Ice Data Center ftp://sidads.colorado.edu/DATASETS/NOAA/G02135/Sep/N_09_area.txt (last accessed 2 April 2007)

OSPAR (2005). *2005 Assessment of data collected under the Riverine Inputs and Direct Discharges (RID) for the period 1990 – 2002*. OSPAR Commission http://www.ospar.org/eng/html/welcome.html (last accessed 3 May 2007)

Pauly, D., Alder, J., Bennett, E., Christensen, V., Tyedmers, P. and Watson, R. (2003). The future for fisheries. In *Science* 302(5649):1359-1361

Philander, S.G.H. (1990). *El Niño, La Niña and the Southern Oscillation*. Academic Press, San Diego, California

Piersma, T., Koolhaas, A., Dekinga, A., Beukema, J.J., Dekker, R. and Essink, K. (2001). Long-term indirect effects of mechanical cockle-dredging on intertidal bivalve stocks in the Wadden Sea. In *Journal of Applied Ecology* 38:976-990

PLUARG (1978). *Environmental Management Strategy for the Great Lakes Basin System*. Pollution from Land Use Activities Reference Group, Great Lakes Regional Office, International Joint Commission, Windsor, Ontario

Postel, S. and Richter, B. (2003). *Rivers for Life: Managing Water for People and Nature*. Island Press, Washington, DC

Prowse, T. D., Wrona, F. J. and Power, G. (2004). Dams, reservoirs and flow regulation. In *Threats to Water Availability in Canada*. National Water Research Institute, Meteorological Service of Canada, Environmental Conservation Service of Environment Canada, Burlington, ON, 9-18 http://www.nwri.ca/threats2full/intro-e.html (last accessed 2 April 2007)

Richardson, C.J., Reiss, P., Hussain, N.A., Alwash, A.J. and Pool, D.J. (2005). The restoration potential of the Mesopotamian Marshes of Iraq. In *Science* 307:1307-1311

Royal Society (2005). *Ocean acidification due to increasing atmospheric carbon dioxide*. Policy Document 12/05, The Royal Society, London http://www.royalsoc.ac.uk/displaypagedoc.asp?id=13539 (last accessed 31 March 2007)

SAUP (2006). Sea Around Us Project. http://www.seaaroundus.org (last accessed 26 March 2007)

Schorr, D. (2004). *Healthy fisheries, sustainable trade: crafting new rules on fishing subsidies in the World Trade Organization*. World Wide Fund for Nature, Gland http://www.wto.org/english/forums_e/ngo_e/posp43_wwf_e.pdf (last accessed 3 May 2007)

SIWI, IFPRI, IUCN and IWMI (2005). *Let it reign: The new water paradigm for global water security*. Final Report to CSD-13. Stockholm International Water Institute, Stockholm

Shiklomanov, I.A. and UNESCO (1999). World Water Resources: Modern Assessment and Outlook for the 21st Century. Summary of World Water Resources at the Beginning of the 21st Century. Prepared in the framework of IHP-UNESCO

Shiklomanov, I. A. and Rodda, J. C. (2003). *World Water Resources at the Beginning of the 21st Century*. Cambridge University Press, Cambridge

Shuval, H. (2003). Estimating the global burden of thalassogenic diseases: Human infectious diseases caused by wastewater pollution of the marine environment. In *Journal of Water and Health* 1(2):53–64

Snoussi, M. (2004). Review of certain basic elements for the assessment of environmental flows in the Lower Moulouya. In *Assessment and Provision of Environmental Flows in the Mediterranean Watercourses: Basic Concepts, Methodologies and Emerging Practice*. The World Conservation Union, Gland

Syvitski, J., Vörösmarty, C., Kettner, A. and Green, P. (2005). Impact of humans on the flux of terrestrial sediment to the global coastal ocean. In *Science* 308:376-380

Turner, K., Georgiou, S., Clark, R., Brouwer, R. and Burke, J. (2004). *Economic valuation of water resources in agriculture: From the sectoral to a functional perspective of natural resource management*. Water Report 27, Food and Agriculture Organization of the United Nations, Rome http://www.fao.org/docrep/007/y5582e/y5582e00.HTM (last accessed 31 March 2007)

UN (2004). International Decade for Action, "Water for Life", 2005-2015. UN Resolution 58/217 of 9 February 2004. United Nations General Assembly, New York, NY http://www.unesco.org/water/water_celebrations/decades/water_for_life.pdf (last accessed 3 May 2007)

UNECE (2000). Guidelines on Sustainable Flood Prevention. In *UN ECE Meeting of the Parties to the Convention on the Protection and Use of Transboundary Watercourses and International Lakes, Second Meeting, 23-25 March 2000, The Hague* http://www.unece.org/env/water/publications/documents/guidelinesfloode.pdf (last accessed 2 April 2007)

UNEP (2001). *The Mesopotamian Marshlands: Demise of an Ecosystem*. UNEP/DEWA/TR.01-3. UNEP Division of Early Warning and Assessment/GRID-Europe, Geneva in cooperation with GRID-Sioux Falls and the Regional Office for West Asia (ROWA), Geneva http://www.grid.unep.ch/activities/sustainable/tigris/mesopotamia.pdf (last accessed 11 April 2007)

UNEP (2005a). *Marine litter. An analytical overview*. United Nations Environment Programme, Nairobi

UNEP (2005b). *One Planet Many People: Atlas of Our Changing Environment*. UNEP Division of Early Warning and Assessment, Nairobi

UNEP (2006). *Iraq Marshlands Observation System*. UNEP Division of Early Warning and Assessment/GRID-Europe http://www.grid.unep.ch/activities/sustainable/tigris/mmos.php (last accessed 31 March 2007)

UNEP-GEMS/Water Programme (2006). UNEP Global Environment Monitoring System, Water Programme www.gemswater.org and www.gemstat.org (last accessed 31 March 2007)

UNEP-GEMS/Water Programme (2007). *Water Quality Outlook*. UNEP Global Environment Monitoring System, Water Programme, National Water Research Institute, Burlington, ON http://www.gemswater.org/common/pdfs/water_quality_outlook.pdf (last accessed 3 May 2007)

UNEP-GIWA (2006a). *Challenges to International Waters – Regional Assessments in a Global Perspective*. Global International Waters Assessment Final Report. United Nations Environment Programme, Nairobi http://www.giwa.net/publications/finalreport/ (last accessed 31 March 2007)

UNEP-GIWA (2006b). *Mekong River, GIWA Regional Assessment 55*. Global International Waters Assessment, University of Kalmar on behalf of United Nations Environment Programme, Kalmar http://www.unep.org/dewa/giwa/publications/r55.asp (last accessed 31March 2007)

UNEP/GPA (2006a). *The State of the Marine environment: Trends and processes*. UNEP-Global Programme of Action for the Protection of the Marine Environment from Land-based Activities, The Hague

UNEP/GPA (2006b). *Ecosystem-based management: Markers for assessing progress*. UNEP-Global Programme of Action for the Protection of the Marine Environment from Land-based Activities, The Hague

UNEP/GRID-Arendal (2002). *Vital Water Graphics. An overview of the State of the World's Fresh and Marine Waters*. United Nations Environment Programme, Nairobi http://www.unep.org/vitalwater/ (last accessed 31 March 2007)

UNEP/GRID-Arendal (2005). *Vital Climate GraphicsUpdate*. United Nations Environment Programme, Nairobi and GRID-Arendal, Arendal http://www.vitalgraphics.net/climate2.cfm (last accessed 2 April 2007)

UNEP-WCMC (2006). *In the front line. Shoreline protection and other ecosystem services from mangroves and coral reefs*. UNEP-World Conservation Monitoring Centre, Cambridge

UNESCO (2006). *Groundwater Resources of the World: Transboundary Aquifer Systems*. WHYMAP 1:50 000 000 Special Edition for the 4th World Water forum, Mexico, DF

UN Secretary General's Advisory Board on Water and Sanitation (2006). Compendium of Actions, March 2006. United Nation, New York, NY http://www.unsgab.org/top_page.htm (last accessed 2 April 2007)

UN Water (2007). *Coping with water scarcity: challenge of the twenty-first century*. Prepared for World Water Day 2007 http://www.unwater.org/wwd07/downloads/documents/escarcity.pdf (last accessed 23 March 2007)

USEPA (2006). *Nonpoint Source Pollution: The Nation's Largest Water Quality Problem*. US Environmental Protection Agency, Washington, D.C. http://www.epa.gov/nps/facts/point1.htm (last accessed 2 April 2007)

Vörösmarty, C. J. and Sahagian, D. (2000). Anthropogenic disturbance of the terrestrial water cycle. In *Bioscience* 50:753-765

Vörösmarty, C. J., Sharma, K., Fekete, B., Copeland, A. H., Holden, J. and others (1997). The storage and ageing of continental runoff in large reservoir systems of the world. In *Ambio* 26:210-219

Watson, R. and Pauly, D. (2001). Systematic distortions in world fisheries catch trends. In *Nature* 414(6863):534-536

WCD (2000). *Dams and Development – A New Framework for Decision-Making: the Report of the World Commission on Dams*. Earthscan Publications Ltd., London http://www.dams.org/report/contents.htm (last accessed 2 April 2007)

WFD (2000). Directive 2000/60/EC of the European Parliament and of the Council establishing a framework for Community action in the field of water policy. OJ (L 327). European Commission, Brussels http://ec.europa.eu/environment/water/water-framework/index_en.html (last accessed 2 April 2007)

WHO (2000). *WHO Report on Global Surveillance of Epidemic-prone Infectious Diseases: Chapter 4, Cholera*. World Health Organization, Geneva http://www.who.int/csr/resources/publications/surveillance/en/cholera.pdf (last accessed 31 March 2007)

WHO (2004). *Global burden of disease in 2002: data sources, methods and results*. February 2004 update. World Health Organization, Geneva http://www.who.int/healthinfo/paper54.pdf (last accessed 2 May 2007)

WHO and UNICEF (2004). *Meeting the MDG Drinking Water and Sanitation Target: A Mid-term Assessment of Progress*. Joint Monitoring Programme for Water Supply and Sanitation. World Health Organization, Geneva and United Nations Children's Fund, New York, NY

WHO and UNICEF (2006). Joint Monitoring Programme for Water Supply and Sanitation (in GEO Data Portal). World Health Organization, Geneva and United Nations Children's Fund, New York, NY

Wilkinson, C. ed. (2004). *Status of coral reefs of the world: 2004*. Australian Institute of Marine Science, Townsville

World Bank (2005). *Towards a More Effective Operational Response: Arsenic Contamination of Groundwater in South and East Asian Countries*. Environment and Social Unit, South Asia Region, and Water and Sanitation Program, South and East Asia, Vol. II, Technical Report. International Bank for Reconstruction and Development, Washington, DC

World Bank (2006). *Measuring Coral Reef Ecosystem Health: Integrating Social Dimensions*. World Bank Report No. 36623 – GLB. The World Bank, Washington, DC

WRI (2005). *World Resources 2005: The Wealth of the Poor-Managing Ecosystems to Fight Poverty*. World Resources Institute, in collaboration with United Nations Development Programme, United Nations Environment Programme and The World Bank, Washington, DC

WWAP (2006). *The State of the Resource, World Water Development Report 2, Chapter 4*. World Water Assessment Programme, United Nations Educational, Scientific and Cultural Organization, Paris http://www.unesco.org/water/wwap/wwdr2/pdf/wwdr2_ch_4.pdf (last accessed 31 March 2007)

WWDR (2006). *Water a shared responsibility*. The United Nations World Water Development Report 2. UN-Water/WWAP/2006/3. World Water Assessment Programme, United Nations Educational, Scientific and Cultural Organization, Paris and Berghahn Books, New York, NY

WWF (2006). *The Best of Texts, the Worst of Texts*. World Wide Fund for Nature, Gland

WWF (2007). *World's top 10 rivers at risk*. World Wide Fund for Nature, Gland http://assets.panda.org/downloads/worldstop10riversatriskfinalmarch13.pdf (last accessed 31 March 2007)

Chapter 5

Biodiversity

Coordinating lead authors: Neville Ash and Asghar Fazel

Lead authors: Yoseph Assefa, Jonathan Baillie, Mohammed Bakarr, Souvik Bhattacharjya, Zoe Cokeliss, Andres Guhl, Pascal Girot, Simon Hales, Leonard Hirsch, Anastasia Idrisova, Georgina Mace, Luisa Maffi, Sue Mainka, Elizabeth Migongo-Bake, José Gerhartz Muro, Maria Pena, Ellen Woodley, and Kaveh Zahedi

Contributing authors: Barbara Gemmill, Jonathan Loh, Jonathan Patz, Jameson Seyani, Jorge Soberon, Rick Stepp, Jean-Christophe Vie, Dayuan Xue, David Morgan, David Harmon, Stanford Zent, and Toby Hodgkin

Chapter review editors: Jeffrey A. McNeely and João B. D. Camara

Chapter coordinator: Elizabeth Migongo-Bake

Main messages

Biodiversity provides the basis for ecosystems and the services they provide, upon which all people fundamentally depend. The following are the main messages of this chapter:

People rely on biodiversity in their daily lives, often without realizing it. Biodiversity contributes to many aspects of people's livelihoods and well-being, providing products, such as food and fibres, whose values are widely recognized. However, biodiversity underpins a much wider range of services, many of which are currently undervalued. The bacteria and microbes that transform waste into usable products, insects that pollinate crops and flowers, coral reefs and mangroves that protect coastlines, and the biologically-rich landscapes and seascapes that provide enjoyment are only a few. Although much more remains to be understood about the relationships between biodiversity and ecosystem services, it is well established that if the products and services that are provided by biodiversity are not managed effectively, future options will become ever more restricted, for rich and poor people alike. However, poor people tend to be the most directly affected by the deterioration or loss of ecosystem services, as they are the most dependent on local ecosystems, and often live in places most vulnerable to ecosystem change.

Current losses of biodiversity are restricting future development options. Ecosystems are being transformed, and, in some cases, irreversibly degraded, a large number of species have gone extinct in recent history or are threatened with extinction, reductions in populations are widespread and genetic diversity is widely considered to be in decline. It is well established that changes to biodiversity currently underway on land and in the world's fresh and marine waters are more rapid than at any time in human history, and have led to a degradation in many of the world's ecosystem services.

Reducing the rate of loss of biodiversity, and ensuring that decisions made incorporate the full values of goods-and-services provided by biodiversity will contribute substantially towards achieving sustainable development as described in the report of the World Commission on Environment and Development (Brundtland Commission report).

- **Biodiversity plays a critical role in providing livelihood security for people.** It is particularly important for the livelihoods of the rural poor, and for regulating local environmental conditions. Functioning ecosystems are crucial as buffers against extreme climate events, as carbon sinks, and as filters for water-borne and airborne pollutants.

- **From the use of genetic resources to harnessing other ecosystem services, agriculture throughout the world is dependent on biodiversity.** Agriculture is also the largest driver of genetic erosion, species loss and conversion of natural habitats. Meeting increasing global food needs will require one or both of two approaches: intensification and extensification. Intensification is based on higher or more efficient use of inputs, such as more efficient breeds and crops, agrochemicals, energy and water. Extensification requires converting increasing additional areas of land to cultivation. Both approaches have the potential to dramatically and negatively affect biodiversity. In addition, the loss of diversity in agricultural ecosystems may undermine the ecosystem services necessary to sustain agriculture, such as pollination and soil nutrient cycling.

- **Many of the factors leading to the accelerating loss of biodiversity are linked to the increasing use of energy**

by society. Dependence on and growing requirements for energy are resulting in significant changes in species and ecosystems, as a result of the search for energy sources and of current energy use patterns. The consequences can be seen at all levels: locally, where the availability of traditional biomass energy is under threat, nationally, where energy prices affect government policies, and globally, where climate change driven by fossil-fuel use is changing species ranges and behaviour. The latter is likely to have very significant consequences for livelihoods, including changing patterns of human infectious disease distribution, and increased opportunities for invasive alien species.

- **Human health is affected by changes in biodiversity and ecosystem services.** Changes to the environment have altered disease patterns and human exposure to disease outbreaks. In addition, current patterns of farming, based on high resource inputs (such as water and fertilizers) and agricultural intensification, are putting great strains on ecosystems, contributing to nutritional imbalances and reduced access to wild foods.

- **Human societies everywhere have depended on biodiversity for cultural identity, spirituality, inspiration, aesthetic enjoyment and recreation.** Culture can also play a key role in the conservation and sustainable use of biodiversity. Loss of biodiversity affects both material and non-material human well-being. Both the continued loss of biodiversity and the disruption of cultural integrity represent obstacles towards the attainment of the Millennium Development Goals (MDGs).

Biodiversity loss continues because current policies and economic systems do not incorporate the values of biodiversity effectively in either the political or the market systems, and many current policies are not fully implemented. Although many losses of biodiversity, including the degradation of ecosystems, are slow or gradual, they can lead to sudden and dramatic declines in the capacity of biodiversity to contribute to human well-being. Modern societies can continue to develop without further loss of biodiversity only if market and policy failures are rectified. These failures include perverse production subsidies, undervaluation of biological resources, failure to internalize environmental costs into prices and failure to appreciate global values at the local level. Reducing the rate of biodiversity loss by 2010 or beyond will require multiple and mutually supportive policies of conservation, sustainable use and the effective recognition of value for the benefits derived from the wide variety of life on Earth. Some such policies are already in place at local, national and international scales, but their full implementation remains elusive.

INTRODUCTION

The understanding of the importance of biodiversity has developed in the 20 years since the report of the World Commission on Environment and Development (Brundtland Commission). There is increased recognition that people are part of, not separate from, the ecosystems in which they live, and are affected by changes in ecosystems, populations of species and genetic changes. Along with human health and wealth, human security and culture are strongly affected by changes in biodiversity, and associated impacts on ecosystem services.

As the basis for all ecosystem services, and the foundation for truly sustainable development, biodiversity plays fundamental roles in maintaining and enhancing the well-being of the world's more than 6.7 billion people, rich and poor, rural and urban alike. Biodiversity comprises much of the renewable natural capital on which livelihoods and development are grounded. However, ongoing, and in many cases, accelerating declines and losses in biodiversity over the past 20 years have decreased the capacity of many ecosystems to provide services, and have had profound negative impacts on opportunities for sustainable development around the planet. These impacts are particularly pronounced in the developing world, in large part due to the patterns of consumption and trade in the industrial world, which themselves are not sustainable.

If future concerns are not taken into account, and the products and services provided by biodiversity are not managed effectively, future options become limited or are eliminated, for rich and poor people alike. While technological alternatives to some of the services provided by biodiversity are available, they are typically more costly, compared to the benefits derived from well-managed ecosystems. Biodiversity loss particularly affects the poor, who are most directly dependent on ecosystem services at the local scale, and are unable to pay for alternatives. Although the private, more restricted, financial benefits of activities that result in the loss of biodiversity, such as the conversion of mangroves to aquaculture enterprises, are usually high, they often externalize many of the social and environmental costs. The overall benefits are frequently considerably less than the societal, more distributed, benefits that are lost along with the biodiversity, but for which the monetary value is often not known. For example, the loss of mangrove ecosystems contributes to declining fisheries, timber and fuel, the reduction of storm protection, and increased vulnerability to the impacts of extreme events.

In addition to the values of biodiversity for the supply of particular ecosystem services, biodiversity also has intrinsic value, independent from its functions and other benefits to people (see Box 5.1). The challenge is to balance the cultural, economic, social and environmental values so that the biodiversity of today is conserved and used in a manner that will allow it to be available for and to sustain the generations of the future. Biodiversity management and policies have an impact upon all sectors of society, and have strong cross-cultural and cross-boundary implications. Policies relating to issues such as trade, transport, development, security, health care and education all have impacts on biodiversity. Discussions on access and benefit sharing relating to genetic resources, one of the provisions of the UN Convention on Biological Diversity (CBD), show that understanding the full value of biodiversity is not simple. In addition to the gaps remaining in the understanding of biodiversity and ecosystem functioning, each individual stakeholder may hold different values for the same attribute of biodiversity. Building a fuller understanding of these values will require considerable additional research, and increasingly comprehensive, interdisciplinary and quantified assessments of the benefits that biodiversity provides to people's health, wealth and security.

Box 5.1 Life on Earth

Biodiversity is the variety of life on Earth. It includes diversity at the genetic level, such as that between individuals in a population or between plant varieties, the diversity of species, and the diversity of ecosystems and habitats. Biodiversity encompasses more than just variation in appearance and composition. It includes diversity in abundance (such as the number of genes, individuals, populations or habitats in a particular location), distribution (across locations and through time) and in behaviour, including interactions among the components of biodiversity, such as between pollinator species and plants, or between predators and prey. Biodiversity also incorporates human cultural diversity, which can be affected by the same drivers as biodiversity, and which has impacts on the diversity of genes, other species and ecosystems.

Biodiversity has evolved over the last 3.8 billion years or so of the planet's approximately 5 billion-year history. Although five major extinction events have been recorded over this period, the large number and variety of genes, species and ecosystems in existence today are the ones with which human societies have developed, and on which people depend.

The relationships among biodiversity and the five main themes assessed in this chapter – livelihood security, agriculture, energy, health and culture – clearly demonstrate the importance of biodiversity to these aspects of human well-being. Biodiversity forms the basis of agriculture, and enables the production of foods, both wild and cultivated, contributing to the health and nutrition of all people. Genetic resources have enabled past and current crop and livestock improvements, and will enable future ones, and allow for flexibility according to market demand and adaptation according to changing environmental conditions. Wild biodiversity is perhaps of greatest direct importance to the one billion people around the world who live a subsistence lifestyle. The decline of this diversity has considerable implications for their health, culture and livelihoods. Supporting services, such as nutrient cycling and soil formation, and regulating services, such as pest and disease control, flood regulation and pollination, underpin successful agricultural systems, and contribute to livelihood security.

Cultural ecosystem services are being increasingly recognized as key determinants of human well-being, including through the maintenance of cultural traditions, cultural identity and spirituality. Among the wide range of other benefits from biodiversity, it has enabled the production of energy from biomass and fossil fuels. Such use of biodiversity has brought tremendous benefit to many people (see Box 5.2), but has had

Box 5.2 Value of biodiversity and ecosystem services

The supply of ecosystem services depends on many attributes of biodiversity. The variety, quantity, quality, dynamics and distribution of biodiversity that is required to enable ecosystems to function, and the supplying benefits to people, vary between services. The roles of biodiversity in the supply of ecosystem services can be categorized as provisioning, regulating, cultural and supporting (see Chapter 1), and biodiversity may play multiple roles in the supply of these types of services. For example, in agriculture, biodiversity is the basis for a provisioning service (food, fuel or fibre is the end product), a supporting service (such as micro-organisms cycling nutrients and soil formation), a regulatory service (such as through pollination), and potentially, a cultural service in terms of spiritual or aesthetic benefits, or cultural identity.

The contributions of biodiversity-dependent ecosystem services to national economies are substantial. The science of valuation of ecosystem services is new, and still developing basic conceptual and methodological rigour and agreement, but it has already been very instructive, since the value of such services is generally ignored or underestimated at decision and policy making levels. Identifying economic values of ecosystem services, together with the notions of intrinsic value and other factors, will assist significantly in future decisions relating to trade-offs in ecosystem management.

Value of:
- Annual world fish catch – US$58 billion (provisioning service).
- Anti-cancer agents from marine organisms – up to US$1 billion/year (provisioning service).
- Global herbal medicine market – roughly US$43 billion in 2001 (provisioning service).
- Honeybees as pollinators for agriculture crops – US$2–8 billion/year (regulating service).
- Coral reefs for fisheries and tourism – US$30 billion/year (see Box 5.5) (cultural service).

Honey bees (*Apis mellifera, Apis mellifica*) provide regulatory services through pollination.

Credit: J. Kottmann/WILDLIFE/Still Pictures

Cost of:
- Mangrove degradation in Pakistan – US$20 million in fishing losses, US$500 000 in timber losses, US$1.5 million in feed and pasture losses (regulating provisioning services).
- Newfoundland cod fishery collapse – US$2 billion and tens of thousands of jobs (provisioning service).

Of those ecosystem services that have been assessed, about 60 per cent are degraded or used unsustainably, including fisheries, waste treatment and detoxification, water purification, natural hazard protection, regulation of air quality, regulation of regional and local climate, and erosion control (see Chapters 2, 3, 4 and 6). Most have been directly affected by an increase in demand for specific provisioning services, such as fisheries, wild meat, water, timber, fibre and fuel.

Sources: Emerton and Bos 2004, FAO 2004, MA 2005, Nabhan and Buchmann 1997, UNEP 2006a, WHO 2001

some significant negative knock-on effects in the form of human-induced climate change, and habitat conversion. These trade-offs, inherent in so much of biodiversity use, are becoming increasingly apparent, as there are greater demands for ecosystem services.

People directly use only a very small percentage of biodiversity. Agriculture reduces diversity to increase productivity for a component of biodiversity of particular interest. However, people rely indirectly on a much larger amount of biodiversity without realizing it. There are bacteria and microbes that transform waste into usable products, insects that pollinate crops and flowers, and biologically diverse landscapes that provide inspiration and enjoyment around the world. Such ecosystem services, or the benefits derived from biodiversity, are ultimately dependent on functioning ecosystems. However, the amount of biodiversity required to enable ecosystems to function effectively varies enormously, and how much biodiversity is needed for the sustainable supply of ecosystem services in the present, and into the future, remains largely unknown.

Despite the critical need for more effective conservation and sustainable use, the loss of biodiversity continues, and in many areas is currently increasing in magnitude. Rates of species extinction are 100 times higher than the baseline rate shown by the fossil record (see Box 5.3). The losses are due to a range of pressures, including land-use change and habitat degradation, overexploitation of resources, pollution and the spread of invasive alien species. These pressures are themselves driven by a range of socio-economic drivers, chiefly the growing human population and associated increases in global consumption of resources and energy, and the inequity associated with high levels of per capita consumption in developed countries.

Responses to the continuing loss of biodiversity are varied, and include further designation of protected areas, and, increasingly, the improved management for biodiversity in production landscapes and seascapes. There are recent signs of an emerging consensus that biodiversity conservation and sustainable development are inextricably linked, as for example illustrated by the endorsement by the 2002 Johannesburg World Summit on Sustainable Development (WSSD) of the CBD's 2010 target, and its subsequent incorporation into the Millennium Development Goals.

GLOBAL OVERVIEW OF THE STATUS OF BIODIVERSITY

Ecosystems

Ecosystems vary greatly in size and composition, ranging from a small community of microbes in a drop of water, to the entire Amazon rain forest. The very existence of people, and that of the millions of species with which the planet is shared, is dependent on the health of our ecosystems. People are putting increasing strain on the world's terrestrial and aquatic ecosystems (see Chapters 3 and 4). Despite the importance of ecosystems, they are being modified in extent and composition by people at an unprecedented rate, with little understanding of the implications this will have in terms of their ability to function and provide services in the future (MA 2005). Figure 5.1 depicts an analysis of the status of terrestrial ecosystems.

For more than half of the world's 14 biomes, 20–50 per cent of their surface areas have already been converted to croplands (Olson and others 2001). Tropical dry broadleaf forests have undergone the most rapid conversion since 1950, followed by temperate grasslands, flooded grasslands and savannahs. Approximately 50 per cent of inland water habitats are speculated to have been transformed for human use during the twentieth century (Finlayson and D'Cruz 2005) (see Chapter 4). Some 60 per cent of the world's major rivers have been fragmented by dams and diversions (Revenga and others 2000), reducing biodiversity as a result of flooding of habitats, disruption of flow patterns, isolation of animal populations and blocking of migration routes. River systems are also being significantly affected by water withdrawals, leaving some major rivers nearly or completely dry. In the marine realm, particularly threatened ecosystems include coral reefs and seamounts (see Box 5.4).

Box 5.3 The sixth extinction

All available evidence points to a sixth major extinction event currently underway. Unlike the previous five events, which were due to natural disasters and planetary change (see Box 5.1), the current loss of biodiversity is mainly due to human activities. The current rapid rates of habitat and landscape changes and modifications, increased rates of species extinction, and the reduction in genetic variability due to population declines, are having impacts on natural processes and on the needs of people. The details of many of these impacts remain uncertain, but their major negative influences can be foreseen and avoided or mitigated.

Figure 5.1 Status of terrestrial ecoregions

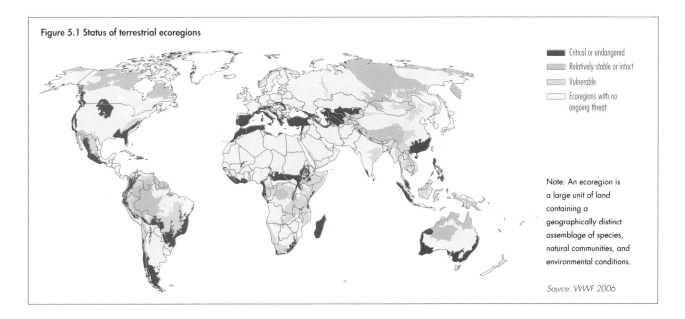

■ Critical or endangered
▨ Relatively stable or intact
▨ Vulnerable
☐ Ecoregions with no ongoing threat

Note: An ecoregion is a large unit of land containing a geographically distinct assemblage of species, natural communities, and environmental conditions.

Source: WWF 2006

Box 5.4 Deep-sea biodiversity

The deep sea is increasingly recognized as a major reservoir of biodiversity, comparable to the biodiversity associated with tropical rain forests and shallow-water coral reefs. The wealth of diverse deep-sea habitats – hydrothermal vents, cold seeps, seamounts, submarine canyons, abyssal plains, oceanic trenches and recently-discovered asphalt volcanoes – contain a vast array of unique ecosystems and endemic species. Although the magnitude of deep-sea diversity is not yet understood (only 0.0001 per cent of the deep seabed has been subject to biological investigations), it has been estimated that the number of species inhabiting the deep sea may be as high as 10 million. It is believed that the deep seabed supports more species than all other marine environments. Marine biodiversity and ecosystems are threatened by pollution, shipping, military activities and climate change, but today fishing presents the greatest threat. The emergence of new fishing technologies and markets for deep-sea fish products has enabled fishing vessels to begin exploiting these diverse, but poorly understood deep-sea ecosystems.

The greatest threat to biodiversity in the deep sea is bottom trawling. This type of high seas fishing is most damaging to seamounts and the coldwater corals they sustain. These habitats are home for several commercial bottom-dwelling fish species. Seamounts are also important

spawning and feeding grounds for species, such as marine mammals, sharks and tuna, which make them very attractive fishing grounds. The long life cycles and slow sexual maturation of deep-sea fish make them particularly vulnerable to large-scale fishing activities. The lack of data on deep-sea ecosystems and associated biodiversity makes it difficult to predict and control the impacts of human activities, but current levels of bottom trawling on the high seas is unlikely to be sustainable, and may even be unsustainable at greatly reduced levels.

Effective management measures for deep-sea fisheries and biodiversity need to be established. Conservation of marine ecosystems has recently extended to the deep sea with the designation in 2003 of the Juan de Fuca Ridge system and associated Endeavour Hydrothermal Vents (2 250 metres deep and 250 kilometres south of Vancouver Island, Canada) as a marine protected area. There are several mechanisms to conserve deep seas, such as the 1982 UN Convention of the Law of the Sea (UNCLOS), 1995 UN Fish Stocks Agreement (UNFSA), International Seabed Authority (ISA), 1992 Convention on Biological Diversity (CBD) and the 1973 Convention on Trade in Endangered Species (CITES). However, these mechanisms need more effective implementation if deep-sea ecosystems are to be conserved and sustainably used.

Sources: Gianni 2004, UNEP 2006b, WWF and IUCN 2001

Examples of species inhabiting the deep sea. False boarfish, *Neocytlus helgae* (left) and cold water coral, *Lophelia* (right).

Credit: Deep Atlantic Stepping Stones Science Party, IFE, URHAO and NOAA (left), UNEP 2006b (right)

The seafloor off Northwest Australia showing dense populations of corals and sponges before trawling (left) and after trawling (right).

Credit: Keith Sainsbury, CSIRO

The fragmentation of ecosystems is increasingly affecting species, particularly migratory species that need a contiguous network of sites for their migratory journeys, species that rely on particular microhabitats and those that require multiple types of habitats during different life cycle stages.

Species

Although about 2 million species have been described, the total number of species range between 5 and 30 million (IUCN 2006, May 1992). Much of this uncertainty relates to the most species-rich groups such as invertebrates.

Current documented rates of extinction are estimated to be roughly 100 times higher than typical rates in the fossil record (MA 2005). Although conservation success in the recovery of several threatened species has been noted (IUCN 2006), and a few species that were presumed extinct have been rediscovered (Baillie and others 2004), it is feasible that extinction rates will increase to the order of 1 000–10 000 times background rates over the coming decades (MA 2005).

Fewer than 10 per cent of the world's described species have thus far been assessed to determine their conservation status. Of these, over 16 000 species have been identified as threatened with extinction. Of the major vertebrate groups that have been comprehensively assessed, over 30 per cent of amphibians, 23 per cent of mammals and 12 per cent of birds are threatened (IUCN 2006).

To understand trends in extinction risk, the conservation status of an entire species group must be assessed at regular intervals. Currently, this information is only available for birds and amphibians, both of which indicate a continuing increase in the risk of extinction from the 1980s to 2004 (Baillie and others 2004, Butchart and others 2005, IUCN 2006).

The threat status of species is not evenly distributed. Tropical moist forests contain by far the highest number of threatened species, followed by tropical dry forests, montane grasslands and dry shrublands. The distribution of threatened species in freshwater habitats is poorly known, but regional assessments from the United States, the Mediterranean Basin and elsewhere indicate that freshwater species are, in general, at much greater risk of extinction than terrestrial taxa (Smith and Darwall 2006, Stein and others 2000). Fisheries have also been greatly depleted, with 75 per cent of the world's fish stocks fully or overexploited (see Chapter 4).

The Living Planet Index measures trends in the abundance of species for which data is available around the world (Loh and Wackernagel 2004). Despite the fact that invertebrates comprise the vast majority of species, trend indices for invertebrate groups only exist for a very small number of species groups, such as butterflies in Europe (Van Swaay 1990, Thomas and others 2004a). The existing limited information suggests that invertebrate and vertebrate population declines may be similar, but further studies are required (Thomas and others 2004b).

Invertebrates, including butterflies, comprise the vast majority of species.

Credit: Ngoma Photos

Genes

Genetic diversity provides the basis for adaptation, allowing living organisms to respond to natural selection, and adapt to their environment. Genes therefore play a strong role in the resilience of biodiversity to global changes, such as climate change or novel diseases. Genes also provide direct benefits to people, such as the genetic material needed for improving yield and disease resistance of crops (see the Agriculture section) or for developing medicines and other products (see the Health and Energy sections).

Over the past two decades, many of the world's most important agricultural crops have lost genetic diversity due to changes in agricultural practices (Heal and others 2002). The continued loss of genetic diversity of such crops may have major implications on food security (see Agriculture section). The amount or rate of loss of genetic diversity is poorly known, but inferences can be made from documented extinctions and population declines, which suggest that substantial genetic loss is occurring (IUCN 2006).

Global responses to curb biodiversity loss

In 2002, parties to the CBD committed themselves to actions to "achieve, by 2010, a significant reduction of the current rate of biodiversity loss at the global, regional and national levels as a contribution to poverty alleviation and to the benefit of all life on earth" (Decision VI/26, CBD Strategic Plan). Setting this target has helped to highlight the need for improved biodiversity indicators, capable of measuring trends in a range of aspects of global biodiversity. It has also helped to galvanize the scientific community to try to develop indicators capable of measuring trends in the various aspects or levels of biodiversity. Figure 5.2 provides a sample of global biodiversity indicators that will be used to measure progress towards the 2010 target. They measure trends in vertebrate populations, extinction risks for birds, global consumption and the establishment of protected areas (SCBD 2006).

The population and extinction risk indices demonstrate a continuing decline in biodiversity, and the ecological footprint indicates that consumption is rapidly and unsustainably increasing. These trends do not bode well for meeting the 2010 biodiversity target at

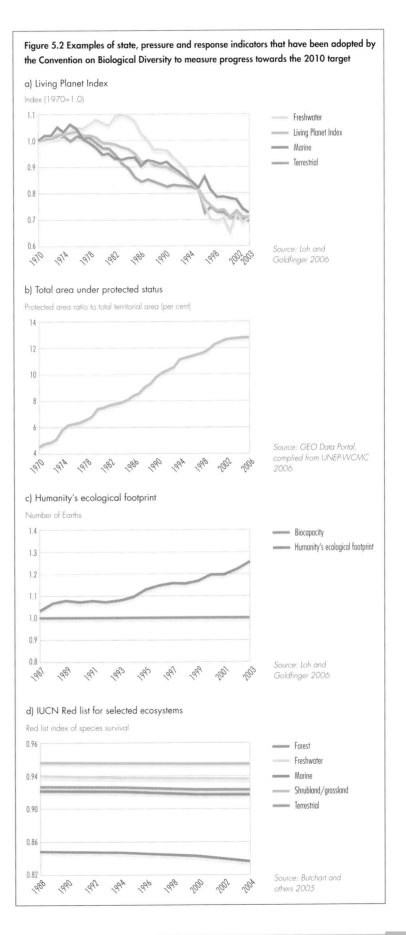

Figure 5.2 Examples of state, pressure and response indicators that have been adopted by the Convention on Biological Diversity to measure progress towards the 2010 target

a) Living Planet Index

Index (1970=1.0)

Freshwater
Living Planet Index
Marine
Terrestrial

Source: Loh and Goldfinger 2006

b) Total area under protected status

Protected area ratio to total territorial area (per cent)

Source: GEO Data Portal, complied from UNEP-WCMC 2006

c) Humanity's ecological footprint

Number of Earths

Biocapacity
Humanity's ecological footprint

Source: Loh and Goldfinger 2006

d) IUCN Red list for selected ecosystems

Red list index of species survival

Forest
Freshwater
Marine
Shrubland/grassland
Terrestrial

Source: Butchart and others 2005

a global scale. Responses to the continuing loss of biodiversity are varied, and include further designation of land and areas of water within protected areas, and increasingly, the improved management for biodiversity in production landscapes and seascapes. The protected areas coverage indicator demonstrates a promising trend in the form of a steady increase in the area under protection.

During the past 20 years, the number of protected areas grew by over 22 000 (Chape and others 2005) and currently stands at more than 115 000 (WDPA 2006). However, the number of protected areas and their coverage can be misleading indicators of conservation (especially for marine areas), as their establishment is not necessarily followed by effective management and enforcement of regulations (Mora and others 2006, Rodrigues and others 2004). Also the percentage and degree to which each ecosystem is protected varies greatly. Roughly 12 per cent of the world's land surface is included within some kind of protected area, but less than one per cent of the world's marine ecosystems are protected, with the Great Barrier Reef and the northwestern Hawaiian islands making up one-third the area of all marine protected areas (Figure 5.3) (Chape and others 2005, SCBD 2006).

In addition to ensuring the effective management of protected areas, emphasis will increasingly need to be placed on the conservation of biodiversity outside protected areas, and in conjunction with other land uses if the rate of loss of biodiversity is to be reduced. The establishment of new policies and processes at all scales, the re-emergence of sustainable agricultural practices, the further development of collaboration among sectors, including corporate partnerships between conservation organizations and extractive industries, and the mainstreaming of biodiversity issues into all areas of decision making, will all contribute to a more secure future for biodiversity, and for sustainable development.

Over the last 20 years environmental issues have increasingly been recognized as important in the development sector at a global scale. The commitment by parties to the CBD to achieve a significant reduction in the rate of biodiversity loss by 2010 as a contribution to poverty alleviation and the benefit of all life on Earth, the endorsement by the 2002 Johannesburg World Summit on Sustainable Development (WSSD) of the CBD's 2010 target, and the incorporation of the 2010 biodiversity target into the Millennium Development Goals as a new target under Goal 7 on environmental sustainability are some examples. A framework for action was proposed at WSSD to implement sustainable development policies, which covered five key areas (water, energy, health, agriculture and biodiversity), This "WEHAB" framework provided a focus, and confirmed the recognition of biodiversity as a key component of the sustainable development agenda.

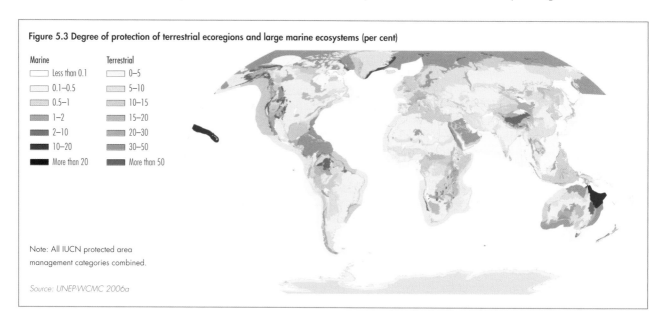

Figure 5.3 Degree of protection of terrestrial ecoregions and large marine ecosystems (per cent)

Marine
- Less than 0.1
- 0.1–0.5
- 0.5–1
- 1–2
- 2–10
- 10–20
- More than 20

Terrestrial
- 0–5
- 5–10
- 10–15
- 15–20
- 20–30
- 30–50
- More than 50

Note: All IUCN protected area management categories combined.

Source: UNEP-WCMC 2006a

DRIVERS OF CHANGE AND PRESSURES

Currently, population growth and patterns of consumption, which lead to increased demand for ecosystem services and energy, are the most important drivers affecting biodiversity. These drivers result in pressures that have direct impacts on ecosystems, species and genetic resources (see Table 5.1). Human activities cause changes in both the living and non-living components of ecosystems, and these pressures have increased dramatically over the past few decades.

Drivers and pressures seldom act in isolation. They tend to interact in synergistic ways, and their impacts on biodiversity are more than the sum of the effects of the individual drivers and pressures themselves (MA 2005). Additionally, the interaction shows considerable regional variation (see Chapter 6). Drivers and pressures act at different temporal and spatial scales. For example, sediments from deforestation in the headwaters of the Orinoco River, deep in South America, have impacts far out in the Wider Caribbean Sea basin, changing the nutrient availability and turbidity of the waters (Hu and others 2004).

Since the Brundtland Commission report, the globalization of agriculture and inappropriate agricultural policies have emerged as leading drivers influencing the loss of species and ecosystem services. Globalization is leading to major changes in where, how and who produces food and other agricultural commodities. Global market demand for high value commodities such as soybeans, coffee, cotton, oil palm, horticultural crops and biofuels has resulted in substantial habitat conversion and ecosystem degradation. This has replaced diverse smallholder farms with larger monoculture enterprises. In other cases, globalization has concentrated and intensified production on the most productive lands, reducing net deforestation rates.

Virtually all of the factors leading to the accelerating loss of biodiversity are linked to the development of and increasing demand for energy by society. Of particular importance are the high levels of per capita energy use in the developed world, and the potential growth in energy use in the large emerging economies. The rapid increase in demand for energy has profound impacts on biodiversity at two levels (Guruswamy and McNeely 1998, Wilson 2002): impacts from the production and distribution of energy, and those resulting from the use of energy. Exploration for hydrocarbons,

pipeline construction, uranium and coal mining, hydroelectric dam construction, harvesting for fuelwood and, increasingly, biofuel plantations can all lead to significant biodiversity loss, both on land and at sea.

The widespread anthropogenic changes to the environment have altered patterns of human disease, and increased pressures on human well-being. The loss of genetic diversity, overcrowding and habitat fragmentation all increase susceptibility to disease outbreaks (Lafferty and Gerber 2002). Some ecosystem changes create new habitat niches for disease vectors, for example, increasing the risk of malaria in Africa and the Amazon Basin (Vittor and others 2006).

Deforestation in Serra Parima, Orinoco River basin..

Credit: Mark Edwards/Still Pictures

The Orinoco River carries sediment that originates from land degradation far away in the Andes all the way to the Caribbean. By contrast, the Caroní river water is clear blue, as it drains the ancient landscapes of the Guyana Highlands, where erosion is much slower.

Credit: NASA 2005

Trends in biodiversity over the next few decades will largely depend on human actions, especially those relating to land-use changes, energy production and conservation. These actions will, in turn, be affected by various factors including advances in our understanding of ecosystem services, development of viable alternatives to natural resources (especially fossil fuels), and the emphasis placed on the environment and conservation by developed and developing country governments alike. Efforts made to predict the prospects for species-level biodiversity have indicated that extinctions are likely to continue at a pace well above the background rate, with as many as 3.5 per cent of the world's birds (BirdLife International 2000), and perhaps a greater proportion of amphibians and freshwater fish, being lost or committed to extinction by the middle of the century.

Climate change is likely to play an increasing role in driving changes in biodiversity, with species' distributions and relative abundances shifting as their preferred climates move towards the poles and higher altitudes, leaving those endemic to polar and high mountain regions most at risk. In addition, changes in the ranges of vector species may facilitate the spread of diseases affecting humans and other species, for example, malaria and the amphibian fungal disease, chytridiomycosis.

Further pressure on biodiversity will result from the continuing increase in the global human population, which is predicted to reach 8 billion by 2025 (GEO Data Portal, from UNPD 2007). All will require access to food and water, leading to an unavoidable increase in stresses on natural resources. The increased infrastructure required to support such a global population of more than 8 billion people will likely have particular effects on

biodiversity in the future (see Chapter 9). The increased need for agricultural production to feed the population will likely be met largely by commercial intensification, with negative consequences for the genetic diversity of agricultural crops and livestock. Extensification will also help to meet the need, with a predicted additional 120 million hectares required by 2030 in developing countries, including lands of high biodiversity value (Bruinsma 2003).

Tropical forests are the terrestrial system likely to be the most affected by human actions in the first half of this century, largely through habitat conversion for agricultural expansion (including the growth of biofuel plantations). Ongoing fragmentation will result in the degradation of the largest remaining areas of species-rich forest blocks in Amazonia and the Congo basin. Marine and coastal ecosystems are also expected to continue to be degraded, with existing impacts, such as fishing, eutrophication from terrestrial activities and coastal conversion for aquaculture, increasing (Jenkins 2003). Large species, including top predators, will be particularly affected, with considerable declines and some extinctions likely.

Changes, both positive and negative, in biodiversity trends over the next few decades are inevitable, yet the details of these changes are not yet set in stone. Their magnitude can be somewhat reduced and mitigated by the further integration of biodiversity considerations into national policies, increasing corporate social responsibility activities and conservation actions. With commitment from governments, the private sector, scientific institutions and civil society, action can be taken to ensure progress towards the CBD 2010 target, the Millennium Development Goals, and beyond.

The *Telestes polylepis*, a critically endangered freshwater species found in Croatia.

Credit: Jörg Freyhof

ENVIRONMENTAL TRENDS AND RESPONSES

Biodiversity is closely linked to livelihood security, agriculture, energy, health and culture, the five themes analysed in this chapter. Of these themes, agriculture (in terms of food security) and energy were explicitly considered in the Brundtland Commission report, and with a focus on water and health, tie together the WEHAB framework for action arising from WSSD. These linkages are likely to emerge as the most critical in implementing actions that will result in truly sustainable development. Table 5.1 summarizes some of the impacts of major drivers on biodiversity, ecosystems and human well-being.

LIVELIHOOD SECURITY

Ecosystems provide critical services

Biodiversity contributes directly and indirectly to livelihood security (MA 2005). Functioning ecosystems are crucial buffers against extreme climate events, and act as carbon sinks and filters for water-borne and airborne pollutants. For example, the frequency of shallow landslides appears to be strongly related to vegetation cover, as roots play an important role in slope stability, and can give the soil mechanical support at shallow depth. In coastal areas, mangroves and other wetlands are particularly effective in providing shoreline stability, reducing erosion, trapping sediments, toxins and

Table 5.1 Impacts on biodiversity of major pressures and associated effects on ecosystem services and human well-being

Pressures	Impacts on biodiversity	Potential implications for ecosystem services and human well-being	Examples
Habitat conversion	■ Decrease in natural habitat ■ Homogenization of species composition ■ Fragmentation of landscapes ■ Soil degradation	■ Increased agricultural production ■ Loss of water regulation potential ■ Reliance on fewer species ■ Decreased fisheries ■ Decreased coastal protection ■ Loss of traditional knowledge	Between 1990 and 1997, about 6 million hectares of tropical humid forest were lost annually. Deforestation trends differ from region to region, with the highest rates in Southeast Asia, followed by Africa and Latin America. Additionally, about 2 million ha of forest are visibly degraded each year (Achard and others 2002). (See Chapter 3.)
Invasive alien species	■ Competition with and predation on native species ■ Changes in ecosystem function ■ Extinctions ■ Homogenization ■ Genetic contamination	■ Loss of traditionally available resources ■ Loss of potentially useful species ■ Losses in food production ■ Increased costs for agriculture, forestry, fisheries, water management and human health ■ Disruption of water transport	The comb jelly, *Mnemiopsis leidyi*, accidentally introduced in 1982 by ships from the US Atlantic coast, dominated the entire marine ecosystem in the Black Sea, directly competing with native fish for food, and resulting in the destruction of 26 commercial fisheries by 1992 (Shiganova and Vadim 2002).
Overexploitation	■ Extinctions and decreased populations ■ Alien species introduced after resource depletion ■ Homogenization and changes in ecosystem functioning	■ Decreased availability of resources ■ Decreased income earning potential ■ Increased environmental risk (decreased resilience) ■ Spread of diseases from animals to people	An estimated 1–3.4 million tonnes of wild meat (bushmeat) are harvested annually from the Congo Basin. This is believed to be six times the sustainable rate. The wild meat trade is a large, but often invisible contributor to the national economies dependent on this resource. It was recently estimated that the value of the trade in Côte d'Ivoire was US$150 million/year, representing 1.4 per cent of the GNP (POST 2005). (For more on overexploitation of fish stocks, see Chapter 4.)
Climate change	■ Extinctions ■ Expansion or contraction of species ranges ■ Changes in species compositions and interactions	■ Changes in resource availability ■ Spread of diseases to new ranges ■ Changes in the characteristics of protected areas ■ Changes in resilience of ecosystems	Polar marine ecosystems are very sensitive to climate change, because a small increase in temperature changes the thickness and amount of sea ice on which many species depend. The livelihoods of indigenous populations living in sub-arctic environments and subsisting on marine mammals are threatened, since the exploitation of marine resources is directly linked to the seasonality of sea ice (Smetacek and Nicol 2005). (For more on climate change, see Chapter 2.)
Pollution	■ Higher mortality rates ■ Nutrient loading ■ Acidification	■ Decreased resilience of service ■ Decrease in productivity of service ■ Loss of coastal protection, with the degradation of reefs and mangroves ■ Eutrophication, anoxic waterbodies leading to loss of fisheries	Over 90 per cent of land in the EU-25 countries in Europe is affected by nitrogen pollution greater than the calculated critical loads. This triggers eutrophication, and the associated increases in algal blooms and impacts on biodiversity, fisheries and aquaculture (De Jonge and others 2002). (See Chapters 4 and 6.)

Source: Adapted from MA 2005

Pita, *Aechmea magdalane*, a thorny-leaved terrestrial bromeliad, grows naturally in lowland forests of southeast Mexico. It is harvested for the commercial extraction of fibre used in the stiching and embroidering of leatherwork. One hectare of forest can provide up to 20 kilogrammes of pita fibre per year, generating an average cash income of US$1 000/ha.

Credit: Elaine Marshall

nutrients, and acting as wind and wave breaks to buffer against storms. The role of inland wetlands in storing water and regulating stream-flow is both a function of their vegetative composition, which helps to maintain soil structure, and their characteristic gentle slopes.

Current trends in land degradation and habitat loss continue to contribute to reducing livelihood options while heightening risks. Changes in land management, particularly the replacement of fire-adapted systems with other forms of land cover, can increase the intensity and extent of fires, increasing the hazard to people. Land-use change also influences climate at local, regional and global scales. Forests, shrub and grasslands, freshwater and coastal ecosystems provide critical sources of food and complementary sources of income (see Box 5.2). Fish and wild meat provide animal protein, while other forest resources provide dietary supplements. These ecosystem goods act as critical safety nets for millions of rural poor. Traditionally, access rights and tenure arrangements for these public goods have evolved to enable equitable distribution of such extractive activities. More recently, due to increased population densities and the introduction of market models, access to these common property resources has been increasingly

restricted, with resulting impacts on rural livelihoods. With reliable access to markets, the commercialization of many wild-harvested products can be extremely successful in contributing to sustaining rural livelihoods (Marshall and others 2006).

Box 5.5 Coral reefs in the Caribbean

The global net value of coral reefs relating to fisheries, coastal protection, tourism and biodiversity, is estimated to total US$29.8 billion/year. However, nearly two-thirds of Caribbean coral reefs are reported to be threatened by human activities. The predominant pressure in the region is overfishing, which affects approximately 60 per cent of Caribbean reefs. Other pressures include large quantities of dust originating from deserts in Africa, which are blown across the Atlantic Ocean and settle on reefs in the Caribbean, leading to significant coral mortality. It has been proposed that this phenomenon led to a coral bleaching event that began in 1987, correlating with one of the years of maximum dust flux into the Caribbean. Coral degradation has negative impacts on coastal communities, including the loss of fishing livelihoods, protein deficiencies, loss of tourism revenue and increased coastal erosion.

Sources: Burke and Maidens 2004, Cesar and Chong 2004, Griffin and others 2002, MA 2005, Shinn and others 2000

Environmental degradation, combined with heightened exposure and vulnerability of human settlements to risk, contributes to vulnerability to disasters. Almost 2 billion people were affected by disasters in the last decade of the 20th century, 86 per cent of them due to floods and droughts (EM-DAT). Long spells of drought associated with the El Niño Southern Oscillation phenomenon (ENSO) contributed to forest fires in the Amazon Basin, Indonesia and Central America in 1997–1998. In Indonesia alone, an estimated 45 600 square kilometres of forest were destroyed (UNEP 1999). In Central America, the loss of over 15 000 km² of forests due to wildfires reduced the capacities of natural forests to buffer the impacts of heavy rainfall and hurricanes, and contributed to the devastating impact of Hurricane Mitch in 1998 (Girot 2001). These impacts spread beyond the tropics, as the large forest fires of California, Spain, Portugal and other Mediterranean countries in 2005 illustrated (EFFIS 2005). Furthermore, coral degradation has negative impacts on coastal communities (see Box 5.5).

The clustering of climate-related and biological risks will also contribute to impacts on human well-being through events such as heat waves and crop failures. The impact on human health has been addressed in greater detail in the Health section.

Ecosystems minimize risks

The linkages between biodiversity and livelihood security are complex, and based on the intrinsic relationship between societies and their environment. Policies that can address both the risks and opportunities posed by rapid environmental changes will require a combined focus on ecosystem management, sustainable livelihoods and local risk management. For example, policies aimed at the improved management of water resources and the non-structural mitigation of weather-related hazards can contribute to the reduction of disaster risks by enhancing landscape restoration, coastal forest management and local conservation and sustainable use initiatives. In coastal ecosystems, restoring mangroves in cyclone-prone areas increases physical protection against storms, creates a reservoir for carbon sequestration and increases livelihood options by generating much-needed income for local communities (MA 2005). Although the evidence

---

Box 5.6 Mangrove restoration for buffering storm surges in Viet Nam

In Viet Nam, tropical cyclones have caused a considerable loss of livelihood resources, particularly in coastal communities. Mangrove ecosystem rehabilitation along much of Viet Nam's coastline is an example of a cost-effective approach to improving coastal defences while generating local livelihoods. Since 1994, the Viet Nam National Chapter of the Red Cross has worked with local communities to plant and protect mangrove forests in northern Viet Nam. Nearly 120 km² of mangroves have been planted, with substantial resulting benefits. Although planting and protecting the mangroves cost approximately US$1.1 million, it saved US$7.3 million/year in dyke maintenance.

During the devastating typhoon Wukong in 2000, project areas remained unharmed, while neighbouring provinces suffered huge losses in lives, property and livelihoods. The Viet Nam Red Cross has estimated that some 7 750 families have benefited from mangrove rehabilitation. Family members can now earn additional income from selling crabs, shrimp and molluscs, while increasing the protein in their diets.

Source: IIED 2003

---

base is varied, communities hit by the 2004 tsunami in South Asia reported less damage in areas with healthy mangrove forests than those with few natural sea defences (Dahdouh-Guebas and others 2005). India and Bangladesh have come to recognize the importance of the Sunderbans mangrove forest in the Gulf of Bengal, not only as a source of livelihoods for fishing communities, but also as an effective mechanism for coastal protection. Viet Nam is also investing in mangrove restoration as a cost-effective means for increased coastal protection (see Box 5.6). Similar benefits can be derived from coral reefs (UNEP-WCMC 2006b).

AGRICULTURE

Links between biodiversity and agriculture

Agriculture is defined broadly here to include crops and agroforestry products, livestock and fisheries production. Of some 270 000 known species of higher plants about 10 000–15 000 are edible, and about 7 000 of them are used in agriculture. However, increased globalization threatens to diminish the varieties that are traditionally used in most agricultural systems. For example, only 14 animal species currently account for 90 per cent of all livestock production, and only 30 crops dominate global agriculture, providing an estimated 90 per cent of the calories consumed by the world's population (FAO 1998). Despite its crucial importance in supporting societies, agriculture remains the largest driver of genetic erosion, species loss and conversion of natural habitats around the world (MA 2005) (see Figure 5.4).

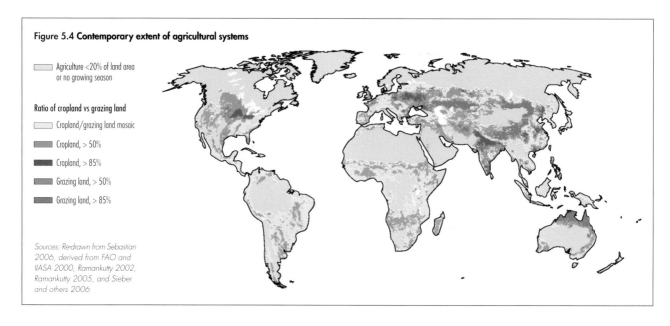

Figure 5.4 Contemporary extent of agricultural systems

Agriculture <20% of land area
or no growing season

Ratio of cropland vs grazing land

Cropland/grazing land mosaic

Cropland, > 50%

Cropland, > 85%

Grazing land, > 50%

Grazing land, > 85%

Sources: Re-drawn from Sebastian 2006, derived from FAO and IIASA 2000, Ramankutty 2002, Ramankutty 2005, and Sieber and others 2006

Both cultivated and wild biodiversity provide services necessary for agriculture (see Table 5.2). Although seldom valued in economic terms, these services play a very significant role in national and regional economies. Different types of agricultural production systems (such as commercial intensive, smallholder, pastoralism and agroforestry systems) use these services to varying degrees and intensity. For example, the use of nitrogen-fixing legume trees in maize-based systems of Eastern and Southern Africa is helping local farming populations to increase per hectare production of maize without otherwise investing in inorganic fertilizers (Sanchez 2002). In addition, environmental benefits are gained through carbon sequestration and provision of fuelwood.

Habitat conversion is often justified as essential to increasing agricultural production, and trends in agricultural land use over the past 20 years are presented in Chapters 3 and 6. Although more than 300 000 km² of land have been converted to agricultural use in the tropics alone (Wood and others 2000), much of this is of marginal use for agriculture or particular crops. This has led to inefficient use of resources, often resulting in degradation of land and ecosystem services (see Chapter 3). Some 1.5 billion people, about half of the world's total labour force and nearly one-quarter of the global population, are employed in agriculture, or their livelihoods are directly linked to it (MA 2005), and women make up the majority of agricultural workers. When agriculture on marginal lands is reduced and these lands are appropriately managed, ecosystems can recover, as demonstrated by the expansion of forests in parts of Europe, North America, Japan, China, India, Viet Nam, New Zealand and Latin America (Aide and Grau 2004, Mather and Needle 1998).

Table 5.2 Biodiversity benefits to agriculture through ecosystem services

Provisioning	Regulating	Supporting	Cultural
■ Food and nutrients ■ Fuel ■ Animal feed ■ Medicines ■ Fibres and cloth ■ Materials for industry ■ Genetic material for improved varieties and yields ■ Pollination ■ Pest resistance	■ Pest regulation ■ Erosion control ■ Climate regulation ■ Natural hazard regulation (droughts, floods and fire)	■ Soil formation ■ Soil protection ■ Nutrient cycling ■ Water cycling	■ Sacred groves as food and water sources ■ Agricultural lifestyle varieties ■ Genetic material reservoirs for improved varieties and yields ■ Pollinator sanctuaries ■ Erosion control

Source: MA 2005

Meeting global food needs poses increasing challenges, and will require either intensification or extensification to increase agricultural productivity (Tillman and others 2002). Intensified systems tend to be dominated by only a few varieties. This approach is usually associated with higher levels of inputs, including technology, agrochemicals, energy and water use. The latter three, at least, have serious negative impacts on biodiversity.

Extensification relies on lower inputs, and generally on more land being used, often through habitat conversion. In many parts of the world, agricultural extensification involves converting more land for the cultivation of major commodities such as soybeans (Latin America and the Caribbean), oil palm and rubber (Asia and the Pacific), and coffee (Africa, Latin America and Asia), and it is exacerbated by the emergence of new markets for export. In Brazil, for example, the area of land used for growing soybeans (most of which are exported to China) grew from 117 000 km^2 in 1994 to 210 000 km^2 in 2003. This was driven by a 52 per cent increase in world consumption of soybeans and soybean products (USDA 2004), and these figures continue to rise dramatically.

A major agricultural biotechnology innovation during the past two decades is the use of "transgenic" or living modified organisms (LMOs) to provide new attributes in different crops and breeds (FAO 2004, IAASTD 2007). The technology is very young, and major investments are being made to enhance its contributions to human well-being and business stability. Research on LMOs has focused mainly on mitigating the impacts of pests and diseases, and there is evidence of reduced needs for pesticides and herbicides in some crops, such as cotton and maize, through genetic modification (FAO 2004). The global production of genetically modified crops (mainly maize, soybean and cotton) was estimated to cover more than 900 000 km^2 in 2005 (James 2003). The use of LMOs is, as for many new technologies, highly controversial, specifically in relation to the uncertain impacts on ecosystems (through escape and naturalization in the landscape), human health and social structures. There are concerns about how its introduction will affect poor people, whose livelihoods depend primarily on traditional low input agricultural practices. Increased research, monitoring and regulation are needed to ensure these negative impacts are avoided as this technology is developed

(see Chapter 3). The Cartagena Protocol on Biosafety was negotiated and adopted under the CBD to develop a global framework for managing and regulating LMOs (FAO 2004, Kormos 2000).

More recently, increasing attention is being given to the existing and potential impacts of climate change on agriculture. Issues include the timing of growth, flowering and maturing of crops, and the impacts of (and on) pollinators, water resources and the distribution of rainfall. There are also issues of changes in market structures, yields for different crops and strains, and the impacts of extreme weather events on traditional methods and livelihoods (Stige and others 2005). Models show that in some areas, specifically where low temperature is a growth-limiting factor, agricultural productivity may increase with climate change. In other areas, where water and heat are limiting factors, productivity may be severely curtailed (IPCC 2007).

Changes in production practices and loss of diversity in agro-ecosystems can undermine the ecosystem services necessary to sustain agriculture. For example, pollinator diversity and numbers are affected by habitat fragmentation (Aizen and Feinsinger 1994, Aizen and others 2002), agricultural practices (Kremen and others 2002, Partap 2002), the land-use matrix surrounding agricultural areas (De Marco and Coelho 2004, Klein and others 2003) and other land-use changes (Joshi and others 2004). Although some of the crops that supply a significant proportion of the world's major staples do not require animal pollination (such as rice and maize), the decline of pollinators has long-term consequences for those crop species that serve as crucial sources of micronutrients and minerals (such as fruit trees and vegetables) in many parts of the world.

Genetic erosion, loss of local populations of species, and loss of cultural traditions are often intimately intertwined. While rates of genetic erosion are poorly known, they generally accompany the transition from traditional to commercially developed varieties (FAO 1998). In crop and livestock production systems throughout the developing world, genetic erosion reduces smallholder farmer options for mitigating impacts of environmental change and reducing vulnerability, especially in marginal habitats or agricultural systems that are predisposed to extreme weather conditions (such as arid and semi-arid lands of Africa and India).

Implications for agricultural technologies and policy
Methodological and technological innovation

Since the Brundtland Commission report, agricultural research and development has made major advances in integrating conservation and development to mitigate loss of biodiversity, reverse land degradation and foster environmental sustainability. Much remains to be done to create the appropriate enabling environment in many countries, rich and poor alike, especially in eliminating anti-conservation regulations and inappropriate agricultural production subsidies.

A particular area of advancement is the use of innovative agricultural practices to enhance production while conserving native biodiversity (Collins and Qualset 1999, McNeely and Scherr 2001, McNeely and Scherr 2003, Pretty 2002). Efforts to foster biodiversity-friendly practices by integrating trees on farms (agroforestry), conservation agriculture, organic agriculture and integrated pest management are all contributing towards the sustainability of production landscapes (see Chapter 3). Agroforestry, for example, has emerged as a major opportunity for achieving biodiversity conservation and sustainability in production landscapes (Buck and others 1999, McNeely 2004, Schroth and others 2004), through three major pathways: reducing pressure on natural forests, providing habitat for native plant and animal

species and serving as an effective land use in fragmented landscapes (see Box 5.7).

Integrated land management approaches are also helping to enhance ecosystem resilience through participatory processes that engage and empower farmers, strengthen local institutions and create options for value-added income generation. These approaches offer significant prospects for restoring degraded lands to enhance habitat connectivity and ecosystem processes. In the tropical forest margins, where slash-and-burn farming is a major cause of deforestation, knowledge of land-use dynamics has helped to identify practical options that are profitable for small-scale farmers and at the same time environmentally sustainable (Palm and others 2005). However, a major challenge to wide-scale implementation of these approaches is the lack of appropriate policy frameworks that align rural and agricultural policies with the protection of biodiversity and ecosystem services. Without such links, the value of integrated natural resource management (Sayer and Campbell 2004) and ecoagriculture (McNeely and Scherr 2003) innovations will remain marginal in ensuring the long-term viability of biodiversity.

Very substantial collections of plant genetic resources for food and agriculture are now maintained

Agriculture in a rain forest in Ghana, growing cassava and fruits such as bananas and papayas.

Credit: Ron Giling/Still Pictures

around the world through the Consultative Group on International Agricultural Research (CGIAR) system. These institutional gene banks are vital for safeguarding germplasm. Farmers have much to contribute at the local level in maintaining the viability of different varieties, such as is being done in an innovative partnership between the International Potato Center and local communities in Peru, an approach that produces income for the farmers while conserving genetic variability. This also helps maintain local ecological knowledge.

Policy options and governance mechanisms

Local and community initiatives remain crucial for supporting agricultural approaches that maintain biodiversity. It is challenging to expand these initiatives, since they are based on local differentiation and diversity, rather than homogenization and mass production. The development of recognized standards and certification for production methods can help give producers in these initiatives greater weight and value in the global market.

However, little progress has been made overall on institutionalizing a more diverse approach to production systems, and in monitoring its effects. The techniques that would support reduced pesticide or herbicide use, for example, have yet to be adopted in most countries, and the full value of the ecosystem services provided by ecologically oriented agricultural systems are only very slowly being recognized. Increased research, and the adoption of techniques, such as integrated pest management, can reduce the use of chemicals while providing important biodiversity conservation services. Similarly, remedial measures required to restore productivity to degraded lands are not being implemented on the scale required. The ecosystem approach can provide a framework for developing practices, such as riparian buffer systems, to both support biodiversity conservation, and assist in water management and purification.

National level legislative and policy measures on land tenure and land-use practices will be key to facilitating wide-scale adoption of proven biodiversity-supporting methodologies and technological options in agriculture. The options offer practical solutions that reduce the impacts of agriculture on biodiversity, but need to be considered within a supportive policy framework that encompasses both commercial and small-scale agricultural production landscapes.

Box 5.7 Serenading sustainability: rewarding coffee farmers in Central America for biodiversity-friendly practices

Research into the disappearance of songbirds in the US Midwest is leading to innovations in the production practices and marketing of high-value coffee produced in Central America. Smithsonian Institution researchers found that conversion of forests in Central America for coffee plantations substantially reduced the winter habitats for many migratory birds, reducing their breeding success and their numbers. They worked with coffee producers to test methods of "bird friendly" planting, using intact or minimally-thinned forests for coffee tree planting. This method of planting produces somewhat fewer coffee beans, but they are of higher quality, and require fewer pesticide and fertilizer inputs. Additionally, the coffee can be marketed as coming from environmentally-friendly sources, potentially bringing in higher prices. Different certification systems, for example for Bird Friendly® and Shade Grown coffee, show the development and limitations of markets for more sustainably grown crops.

Sources: Mas and Dietsch 2004, Perfecto and others 2005

Box 5.8 Initiatives for implementation by biodiversity Multilateral Environmental Agreements

In 1996, parties to the CBD adopted a programme of work on conservation and sustainable use of agricultural biological diversity. In addition, the CBD has established the International Initiative for the Conservation and Sustainable Use of Pollinators, and the International Initiative for the Conservation and Sustainable Use of Soil Biodiversity, both to be implemented in cooperation with FAO and the Global Strategy for Plant Conservation. Although much remains to be done, global policy processes are helping national governments, particularly in developing countries, to better understand the implications of globalization in agriculture for national policies and development priorities. The entry into force in June 2004 of the International Treaty on Plant Genetic Resources for Food and Agriculture represents another step in governance of the conservation and use of crop genetic resources, especially for large-scale commercial agriculture. This provides for a multilateral system of exchange for some 30 crops and 40 forage species, and should greatly facilitate use and stimulate the development of effective benefit sharing mechanisms.

At the global scale, ongoing international negotiations are addressing imbalances in markets, subsidies and property rights, all of which have direct links to land use in agriculture (see Box 5.8). However, there are still major challenges to the conclusion and implementation of the kind of agreements that would generate tangible impacts on biodiversity and agriculture, particularly in the developing world.

ENERGY

Links between biodiversity and energy

Many forms of energy are the result of a service provided by ecosystems, now or laid down in the form of fossil fuels far in the past. Conversely, society's growing requirements for energy are resulting in significant changes in those same ecosystems, both in the search for energy sources, and as a result of energy use patterns. Given that energy is a fundamental requirement for supporting development in all economies, the challenge is to sustainably provide it without driving further loss of biodiversity. It is necessary to define the trade-offs required, and develop appropriate mitigation and adaptation strategies.

Demand for energy is projected to grow at least 53 per cent by 2030 (IEA 2006). Energy from biomass and waste is projected to supply about 10 per cent of global demand until 2030 (see Figure 5.5). However, this assumes that adequate fossil fuels will be available to address the majority of the increase in demand, and some have suggested this may not be realistic (Campbell 2005). Energy-related carbon dioxide emissions are expected to increase slightly faster than energy use by 2030 (see Chapter 2).

Energy use has impacts at local, national and global levels. Pollution from burning fossil fuels, and the associated effects of acid rain have been a problem for European and North American forests, lakes and soils, although the impacts on biodiversity have not been as significant or widespread as cautioned in the Brundtland Commission report. While emission controls in Europe and North America led to a reversal of acidification trends, there is now a risk of acidification in other areas of the world, particularly Asia (see Chapters 2 and 3). Use of thermal and nuclear power results in waste disposal problems, as do solar cells, which can result in soil contamination by heavy metals. Desertification in the Sahel and elsewhere in sub-Saharan Africa has been linked in part to fuel demand from biomass (see Box 5.9) (Goldemberg and Johansen 2004). Indirect effects of energy use include both overexploitation of natural resources and greatly facilitated spread of invasive alien species through global trade, both made possible through cheap and easily-available energy for transport.

The impacts noted above are relatively localized and small in comparison to the potential impacts of climate change, which results largely from energy use (see Chapters 2, 3 and 4). As a result of climate change, species ranges and behaviour are changing (see Box 5.10 and Chapter 6), with consequences for human well-being, including changing patterns of human disease distribution, and increased opportunities for invasive alien species. Species most likely to be affected include those that already are rare or threatened, migratory species, polar species, genetically impoverished species, peripheral populations and specialized species, including those restricted to alpine areas and islands. Some amphibian species extinctions have already been linked with climate change (Ron and others 2003, Pounds and others 2006), and a recent global study estimated that 15–37 per cent of regional endemic species could be committed to extinction by 2050 (Thomas and others 2004 b).

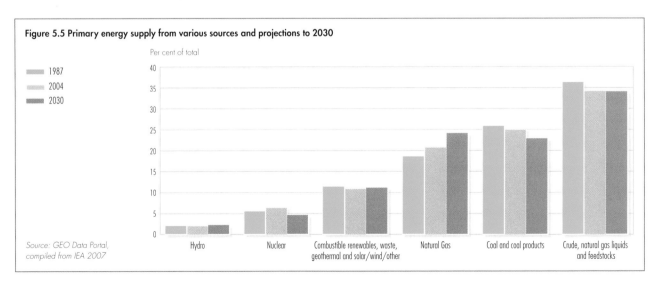

Figure 5.5 Primary energy supply from various sources and projections to 2030

Per cent of total

Legend: 1987, 2004, 2030

Categories: Hydro | Nuclear | Combustible renewables, waste, geothermal and solar/wind/other | Natural Gas | Coal and coal products | Crude, natural gas liquids and feedstocks

Source: GEO Data Portal, compiled from IEA 2007

Biodiversity-based energy sources include both traditional biomass and modern biofuels. Ecosystems provide relatively inexpensive and accessible sources of traditional biomass energy, and therefore have a vital role to play in supporting poor populations (see Figure 5.6). If these resources are threatened, as is the case in some countries with extreme deforestation, poverty reduction will be an even greater challenge. Use of fuelwood can cause deforestation, but demand for fuelwood can also encourage tree planting, as occurs, for example, in Kenya, Mali and several other developing countries.

Source: Barnes and others 2002, FAO 2004

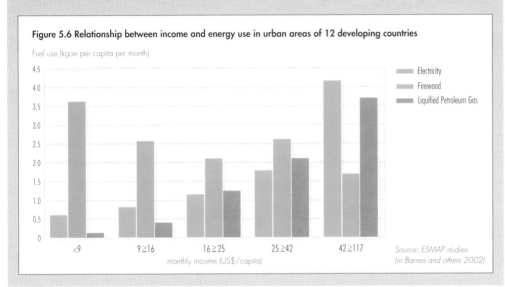

Figure 5.6 Relationship between income and energy use in urban areas of 12 developing countries

Fuel use (kgoe per capita per month)

Legend: Electricity, Firewood, Liquified Petroleum Gas

monthly income (US$/capita)

Source: ESMAP studies (in Barnes and others 2002)

Climate change is also having impacts at ecosystem scales. By 2000, 27 per cent of the world's coral reefs had been degraded in part by increased water temperatures, with the largest single cause being the climate-related coral bleaching event of 1998. For some reefs recovery is already being reported (Wilkinson 2002). Mediterranean-type ecosystems found in the Mediterranean basin, California, Chile, South Africa and Western Australia are expected to be strongly affected by climate change (Lavorel 1998, Sala and others 2000).

Managing energy demand and biodiversity impacts

Few energy sources are completely biodiversity neutral, and energy choices need to be made with an understanding of the trade-offs involved in any specific situation, and the subsequent impacts on biodiversity and human well-being (see Table 5.3). Biodiversity management is emerging as a key tool for the mitigation of and adaptation to the impacts of climate change – from avoided deforestation to biodiversity offsets – while contributing to the conservation of a wide range of ecosystem services.

Reports of extinctions
- Amphibians (Pounds and others 2006)

Reports of changes in species distribution
- Arctic foxes (Hersteinsson and MacDonald 1992)
- Mountain plants (Grabbherr and others 1994)
- Intertidal organisms (Sagarin and others 1999)
- Northern temperate butterflies (Parmesan and others 1999)
- Tropical amphibians and birds (Pounds and others 1999)
- British birds (Thomas and Lennon 1999)
- Tree distributions in Europe (Thuiller 2006)

Reports of changes in species behaviour
- Earlier flight times in insects (Ellis and others 1997, Woiwod 1997)
- Earlier egg laying in birds (Brown and others 1999, Crick and Sparks 1999)
- Breeding in amphibians (Beebee 1995)
- Flowering of trees (Walkovsky 1998)
- Ant assemblages (Botes and others 2006)
- Salamanders (Bernardo and Spotila 2006)

Reports of changes in population demography
- Changes in population sex ratios in reptiles (Carthy and others 2003, Hays and others 2003, Janzen 1994)

There are a number of management and policy responses to the increasing demand for energy and the impacts on biodiversity. One important response to the rising price of oil is increasing interest in other energy sources. Prime among these are biofuels, with several countries investing significant resources in this field (see Box 5.11). The world output of biofuels, assuming current practice and policy, is projected to increase almost fivefold, from 20 million tonnes of oil equivalent (Mtoe) in 2005 to 92 Mtoe in 2030. Biofuels, which are produced on 1 per cent of the world's arable land, support 1 per cent of road transport demand, but that is projected to increase to 4 per cent by 2030, with the biggest increases in United States and Europe. Without significant improvement in productivity of biofuel crops, along with similar progress in food crop agricultural productivity, achieving 100 per cent of transport fuel demand from biofuels is clearly impossible (IEA 2006). In addition, large-scale biofuel production will also create vast areas of biodiversity-poor monocultures, replacing ecosystems such as low-productivity agricultural areas, which are currently of high biodiversity value.

Current actions to address the impacts of climate change can be both beneficial and harmful to biodiversity. For example, some carbon sequestration programmes, designed to mitigate impacts of greenhouse gases, can lead to adverse impacts on biodiversity through the establishment of monoculture forestry on areas of otherwise

high biodiversity value. Avoiding deforestation, primarily through forest conservation projects, is an adaptation strategy that may be beneficial, with multiple benefits for climate change mitigation, forest biodiversity conservation, reducing desertification and enhancing livelihoods. It must be recognized that some "leakage" in the form of emissions resulting from those conservation efforts can occur (Aukland and others 2003). Climate change will also affect current biodiversity conservation strategies (Bomhard and Midgley 2005). For example, shifts from one climate zone to another could occur in about half of the world's protected areas (Halpin 1997), with the effects more pronounced in those at higher latitudes and altitudes. Some protected area boundaries will need to be flexible if they are to continue to achieve their conservation goals.

The impacts of energy production and use on biodiversity have been addressed as a by-product of several policy responses in the past few decades. Examples include Germany's effort to reduce subsidies in the energy and transport sectors, promoting increases in the proportion of organic farming and reducing nitrogen use in agriculture (BMU 1997, OECD 2001). However, responses have not been comprehensive, coordinated or universal. Commitments, including shared plans of action, have been made in various fora, but implementation has proved to be extremely challenging, due both to problems of securing required finance and lack of political will or vision.

Box 5.11 Top biofuel producers in 2005 (million litres)	
Biodiesel	
Germany	1 920
France	511
United States	290
Italy	270
Austria	83
Bio-ethanol	
Brazil	16 500
United States	16 230
China	2 000
European Union	950
India	300

Source: Worldwatch Institute 2006

The world output of biofuels, assuming current practice and policy, is projected to increase almost fivefold. Above, an experimental farm for the production of biodiesel in Gujarat, India.

Credit: Joerg Boethling/Still Pictures

Table 5.3 Energy sources and their impacts on biodiversity

Energy source *	Impacts on biodiversity	Subsequent impact on human well-being
Fossil fuels Crude oil Coal Natural gas	■ Global **climate change** and associated disturbances, particularly when coupled with human population growth and accelerating rates of resource use, will bring losses in biological diversity. ■ **Air pollution** (including acid rain) has led to damage to forests in southern China amounting to US$14 billion/year. Losses from air pollution impacts on agriculture are also substantial, amounting to US$4.7 billion in Germany, US$2.7 billion in Poland and US$1.5 billion in Sweden (Myers and Kent 2001). ■ The direct impact of **oil spills** on aquatic and marine ecosystems are widely reported. The most infamous case is the Exxon Valdez, which ran aground in 1989, spilling 37 000 tonnes of crude oil into Alaska's Prince William Sound (ITOPF 2006). ■ Impacts also come through the **development of oil fields** and their associated infrastructure, and human activities in remote areas that are valuable for conserving biodiversity (such as Alaska's Arctic National Wildlife Refuge that may be threatened by proposed oil development).	■ Changes in distribution of and loss of natural resources that support livelihoods. ■ Respiratory disease due to poor air quality.
Biomass Combustibles, renewables and waste	■ Decreased amount of land available for food crops or other needs due to greatly **expanded use of land** to produce biofuels, such as sugar cane or fast-growing trees, resulting in possible natural habitat conversion to agriculture, and intensification of formerly extensively developed or fallow land. ■ Can contribute **chemical pollutants** into the atmosphere that affect biodiversity (Pimentel and others 1994). ■ Burning crop residues as a fuel also **removes essential soil nutrients**, reducing soil organic matter and the water-holding capacity of the soil. ■ Intensively managing a biofuel plantation may require **additional inputs of fossil fuel** for machinery, fertilizers and pesticides, with subsequent fossil fuel related impacts. ■ Monoculture of biomass fuel plants can increase **soil and water pollution** from fertilizer and pesticide use, **soil erosion and water run-off**, with subsequent loss of biodiversity.	■ Cardiovascular and respiratory disease from reduced indoor air quality, due to wood-burning stoves, especially among poor women and children. ■ Decreased food availability.
Nuclear energy	■ Water used to cool reactors is released to environment at **significantly above ambient temperatures**, and accentuates ecological impacts of climatic extremes, such as heat waves, on riverine fauna. ■ Produces relatively small **amounts of greenhouse gases** during construction. ■ Because of the potential risks posed by nuclear energy, some nuclear plants are surrounded by **protected areas**. For example, the Hanford Site occupies 145 000 ha in southeastern Washington State. It encompasses several protected areas and sites of long-term research (Gray and Rickard 1989), and provides an important sanctuary for plant and animal populations. ■ A nuclear accident would have grave implications for people and biodiversity.	■ Health impacts of ionising radiation include deaths and diseases due to genetic damage (including cancers and reproductive abnormalities).
Hydroelectricity	■ Building large dams leads to **loss of forests, wildlife habitat and species populations, disruption of natural river cycles, and the degradation of upstream catchment areas** due to inundation of the reservoir area (WCD 2000). ■ Dam reservoirs also emit **greenhouse gases** due to the rotting of vegetation and carbon inflows from the basin. ■ On the positive side, some dam reservoirs provide **productive fringing wetland ecosystems** with fish and waterfowl habitat opportunities.	■ Building large dams can result in displacement of people. ■ Alterations in availability of freshwater resources (both improved and declining, depending on the situation) for human use.
Alternative energy sources Geothermal Solar, wind, tidal and wave	■ **Ecosystem disruption** in terms of desiccation, habitat losses at large wind farm sites and undersea noise pollution. ■ Tidal power plants may **disrupt migratory patterns** of fish, **reduce feeding areas** for waterfowl, **disrupt flows of suspended sediments** and result in various other changes at the ecosystem level. ■ Large photovoltaic farms **compete for land** with agriculture, forestry and protected areas. ■ Use of **toxic chemicals** in the manufacture of solar energy cells presents a problem both during use and disposal (Pimentel and others 1994). ■ Disposal of water and wastewater from geothermal plants may cause significant **pollution** of surface waters and groundwater supplies. ■ Rotors for wind and tidal power can cause some **mortality** for migratory species, both terrestrial and marine (Dolman and others 2002). ■ Strong **visual impact** of wind farms.	■ Decreased species populations to provide basic materials of life. ■ Toxins released to the environment may cause public health problems. ■ Decreased economic value of lands near wind farms, due to strong visual impacts.

* See Figure 5.5 for percentage of total primary energy supply

There are also attempts to address this issue through impact management within the private sector, and especially in the energy industry. The private sector is increasingly accepting its responsibilities as a steward of the environment. It is collaborating with non-governmental organizations, through fora such as the Energy and Biodiversity Initiative (EBI 2007), to better understand impacts and possible mitigation and adaptation strategies that make business sense. Beyond legislation and regulation, the use of payments for ecosystem services, as exemplified by the emerging carbon market, represents an innovative though somewhat controversial approach to addressing the impacts of energy use on the environment. The *State of the Carbon Market 2006*, which covers the period from 1 January, 2005 to 31 March, 2006, records a burgeoning global carbon market, worth over US$10 billion in 2005, 10 times the value of the previous year, and more than the value (US$7.1 billion) of the entire US wheat crop in 2005 (World Bank 2006).

Ensuring access to energy while maintaining biodiversity and vital ecosystem services will require an integrated multi-sectoral approach (see Chapters 2 and 10) that includes:

- an ecosystem approach to management of biodiversity and natural resources that ensures inclusion of lessons learned in ongoing management of natural resources affected by energy production and use;
- a major shift in environmental governance to incorporate policies and incentives promoting energy production and use that mainstreams action to address biodiversity concerns, especially with respect to climate change; and
- increasing partnership with the private sector, including extractive industries and the financial sector, to promote energy programmes that internalize the full costs on biodiversity and livelihoods.

HEALTH
Biodiversity change affects human health
Although there is limited understanding of the consequences of many specific changes in biodiversity for health and the incidence of disease in people and other species, the conceptual links between broader environmental changes and human health are well understood, as seen in Figure 5.7. Emerging diseases resulting from the destruction and fragmentation of tropical forests and other ecosystems, wildlife-human

disease linkages (for example, Lyme disease, West Nile virus and avian influenza), the many known and as yet undiscovered pharmaceutical products found in nature, the contribution of ecosystem services to human health and the increasing recognition of the impacts of endocrine disrupters on both animal and human health, all underline the links between biodiversity and human health (Chivian 2002, Osofsky and others 2005).

About 1 billion people live a subsistence lifestyle, and loss of ecosystem productivity (for example through loss of soil fertility, drought or overfishing) can rapidly lead to malnutrition, stunted childhood growth and development, and increased susceptibility to other diseases. There is a profound global nutritional imbalance, with a billion overnourished (mainly rich) people and a similar number undernourished (mainly poor people). Historically, this imbalance has been driven primarily by social and economic factors, but ecological factors will probably play an increasingly important role in the future. Some 70 per cent of infectious diseases originate in animals, and conservation issues are central to their epidemiology. Increased risks of infectious disease spread and crossover can result from land-use changes, many forms of intensive animal production, invasive alien species and the international wildlife trade. Climate change is expanding the range and activity level of disease vectors, particularly insect-borne vectors. The recent international scares over Sudden Acute Respiratory Syndrome (SARS) and avian influenza have brought a dramatic new dimension to the global health debate.

Along with biodiversity changes, there are a number of other factors that are increasing the exposure to and risk of disease. An increasing human population provides an increased number of hosts for disease agents; climate change raises temperatures, altering the wider distribution of disease vectors, such as mosquitoes; drug resistance to conventional treatments is increasing; and continuing poverty and malnutrition make many people more susceptible to disease. Recent experiences with West Nile virus, hantavirus, avian influenza and tuberculosis provide evidence that disease causing micro-organisms are rapidly adapting to changing circumstances, and emerging or increasing rates of infectious diseases are the result (Ayele and others 2004, Campbell and others 2002, Harvell and others 2002, Zeier and others 2005). However, changes to ecosystems

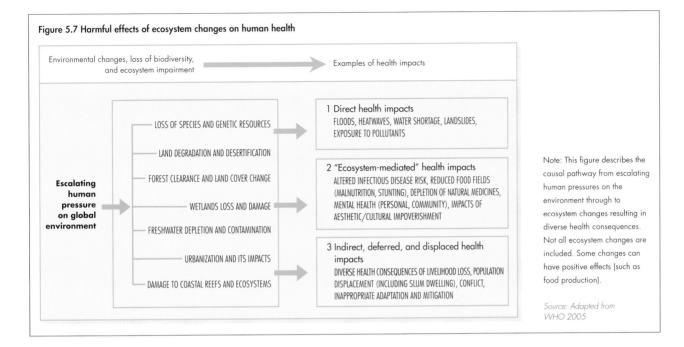

Figure 5.7 Harmful effects of ecosystem changes on human health

Environmental changes, loss of biodiversity, and ecosystem impairment → Examples of health impacts

Escalating human pressure on global environment

- LOSS OF SPECIES AND GENETIC RESOURCES
- LAND DEGRADATION AND DESERTIFICATION
- FOREST CLEARANCE AND LAND COVER CHANGE
- WETLANDS LOSS AND DAMAGE
- FRESHWATER DEPLETION AND CONTAMINATION
- URBANIZATION AND ITS IMPACTS
- DAMAGE TO COASTAL REEFS AND ECOSYSTEMS

1 Direct health impacts
FLOODS, HEATWAVES, WATER SHORTAGE, LANDSLIDES, EXPOSURE TO POLLUTANTS

2 "Ecosystem-mediated" health impacts
ALTERED INFECTIOUS DISEASE RISK, REDUCED FOOD FIELDS (MALNUTRITION, STUNTING), DEPLETION OF NATURAL MEDICINES, MENTAL HEALTH (PERSONAL, COMMUNITY), IMPACTS OF AESTHETIC/CULTURAL IMPOVERISHMENT

3 Indirect, deferred, and displaced health impacts
DIVERSE HEALTH CONSEQUENCES OF LIVELIHOOD LOSS, POPULATION DISPLACEMENT (INCLUDING SLUM DWELLING), CONFLICT, INAPPROPRIATE ADAPTATION AND MITIGATION

Note: This figure describes the causal pathway from escalating human pressures on the environment through to ecosystem changes resulting in diverse health consequences. Not all ecosystem changes are included. Some changes can have positive effects (such as food production).

Source: Adapted from WHO 2005

and their services, especially freshwater sources, food-producing systems and climatic stability, have been responsible for significant adverse impacts on human health in the past 20 years, predominantly in poor countries. Wealthy communities are often able to avoid the effects of local ecosystem degradation by migration, substitution or by appropriation of resources from less-affected regions.

Biodiversity is also the source for many cures. In 2002–2003, 80 per cent of new chemicals introduced globally as drugs could be traced to or were inspired by natural products. Profits from such developments can be enormous. For example, a compound derived from a sea sponge to treat herpes was estimated to be worth US$50–100 million annually, and estimates of the value of anti-cancer agents from marine organisms are up to US$1 billion a year (UNEP 2006a).

Traditional medicines, mainly derived from plants, are a mainstay of primary health care for a significant proportion of the population in developing countries. It is speculated that some 80 per cent of people in developing countries rely on traditional medicines, mostly derived from plants, and more than half of the most frequently prescribed drugs in developed countries derive from natural sources.

Loss of biodiversity may decrease our options for new treatments in the future. WHO has identified

20 000 species of medicinal plants for screening, and there are many more species whose medicinal values are only just being discovered, or may prove important in the future. The value of the global herbal medicine market was estimated at roughly US$43 billion in 2001 (WHO 2001).

The capacity of ecosystems to remove wastes from the environment is being degraded, due to both increased loading of wastes and degradation of ecosystems, leading to local and sometimes global waste accumulation (MA 2005). Examples include the accumulation of particles and gases in the air, and of microbial contaminants, inorganic chemicals, heavy metals, radioisotopes and persistent organic pollutants in water, soil and food. Such wastes have a wide range of negative health impacts.

Managing biodiversity change and human health impacts

Access to ecosystem services is not equitably distributed, and far from optimal from a population health perspective. Essential resources, such as shelter, nutritious food, clean water and energy supplies, are top priorities in effective health policies. Where ill health is directly or indirectly a result of excessive consumption of ecosystem services, substantial reductions in consumption would have major health benefits, and simultaneously reduce pressure on ecosystems (WHO 2005). For example, in rich countries, where overconsumption is causing increasing health impacts, the reduced consumption of animal products and refined carbohydrates would have significant benefits for both human health and for ecosystems globally (WHO 2005). Integration of national agricultural and food security policies with the economic, social and environmental goals of sustainable development could be achieved, in part, by ensuring that the environmental and social costs of production and consumption are reflected more fully in the price of food and water.

Responses that mitigate the impacts of ecosystem changes on human health often involve policies and actions outside the health sector. Action to mitigate impacts of climate change will require cooperation across multiple sectors. However, the health sector bears responsibility for communicating the health impacts of ecosystem changes, and of effective and innovative interventions. Where there are trade-offs, such as between mitigation of negative health impacts and economic growth in other sectors, it is important that the health consequences are well understood, so that they can be included when setting priorities and determining trade-offs.

CULTURE

Interactions between biodiversity and culture

Over the past two decades, there has been growing recognition of the relevance of culture and cultural diversity for the conservation of biodiversity and for sustainable development, as made explicit during the WSSD in 2002 (Berkes and Folke 1998, Borrini-Feyerabend and others 2004, Oviedo and others 2000, Posey 1999, Skutnabb-Kangas and others 2003, UNDP 2004, UNEP and UNESCO 2003).

In each society, culture is influenced by locally specific relationships between people and the environment, resulting in varied values, knowledge and practices related to biodiversity (Selin 2003). Cultural knowledge and practices have often contributed specific strategies for the sustainable use and management of biodiversity (see Anderson and Posey 1989, Carlson and Maffi 2004, Meilleur 1994, for examples). The diversity of cultures that have developed globally provides a vast array of responses to different ecosystems, and to variation and change in environmental conditions within them. This cultural diversity forms an essential part

of the global pool of resources available to address the conservation of biodiversity (ICSU 2002, UNESCO 2000). However, cultural diversity is being rapidly lost, in parallel to biological diversity, and largely in response to the same drivers (Harmon 2002, Maffi 2001). Taking linguistic diversity as an indicator of cultural diversity, over 50 per cent of the world's 6 000 languages are currently endangered (UNESCO 2001), and it has been speculated that up to 90 per cent of existing languages may not survive beyond 2100 (Krauss 1992). With loss of languages comes the loss of cultural values, knowledge, innovations and practices, including those related to biodiversity (Zent and López-Zent 2004).

In addition to the importance of culture for the conservation and sustainable use of biodiversity, human societies everywhere are themselves dependent on biodiversity for their livelihoods, as well as for cultural identity, spirituality, inspiration, aesthetic enjoyment and recreation (MA 2005). Loss of biodiversity thus affects both material and non-material human well-being.

Although societies in industrialized countries may be further removed from the immediate impacts of biodiversity loss, they are nevertheless adversely affected by loss or decline in ecosystem services. Certain categories of people are especially vulnerable to drastic environmental and social change. They include the poor, women, children and youth, rural communities, and indigenous and tribal peoples. The latter constitute the majority of the world's cultural diversity (Posey 1999).

Correlations have been identified between the respective geographic distributions of cultural and biological globally and regionally (Harmon 2002, Oviedo and others 2000, Stepp and others 2004, Stepp and others 2005). Figure 5.8 highlights this, showing the worldwide distributions of plant diversity and linguistic diversity. Areas of high biodiversity tend to be areas of a higher concentration of distinct cultures. Meso-America, the Andes, Western Africa, the Himalayas, and South Asia and the Pacific, in particular, present this pattern of high "biocultural" diversity. This pattern is supported by research that combines indicators of cultural diversity with indicators of biodiversity into a global biocultural diversity index (Loh & Harmon 2005).

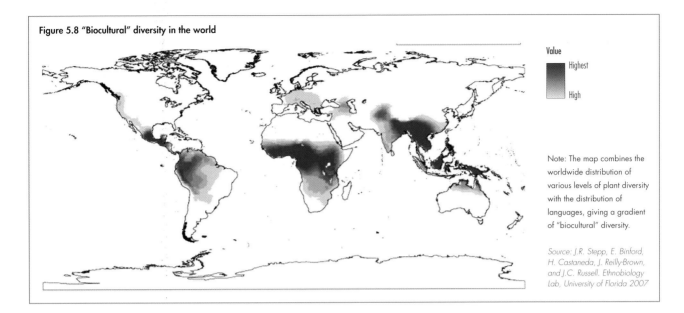

Figure 5.8 "Biocultural" diversity in the world

Value

Highest

High

Note: The map combines the worldwide distribution of various levels of plant diversity with the distribution of languages, giving a gradient of "biocultural" diversity.

Source: J.R. Stepp, E. Binford, H. Castaneda, J. Reilly-Brown, and J.C. Russell. Ethnobiology Lab, University of Florida 2007

While correlations are evident at the global level, the identification of any causal links between biodiversity and cultural diversity requires research at the local level. Empirical evidence that supports the interrelationships between cultures and biodiversity includes:

- anthropogenic creation and maintenance of biodiverse landscapes through traditional low-impact resource management practices (Baleé 1993, Posey 1998, Zent 1998);

- large contribution of traditional farmers to the global stock of plant crop varieties and animal breeds (Oldfield and Alcorn 1987, Thrupp 1998);

- customary beliefs and behaviours that contribute directly or indirectly to biodiversity conservation, such as sustainable resource extraction techniques, sacred groves, ritual regulation of resource harvests and buffer zone maintenance (Moock and Rhoades 1992, Posey 1999); and

- dependence of socio-cultural integrity and survival of local communities on access to and tenure of

traditional territories, habitats and resources, which also importantly affect food security (Maffi 2001).

These findings point to significant ecological and societal implications of the increasing threats to the world's cultural diversity. Global social and economic change (see Chapter 1), is driving the loss of biodiversity, and disrupting local ways of life by promoting cultural assimilation and homogenization. Cultural change, such as loss of cultural and spiritual values, languages, and traditional knowledge and practices, is a driver that can cause increasing pressures on biodiversity, including overharvesting, widespread land-use conversion, overuse of fertilizers, reliance on monocultures that replace wild foods and traditional cultivars, and the increase and spread of invasive alien species that displace native species (MA 2005). In turn, these pressures impact human well-being. The disruption of cultural integrity also impedes the attainment of the Millennium Development Goals (MDGs) (see Table 5.4).

Table 5.4 Impacts of loss of cultural diversity	
Impact on vulnerable groups dependent on local resources	**Relevance to MDGs**
■ Local food insecurity due to reduction of traditional varieties of crops and access to wild foods (IUCN 1997)	■ Goal 1 Eradicate extreme poverty and hunger
■ Devaluation of gender-specific knowledge of biodiversity, especially women's knowledge of medicines and food sources (Sowerwine 2004)	■ Goal 3 Promote gender equality and empower women
■ Loss of traditional and local knowledge, practices and language relevant to conservation and sustainable use of biodiversity (Zent and Lopez-Zent 2004)	■ Goal 7 Ensure environmental sustainability

The spread of invasive alien
species such as the water
hyacinth can have adverse
impacts on biodiversity.

Credit: Ngoma Photos

Managing biological and cultural diversity

The growing recognition over the past two decades
of the importance of culture and cultural diversity to
the environment and human well-being has led to
significant developments in terms of policy and other
responses relevant to sustainable development and
biodiversity conservation at international, national and
local levels (see Chapter 6, Arctic). The policies and
activities of UNEP, UNESCO, IUCN and the CBD
now include a focus on the interrelationships between
biodiversity and cultural diversity, and the indicators for
measuring progress towards meeting the CBD's 2010
target include a focus on trends in cultural diversity.
In 2006, the UN Human Rights Council adopted the
UN Declaration on the Rights of Indigenous Peoples,
recognizing that "respect for indigenous knowledge,
cultures and traditional practices contributes to
sustainable and equitable development and proper
management of the environment."

National policies have also taken the initiative to
strengthen the links between biodiversity and cultures
in accord with the CBD. For example, the Biological
Diversity Act of India (2002) stipulates that central
government shall endeavour to respect and protect
the knowledge of local people relating to biodiversity.
In doing so, the act provides that forests protected
as sacred groves in the context of local communities'

belief systems may be recognized as heritage sites. In
Panama, legal recognition has been given in the form
of sovereignty to the seven major groups of indigenous
peoples in that country. Panama was the first
government in Latin America to recognize this class of
rights for its indigenous populations, and 22 per cent
of the national territory is now designated as sovereign
indigenous reserves.

Effective biodiversity conservation, particularly that
outside of protected areas, relies on integrating local
participation, knowledge and values in land-use
planning, for example in the co-management of forests,
watersheds, wetlands, coastal areas, agricultural lands
and rangelands, fisheries, and migratory bird habitats
(Borrini-Feyerabend and others 2004). Successful co-
management often involves partnerships between local
communities and governments, international and local
organizations (see Chapter 6, the Polar Regions) and
the private sector, including ecotourism ventures.

Incorporating local and traditional knowledge in
policy decisions and on-the-ground action calls for
mainstreaming the links between biodiversity and
culture into social and sectoral plans and policies
(UNESCO 2000). This approach involves developing
and strengthening institutions at all scales, so that local
knowledge for the conservation and sustainable use of

biodiversity can be successfully transferred to landscape and national scales. It also involves strengthening the retention of traditional knowledge through education, conservation of languages and support for passing on knowledge between generations.

An integrative approach to biodiversity conservation for sustainable development takes into account the importance of maintaining the diversity of culturally-based knowledge, practices, beliefs and languages that have contributed to the conservation and sustainable use of local biodiversity. The adoption of this integrative approach in international and national policy directives and on-the-ground interventions signals positive change. Further recognition of impacts on the most vulnerable societies and social categories of people, and efforts to strengthen the contribution of local and traditional ecological knowledge to policy recommendations (Ericksen and Woodley 2005), will assist in the maintenance of sustainable relationships between people and biodiversity.

CHALLENGES AND OPPORTUNITIES
CHALLENGES
Undervaluation of biodiversity
Biodiversity loss continues because the values of biodiversity are insufficiently recognized by political and market systems. In part this is due to the costs of biodiversity loss not being borne solely by those responsible for its loss. An added complexity is that the global nature of many biodiversity values results in the impact of biodiversity loss being felt far beyond national boundaries. Losses of biodiversity, such as the erosion of genetic variability in a population, are often slow or gradual, and are often not seen or fully recognized until it is too late. The dramatic and immediate problems typically receive greater policy attention and budgetary support, so funding is often more available for charismatic megafauna, such as tigers or elephants, than for the wider, yet less celebrated variety of biodiversity that forms key components of the planet's infrastructure, and makes the most substantial contribution to delivering the wide range of ecosystem services from which people benefit.

Many of the attempts to calculate the values of biodiversity consider transaction values of the individual components of biodiversity, the price paid for particular goods-and-services. Although

this incorporates some of biodiversity's values, it consistently undervalues many ecosystem functions that are essential for the delivery of ecosystem services. In addition, some elements of biodiversity are irreplaceable when lost, for example through species extinction, or gene loss. Economic valuation and new market mechanisms need to be part of a larger policy toolbox, to take account of such irreversible changes to biodiversity, and although more complete economic valuation is necessary to help create important incentives and opportunities for conservation, it will be insufficient to fully conserve biodiversity for future generations. Traditional conservation programmes, focused on protecting components of biodiversity from exploitation and other drivers, will remain an important policy tool to protect the irreplaceable and many of the intangible values of biodiversity (see Box 5.12).

Society can only develop without further loss of biodiversity if market and policy failures are corrected, including perverse production subsidies, undervaluation of biological resources, failure to internalize environmental costs into prices, and failure to recognize global values at the local level. Most policy sectors have impacts on biodiversity, and biodiversity change has significant implications for those sectors. However, biodiversity concerns are rarely given sufficient standing when industrial, health, agricultural, development or security policies are developed. Although any society or economy that continues to deplete biodiversity is, by definition, unsustainable,

Box 5.12 Payments for ecosystem services: reforesting the Panama Canal Watershed

An April 2005 cover article in *The Economist* entitled "Rescuing Environmentalism" led with a analysis of the work by PRORENA, a Panamanian NGO, to establish a diverse native forest cover across extensive areas of deforested lands in the Panama Canal watershed. There has been heavy support from the reinsurance industry, which sees that a regular water flow is necessary for the long-term working of the canal. The project works with local communities to identify a mix of useful tree species, and to research optimal rearing and planting options. It provides income streams for the communities, while improving water retention and flow dynamics for the canal region. It has demonstrated that large-scale ecological restoration in the tropics is technically feasible, socially attractive and financially viable.

Source: The Economist 2005

mainstreaming biodiversity concerns effectively into broader policy making so that all policy supports environmental sustainability remains a key challenge.

Reducing the rate of biodiversity loss will require multiple and mutually supportive policies of conservation and sustainable use, and the recognition of biodiversity values. New policies of integrated landscape and watershed management and sustainable use – the ecosystem approach – can be effective in reducing biodiversity loss (see Box 4.9). In recent years, legal structures such as "biodiversity easements" and "payments for biodiversity services" have been developed to use market mechanisms to provide additional financial resources, and new markets for biodiversity-friendly products are developing new options for producers. These present new opportunities to recognize and mainstream the value of biodiversity, and can address many of the drivers of biodiversity loss. With a supportive policy framework, such changes will initiate market and behavioural corrections that will move society towards increased sustainability. Although they only make up a small fraction of total market share, organic and sustainably produced agricultural products, such as

"bird-friendly" coffee and cocoa, are clear examples of this. However, each of these attempts also has to be cost-effective in the local or global marketplace, and comply with other obligations, such as international trade rules, which often remain perversely disconnected from environmental needs and policies.

Ineffective governance systems

Political authority and power often reside far from where decisions that affect biodiversity conservation and sustainable use are taken. This includes disjuncture between and within countries, where different ministries frequently take different approaches to the issue of biodiversity management. Biodiversity concerns are dealt with in numerous international and regional agreements, many of which have come into force in the past 20 years. In 2004, five of the key global biodiversity-related conventions (CBD, CITES, CMS, Ramsar and the World Heritage Convention) created the Biodiversity Liaison Group to help facilitate a more coordinated approach to policy development and implementation. UNEP has created the Issues-based Modules project, which aims to assist countries and other stakeholders to understand the intersections of obligations coming from the various conventions. Such actions and projects epitomize the call from the WSSD to shift from policy development to implementation, and provide a start towards an integrated approach to biodiversity management.

Biodiversity governance involves multiple stakeholders, including landholders, community and political jurisdictions (local, national and regional), the private sector, specific arrangements such as fisheries management councils, species protection agreements and the global agreements. Most of these suffer from a lack of financial and human capacity to effectively manage biodiversity. Even very clear policies do not ensure compliance or enforcement, as is evident from the ongoing illegal international trade in species and their parts, in contravention of CITES.

The proliferation of authorities has, in many instances, created confusion, dispersed resources, and slowed policy development and implementation. This has led to coordination problems between and within scales: local to national, inter-ministerial, regional and international. In most countries, biodiversity concerns are the responsibility of relatively weak, underfunded and understaffed environmental ministries. Decisions that severely threaten biodiversity, such as land-use changes

---

Box 5.13 Key questions to assist a fuller consideration of biodiversity and governance in policy development and implementation

Nations, communities, public and private organizations, and international processes have been grappling with how to implement policies that take biodiversity concerns into account. A list of questions best indicates the kinds of information that are useful to collate and consider with stakeholders:

- What are the local, national and global values of biodiversity?
- How can biodiversity concerns be integrated into all sectoral decision making?
- How does the ecosystem approach at the landscape level that is necessary to protect biodiversity and ecosystem services fit with existing land tenure and governmental jurisdictions?
- What does sovereignty over genetic resources actually mean? Because many, if not most, genetic resources occur in multiple jurisdictions, how can potential (and likely) multiple claims over the same or related resources be addressed?
- How can biodiversity effectively be both used and conserved?
- What are the potential and plausible environmental impacts of living modified organisms, and what are the appropriate regulatory regimes for them?
- How should the standards of invention, usefulness and non-obviousness be applied in terms of patenting genes, gene expressions and life forms?
- Will the benefits from use of genetic resources justify the costs and the restrictions on research and access?
- How does the enclosure of biodiversity fit within national legal and property rights systems? And, how does this affect the rights of traditional and indigenous communities who may have more communal approaches and traditions to resource management and appropriation?
- Who should be the beneficiaries of such benefits: governments, communities, patent holders, inventors, local people or biodiversity itself?

and the introduction of potentially invasive species (either by design or accident), are most frequently taken by agriculture, fisheries, commerce or mining ministries. Often this is done without effective consultation with the authorities responsible for the environment, or recognition of the costs of such impacts.

Biodiversity governance is in a major period of flux. Historically, biodiversity was largely considered as a common heritage, and a public good. The late 20th century saw an unprecedented "enclosing" of genetic resources, a shift from considering them common heritage to seeing them as products to be owned in whole or in part. Two components of this recent enclosure movement are the patenting of genes, gene expressions and derived life forms on the one hand, and the fundamental shift to the concept of ownership of genetic resources that arose through the CBD and the FAO International Treaty on Plant Genetic Resources, in terms of national sovereignty over biological diversity (Safrin 2004). At the same time, the importance of biodiversity is better recognized, not only as a source of new products, but also as fundamental for the supply of the full range of ecosystem services (see Box 5.13).

In 2002, the CBD adopted the Bonn Guidelines on Access to Genetic Resources and Fair and Equitable Sharing of the Benefits Arising out of their Utilization (ABS), and the WSSD subsequently called for further elaboration of the international regime on access and benefit sharing. Although the resulting negotiations have since dominated much of the international biodiversity discourse, the "green gold" predicted by the early advocates of the issue within the CBD, and the "gene gold" predicted by the rush to patent genetic information have not materialized. Whether this reflects an early market or overinflated predictions is not clear. However, these ABS discussions are likely to continue to dominate international negotiations, not only on biodiversity, but also on trade and intellectual property, distracting discussion from other fundamental issues of greater importance to the sustainable supply of ecosystem services for development. Further research and understanding on how to capture and distribute the benefits arising from the use of biodiversity will contribute to these discussions, as illustrated by the Indian case outlined in Box 5.14.

The CBD has taken a novel and progressive approach to identify a mechanism to respect the breadth of

> **Box 5.14 Access and benefit sharing in India**
>
> The Kani-TBGRI model of benefit sharing with local communities relates to an arrangement between the Kani tribe from the southern and western Ghat region of Kerala State, India and the Tropical Botanic Garden and Research Institute (TBGRI). Under this agreement the Kani tribe receives 50 per cent of the licence fee and royalties resulting from the sale of the manufacturing licence for Jeevni, an anti-fatigue drug, by TBGRI to the pharmaceutical company Aryavaidya Pharmacy Coimbatore Ltd. Jeevni is a formulation based on molecules found in the leaves of a wild plant, *Trichophus zeylanicus*, used by the Kani to keep them energetic and agile. In 1997 a group of Kani tribal members, with assistance from TBGRI, developed the Kerala Kani Samudaya Kshema Trust. The trust's objectives include welfare and development activities for the Kani, preparation of a biodiversity register to document the Kani's knowledge base, and the evolution of and support for methods to promote the sustainable use and conservation of biological resources.
>
> *Source: Anuradha 2000*

traditional knowledge on the uses for biodiversity. The strengthened voices of indigenous communities have brought forward important and as yet unresolved issues, including tensions between different ways of knowing (western science and community cosmology), valuing (economically based and culturally based) and governing (formalized written and customary law). Local and indigenous communities, and women within them, have been and will continue to be important stewards of biodiversity, and national systems of land tenure and respect for indigenous communities are intertwined with biodiversity policy making at the local and international levels.

OPPORTUNITIES

New and evolving concepts of ownership over biodiversity and genetic resources, protection of traditional knowledge, the ecosystem approach, ecosystem services and valuation, have created policy challenges for all of the actors. Governments at all levels, communities and businesses are grappling with how to incorporate environmental, social and cultural concerns more effectively into their decision making processes. In order to achieve sustainable development, biodiversity needs to be mainstreamed into energy, health, security, agricultural, land use, urban planning and development policies.

Management interlinkages

At the international level, the biodiversity-related conventions have increased their collaboration, and are attempting to link more closely with economic instruments such as World Intellectual Property Organization and the World Trade Organization. Each of these processes has developed strategies and action plans that need to be implemented nationally, and there is a clear need to find which approaches work best, under which circumstances, and to deliver more effective advice at each level.

Private sector interventions

Some private corporations have started to build biodiversity concerns into their planning and implementation, but many more still need to analyse and minimize the negative impacts of infrastructure development, and operations, such as processing and transportation, on biodiversity. Seemingly good policy may mask environmental degradation elsewhere, such as the movement of polluting industries to, or sourcing of wood products from, less-regulated areas. Codes of conduct, certification schemes, transparency through triple-bottom line accounting and international regulatory standards are key policy options for creating incentives and level playing fields that will minimize these cost-shifting behaviours. Regional organizations, such as the European Community, NAFTA and SADC, play important roles in creating such level playing fields, and collaboration across sectors within government is also required. Interagency coordination is needed to bring coherence to international negotiations, and to bring biodiversity concerns into national policy development.

Market mechanisms

Appropriately recognizing the multiple values of biodiversity in national policies is likely to require new regulatory and market mechanisms, such as:

- better valuation and the creation of markets for ecosystem services;
- more widespread certification systems;
- payment programmes to increase incentives for conservation and protection of biodiversity and ecosystems;
- new policies providing tax incentives for low biodiversity impact operations;
- reducing and eliminating perverse incentives for biodiversity loss;
- developing conservation easements; and
- mechanisms for upstream-downstream transfers.

Pro-poor policies

Implementing policies that benefit the poorest in society will be challenging, but necessary. Raising the profile and representation of direct biodiversity users and stewards, especially smallholders, will be key in developing effective implementation mechanisms. Recognizing the role women play in protecting, using and understanding biodiversity in many parts of the world can lead to the mutual benefit of empowering communities and ensuring sustainable use of biodiversity. Including all stakeholders in the shaping and testing of policies will be necessary to ensure long-term viability and acceptance of the policy changes. Generalizing and scaling up inclusive projects is a key challenge and opportunity for the international community.

Conservation measures

Natural disasters in recent years – tsunamis, hurricanes and earthquakes – have highlighted a range of environmental and biodiversity concerns. Preservation and restoration of coastal mangroves, seagrasses, coastal wetlands and reef systems protect shorelines from the power of storms. Forests regulate water flow, and soil structure and stability. Policies that help protect biodiversity also protect people and infrastructure. Taking the range of biodiversity and environmental concerns into account in land-use planning, and enforcing rules and regulations are key to success.

New governance structures

The understanding of biodiversity, its role and uses, and the governance structure of enclosure is all in its infancy, as nations and localities are testing options, and finding opportunities and obstacles (see Box 5.15). Further analysis and assessment of valuation programmes, mainstreaming attempts, and new governance structures are needed to develop best practices and share lessons learned. As more policy tools and mechanisms based on success are developed, new ways will emerge to conserve and use the world's biodiversity. However, enough is already known to make better decisions on the conservation and wise use of biodiversity. Given the documented rate of habitat conversion and degradation, and declines in populations and genetic resources, much more action is needed immediately to conserve biodiversity so that future generations will have the full range of opportunities to benefit from its use.

Box 5.15 Information gaps and research needs

Due to the complexity of the concept of biodiversity there is no simple list of information gaps that, if filled, would answer the majority of the questions this chapter has raised. However, each level has some significant information needs, and addressing these would provide multiple benefits:

What exists on Earth and where?
These fundamental questions of description and biogeography underpin all biodiversity and ecosystem research. The discovering, naming, describing and ordering of the different species on Earth is a science called taxonomy. It is needed, for example, to identify invasive species, differentiate between different disease vectors and reservoirs, and identify likely candidates for new medicines and other useful chemicals and enzymes. However, the majority of the world's species have not been identified, and some key groups, such as invertebrates and micro-organisms, are especially poorly understood. The CBD has created the Global Taxonomy Initiative (GTI) to try to overcome this impediment, and the Global Biodiversity Information Facility (GBIF) has been created to pull together the disparate data from taxonomic institutions around the world for integrated use, leveraging each country's investment for the common good. However, greater financial and collaborative support from governments and civil society is needed for these efforts.

How do biological resources function?
From the genetic level through to research on how different organisms move, and process food, water, salt and other inputs (including pollutants), there is an increasing understanding of the range of processes that nature has developed, and that can be used to move towards a more sustainable development path. Examples include:
- the increasing understanding of the genetics of key agricultural organisms, such as rice and potato, which should contribute to the development of more hardy and prolific strains;
- the study of the ability of different classes of microbes to perform a range of functions, from breaking down pollutants to isolating and purifying metals; and
- the identification of processes that will allow people to most effectively develop technologies, such as biofuels, without further damaging the environment or harming food security.

Considerable resources are going into this range of research, frequently driven by specific economic interests, but the work is often hampered by a lack of taxonomic and biogeographic understanding.

How does the system interact?
The multitude of questions about ecology range from the very local (how do soil microbes support plant growth) to global (how do forest and ocean organisms sequester carbon and regulate climate systems). Answering these questions, and understanding the dynamics within them frequently takes many years' research with repeated observations. In many areas increased research is needed, for example, on:
- the impacts of fragmentation on biodiversity structure and functioning, resiliency of ecosystems to change (such as from climate change and human interventions);
- the role of biodiversity in mitigating and responding to climate change;
- the role of restoration ecology in remediating changed and degraded lands; and
- reservoirs and vectors of pathogens and zoonotic diseases.

New mechanisms are also needed to bring together the vast research results in a way to use the data for new modelling and research questions.

How do people use and understand biodiversity?
The vast array of different cultures, and the associated range of knowledge about biodiversity, contributes key understanding for conservation and sustainable use of biodiversity. Many new governance structures and techniques are being developed, and these need to be understood more clearly if their effectiveness and synergies are to be maximized, and the spread of perverse incentives is to be avoided. There is a need for increased capacity building, to convert knowledge into practice in many parts of the world. Increased understanding of how people relate to biodiversity, and how to move towards greater stewardship of biodiversity may be the biggest question the world still must answer.

How can biodiversity be valued?
Substantial research on internalizing the values of biodiversity, and the adoption of new indices of global and national wealth based on functioning ecosystems are required, including clear and consistent rules and processes that cross economic and political jurisdictions, such as are emerging in areas of forest and organic certification.

References

Achard, F., Eva, H.D., Stibig, H.J., Mayaux, P., Gallego, J., Richards, T. and Malingreau, J.P. (2002). Determination of Deforestation Rates of the World's Humid Tropical Forests. In *Science* 297(5583):999-1002

Aide, T. M. and Grau H.R. (2004). Globalization, migration and Latin American ecosystems. In *Science* 305:1915-1916

Aizen, M. A., Ashworth L. and Galetto, L. (2002). Reproductive success in fragmented habitats: do compatibility systems and pollination specialization matter? In *Journal of Vegetation Science* 13(6):885-892

Aizen, M. A. and Feinsinger, P. (1994). Forest Fragmentation, Pollination, and Plant Reproduction in a Chaco Dry Forest, Argentina. In *Ecology* 75(2):330-351

Anderson, A. and Posey, D. (1989). Management of a sub-tropical scrub savanna of the Gorotire Kayapo of Brazil. In *Advances in Economic Botany* 7:159-173

Anuradha, R V. (2000). *Sharing the Benefits of Biodiversity: the Kani-TBGRI Deal in Kerala, India.* Kalpavriksh Environment Action Group, Pune, Maharashtra

Aukland, L, Costal, P. M. and Brown. S. (2003). A conceptual framework and its application for addressing leakage: the case of avoided deforestation. In *Climate Policy* 3(2):123-136

Ayele, W.Y., Neill, S.D., Zinsstag, J., Weiss, M.G. and Pavlik, I. (2004). Bovine tuberculosis; an old disease but a new threat to Africa. In *The International Journal of Tuberculosis and Lung Disease* 8:924-937

Baillie, J.E.M., Hilton-Taylor, C. and Stuart, S.N. (2004). *2004 IUCN Red List of Threatened Species. A Global Species Assessment.* World Conservation Union (IUCN), Gland and Cambridge http://www.iucn.org/themes/ssc/red_list_2004/main_EN.htm (last accessed 8 May 2007)

Baleé, W. (1993). Indigenous Transformation of Amazonian Forests. In *L'Homme* 126-128:231-254

Barnes, D.F., Krutilla, K. and Hyde, W. (2002). *The Urban Energy Transition? Energy, Poverty, and the Environment.* The World Bank, Washington, DC

Barthlott, W., Biedinger, N., Braun, G., Feig, F., Kier, G. and Mutke, J. (1999). Terminological and methodological aspects of the mapping and analysis of global biodiversity. In *Acta Botanica Finnica* 162:103-110

Beebee, T.J.C. (1995). Amphibian breeding and climate. In *Nature* 374:219-220

Berkes, F. and Folke, C. (1994). Investing in cultural capital for sustainable use of natural resources. In Koskoff, S. (ed.) *Investing in Natural Capital: The Ecological Economics Approach to Sustainability.* Island Press, Washington, DC

Berkes, F. and Folke, C. (eds.) (1998). *Linking Social and Ecological Systems: Management Practices and Social Mechanisms for Building Resilience.* Cambridge University Press, Cambridge

Bernardo, J. and Spotila, J. R. (2006). Physiological constraints on organismal response to global warming: mechanistic insights from clinally varying populations and implications for assessing endangerment. In *Biology Letters* 2(1):135-139 http://www.cofc.edu/~bernardoj/bernardo&spotila.2005.pdf (last accessed 8 May 2007)

BirdLife International (2000). *Threatened Birds of the World.* Lynx Edicions and BirdLife International, Barcelona and Cambridge

BMU (1997). *Sustainable Germany – towards an environmentally sound development.* Federal Environmental Agency of Germany http://www.umweltdaten.de/publikationen/fpdf-l/2537.pdf (last accessed 8 May 2007)

Bomhard, B. and Midgley, G. (2005). Securing Protected Areas in the Face of Global Change: Lessons Learned from the South African Cape Floristic Region. A Report by the Ecosystems, Protected Areas, and People Project. IUCN, Bangkok and SANBI, Cape Town http://www.iucn.org/themes/wcpa/pubs/pdfs/pasclimatechange.pdf (last accessed 30 June 2007)

Borrini-Feyerabend, G., MacDonald, K. and Maffi, L. (eds.) (2004). History, Culture and Conservation. In *Policy Matters* 13 Spec. Issue

Botes, A., McGeoch, M. A., Robertson, H. G., Van Niekerk, A., Davids, H. P., Chown, S. L. (2006). Ants, altitude and change in the northern Cape Floristic Region. In *Journal of biogeography* 33(1):71-90

Brown J.L.., Li, S.H. and Bhagabati, N. (1999). Long-term trend toward earlier breeding in an American bird: a response to global warming? In *Proceedings of the National Academy of Science* 96:5565-5569

Bruinsma, J. (ed.) (2003). *World Agriculture: Towards 2015 / 2030, an FAO Perspective.* Food and Agriculture Organization of the United Nations, Rome and Earthscan, London

Buck, L.E., Lassoie, J.P. and Fernades, E.C.M. (eds.) (1999). *Agroforestry in Sustainable Agricultural Systems.* CRC Press LLC, Boca Raton, FL

Burke, L. and Maidens, J., (2004). *Reefs at Risk in the Caribbean.* World Resources Institute, Washington, DC

Butchart, S.H.M., Stattersfield, A.J., Baillie, J., Bennun, L.A., Stuart, S.N., Akcakaya, H.R., Hilton-Taylor, C. and Mace, G.M. (2005) Using Red List Indices to measure progress towards the 2010 target and beyond. In *Philosophical Transactions of the Royal Society B* 360:255–268

Campbell, C J. (2005). *Oil Crisis.* Multi Science Publishing Co. Ltd., Essex

Campbell, G.L., Martin, A.A., Lanciotti, R.S., and Gubler, D.J. (2002). West Nile Virus. A review. In *The Lancet Infectious Diseases* 2:519-529

Carlson, T.J.S. and Maffi, L. (eds.) (2004). Ethnobotany and Conservation of Biocultural Diversity. *Advances in Economic Botany Series* Vol. 15. New York Botanical Garden Press, Bronx, NY

Carthy, R.R., Foley, A.M., Matsuzawa, Y. (2003). Incubation environment of loggerhead turtle nests: effects on hatching success and hatchling characteristics. In Bolten, A.B. and Witherington, B.E. (eds.) *Loggerhead Sea Turtles.* Smithsonian Institution Press, Washington, DC

Cesar, H. and Chong, C.K. (2004). Economic Valuation and Socioeconomics of Coral Reefs: Methodological Issues and Three Case Studies. In *Economic Valuation and Policy Priorities for Sustainable Management of Coral Reefs.* WorldFish Centre, Penang

Chape, S., Harrison, J., Spalding, M. and Lysenko, I. (2005). Measuring the extent and effectiveness of protected areas as an indicator for meeting global biodiversity targets. In *Philosophical Transactions of the Royal Society B* 360:443-455

Chivian, E. (2002). *Biodiversity: Its Importance to Human Health.* Center for Health and the Global Environment, Cambridge, MA

Collins, W.W. and Qualset, C.Q. (eds.) (1999). *Biodiversity in Agroecosystems.* CRC Press, Boca Raton, FL

Crick H.Q.P. and Sparks, T.H. (1999). Climate change related to egg-laying trends. In *Nature* 399:423-424

Dahdouh-Guebas F., Jayatissa, L.P. Di Nitto, D., Bosire, J.O., Lo Seen, D. and Koedam, N.(2005). How effective were mangroves as a defence against the recent tsunami? In *Current Biology* 15(12):R443-R447

De Jonge, V.N., Elliot, M. and Orive, E. (2002). Causes, historical development, effects and future challenges of a common environmental problem: eutrophication. In *Hydrobiologia* 475/476:1-19

De Marco, P. and F. M. Coelho (2004). Services performed by the ecosystem: forest remnants influence agricultural cultures' pollination and production. In *Biodiversity and Conservation* 13(7):1245-1255

Dolman, S. J., Simmonds, M.P., and Keith, S. (2003). *Marine wind farms and cetaceans.* IWC/SC/55/E4. International Whaling Commission, Cambridge

EBI (2007). *The Energy and Biodiversity Initiative* http://www.theebi.org/pdfs/practice.pdf (last accessed 8 May 2007)

EFFIS (2005). *Forest Fires in Europe 2005.* Report number 6. European Forest Fire Information System, European Commission Joint Research Centre, Ispra

Ellis, W.N., Donner, J.H. and Kuchlein, J.H. (1997). Recent shifts in phenology of Microlepidoptera, related to climatic change (Lepidoptera). In *Entomologische Berichten* 57(4):66-72

EM-DAT (undated). *Emergency Events Database: The OFDA/CRED International Disaster Database* (in GEO Data Portal). Université Catholique de Louvain, Brussels

Emerton, L. and Bos, E. (2004). *Value: Counting ecosystems as an economic part of water.* World Conservation Union (IUCN), Gland

Ericksen, P. and Woodley, E. (2005). Using Multiple Knowledge Systems: Benefits and Challenges. In *Millennium Ecosystem Assessment.* Volume 4, Multiscale Assessments. Island Press, Washington, DC

FAO (1998). *The State of the World's Plant Genetic Resources for Food and Agriculture.* Food and Agriculture Organization of the United Nations, Rome

FAO (2004). *The State of the World's Fisheries and Aquaculture 2004.* Food and Agriculture Organization of the United Nations, Rome

FAO and IIASA (2000). *Global Agro-ecological Zoning (CD-ROM).* FAO Land and Water Digital Media Series Nr. 11. Food and Agriculture Organization of the United Nations and International Institute for Applied Systems Analysis, Rome

Finlayson, C.M. and D'Cruz, R. (CLAs) (2005). Inland Water Systems. Chapter 20. In *Ecosystems and Human Well-being: Current Status and Trends.* Millennium Ecosystem Assessment. Island Press, Washington, DC

GEO Data Portal. *UNEP's online core database with national, sub-regional, regional and global statistics and maps, covering environmental and socio-economic data and indicators.* United Nations Environment Programme, Geneva http://www.unep.org/geo/data or http://geodata.grid.unep.ch (last accessed 7 June 2007)

Gianni, M. (2004). *High seas bottom trawl fisheries and their impacts on the biodiversity of vulnerable deep-sea ecosystems: Options for international action.* World Conservation Union (IUCN), Gland

Girot, P.O. (2001). Vulnerability, Risk and Environmental Security in Central America: Lessons from Hurricane Mitch. In *Conserving the Peace: Resources, Livelihoods and Security.* IUCN/IISD Task Force on Environment and Security. International Institute for Sustainable Development, Geneva

Goldemberg, J. and Johansson, T. B. (2004). *World Energy Assessment. Overview: 2004 update.* United Nations Development Programme and United Nations Department of Economic and Social Affairs, New York, NY

Grabbherr, G., Gottfried, M. and Pauli, H. (1994). Climate effects on mountain plants. In *Nature* 369(6480):448

Gray, R.H. and Rickard, W.H. (1989). The protected area of Hanford as a refugium for native plants and animals. In *Environmental Conservation* 16(3):251-260

Griffin, D.W., Kellogg, C.A., Garrison, V.H. and Shinn, E.A. (2002). The Global Transport of Dust. In *American Scientist* 90:230-237

Grin, F. (2005). The Economics of Language Policy Implementation: Identifying and Measuring Costs. In N. Alexander (ed.). *Proceedings of the Symposium "Mother-Tongue Based Education in Southern Africa: The Dynamics of Implementation.").* Multilingualism Network, University of Cape Town, Cape Town

Guruswamy, L.D. and McNeely, J.A. (eds.) (1998). *Protection of Global Biodiversity: Converging Strategies.* Duke University Press, Durham and London

Halpin, P.N. (1997). Global climate change and natural-area protection: management responses and research directions. In *Ecological Applications* 7:828-843

Harmon, D. (2002). *In Light of Our Differences: How Diversity in Nature and Culture Makes Us Human.* Smithsonian Institution Press, Washington, DC

Harvell, C.D., Mitchel, C.E., Ward, J.R., Altizer, S., Dobson, A.P., Ostfeld, R.S. and Samuel, M.D. (2002). Climate warming and disease risks for terrestrial and marine biota. In *Science* 296(5576):2158-62

Hays, G.C., Broderick, A.C., Glen, F. and Godley, B.J. (2003). Climate change and sea turtles: 150-year reconstruction of incubation temperatures at a major marine turtle rookery. In *Global Change Biology* 9:642-646

Heal G., Dasgupta, P., Walker, B., Ehrlich, P., Levin, S., Daily, G., Maler, K.G., Arrow, K., Kautsky, N., Lubchenco, J., Schneider, S., and Starrett, D. (2002). *Genetic Diversity and Interdependent Crop Choices in Agriculture.* Beijer Discussion Paper 170. The Beijer Institute, The Royal Swedish Academy of Sciences, Stockholm

Hersteinsson, P. and MacDonald, D.W. (1992). Interspecific competition and the geographical distribution of red and arctic foxes Vulpes vulpes and Alopex lagopus. In *Oikos* 64:505-515

Hu, C., Montgomery, E.T., Schmitt, R.W. and Muller-Karger, F.E. (2004). The dispersal of the Amazon and Orinoco River water in the tropical Atlantic and Caribbean Sea: Observation from space and S-PALACE floats. Deep Sea Research II. In *Topical Studies in Oceanography* 51(10-11):1151-1171

IAASTD (2007). *The International Assessment of Agricultural Science and Technology for Development.* http://www.agassessment.org (last accessed 8 May 2007)

IEA (2006). *World Energy Outlook 2006.* International Energy Agency, Paris

IEA (2007). *Energy Balances of OECD Countries and Non-OECD Countries 2006 edition.* International Energy Agency, Paris (in GEO Data Portal)

IIED (2003). Climate Change – Biodiversity and Livelihood Impacts. Chapter 3. In *The Millennium Development Goals and Conservation; Managing Natures Wealth for Society's Health.* International Institute for Environment and Development, London

ICSU (2002). *Science and Traditional Knowledge.* Report from the ICSU Study Group on Science and Traditional Knowledge http://www.icsu.org/Gestion/img/ICSU_DOC_DOWNLOAD/220_DD_FILE_Traitional_Knowledge_report.pdf (last accessed 30 June 2007)

IPCC (2007). *Climate Change 2007: Impacts, Adaptation and Vulnerability.* Working Group II Contribution to the Intergovernmental Panel on Climate Change Fourth Assessment Report. Cambridge University Press, Cambridge

ITOPF (2006). Summaries of major tanker spills from 1967 to the present day. http://www.itopf.com/casehistories.html#exxonvaldez (last accessed 30 June 2007)

IUCN (1997). *Indigenous Peoples and Sustainablility. Cases and Actions,* IUCN Inter-Commission Task Force on Indigenous Peoples. International Books, Utrecht

IUCN (2006). *2006 IUCN Red List of Threatened Species.* http://www.iucnredlist.org/ (last accessed 30 June 2007)

James, C. (2003). *Preview: Global status of commercialized transgenic crops.* Briefs No. 30. International Service for the Acquisition of Agri-biotech Applications (ISAAA). Ithaca, NY

Janzen, F.J. (1994). Climate Change and Temperature-Dependent Sex Determination in Reptiles. In *Proceedings of the National Academy of Sciences* 91:7487-7490

Jenkins, M. (2003). 'Prospects for Biodiversity'. In *Science* 302:1175-1177

Joshi, S. R., Ahmad, F. and Gurung, M.B. (2004). Status of *Apis laboriosa* populations in Kaski district, western Nepal. In *Journal of Apicultural Research* 43(4):176-180

Klein, A. M., Steffan-Dewenter, I. and Tscharntke, T. (2003). Pollination of *Coffea canephora* in relation to local and regional agroforestry management. In *Journal of Applied Ecology* 40:837-845

Kormos, C. and Hughes, L. (2000). *Regulating Genetically Modified Organisms: Striking a Balance between Progress and Safety.* Advances in Applied Biodiversity Science, Number 1. Conservation International, Washington, DC

Krauss, M. (1992). The world's languages in crisis. In *Language* 68:4-10

Kremen, C., Williams, N. M. and Thorp, R.W. (2002). Crop pollination from native bees at risk from agricultural intensification. In *Proceedings of the National Academy of Sciences* 99(26):16812-16816

Lafferty, K. D. and Gerber, L. (2002). Good medicine for conservation biology: The intersection of epidemiology and conservation theory. In *Conservation Biology* 16:593-604

Lavorel, S. (1998). Mediterranean terrestrial ecosystems: research priorities on global change effects. In *Global Ecology and Biogeography* 7:157-166

Loh, J. and Wackernagel, M. (2004). Living planet report 2004. World Wide Fund for Nature, Gland

Loh, J. and D. Harmon (2005). A global index of biocultural diversity. In *Ecological Indicators* 5, 231-241.

Loh, J. and Goldfinger, S. (eds.) (2006). *Living planet report 2006.* World Wide Fund for Nature, Gland

MA (2005). *Ecosystems and Human well-being: Biodiversity Synthesis.* Millennium Ecosystem Assessment. World Resources Institute, Washington, DC

Maffi, L. (ed.) (2001). *On Biocultural Diversity: Linking Language, Knowledge, and the Environment.* Smithsonian Institution Press, Washington, DC

Marshall, E., K. Schreckenberg, and Newton, A. (2006). *Commercialization of Non-timber Forest Products; Factors Influencing Success.* UNEP- World Conservation Monitoring Centre Biodiversity Series No 23, Cambridge

Mas, A.H. and Dietsch, T.V. (2004). Linking shade coffee certification to biodiversity conservation: butterflies and birds in Chiapas, Mexico. In *Ecological Applications* 14(3):642-654

Mather, A. and Needle, C.L. (1998). The Forest transition: a theoretical basis. In *Area* 30:117-124

May, R.M. (1992). How many species inhabit the earth? In *Scientific American* 267(4):42

McNeely, J.A. (2004). Nature vs. Nurture: managing relationships between forests, agroforestry and wild biodiversity. In *Agroforestry Systems* 61:155-165

McNeely, J.A. and Scherr, S.J. (2001). *Common Ground, Common Future: How Ecoagriculture Can Help Feed the World and Save Wild Biodiversity.* World Conservation Union and Future Harvest, Washington, DC

McNeely, J.A. and Scherr, S.J. (2003). *Ecoagriculture: Strategies to Feed the World and Save Wild Biodiversity.* Island Press, Washington, DC

Meilleur, B. (1994). In Search of "Keystone Societies." In Etkin, N.L. (ed.) *Eating on the Wild Side: The Pharmacologic, Ecologic, and Social Implications of Using Noncultigens.* University of Arizona Press, Tucson, AZ

Moock, J. and Rhoades, R. (1992). *Diversity, Farmer Knowledge and Sustainability.* Cornell University Press, Ithaca, NY

Mora, C., Andréfouët, S., Costello, M.J., Kranenburg, C., Rollo, A., Veron, J., Gaston, K. J., Myers, R. A. (2006). Coral Reefs and the Global Network of Marine Protected Areas. In *Science* 312:1750-1751

Myers, N. and Kent. J. (2001). *Perverse Subsidies: How Tax Dollars Can Undercut the Environment and Economy.* Island Press, Washington, DC

Nabhan, G. P. and S. L. Buchman. 1997. Services provided by pollinators. In Daily G. E. (ed.) *Nature's Services – Societal Dependence on Natural Ecosystems.* Island Press, Washington, DC

NASA (2005). Earth Observatory. Astronaut image ISS012-E-11779 http://earthobservatory.nasa.gov/Newsroom/NewImages/images.php3?img_id=17161 (last accessed 8 May 2007)

OECD (2001). *Environmental Performance Reviews Germany.* Organisation for Economic Co-operation and Development, Paris

Oldfield, M.L. and Alcorn, J. (1987). Conservation of Traditional Agroecosystems. In *BioScience* 37(3):199-208

Olson, D.M., Dinerstein, E., Wikramanayake, E.D., Burgess, N.D., Powell, G.V.N., Underwood, E.C., D'Amico, J.A., Itoua, I., Strand, H.E., Morrison, J.C., Loucks, C.J., Allnutt, T.F., Ricketts, T.H., Kura, Y., Lamoreux, J.F., Wettengel, W.W., Hedao, P. and Kassem, K.R. (2001). Terrestrial ecoregions of the world: a new map of life on earth. In *BioScience* 51:933-8

Osofsky, S. A., Kock, R. A., Kock, M.D., Kalema-Zikusoka, G., Grahn, R., Leyland, T. and Karesh, W. B. (2005). Building support for protected areas using a "one health" perspective. In McNeely, J. A. (ed.) *Friends for Life: New Partners in Support of Protected Areas.* World Conservation Union, Gland

Oviedo, G., Maffi, L. and Larsen, P. B. (2000). *Indigenous and Traditional Peoples of the World and Ecoregion Conservation: An Integrated Approach to Conserving the World's Biological and Cultural Diversity.* World Wide Fund for Nature and Terralingua, Gland

Palm, C.A., Vosti, S.A., Sanchez, P.A. and Ericksen, P.J. (eds.) (2005). *Slash-and-Burn Agriculture: The Search for Alternatives.* Columbia University Press, New York, NY

Parmesan, C., Ryrholm, N., Steganescu, C., Hill, J.K., Thomas, C.D., Descimon, H., Huntley, B., Kaila, L., Kullberg, J., Tammaru, T., Tennent, W.J., Thomas, J.A. and Warren, M. (1999). Poleward shifts in geographical ranges of butterfly species associated with regional warming. In *Nature* 399:579-83

Partap, U. (2002). *Cash crop farming in the Himalayas: the importance of pollinator management and managed pollination.* Biodiversity and the Ecosystem Approach in Agriculture. Forestry and Fisheries, Food and Agriculture Organization of the United Nations, Rome

Perfecto, I., Vandermeer, J., Mas, A.H. and Soto Pinto, L. (2005). Biodiversity, yield, and shade coffee certification. In *Ecological Economics* 54:435-446

Pimentel, D., Rodrigues, G., Wang, T., Abrams, R., Goldberg, K., Staecker, H., Ma, E., Brueckner, L., Trovato, L., Chow, C., Govindarajulu, U. and Boerke, S. (1994). Renewable energy: economic and environmental issues. In *BioScience* 44:536-547

Posey, D.A. (1998). Diachronic Ecotones and Anthropogenic Landscapes in Amazonia: Contesting the Consciousness of Conservation. In Baleé, W. (ed.) *Advances in Historical Ecology.* Columbia University Press, New York, NY

Posey, D. A. (ed.) (1999). *Cultural and Spiritual Values of Biodiversity.* Intermediate Technology Publications and United Nations Environment Programme, London and Nairobi

POST (2005). *The bushmeat trade.* POSTNOTE, February 2005, No. 236. Parliamentary Office of Science and Technology, London

Pounds J.A., Fogden, M. P. L. and Campbell, J. H. (1999). Biological response to climate change on a tropical mountain. In *Nature* 398:611-615

Pounds, J.A., Bustamante, M. R., Coloma, L. A., Consuegra, J. A., Fogden, M. P. L., Foster, P. N., La Marca, E., Masters, K. L., Merino-Viteri, A., Puschendorf, R., Ron, S. R., Sa´nchez-Azofeifa, G. A., Still, C. J. and Young, B. E. (2006). Widespread amphibian extinctions from epidemic disease driven by global warming. In *Nature* 469:161-167

Pretty, J. (2002). *Agri-Culture: Reconnecting People, Land and Nature.* Earthscan, London

Ramankutty, N. (2002). Global distribution of croplands. Center for Sustainability and the Global Environment, University of Wisconsin- Madison. Unpublished data obtained through personal communication

Ramankutty, N. (2005). Global distribution of grazing lands. Center for Sustainability and the Global Environment, University of Wisconsin- Madison. Unpublished data obtained through personal communication

Revenga, C., Brunner, J., Henninger, N., Kassem, K. and Payne, R. (2000). *Pilot Analysis of Global Ecosystems: Freshwater Systems.* World Resources Institute, Washington, DC

Rodrigues, A.S.L., Andelman, S.J., Bakarr, M.I., Boitani, L., Brooks, T.M., Cowling, R.M., Fishpool, L.D.C., da Fonseca, G.A.B., Gaston, K.J., Hoffmann, M., Long, J.S., Marquet, P.A., Pilgrim, J.D., Pressey, R.L., Schipper, J., Sechrest, W., Stuart, S.N., Underhill, L.G., Waller, R.W., Watts, M.E.J. and Yan, X. (2004). Effectiveness of the global protected area network in representing species diversity. In *Nature* 428(6983):640-643

Ron, S. R., Duellman, W. E., Coloma, L. A. and Bustamante, M.R. (2003). Population declines of the Jambato toad *Atelopus ignescens* (Anura:Bufonidae) in the Andes of Ecuador. In *J. Herpetol* 37:116-126

Safrin, S. (2004). Hyperownership in a time of biotechnical promise: the international conflict to control the building blocks of life. In *American Journal of International Law* 641

Sagarin, R.D., Barry, J.P., Gilman, S.E. and Baxter, C.H. (1999). Climate-related change in an intertidal community over short and long time scales. In *Ecological Monographs* 69:465-490

Sala, O.E., Chapin, F.S. and Armesto, J.J. (2000). Global biodiversity scenarios for the year 2100. In *Science* 287:1770–1774

Sanchez, P.A. (2002). Soil Fertility and Hunger in Africa. In *Science* 295:2019-2020

Sayer, J. and Campbell, B. (2004). *The Science of Sustainable Development: Local Livelihoods and the Global Environment.* Cambridge University Press, Cambridge

SCBD (2006). *Global Biodiversity Outlook 2.* Secretariat of the Convention on Biological Diversity, Montreal

Schroth, G., Da Fonseca, G.A.B., Harvey, C.A., Gascon, C., Vasconcelos, H.L. and Izac, A-M. (2004). *Agroforestry and Biodiversity Conservation in Tropical Landscapes.* Island Press, Washington, DC

Sebastian, K. (2006). *Global Extent of Agriculture.* International Food Policy Research Institute, Washington, DC

Selin, H. (ed.) (2003). *Nature Across Cultures: Views of Nature and the Environment in Non-Western Cultures.* Kluwer Academic Publishers, Dordrecht

Shiganova, T. and Vadim, P. (2002). Invasive species *Mnemiopsis leidyi.* Prepared for the Group on Aquatic Alien Species (GAAS). www.zin.ru/projects/invasions/gaas/mnelei.htm (last accessed 8 May 2007)

Shinn, E.A., Smith, G.W., Prospero, J.M., Betzer, P., Hayes, M.L., Garrison, V. and Barber, R.T. (2000). African dust and the demise of Caribbean coral reefs. In *Geophysical Research Letters* 27(19):3029-3032

Siebert, S., Doell, P., Feick, S. and Hoogeveen, J. (2006). *Global map of irrigated areas version 4.0.* Johann Wolfgang Goethe University, Frankfurt am Main and Food and Agriculture Organization of the United Nations, Rome

Skutnabb-Kangas, T., Maffi, L. and Harmon, D. (2003). *Sharing a World of Difference: The Earth's Linguistic, Cultural, and Biological Diversity.* United Nations Educational, Scientific and Cultural Organization, Paris

Smetacek, V. and Nicol, S. (2005). Polar ocean ecosystems in a changing world. In *Nature* 437(7057):362-368

Smith, K. and Darwall, W. (compilers) 2006. *The Status and Distribuiton of Freshwater Fish Endemic to the Mediterranean Basin.* World Conservation Union, Gland and Cambridge

Sowerwine, J.C. (2004). Effects of economic liberalization on Dao women's traditional knowledge, ecology, and trade of medicinal plants in Northern Vietnam. In Carlson, T.J.S. and Maffi, L. (eds.) (2004)

Stein, B.A., Kutner, L.S. and Adams, J.S. (2000). *Precious Heritage: The Status of Biodiversity in the United States.* Oxford University Press, New York, NY

Stepp, J. R., Cervone, S., Castaneda, H., Lasseter, A., Stocks, G., and Gichon, Y. (2004). Development of a GIS for global biocultural diversity. In Borrini-Feyerabend, G., MacDonald, K. and Maffi, L. (eds.) (2004)

Stepp, J. R., Castaneda, H. and Cervone, S. (2005). Mountains and biocultural diversity. In *Mountain Research and Development* 25(3):223-227

Stige, L.C., Stave, J., Chan Kung-Sik, Ciannelli, L., Pettorelli, N., Glantz, M., Herren, H.R. and Stenseth, N.C. (2005). The effect of climate variation on agro-pastoral production in Africa. In *Proceedings of the National Academy of Sciences* 103(9):3049-3053

The Economist (2005). Saving Environmentalism and Are you being served. In *The Economist* 23 April 2005: 75-78

Thomas, J.A., Telfer, M.G., Roy, D.B., Preston, C.D., Greenwood, J.J.D., Asher, J., Fox, R., Clarke, R.T. and Lawton, J.H. (2004a). Comparative losses of British butterflies, birds, and plants and the global extinction crisis. In *Science* 303(5665):1879-81

Thomas, C.D., Cameron, A., Green, R.A., Bakkenes, M., Beaumont, L.J., Collingham, Y.C., Erasmus, B.F.N., de Siquera, M.F., Grainger, A., Hannah, L., Hughes, L., Huntley, B., van Jaarsveld, A.S., Midgley, G.F., Miles, L., Ortega-Huerta, M.A., Townsend Peterson, A., Phillips, O.L. and Williams, S.E. (2004b). Extinction risk from climate change. In *Nature* 427:145-8

Thomas, C.D. and Lennon, J.J. (1999). Birds extend their ranges northwards. In *Nature* 399:213

Thuiller, W. (2006). Patterns and uncertainties of species' range shifts under climate change. In *Global Change Biology* 10(12):2020

Thrupp, L.A. (1998). *Cultivating Diversity.* World Resources Institute, Washington, DC

Tillman, D., Cassman, K.G., Matson, P.A., Naylor, R. and Polasky, S. (2002). Agricultural sustainability and intensive production practices. In *Nature* 418:671-677

UNDP (2004). Human Development Report 2004: *Cultural Liberty in Today's Diverse World.* United Nations Development Programme. New York, NY

UNEP (1999). *Wildland Fires and the Environment: a Global Synthesis.* UNEP/DEIA&EW/TR.99-1. United Nations Environment Programme, Nairobi

UNEP (2006a). *Marine and coastal ecosystems and human well-being: A synthesis report based on the findings of the Millennium Ecosystem Assessment.* DEW/0785/NA. United Nations Environment Programme, Nairobi

UNEP (2006b). *Ecosystems and biodiversity in deep waters and high seas.* UNEP Regional Seas Reports and Studies No. 178. United Nations Environment Programme and World Conservation Union (IUCN), Nairobi

UNEP-WCMC (2006). *World Database of Protected Areas.* World Conservation Monitoring Centre, UK (in GEO Data Portal)

UNEP-WCMC (2006a). *World Database on Protected Areas.* www.unep-wcmc.org/wdpa/index.htm (last accessed 8 May 2007)

UNEP-WCMC (2006b). *In the front line: shoreline protection and other ecosystem services from mangroves and coral reefs.* UNEP-World Conservation Monitoring Centre, Cambridge

UNEP and UNESCO (2003). *Cultural Diversity and Biodiversity for Sustainable Development.* United Nations Environment Programme, Nairobi

UNESCO (2000). *Science for the Twenty-First Century: A New Commitment.* World Conference on Science, United Nations Educational, Scientific and Cultural Organization, Paris

UNESCO (2001). *Atlas of the World's Languages in Danger of Disappering.* United Nations Educational, Scientific and Cultural Organization Publishing, Paris

UNPD (2007). *World Population Prospects: The 2006 Revision.* UN Population Division, New York, NY (in GEO Data Portal)

USDA (2004). *The Amazon: Brazil's Final Soybean Frontier.* Washington, DC http://www.fas.usda.gov/pecad/highlights/2004/01/Amazon/Amazon_soybeans.htm (last accessed 8 May 2007)

Van Swaay, C.A.M. (1990). An assessment of the changes in butterfly abundance in the Netherlands during the 20th Century. In *Biological Conservation* 52:287-302

Vittor, A.Y., Gilman, R.H., Tielsch, J., Glass, G.E., Shields, T.M., Sanchez-Lozano W, Pinedo, V.V. and Patz, J.A. (corresponding author) (2006). The effects of deforestation on the human-biting rate of Anopheles darlingi, the primary vector of falciparum malaria in the Peruvian Amazon. In *American Journal of Tropical Medicine and Hygiene* 74:3-11

Walkovsky, A. (1998). Changes in phenology of the locust tree (*Robinia pseudoacacia L.*) in Hungary. In *International Journal of Biometeorology* 41:155-160

WCD (2000). *Dams and Development: A New Framework for Decision-Making.* World Commission on Dams. Earthscan, London

WDPA (2006). World Database on Protected Areas. IUCN-WCPA and UNEP-WCMC, Washington, DC

Wetlands International (2002). *Waterbird Population Estimates.* Third Edition. Wetlands International Global Series No. 12, Wageningen

WHO (2001). Herbs For Health, But How Safe Are They? In *WHO News. Bulletin of the World Health Organization* 79(7):691

WHO (2005). *Ecosystems and Human Well-being: Health Synthesis.* Millennium Ecosystem Assessment. World Resources Institute, Washington, DC

Wilkinson, C. (2002). Coral bleaching and mortality — The 1998 event 4 years later and bleaching to 2002. In Wilkinson, C. (ed). *Status of coral reefs of the world: 2002.* Australian Institute of Marine Science http://www.aims.gov.au/pages/research/coral-bleaching/scr2002/scr-00.html (last accessed 8 May 2007)

Wilson, E.O. (2002). *The Future of Life.* Alfred A. Knopf, New York, NY

Woiwod, I.P. (1997). Review. In *Journal of Insect Conservation* 1:149-158

Wood, S., Sebastian, K. and Scherr, S.J. (2000). *Pilot Analysis of Global Ecosystems: Agroecosystems.* Report Prepared for the Millennium Assessment of the State of the World's Ecosystems. International Food Policy Research Institute and World Resources Institute, Washington, DC

World Bank (2006). *World Development Indicators.* The World Bank, Washington, DC

Worldwatch Institute (2006). *Biofuels for transportation, global potential and implications for sustainable agriculture and energy in the 21st century.* Worldwatch Institute, Washington, DC

WWF and IUCN (2001). *The status of natural resources on the high-seas.* http://www.iucn.org/THEMES/MARINE/pdf/highseas.pdf (last accessed 8 May 2007)

WWF (2006). *Conservation Status of Terrestrial Ecoregions.* World Wide Fund for Nature, Gland http://www.panda.org/about_wwf/where_we_work/ecoregions/maps/index.cfm (last accessed 8 May 2007)

Zeier, M., Handermann, M., Bahr, U., Rensch, B., Muller, S., Kehm, R., Muranyi, W. and Darai, G. (2005). New ecological aspects of hantavirus infection: a change of a paradigm and a challenge of prevention — a review. In *Virus Genes* 30(2):157-80

Zent, E.L. (1998). A Creative Perspective of Environmental Impacts by Native Amazonian Human Populations. In *Interciencia* 23(4):232-240

Zent, S. and Lopez-Zent, E. (2004). Ethnobotanical Convergence, Divergence, and Change among the Hoti of the Venezuelan Guayana. In Carlson, T.J.S. and Maffi, L. (eds.) (2004)

Section C

Regional Perspectives: 1987–2007

Chapter 6 **Sustaining a Common Future**

*Continued environmental degradation in
all regions is unfairly shifting burdens onto
future generations, and contradicts the
principle of intergenerational equity.*

Sustaining a Common Future

Coordinating lead authors: Jane Barr and Clever Mafuta

Lead authors:
Africa: Clever Mafuta
Asia and the Pacific: Murari Lal and Huang Yi
Europe: David Stanners
Latin America and the Caribbean: Álvaro Fernández-González, Irene Pisanty-Baruch, and Salvador Sánchez-Colón
North America: Jane Barr
West Asia: Waleed K. Al-Zubari and Ahmed Fares Asfary
Polar Regions: Joan Eamer and Michelle Rogan-Finnemore

Contributing authors:
Africa: Washington Ochola, Ahmed Abdelrehim, Charles Sebukeera, and Munyaradzi Chenje
Asia and the Pacific: Jinhua Zhang, Tunnie Srisakulchairak Sithimolada, Sansana Malaiarisoon, and Peter Kouwenhoven
Europe: Gulaiym Ashakeeva, Peter Bosch, Barbara Clark, Francois Dejean, Nikolay Dronin, Jaroslav Fiala, Anna Rita Gentile, Adriana Gheorghe, Ivonne Higuero, Ybele Hoogeveen, Dorota Jarosinska, Peder Jensen, Andre Jol, Jan Karlsson, Pawel Kazmierczyk, Peter Kristensen, Tor-Björn Larsson, Ruben Mnatsakanian, Nicolas Perritaz, Gabriele Schöning, Rania Spyropoulou, Daniel Puig, Louise Rickard, Gunnar Sander, Martin Schäfer, Mirjam Schomaker, Jerome Simpson, Anastasiya Timoshyna, and Edina Vadovics
Latin America and the Caribbean: Paola M. García-Meneses, Elsa Patricia Galarza Contreras, Sherry Heilemann, Thelma Krug, Ana Rosa Moreno, Bárbara Garea, José Gerhartz Muro, Stella Navone, Joana Kamiche-Zegarra, and Farahnaz Solomon
North America: Bruce Pengra and Marc Sydnor
West Asia: Asma Ali Abahussain, Mohammed Abido, Rami Zurayk, Abdullah Al-Droubi, Ibrahim Abdul Gelil Al-Said, Saeed Abdulla Mohamed, Sabah Al-Jenaid, Mustafa Babiker, Maha Yahya, Hratch Kouyoumjian, Anwar Shaikheldin Abdo, Dhari Al-Amji, Samira Asem Omar, Asadullah Al-Ajmi, Yousef Meslmani, Gilani Abdelgawad, Sami Sabry, Mohamed Ait Belaid, Sahar Al-Barari, Fatima Haj Mousa, Ahlam Al-Marzouqi, Elham Tomeh, Omar Jouzdan, Said Jalala, Mohammed Eila, and Nahida Butayban
Polar Regions: Alan Hemmings, Christoph Zöckler, and Christian Nellemann

Chapter review editors: Rudi Pretorius and Fabrice Renaud

Chapter coordinator: Ron Witt

Main messages

Multistakeholder consultations organized in the seven GEO regions as part of the GEO-4 assessment show that the regions share common concerns about a number of critical environmental and sustainability issues as well as face tremendous differences in their environmental challenges. The assessment highlights strong interdependencies reinforced by globalization and trade, with growing demand on resources in and across the regions. Some of the main common messages arising out of regional analyses include the following:

Population and economic growth are major factors fuelling increased demand on resources, and contributing to global environmental change in terms of the atmosphere, land, water and biodiversity. Four of the regions identified climate change as a key priority issue (Europe, Latin America and the Caribbean, North America, and the Polar Regions). The other regions have also highlighted climate change as a major issue. Developed regions have higher per capita emissions of greenhouse gases, while climate change impacts will have greater effects on the poor and other vulnerable people and countries.

There has been an encouraging decoupling of environmental pressures from economic growth in some areas. However, globalization has contributed to the achievement of environmental progress in some developed regions at the expense of developing countries through the outsourcing of energy, food and industrial production, and the subsequent relocation of related environmental and social impacts. Disparities in ecological impacts prevail, and environmental inequities continue to grow. Gender inequities continue in many regions where women often have limited access to natural resources, and are exposed to the health risks of indoor air pollution.

There are examples of good environmental governance and investments in new technologies that provide models for other regions. Economic, political and social integration, combined with good governance, is making Europe a leader in transboundary environmental decision making. North America is a model in providing access to superior quality environmental information, and investments in research and development. Africa, Asia and the Pacific, Latin America and the Caribbean and West Asia have also made great strides in tackling some of their environment and development challenges. Integrated watershed management is increasing in many regions, helping to protect and restore ecosystems.

Unique regional concerns emphasize the diversity of environmental issues across the globe. The variety of the main regional environmental messages is reflected in the following:

In **Africa**, land degradation is the overarching environmental issue of concern, affecting some 5 million square kilometres of land by 1990 and contributing to loss of livelihoods. Poverty is both a cause and a consequence of land degradation: poor people are forced to put immediate needs before the long-term quality of the land, while degraded farmland and poor yields contribute to food and income insecurity. Per capita food production in Africa has declined by 12 per cent since 1981. Drought and climate change and variability exacerbate land degradation. In addition to threatening the livelihoods of the rural poor, land degradation has widespread effects on Africa's river catchments, forests, the expansion of deserts, and it diminishes ecosystem services. Regional efforts to stop land degradation include integrated crop and land management programmes that also seek to improve yields. Policy shortcomings in addressing the issue still exist, as do unfair agricultural subsidies in developed regions.

In **Asia and the Pacific**, rapid population growth, higher incomes, and burgeoning industrial and urban development are causing a number of environmental problems that have implications for human health and well-being. The priority issues are urban air quality, stress on freshwater, degraded ecosystems, agricultural land use and increased waste. A number of factors have led to an increase in urban air pollution: a highly urbanized population; poorly planned municipal development; a lack of affordable and clean mass transport services; the massive increase in motorized vehicles, with the use of passenger cars in the region increasing about 2.5 times over the last two decades; and haze pollution from forest fires in South East Asia. Air pollution causes the premature deaths of about 500 000 people annually in Asia. Excessive withdrawals from surface waters and aquifers, industrial pollution, inefficient use, climate change and variability, and natural disasters are major causes of water stress, threatening human well-being and ecological health. There has been remarkable progress in the provision of improved drinking water over the last decade, but some 655 million people in the region (17.6 per cent) still lack access to safe water. Valuable ecosystems continue to be degraded. Except for Central Asia, most sub-regions have applied sufficient counter measures to successfully overcome the impacts of land degradation on agricultural production. Rapid economic growth, together with new lifestyles associated with greater affluence, have led to rapid changes in consumption patterns. This has contributed to the generation of large quantities of waste, and changes in waste composition. The illegal traffic in electronic and hazardous waste and their effects on human health and the environment pose new and growing challenges. Most countries have developed extensive domestic laws, regulations and standards related to the environment, and participate in global action through multilateral and bilateral agreements.

In **Europe**, rising incomes and growing numbers of households contribute to unsustainable production and consumption, increased energy use and emissions of greenhouse gases, poor urban air quality and transportation challenges. Biodiversity loss, land-use change and freshwater stresses are the other priority issues. The region has made progress in decoupling economic growth from resource use and environmental pressures, although per capita household consumption is steadily increasing. Since 1987, greenhouse gas (GHG) emissions from the energy sector have been reduced in Western Europe, but have increased since the end of the 1990s across the whole region, partly because increasing natural gas prices have re-established coal as a key energy source. Recently, growing public awareness, underpinned by rising energy prices, has given a new political momentum to climate change policies. Despite much progress, poor water and urban air quality still cause substantial problems in parts of the region, affecting the health and quality of life of many people. However, in most parts of the region, water quality has improved since 1990 due to reductions in contaminant loads from wastewater treatment and industries, as well as a decline in industrial and agricultural activity. Emissions of air pollutants are largely driven by the demand for greater mobility. The EU has been imposing progressively stricter pollution controls on vehicles. Farming in marginal areas is under pressure, and is subject to both land abandonment and intensification, both of which have impacts on biodiversity. With its many action plans and legal instruments at different levels, the region has a unique experience of environmental cooperation.

In **Latin America and the Caribbean**, the priority environmental issues are growing cities, threats to biodiversity and ecosystems, degraded coasts and polluted seas, and regional vulnerability to climate change. Regionalization and globalization have triggered an increase in oil and gas extraction, expanded the use of arable land for monoculture exports and intensified tourism. As a result, decreased access to rural livelihoods has helped fuel the continued unplanned growth of urban areas. The region is the most urbanized in the developing world, with 77 per cent of the total population living in cities. The quality of fuels (both gasoline and diesel) has gradually improved throughout the region, but urban air pollution and associated health impacts are high and increasing. Untreated domestic and industrial wastewaters are on the rise, affecting coastal areas, where 50 per cent of the people live. Domestic waste is generally insufficiently treated. Land use change has had impacts on biodiversity and cultural diversity. The conversion of forest land to pastures, monoculture planted forests, infrastructure and urban areas is causing habitat loss and fragmentation, as well as the loss of indigenous knowledge and cultures. Other pressures are from wood harvesting, forest fires and the extraction of fossil fuels. Integrated prevention and control programmes are helping to decrease annual deforestation rates in the Amazon. Land degradation affects 15.7 per cent of the region, due to deforestation, overgrazing and inappropriate irrigation. Protected areas now cover 11 per cent of the land base, and new efforts are being made to conserve corridors and the Amazon, but more efforts are needed to protect hot spots. Declining water quality, climate change and algal blooms have contributed to the rise in water-borne diseases in coastal regions. To address the pressures, integrated marine and coastal areas management is increasing. Extreme climatic events have increased over the past 20 years, and the region is subject to climate change impacts, such as retreating glaciers.

In **North America**, energy use, urban sprawl and freshwater stresses can all be related to climate change, an issue the region is struggling to address. With only 5.1 per cent of world population, North America consumes just over 24 per cent of global primary energy. The consumption of energy is responsible for much of the region's high and increasing GHG emissions that contribute to climate change. Canada signed the Kyoto Protocol, and has produced a plan to become more energy efficient, while some US states

are showing impressive action to reduce energy use and emissions in the absence of mandatory federal caps. Further gains in energy efficiency have been hampered by increased use of larger and less fuel-efficient vehicles, low fuel economy standards, and increases in distances travelled and the number of cars. Sprawling suburbs and a growing trend towards exurban settlement patterns are fragmenting ecosystems, increasing the urban-wildlife interface and paving over prime agricultural land. Although there are policies to contain sprawl, suburban life is deeply imbedded in the culture and landscape. The past 20 years have seen important regional water shortages, and climate change is expected to exacerbate water deficits. Agriculture is the major water user, and irrigation continues to increase, competing with urban centres for limited supplies. In response, water restrictions and conservation strategies have become widespread. The human health impacts of environmental change are an emerging issue, as it becomes more evident that air pollution is linked to respiratory diseases, and there are significant economic costs.

In **West Asia**, freshwater stresses, land degradation, coastal and marine ecosystems degradation, urban management, and peace and security are the priority issues. The region's environment is predominantly dryland, with great variability in rainfall within and between seasons, and frequent spells of drought. West Asia is one of the most water-stressed regions in the world. Reflecting rapid population growth and socio-economic development, overall per capita freshwater availability has fallen, while consumption has risen. Agriculture uses 80 per cent of the region's available water. Aided by subsidies, irrigated agriculture has expanded to achieve food security, but inefficient methods and poor planning have put immense pressure on the limited resource. Desalination of seawater provides the bulk of municipal water in the GCC countries, but sustainability is hampered by a lack of demand management and price-signalling mechanisms. The level of sewage treatment is low, so shallow aquifers are polluted, and have high levels of nitrates, a health hazard. In the Mashriq sub-region, water-borne diseases are a major concern. Since more than 60 per cent of surface water resources originate from outside the region, the sharing of international water resources is another major challenge. With increased urbanization and growing economies, the number of vehicles has risen enormously. Although unleaded gasoline has been introduced in most of the countries, continued use of leaded gas contributes to air pollution, and poor human health and economic performance. In some countries, growing economic disparities, rising rural-urban migration and/or military conflicts have led to the expansion of slum areas, and increased human suffering, often related to deteriorating environmental conditions. Land degradation is a key issue, especially since fragile drylands constitute about 64 per cent of the total area. A number of factors have led to the deterioration of marine and coastal areas, including fisheries, mangroves and coral reefs. They include rapid development of urban and tourism infrastructure, and of refineries, petrochemical complexes, power and desalination plants, as well as oil spills from ship ballast. Vast areas of terrestrial and marine ecosystems have been severely affected by wars, which led to the discharge of millions of barrels of crude oil into coastal waters. They have also been harmed by the infiltration of oil and seawater into aquifers, and by hazardous waste disposal. Environmental impact assessment requirements were introduced recently. Other responses include programmes to conserve biodiversity, manage coastal zones and develop marine protected areas.

The **Polar Regions** influence major environmental processes, and have direct impacts on global biodiversity and human well-being. The priority issues are climate change, persistent pollutants, the depletion of the ozone layer, and commercial activity. Even though their GHG emissions are negligible, the Polar Regions are part of a cycle of global climate change impacts, such as altered ocean currents and rising sea levels. Evidence shows that circulation of the deep, cold water of the North Atlantic conveyor belt may have slowed. Its breakdown could precipitate an abrupt change in global climate regimes. As a result of climate change, the Arctic is warming twice as fast as the world average, causing shrinking sea ice, melting glaciers and changes in vegetation. The Greenland and Antarctic ice sheets are the largest contributors to the sea level rise from melting land ice. Observed climate change has wide-ranging impacts on plants, animals and human well-being in the Arctic. Although the manufacture and use of many persistent organic pollutants (POPs) have been banned in most industrialized countries, they persist in the environment and accumulate in cold regions where they enter marine and terrestrial ecosystems, and build up in food chains. Mercury from industrial emissions is also increasing in the environment. These toxic substances pose a threat to the integrity of the traditional food system and the health of indigenous peoples. Action by scientists and northern indigenous peoples resulted in several important treaties to address toxic chemicals. Stratospheric ozone depletion in the Polar Regions has resulted in seasonal increase in ultraviolet radiation, with impacts on ecosystems and increased human health risks. Despite the success of the Montreal Protocol, recovery of the stratospheric ozone layer is expected to take another half century, or more.

REGIONAL DEVELOPMENTS

Since the World Commission on Envionment and Development (Brundtland Commission), international and national environmental policies have invoked sustainable development to address the impacts of economic growth, ensure a clean environment today and in the future, and reduce the cumulative effects of poverty. The 2005 United Nations Summit was one of the largest gatherings of world leaders in history. It emphasized the urgency and relevance of achieving more sustainable development. According to the World Business Council for Sustainable Development, "the planet seems at least as unsustainable as in 1987" (WBCSD 2007). Sustainable development is particularly critical in a world where pressures on the environment are increasing, with a wide range of ensuing impacts on the environment and human health. Some of the impacts, such as climate change, long-range air pollution and upstream-to-downstream water pollution can have far-reaching effects.

Economic trends

Economic trends over the past two decades have played a significant role in shaping the state of the world's environment (see Chapter 1). In 1987, many developing countries were in an economic downturn, characterized by falling prices for their exports, which consist mainly of raw materials, such as mineral ores and agricultural products. The prices of such commodities have often not increased significantly since the 1980s, and the current economic order has been worsened by growing loan repayment burdens. Africa, for example, enjoys only 5 per cent of the developing world's income, but it carries about two-thirds of the global debt (AFRODAD 2005), with sub-Saharan Africa spending US$14.5 billion yearly on debt repayments (Christian Reformed Church 2005). Despite some recent efforts towards debt relief, developing countries in Africa and other regions are still forced to exploit the limited capital tied up in their natural resources.

Cotton bales ready for export from Cameroon. Farmers in developing regions face many challenges due to unfair global markets.

Credit: Mark Edwards/Still Pictures

The Brundtland Commission highlighted poverty alleviation as one of the key responses needed to address the world's environmental problems, and this is still valid today. Poverty and environmental degradation have a cause-and-effect relationship, and can fall into a cycle that is difficult to reverse (UNEP 2002a). Some argue that the past two decades have seen too modest and uneven economic growth rates (see Figure 6.1 and Figure 1.7 in Chapter 1, which shows 20-year averages at a country level) to have influenced a significant positive effect on the state of the environment. However, a contrary argument is that economic growth is the cause of current environmental degradation. The conundrum is illustrated by the case of nutrient loading, which was highlighted in GEO-3 as a priority environmental issue (UNEP 2002a). Large-scale fertilizer applications boost yields of hybrid crops, which the Brundtland Commission projected would increase food production through a green revolution. While fertilizers have made a positive contribution to growth of the agricultural sector, and ultimately to the economy, excessive nutrients from agricultural inputs have also contributed to soil degradation, and affected freshwater quality and marine ecosystems, endangering the ecosystem services that are the basis of long-term economic prosperity (MA 2005).

Livelihood conditions

Across the world, there is a clear trend of increasing food production to keep pace with growing populations and rising incomes. The world's population is currently estimated at 6.7 billion, having grown by some 1.7 billion since 1987 (GEO Data Portal, from UNPD 2007). The Brundtland Commission warned against attributing environmental problems to population growth alone, since global environmental problems can also be accounted for by inequalities in access to resources and their unsustainable use. Prior to 1987, developed countries, with one-quarter of the world's population, consumed about 80 per cent of commercial energy and metals, 85 per cent of paper, and more than half of the fat contained in food (Court 1990). The situation virtually remains the same today, with North America, for example, consuming over 24 per cent of the total global primary energy despite having only 5.1 per cent of the world's population (GEO Data Portal, from IEA 2007 and UNPD 2007).

The world continues to undergo regional and national economic changes that have global consequences. These include trade and subsidies. For example, the World Trade Organization (WTO) relies on regional trade agreements to settle disputes between member countries. While

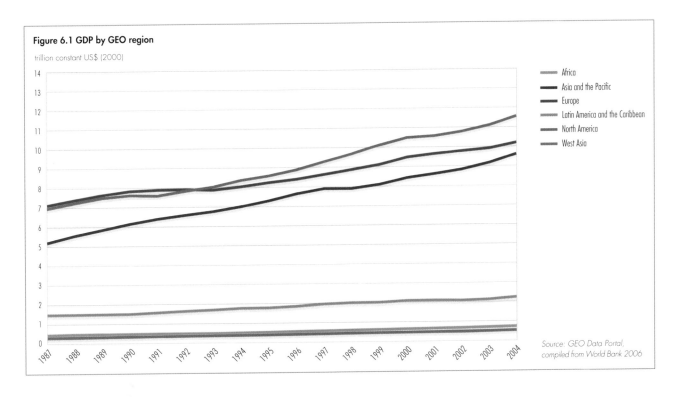

Figure 6.1 GDP by GEO region

trillion constant US$ (2000)

Legend:
- Africa
- Asia and the Pacific
- Europe
- Latin America and the Caribbean
- North America
- West Asia

Source: GEO Data Portal, compiled from World Bank 2006

Global environmental problems are more evident today than they were two decades ago. For example, by 2003 global CO_2 emissions had increased by 17 per cent compared to 1990 levels. The rapidly-expanding economies of China and India are contributing significantly to this increase. China is already the second largest CO_2 emitter after the United States.

Most of these emissions come from energy generation. The resulting air pollution has a significant impact not only on local air quality and human health, but also on the global climate (see Chapter 2). Despite the Brundtland Commission's recommendation for the introduction of fuel-efficient modern technologies, and the commitment by the World Summit on Sustainable Development in 2002 to diversify energy supply and substantially increase the global share of renewable energy sources, it is projected that fossil fuels will remain the dominant energy source to 2025, accounting for over 80 per cent of the energy demand. Therefore, the world continues to be locked into unsustainable energy patterns that are associated with climate change and other environmental and human health threats.

This situation is compounded by disparities in regional energy consumption patterns (see Figure 1.8 in Chapter 1). It is projected that over 70 per cent of the increase in energy demand up to 2025 will come from developing countries, with China alone accounting for 30 per cent, implying that both developed and developing regions will have major impacts on both air quality and global climate change.

The unsustainability of the way the Earth's natural resources are being used is increasingly evident. As a result of the growing competition and demand for global resources, the world's population has reached a stage where the amount of resources needed to sustain it exceeds what is available. An example of ecological overshoot is seen in attempts to increase food production that result in increased levels of environmental degradation, such as deforestation of marginal lands, including wetlands, upper watersheds and protected areas that have been converted to farmlands. According to the *2005 Footprint of Nations* report, humanity's footprint is 21.9 ha/person, while the Earth's biological capacity is, on average, only 15.7 ha/person, with the ultimate result that there is net environmental degradation and loss. On a regional level, the differences in footprint are profound, as illustrated in the Living Planet Report 2006 (see Figure 6.2).

Sources: IEA 2007, UNFCCC-CDIAC 2006, Venetoulis and Talberth 2005, World Bank 2006, WWF 2006a

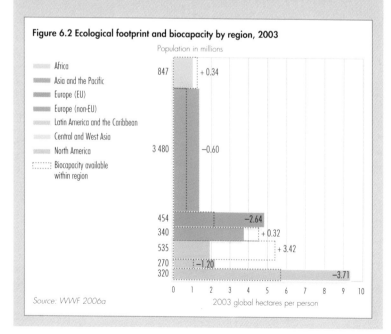

Figure 6.2 Ecological footprint and biocapacity by region, 2003

Population in millions

- Africa
- Asia and the Pacific
- Europe (EU)
- Europe (non-EU)
- Latin America and the Caribbean
- Central and West Asia
- North America
- Biocapacity available within region

847	+ 0.34
3 480	−0.60
454	−2.64
340	+ 0.32
535	+ 3.42
270	−1.20
320	−3.71

0   1   2   3   4   5   6   7   8   9   10

2003 global hectares per person

Source: WWF 2006a

many of these disputes deal exclusively with trade issues, some focus on the use of environmental or social measures taken by countries that allegedly affect foreign competition. Examples include efforts by the United States to protect dolphins and sea turtles from harmful fishing, efforts that have been challenged under the General Agreement on Tariffs and Trade and the WTO. These were called the tuna-dolphin and shrimp-turtle disputes. Other well-known cases include beef and hormones (United States vs. European Community), gasoline and air quality (Venezuela and Brazil vs. United States), softwood lumber (Canada vs. United States), asbestos (Canada vs. France and the European Community), and most recently, genetically modified organisms (United States vs. European Community) (Defenders of Wildlife 2006).

In the United States and Europe, food surpluses are partly a result of subsidies and other incentives that stimulate production, even where there is little or no demand. In the 10-year period from 1995 to 2004, the US government provided some US$143.8 billion in subsidies to its farmers (EWG 2005). Although this averages about half the annual cost of food aid in 1986, which stood at US$25.8 billion (Court 1990), the impact on developing countries is significant. Many of them find it cheaper to import food than to produce their own, and are forced to focus on producing export crops such as cotton, tobacco, tea and coffee. This reduces agricultural opportunities for smallholder subsistence farmers, leading to food insecurity, particularly in rural areas, or unsustainable urban growth as a result of rural to urban migration.

In theory, global natural resources have the capacity to produce enough food, medicine, shelter and other life-supporting services for an even larger population (see Box 6.1). In reality, this is not the case, due to the uneven distribution of such resources, including fertile and well-watered land, forests, wetlands and genetic resources. The capacity of these natural resources to support life is diminished by land degradation, air and water pollution, climate variability and change, deforestation, and loss of habitats and biodiversity. As a result of the uneven access to, and unbalanced production levels of natural resources, the world continues to suffer disparities in food

Table 6.1 Key regional priority issues selected for GEO-4	
Africa	Land degradation and its cross-cutting impacts on forests, freshwater, marine and coastal resources, as well as pressures such as drought, climate variability and change, and urbanization
Asia and the Pacific	Transport and urban air quality, freshwater stress, valuable ecosystems, agricultural land use, and waste management
Europe	Climate change and energy, unsustainable production and consumption, air quality and transport, biodiversity loss and land-use change, and freshwater stress
Latin America and the Caribbean	Growing cities, biodiversity and ecosystems, degrading coasts and polluted seas, and regional vulnerability to climate change
North America	Energy and climate change, urban sprawl and freshwater stress
West Asia	Freshwater stress, land degradation, degrading coasts and marine ecosystems, urban management, and peace and security
Polar Regions	Climate change, persistent pollutants, the ozone layer, and development and commercial activity

production levels, with both food surpluses in some regions and widespread food shortages in others.

Selected environmental issues

The following sections of this chapter examine the most significant environmental issues in the seven UNEP regions: Africa, Asia and the Pacific, Europe, Latin America and the Caribbean, North America, West Asia, and the Polar Regions (see the regional maps in the introductory section of this report). Regional overlaps exist across some of the regions, due to historical links and biophysical ties, which make it difficult to strictly disaggregate data. Examples of overlaps among regions include the case of Africa, Europe and West Asia, where the Mediterranean provides a quasi border (see Box 6.46), and that of Latin America and Caribbean and the North American regions, with their overlaps.

Each region held consultations to identify its regional issues of global significance. From these consultations, between one and five key environmental priority issues were selected for focused analyses in each regional section (see Table 6.1).

All regions report progress over the past 20 years in making environmental matters part of mainstream politics. In most regions, sustainable development strategies have been formulated, and are being integrated into national policies. The public, including indigenous peoples, participates to a much greater degree in environmental decision making (see the Polar section).

A more holistic approach to environmental management is being taken, with ecosystem approaches becoming common. For example, promising new integrated management strategies that involve public participation are being introduced in both freshwater and marine systems to protect valuable resources and livelihoods. The economic value of ecosystem services is now recognized, and some payment schemes are emerging. In many regions, proposed projects now require environmental impact assessments. Recycling and other waste management strategies are evolving in many areas, and sustainable consumption is increasingly promoted. In recognition of the transboundary nature of environmental pressures and impacts, better models of managing shared environments have emerged, such as regional seas.

AFRICA
DRIVERS OF CHANGE
Socio-economic trends

Africa's social and economic performance has improved in recent years. Between 1995 and 2004, African economies have grown (Figure 6.3). In 2004, the economic growth rate in terms of purchasing power was 5.8 per cent, up from 4 per cent in 2003 (GEO Data Portal from World Bank 2006). Sub-Saharan Africa's economies must grow at an average annual rate of 7 per cent to reduce income poverty by half by 2015 (AfDB 2004). Improved economic growth since the mid-1990s has increased the region's chances to meet key MDG targets, and this may have a positive

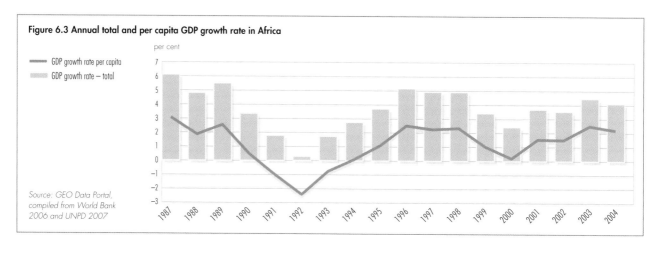

Figure 6.3 Annual total and per capita GDP growth rate in Africa

per cent

— GDP growth rate per capita
▒ GDP growth rate – total

Source: GEO Data Portal, compiled from World Bank 2006 and UNPD 2007

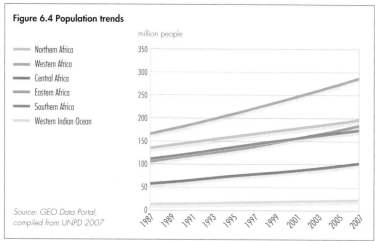

Figure 6.4 Population trends

million people

— Northern Africa
— Western Africa
— Central Africa
— Eastern Africa
— Southern Africa
— Western Indian Ocean

Source: GEO Data Portal, compiled from UNPD 2007

effect on the environment (UNEP 2006a). However, the demand on the region's resources is growing as a result of the increasing human population (see Figure 6.4) and economic activities.

Environmental governance

Since 1987, several major regional developments have resulted in significant changes in the way environmental issues are managed in Africa. They include political reforms, institution building and new policy measures that reinforce the Brundtland Commission's messages, and seek to promote sustainable development.

Among key political reforms since 1987 was the transformation of the Organization of African Unity (OAU) to the African Union (AU) in 2002 to focus greater attention on accelerated regional political and socio-economic development. In this context, African leaders launched a major regional socio-economic and development plan in 2003, the New Partnership for Africa's Development (NEPAD).

The UN General Assembly adopted NEPAD as the framework for Africa's development, under which the 2003 Action Plan of the Environment Initiative (EAP), Africa's most recent regional environmental policy, was developed. The EAP seeks to address Africa's environmental challenges, while combating poverty and promoting socio-economic development. Prepared under the leadership of the African Ministerial Conference on the Environment (AMCEN), a pan-African forum for environment ministers established in 1985, it strengthens cooperation in halting the degradation of Africa's environment, and in satisfying the region's food and energy needs (UNEP 2003a). AMCEN has since matured into a forum that provides a framework for environmental policy orientation while defending Africa's stake and interests on the international stage.

Although still weak, there have been a number of policy initiatives since the Brundtland Commission, including landmark multilateral agreements, such as the 1991 Bamako Convention on the Ban of the Import into Africa and the Control of Transboundary Movement and Management of Hazardous Wastes Within Africa, and the 1994 Lusaka Agreement on Co-operative Enforcement Operations Directed at Illegal Trade in Wild Fauna and Flora.

Some policies were already in place before 1987. These include the African Convention on the Conservation of Nature and Natural Resources (Algiers Convention), the first Africa-wide environmental convention for the conservation, use and development of soil, water, flora and fauna in accordance with scientific principles, and with due regard to the best interests of the people. The treaty has been revised and was adopted by the AU Assembly in July 2003. The new text makes the convention comprehensive and modern, and the first regional treaty to deal with a wide spectrum of sustainable development issues (UNEP 2003b). Other earlier regional conventions include the 1981 Convention for Cooperation in the Protection and Development of the Marine and Coastal Environment of the West and Central African Region (Abidjan Convention), and the 1985 Nairobi Convention for the Protection, Management and Development of the Marine and Coastal Environment of the Eastern African Region.

People in Africa recognize that land use and degradation have cross-cutting impacts on other resources, including forests, freshwater, marine and coastal resources. Similarly, issues such as drought, climate variability and change, and urbanization act as pressures that exacerbate land degradation.

SELECTED ISSUE: LAND DEGRADATION

Land resources: endowments and opportunities

Africa's 53 countries have a total land area of about 30 million square kilometres, comprised of a variety of ecosystems, including forests and woodlands, drylands, grasslands, wetlands, arable lands, coastal zones, freshwater, mountain and urban areas. The 8.7 million km² of Africa's land that is considered suitable for agricultural production has the potential to support the majority of the region's people (FAO 2002). Forest land covers 6.4 million km², representing 16 per cent of the global forest cover (GEO Data portal, from FAO 2005). The Congo River basin has Africa's largest forest reserve, and is the second largest contiguous block of tropical rain forest in the world after the Amazon (FAO 2003a).

Covering about 1 per cent of Africa's total land area, wetlands are found in virtually all countries, and are a key land feature (UNEP 2006a). Some of the more prominent wetlands include the Congo Swamps, the Chad Basin, the Okavango Delta, the Bangweulu swamps, Lake George, the floodplains and deltas of the Niger and Zambezi rivers, and South Africa's Greater St Lucia Park wetlands.

About 43 per cent of Africa's land is "susceptible" dryland (UNEP 1992) (see Chapter 3). This figure excludes hyper-arid areas, such as about two-thirds of the Sahara in Northern Africa, which, at over 9 million km², is the largest desert in the world (Columbia Encyclopedia 2003). Together with

Namibia's Skeleton Coast, the Kgalagadi (Kalahari) desert in Southern Africa (mainly arid land) is the world's largest body of sand (Linacre and Geerts 1998).

Mountains are also important land features in Africa, especially for smaller countries, including Swaziland, Lesotho and Rwanda, which rank among the world's top 20 mountainous countries (Mountain Partnership 2001). Kilimanjaro (Tanzania), Mount Kenya and Ruwenzori (Uganda and the Democratic Republic of Congo) are Africa's three highest mountains (UNEP 2006a).

Rolling grasslands dotted with trees, often called savannah, are extensive in Africa. Savannah grasslands occur in areas where rainfall is sufficient to prevent the establishment of desert vegetation but too low to support rain forests. They are held between these two extremes by climate, grazing and fire. Savannahs are among the most spectacular biomes from both landscape and wildlife perspectives. Savannah grasslands mainly cover parts of most sub-Saharan countries (Maya 2003).

Clearly, land is an environmental, social and economic good that is critical to the realization of opportunities for the people of Africa. Figure 6.5 shows the share of Africa's main land uses, including pastures, cropland, forests and woodlands. Agriculture is the dominant land use in Africa, and the biggest employer, although trends since 1996 show a small decline in its importance relative to other employment sectors (see Figure 6.6). Other economic activities that African people depend on include fisheries, forestry, mining and tourism.

Some of the world's major tea, coffee and cocoa-producing countries are in Africa. For example, Kenya is the fourth largest tea producer in the world with

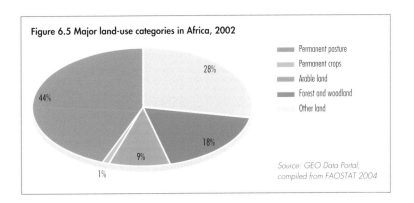

Figure 6.5 Major land-use categories in Africa, 2002

- Permanent pasture
- Permanent crops
- Arable land
- Forest and woodland
- Other land

28%
44%
18%
9%
1%

Source: GEO Data Portal, compiled from FAOSTAT 2004

Figure 6.6 Changes in sectoral shares in employment in sub-Saharan Africa

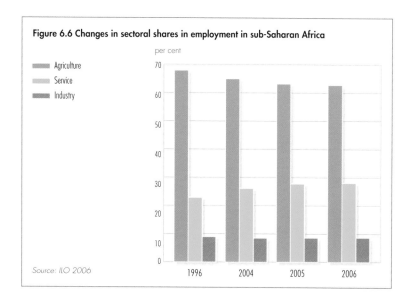

per cent

Legend:
- Agriculture
- Service
- Industry

Source: ILO 2006

(x-axis: 1996, 2004, 2005, 2006; y-axis: 0 to 70)

324 600 tonnes in 2004, up from 236 290 tonnes in 2000 (Export Processing Zones Authority 2005).

Horticulture, which accounts for 20 per cent of global agricultural trade, and is the fastest-growing agricultural sector, has significant potential in Africa. According to the *Africa Environment Outlook 2* report (UNEP 2006a), horticultural exports in sub-Saharan Africa exceed US$2 billion/year. Africa could benefit more if it were to fully utilize its

irrigation potential: a mere 7 per cent of all arable land in Africa is under irrigation (GEO Data Portal, from FAOSTAT 2005).

In addition to agriculture, African peoples rely on fisheries to provide some of their food needs. Nearly 10 million people depend on fishing, fish farming, and fish processing and trade. Africa produces 7.3 million tonnes of fish/year, 90 per cent of which is caught by small-scale fishers. In 2005, the region's fish exports were worth US$2.7 billion (New Agriculturalist 2005).

Electricity, mainly hydropower, is critical to the growth of the economy. Africa's hydropower potential is not fully used; only five per cent of the economically feasible hydropower potential of 1 million gigawatt hours/year is utilized (UNECA 2000).

Resources such as forests and woodlands provide a wide range of goods-and-services, including firewood and construction timber. Although less evident, they also provide ecosystem functions, such as protecting the soil from erosion, protecting watersheds and regulating water flows. Through habitat provision, land resources are vital to the growth of wildlife-based tourism in Africa (see Box 6.2). Africa is also endowed with different minerals, including 70 per cent of the

Box 6.2 Nature-based tourism

Nature-based tourism is one of the fastest growing tourism sectors worldwide, representing 7 per cent of the total worldwide export of goods-and-services. Nature-based tourism depends on the conservation of natural landscapes and wildlife, and using ecosystems in this way promotes both human well-being and biodiversity conservation (see Chapter 7).

Sources: Christ and others 2003, Scholes and Biggs 2004

Nature-based tourism is a major growth industry.

Credit: Ngoma Photos

world diamonds, 55 per cent of its gold and at least 25 per cent of chromites (UNEP 2006a). Many minerals have yet to be exploited.

Land pressures

Africa's land is under pressure from increased resource demand due to a growing population, natural disasters, climate change and extreme weather events such as drought and floods, and the inappropriate use of technology and chemicals. Drought can exacerbate land degradation in the drylands (see Chapter 3 and Box 6.3). Land is also degraded through poorly planned and managed activities related to agriculture, forestry and industry, as well as from the impacts of urban slums and infrastructure development (see Chapter 3).

Africa is one of the most vulnerable regions to climate variability and change because of multiple stresses and low adaptive capacity, according to some new studies (see Figure 6.7). Some adaptation to current climate variability is taking place, however, this may be insufficient for future changes in climate (Boko and others 2007).

With a growing population, Africa faces declining per capita access to arable land (Figure 6.8) even as the region struggles to increase food production per unit area. Per capita agricultural production declined by 0.4 per cent between 2000 and 2004 (AfDB 2006b). Land degradation exacerbates poor food production, increasing food insecurity.

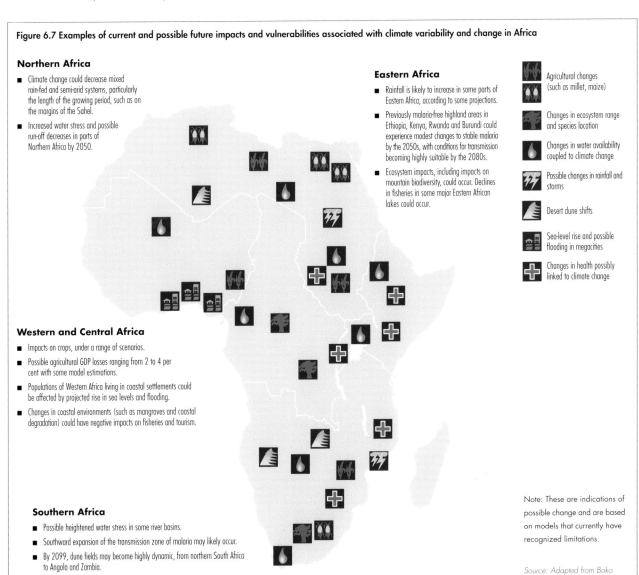

Figure 6.7 Examples of current and possible future impacts and vulnerabilities associated with climate variability and change in Africa

Northern Africa
- Climate change could decrease mixed rain-fed and semi-arid systems, particularly the length of the growing period, such as on the margins of the Sahel.
- Increased water stress and possible run-off decreases in parts of Northern Africa by 2050.

Eastern Africa
- Rainfall is likely to increase in some parts of Eastern Africa, according to some projections.
- Previously malaria-free highland areas in Ethiopia, Kenya, Rwanda and Burundi could experience modest changes to stable malaria by the 2050s, with conditions for transmission becoming highly suitable by the 2080s.
- Ecosystem impacts, including impacts on mountain biodiversity, could occur. Declines in fisheries in some major Eastern African lakes could occur.

Agricultural changes (such as millet, maize)

Changes in ecosystem range and species location

Changes in water availability coupled to climate change

Possible changes in rainfall and storms

Desert dune shifts

Sea-level rise and possible flooding in megacities

Changes in health possibly linked to climate change

Western and Central Africa
- Impacts on crops, under a range of scenarios.
- Possible agricultural GDP losses ranging from 2 to 4 per cent with some model estimations.
- Populations of Western Africa living in coastal settlements could be affected by projected rise in sea levels and flooding.
- Changes in coastal environments (such as mangroves and coastal degradation) could have negative impacts on fisheries and tourism.

Southern Africa
- Possible heightened water stress in some river basins.
- Southward expansion of the transmission zone of malaria may likely occur.
- By 2099, dune fields may become highly dynamic, from northern South Africa to Angola and Zambia.
- Food security is likely to be further aggravated by climate variability and change.

Note: These are indications of possible change and are based on models that currently have recognized limitations.

Source: Adapted from Boko and others 2007

Drought occurs in some parts of sub-Saharan Africa virtually every year. Some of the major droughts in the past two decades include those of 1990–92 and 2004–05. Widespread drying was observed in Western and Southern Africa between the 1970s and early 2000s. Poor rains were the main factor behind the expansion of dry soils in Africa's Sahel region and in Southern Africa, where El Niño-related episodes have become more frequent since the 1970s (see Figure 4.5 in Chapter 4, showing global precipitation trends in the 20th century).

The 2004–2005 drought was the most widespread in Africa in recent times. It was not limited to the Sahel and Southern Africa, but extended up the eastern coast, where many countries were subject to a multi-year drought that caused food shortages from Tanzania in the south to Ethiopia, Kenya and Eritrea in the north. In the Horn of Africa (Somalia, Ethiopia, Eritrea and Djibouti), it was the sixth consecutive year of severe drought.

Sources: Darkoh 1993, FEWSNET 2005, Stafford 2005

Figure 6.8 Per capita arable land

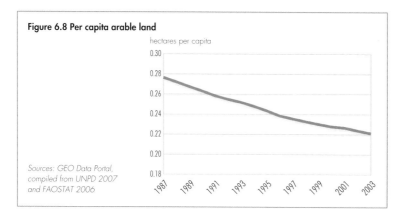

hectares per capita

Sources: GEO Data Portal, compiled from UNPD 2007 and FAOSTAT 2006

Forest conversion

Africa has the highest deforestation rate of the world's regions. The region loses an estimated 40 000 km², or 0.62 per cent of its forests annually, compared to the global average deforestation rate of 0.18 per cent (FAO 2005). Africa's pristine natural forests are being replaced by extensive areas of secondary forests, grasslands and degraded lands. Variation within the region is large. Reported net losses are most significant in those countries with the greatest extent of forests, such as Angola, Tanzania and Zambia in Eastern and Southern Africa, but the rate of loss has shown signs of declining slightly since 2000 (FAO 2007a).

Land tenure

The communal land tenure system, in which property is collectively owned, is often cited as the reason for overexploitation of land, contributing to land degradation and deforestation. Under such a system, the costs of impacts such as land degradation, siltation and water pollution are borne by the community as a whole, while potential benefits accrue to the individual. Poor land tenure regimes that precipitate ineffective land-use planning and management can only lead to overexploitation of the resource, contributing to increased land degradation, salinization, pollution, soil erosion and conversion of fragile lands (UNEP 2006a).

Urbanization

Although by far the least urbanized region of the world (see Chapter 1, Figure 1.6), at 3.3 per cent annual growth between 2000 and 2005, Africa has the world's highest rate of urbanization, with the urban population doubling every 20 years, and an estimated 347 million people (38 per cent of Africans) living in urban settlements by 2005 (GEO Data Portal, from UNPD 2005). While urban areas are centres of economic activity, innovation and development, the rapidly expanding urban centres are encroaching on rural and agriculturally productive land. In addition some of Africa's urban centres are increasingly characterized by rising poverty levels. Over 72 per cent of the urban population in sub-Saharan Africa lives in slums, without adequate housing, drinking water and sanitation facilities (UN-HABITAT 2006). Informal settlements pose a threat to environmental integrity through practices such as illegal and uncontrolled waste disposal. Poverty is forcing urban dwellers to adopt alternative livelihood strategies such as urban agriculture, which supplements food needs and generates household income.

Land degradation trends

Land degradation is a serious problem in Africa, especially in drylands (see Chapter 3). By 1990, land degradation affected an estimated 5 million km² of the continent (Oldeman and others 1991). In 1993, 65 per cent of agricultural land was degraded, including 3.2 million km² (25 per cent) of Africa's susceptible drylands (arid, semi-arid and dry sub-humid areas) (WRI 2000). Chapter 3 presents a recent assessment of land degradation, based on the last 25-year trend in biomass production (from satellite measurements) per unit of rainfall (see Figure 3.6 in Chapter 3). The most common processes of land degradation in Africa are soil erosion, soil nutrient depletion, contamination of soils and salinization.

Soil erosion

The Brundtland Commission warned that 5.4 million km² of fertile land would be affected by soil erosion in Africa and Asia unless adequate conservation measures were taken (WCED 1987). Soil erosion is now widespread in Africa (see Chapter 3). For example, half of Rwanda's farmland is moderately to severely eroded, with two-thirds of the soil classified as acidic and exhausted (IFAD and GEF 2002).

Despite the reduced productivity of eroded soil, many African farmers are forced to continuously use the same land because of factors such as population pressure, inequitable land ownership and poor land-use planning. There is a strong relationship between population density and soil erosion. The estimated per capita productive land available in Central and Eastern Africa varies from a low of 0.69 ha in the Democratic Republic of Congo, to 0.75 ha in Burundi, 0.85 in Ethiopia, 0.88 in Uganda, 0.89 in Cameroon, 0.90 in Rwanda, 1.12 ha in the Central African Republic, 1.15 in Congo and 2.06 in Gabon (UNEP 2006a).

Coastal erosion, resulting from beachfront developments, and the mining of sand, coral and lime, is also worsening, with erosion rates as high as 30 metres/year in Western Africa, mainly in Togo and Benin (UNEP 2002b).

Salinization

While irrigation could provide some of the impetus towards a Green Revolution in Africa, inefficient application could lead to land degradation due to salinization. About 647 000 km^2, or 2.7 per cent of Africa's total land area is affected by salinization, representing over 26 per cent of the world's salinized land area (see Table 6.2) (FAO TERRASTAT 2003).

Desertification

At present, almost half of Africa's land area is vulnerable to desertification. Africa's drylands are unevenly distributed across the region, and some are even found in the usually wet tropical zones of Central and Eastern Africa (see Chapter 3). Across Africa, drylands occupy 43 per cent of the region (CIFOR 2007). Areas most affected by desertification (defined as land degradation in susceptible drylands) are located in the Sudano-Sahelian region and Southern Africa. The area along the desert margins, which occupies about 5 per cent of Africa's land, is at the highest risk of desertification (Reich and others 2001). Areas particularly at risk include the Sahel, a band of semi-arid lands stretching along the southern margin

Soil erosion is now widespread in Africa, affecting food production and food security.

Credit: Christian Lambrechts

of the Sahara Desert, and some nations that consist entirely of drylands, such as Botswana and Eritrea.

Land degradation impacts

Land degradation is the biggest threat to realizing the region's full potential from land. It undermines soil fertility, and, especially in the drylands, it can cause productivity losses of as much as 50 per cent (UNCCD Secretariat 2004). The decline in land quality causes economic stresses, and affects biodiversity through impacts on terrestrial and aquatic ecosystems, and on fishery resources. The degradation also reduces water availability and quality, and can alter the flows of rivers, all leading to serious downstream consequences. The process is closely linked to poverty, which is both a cause and a consequence of land degradation. Poor people are forced to put immediate needs before the long-term quality of the land. The ensuing social, economic and political tensions can create conflicts, more impoverishment and increased land degradation,

Table 6.2 African countries with 5 per cent or more of their land affected by salinization	
Country	Salinity thousand km^2
Botswana	63
Egypt	87
Ethiopia	51
Morocco	23
Somalia	57

Source: FAO TERRASTAT 2003

and force people to seek new homes and livelihoods (UNEP 2006b). Also dust storms are considered by some as an impact of land degradation, while in fact such storms are mainly natural processes that build up in desert areas (see Box 6.4).

Food security and poverty

In Africa, the proportion of people living below the poverty line increased from 47.6 per cent in 1985 to 59 per cent in 2000 (UNECA 2004). Some 313 million Africans lived on less than US$1 per day in 2005 (UNDP 2005a). As a result of poverty, more people in Africa have limited access not only to food, but also to potable water, minimum health care and education. Poverty is exacerbated by extensive use of degraded land, or soils of poor fertility. Unless the land is rehabilitated, both degradation and poverty deepen.

Food insecurity and reduced caloric intake are the major socio-economic impacts of land degradation. Declining soil fertility causes average yield losses of as much as 8 per cent (FAO 2002). With the relatively high share of the agricultural sector in Africa's GPD, as much as 34 per cent in Eastern Africa, it is estimated that land degradation can lead to an annual loss of 3 per cent of agriculture's contribution to GDP in sub-Saharan Africa. In Ethiopia alone, GDP loss from reduced agricultural productivity is estimated at US$130 million/year (TerrAfrica 2004). While

global per capita food production has risen by over 20 per cent since 1960, it has been falling steadily in Africa, declining by 12 per cent since 1981 (Peopleandplanet.net 2003).

The region's food insecurity is due to a number of factors, including unfavourable weather, land degradation, poverty, conflict and civil strife, HIV/AIDS, low soil fertility, and pests. The proportion of undernourished individuals in sub-Saharan Africa has, on average, fallen from 35 per cent in 1990 to 32 per cent in 2003, but the absolute number of undernourished people increased, from about 120 million around 1980 to some 180 million around 1990 to 206 million in 2003 (FAO 2007b). As such, Africa is the only region in the world where the need for food aid is increasing (see Figure 6.9). In 2004, 40 countries in sub-Saharan Africa received almost 3.9 million tonnes of food aid (52 per cent of global aid) (WFP 2005), compared to an annual average of just over 2 million tonnes received during the period 1995–1997 (FAOSTAT 2005) (see Box 6.5).

Genetic modification (GM) technology has the potential to improve the yields and quality of food crops, as well as to create resistance to diseases, such as the damaging cassava virus in Western Africa. However, GM technology is controversial, since genetically modified organisms (GMOs) have not been fully tested for environmental or health effects. Many African countries have been declining GM food aid because of such concerns, despite the fact that many of them experience food shortages. The region has 810 000 km² under GMOs (James 2004), mainly in South Africa.

Agricultural production in many African countries is projected to be severely compromised due to climate variability and change. The area suitable for agriculture, the length of growing seasons and yield potential, particularly along the margins of semi-arid and arid areas, are expected to decrease. This would further adversely affect food security and exacerbate malnutrition in the region. In some countries, yields from rain-fed agriculture could be reduced by up to 50 per cent by 2020 (Boko and others 2007).

Environmental impacts

Land degradation threatens tropical forests, rangelands and other ecosystems. For example, the drylands of Eastern and Southern Africa

Box 6.4 Deserts and dust

Storms can transport fine sand and dust over large areas, having both positive (fertilizer) and negative (small particles) impacts on ecosystems and human health regionally and globally. As described in Chapter 3, some 90 per cent of such dust originates from natural processes in true deserts in Africa and Asia.

Sandstorm in Gao, Mali.

Credit: BIOS Crocetta Tony/Still Pictires

As a result of inadequate food production, Africa spends US$15–20 billion on food imports annually, in addition to receiving US$2 billion/year in food aid. The World Food Programme, which accounts for 40 per cent of international food aid, has spent US$12.5 billion, 45 per cent of its total investment since its establishment, in Africa. These are vast amounts of money, which could be used to revitalize agriculture through measures such as the provision of agricultural inputs and the rehabilitation of degraded land.

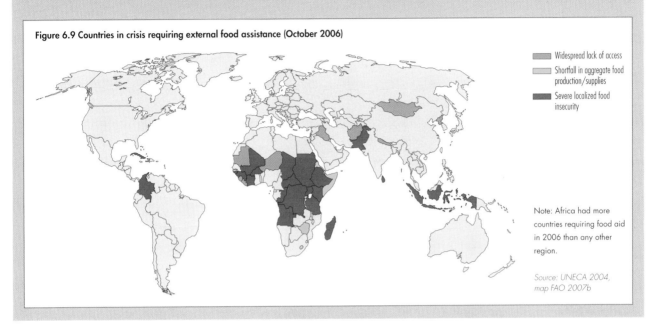

Figure 6.9 Countries in crisis requiring external food assistance (October 2006)

Widespread lack of access

Shortfall in aggregate food production/supplies

Severe localized food insecurity

Note: Africa had more countries requiring food aid in 2006 than any other region.

Source: UNECA 2004, map FAO 2007b

are particularly vulnerable to vegetation loss, and savannahs are at a very high risk of land degradation. The impacts include loss of biodiversity, rapid loss of land cover, and depletion of water availability through destruction of catchments and aquifers. Increased siltation fills up dams, and leads to flooding in rivers and estuaries. In Sudan, for example, the total capacity of the Roseires reservoir, which generates 80 per cent of the country's electricity, fell by 40 per cent in 30 years, due to siltation of the Blue Nile (UNEP 2002b).

As a result of habitat loss due to land degradation, four antelope species in Lesotho and Swaziland, the blue wildebeest in Malawi, the Tssessebe in Mozambique, the blue buck from the southwestern Cape in South Africa and the kob in Tanzania are threatened with extinction. In Mauritania, an estimated 23 per cent of the mammals are at risk of extinction (UNEP 2006a). In Western and Central Africa, endangered timber trees and plants include the rock elm (*Milicia excelsa*), prickly ash bark (*Zanthoxylum americanum*) and African oil palm (*Brucea guineensis*). Endangered mammals include the chimpanzee (*Pan troglodytes*), Senegal hartebeest (*Alcelaphus bucelaphus*), elephants (*Loxodanta africana*) and one of the three manatee species (*Trichechus senegalensis*). In Eritrea alone, 22 plant species are reportedly threatened with extinction (UNEP 2006a).

Land degradation affects important ecosystems, such as wetlands, causing loss of habitat for birds (see Box 6.6). Wetland degradation also reduces ecosystem functions, such as flood regulation. Wetland loss in Africa is significant and critical but not well documented; 90 per cent of wetlands in the Tugela Basin in South Africa have reportedly been lost, while in the Mfolozi catchment, also in South Africa, 58 per cent (502 km²) of the wetland area has been degraded. In Tunisia's Medjerdah catchment, 84 per cent of the wetland area has vanished (Moser and others 1996).

Land degradation is also rampant along Africa's 40 000 km coastline (UNEP 2002b). Mining of sand, gravel and limestone from estuaries, beaches or the nearshore continental shelf is common in Africa's coastal states and islands. Sand and gravel mining in coastal rivers and estuaries in particular tends to diminish the amount of fluvial sediment input to the coastline, accelerating shoreline retreat. Dredging of

Degradation and loss of wetland habitats constitute the most significant threat to the endangered wattled crane, which is endemic to Africa, and ranges across 11 countries, from Ethiopia to South Africa. It is the most wetland-dependent of Africa's cranes, occurring in the extensive floodplain systems of Southern Africa's large river basins, especially the Zambezi and Okavango. Intensified agriculture, overgrazing, industrialization and other pressures on wetlands have contributed to its decline, especially in South Africa and Zimbabwe.

Source: International Crane Foundation 2003

Credit: BIOS Courteau Christophe (B)/Still Pictures

sand from the inner continental shelf is an obvious cause of beach erosion in Africa. This problem has been documented in Benin, Liberia, Sierra Leone, Cote d'Ivoire, Ghana, Nigeria, Mauritius, Tanzania, Togo, Kenya, the Seychelles and Mozambique (Bryceson and others 1990). Coastal erosion is also influenced by the modification of stream flows through river impoundments, which, in turn, cause habitat change in estuaries (See Box 6.7).

Land degradation in coastal zones is associated with the development of coastal settlements. Coastal towns are by far the most developed of Africa's urban areas and, by implication, the concentration of residential, industrial, commercial, agricultural, educational and military facilities in coastal zones is high. Major coastal cities include Abidjan, Accra, Alexandria, Algiers, Cape Town, Casablanca, Dakar, Dar es Salaam, Djibouti, Durban, Freetown, Lagos, Libreville, Lome, Luanda, Maputo, Mombasa, Port Louis and Tunis.

Changes in a variety of ecosystems are already being detected at a faster rate than anticipated due to climate change. This is particularly the case in Southern Africa. Climate change, interacting with human drivers such as deforestation and forest fires are a threat to Africa's ecosystems. It is estimated that, by the 2080s, the proportion of arid and semi-arid lands in Africa is likely to increase by 5-8 per cent (Boko and others 2007). Climate change will also aggravate the water stress currently faced by some countries, while some that currently do not experience water stress will be at greater risk of water stress.

Conflict

Land degradation in Africa is also linked to civil conflicts, such as in the Darfur region of Sudan, where the clearing of tree cover around water points has degraded the land since 1986 (Huggins 2004). In Darfur, rainfall has declined steadily over the last 30 years, with negative impacts on farming communities and pastoralists. The resultant land degradation has forced many to migrate southwards, leading to conflict with farming communities where they settle (UNEP 2006a). In countries that recently emerged from wars, such as Angola, land mines prevent the use of land for productive purposes, such as agriculture.

Addressing land degradation

Addressing the issue of land degradation is key to helping Africa reduce poverty, and achieve some of its targets as set out under the Millennium Development Goals. Although policy shortcomings still exist, Box 6.8 lists some of the promising regional policy initiatives that address land degradation.

Efforts to stop land degradation include integrated crop and land management programmes that seek to provide tangible, short-term benefits to farmers, such as increased yields and reduced risks. Efforts, though localized, include water harvesting, agroforestry, and a variety of new and traditional grazing strategies. There are opportunities to expand these methods, which focus not only on increasing yields, but also on building healthy soils, maintaining crop diversity and avoiding the use of expensive chemical fertilizers and pesticides that pollute water sources, and are a human health risk (see the sections on soil erosion and desertification in Chapter 3). These strategies are particularly adapted to ecological constraints faced by poor farmers on marginal or less favoured lands, because they address problems of soil fertility and water availability that biotechnology or more conventional means for the intensification of production cannot readily overcome (Halweil 2002).

NEPAD's Comprehensive Agricultural Development Programme seeks to promote irrigated agriculture (UNEP 2006a) through extending the area under sustainable land management and reliable water control systems. This would include rapidly increasing the area under irrigation, especially smallholder irrigation, improving rural infrastructure and trade-related capacities for markets, and increasing food supplies. All this would help in reducing hunger.

Land degradation is partly blamed on the failures of the "Western" land administration system of land title, which has often not benefited the poor. More

attention is now being given to the inclusion of customary tenure in national land administration laws to protect people's customary land rights. Innovative tools to both improve tenure security for the poor as well as to address land degradation problems include occupancy licences, customary leases and certificates. However, such tools also present problems. For example, in Zambia, registration of customary land often leads to denial of other customary rights, while in Uganda the pace of issuing certificates has been slow, with no certificates issued since 1998. In Mozambique, certificates are successfully issued, although it is unclear if the innovative tools have been fully embedded in society (Asperen and Zevenbergen 2006).

ASIA AND THE PACIFIC
DRIVERS OF CHANGE
Socio-economic trends

The Asia and the Pacific region is comprised of 43 countries and a number of territories, and is, for the purposes of this report, divided into six sub-regions. It is endowed with a rich diversity of natural, social and economic resources. The length of its coastline is two-thirds of the global total, and it has the world's largest mountain chain. The region includes some of the poorest nations in the world, several highly advanced economies, and a number of rapidly growing ones, notably China and India. From 1987 to 2007, the population increased from almost 3 billion to almost 4 billion people, and the region is now home of 60 per cent of the world's people (GEO Data Portal, from UNPD 2007), representing a wide range of different ethnicities, cultures and languages.

In most nations, central governments have played a key role in economic planning to achieve development goals, and have been instrumental in formulating environmental policies. For the region as a whole, GDP (purchasing power parity, in constant 2000 US dollars) increased from US$7.5 trillion in 1987 to US$18.8 trillion in 2004 (GEO Data Portal, from World Bank 2006).

Many countries have made considerable progress towards attaining the Millennium Development Goals (MDGs), although achievements are marked by wide disparities and stark contrasts (see Box 6.9). Since several countries have already achieved many of the MDG targets, they have raised their targets, setting new goals, called MDG Plus.

Since 2000, Asia and the Pacific's GDP growth has surpassed the 5 per cent rate suggested by the Brundtland Commission in 1987 (ADB 2005), but ecosystems and human health continue to deteriorate. Population increases and fast economic development have driven significant environmental degradation and natural capital losses during the last two decades. In turn, deteriorating environmental conditions are threatening and diminishing the quality of life for millions of people.

Rapid population growth, fast economic development and urbanization have led to increased energy needs. Between 1987 and 2004 energy use in this region increased by 88 per cent, compared to a global average rise of 36 per cent (GEO Data Portal, from IEA 2007). Presently, Asia and the Pacific is responsible for only about 34 per cent of total global energy consumption, and per capita energy consumption is much lower than the world average (see Chapter 2). There are strong signs that regional energy demands will continue to increase (IEA 2006) (see Figure 6.10). Asia and the Pacific's share of global CO_2 emissions increased from 31 per cent 1990 to 36 per cent in 2003, with considerable variation within the region (see Figure 6.11). These energy and related CO_2 emission trends are part of a pattern of global increases that are contributing to climate change (see Chapter 2).

Box 6.9 Progress towards the Millennium Development Goals

Remarkable progress has been made in overall poverty reduction in this region. Between 1990 and 2001, the number of people living on less than US$1/day dropped by nearly 250 million. Sustained growth in China and the acceleration of India's economy contributed to such progress. Efforts at reducing malnutrition, however, have been less successful. The most severe problems are evident in South Asia, where nearly half the children five years old and under are malnourished.

The region has also made progress towards MDG 7 on the environment. Environmental protection is considered to be a fundamental element in achieving several MDGs, and is a powerful engine for economic growth and poverty eradication. South Asia made the most impressive achievements in providing safe sources of drinking water, and India contributed substantially to the positive trend. Another encouraging sign is the significant progress in improving energy efficiency, and in providing access to clean technology and fuels in eastern and southern Asia. However, energy efficiency continues to decline in South East Asia.

Source: UN 2005a

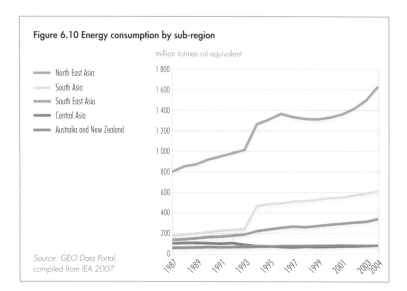

Figure 6.10 Energy consumption by sub-region

million tonnes oil equivalent

- North East Asia
- South Asia
- South East Asia
- Central Asia
- Australia and New Zealand

1 800
1 600
1 400
1 200
1 000
800
600
400
200
0

1987 1989 1991 1993 1995 1997 1999 2001 2003 2004

Source: GEO Data Portal, compiled from IEA 2007

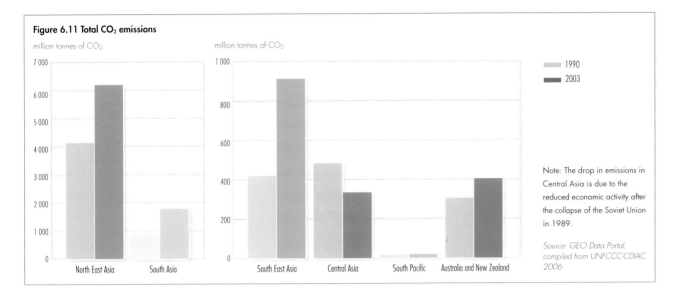

Figure 6.11 Total CO₂ emissions

million tonnes of CO₂

million tonnes of CO₂

Legend:
1990
2003

Note: The drop in emissions in Central Asia is due to the reduced economic activity after the collapse of the Soviet Union in 1989.

Source: GEO Data Portal, compiled from UNFCCC-CDIAC 2006

Environmental governance

These problems are not new, although many are intractable, and some are getting worse. Most countries in Asia and the Pacific have developed extensive domestic laws, regulations and standards related to the environment, and participate in global action through multilateral and bilateral agreements. However, the implementation of laws and agreements has been hampered by a wide variety of factors. They include: inadequate implementation, enforcement and monitoring; a lack of capacity, expertise, know-how and coordination among different government agencies; and insufficient public participation, environmental awareness and education. Most importantly, the lack of integration of environmental and economic policies has been the major constraint in establishing an effective system of environmental management. All of these factors undermine efforts to alleviate pressure on environmental quality and ecosystem health.

Furthermore, the region is highly vulnerable to natural hazards. Notable examples include the 2004 Indian Ocean Tsunami and the 2005 earthquake in Pakistan. Evidence exists of significant increases in the intensity and/or frequency of extreme weather events, such as heat waves, tropical cyclones, prolonged dry spells, intense rainfall, tornadoes, snow avalanches, thunderstorms and severe dust storms since the 1990s (IPCC 2007a). Impacts of such disasters range from hunger and susceptibility to disease, to loss of income and livelihoods, affecting the survival and human well-being of both present and future generations.

Clearly, the region still faces some formidable environmental governance challenges to protect valuable natural resources and the environment while alleviating poverty and improving living standards with limited natural resources.

SELECTED ISSUES

Increases in consumption and associated waste have contributed to the exponential growth in existing environmental problems, including deteriorating water and air quality. Land and ecosystems are being degraded, threatening to undermine food security. Climate change is likely to affect the region with thermal stress, and more severe droughts and floods, as well as soil degradation, coastal inundation and salt water intrusion due to sea-level rise (IPCC 2007b). Agricultural productivity is likely to decline substantially, due to projected warmer temperatures and shifts in rainfall patterns in most countries. Major trends and responses are described for five such environmental issues that are key priorities in the region: transport and urban air quality, freshwater stress, valuable ecosystems, agricultural land use and waste management.

TRANSPORT AND URBAN AIR QUALITY
Air pollution

The growing energy needs and the associated increase in resource mixes and fuel types have resulted in the intensification of urban air pollution and the serious degradation of air quality in many Asian cities. This has been further complicated by the region's relatively poor energy intensity and fuel efficiency. Increased energy consumption has

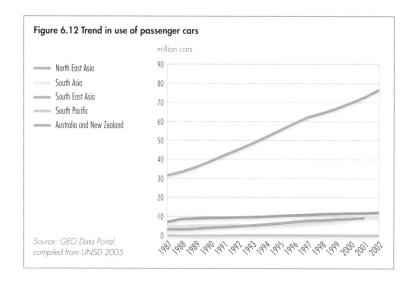

Figure 6.12 Trend in use of passenger cars

million cars

- North East Asia
- South Asia
- South East Asia
- South Pacific
- Australia and New Zealand

Source: GEO Data Portal, compiled from UNSD 2005

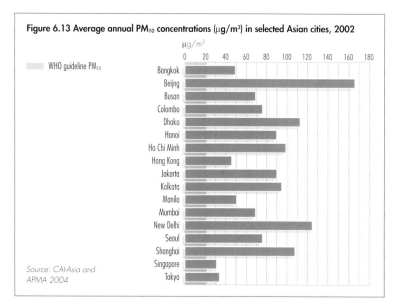

Figure 6.13 Average annual PM₁₀ concentrations (µg/m³) in selected Asian cities, 2002

µg/m³

WHO guideline PM₁₀

Bangkok, Beijing, Busan, Colombo, Dhaka, Hanoi, Ho Chi Minh, Hong Kong, Jakarta, Kolkata, Manila, Mumbai, New Delhi, Seoul, Shanghai, Singapore, Tokyo

Source: CAI-Asia and APMA 2004

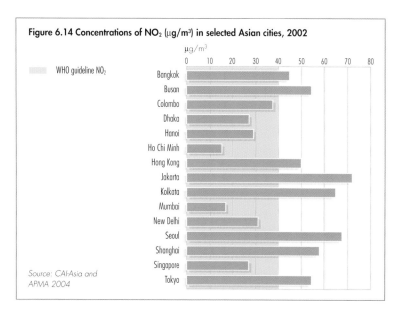

Figure 6.14 Concentrations of NO₂ (µg/m³) in selected Asian cities, 2002

µg/m³

WHO guideline NO₂

Bangkok, Busan, Colombo, Dhaka, Hanoi, Ho Chi Minh, Hong Kong, Jakarta, Kolkata, Mumbai, New Delhi, Seoul, Shanghai, Singapore, Tokyo

Source: CAI-Asia and APMA 2004

also led to a rise in greenhouse gas emissions, contributing to climate change (see Box 6.11 and Figure 6.11), which has major impacts on ecosystems and human well-being.

The exploding growth in motorized vehicles (see Figure 6.12) is the key factor in both traffic congestion and the levels of urban air pollution in many cities. Between 1987 and 2003, the use of passenger cars has increased about 2.5 times (GEO Data Portal, from UNSD 2005). During the 1990s, the number of cars and two-wheeled motorcycles in China and India rose by more than 10 per cent/year (Sperling and Kurani 2003). China had some 27.5 million passenger cars and 79 million motorcycles in use by 2004 (CSB 1987–2004). In India, passenger car ownership nearly tripled from 2.5 per 1 000 people in 1987 to 7.2 per 1 000 people in 2002 (GEO Data Portal, from UNSD 2005). Other factors contribute to a sharp deterioration in urban air quality. There is a higher concentration of people living in large cities than in other regions. With the exception of a few cities, municipal development is poorly planned. There is a lack of affordable and clean mass transit services. In addition, there is haze pollution caused by forest fires in South East Asia.

The most common urban air pollutants are nitrogen oxides, sulphur dioxide, particulate matter, lead and ozone. Levels of PM₁₀ (particulate matter less than 10µ in diameter) remain high in many Asian cities, far exceeding standards prescribed by the World Health Organization (see Figure 6.13) (see Chapter 2). In particular, South Asian cities continue to record the highest levels of outdoor particulate pollution worldwide (World Bank 2003a). While there are indications that the concentrations of sulphur dioxide in selected Asian cities have declined in recent years, large and growing motor vehicle fleets in mega-cities continue to contribute to high nitrogen dioxide levels (see Figure 6.14).

Recent assessments suggest that outdoor and indoor urban air pollution, especially from particulates, has considerable impacts on public health. A WHO study estimates that more than 1 billion people in Asian countries are exposed to outdoor air pollutant levels exceeding WHO guidelines (WHO 2000a), and this causes the premature death of about 500 000 people annually in Asia (Ezzati and others 2004a, Ezzati and others 2004b, Cohen

Table 6.3 Health and economic costs of PM_{10} for selected cities	
Manila	About 8 400 cases of chronic bronchitis and about 1 900 cases of excess deaths associated with PM_{10} resulted in a cost of US$392 million in 2001 (World Bank 2002a).
Bangkok	About 1 000 cases of chronic bronchitis and about 4 500 cases of excess deaths associated with PM_{10} resulted in a cost of US$424 million in 2000 (World Bank 2002b).
Shanghai	About 15 100 cases of chronic bronchitis and about 7 200 cases of premature deaths associated with PM_{10} resulted in a cost of US$880 million in 2000 (Chen and others 2000).
India (for 25 most polluted cities)	Estimated annual health damage of pre-Euro standards for vehicle exhaust emissions is estimated between US$14 million and US$191.6 million per city (GOI 2002).

and others 2005). The region has the world's highest burden of disease attributable to indoor air pollution (see Chapter 2). In addition, air pollution leads to substantial financial and economic costs to households, industry and governments in Asia. Limited studies have been done, but some show the health and economic costs of particulates (PM_{10}) in selected cities and groups of cities in Asia (see Table 6.3).

Addressing urban air pollution

Most countries in Asia and the Pacific have established a legislative and policy framework to address air pollution, and there are a number of institutional arrangements at national and city levels. Most countries have phased out leaded fuels (UNEP 2006c). Many cities, including Bangkok, Beijing, Jakarta, Manila, New Delhi and Singapore, are noteworthy for their recent implementation of such actions. To address haze pollution, members of the Association of Southeast Asian Nations (ASEAN) agreed to a regional plan of action, and created a Haze Fund to implement the ASEAN Agreement on Transboundary Haze Pollution (ASEAN 2006).

Monitoring air pollutants is a key tool for informed policy making, regulation and enforcement, and for assessing impacts, but only some cities conduct regular monitoring. The region needs to accelerate the switch from fossil fuels to cleaner and renewable forms of energy. It also needs to promote a reduction in private vehicle use, as well as drastically improve the efficiency and availability of mass transit systems, with approaches such as those envisaged under the Regional Environmentally Sustainable Transport (EST) Forum, launched in 2005 in the North East and South East Asian sub-regions (Ministry of the Environment of Japan 2005). Sustainable city planning is another long-term measure that should be undertaken.

FRESHWATER STRESS

Water quantity and quality

Of all freshwater-related issues, adequate water supply is the major challenge to all the Asia and the Pacific nations. The region has 32 per cent of the world's freshwater resources (Shiklomanov 2004), but is home to about 58 per cent of the world population. The South Pacific (along with many African countries) has the lowest per capita freshwater availability in the world.

Since Asian economies depend heavily on agriculture and irrigation, agriculture puts the greatest demands on water resources (see Figure 6.15). Excessive withdrawals from surface waters and underground aquifers, pollution of freshwater resources by industrial sectors, and inefficient use of freshwater are major causes of water stress (WBCSD 2005). Climate change has the potential to exacerbate

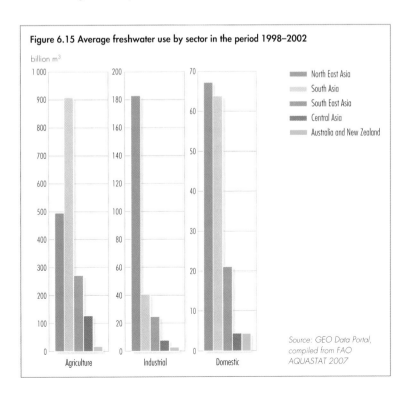

Figure 6.15 Average freshwater use by sector in the period 1998–2002

Source: GEO Data Portal, compiled from FAO AQUASTAT 2007

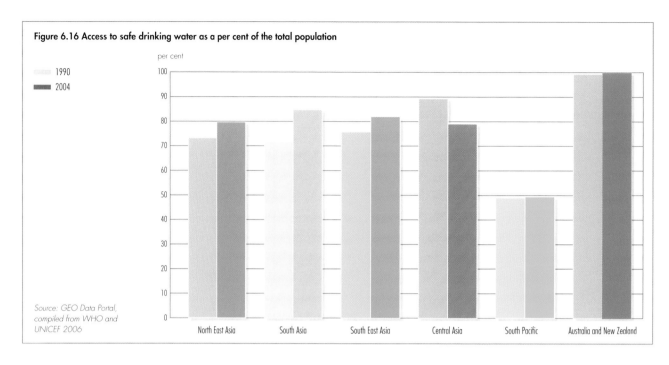

Figure 6.16 Access to safe drinking water as a per cent of the total population

1990
2004

per cent

Source: GEO Data Portal,
compiled from WHO and
UNICEF 2006

North East Asia | South Asia | South East Asia | Central Asia | South Pacific | Australia and New Zealand

water resource stress in many countries of Asia and the Pacific (IPCC 2007b). There are reports of unprecedented glacier retreats in the Himalayan Highlands over the past decade (WWF 2005). Furthermore, climatic variability and natural disasters have threatened watershed quality in recent years, causing damage to sanitation facilities and the contamination of groundwater (UNEP 2005a) (see Chapter 4).

Human activities, such as land-use change, water storage, interbasin transfers, and irrigation and drainage, influence the hydrological cycle in many

river basins (see Chapter 4) (Mirza and others 2005). Changes in recent years in continuity and withdrawal patterns in the summer monsoon have led to considerable spatial and temporal variations in rainwater availability (Lal 2005). Southwest Bangladesh suffers from extreme water shortages as well as acute moisture stress during the dry months, adversely affecting both ecological functions and agricultural production. Floods during the monsoon season inundate an average of 20.5 per cent of Bangladesh, and can flood as much as 70 per cent of the country during an extreme flood event (Mirza 2002). Furthermore, the influx of saline water is a major hazard in South Asia and South East Asia, and in the atoll islands of the Pacific.

Although remarkable progress in the provision of improved drinking water has been made over the last decade (see Figure 6.16), especially in South Asia, some 655 million people in the entire region (or 17.6 per cent) still lack access to safe water (GEO Data Portal, from WHO and UNICEF 2006). While South Pacific states have not made any progress, conditions in Central Asian countries actually deteriorated. In many mega-cities, up to 70 per cent of citizens live in slums, and generally lack access to improved water and sanitation.

Water pollution and inadequate access to improved drinking water are severe threats to human well-being and ecological health. The expansion of

---

Box 6.10 Water pollution and human health in South Asia and South East Asia

High natural concentrations of arsenic and fluoride in water have resulted in acute health problems in parts of India and Bangladesh. More than 7 000 wells in West Bengal have high levels of dissolved arsenic, reaching over 50 mg/litre, five times the WHO guideline. Water-borne diseases are associated with degraded water quality, and in developing countries they cause 80 per cent of all illnesses. With two-thirds of the South Asian population lacking adequate sanitation, water-borne diseases are prevalent, including diarrhoea, which kills 500 000 children each year.

There have been attempts to reform the water and sanitation sector in South Asia and South East Asia, including large-scale subsidization of water for the poor. For example, under its National Growth and Poverty Eradication Strategy (NGPES), Laos is developing the infrastructure to ensure greater access to safe water and sanitation, especially for the rural population. Singapore is recycling wastewater, bringing it up to drinking quality standards by using a new filtration technology.

Sources: CPCB 1996, OECD 2006a, OECD 2006b, Suresh 2000, WBCSD 2005, WHO and UNICEF 2006

agriculture, with increased use of agrochemicals, will cause more serious water pollution, as chemicals get into rivers and coastal waters. An increase in the volume of domestic wastewater is also degrading water quality in urban areas. Although discharges of organic water pollutants have declined in a number of Asian countries in recent years (Basheer and others 2003), the cumulative amount of discharges is greater than natural recovery capacity, and this continues to degrade water quality. Human health is threatened by unsafe water (see Box 6.10).

Balancing freshwater supply and quality with increasing demand

Nations in the region are taking numerous steps to address the high demand for safe water. North East Asia relies on command-and-control policies, specifically the "polluter-pays-principle," to target individual polluting sources. These measures have achieved significant water quality improvements. They now show diminishing returns, however, due to continued population growth and rapid urbanization. China introduced a series of policy measures promoting small-scale projects, and invested more than US$2.5 billion between 2000 and 2004, increasing the number of people with access to

safe drinking water by 60 million (Wang Shu-cheng 2005). The Three Gorges Dam in China is expected to provide a source of water, renewable energy (annual generation of electricity up to some 85 billion kWh) and flood control (upgrading the flood control standard from 10-year floods to 100-year floods), but is also expected to have social and environmental impacts such as loss of livelihoods in areas that will be submerged and loss of some biodiversity and ecosystem functioning. The scale and magnitude of these impacts, however, will have to be further investigated (Huang and others 2006). Mongolia and China adopted demand side management and watershed management policies to complement existing supply side management. Efforts are also underway in some Central Asian countries to use water and wastewater more efficiently, especially in agriculture.

Improvements in water use efficiency, especially in the irrigation sector, will have immediate positive impacts on water availability. Cooperation among governments, industries and public utility services would lead to a better appreciation for the need to use market-based instruments (MBIs) to lower some of the implementation costs in designing and applying such changes.

The Three Gorges Dam in China: the 1987 image on the left shows the river and surrounding landscape (overview and detail) before the dam was constructed; the 2000 image (top right) shows the dam under construction and in the 2006 image (bottom right) the dam is operational.

Credit: Landsat and ASTER images from NASA/USGS compiled by UNEP/GRID–Sioux Falls

VALUABLE ECOSYSTEMS

Biodiversity at risk

Over the last two decades, as Asia and the Pacific has become the world's fastest developing region, enormous pressures have been put on its ecosystems to support the ever-growing demand for natural resources and energy.

Coastal ecosystems, the locus of land-ocean interaction, play an important role. The region has an extremely long coastline, and more than half of its inhabitants live on or near the sea. They depend directly on coastal resources, such as mangroves and coral reefs, for part of their livelihoods (Middleton 1999). Due to large-scale exploitation of natural resources, most of the inland ecosystems in Central Asia have been severely depleted. Factors that threaten biodiversity and ecosystem functions include rapid changes in land use, extensive but poorly managed irrigation, more intensive use of rangelands, medicinal and food plant collection, construction of dams and fuelwood collection.

Asia and the Pacific has about 50 per cent of the world's remaining mangrove forests, although they have been extensively damaged or destroyed by industrial and infrastructure development (see Table 6.4) (FAO 2003b, UNESCAP 2005a). The most significant degradation of mangroves in South East Asia can be attributed to extensive coastal development. In addition, the mangroves are affected by sedimentation and pollutants from inland sources. Mangroves are vital to coastal ecosystems. They fulfil important functions in providing wood and non-wood forest products, coastal protection, habitat, spawning grounds and nutrients for a variety of fish and shellfish species. They are important for biodiversity conservation.

Box 6.11 Climate change and its potential impacts

A progressive and accelerated long-term warming trend has been reported for Asia for the period 1860–2004. Australia is suffering severe drought in recent years and had its warmest year on record, as well as its hottest April, in 2005.

Both ecosystems and human well-being are very vulnerable to climate change. Coasts and rapidly growing coastal settlements and infrastructure in countries such as Bangladesh, China, India, Myanmar and Thailand are at risk from any increase in coastal flooding and erosion due to sea-level rise and meteorological changes.

South Pacific island states are extremely vulnerable to global climate change and global sea-level rise. In a number of islands, vital infrastructure and major concentrations of settlements are very likely to be at risk. In some extreme cases, migration and resettlement outside national boundaries might have to be considered. In addition, climate change is projected to exacerbate health problems, such as heat-related illness, cholera, dengue fever and biotoxin poisoning, placing additional stress on the already overextended health systems of most small island states (see Chapter 2).

Sources: Greenpeace 2007, Huang 2006, IPCC 2007a

Boy (left) runs to catch the school boat in Pramukha island of Kepulaun Seribu (thousand islands) north of Jakarta, Indonesia and children (right) play at the wooden quay of Panggang island of Kepulaun Seribu. It is believed that about 2 000 islands are threatened with coastal flooding in this archipelagic nation due to climate change-induced sea level rise.

Credit: Greenpeace/Shailendra Yashwant

Coral reefs are fragile ecosystems, sensitive to climate change, human activities, such as tourism, and natural threats and disasters. Asia and the Pacific has some 206 000 km² of coral reefs, 72.5 per cent of the world's total (Wilkinson 2000, Wilkinson 2004). Heavy reliance on marine resources across the region has resulted in the degradation of many coral reefs, particularly those near major population centres. Moreover, higher sea surface temperatures have led to severe bleaching of the corals in coastal regions. About 60 per cent of the region's coral reefs are estimated to be at risk, with mining and destructive fishing the greatest threats (see Figure 6.17) (UNESCAP 2005b). The ultimate impacts are habitat degradation and destruction, which threaten important and valuable species, and increase the loss of biodiversity (see Table 6.5).

The destruction and reduction of ecosystem services and functions in turn reduce their contribution to human well-being. Deforestation, for example, has caused the rapid reduction of timber production, especially of the valuable timber only found in natural forests, affecting the livelihoods of people who depend on those forests (SEPA 2004). However, well conserved and managed valuable ecosystems continue to support human well-being. For example, large mangrove forests in the north and south of Phang Nga, the most tsunami-affected region in Thailand, significantly mitigated the impact of the 2004 Indian Ocean Tsunami (UNEP 2005a).

Alleviating pressures on ecosystems

The common policy response to ecosystem destruction is the establishment of protected areas. South East Asia, where coastal ecosystems are abundant, set aside 14.8 per cent of its land for protection, a higher

Table 6.4 Change in mangrove area by sub-region

Sub-region	1990 (km²)	2000 (km²)	Annual change 1990–2000 (per cent)
North East Asia	452	241	8.0
South Asia	13 389	13 052	0.2
South East Asia	52 740	44 726	1.6
South Pacific	6 320	5 520	1.3
Australia and New Zealand	10 720	9 749	0.9
Total	83 621	73 288	1.3

Source: based on FAO 2003b

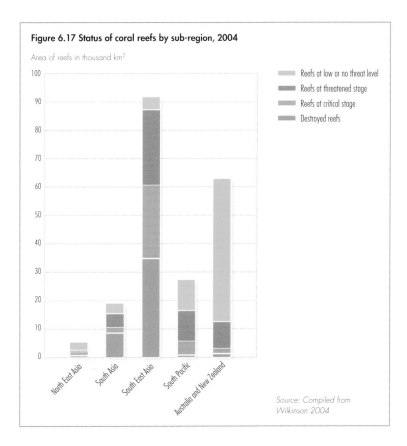

Figure 6.17 Status of coral reefs by sub-region, 2004

Area of reefs in thousand km²

Legend:
- Reefs at low or no threat level
- Reefs at threatened stage
- Reefs at critical stage
- Destroyed reefs

Source: Compiled from Wilkinson 2004

Table 6.5 Threatened species by sub-region

Sub-region	Mammals	Birds	Reptiles	Amphibians	Fishes	Molluscs	Other invertebrates	Plants
North East Asia	175	274	55	125	153	28	32	541
South Asia	207	204	64	128	110	2	78	538
South East Asia	455	466	171	192	350	27	49	1 772
Central Asia	45	46	6	0	19	0	11	4
South Pacific	119	270	63	13	186	99	15	534
Australia and New Zealand	72	145	51	51	101	181	116	77
Total	1 073	1 405	410	509	919	337	301	3 466

Source: IUCN 2006

proportion than the 2003 world average of 12 per cent. In the other Asia and the Pacific sub-regions, less than 10 per cent of their land is protected (UN 2005a). Countries cooperate in protecting marine and coastal ecosystems through four Regional Seas Action Plans: East Asian Seas, North-West Pacific, South Asian Seas and the Pacific Plan (UNEP 2006d). However, a recent study reveals that East Asia and South Asia discharge 89 per cent and 85 per cent respectively of their untreated wastewater directly into the sea (UNEP 2006d). This indicates that concrete measures are needed to achieve action plan goals.

In the South Pacific, as well as in Indonesia and the Philippines, local communities or land-owning groups, together with local governments and/or other partners, collaboratively manage 244 designated coastal areas, which include 276 smaller protected areas. Many are truly locally-managed marine areas (LMMA), a rapidly expanding approach, using traditional knowledge-based practices (see Chapters 1 and 7) (LMMA 2006). The LMMA strategy offers an alternative approach to more central systems managed by formal government institutions.

Along with sound policies and legislation, the nations of Asia and the Pacific need to raise public awareness of biodiversity and ecosystem service values, and to reduce human demands on ecosystems in order to alleviate pressures on them.

AGRICULTURAL LAND USE
Land quality
Human activities can have a negative impact on the quality of land. Poor land management can cause soil erosion, overgrazing can result in degradation of grasslands, overuse of fertilizers and pesticides reduces soil quality, and, in some areas, landfills, industrial activities and military activities cause contamination (see Chapter 3).

Agricultural land use is expanding in all countries and sub-regions, except for Australia and New Zealand, and Central Asia. In these sub-regions, agricultural land represents about 60 per cent of total land. The agricultural area in the six sub-regions of Asia and the Pacific, comparing changes over time, is illustrated in Figure 6.18.

Systematic data are lacking, but experts agree that land is being degraded in all sub-regions (IFAD 2000, Scherr and Yadev 2001, UNCCD 2001, ADB and GEF 2005). This degradation can have serious consequences for agriculture and ecosystem integrity, threatening food security and human well-being.

As food security has a very high priority in the region, land degradation is being tackled by countermeasures, such as substituting new arable land for degraded land. Although these shifts do not register in national figures of agricultural area, local people living in degraded areas feel the effect in terms of their well-being.

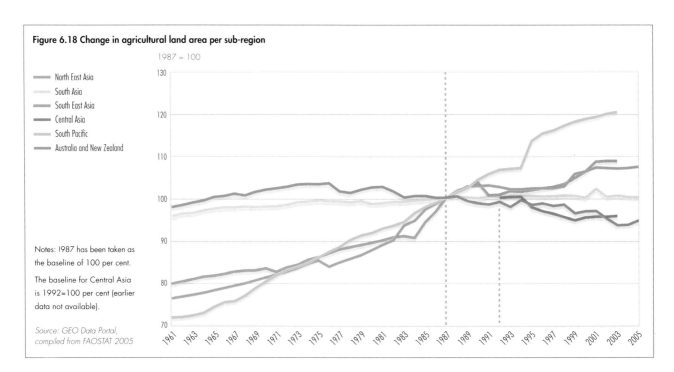

Figure 6.18 Change in agricultural land area per sub-region

1987 = 100

North East Asia
South Asia
South East Asia
Central Asia
South Pacific
Australia and New Zealand

Notes: 1987 has been taken as the baseline of 100 per cent.

The baseline for Central Asia is 1992=100 per cent (earlier data not available).

Source: GEO Data Portal, compiled from FAOSTAT 2005

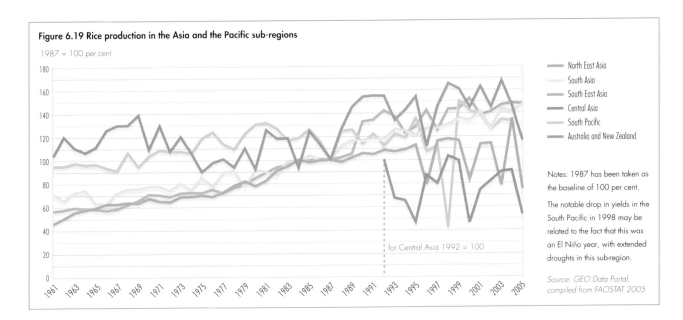

Figure 6.19 Rice production in the Asia and the Pacific sub-regions

1987 = 100 per cent

Legend:
- North East Asia
- South Asia
- South East Asia
- Central Asia
- South Pacific
- Australia and New Zealand

for Central Asia 1992 = 100

Notes: 1987 has been taken as the baseline of 100 per cent.

The notable drop in yields in the South Pacific in 1998 may be related to the fact that this was an El Niño year, with extended droughts in this sub-region.

Source: GEO Data Portal, compiled from FAOSTAT 2005

From the 1960s to 1987, most parts of this region achieved remarkable increases in rice production, the dominant food crop, and most sub-regions were able to prolong this trend (see Figure 6.19). Declines in fertility were more than compensated for by such factors as the use of additional fertilizers and pesticides, increasing yields.

It appears that most countries applied sufficient countermeasures to successfully overcome the impacts of land degradation on agricultural production (Ballance and Pant 2003). The five Central Asian countries are the exception, with deepening declines after the collapse of the Soviet Union in 1991. Land degradation in the form of salinization from poor irrigation practices continued, especially since energy supply was insufficient to allow for pumping to drain accumulated salty water. At the same time, the use of costly fertilizers and pesticides dropped sharply.

Towards more sustainable land management

Since agriculture is the main land use in Asia and the Pacific, land conservation as a tool of sustainable agriculture has been heavily

Poor land management can cause soil erosion. Terracing is one countermeasure that overcomes the impacts of land degradation.

Credit: Christian Lambrechts

emphasized. Sustainable agriculture can promote rural development, as well as increase food security and ecosystem vitality. Immediate responses include reforestation, redefining protected areas and using integrated approaches, such as integrated pest management, organic farming and integrated watershed management. Proper management of fertilizers and pesticides in agricultural activities is also crucial to protecting human health. Good governance is the basic foundation of any land conservation and management strategy. Besides providing appropriate legal and policy mechanisms for administering land ownership, it can foster the active participation of civil society in land reform efforts, and ensure the equitable distribution of agrarian development benefits.

Many farmers in South Asia and South East Asia are women, but their contribution tends to go unnoticed because they lack access to resources; men are inclined to have better access to land for farming or forestry. Land management and conservation schemes should recognize and protect the rights of female participants in agriculture, and they should share the benefits (FAO 2003c).

WASTE MANAGEMENT
Consumption and waste generation

The industrial model of development has driven the region's economy into a stage of rapid growth, accompanied by increased environmental pollution. This pattern follows the general trend in early economic growth described by the environmental Kuznets Curve (Kuznets 1995, Barbier 1997). This development model, together with new lifestyles associated with greater affluence, has led to rapid changes in consumption patterns, the generation of large quantities of waste and changes in waste composition. These are the drivers behind exponentially growing waste management problems in Asia and the Pacific.

The region currently generates 0.5–1.4 kilogrammes of municipal waste per person daily (Terazono and others 2005, UNEP 2002c). This trend shows no sign of abating, as illustrated in Figure 6.20, which extends the trend until 2025. Compostable wastes, such as vegetable and fruit peels and other leftover foodstuffs, represent 50–60 per cent of the waste stream (World Bank 1999).

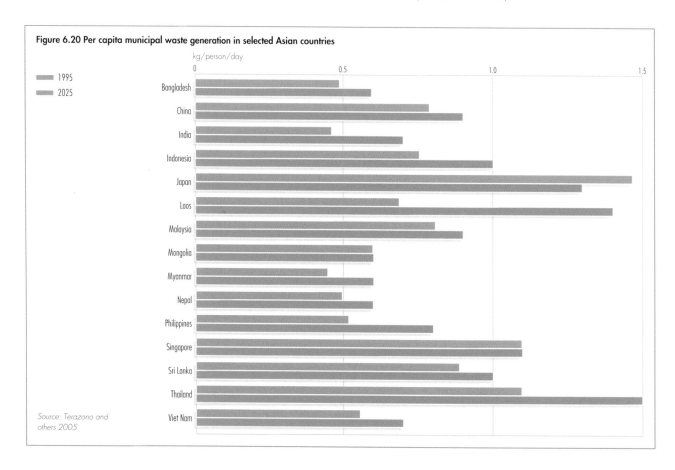

Figure 6.20 Per capita municipal waste generation in selected Asian countries

Source: Terazono and others 2005

The use of unsanitary landfills is becoming problematic, because they contaminate land and groundwater. Poor people, especially those who depend on local resources for their food supply, or who earn their livelihoods from recycling, are highly vulnerable to such impacts. The Japan Environmental Council (2005) found that in the Philippines, people who collect recyclable materials from landfills frequently give birth to deformed children. The illegal traffic in electronic and hazardous waste, and the effects on human health and the environment pose new and growing challenges for Asia and the Pacific (see Box 6.12).

Although most countries in Asia and the Pacific have ratified the Basel Convention on the Control of Transboundary Movements of Hazardous Wastes and their Disposal, the region as a whole lacks a common approach to the import of hazardous wastes.

Sustainable waste management

Recently, several countries have initiated a variety of policy responses to address the growing waste problem. For example, Dhaka has been implementing community-based solid waste management and composting projects. They benefit the municipality by saving transport and collection costs, and reducing the amount of land needed for landfills. They also contribute to progress in achieving some Millennium Development Goals (MDGs), including reducing poverty, as well as unemployment, pollution, soil degradation, hunger and illness (UNDP 2005b). The proper reuse and recycling of waste (its collection, sorting and processing) is labour intensive, and can provide employment for the poor and unskilled. Substantial numbers of people in developing countries earn their living through well-organized systems of waste collection, such as rag picking and recycling. In India alone, more than 1 million people find livelihood opportunities dealing with waste (Gupta 2001). Although there are examples of policies and strategies to tackle waste problems, effective waste management strategies and systems are still lacking or inadequate in many countries, posing a serious threat to human health and the environmental.

Many countries are starting to implement cleaner production policies and practices. Market-based tools, such as eco-labelling, have gained ground in the Philippines, Thailand, Singapore and Indonesia. For example, in cooperation with the government, business and other stakeholders, the Thailand Business Council for Sustainable Development launched its Green Label

Effective waste management strategies are lacking or are inadequate in many countries.

Credit: Ngoma Photos

project in 1994. By August 2006, 31 companies had submitted applications to use the label for 148 brands or models in 39 product categories (TEI 2006). The Thai Green Label, recognized by both companies and environmentally aware consumers, is gradually becoming a trademark for environmentally friendly products (Lebel and others 2006).

Several countries, such as Japan and South Korea, are adopting the "reduce, reuse and recycle" (3R) approach (see Chapter 10), and governments are integrating policies aimed at more efficient natural resource use into their agendas. The goal is to move towards a Sound Material-Cycle Society, characterized by preventing waste generation in the first place through lower input of natural resources, smarter product design, more efficient manufacturing and more sustainable consumption. It also involves reuse, recycling and proper treatment of materials that would otherwise enter the waste stream. In the Pacific, Fiji introduced in 2007 new measures to integrate air pollution, and solid and liquid waste management into a National Waste Management Strategy. Some countries lag behind. Mongolia has not developed comprehensive waste management laws, and South Asian countries have not yet instituted policy measures to promote more sustainable consumption.

EUROPE
DRIVERS OF CHANGE
Socio-economic and consumption trends
The past two decades have seen substantial changes across the European Region. Within this broader region, the European Union (EU) has gradually expanded to include 27 countries, while 32 European countries now participate in the activities of the European Environment Agency (EEA) and its information network, Eionet (see Box 6.13).

About 830 million people (less than one-sixth of the world's population) live in the European Region, of which over half (489 million) live in the EU-27 (GEO Data Portal, from UNPD 2007). The diversity of the European Region can be seen in the various countries' socio-economic systems, environmental governance and the priority given to environmental issues on their policy agendas. The nature of environmental challenges in Europe has been changing. While industrial pollution is still a major problem in many non-EU countries, environmental problems now also include more complex problems related to lifestyle issues.

Rising standards of wealth (leading to rising consumption of energy, transport and consumer goods) and growing numbers of households are driving greenhouse gas emissions from human activities (see Figure 6.22). A reliable and affordable energy supply and an effective transport system are preconditions for economic growth, but are also major sources of greenhouse gas emissions, and other environmental pressures.

Environmental governance: an evolution of ideas
At the time of the report of the Brundtland Commission in 1987, the region was just waking up to the potential transnational consequences of its industrial activities. Today, Europe, particularly the European Union, recognizes responsibility for its contribution to global environmental problems. The European region, and the EU consumer society in particular, leave an "ecological footprint" on other parts of the world. Shrinking the footprint, and tackling environmental issues will, at least in the case of the European Union, require managing and stabilizing demand, as rising consumption may offset even the best technological and efficiency improvements.

The Brundtland Commission report, *Our Common Future*, was a milestone in integrating the objectives of sustainable and equitable environmental development into the heart of European policy. In the two decades since then, substantial progress in environmental protection has been achieved across Europe, especially in the EU member states.

The Brundtland messages resonated in a Europe scarred by two serious environmental accidents during the previous year. An incident at the Chernobyl nuclear power plant in Ukraine led to radioactive fallout in many parts of Europe, and the Sandoz chemical fire in Basel sent toxic materials into the Rhine. Both of these industrial accidents led to severe, transnational, long-term human and environmental repercussions, some of which are still felt today. The accidents perhaps helped to set the scene for the broad acceptance of the Brundtland Commission report, by focusing public attention on the need for increased international action and cooperation to protect human life, and to safeguard the environment for future generations.

The European Region comprises the countries of Eastern, Central and Western Europe. The country groupings in this report are different from the divisions used in earlier GEO reports, to better describe groupings based on various socio-political characteristics (for a full list of countries in the European Region, refer to the introductory section of this report). Even though Central Asia is also considered as part of the wider European region, its environmental issues are analysed under the Asia and the Pacific region to avoid overlap.

Region (Group)	Sub-groups		Countries
Western and Central Europe (WCE)	EU-27	EU-15	The pre-2004 European Union member states: Austria, Belgium, Denmark, Finland, France, Germany, Greece, Ireland, Italy, Luxembourg, Netherlands, Portugal, Spain, Sweden and the United Kingdom
		New EU	Bulgaria*, Cyprus, Czech Republic, Estonia, Hungary, Latvia, Lithuania, Malta, Poland, Slovakia, Slovenia and Romania*
	European Free Trade Association (EFTA)		Iceland, Liechtenstein, Norway and Switzerland
	Other WCE countries		Andorra, Monaco and San Marino
Eastern Europe, and Caucasus (EE&C)	Caucasus		Armenia, Azerbaijan and Georgia
	Other EE&C countries		Belarus, Republic of Moldova, The Russian Federation and Ukraine
Southeastern Europe (SEE)	Western Balkans		Albania, Bosnia-Herzegovina, Croatia, Montenegro**, The Former Yugoslav Republic of Macedonia and Serbia**
	Other SEE countries		Bulgaria, Romania and Turkey
European Environment Agency (EEA-32)			Austria, Belgium, Bulgaria, Cyprus, Czech Republic, Denmark, Estonia, Finland, France, Germany, Greece, Hungary, Iceland, Ireland, Italy, Latvia, Lithuania, Liechtenstein, Luxembourg, Malta, the Netherlands, Norway, Poland, Portugal, Romania, Slovenia, Slovakia, Spain, Sweden, Switzerland, Turkey and the United Kingdom

*   Bulgaria and Romania joined the EU on 1 January 2007; data are not always integrated yet. In such cases the text refers to EU-25.

** Montenegro and Serbia were proclaimed independent republics on 3 and 5 June 2006, respectively: data are still shown for Serbia and Montenegro together.

The European Union is developing as a global leader in environmental governance, and the whole European region has a unique experience of environmental cooperation, with its many action plans and legal instruments acting at a variety of levels. The prospect of EU accession has been, and remains, the main driver of change in environmental policy in the candidate and pre-candidate countries. The focus of EU environmental policy moved from the use of remedial measures in the 1970s to end-of-pipe pollution reduction solutions in the 1980s, then to integrated pollution prevention and control, using best available techniques, in the 1990s. Today, policies are moving beyond these technical solutions to also address the patterns and drivers of unsustainable demand and consumption, and towards an integrated approach of the issues focusing on prevention. Policy changes in the New EU countries are following a similar broad sequence, but they have some opportunities for "leapfrogging," based on EU experience, which should result in cost savings and improved effectiveness.

There are many opportunities for further improving cooperation at all levels in Europe, for example, in establishing sustainable systems of energy, transport and agriculture. Air quality is an area where environmental policy has been effective, but where much still needs to be done. Some issues faced by the European Union in the 1980s are now, more than 20 years later, being tackled in Eastern Europe. More could be done to maximize the learning experience from the Western European countries, and to disseminate it elsewhere.

To complement its relatively good domestic environmental progress, Europe also has responsibility for sustainable resource management outside its borders. This is the next step towards the equitable and sustainable environmental future envisaged by the Brundtland Commission in 1987.

SELECTED ISSUES

Despite much progress, poor water and urban air quality still cause substantial problems in parts of the European region, affecting the health and quality of life of many people. Emissions of air pollutants are largely driven by the demand for greater mobility. Water pollution and scarcity problems are caused by the impacts of industrial and agricultural activities, poor management of water as a resource, and the disposal of sewage wastes, all also threatening biodiversity. Changing climatic conditions further compound these issues.

CLIMATE CHANGE AND ENERGY

The climate on Earth is changing, and in Europe the average temperature has increased by about 1.4°C compared to pre-industrial levels. Annual mean deviations in Europe tend to be larger than global deviations (see Figure 6.21 and Figure 2.18). In the Arctic regions of Russia, it has increased by up to 3°C over the past 90 years (Russian 3rd Nat. Comm 2002, ACIA 2004). The European mean temperature is projected to increase by 2.1–4.4°C by 2080. Sea levels are rising and glacier melting is accelerating; during the 20th century the global mean sea-level rise was 1.7 mm/year and is projected to rise by 0.18–0.59 metres by 2100 (IPCC 2007a).

Energy emission and efficiency trends

Since 1987, greenhouse gas emissions from the energy sector have been reduced in Western Europe, but since the end of the 1990s, these emissions have increased across the European region, partly because increasing natural gas prices have re-established coal as a key fuel (see Figure 6.22). While energy use grew at a slightly lower pace than economic activity over the past 15 years, Europe as a whole has not succeeded in stabilizing its energy consumption levels. There is a clear energy efficiency gap between the EU-15 and the New EU, due to both technological and structural issues (see Box 6.14).

Emissions for the New EU are projected to remain well below their 1990 levels, even allowing for a doubling of economic output. This is not so for Israel, which has no obligations under the Kyoto Protocol, but forecasts significant increases compared to its 1996 levels. Current trends and prospects for the EU-15 are worrying. With existing domestic policies and measures, total EU-15 greenhouse gas emissions will only be 0.6 per cent below base year levels in 2010. Taking into account additional domestic policies and measures being planned by member states, a total EU-15 emissions reduction of 4.6 per cent is projected. This relies on the assumption that several member states will cut emissions by more than is required to meet their national targets. The projected use of Kyoto mechanisms by 10 member states will reduce emissions by 2010 by a further 2.6 per cent. Finally, the use of carbon sinks, under Articles 3.3 and 3.4 of the Kyoto Protocol, would contribute an additional 0.8 per cent (EEA2006a).

Towards a more sustainable energy system

A number of pan-regional plans have been initiated to develop common energy policy objectives, promote

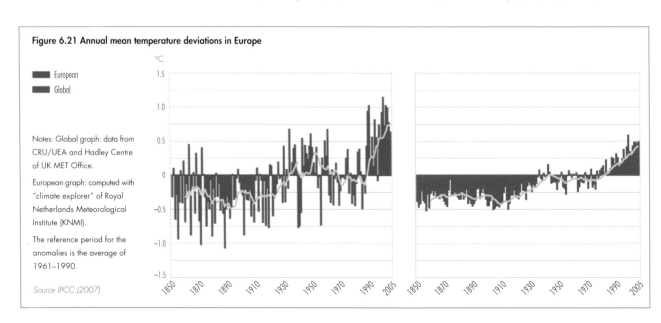

Figure 6.21 Annual mean temperature deviations in Europe

European
Global

Notes: Global graph: data from CRU/UEA and Hadley Centre of UK MET Office.

European graph: computed with "climate explorer" of Royal Netherlands Meteorological Institute (KNMI).

The reference period for the anomalies is the average of 1961–1990.

Source IPCC (2007)

more sustainable energy production and consumption, and ensure stability of supply. For example, in November 2006, the European Union and the countries of the Black Sea and Caspian Sea regions agreed on a common energy strategy, based on four areas: converging energy markets, enhancing energy security, supporting sustainable energy development, and attracting investment for common projects (EC 2006a). In March 2007, the European Union agreed on an integrated climate change and energy action plan (EC 2007a), based on a comprehensive package of proposals from the European Commission (EC 2007b). Figure 6.23 illustrates some of the impacts of CO_2 reduction initiatives. The charts show the estimated contributions of various factors that have affected emissions from public electricity and heat production.

The capital investment needed to meet forecast energy growth is an important incentive for energy savings and energy efficiency measures, as well as for changes in fuel mixes. There is a particular need for investment in energy infrastructure in some Southeastern European countries. Harnessing renewable energy sources would also make a major contribution to a more sustainable energy system (EEA 2005b). In this respect, the use of the Clean Development Mechanism may offer win-win situations by helping industrialized countries to meet their Kyoto targets, and simultaneously offering investment in new technologies for developing countries.

The Kyoto targets are only a first step towards the more substantial global emission reductions that

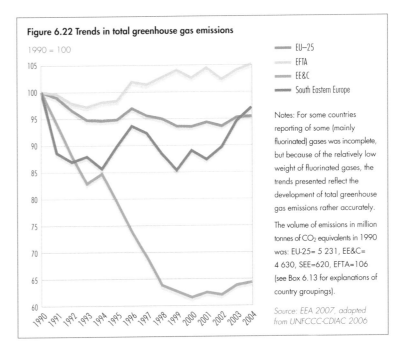

Figure 6.22 Trends in total greenhouse gas emissions

1990 = 100

— EU–25
— EFTA
— EE&C
— South Eastern Europe

Notes: For some countries reporting of some (mainly fluorinated) gases was incomplete, but because of the relatively low weight of fluorinated gases, the trends presented reflect the development of total greenhouse gas emissions rather accurately.

The volume of emissions in million tonnes of CO_2 equivalents in 1990 was: EU-25= 5 231, EE&C= 4 630, SEE=620, EFTA=106 (see Box 6.13 for explanations of country groupings).

Source: EEA 2007, adapted from UNFCCC-CDIAC 2006

Box 6.14 Energy efficiency and industrial restructuring in Central and Eastern Europe

Energy intensity in non-OECD Europe is expected to decline at an annual rate of 2.5 per cent between 2003 and 2030. The energy efficiency gap between Eastern and Western Europe is due to both technological and structural aspects, with the latter playing a more crucial role than is often acknowledged.

Energy-intensive industries make up an increasing share of the industrial fabric in Eastern Europe, while the reverse is true for Western Europe. Sector-specific statistics show that the energy efficiency of energy-intensive industries in Western Europe has not improved dramatically over the past few years.

Source: EIA 2006a

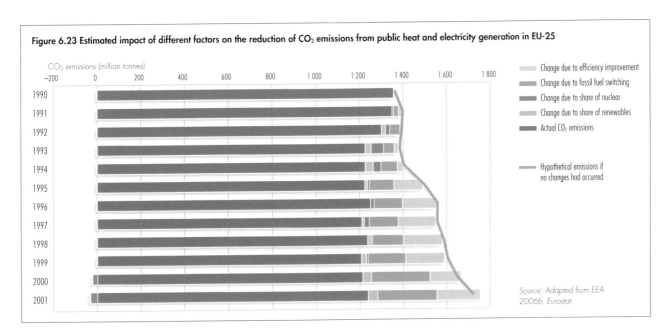

Figure 6.23 Estimated impact of different factors on the reduction of CO_2 emissions from public heat and electricity generation in EU-25

CO_2 emissions (million tonnes)

Change due to efficiency improvement
Change due to fossil fuel switching
Change due to share of nuclear
Change due to share of renewables
Actual CO_2 emissions

Hypothetical emissions if no changes had occurred

Source: Adapted from EEA 2006b, Eurostat

will be needed to reach the long-term objective of the UN Framework Convention on Climate Change (UNFCCC). No agreement has yet been reached on new UNFCCC targets for industrialized countries, or on possible new emission reduction strategies for other countries. Growing public awareness, underpinned by rising energy prices, has given a new political momentum to climate change policies in Europe. This has been stimulated by extreme weather conditions, although those are not necessarily a consequence of climate change. To limit the impacts of climate change to a manageable level, the European Union has proposed that the global temperature should not exceed an average of 2°C above pre-industrial temperatures. To achieve this target, worldwide greenhouse gas emissions will need to peak before 2025, and by 2050 they should fall by up to 50 per cent, compared to 1990 levels. This implies emission reductions of 60–80 per cent by 2050 in developed countries. If developing countries accept emissions reduction commitments, they will need to significantly reduce their emissions (EC 2007a, EC 2007b).

SUSTAINABLE CONSUMPTION AND PRODUCTION

Unsustainable resource use

European consumption and production contribute to the high (and often unsustainable) use of resources, increasing environmental degradation, depletion of natural resources and growing amounts of waste inside as well as outside Europe. The wealthier the society, the more resources it tends to use and the more waste it generates. Household consumption expenditure is steadily increasing throughout Europe (see Figure 6.24), with Western European households having some of the highest consumption levels in the world. At the same time, patterns of consumption are changing, with the food component decreasing and the shares for transport, communication, housing, recreation and health on the rise.

The goods-and-services that cause the highest environmental impacts through their life cycles have been identified as housing, food and mobility (EEA 2005b, EEA 2007, EC 2006b). The dominant stage with respect to impacts differs significantly between different goods-and-services. For food and beverages, the majority of environmental impacts

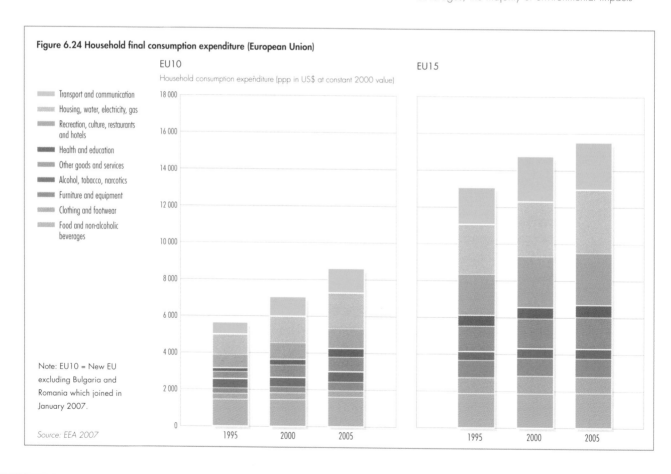

Figure 6.24 Household final consumption expenditure (European Union)

EU10

Household consumption expenditure (ppp in US$ at constant 2000 value)

Transport and communication
Housing, water, electricity, gas
Recreation, culture, restaurants and hotels
Health and education
Other goods and services
Alcohol, tobacco, narcotics
Furniture and equipment
Clothing and footwear
Food and non-alcoholic beverages

EU15

Note: EU10 = New EU excluding Bulgaria and Romania which joined in January 2007.

Source: EEA 2007

are related to agricultural or industrial production activities, while for personal transport the majority of the impacts are in the use phase, when driving the car or flying in an airplane.

Decoupling resource use from economic growth

The European Union has made important progress in decoupling resource use from economic growth, as has the wider European region, albeit at a slower pace (see Box 6.15). However, absolute reduction in resource use has not been achieved. Improvements have also been made in eco-efficiency, but attempts to change consumption patterns have had limited success. Over the past four decades productivity in the use of raw materials and energy has increased by 100 per cent and 20 per cent respectively, but there is still much room for improvement in how Europeans use energy and resources.

In the new EU member states, a number of factors have contributed to stabilization or even a decrease in use of natural resources over the past few years (EEA 2007). They include changes in the structure of the economy and production, particularly a reduction in the level of industrial production and agricultural intensity, together with the modernization of technologies and improvements in efficiency. In Western Europe, achieving an absolute decoupling of environmental impacts, material use and waste generation from economic growth remains a challenge.

Products are being redesigned to meet this challenge, but it remains to be seen if this will eventually lead to absolute decoupling. Voluntary measures have also been introduced to stimulate sustainable production and consumption, including eco-labelling, corporate social responsibility, the European Union's Eco-Management and Audit Scheme (EMAS) and voluntary agreements with various industries.

Nevertheless, increasing consumption and production, coupled with a lack of prevention, often outstrip efficiency gains (see Box 6.16). To make consumption and production patterns more sustainable, economic instruments that reflect the real environmental and social costs of materials and energy are needed, and should be combined with legal instruments, information-based and other instruments.

Box 6.15 Sustainable Consumption and Production (SCP) and the environmental policy agenda

The UN Conference on Environment and Development (UNCED) in Rio de Janeiro in 1992 highlighted the problem of unsustainable consumption. Ten years later, the World Summit on Sustainable Development (WSSD) in Johannesburg resulted in the agreement to develop "a framework of programmes on sustainable consumption and production." The global Marrakech process on sustainable consumption and production, including its seven task forces led by countries, was established after the Johannesburg summit, and aims to prepare a framework of programmes for the UN Commission on Sustainable Development (CSD) in 2010–2011.

In the European region, sustainable consumption and production, decoupling of environmental impacts from economic growth, increasing eco-efficiency, and sustainable management of resources are now increasingly visible on the policy agenda. The EU thematic strategies on sustainable use of natural resources, and on prevention and recycling of waste, and the renewed EU Sustainable Development Strategy specifically refer to tackling unsustainable consumption and production. An EU Action Plan on Sustainable Consumption and Production is being prepared by the Commission.

National strategies related to sustainable consumption and production have been prepared in, for example, the Czech Republic, Finland, Sweden and the United Kingdom. In some regions of Europe, including in the EE&C and the Balkan countries, work on those issues remains at an early stage.

Source: EEA 2007

While for Western Europe the challenge is to achieve decoupling, in some sub-regions of Europe, lack of efficient waste collection and safe disposal remains a major problem, as it causes contamination of land and groundwater (EEA 2007). Some EE&C countries face yet another threat – accumulated hazardous waste generated during the Soviet era. It includes mainly radioactive, military and mining wastes, but there are also large stockpiles of obsolete pesticides containing persistent organic pollutants (POPs). The lack of funds for proper disposal makes them a large risk to the environment (UNEP 2006e).

AIR QUALITY AND TRANSPORT
Air pollutants

Despite progress in reducing emissions, air pollution still poses risks for both human health and the environment. The main public health impact is caused by small airborne particles (particulate matter), their toxic constituents, such as heavy metals and polyaromatic hydrocarbons, as well as by tropospheric ozone. Growth in the number of motor vehicles, along with emissions from industry, power production and households all contribute to air pollution (see Box 6.16).

Emissions of air pollutants in Western Europe have declined by 2 per cent/year since 2000, as a

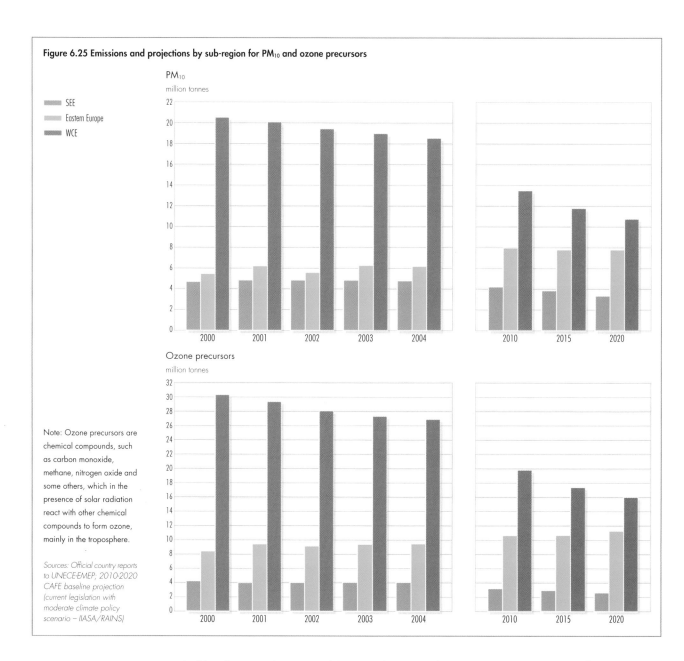

Figure 6.25 Emissions and projections by sub-region for PM$_{10}$ and ozone precursors

SEE
Eastern Europe
WCE

PM$_{10}$
million tonnes

Ozone precursors
million tonnes

Note: Ozone precursors are chemical compounds, such as carbon monoxide, methane, nitrogen oxide and some others, which in the presence of solar radiation react with other chemical compounds to form ozone, mainly in the troposphere.

Sources: Official country reports to UNECE-EMEP; 2010-2020 CAFE baseline projection (current legislation with moderate climate policy scenario – IIASA/RAINS)

result of the effective implementation of EU air quality policies, a trend that is expected to continue to 2020 (see Figure 6.25). In Southeastern Europe, emissions stabilized between 2000 and 2004, and reductions of some 25 per cent are expected by about 2020. In Eastern Europe, economic recovery since 1999 has led to a 10 per cent increase in air emissions, and projections to 2020 are for further emission increases, except for sulphur dioxide (Vestreng and others 2005). Stronger efforts will be needed to achieve safer levels of air quality. In Western Europe and Southeastern Europe, the expected reduction in emissions will reduce impacts on public health and ecosystems significantly by 2020, but not enough to reach safe levels.

In the year 2000, exposure to particulate matter was estimated to reduce average statistical life expectancy by approximately nine months in the EU-25. This is comparable to the impacts of traffic accidents (EC 2005a, Amann and others 2005).

Sulphur deposition, the main acidifying factor, has fallen over the past 20 years (CHMI 2003). In 2000, acidifying deposition was still above critical loads in parts of Western Europe, but the percentage of EU-25 forest areas affected is projected to decrease from 23 per cent in 2000 to 13 per cent in 2020. Ammonia is projected to be the dominant source of acidification in the future.

Box 6.16 Rising transport demand outstrips technical improvements

Catalytic converters, required since 1993, contribute to improved air quality in Western Europe (EEA 2006c), but the gains have been partly offset by increased road traffic, and higher numbers of diesel cars. In Central and Eastern Europe, the public transport systems have been deteriorating since the early 1990s, and car ownership has risen (see Figure 6.26). In Western and Central Europe, car ownership in 2003 ranged from 252 per 1000 people (Slovakia) to 641 (Luxembourg). The number for Belarus is from 1998, and assuming that continued growth since then would put that country's car ownership at Russian levels. The number for Armenia is from 1997, and its fleet size was stable between 1993 and 1997.

In Central and Eastern Europe, total emissions from vehicles are lower than those in Western and Central Europe, but emissions per vehicle are much higher because of the poorer quality of roads and vehicles, and the lack of effective traffic management, which contribute to higher fuel consumption. In addition, there is poorer quality fuel in some parts of Central and Eastern Europe.

In Western and Central Europe, road freight transport continues to grow faster than the economy, driven by EU expansion and the growing internationalization of markets. In Central and Eastern Europe, freight transport has been increasing since the early 1990s. In addition, e-business and the comparatively low costs of road transport – resulting from the lack of liability for costs to infrastructure and environmental externalities – are changing the freight sector through processes such as outsourcing, low-storage production, decentralized distribution and "just in time" delivery.

Source: EEA 2006c, EEA 2007

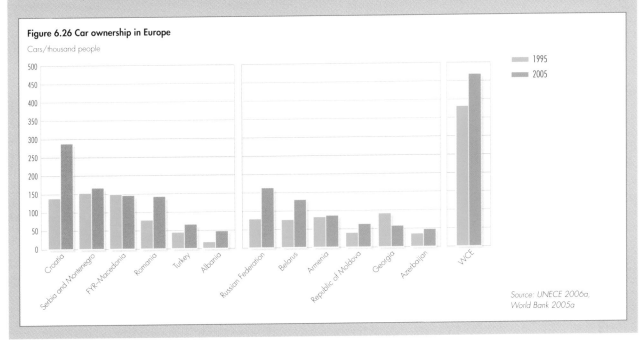

Figure 6.26 Car ownership in Europe

Cars/thousand people

Legend: 1995, 2005

Source: UNECE 2006a, World Bank 2005a

Reducing air pollution

Between 1990 and 2004, there has been progress in reducing air pollution. Most of the reductions of particulates came from the energy supply sector and industry, and emissions are expected to decrease further as cleaner vehicle engine technologies are adopted, and stationary fuel combustion emissions are controlled through abatement or use of low-sulphur fuels, such as natural gas or unleaded gasoline (see Box 6.17). From 1993 through 2007, the European Union has been imposing progressively stricter pollution controls on vehicles. This has controlled such pollutants as CO, HC, NO_X and PM_{10}, using technologies such as catalytic converters and better engine controls. A Euro 5 norm will come into place in 2009, and will further reduce the emissions of regulated pollutants.

Eastern and Southeastern Europe have their own car industries, which have not automatically adopted Western European vehicle technologies such as catalytic converters. However, the technologies are widely available in Western Europe at low cost, so the introduction of emission regulation may be a cost-effective means of reducing pollutant emissions from transport in Eastern Europe. The adoption of EURO vehicle emission norms (Table 6.6) by Russia and Ukraine, for example, would affect a majority of the population of the entire EE&C region, and an even larger share of the economy and vehicle fleet.

It would also have an impact in countries that have not introduced the standards because most manufacturers would meet the new standards.

These are all promising developments, but people are still exposed to levels of air pollution that exceed the air quality standards set by the European Union and the World Health Organization (WHO). In the period 1997–2004, 23–45 per cent of the urban population was still potentially exposed to ambient air concentrations of particulate matter (PM_{10}) in excess of the EU limit value set to protect human health. There was no discernible trend over the period (see Figure 6.27). For ozone, there is considerable variation from year to year. Over the period, 20–25 per cent of the urban population was exposed to concentrations above the ozone target value. In 2003, a year with extremely high ozone concentrations due to high temperatures related to meteorological conditions (EEA 2004a), this increased to about 60 per cent.

The situation for NO_2 is improving, but about 25 per cent of the European urban population is still potentially exposed to concentrations above the limit value. The share of urban population exposed to SO_2 concentrations above the short-term limit value decreased to less than 1 per cent, and the EU limit value is close to being met.

The level of air pollution in the largest cities of Russia, Ukraine and Moldova has increased over the last years, and frequently exceeds WHO air quality standards. This increase is mainly caused by an increase in particulates, nitrogen dioxide and

Table 6.6 Adoption of EURO vehicle emission standards by non-EU countries

	EURO 1	EURO 2	EURO 3	EURO 4
EU passenger cars/light commercial vehicles	1993/1993	1997/1997	2001/2002	2006/2007
Bulgaria				2007/2007
Romania				2006/2007
Turkey				2006/2007
Croatia		2000		
Albania	National limits for CO and HC			
FYROM	National limits for CO			
Bosnia and Herzegovina	No regulation			
Serbia and Montenegro	No regulation			
Belarus		2002	2006	Q4-2006
Russia		2006	2008	2010
Ukraine (only on imported vehicles)		2005	2008	2010

Notes:

Belarus: unclear if information indicates an obligatory norm or simply availability of vehicles conforming to the norm.

Russia: unconfirmed press reports indicate that introduction of norms may be delayed.

The recent EURO 5 norm is not included, as non-EU countries have not started to implement it.

Years indicate when norms are/will be introduced: passenger cars/light commercial vehicles.

Source: based on information received from EEA contact points

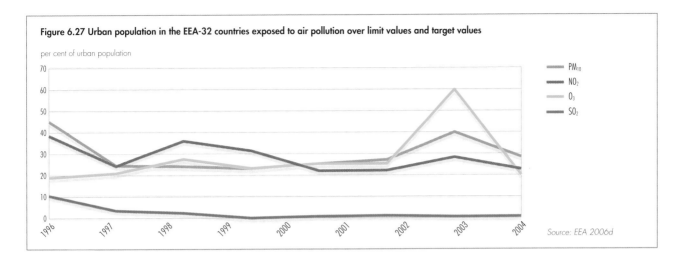

Figure 6.27 Urban population in the EEA-32 countries exposed to air pollution over limit values and target values

per cent of urban population

Legend: PM₁₀, NO₂, O₃, SO₂

Source: EEA 2006d

benzo(a)pyrene. In Russia, the number of cities with concentrations of benzo(a)pyrene over maximum allowable concentration has increased in the last five years, reaching 47 per cent in 2004.

The EU 6th Environmental Action Programme (6EAP) has the objective of achieving levels of air quality that do not give rise to significant negative impacts on and risks to human health and the environment. The Thematic Strategy on Air Pollution (EC 2005b) was adopted in September 2005, and sets air quality goals in the European Union up to 2020. Table 6.7 summarizes the anticipated benefits of the strategy, compared to the 2000 situation.

LAND-USE CHANGE AND BIODIVERSITY LOSS

Land-use threats

Agriculture in Europe has two trends that threaten biodiversity: intensification and abandonment. In socio-economic terms, farming in marginal areas is under pressure, and is subject to both land abandonment

and intensification (EEA 2004b, EEA 2004c, Baldock and others 1995). Urban sprawl, infrastructure development, illegal logging and human-induced fires are other increasingly significant problems for biodiversity in the European region.

The most intensive farm systems result in highly-productive monocultures, with very low biodiversity. At the other end of the scale are the species-rich traditional farming systems that have shaped the European landscape and created habitats rich in species. They have low stocking densities, little or no chemical inputs and labour-intensive management, such as shepherding. Their ecosystems include semi-natural pastures, such as steppes, dehesas (grasslands with scattered oaks, typical of Portugal and Spain) and mountain pastures. Conservation of these habitats requires the continuation of traditional land management practices.

The agricultural sector suffers from a lack of follow-through in the liberalization process, and in the building

Table 6.7 Anticipated benefits of the EU Thematic Strategy on Air Pollution

	Human health benefits in the EU-27			Natural environment benefits in the EU-27 (km²)		
	Monetized health benefits (Euro/year)	Years of life lost due to PM$_{2.5}$	Premature deaths due to PM$_{2.5}$ and O$_3$	Acidification (forest area exceeded)	Eutrophication (ecosystem area exceeded)	Ozone (forest area exceeded)
2000	–	3.62 million	370 000	243 000	733 000	827 000
Strategy 2020	42–135 billion	1.91 million	230 000	63 000	416 000	699 000

Notes:

Ecosystem benefits have not been monetized, but are expected to be significant.

Ecosystem benefits for the strategy scenario have been interpolated from existing analyses.

Source: EC 2005b

of market institutions to support the development of competitive food markets. As a result, subsistence agriculture is now spreading in Eastern Europe. Socio-economic conditions in rural areas with small-scale farming are generally unfavourable, leading to low incomes, difficult working conditions and a lack of social services, all of which makes farming an unattractive option for young people. The result is population loss from rural areas and land abandonment. Over 200 000 km2 of arable land have already been deserted in European Russia for example, and this trend is expected to continue (Prishchepov and others 2006) (see Box 6.18). With such abandonment traditional management practices also disappear due to which high-nature value farmlands are degraded, for example, through encroachment when sheep and controlled winter fires no longer keep grasses short.

Lack of good farm management, involving improper drainage, overgrazing and irrigation, contribute to land degradation in the form of falling organic soil carbon content, increased erosion rates, salinization, lowered productivity and vegetation loss.

Forestry in Europe is sustainable, but regional problems exist, notably illegal logging in Eastern Europe and human-induced forest fires. In recent years, the magnitude and frequency of forest fires has increased (Goldammer and others 2003, Yefremov and Shvidenko 2004). The Balkan Region, Croatia, Turkey, the FYROM and Bulgaria had a strong peak in the year 2000 (FAO 2006a), while the summer of 2003 saw one of the most severe fire seasons in recent decades in Southeastern Europe, France and Portugal (EC 2004). Currently, the forest area of Europe is 10.3 million km2 (79 per cent in Russia). About one-quarter is primary forest, with no clearly visible indications of human activities, 50 per cent is modified natural forest with little human influence and the rest is heavily modified.

Climate change is an overarching pressure that is expected to become the main driver of biodiversity loss in the future, affecting productivity, the growth cycle of plants and animals and species distribution (Ciais and others 2005, Thomas and others 2004). Table 6.8 summarizes the main threats to biodiversity in Europe.

Managing biodiversity

The wider countryside covers a large part of the European and global terrestrial landscape, and a considerable part of biodiversity depends on the adequate management of the wider countryside. The goal for the wider countryside must be to maintain or restore robust functional ecosystems as a basis for sustainable development, securing long-term ecologically favourable conditions. Only then can the loss of biodiversity be halted and the social, economic, and cultural value for people living in and depending on the wider countryside be secured.

The EU target to halt biodiversity loss by 2010 (EC 2006c, UNECE 2003a), is more stringent than the global-level CBD target, which aims to significantly reduce the current rate of biodiversity loss. Although much has been achieved, the target will not be reached for all ecosystems, species and habitats in Europe.

The Pan-European Ecological Network (PEEN) is a non-binding framework that aims to enhance ecological connectivity across Europe by promoting synergy between policies, land-use planning and rural and urban development at all scales (Council of Europe 2003a). The establishment of the PEEN is supported by legal provisions and instruments under various conventions and international agreements.

The European Commission Communication on a European Biodiversity Strategy (EC 2006c) calls for EU countries to reinforce coherence and connectivity of the NATURA 2000 network. It also highlights the need to restore biodiversity and ecosystem services in non-protected rural areas of the European Union. Compliance with these objectives by EU countries is key to the implementation of the PEEN. The main instrument for nature conservation is the Habitat and Birds Directives, with a network of Natura 2000 areas that covers 16 per cent of the European Union.

Table 6.8 Main threats to biodiversity reported in the pan-European Region				
Threat	Northwestern Europe	Caucasus	Eastern Europe	Southeastern Europe
Climate change	**	***	**	**
Urbanization and infrastructure	**	*	**	**
Agricultural intensification	**	*	**	**
Land abandonment	**		**	***
Desertification	*	**	*	**
Acidification	*		***	*
Eutrophication	***	*	**	**
Radioactive contamination			**	
Forest fires	*		**	**
Illegal logging		**	**	*
Illegal hunting and wildlife trade		***	*	
Invasive alien species	**	*	**	*

* minor threat   ** moderate threat   *** serious threat

Source: EEA 2007

In 1987, the Brundtland Commission report recommended removing subsidies for intensive agriculture, and decoupling production from subsidies. In 2003, the EU Common Agricultural Policy (CAP) was reformed, with more attention given to rural development. Intensive agriculture still receives a larger share of agricultural subsidies under the CAP, but the range of agri-environment policy tools has been widened.

Agri-environmental schemes include support for the conservation of high nature-value farmland. These areas comprise the "hot spots" of biodiversity in rural areas (EEA 2004b, EEA 2004c). Subsidies are also offered to farmers who comply with good agricultural practices, such as reducing erosion and nitrate leaching. New EU countries have been slow to implement some of these environmental instruments.

FRESHWATER STRESS

Water quality and quantity

Although most people in the European region are well served, some still do not have access to improved drinking water, and even more have no access to sanitation. WHO estimates that for the European region unsafe water, sanitation and hygiene annually results in 18 000 premature deaths, 736 000 disability adjusted life years (DALYs), and the loss of 1.18 million years of life.

Generally, the population in Western Europe has continuous access to good quality drinking water. However, in the Balkan countries and some areas of Central Europe, water supplies are often intermittent, and of poor quality (see Box 6.19). Where users receive water intermittently, there is more risk that pollutants will contaminate the network, and there is increased infrastructure deterioration. Leakage losses from distribution networks are often high, and it is not unusual for more than one-third of the water to be lost before delivery in many Central and Eastern European countries (see Figure 6.28).

In most parts of the region, water quality has improved since 1990, due to reductions in contaminant loads from wastewater treatment and industries, as well as a decline in industrial and agricultural activity (see Figure 6.29) (EEA 2003, UNEP 2004a, EEA 2007). Some large rivers, such as the Kura and the Volga, are still heavily polluted (EEA 2007).

Agriculture is the main contributor to water pollution in Western Europe. Most nitrate pollution, which also causes eutrophication, is due to farm fertilizer and manure run-off. In the New EU, the financial crisis of the 1990s in the farming sector led to a sharp decline in fertilizer use and cattle numbers, but these are now increasing.

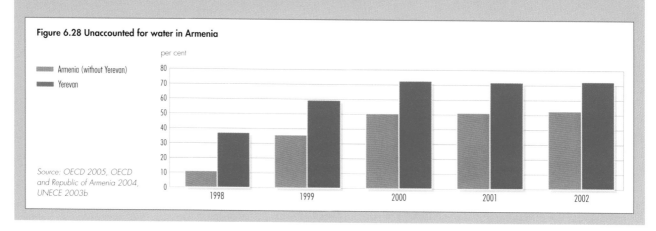
There is a high nitrogen surplus in soils in many
countries, and concentrations remain highest where
agriculture is most intensive (EEA 2005a, UNEP
2004a). Together with the use of agricultural
pesticides, this threatens groundwater sources,
and many groundwater bodies now exceed limits
for nitrate and other contaminants (EEA 2003,
EEA 2007).

Decreasing inputs of oxygen consuming substances,
such as ammonia and phosphorus, from urban

wastewater to surface waters have led to improved
oxygen and nutrient conditions in rivers and lakes.
For nitrogen coming mainly from agriculture, there
has been little or no improvement.

Agriculture is not only responsible for a large share
of water pollution, but also for about one-third of the
water use across Europe, especially in the south.
The proportion of water used for agriculture varies
from none to 80 per cent of the total water demand,
depending on the country.

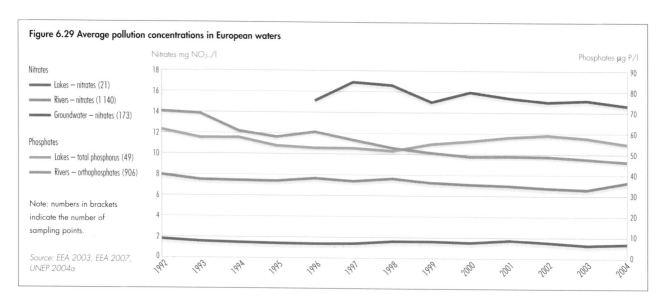

Figure 6.29 Average pollution concentrations in European waters

Nitrates

Lakes – nitrates (21)
Rivers – nitrates (1 140)
Groundwater – nitrates (173)

Phosphates

Lakes – total phosphorus (49)
Rivers – orthophosphates (906)

Note: numbers in brackets
indicate the number of
sampling points.

Source: EEA 2003, EEA 2007,
UNEP 2004a

Industrial water use declined through the 1980s and 1990s, because of water recycling, the closure of industries and a decline in industrial production (EEA 1999). Household water use also decreased in Western Europe, as a result of higher prices.

Managing water and sanitation

A number of conventions set out responses to Europe's water quality problems. The EU Water Framework Directive, introduced in 2000, takes an integrated water resource management approach, with the goal that all water bodies attain good ecological status by 2015.

Partnerships for water management have a long tradition in the European Union, and there are many international agreements, such as the Danube Commission and the International Commission for the Protection of the Rhine. There has, however, been a significant decline in the level of water quality monitoring in parts of Central and Eastern Europe in the 1990s. Since then, improvements have been observed, but in several countries monitoring is still insufficient to obtain a clear picture of water quality (EEA 2007).

The EU focus has shifted from point sources of water pollution to diffuse or non-point sources of contaminants, such as agricultural run-off (EEA 2003, EEA 2005a). Non-point sources are difficult to measure or estimate, and thus are hard to manage (UNEP 2004a).

Modernizing water networks would improve water availability, and tackling leaks would prevent the loss of substantial amounts of piped water in some New EU countries (EEA 2003). Water metering and appropriate pricing would create an incentive to conserve water, and could lead to estimated savings of 10–20 per cent (EEA 2001), but care must be taken to ensure that prices are not prohibitive. The EU Water Framework Directive stipulates that water users should contribute adequately to the full cost of the water supply.

Water demand management can be applied in the agriculture sector, for example through the substitution of crops with plants that have a lower water demand (UNEP 2004a), and more water-efficient irrigation technology. Water recycling is also a good means of using water more sustainably.

LATIN AMERICA AND THE CARIBBEAN
DRIVERS OF CHANGE
Socio-economic trends

Latin American and the Caribbean region (LAC) comprises 33 countries, and can be sub-divided into three sub-regions: the Caribbean, Meso-America (Mexico and Central America) and South America. Over 560 million people, representing over 8 per cent of the world's population, live in the region, with over half of them concentrated in Brazil and Mexico. Between the release of the Brundtland Commission report in 1987 and 2005, the region's population grew by almost 34 per cent. While the annual population growth rate for the region fell from 1.93 to 1.42 per cent, growth rates are still well above 2 per cent in several Central American countries. Over the same period, regional life expectancy increased from about 66 to 71 years (GEO Data Portal, from UNPD 2007).

For the region as a whole, human development, as measured by the UNDP Human Development Index (HDI), is at an intermediate level. Compared to 1985, all nations in the region have climbed in the ranking, indicating that, on average, people have become healthier, better educated and less impoverished (UNDP 2006). However, only 33 per cent of the region's population lives in countries with a high level of human development. Haiti was ranked 154th among 177 countries on the HDI in 2004.

Poverty and inequality persist as serious challenges. While the proportion of poor people fell from 48.3 per cent in 1990 to 43.5 per cent in 1997, it was still 42.9 per cent (222 million people) in 2004, of which 96 million lived in extreme poverty (CEPAL 2005). Of the world's regions, Latin America and the Caribbean has the worst income inequality. The poorest 20 per cent of households get between 2.2 per cent of national income in Bolivia, and 8.8 per cent in Uruguay. The wealthiest 20 per cent of households enjoy 41.8 per cent of national income in Uruguay, and 62.4 per cent in Brazil (CEPAL 2005).

After the "lost decade" of the 1980s, when per capita GDP decreased by 3.1 per cent/year mainly due to a general economic crisis, GDP grew by some 53 per cent between 1990 and 2004, or 2.9 per

Above, children and their parents work all day sifting through rubbish for scraps to sell. The photo below shows the disparity between the rich and the poor.

Credit: Mark Edwards/Still Pictures (top) and Ron Giling/LINEAIR/Still Pictures (bottom)

the Caribbean, which are closely related to energy use, increased by some 24 per cent from 1990 to 2003. However, at 2.4 tonnes/person/year they are still well below those in developed countries (19.8 tonnes/person/year in North America and 8.3 in Europe for 2003). In fact, the region as a whole today only contributes just over 5 per cent of global anthropogenic CO_2 emissions (GEO Data Portal, from UNFCCC-CDIAC 2006).

Between 1980 and 2004, energy intensity (energy consumption per unit of GDP) stagnated in Latin America and the Caribbean (CEPAL 2006), with associated negative economic and environmental impacts. In industrialized countries, it fell by 24 per cent. This lack of progress in energy intensity in Latin America and the Caribbean can be explained by a lack of more efficient technologies, outdated industries, subsidized fuel prices (with respect to international market prices) and the transport sector's high and inefficient use of energy (see Chapter 2).

Science and technology

Latin America and the Caribbean has traditionally lacked competitiveness in scientific development and technological innovation (Philippi and others 2002). But countries have taken a variety of steps towards investing in environmental science and technology relevant to the promotion of sustainability (Toledo and Castillo 1999, Philippi and others 2002). However, very few countries in the region have reached the international goal of investing at least one per cent of GDP in science and technology (RICYT 2003). In addition, there is a significant emigration of highly educated people, creating a "brain drain" to industrialized countries (Carrington and Detragiache 1999).

Governance

Environmental governance is a complicated issue, since the environment has not yet been given the high-priority status it requires (Gabaldón and Rodríguez 2002). Regional participation in global multilateral environmental agreements (MEAs) is generally high (see Box 6.21), and governmental institutions formally devoted to environmental matters were created in most countries over the last 15 years. However, the profile and budgets of environmental institutions are often lower than those of other ministries or departments, which have so far failed to mainstream environmental criteria.

cent/year on average (GEO Data Portal, from World Bank 2006). However, this rate is noticeably lower than that experienced by other developing sub-regions (particularly South East Asia), and well below the 4.3 per cent needed to meet the Millennium Development Goal to reduce extreme poverty (CEPAL 2005). Regionalization and globalization have triggered an increase in oil and gas extraction, expanded the use of arable land for monoculture exports and intensified tourism in the Caribbean (UNEP 2004b).

Energy consumption

Energy consumption is still low, and its use relatively inefficient (see Box 6.20). Anthropogenic carbon dioxide (CO_2) emissions from Latin America and

Box 6.20 Energy supply and consumption patterns

Uneven access to energy, as well as inefficiency in energy use are still challenging sustainable development. Latin America and the Caribbean holds 22 per cent of the world's hydropower potential, 14 per cent of the capacity of the geothermal power systems installed worldwide, 11 per cent of global petroleum reserves, 6 per cent of natural gas and 1.6 per cent of coal. Countries like Argentina, Chile, Ecuador, Mexico and Venezuela depend mostly on fossil fuels, while Costa Rica and Paraguay use more renewable energy, Brazil is the world's foremost producer of biofuels (from sugar cane and soy).

Despite this abundance of energy sources, annual per capita energy consumption, at 0.88 tonnes of oil equivalent, increased only slightly over the 1987–2004 period. It is still below the world average (1.2 tonnes), and much lower than in developed regions (2.4 tonnes in Europe and 5.5 in North America). Transportation and industry are the major energy consumers. The former accounted for 37 per cent of total energy consumption, followed by 34 per cent for the industrial sector during the 1980–2004 period. Fuelwood is still an important energy source, especially in the domestic sector, although its use decreased between 1990 and 2000.

Sources: CEPAL 2005, GEO Data Portal from IEA2007, OLADE 2005, UNECLAC 2002

Brazil is the world's foremost producer of biofuels. Above, distillery in Brazil, for sugar and ethanol production.

Credit: Joerg Boethling/Still Pictures

Current environmental challenges in Latin America and the Caribbean, as well as environmental policies in many of its countries, point clearly to the fact that good governance, and land-use planning in particular, are crucial, cross-cutting issues for the 21st century.

The region has emphasized manufactured and human capital as the basis for development, disregarding natural capital (both natural resources and environmental services) as the physical basis of economic and social activities. This has led to poor urban and rural development planning, an increasing rural to urban influx, the development of social and spatial inequities, and limited institutional capacities to enforce environmental policies and regulations.

Despite these difficulties, governmental, academic and social institutions increasingly ensure that environmental issues are taken into account. Governments increasingly recognize that environmental management is closely related to issues of poverty and inequity, and that good governance should include a stable economy as an instrument for sustainable development, and not only as a goal in itself (Guimaraes and Bárcena 2002).

harbouring examples of all biomes, except tundra and taiga (although alpine tundra occurs in isolated spots). It also has the largest species diversity of the world's regions, and many of its species are endemic and hosts several of the world's greatest river basins, including the Amazon, Orinoco, Paraná, Tocantins, São Francisco and Grijalva-Usumacinta (FAO AQUASTAT 2006).

With nearly 28 000 cubic metres/person/year, per capita freshwater availability is much higher than the world average, but freshwater resources are unevenly distributed. Brazil alone has nearly 40 per cent, and areas such as the Colombian chocó receive over 9 000 mm of rainfall/year. On the other hand, almost 6 per cent of the region's land is desert, and, in some places, such as the Chihuahua or Atacama deserts, there is no appreciable precipitation. Increasing water demand and contamination, especially in and around the growing urban areas, along with inefficient water use, have progressively diminished water availability and quality. For the first time in the last 30 years, water availability has become a limiting factor for the socio-economic development of some Latin America and the Caribbean areas, particularly in the Caribbean (UNECLAC 2002).

Free access to environmental information, and widespread provision of environmental education could foster the necessary impetus and political will for improved environmental policies. Research on the environmental, social and economic dimensions of sustainability is urgently needed to support the design of policies that focus on the sustainable management of land assets, both natural and social. This is perhaps the greatest challenge facing the region.

The Tocantins River hosts a diversity of species, many of which are endemic.

Credit: Mark Edwards/Still Pictures

SELECTED ISSUES

Latin America and the Caribbean represents about 15 per cent of the world's total land area, yet it holds the largest variety of WWF defined ecoregions,

Extensive, unplanned urbanization, threats to terrestrial biodiversity and ecosystems, coastal degradation and marine pollution, and regional vulnerability to climate change are key priorities among the major

environmental issues in the region. These four selected issues are crucial to both regional and global environmental sustainability.

GROWING CITIES

Urbanization

Latin America and the Caribbean is the most urbanized region in the developing world, with an urbanization level similar to that of developed regions. Between 1987 and 2005, the urban population increased from 69 to 77 per cent of the total population (see Figure 6.30). In Guyana and St. Lucia, the urban population represents less than 28 per cent of the total, while in Argentina, Puerto Rico and Uruguay, it is over 90 per cent. The growth rate of the urban population in the region slowed from 2.8 per cent annually between 1985 and 1990 to 1.9 per cent between 2000 and 2005, (GEO Data Portal, from UNPD 2005). The mega-cities of Mexico City, Sao Paulo and Buenos Aires have about 20, 18 and 13 million inhabitants respectively. Between 1980 and 2000, they had annual population growth rates of 2, 4 and 1 per cent, respectively (WRI 2000, Ezcurra and others 2006).

Rural to urban migration

Population growth and rural to urban migration, driven by impoverishment of rural areas and lack of jobs, have transformed the settlement pattern from rural to predominantly urban in less than 50 years (Dufour and Piperata 2004). Migration has occurred at varying rates in different countries. Argentina, Chile and Venezuela urbanized relatively quickly, while the rate has been slowest in Paraguay, Ecuador and Bolivia. Today, these least-urbanized countries show the highest urbanization rates (3.3–3.5 per cent) (Galafassi 2002, Anderson 2002, Dufour and Piperata 2004). The growth rate of many major metropolitan areas, such as Mexico City, Monterrey and Guadalajara in Mexico, has declined, but intermediate-size cities are still growing, especially those with tourism and manufacturing industries (Garza 2002, Dufour and Piperata 2004, CONAPO 2004). In Peru, for example, Cuzco, Juliaca, Ayacucho and Abancay have higher growth rates than Lima (Altamirano 2003). In Brazil, more than half of the Amazon's urban dwellers do not live in the bigger cities of Manaos and Belem (Browder and Godfrey 1997). In some cases, migration from rural areas has prompted the recovery of natural forest ecosystems, a process known as forest transition (Anderson 2002, Mitchell and Grau 2004).

Urban air pollution

Urban air pollution, mainly caused by intensive fossil fuel use in the transport and industrial sectors, remains an issue (see Chapter 2). Only one-third of the region's countries have established air quality standards or emission limits. It is being monitored and better managed in some of the region's biggest cities, such as Mexico City (Molina and Molina 2002) and Sao Paulo, where conditions had long been the worst. Mexico City has completely eliminated atmospheric lead pollution, but still faces serious problems with tropospheric ozone, sulphur compounds and particulates (Bravo and others 1992, Ezcurra and others 2006). The quality of fuels (both gasoline and diesel) has gradually improved throughout the region; unleaded gasoline and diesel with lower sulphur levels are used increasingly (IPCC 2001). Bogota has reduced pollution from motor vehicles, but it still struggles to control emissions from industries in urban areas. However, air pollution is increasing in medium and smaller cities, where resources and control technologies are less available, and where urban growth management is still inadequate (UNECLAC 2002). Indoor air pollution, mainly affecting poor people who use traditional biomass for cooking and heating, has a lower profile on the urban environmental agenda.

Water, sanitation and waste collection services

Production and consumption are centralized in urban areas, affecting their surroundings through

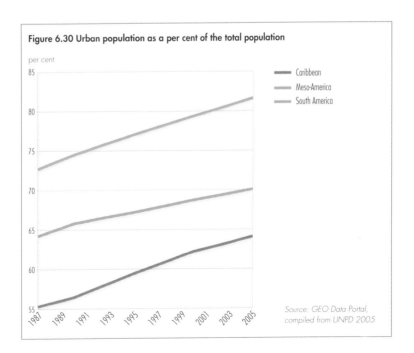

Figure 6.30 Urban population as a per cent of the total population

per cent

- Caribbean
- Meso-America
- South America

Source: GEO Data Portal, compiled from UNPD 2005

deforestation, land degradation, loss of biodiversity, soil, air and water contamination, and extraction of building materials. The generally better supply of services (such as water, energy and sewage) in urban areas contrasts with the undersupply of health, education and other social services that many (particularly poor) urban dwellers face, increasing the toll on their well-being. Urban poverty is a key issue: 39 per cent of urban families live below the poverty line, and 54 per cent of the extremely poor are urban (CEPAL 2005).

Over the course of the 20th century, water extraction (see Chapter 4) has increased by a factor of 10, and now amounts to some 263 cubic kilometres per year, with Mexico and Brazil together accounting for 51 per cent (UNECLAC 2002). Between 1998 and 2002, 71 per cent of the region's water use was for agriculture (GEO Data Portal, from FAO AQUASTAT 2007). The region's population with access to improved drinking water increased from 82.5 per cent in 1990 to 91 per cent in 2004. In the same period, access to safe water in urban areas increased from 93 to 96 per cent, and in rural areas from 60 per cent to 73 per cent. However, by 2005 nearly 50 million inhabitants of the region still lacked access to improved drinking water (GEO Data Portal, from WHO and UNICEF 2006, and UNPD 2007), with 34 million of them in rural areas (OPS 2006). The cost of water supply is rising, due to increasing demand and decreasing accessibility. In Mexico City, water imported from the Cutzamala watershed must be pumped upto 1 100 metre to reach the high altitude Basin of Mexico (Ezcurra and others 2006).

Provision of sanitation services (see Chapter 4) increased from 67.9 per cent of the region's population in 1990 to 77.2 per cent in 2004 (85.7 and 32.3 per cent in urban and rural areas, respectively). However, only 14 per cent of the sewage is adequately treated (CEPAL 2005), and in 2004 some 127 million people still lacked access to sanitation services (GEO Data Portal, from WHO and UNICEF 2006 and OPS 2006). Surface- and groundwater resources are frequently polluted with a variety of substances, including nitrates and heavy metals, but the region still lacks adequate systematic monitoring and protection of water sources. Water pollution has significant impacts on coastal areas, where some 50 per cent of the population is located (GEO Data Portal, from UNEP/DEWA/GRID-Europe 2006).

Urbanization has prompted a rapid increase in solid waste generation in the region. Municipal solid waste production increased from an estimated 0.77 kilogrammes/person/day in 1995 to 0.91 kg/person/day in 2001. On average, municipal waste still contains a high level of organic (putrescible) residues (about 56 per cent) and a moderate amount (about 25 per cent) of such materials as paper, plastics, fabric, leather and wood (OPS 2005). Formal recycling efforts are still incipient. Although 81 per cent of all municipal solid waste generated is collected, only 23 per cent is adequately disposed of. The rest is discarded in an uncontrolled manner at unofficial dump sites, in watercourses and along roadsides or is burned, polluting land, air, and waterbodies (OPS 2005).

Improving urban planning and management

During the past decade, an emerging policy response to environmental issues combines command-and-control approaches, such as regulations and standards, with economic instruments, such as those implementing the polluter-pays-principle and payment for environmental services. Several recent examples have shown, however, that privatization is not by definition the best approach to introduce such concepts as payment for water services, as it does not necessarily lead to a more sustainable and equitable use of the resource (Ruiz Marrero 2005). The potential for these policies to improve ecosystem and human well-being should be carefully evaluated. Payment schemes have no power to reverse the damage if careful urban planning is neglected.

The chaotic growth of cities, their demand for resources, and the pressure created by current production and consumption patterns should give way to a sustainable use of the resource base so as to improve people's quality of life, and meet long-sought development goals. To achieve this, the use of economic instruments and effective compliance with environmental law need to be coupled with participatory and ecologically oriented urban planning as the strategic basis for sustainability.

Several successful examples clearly demonstrate the feasibility of developing and implementing policies that address at least some of these pressing environmental problems in cities, such as urban air pollution. All are based on sounder urban environmental planning and management. For example, the integrated public

The integrated public transportation system in Curitiba, Brazil.

Credit: Ron Giling/Still Pictures

transport systems developed in Curitiba (Brazil) and Bogota (Colombia) have become a model for other large cities in the region (Mexico City, Sao Paulo and Santiago de Chile) and in Europe (Bilbao and Seville), as has the integrated programmes for air quality management implemented in major Mexican cities since the 1990s (Molina and Molina 2002). Other examples include the urban agriculture and restoration of the waterfront of Havana (a UNESCO World Heritage City), water law reforms in Chile that have improved water efficiency and wastewater treatment (Winchester 2005, PNUMA 2004, UN-HABITAT 2001), and the community-based solid waste management scheme adopted in Curitiba (Braga and Bonetto 1993).

TERRESTRIAL BIODIVERSITY

Damage to biodiversity

Latin America and the Caribbean is characterized by an extremely high biological diversity, at ecosystem, species and genetic levels. Amazonia alone is considered to have about 50 per cent of the world's biodiversity (UNECLAC 2002). Six of its countries (Brazil, Colombia, Ecuador, Mexico, Peru and Venezuela) are considered mega-diverse. Each of these countries has more species of plants, vertebrates and invertebrates than most of the nations on the planet together (Rodriguez and others 2005). The ecoregions together form a huge terrestrial corridor of 20 million square kilometres (Toledo and Castillo 1999).

This immense biodiversity is under threat due to habitat loss, land degradation, land-use change, deforestation and marine pollution (Dinerstein and others 1995, UNECLAC 2002). Eleven per cent of the region is currently under formal protection (GEO Data Portal, from UNEP-WCMC 2007). Of 178 ecoregions recognized in the region by the World Wide Fund for Nature (WWF) (Dinerstein and others 1995, Olson and others 2001), only eight are relatively intact, 27 are relatively stable, 31 are critically endangered, 51 are endangered, 55 are vulnerable, and the remaining six are unclassified. Around one-sixth of the world's endemic plants and vertebrates are threatened by habitat loss in seven regional "hot spots." Forty-one per cent of the threatened endemic plants are in the tropical Andes, some 30 per cent are in Meso-America (including the Chocó-Darién-Esmeraldas area between Panama and Colombia) and the Caribbean, and 26 per cent are in the Brazilian Atlantic Forest and Cerrado (savannah) (UNEP 2004b).

High ecological diversity is accompanied by rich cultural diversity (see Chapter 5). Over 400 different indigenous groups are estimated to live in the region – roughly 10 per cent of the total population. Frequently, they live on society's margins, and have no role in decision making at the national level. Many indigenous cultures have already disappeared, and others are on their way to extinction (Montenegro and Stephens 2006). As economics turn towards market homogeneity, cultural heterogeneity and traditional management knowledge is increasingly threatened (see Chapter 5) (see Box 6.22).

Box 6.22 Cultural diversity, traditional knowledge and trade

Indigenous and *campesino* communities have a long history of environmental management, developed in close relationship with the huge biodiversity of the region. This has led to both successes and failures in protecting environmental resources. Common property is a widespread land tenure system that offers both challenges and opportunities.

In many cases, the domestication and diversification of resources that are currently highly valued originated in indigenous communities. This knowledge is transmitted orally from one generation to the next, but as indigenous groups tend to be marginalized by migration and land-use change, it is being progressively lost. Traditional knowledge has proven to be enormously valuable, as, for example, in bioprospecting and biotechnology in recent times. Many modern drugs derive from traditional uses of plants by indigenous groups. In some cases, traditional knowledge has led to what is now recognized as sustainable environmental management. A deep understanding of this type of knowledge, and an adequate system of intellectual property rights is badly needed in the region.

Sources: Carabias 2002, Cunningham 2001, Maffi 2001, Peters 1996, Peters 1997, Toledo 2002

Indigenous people in Brazil, collecting medicinal plants. Traditional knowledge has proven to be enormously valuable in supporting livelihoods.

Credit: Mark Edwards/Still Pictures

The region contains 23.4 per cent of the world's forest cover, but is losing it rapidly. Trade, unplanned urbanization and lack of land-use planning are driving the conversion of forests to pastures for livestock production, and monoculture planted forests for crops such as corn, wheat, rice, coca and soy, for export and to produce biofuel. Forests are also cleared for infrastructure, such as roads and large dams, and the growth of urban areas. Other pressures include land speculation, wood harvesting, timber demand and forest fires (UNEP 2004b).

Some 66 per cent of the global forest cover loss from 2000 to 2005 occurred in Latin America (see Figure 6.31), while the region contains over 23 per cent of the world's forest cover (FAO 2005). South America suffered the largest net loss (almost 43 000 km²/year), of which 73 per cent occurred in Brazil (FAO 2005). Deforestation can reduce the quantity and quality of water resources, result in increased soil erosion and sedimentation of water bodies, and cause severe degradation or loss of biodiversity (McNeill 2000, UNEP 2006i). It is also an important cause of greenhouse gas emissions. Deforestation in the region is responsible for an estimated 48.3 per cent of the total global CO_2 emissions from land-use change (see Chapter 2), with nearly half of this coming from deforestation in Brazil, particularly in the Amazon

Figure 6.31 Average annual forest change

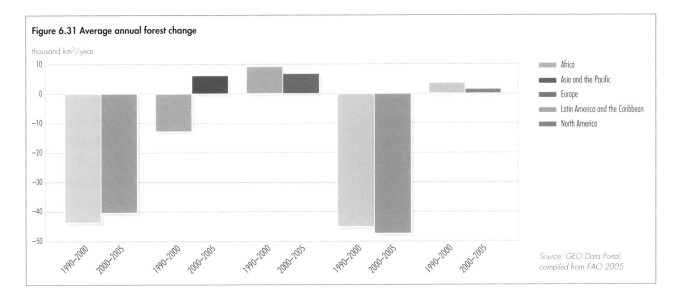

thousand km²/year

Legend:
- Africa
- Asia and the Pacific
- Europe
- Latin America and the Caribbean
- North America

Source: GEO Data Portal, compiled from FAO 2005

Basin. New efforts are being made to ameliorate this devastation. As a result of integrated prevention and control programmes, annual deforestation in the Amazon decreased from some 26 100 km² in 2004, to 13 100 km² in 2006 (INPE 2006). The so-called "Zero Deforestation Law" passed in 2004 by the Paraguayan Congress has helped to reduce the deforestation rate in Paraguay's Eastern Region by 85 per cent. Until 2004, Paraguay had one of the highest deforestation rates in the world (WWF 2006b).

Land degradation is another major environmental issue in this region (also see Chapter 3). Some 3.1 million km², or 15.7 per cent, is degraded land. The problem is more severe in Meso-America, where it affects 26 per cent of the territory, while 14 per cent of South America is affected (UNEP 2004b). Water erosion is the main cause of land degradation, while wind erosion is significant in some locations, such as the area bordering Bolivia, Chile and Argentina (WRI 1995). The mountain regions of Meso-America and the Andes are among the most seriously eroded areas in the world (WRI 1995).

Desertification affects 25 per cent of the territory due to deforestation, overgrazing and inadequate irrigation (see Chapter 3) (UNEP 2004b). Salinization of agricultural soils due to irrigation is particularly significant in Argentina, Cuba, Mexico and Peru, which have extensive dryland areas that are often subject to inappropriate use or protracted droughts (UNEP 2004b). Furthermore, the agricultural intensification is causing nutrient depletion (see Box 6.23).

Protecting terrestrial biodiversity

The area under protection (both marine and terrestrial IUCN Categories I-VI) almost doubled from 1985 to 2006, and now shields 10.5 per cent of total territory, with greater relative coverage in South America (10.6 per cent) and Meso-America (10.1 per cent) than in the Caribbean (7.8 per cent) (GEO Data Portal, from UNEP-WCMC). New efforts are being made, such as the creation of the Meso-American Biological Corridor, which extends from southern Mexico to Panama, and the Pilot Programme to Conserve the Brazilian Rain Forest. In the Amazon, seven new conservation areas have been created, totalling about 150 000 km² and including the largest (42 500 km²) strictly-protected area ever created in a tropical forest, the Grão-Pará Ecological

---

Box 6.23 Agricultural intensification in Latin America and the Caribbean

In South America, an estimated 682 000 km² are affected by nutrient loss, with about 450 000 km² affected to a moderate or severe degree. Fertility is decreasing in northeastern Brazil and northern Argentina, while other critical areas are found in Mexico, Colombia, Bolivia and Paraguay. Only 12.4 per cent of the region's agricultural land has no fertility limitations; 40 per cent of the territory has low potassium, and nearly one-third has aluminium toxicity, a condition found especially in the tropics.

In 2002, the region consumed approximately 5 million tonnes of nitrogen fertilizers, equivalent to 5.9 per cent of global consumption, of which 68 per cent was consumed by Argentina, Brazil and Mexico alone. The major environmental impact of excessive use of such fertilizers is increased nitrification of waterbodies and soil (see Chapter 3), which also affects coastal zones (see section below), drinking water supplies (see Chapter 4) and biodiversity (see Chapter 5).

Sources: FAOSTAT 2004, Martinelli and others 2006, UNECLAC 2002, Wood and others 2000

Station (Conservation International 2006, PPG7 2004). In general, "biodiversity hot spots" are poorly protected throughout the region. Protective actions and continuous efforts are needed in most hot spots, as well as in other areas rich in biodiversity.

The policy environment has changed dramatically in recent years, with an increasing mobilization of civil society to address issues such as extraction of oil and gas, water access and protection of regional biodiversity (see Chapters 3, 4 and 5). Some recent examples include the geopolitical alert over the Guarani Aquifer (one of the world's largest, encompassing 1.2 million km[2] in Brazil, Paraguay, Uruguay and Argentina) (Carius and others 2006), and the debates over the Pascua-Lama gold mining project in Chile (Universidad de Chile 2006), a new law on protected natural areas in the Dominican Republic, and the construction of pulp mills on the Uruguay River.

Biodiversity conservation and the effective enforcement of environmental laws remain policy challenges in the protection of biological resources. Current policies can impose restrictions for conservation efforts, and they should be revised at the local, national and regional levels. Local institutions and common property approaches should be considered in planning for conservation and sustainable management, while adequate funding and revenue strategies are still needed. Payment for environmental services (MA 2005) may be a crucial instrument for effectively protecting biodiversity (CONABIO 2006), and promising examples are underway in several countries, such as Mexico, Costa Rica and Colombia (Echavarria 2002, Rosa and others 2003).

DEGRADED COASTS AND POLLUTED SEAS
Coastal degradation threats
The effects of coastal degradation, and the deterioration or loss of the wide range of environmental services provided by marine and coastal ecosystems are felt inland, often a long way from the coast (UNEP 2006i) (see Chapters 3 and 4). About half of the regional population lives within 100 kilometres of the coast (GEO Data Portal, from UNPD 2005 and UNEP/DEWA/GRID-Geneva 2006). Nearly one-third of the coastline in North and Central America and about half of that in South America are under moderate to high threats from the impacts of development. As a result,

mangrove losses range from 67.5 per cent in Panama and 36 per cent in Mexico, to 24.5 per cent in Peru, while Costa Rica recorded a gain of 5.9 per cent (Burke and others 2001, FAO 2003b). The growth of aquaculture and shrimp farming has also contributed to mangrove damage (UNEP 2006i). The destruction of these ecosystems has increased risks for coastal populations and infrastructure (Goulder and Kennedy 1997, Ewel and others 1998).

The Caribbean Sea provides many ecosystem services, such as fisheries and recreation opportunities, and attracts about 57 per cent of the world's scuba diving tours (UNEP 2006i). Between 1985 and 1995, 70 per cent of monitored beaches in the eastern Caribbean islands had eroded, indicating a loss of shoreline protection capacity and increased vulnerability to erosion and storm effects (Cambers 1997). In the Caribbean as a whole, 61 per cent of the coral reef area is under medium or high threat from sediment, marine and land-based sources of pollution, as well as from overfishing (Bryant and others 1998). Coastal groundwater contamination (including saltwater intrusion) is occurring throughout the region, at great economic cost (UNEP/GPA 2006a).

The region's oceans face a number of threats, including eutrophication caused by land-based sources of nutrient pollution, unplanned urbanization, lack of sewage treatment, salinization of estuaries due to decreased freshwater flow, unregulated ballast waters from ships and invasive of alien species (UNEP 2006i, Kolowski and Laquintinie 2006).

Specific threats to the region's marine waters include:
- Some 86 per cent of the sewage goes untreated into rivers and oceans; in the Caribbean, the figure rises to 80–90 per cent (OPS 2006, UNEP/GPA 2006a).
- There is elevated oil pollution from refineries in the Greater Caribbean and the Gulf of Mexico, and from deep offshore drilling in the Gulf of Mexico and off Brazil. Oil spills are a serious problem in the Gulf of Mexico (Beltrán and others 2005, Toledo 2005, UNEP/GPA 2006a).
- Agrochemical run-off is also important, and highly toxic concentrations have been found in estuaries in the Caribbean, Colombia and Costa Rica (PNUMA 1999).

- With nearly a tripling in the volume of marine cargo in the region between 1970 and 2004 (UNCTAD 2005), marine transport is an important source of pollution.
- Hazardous waste, including radioactive materials from other regions, is shipped around South America or through the Panama Canal, and heavy metals pollute the Gulf of Mexico (Botello and others 2004).
- Many invasive alienspecies (crustaceans, land molluscs and insects) inadvertently introduced through freight and ballast, have inflicted important economic damage to infrastructure and crops (Global Ballast Water Management Programme 2006).
- Overfishing is a major source of concern, particularly in the Caribbean, where pelagic predator biomass appears to have been depleted (see Box 6.24).

Degraded coastal waters put people's health at risk. Cholera and other water-borne diseases are on the rise in coastal areas, and can be related to declining water quality, climate changes and eutrophication-driven algal blooms. These blooms (including red tides) have caused neurological damage and death in people through consumption of affected seafood (UNEP 2006i). Cholera increases the rates of sickness and death, and has a severe economic impact in coastal regions. For instance, tuna from countries with incidences of cholera has to be quarantined. Human health effects are also caused by nearshore water pollution, as people consume fish or other seafood containing heavy metals and other toxins that have bioaccumulated in the food chain (see Chapter 4) (Vázquez-Botello and others 2005, UNEP 2006i).

There has been nearly a tripling in the volume of marine cargo in the region. Above, container and cranes at the freight port of Panama City.

Credit: Rainer Heubeck/Das Fotoarchiv/Still Pictures

The regional marine catch peaked in 1994, accounting for close to 28 per cent of the global catch. Harvests in Peru and Chile, which account for most of this catch, had doubled or tripled in the preceding decade. Regional fisheries fell to 50 per cent of that level in 1998, but recovered to 85 per cent of the 1994 level by 2000. The greatest impact of these fluctuations has been on small pelagics (anchovy, sardine and mackerel)

that predominate the regional catch in the Humboldt Current Large Marine Ecosystem, along the west coast of South America (see Figure 6.32). In 2001, regional aquaculture represented 2.9 per cent of the world's total volume and 7.1 per cent of the value. It was concentrated in Chile (51 per cent) and Brazil (19 per cent), but was developed at the expense of mangroves, estuaries and salt marshes.

Sources: Sea Around Us 2006, UNEP 2004b

Figure 6.32 Catch of major groups of fish and invertebrates in the Humboldt Current Large Marine Ecosystem

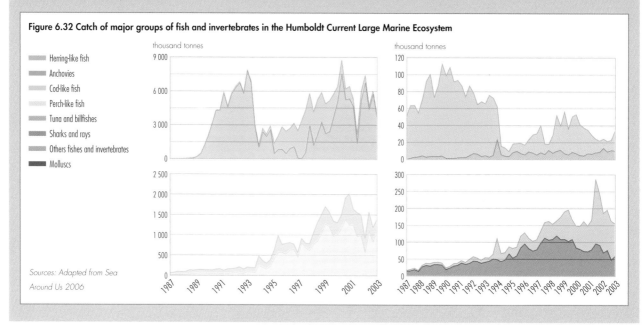

Herring-like fish
Anchovies
Cod-like fish
Perch-like fish
Tuna and billfishes
Sharks and rays
Others fishes and invertebrates
Molluscs

Sources: Adapted from Sea Around Us 2006

Responses to marine and coastal pollution

Most regional and sub-regional responses are related to the UNEP Regional Seas programme, the UN Convention on the Law of the Sea, international conventions on maritime transport and conventions on fisheries (UNEP 2004b, UNECLAC 2005). Only a few countries have ratified global agreements against illegal exploitation of highly migratory fish populations.

Regional Seas Programmes exist for the North East Pacific, the South East Pacific and the Wider Caribbean. All programmes are underpinned by regional conventions: the 1981 Convention for the Protection of the Marine Environment and Coastal Zones of the South-East Pacific (Lima Convention), the 2002 Convention for Cooperation in the Protection and Sustainable Development of the Marine and Coastal Environment of the North-East Pacific (The Antigua Convention) and the 1983 Convention for the Protection and Development of the Marine Environment of the Wider Caribbean Region (Cartagena Convention). Protocols to address specific problems (such as mitigating and preventing pollution from land-based

activities, radioactive pollution, oil spills, and protected areas and wildlife) have been adopted in the South East Pacific and the Wider Caribbean, while the North East Pacific programme is still at an early stage, seeking financial support for implementing its action plan (UNEP/GPA 2006b). The effectiveness of these programmes still remains to be assessed. In general, there is inadequate use of economic instruments, making compliance dependent on limited monitoring resources.

However, integrated marine and coastal areas management is gaining ground, with growing protection for marine areas, and increasing efforts to establish marine protected areas, such as the whale sanctuary in the marine areas of Mexico, established in 2002 (SEMARNAT 2002). But, more focus is needed on the integration of coastal area and inland river basin management (ICARM) as a key response to coastal and marine pollution (see Chapters 4 and 5). The Global Environment Facility (GEF) and the UNEP/GPA Secretariat are supporting this approach, as well as the integrated management of shared living marine resources in the Caribbean.

REGIONAL VULNERABILITY TO CLIMATE CHANGE

Extreme climate related events

Findings by the Intergovernmental Panel on Climate Change (IPCC) indicate that impacts of global warming in Latin America and the Caribbean include rising sea levels, higher rainfall, increased risk of drought, stronger winds and rain linked to hurricanes, more pronounced droughts and floods associated with El Niño events, decline in water supplies stored in glaciers and declines in crop and livestock productivity (IPCC 2007a). Tropical rain forests in Meso-America and the Amazon basin, mangroves and coral reefs in the Caribbean and other tropical areas, mountain ecosystems in the Andes, and coastal wetlands are some of the ecosystems more vulnerable to the effects of climate change (IPCC 2007b). Small island states are particularly extreme cases, as they may be affected by surface warming, droughts and reduced water availability, floods, erosion of beaches and coral bleaching, all of which might affect local resources and tourism (IPCC 2007b). Furthermore, changes in El Niño behaviour may be associated with increasingly severe and more frequent extreme climate events (Holmgren and others 2001).

Extreme climate events affecting the region have already increased over the last 20 years. The number, frequency, duration and intensity of tropical storms and hurricanes in the North Atlantic Basin have increased since 1987 (see Figure 6.33). The 2005 season was the most active and the longest on record, with 27 tropical storms, 15 of which became hurricanes. An unprecedented four of these storms reached category five on the Saffir-Simpson scale, with Wilma being the most intense in history (Bell and others 2005). Hurricanes Jeanne and Ivan in September 2004, and Dennis in July 2005, had severe impacts on Caribbean islands, killing 2 825 people and affecting just over 1 million more (EM-DAT). Hurricane Stan in October 2005 left 1 600 dead and 2.5 million people affected in Haiti, Central America and Mexico (EM-DAT).

In South America, flooding and landslides had huge impacts in the 2000–2005 period, including 250 dead and 417 500 affected in Bolivia (EM-DAT). Economic damage is increasing, partly due to more people being exposed. From 1997–2006, it doubled in Central America and the Caribbean, and grew by

50 per cent in South America (as compared to the preceding decade) (EM-DAT). Poverty and settlement on vulnerable sites, such as coastal areas and marginal zones, expose people to increased risks from flooding, landslides and other hazards. In addition, natural and social conditions throughout the region increase the risk of exposure to infectious diseases, such as malaria or dengue fever, which are in turn exacerbated by climate change (see Box 6.25).

Between 2000 and 2005, droughts caused serious economic losses for more than 1.23 million people in Bolivia, Brazil, Cuba, El Salvador, Guatemala, Honduras, Haiti, Jamaica, Mexico, Nicaragua, Paraguay, Peru and Uruguay (EM-DAT). In 2003 and 2004, the Amazon River experienced its lowest water level in a decade, and Cuba had only 60 per cent of its average rainfall (INSMET 2004, UNEP 2006f).

The loss of Latin American glaciers is particularly dramatic evidence of climate change: the Andean ridge and Patagonia in Argentina are showing evidence of glacier retreat, and a reduction in snow-covered zones (see Figure 6.35). In Peru, the Andean glaciers of Yanamarey, Uruashraju and Broggi are diminishing in size, while the Antisan glacier in Ecuador retreated eight times faster during the 1990s than in previous decades. In Bolivia, the Chacaltaya glacier lost over half of its area since 1990 (CLAES 2003). Glacier loss in the Andes, and saltwater intrusion from sea-level rise will affect the availability of drinking water, and may also affect agricultural production and tourism.

Mitigation and adaptation to climate change

A lack of adaptation capabilities increases the severity

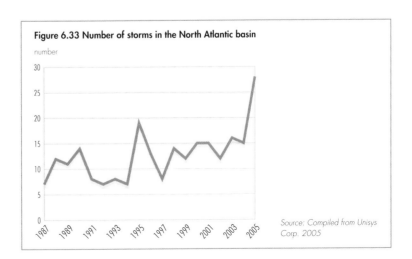

Figure 6.33 Number of storms in the North Atlantic basin

number

Source: Compiled from Unisys Corp. 2005

of climate change impacts (Tompkins and Adger 2003). Latin America and the Caribbean, particularly its small island states, is especially vulnerable to the effects of climate change, such as sea-level rise and extreme events (IPCC 2007b). The region lacks basic information, observation and monitoring systems, capacity building and appropriate political, institutional and technological frameworks. It has generally low income, and many settlements are in vulnerable areas. Under the United Nations Framework Convention on

Box 6.25 Health, climate and land-use change: re-emerging epidemics

Rising temperatures, land cover modification, changing precipitation patterns and shrinking health expenditures are behind the re-emergence of epidemics once under control in Latin America and the Caribbean. Changes related to the El Niño-Southern Oscillation tend to increase the geographic distribution of disease vector organisms, and result in alterations to the life cycle dynamics and seasonal activity of vectors and parasites. This amplifies the risk of transmission for many vector-borne diseases, such as malaria, dengue fever, yellow fever and bubonic plague. A reinfestation of the *Aedes aegypti* mosquito, responsible for

the transmission of yellow fever and dengue fever (see Figure 6.34), and is thought to be related to climate change.

Both too much and too little precipitation can lead to faecal-oral transmission infections, such as cholera (as in Honduras, Nicaragua and Peru in 1997 and 1998), typhoid fever and various diarrhoeas. Flooding can contaminate water with human waste, while a lack of water means less for hygiene. The loss of vegetation and the occurrence of extreme weather events facilitate the contamination of water, and an increase in pests.

Sources: Githeko and others 2000, Hales and others 2002, McMichael and others 2003, UNEP 2004b, WHO 2006

Figure 6.34 Re-infestation by *Aedes aegypti* in Latin America and the Caribbean

1970

2002

Figure 6.35 Retreating glacier zone in the border area between Argentina and Chile:

a) 1973

b) 2000

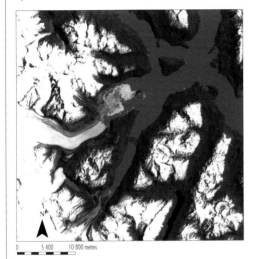

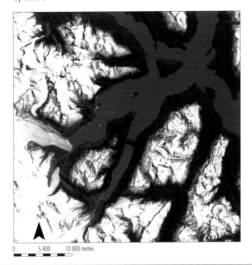

Source: Compiled from
Lansat.org 2006

Climate Change (UNFCCC), countries in the region agreed to mitigation and adaptation activities in the energy, transport, agricultural and waste management sectors, and to increase the capacity of carbon sinks (Krauskopfand Retamales Saavedra 2004, Martínez and Fernández 2004).

NORTH AMERICA
DRIVERS OF CHANGE
Socio-economic trends

Over the past 20 years, North America (Canada and the United States) has continued to enjoy generally high levels of human, economic and environmental well-being. Canada and the United States rank sixth and eighth respectively on the 2006 Human Development Index (UNDP 2006). Since 1987, the region's total population grew, mostly due to immigration, by 23 per cent to almost 339 million in 2007, with 90 per cent in the United States (GEO Data Portal, UNPD 2007). There has been strong growth in per capita GDP (see Figure 6.36). The ratio between energy use and GDP continued a slow but positive decline that began in 1970, reflecting a shift to less resource-intensive production patterns, although the region remains among the most energy-intensive in the industrialized world. The two countries (along with Mexico) became more economically integrated following the 1994 North American Free Trade Agreement (NAFTA). The combination of a growing population and economy has many implications for environment and development. It leads to increases in energy use, anthropogenic greenhouse gas (GHG) emissions and the overuse of the planet's resources.

Energy consumption

With only 5.1 per cent of the world population, North America consumes just over 24 per cent of global primary energy. Per capita energy consumption in both countries is shown in Figure 6.37. Total energy consumption grew 18 per cent. The US transport sector consumes 40 per cent of total energy used (see Figure 6.38), making it the dominant energy user in the region. Total transport energy consumption rose by 30 per cent between 1987 and 2004 (GEO Data Portal, from IEA 2007). More big, less fuel-efficient vehicles and a rise in number of vehicles and distances travelled contributed to this trend.

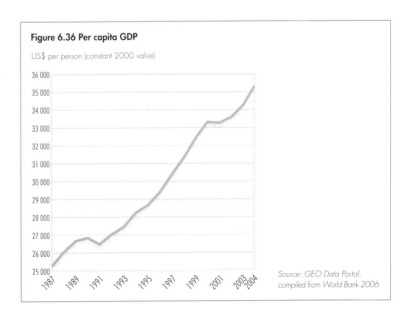

Figure 6.36 Per capita GDP

US$ per person (constant 2000 value)

Source: GEO Data Portal,
compiled from World Bank 2006

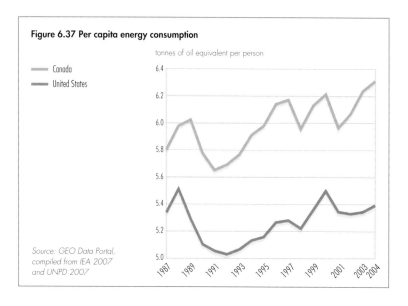

Figure 6.37 Per capita energy consumption

tonnes of oil equivalent per person

Canada

United States

Source: GEO Data Portal, compiled from IEA 2007 and UNPD 2007

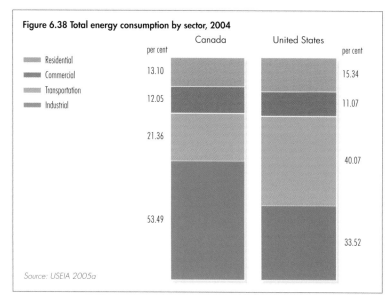

Figure 6.38 Total energy consumption by sector, 2004

Residential
Commercial
Transportation
Industrial

Canada

per cent

13.10
12.05
21.36
53.49

United States

per cent

15.34
11.07
40.07
33.52

Source: USEIA 2005a

Environmental governance

Building on the solid foundation of environmental legislation from the 1970s, the two countries strengthened domestic policies, and entered into important bilateral and multilateral environmental agreements over the past two decades. Following the report of the World Commission on Environment and Development in 1987, sustainable development was adopted in principle, and governments are integrating it into policy and governing structures. Market-based programmes have been effective, especially in controlling SO_x emissions, and, as a result, have been applied widely in North America and elsewhere. New modes of paying for ecosystem services are providing increased incentives for controlling pollution and conserving natural resources.

They have made important strides in transboundary cooperation to address common environmental issues, through organizations such as the Commission for Environmental Cooperation (CEC), the International Joint Commission (IJC) and the Conference of Eastern Canadian Premiers and New England Governors. The CEC, which includes Mexico, has a mechanism for citizen submissions that has enabled the public to play an active whistle-blower role when a government appears to be failing to enforce its environmental laws effectively. Both Canada and the United States are federal systems, where decisions are often made at local or regional levels. States and provinces, municipalities and other local actors have become especially progressive in addressing environmental issues. North America is a leader in research in environmental science and ecology, developing state of environment reporting, integrating the public into environmental decision making and providing access to timely information on environmental conditions.

The United States leads the world in the production of goods-and-services that mitigate or prevent environmental damage (Kennett and Steenblik 2005). The growth engine of North America also helps drive creation of employment and wealth in most parts of the world. As in other regions, the transition from a growth only, heavy polluting model to a sustainable development one is still in process, with much more to be done.

SELECTED ISSUES

North American priority issues identified for this report through regional consultation are energy and climate change, urban sprawl, and freshwater quality and quantity. The analyses illustrate how atmospheric and water pollution, as well as sprawl, have direct impacts on both ecosystems and human health, and can cause economic, social and cultural impacts and disruptions. As in the rest of the world, the most vulnerable populations suffer the greatest impacts.

ENERGY AND CLIMATE CHANGE

Although total energy consumption has increased since 1987 (see above), progress has been made in using energy more efficiently. Energy consumption per unit of GDP dropped since the 1980s, reflecting the rising importance of the service sector, and information and communication technologies, which use less energy to create wealth than do heavy industries. Investments in energy efficiency have proven to be economically and environmentally beneficial (see Box 6.26). Improved energy efficiency can also be attributed in part to the outsourcing of some production activity, relocating energy use and its impacts in other parts of the world (Torras 2003).

Energy production

The United States and Canada exhibit similar energy consumption patterns, which have hardly changed over the past 15 years and are dominated by just over 50 per cent petroleum products. They do differ in energy production, though. While total energy production grew in both countries, oil production declined in the United States (see Figure 6.39), resulting in a growing dependence on imported oil. Driven by demand for transportation fuel, the rising price of crude oil, uncertainty about supplies and a favourable fiscal regime, heavy investments were made in Canada, doubling oil production from oil sands to about 150 000 tonnes/day between 1995 and 2004. Production may reach some 370 000 tonnes/day by 2015, with an associated doubling of GHG emissions (Woynillowicz and others 2005).

Extracting oil from the sands involves the use of large amounts of natural gas and water. It results in substantial releases of GHGs, the disposal of dangerous tailings and wastewater, and the radical alteration of landscapes and damage to boreal forests, which threatens wildlife habitat and requires extensive reclamation. Improved environmental performance of oil sands production is likely to be eroded by a huge increase in development (Woynillowicz and others 2005).

Heavy reliance on imported fossil fuels heightened concerns over US energy security during the last decade (see Chapter 7). Canada's concern focuses on the impacts that US demand for Canadian energy will have on its own energy supply, and on the environment. Canada is the most important source of US oil imports, and over 99 per cent of Canada's crude oil exports go to the United States (USEIA 2005b). To help meet US demand, oil and gas exploration in that country increased dramatically over the past 20 years; from 1999 to 2004, the number of drilling permits more than tripled (GAO 2005).

Coal-bed methane production has increased in both countries since the mid-1990s. It is still in its infancy in Canada, but it accounted for about 7.5 per cent of

Box 6.26 Energy efficiency makes economic sense

A 13 per cent improvement in national energy consumption per unit of GDP between 1990 and 2003 saved Canadians almost US$7.4 billion in energy costs in 2003, and reduced annual GHG emissions by an estimated 52.3 million tonnes.

The US Energy Star programme, a voluntary labelling programme promoting energy-efficient products and practices, prevented 35 million tonnes of GHG emissions, and saved about US$12 billion in 2005 alone.

Sources: NRCan 2005, USEPA 2006a

Oil sand mining has major negative environmental impacts.

Credit: Chris Evans,
The Pembina Institute
http://www.OilSandsWatch.org

Figure 6.39 Energy production by fuel type

1987

Crude oil, natural gas liquids and feedstocks 31.3% | Coal and coal products 28.2% | Natural gas 24.9%

Nuclear 7.9%
Combustible renewables and waste 4.5%
Hydro 2.6%
Geothermal 0.5%

2005

Crude oil, natural gas liquids and feedstocks 23.1% | Coal and coal products 29.3% | Natural gas 28.4% | Nuclear 11.6%

Combustible renewables and waste 4.3%
Hydro 2.7%
Geothermal 0.4%
Solar/wind/other 0.1%

Source: GEO Data Portal, compiled from IEA 2007

total US natural gas production in 2000. The heavy sodium content of water drawn off in this process can pollute surface- and groundwater used for drinking and irrigation (USEIA 2005b). The exploration for underground deposits, drilling of closely-spaced wells to recover coal-bed methane, and the new infrastructure for fossil fuel exploration, production and distribution have significant impacts. The fragmenting and damaging of wilderness areas (USEPA 2003), increasing air pollution, pipeline leaks and oil transport spills all pose important environmental and health threats (Taylor and others 2005). There is increasing recognition of the human health costs of fossil fuel combustion (see Box 6.27).

Greenhouse gas emissions and climate change

The energy sector is a major CO_2 emitter (see Figure 6.40). The United States emits 23 per cent and Canada 2.2 per cent of global energy-related GHG emissions (USEIA 2004). Fossil fuel combustion accounts for 98 per cent of total US CO_2 emissions. From 1987 to 2003, CO_2 emissions from fossil fuels in North America increased 27.8 per cent, and per

capita emissions per unit of GDP remain high relative to other industrialized regions (GEO Data Portal, from UNFCCC-CDIAC 2006). The transport section is a major emitter of GHGs; in 2005, it accounted for 33 per cent of total US energy-related CO_2 emissions (USEIA 2006b). An emerging issue is the rise in emissions from air travel (see Chapter 2).

The IPCC Fourth Assessment in 2007 stated with very high confidence that climate change is human-induced through GHG emissions (see Chapters 2, 4, 5 and the Polar section of this chapter), and this will have important consequences for human health (see Box 6.28) (see Chapter 2). With its large output of GHGs, North America has an impact on the changing climate in other parts of the world, disproportionately affecting poor and more vulnerable countries and people (IPCC 2007b).

Responding to climate change

Since the 1990s, North American governments focused on market-based, voluntary and technological measures to address climate change. When it ratified the Kyoto Protocol in 2002, Canada committed to a 6 per cent reduction in GHG emissions below the 1990 level by between 2008 and 2012. The United States, which has signed but not ratified the protocol, called for an 18 per cent reduction in emissions relative to economic output by 2012 compared to 2002 (The White House 2002). From 1992 to 2003, emissions of CO_2 rose by 24.4 and 13.3 per cent for Canada and the United States respectively (UNFCCC 2005).

In 2006, the Canadian government introduced a new Clean Air Act as the centrepiece of a "green"

Box 6.27 Fossil fuels and human health in North America

Burning fossil fuels in power plants and vehicles is the major source of CO_2, SO_2 and NO_x emissions. There are clear associations between exposure to air pollutants and a range of human health problems. In the early part of this decade, air pollution caused the premature deaths of an estimated 70 000 people a year in the United States and some 5 900 in Canada. It is known to exacerbate asthma, which is on the rise, especially among children. Mercury emitted when coal is burned in power plants enters the food chain, affecting indigenous people in the North more than other North Americans (see Chapter 2 and the Polar Regions section in this chapter). It can have severe health effects.

Sources: CEC 2006, Fischlowitz-Roberts 2002, Judek and others 2005

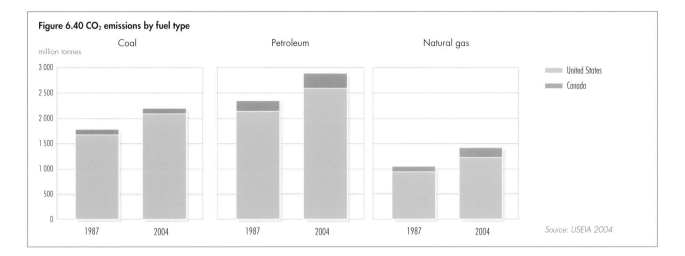

Figure 6.40 CO₂ emissions by fuel type

million tonnes

Coal   Petroleum   Natural gas

United States
Canada

1987 2004 1987 2004 1987 2004

Source: USEIA 2004

agenda. If passed into law, short-term targets for GHG emissions will be set based on intensity (which encourages efficiency, but allows emissions to increase if output grows). In 2007, a new regulatory framework called for an absolute reduction of 150 megatonnes (or 20 per cent below the 2006 level) by 2020 (Environment Canada 2007).

In 2005, the United States initiated a national energy plan. It provides support for the fossil fuel industry, and contains incentives for research and development of cleaner fuels, renewable energy, especially biofuels and hydrogen, and increased energy conservation and efficiency, among other measures (The White House 2005). In 2006, the United States joined five other countries to form the new Asia-Pacific Partnership for Clean Development and Climate, a US-led effort to accelerate the voluntary development and deployment of clean energy technologies. It exemplifies US support for public-private partnerships that replicate successful energy projects globally. Canada has expressed

an interest in participating in this partnership. Both countries' long-term policies include adaptation strategies to cope with climate change impacts (Easterling and others 2004, NRCan 2004). Many states, municipalities, the private sector and other actors, took a variety of significant and innovative steps since the late 1990s to address GHG emissions (see Box 6.29).

Box 6.28 Potential human health impacts of climate change

The major health threat of warmer temperatures is the likely increase in more intense and prolonged heat waves that can cause dehydration, heat stroke and increased mortality. Depending on the location, climate change is expected to increase smog episodes, water- and food-borne contamination, diseases transmitted by insects (such as Lyme disease, West Nile Virus and Hantavirus pulmonary syndrome), and the intensity of extreme weather events (hurricanes such as Katrina, which devastated the northern coast of the Gulf of Mexico in August 2005). Children, the elderly, the poor, disabled people, immigrant populations, aboriginal people, those who work outdoors and people with already compromised health would suffer disproportionately.

Sources: Kalkstein and others 2005, Health Canada 2001

Box 6.29 States, provinces, municipalities and businesses take climate change action

Over the past 20 years, there has been a lack of recognition of the need to protect the environment as the foundation upon which development takes place. The focus has been on development first, often at the expense of the environment. However, leadership in promoting sustainability is emerging from states and provinces, municipalities, cross-border organizations, and the voluntary and private sectors. The following are examples of commitments at different governance levels to mitigate climate change:

■ In 2006, California, the 12th largest carbon emitter in the world, passed the first bill in the United States to cap CO₂ emissions. Many other states have committed to initiatives, including carbon sequestration, GHG-trading, Smart growth, climate action plans and

renewable energy portfolio standards (RPS), that require electric utilities to offer some renewable energy. Well over half of the American public now lives in a state in which an RPS is in operation.

■ At the bilateral level, the New England Governors and Eastern Canadian Premiers adopted a Climate Action Plan in 2001.

■ At the city level, 158 US mayors and the 225-member Federation of Canadian Municipalities agreed to reduce GHG emissions.

■ Numerous leading North American companies have adopted a variety of climate change initiatives.

■ In 2006, a coalition of 86 evangelical leaders committed themselves to influencing their congregations to limit GHGs.

Sources: ECI 2006, FCM 2005, Office of the Governor 2006, Pew Center on Global Climate Change 2006, US Mayors 2005

Concerns about energy security are helping foster an energy transition to replace what has been called America's "addiction to oil" (The White House 2006), but it is not clear if the transition will be to a low-energy economy and lifestyle. Political support and financial incentives led to record growth in wind energy, ethanol and coal production over the past five years (RFA 2005, AWEA 2006, NMA 2006), and there has been renewed interest in nuclear energy. Since 2000, biomass has been the largest renewable energy source in the United States. Sales of hybrid cars also grew over the past several years, although North America is behind many other industrialized regions in promoting and using fuel-efficient alternatives to the internal combustion engine (Lightburn 2004). North America's earlier successes in arresting air pollution and acid rain have become models for other regions.

SPRAWL AND THE URBAN–RURAL INTERFACE
Urban expansion

Sprawl was an issue in GEO-3, and continues to be one of the most daunting challenges to environmental quality in North America. Permissive land-use planning and zoning, and the growth of affluent populations facilitated urban sprawl. Houses and lots have become bigger, while the average number of people per household has fallen (DeCoster 2000). Sprawl has contributed significantly to increases in the number

of cars, vehicle kilometres travelled and the length of paved roads in North America over the past 20 years. Urban development accounts for less than 1 per cent of Canada's land area (OECD 2004) and 3.1 per cent of US land area (Lubowski and others 2006).

Urban sprawl, a settlement pattern and process on the outskirts of urban areas characterized by low-density housing, continued unabated over the past 20 years. By 2000, it was expanding at twice the rate of population growth in the United States (HUD 2000), and Canada now has 3 of the world's 10 most sprawling urban areas (Calgary, Vancouver and Toronto) (Schmidt 2004). In the United States, coastal areas represent only 17 per cent of the US land base, but contain more than half the US population (Beach 2002). Sprawl is increasing, and can stretch 80 km inland.

Rural or exurban sprawl has expanded more than other settlement patterns over the past decade, and is an increasing threat to natural (and protected) areas and their ecosystem services. Exurban sprawl is defined as clusters of large lot, low-density housing developments beyond the urban fringe, separated by natural areas and with extended commuting distance to urban areas (Heimlich and Anderson 2001). Between 1990 and 2000, the exurban population of the 22 states west of the Mississippi River increased by 17.3 per cent (Conner and others 2001). Population growth in

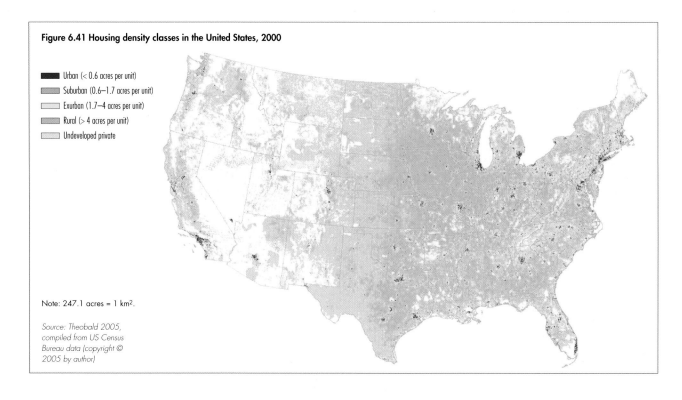

Figure 6.41 Housing density classes in the United States, 2000

- Urban (< 0.6 acres per unit)
- Suburban (0.6–1.7 acres per unit)
- Exurban (1.7–4 acres per unit)
- Rural (> 4 acres per unit)
- Undeveloped private

Note: 247.1 acres = 1 km².

Source: Theobald 2005, compiled from US Census Bureau data (copyright © 2005 by author)

California's Central Valley, which supplies the United States with one-quarter of its foodstuffs, is threatening valuable agricultural land (Hammond 2002).

Exurban expansion, as well as commercial and energy developments close to protected areas, threatens their integrity (Bass and Beamish 2006). In 2000, urban and suburban settlements occupied almost 126 000 square kilometres in the United States, while exurban housing accounted for over seven times that (some 11.8 per cent of the United States mainland) (Theobald 2005) (see Figure 6.41). There was strong growth in rural areas of the Rocky Mountains, southern states and the California interior (OECD 2005) (see Box 1.9 in Chapter 1). Exurban sprawl also characterized most of the increase in Canada's rural population from 1991 to 1996, especially in the west (Azmier and Dobson 2003). Where developments press up against open spaces, they create an urban-rural interface (URI), where social and ecological systems overlap and interact (Wear 2005).

Expansion of exurban sprawl and the URI are associated with the fragmentation and loss of forests, prime agricultural land (see Chapter 3), wetlands and other resources, such as wildlife habitats and biodiversity (see Box 6.30 below). Of the more than 36 400 km² of land developed in the United States between 1997 and 2001, 20 per cent came from cropland, 46 per cent from forest land and 16 per cent from pastureland (NRCS 2003). In Canada, about half the area converted to urban uses over the past 30 years was good agricultural land, meaning land where crop production is not constrained (Hoffmann 2001).

In both countries, grasslands are also being lost and fragmented, leading to drastically altered native prairie landscapes, biodiversity loss and introduced species. The North American Central Grasslands are considered to be among the most threatened ecosystems, both on the continent and in the world (Gauthier and others 2003). Finally, almost half the annual net loss of wetlands from 1982 to 1997 in the United States was due to urban development (NRCS 1999).

Wildfire is a positive agent of natural disturbance in many forest ecosystems, but the increased intermingling of housing with flammable forests and grasslands contributed to a rise in the number of "interface" fire incidents over the past decade (Hermansen 2003, CFS 2004). Interface fires destroy property, threaten human health and wildlife, and can foster invasive species and insect attacks. Such fires are not as severe in Canada, but they still affect thousands of people, and have large economic costs, and the risks are growing (CFS 2004).

Wild elk in the residential area of Gardiner, Montana.

Credit: Jeff and Alexa Henry/ Still Pictures

Increased exurban development affects freshwater in several ways. Impervious surfaces channel water into drains and sewers instead of replenishing groundwater, and suburban run-off contains a host of pollutants (Marsalek and others 2002). Furthermore, the expanding URI is also creating more opportunities for off-road vehicle recreation, which is an emerging source of habitat fragmentation, increased erosion, water degradation, and noise and air pollution, especially in the United States (Bosworth 2003). Although the desire for a healthy environment is a key incentive to move to the suburbs, some health threats occur more in suburbs than in areas with less sprawl (see Box 6.31).

Box 6.31 Sprawl and human health

Traffic deaths and injuries, and illness associated with higher ozone levels occur more in sprawling suburbs than in denser settlements.

Suburban areas are less conducive to walking than are compact neighbourhoods, and lack of exercise can contribute to weight gain and associated health problems, such as diabetes.

Expansion of the URI has led to greater human exposure to diseases and infections transmitted between animals and humans, such as Lyme disease, which is on the rise in the United States.

Sources: Ewing and others 2005, Frumkin and others 2004, Robinson 2005

Smart policy response to sprawl

North America has made important strides in reducing forest, grassland and wetland losses to urban and suburban development over the past 20 years, through public and private preservation, mitigation and restoration programmes.

A number of states, provinces and municipalities have designed and implemented Smart Growth (see UNEP 2002), and other strategies that include a wide variety of policy tools to manage sprawl (Pendall and others 2002). One definition of Smart Growth is about 48 people/hectare, a density deemed conducive to public transit (Theobald 2005). The features of Smart Growth are meant to reduce the environmental impact of human settlements and travel, preserve farmland and green space and their ecosystem services, and increase "liveability." Organizations representing many sectors of society have endorsed the principles of Smart Growth (Otto and others 2002).

Nationally, the United States encourages sustainable urban development through the Smart Growth Network, the Liveability Agenda and the National Award for Smart Growth Achievement (Baker 2000, USEPA2004, SGN 2005). Between 1997 and 2001, 22 US states enacted laws to curb sprawl (El Nasser and Overberg 2001). In Canada, the 30-year plan of the Transportation Association of Canada led to the inclusion of sprawl control in the master plans of most Canadian cities (Raad and Kenworthy 1998). Canada's 2002 Urban Strategy, its 2000 Green Municipal Fund and the 2005 New Deal for Cities and Communities all support sprawl control in various ways (Sgro 2002, Government of Canada 2005). There is a lack of information about the effectiveness of sprawl control, but Bengston and others (2004) found that policy implementation, packages of complementary policy instruments, vertical and horizontal coordination, and stakeholder participation are vital elements for success.

Urban air pollution policy should include such integrated policy packages. Over the past two decades, a number of emissions declined, due to a variety of controls, including clean air legislation, voluntary and regulatory acid rain programmes, and transboundary air quality agreements. Both North American countries now have comparable criteria for air quality (CEC 2004), and real-time air quality information is provided on the Internet. Both countries introduced regulations to reduce emissions from new diesel vehicles, starting in 2007 (Government of Canada 2005, Schneider and Hill 2005), and to reduce mercury emissions from coal-burning power plants (CCME 2005, USEPA2005a). These controls should help to bring down concentrations of traditional urban air pollutants, which remain high compared to other developed regions (OECD 2004).

FRESHWATER

Water supply and demand

North America possesses about 13 per cent of the world's renewable freshwater (GEO Data portal, from FAO AQUASTAT 2007), but despite the apparent abundance, users are not always close to water sources, and some experience periodic water deficits (NRCan 2004). Limited water supplies have led to increased competition for water in parts of western North America (see Box 6.32), the Great Plains (Bails and others 2005) and the Great Lakes basin. Droughts can increase the stress. A severe drought during 2000–2005 affected large areas of North America from the US southwest to Canada's Atlantic provinces (Smith 2005).

Glaciers and snowpacks, a major source of the Canadian Prairies' water, are declining (Donahue and Schindler 2006), and hydrological variability is expected to worsen with climate change, exacerbating competition for water among agriculture, the oil and gas industry, and municipalities. The Prairie Provinces have responded by adopting watershed planning and management strategies (Venema 2006).

The United States and Canada respectively are the two highest per capita water users in the world (see Figure 6.42). One of the key reasons is its low cost, the lowest among the world's industrialized countries, given the subsidies to industry, agriculture and municipalities. Another reason is that North America is a net exporter of food, and thus the world's biggest exporter of "virtual water," which is the water contained in the food (International Year of Freshwater 2003). Since the mid-1990s, some municipalities in both countries introduced limited water metering, and water restrictions in times of shortage. An emerging concern is the loss of municipal water from leaking pipes, which ranges up to 50 per cent in some places, due to ageing infrastructure (Environment Canada 2001, CBO 2002).

Figure 6.43 shows the major categories of water users in both countries. Agriculture accounts for 39 per cent of North America's annual water withdrawal (GEO Data Portal, from FAO 2007a). The United States has over 75 per cent of North America's irrigated cropland, and between 1995 and 2000, this area increased by nearly 7 per cent.

Groundwater demand in North America increased over the past 20 years. Irrigation in drought-prone regions of the United States is responsible for unsustainable withdrawal of water from aquifers, at rates as much as 25 per cent higher than natural replenishment (Pimentel and Pimentel 2004). The impacts of overdrafting aquifers include land subsidence, saltwater intrusion in coastal areas and loss of aquifer capacity (see Chapter 4, Table 4.1). Although groundwater data in Canada is limited, studies suggest that most aquifers are not yet threatened by overdrafting (Nowlan 2005).

Box 6.32 Water shortages in the North American west

With a yearly average rainfall of less than 10.2 centimetres, the western United States is one of Earth's driest regions, but it is home to about one in five US citizens. The Colorado River drains almost 627 000 km², and is completely allocated to providing water for more than 24 million people, irrigating about 8 100 km² of agricultural land and generating 4 000 megawatts of hydropower (see Chapter 4). A trickle now feeds the once-fertile delta at the river's mouth.

In the early 1990s, a water market developed, allowing rapidly growing municipalities to purchase water rights from farmers and ranchers. The United States introduced a number of strategies to forestall conflicts, including water conservation and efficiency, and collaboration.

Sources: Cohn 2004, Harlow 2005, Saunders and Maxwell 2005

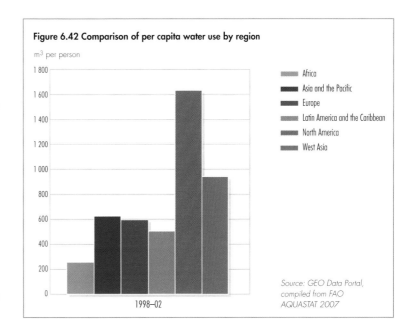

Figure 6.42 Comparison of per capita water use by region

m³ per person

- Africa
- Asia and the Pacific
- Europe
- Latin America and the Caribbean
- North America
- West Asia

1998–02

Source: GEO Data Portal, compiled from FAO AQUASTAT 2007

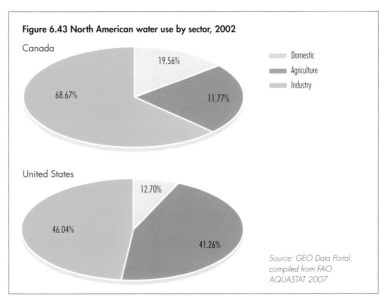

Figure 6.43 North American water use by sector, 2002

Canada

- Domestic
- Agriculture
- Industry

19.56%
68.67%
11.77%

United States

12.70%
46.04%
41.26%

Source: GEO Data Portal, compiled from FAO AQUASTAT 2007

In the United States, water efficiency improved due to water conservation strategies, supported since 2002 by the Farm Bill (NRCS 2005). By 2004, the area irrigated with sprinkler and micro-irrigation systems had grown to more than half the total irrigated land (Hutson and others 2004).

Water quality

Overall, drinking water in North America is the cleanest in the world, but some places in the region have water of lower quality (UNESCO 2003) (see Box 6.33). Measures and definitions of water quality differ between the two countries, making an overall assessment of the region difficult. Preliminary indicators show that Canada's freshwater is "good" or "excellent" at 44 per cent of selected sites, "fair" at 31 per cent, and "marginal" or "poor" at 25 per cent (Statistics Canada 2005). Studies using different measures show that about 36 per cent of US watersheds have moderate water quality problems, 22 per cent have more serious problems and 1 in 15 watersheds is highly vulnerable to further degradation (USEPA 2002). A recent study revealed that 42 per cent of the nation's shallow streams were in poor environmental condition (USEPA 2006b).

The primary causes of degradation are agricultural run-off, sewage treatment plant discharges and hydrologic modifications (see Figure 6.44 and Box 6.33). Significant gains have been made in protecting water quality from point sources of pollution, while non-point contamination, especially from agriculture, the largest source of freshwater impairment, has become a priority in both countries.

Confined (or concentrated) animal feeding operations (CAFOs), which have increased in size, scale and geographic clustering over the past 20 years, are a growing source of non-point nutrient pollution (Naylor and others 2005). When improperly managed, nutrients from manure enter waterbodies and groundwater. Nutrient management plans now require farmers to adhere to certain guidelines to control run-off, but in 2001, only 25 per cent of manure-producing farms had such plans (Beaulieu 2004). The US Clean Water Act regulates waste discharges from livestock systems, while states can impose either stricter or more lenient restrictions (Naylor and others 2005). CAFOs consume large amounts of water, and are under increasing pressure to conserve it (NRCS 2005).

About 40 per cent of major US estuaries are highly eutrophic due to nitrogen enrichment: agricultural fertilizer accounts for about 65 per cent of the nitrogen entering the Gulf of Mexico from the Mississippi Basin (Ribaudo and Johansson 2006). It contributes to the formation of the world's second-largest aquatic "dead zone" (after the Baltic Sea) (Larson 2004). A 2000 Action Plan aims to reduce the average size of the Gulf of Mexico dead zone by half by 2015 (see Chapter 4).

Chesapeake Bay is also subject to nutrient problems and associated large algal blooms that kill fish

---

Box 6.33 Drinking water, wastewater treatment and public health

North American drinking water can contain contaminants from municipal and industrial wastewater effluents, sewer overflows, urban run-off, agricultural wastes and wildlife. Pathogens in drinking water have been responsible for numerous health-related incidents in the region, and its waters also contain pharmaceuticals, hormones and other organic contaminants from residential, industrial and agricultural origins.

Canada
Treatment of effluents in municipalities has improved since 1991, but a number of serious health incidents related to contaminated water in the early part of this decade affected thousands of people, and prompted provinces to increase groundwater monitoring, and adopt better methods to adhere to national guidelines. Many coastal communities still discharge insufficiently treated sewage, and a number of indigenous communities have inferior drinking water and sewage services compared to other Canadians. Overflows from combined sewer and storm-water systems are leading causes of water pollution. Provinces and some municipalities have wastewater and sewage standards, and enforce federal guidelines, but Canada has no nationally enforced drinking water standards. In 2006, with 193 of 750 systems in First Nations communities classified as high risk, the federal government launched a plan of action to address their drinking water concerns.

United States
Concentrations of contaminants rarely exceed US drinking water standards, but guidelines for some compounds have not yet been established, and the interactive effects of complex mixtures are still uncertain. The United States experienced some 250 disease outbreaks and nearly 500 000 cases of water-borne illness from polluted drinking water between 1985 and 2000. In 2005, the United States amended the Safe Drinking Water Act to reduce the entry of microbial contaminants and the health risks from disinfection by-products. Some 3.5 million US residents get sick each year from exposure to pollution from sewer spills and overflows while swimming, boating and fishing. The US Beach Act of 2000 requires states to adopt US Environmental Protection Agency (USEPA) health standards to protect people from harmful pathogens, and information on beach quality is provided in real time. The US Clean Water Act requires all cities to have secondary sewage treatment, and new measures to control storm-water discharge were introduced in the 1990s. However, a 2005 study found that more than half the Great Lakes' municipalities were violating these rules, and ageing infrastructure is an emerging and costly problem.

Sources: American Rivers 2005, Boyd 2006, Environment Canada 2003, EIP 2005, USEPA 2005b, INAC 2006, Kolpin and others 2002, Marsalek and others 2001, OECD 2004, Smith 2003, Surfrider Foundation 2005, Wood 2005

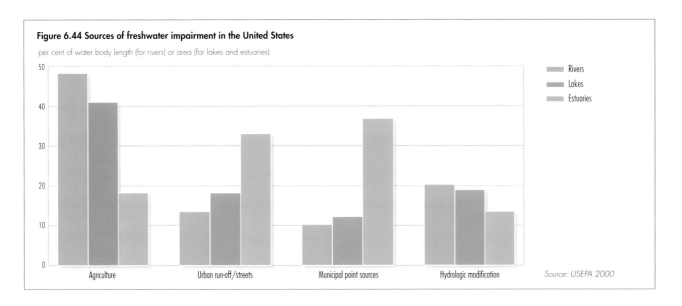

Figure 6.44 Sources of freshwater impairment in the United States

per cent of water body length (for rivers) or area (for lakes and estuaries)

Legend:
- Rivers
- Lakes
- Estuaries

Categories (x-axis): Agriculture, Urban run-off/streets, Municipal point sources, Hydrologic modification

Source: USEPA 2000

and destroy shellfish habitat. Despite programmes in place since 1983, the ecosystem has become seriously degraded as population growth increases pressures (CBP 2007 2004). Dead zones also occur in freshwaters, such as Lake Erie's hypoxic zone, which has expanded since 1998, harming the lake's food web (Dybas 2005). The region has instituted innovative transboundary, multistakeholder and multi-level policy measures to address this and other water problems (see Box 6.34) (see Chapter 4).

In Canada, increased fertilizer use, livestock numbers and manure application also contribute to increased nitrogen contamination in lakes and rivers (Eilers and Lefebvre 2005), including Lake Winnipeg, where phosphorus loadings increased by 10 per cent over the last 30 years, posing a severe threat to the lake's ecological balance (Venema 2006). A 2003 action plan, and Manitoba's 2006 Water Protection Act aim to reduce phosphorus and nitrogen loadings. Nitrate

approx 800 km

Summertime satellite observations of ocean colour in the Gulf of Mexico show highly turbid waters which may include large blooms of phytoplankton extending from the mouth of the Mississippi River all the way to the Texas coast. Reds and oranges represent low oxygen concentrations.

Credit: NASA/Goddard Space Flight Center Scientific Visualization Studio www.gsfc. nasa.gov/topstory/2004/ 0810deadzone.html

concentrations in Canada's agricultural streams have contributed to declines in some amphibian populations (Marsalek and others 2001).

At present, all levels of governments are increasingly attempting to address watersheds as integrated systems through Integrated Water Resource Management (IWRM), comprehensive river basin management and other approaches. The adoption of city and community-based management and restoration strategies is also growing (Sedell and others 2002) (see Chapter 4). For example, New York City's investment in land conservation in the Catskills-Delaware watershed to protect the natural filtration capacity of the city's water source is helping to reduce the cost of its water treatment plants (Postel 2005).

WEST ASIA
DRIVERS OF CHANGE
Socio-economic trends

The 12 countries of West Asia are divided into two sub-regions: the Arabian Peninsula, including the Gulf Cooperation Council (GCC) and Yemen; and the Mashriq, including Iraq, Jordan, Lebanon, Occupied Palestinian Territories (OPT) and Syria.

Although the region made notable progress towards achieving the Millennium Development Goals (MDGs) in health, education and empowerment of women (UNEP, UNESCWA and CAMRE 2001), some 36 million people over 18 years of age (32 per cent of the total population) are still illiterate, including 21.6 million women (UNESCWA 2004). Poverty in the region has been rising since the 1980s, ranging from almost no poverty in Kuwait to

as much as 42 per cent of the population in Yemen (UNESCWA 2004, World Bank 2005a, World Bank 2005b). It is possible for the GCC countries to achieve the MDGs by 2015, but doubtful for the Mashriq and Yemen, and impossible for Iraq and the OPT (UN 2005b).

While the region experienced a large increase in its human development scores between 1960 and 1990, it registered very little progress thereafter (UNDP 2001). People endure low levels of freedom at the family, tribal, traditional, social and political levels, and the majority of countries still lack political institutions and modernized constitutions and laws that protect individual freedoms and human rights (UNDP 2004). There is some indication, however, of a slow and gradual democratization process, which may lead to greater accountability.

Countries in each sub-region have responded differently to socio-economic and geopolitical changes since 1987. The exploitation of natural resources, and sustained population and urban growth remain the major factors in all West Asian economies. Agriculture is the main economic activity in the Mashriq and Yemen, contributing an average 30 per cent to GDP, and employing more than 40 per cent of the work force, whereas oil is the major source of income in the GCC, representing about 40 per cent of GDP, and 70 per cent of government revenues (UNESCWA and API 2002).

West Asia's high dependency on natural resources has made it very vulnerable to economic shocks and fluctuations in international prices, with profound repercussions on growth, employment and economic stability, as well as on the environment. A clear example is the decline of oil prices in the 1980s that plunged the region into a decade of macro-economic instability, characterized by rising debts, high unemployment rates and acute balance of payment difficulties.

With the economic reforms of the late 1980s and early 1990s, and the temporary recovery of oil markets, the region witnessed some economic stabilization during the 1990s, reflected in reduced inflation rates, lower public and foreign deficits and a noticeable increase in investments (World Bank 2003b), but there were limited impacts on economic growth. Population increases likely eroded

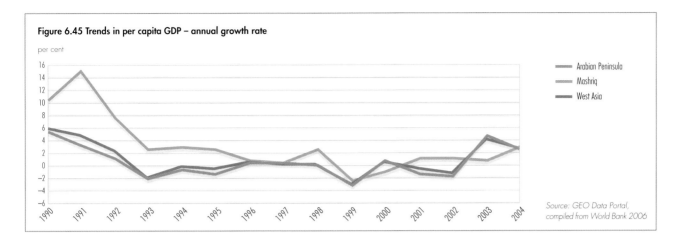

Figure 6.45 Trends in per capita GDP – annual growth rate

per cent

Arabian Peninsula
Mashriq
West Asia

Source: GEO Data Portal,
compiled from World Bank 2006

economic achievements by 2002 (see Figure 6.45). Nevertheless, starting in 2002, and with sharp increases in oil prices, growth picked up significantly, particularly in the GCC countries, which experienced remarkable capital inflows and rising investment levels (World Bank 2005a, UNESCWA 2004).

Recent developments, such as trade agreements and partnerships with the European Union and the United States, are expected to contribute to economic growth and development in the region. Despite these positive developments, demographic and employment pressures will continue to constitute a core development problem, and result in major challenges in the future. Although declining, population growth rates are still close to 3 per cent. On average, the population is 63 per cent urban (GEO Data Portal, from UNPD 2005), and unemployment rates rise above 20 per cent (UNESCWA 2004). Political instability, the shattered economies of Iraq and the OPT, and associated disruptions and sharp declines in growth pose further challenges.

Environmental governance
Significant effort has been made in environmental governance since the World Commission on Environment and Development report. Countries have actively begun enacting environmental regulations, and setting up a variety of local, national and international environmental institutions (UNESCWA 2003a). National environmental strategies and action plans were prepared, and some countries are preparing sustainable development strategies. However, there is still governmental reluctance to implement integrated environmental, economic and social decision making. Governments still routinely conceive and implement economic development

programmes on a sectoral basis, without considering their environmental and social causes, contexts and implications.

The creation of effective institutions, capacity building, and strict environmental legislation and enforcement are urgently required to discourage the relaxation of environmental protection. Regional cooperation and coordination among West Asian countries to manage shared marine and water resources, mitigate the impacts of transboundary environmental problems and enhance regional environmental management capacity should be a priority. Finally, socio-economic integration in the region has the potential to alleviate population pressures on development and the environment.

SELECTED ISSUES
The region's environment is predominantly dryland, with great variability in rainfall within and between seasons, and frequent spells of drought, making water the most precious resource. Poor resource management over several decades has resulted in widespread land and marine degradation. Population growth and changes in consumption patterns have made urbanization a major environmental issue. A protracted history of wars and conflicts has placed peace and security at the centre of environmental concerns. Five regional priority issues were selected: freshwater stress, land degradation, marine and coastal degradation, urban management, and peace and security.

FRESHWATER
Overexploitation of water
West Asia is one of the world's most water-stressed regions. Between 1985 and 2005, its overall per capita freshwater availability fell from 1 700 to

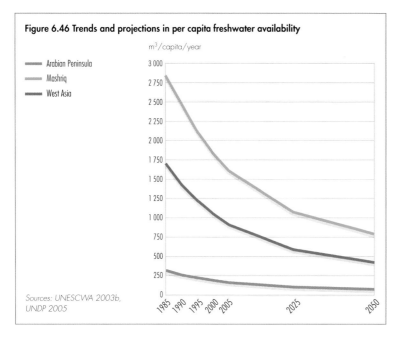

Figure 6.46 Trends and projections in per capita freshwater availability

m³/capita/year

Legend:
- Arabian Peninsula
- Mashriq
- West Asia

y-axis values: 3 000, 2 750, 2 500, 2 250, 2 000, 1 750, 1 500, 1 250, 1 000, 750, 500, 250, 0

x-axis values: 1985, 1990, 1995, 2000, 2005, 2025, 2050

Sources: UNESCWA 2003b, UNDP 2005

907 cubic metres/year (see Figure 6.46). Based on projected population increases, it is expected to decline to 420 m³/year by the year 2050.

The Mashriq relies mainly on surface water, and, to a lesser extent, on groundwater, while the Arabian Peninsula relies on renewable and non-renewable groundwater and desalinated water. Both regions are increasingly using treated wastewater. Since more than 60 per cent of surface water originates outside the region, the issue of shared water resources is a major determinant of regional stability. Riparian countries have not signed agreements on equitable sharing

and management of water resources. Groundwater overexploitation and the continued deterioration in the quality of limited surface- and groundwater, as a result of industrial, domestic and agricultural effluents, aggravate water scarcity, and affect human health and ecological systems (see Chapter 4).

Rapid urbanization, particularly in the Mashriq and Yemen, challenges efforts to meet increasing domestic water demands with scarce public funds. Municipal water consumption escalated from 7.8 billion m³ in 1990 to about 11 billion m³ in 2000, a 40 per cent increase, a trend expected to persist (UNESCWA 2003b). Although most people have access to improved drinking water and sanitation, these services are not always reliable, especially in lower-income areas. Such water shortages are a problem in key cities like Sana'a, Amman and Damascus (Elhadj 2004, UNESCWA 2003b).

In the GCC countries, rapid population growth and urbanization, and the rise in per capita water consumption explain the current alarming increase in urban water demand. With an average consumption range of 300–750 litres/person/day, GCC residents rank among the highest per capita water users in the world (World Bank 2005c). Key reasons include the absence of proper demand management and price signalling mechanisms. Government policies have primarily focused on the supply side of water production from aquifers or desalination plants. Water tariffs are generally quite low, representing an average of no more than 10 per cent of the cost, and therefore there are no incentives for consumers to save water.

Although urban demand is high, the agricultural sector consumes most water, accounting for more than 80 per cent of total water used (see Figure 6.47). During the past few decades, economic policies favouring food self-sufficiency and socio-economic development have prioritized the development and expansion of irrigated agriculture. Agricultural water use increased from about 73.5 billion m³ in 1990 to more than 85 billion m³ in the 1998–2002 period (UNESCWA 2003b), exerting immense pressures on the region's limited water resources (see Box 6.35). Although many countries recently abandoned such policies, agricultural water consumption is expected to increase, and problems in allocating water among agricultural, domestic and industrial sectors will worsen.

Figure 6.47 Current and projected water demand in West Asia

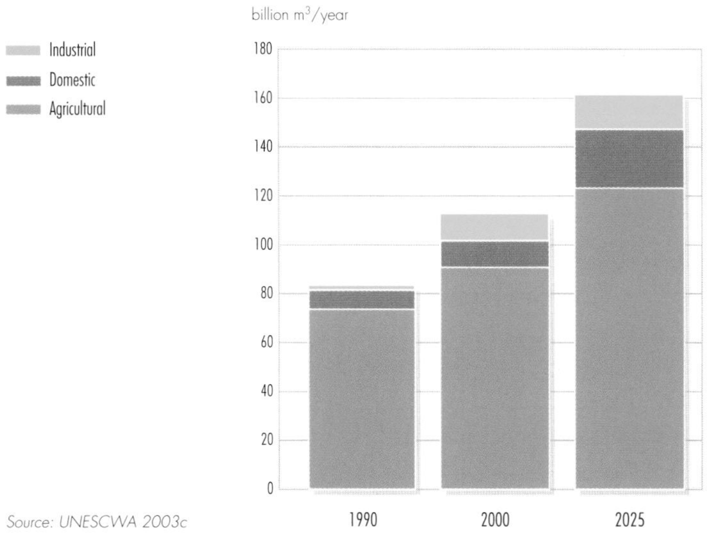

billion m³/year

Legend:
- Industrial
- Domestic
- Agricultural

y-axis values: 180, 160, 140, 120, 100, 80, 60, 40, 20, 0

x-axis values: 1990, 2000, 2025

Source: UNESCWA 2003c

Over the past three decades, economic policies and generous subsidies in most of the GCC countries supported the expansion of irrigated agriculture in an effort to achieve food security. Irrigation water is often used inefficiently, and without considering the economic opportunity cost for potable water and urban or industrial demands. Agriculture contributes less than 2 per cent to GDP in GCC countries, but it overexploits groundwater resources, most of which are non-renewable, resulting in their depletion and quality deterioration due to seawater intrusion and the upflow of saltwater. No clear "exit strategy" exists to address the question of what happens when the water is gone.

Source: Al-Zubari 2005

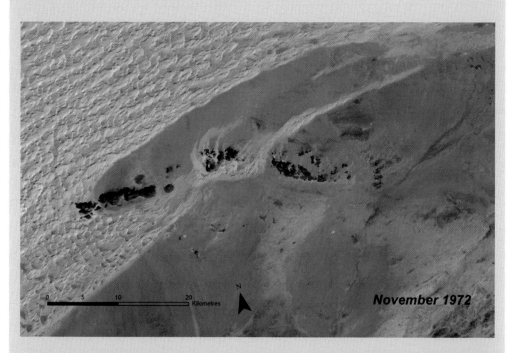

November 1972

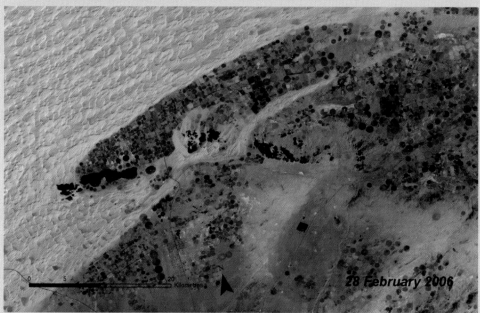

28 February 2006

Agricultural expansion based on fossil groundwater in Saudi Arabia. The bright circles are the areas that are irrigated by large sprinklers.

Credit: UNEP/GRID-Sioux Falls

In the Mashriq, the health impacts of poor water quality are a major concern (see Box 6.36). The main causes are the use of untreated domestic wastewater for irrigation, poor sanitation and inadequate waste management (UNESCWA 2003c). Furthermore, overexploitation of groundwater has caused many natural springs to dry up, resulting in destruction of their surrounding habitats, as well as the loss of their historical and cultural values. An example is the drying up of most of the historical springs in the Palmyra oasis in Syria, including Afka, around which the historical Kingdom of Zanobia was developed (ACSAD 2005).

Box 6.36 The health impacts of water pollution

A 2002–2003 pilot project assessing impacts of groundwater nitrate contamination in the Mashriq countries confirmed it as a serious source of illness in infants. In general, most small villages in the region lack adequate wastewater disposal systems, and rely on individual household cesspits. This contributes to the contamination of groundwater, often a source of untreated drinking water. Extensive use of manure as fertilizer aggravates the problem as run-off seeps into aquifers. Nitrate causes methemoglobaenemia (blue baby syndrome) in infants, a condition that can result in death or retardation.

Source: UNU 2002

Towards sustainable water resources management

The supply driven approach to water management has not delivered a substantial degree of water sustainability or security. Recently, most countries have shifted towards more integrated water management and protection approaches. Water sector policy reforms focus on decentralization, privatization, demand management, conservation and economic efficiency, improved legal and institutional provisions, and public participation (UNESCWA 2005). Very few countries have completed and integrated these strategies into social and economic development frameworks yet because existing institutional capacities are still inadequate (UNESCWA 2001).

In addition, improved population and agriculture policies are crucial to the sustainable management of water resources. The absence of agreements regulating shared surface- and groundwater resources among riparian countries, and lack of financing (mainly in Mashriq countries) constitute major challenges for the region.

LAND DEGRADATION AND DESERTIFICATION
Land quality

Sixty-four per cent of West Asia's 4 million square kilometre land base is drylands (Al-Kassas 1999) on calcareous soils prone to degradation. Just over 8 per cent of the land is cultivated, but historically this has provided the population with ample food, with few adverse environmental impacts. Over the past 20 years, however, a 75 per cent increase in the population (GEO Data Portal, from UNPD 2007) has increased the demand for commodities and land. This was accompanied by the intensive use of inappropriate technology, poor regulation of common property resources, ineffective agricultural policies and rapid, unplanned urban development. These pressures resulted in widespread land-use changes, land degradation and desertification (which is land degradation in drylands, see Chapter 3) in most of the countries.

Wind erosion, salinity and water erosion constitute the major threats, while soil waterlogging, fertility degradation and soil crusting are secondary problems. At the beginning of the 21st century, 79 per cent of the land was degraded, with 98 per cent of that being caused by people (ACSAD and others 2004). The reasons include inadequate land resource policies, centralized governance, lack of public participation, low-profile expertise, and an arbitrary and isolated single discipline approach to planning and management.

Soil degradation and food security

The expansion of cultivated and irrigated land (see Figure 6.48), intensified mechanization, modern technology, the use of herbicides, pesticides and fertilizers, and the expansion of greenhouses and aquaculture resulted in a significant increase in agricultural production. Irrigated lands increased from 4.4 to 7.3 million hectares from 1987 to 2002 (GEO Data Portal, compiled from FAOSTAT 2005). Despite increases in food production, however, the trade deficit continues to increase, threatening food security. Poor management and irrational use of irrigation water has increased salinity and alkalinity (see Box 3.5 in Chapter 3), which affects about 22 per cent of the region's arable land (ACSAD and others 2004). Economic losses from the effects of salinization are expected to be significant (World Bank 2005c).

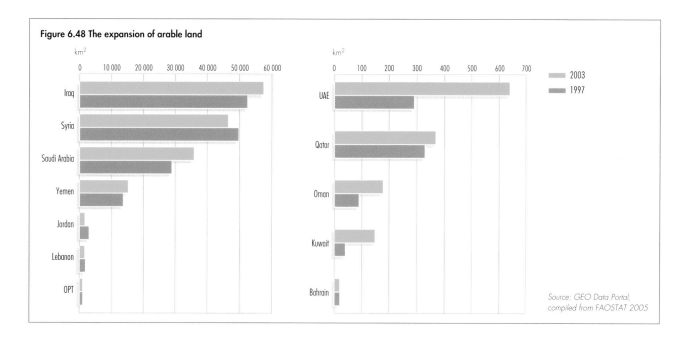

Figure 6.48 The expansion of arable land

Legend:
- 2003
- 1997

Left chart (km², scale 0 to 60 000):
- Iraq
- Syria
- Saudi Arabia
- Yemen
- Jordan
- Lebanon
- OPT

Right chart (km², scale 0 to 700):
- UAE
- Qatar
- Oman
- Kuwait
- Bahrain

Source: GEO Data Portal, compiled from FAOSTAT 2005

Rangelands and soil erosion

Rangelands occupy over 52 per cent of the total land area (GEO Data Portal, compiled from FAOSTAT 2005). The carrying capacity changes annually, depending on the distribution and amount of rainfall. Annual production of animal dry feed is estimated as low, ranging from 47 kilogrammes/ha in Jordan to 1 000 kg/ha in Lebanon (Shorbagy 1986). This indicates a substantial feed gap, although there has been no significant change in the number of standard livestock units (250 kg) since 1987, estimated at 14.6 million (FAOSTAT 2005). West Asia is prone to drought, frost and excessive heat, so vegetation diversity is essential, because it enhances plant cover resilience. Biodiversity is declining, however, due to pressures on forests, woodlands and rangelands.

The impacts of heavy and early grazing, rangeland cultivation and recreational activities have significantly reduced species diversity and density, and increased soil erosion and sand dune encroachment on agricultural lands (Al-Dhabi and others 1997). Observations of plant cover change show that vegetation in drylands may extend approximately 150 kilometres further in a rainy year than in a preceding dry year (Tucker and others 1991). In the period 1985–1993, the land area covered by sand increased by about 375 km² in the Al-Bishri area of Syria (ACSAD 2003), while the size of the dune fields nearly doubled in 15 months north of Jubail, in eastern Saudi Arabia (Barth 1999). Between 1998

and 2001, overgrazing and fuelwood gathering decreased rangeland productivity by 20 per cent in Jordan and 70 per cent in Syria (ACSAD and others 2004) (see Chapter 3).

Forests

Forests occupy 51 000 km² of West Asia, or only 1.34 per cent of the region's total area (GEO Data Portal, from FAO 2005), and account for less than 0.1 per cent of the world's total forested area. Forest degradation occurs widely. Fires, wood cutting, overgrazing, cultivation and urbanization all negatively affect the products and services of the forests (FAOSTAT 2004). There have been no major changes, though, in the total extent of forest area in the last 15 years, because deforestation in some parts was balanced by afforestation in others. Between 1990 and 2000, the forest cover even increased on average by 60 km²/year in the Arabian Peninsula, but remained stable between 2000 and 2005. In the Mashriq, the rate of increase due to afforestation programmes has been 80 km²/year since 1990, and continues to date (GEO Data Portal, from FAO 2005). The main challenges and constraints facing sustainable forest management are weak institutions and law enforcement, unfavourable land tenure practices, climatic and water limitations, lack of technical personnel and agricultural extension services, insufficient financial resources and policy failures (UNEP, UNESCWA and CAMRE 2001).

Mitigating land degradation

National action plans (NAPs) to combat desertification contain well-defined measures to mitigate land degradation and protect threatened areas (ACSAD and others 2004). Countries with completed NAPs (Jordan, Lebanon, Oman, Syria and Yemen), and others still in the process will need to accelerate implementation to stem desertification. The region's countries have joined international efforts to conserve biodiversity, and most have ratified the Convention on Biological Diversity and its biosafety protocol, and have joined FAO's International Treaty on Plant Genetic Resources for Food and Agriculture. More

Measures to combat desertification can transform bare land (top photo, taken in 1995) into areas with good plant cover (bottom photo, taken in 2005). This area in Al-Bishri, Syria received comparable amounts of annual precipitation and spring rain in both years.

Credit: Gofran Kattash, ACSAD

intensive efforts are required, though, to improve the understanding of ecosystem dynamics, and to develop more efficient and sustainable production systems, including integrated forest management programmes.

However, in many countries these plans are not mainstreamed into national development policies. The interaction between land degradation issues and poverty is routinely ignored, leading to irrelevant and ineffective policies. Despite governmental efforts to prevent and reduce land degradation at the national and regional levels, only limited success has been achieved, mostly due to the severity of the problems. More extensive cooperative and participatory efforts are urgently needed.

There already are considerable practical efforts being made to improve degraded lands, such as introducing water-efficient irrigation and agricultural techniques (Al-Rewaee 2003), rehabilitating degraded rangelands (see photos), increasing the area under protection (see Figure 6.49) and afforestation projects. These efforts, however, cover only 2.8 and 13.6 per cent of degraded lands in the Arabian Peninsula and the Mashriq, respectively (ACSAD and others 2004). After a clear increase between 1990 and 1995, the total protected area remained the same, indicating the need for intensified and integrated efforts to enlarge these projects.

Governments have only recently recognized the ecological importance of forests, and are now conserving forest ecosystems and biological diversity, for example through forest reserves and ecotourism. Large water impoundments in Syria, Jordan and Iraq have also created new habitats for resident and migrating species, especially birds. Significant achievements are the restoration of Mesopotamian marshlands of Iraq, in the Eden Again Project 2004 (see Chapter 4, Figure 4.12), and the preservation of local wheat varieties in Jordan and Syria (Charkasi 2000, ICARDA 2002, Iraq Ministry of Environment 2004, UNEP/PCAU 2004).

COASTAL AND MARINE ENVIRONMENTS
Coastal development
Coastal and marine areas in West Asia are threatened by rapid coastal development of residential towns, resorts and recreational projects (see Chapter 4). Land reclamation, oil pollution, chemical contamination and overfishing are also factors.

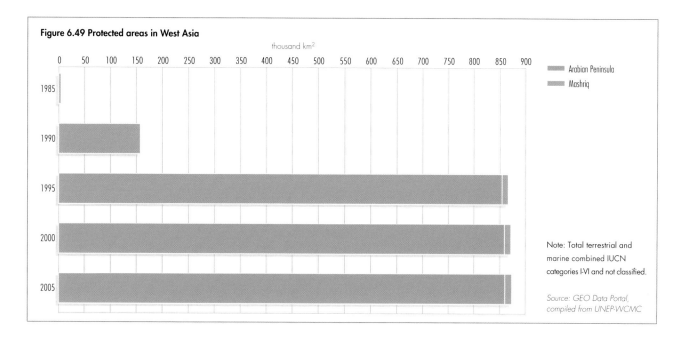

Figure 6.49 Protected areas in West Asia

thousand km²

Legend: Arabian Peninsula, Mashriq

Note: Total terrestrial and marine combined IUCN categories I-VI and not classified.

Source: GEO Data Portal, compiled from UNEP-WCMC

Dredging for urban and transport developments has caused extensive coastline alteration. By the early 1990s, some of the GCC countries had developed 40 per cent of their coastlines (Price and Robinson 1993). The coastal zone in Bahrain increased by about 40 km² in less than 20 years (ROPME 2004). Similarly, since 2001 more than 100 million m³ of rock and sand have been used in the Palm Islands on the coast of Dubai Emirate, United Arab Emirates (UAE) to increase the shoreline by 120 km (DPI 2005, ESA 2004). Over 200 million m³ of dredged sediments were used for Jubail Industrial City, Saudi Arabia (IUCN 1987), and the 25-km causeway connecting Bahrain and Saudi Arabia used about 60 million m³ of dredged mud and sand.

Industry, agriculture, livestock production, and food and beverage processing are the major sources of organic carbon loads and oxygen-demanding compounds discharged to the marine environment in the ROPME Sea Area (RSA), which includes the coasts of the eight Member States of the Regional Organization for the Protection of the Marine Environment (ROPME 2004). Water discharged directly from desalination plants contributes brine, chlorine and thermal pollution, as well as micro-organisms that may include pathogenic bacteria, protozoa and viruses (WHO 2000b).

Oil spills and chemical contamination are other major threats to the region's marine environment, including the Mediterranean countries of the region

(see Box 6.46). Eight refineries and more than 15 petrochemical complexes are located along the coast of the RSA, and more than 25 000 tankers, carrying about 60 per cent of total global oil exports, pass through the Strait of Hormuz annually (ROPME 2004). Ballast waters spill about 272 000 tonnes of oil in the RSA every year (UNEP 1999). Wars and military conflicts contributed additional oil spills and chemical contamination (ROPME 2004).

Jordan's coral reefs, the tidal coral reef terraces in Lebanon and Syria (Kouyoumjian and Nouayhed 2003), and the great diversity of endemic organisms in Yemen and the RSA are at risk unless protected and well managed, while coastal erosion everywhere continues to be a threat. The degradation and loss of coral reefs (see Box 6.37), and the decline in the Dead Sea's water level are additional serious problems affecting marine and coastal environments.

Box 6.37 Coral reef degradation and bleaching

There are more than 200 species of corals in the Red Sea, and 60 species in the RSA. Human activities and other factors account for the continuous degradation of corals in the region. Climate change caused major coral bleaching in the RSA and Red Sea during 1996 and 1998, and mortality of Acropora corals reached 90 per cent.

Sources: PERSGA 2003, Riegl 2003, ROPME 2004, Sheppard 2003, Sheppard and others 1992

Coastal development has exerted significant pressures on fisheries. Contaminants, high temperatures, disease agents and biotoxins accounted for fish mortality in the RSA from 1986 until 2001, resulting in considerable economic losses to the fishing industry and local fishers (ROPME 2004). Furthermore, population growth has led to a gradual decrease in annual per capita fish catches, especially in the RSA (see Figure 6.50), threatening food security. In the RSA, there are more than 120 000 fishers (Siddeek and others 1999). In the past 10 years, fish harvests in Mashriq countries remained at about 5 000–10 000 tonnes/year, while annual fish catches in Yemen alone increased from around 80 000 to 140 000 tonnes. Fishery regulations exist but need better enforcement, particularly in the RSA. The Red Sea has mainly been threatened by reclamation activities (ROPME 2004,

PERSGA 2004), but the emerging and growing shrimp farming industry is expected to seriously threaten remaining mangroves (PERSGA and GEF 2003).

Policy responses

Recently, many countries introduced regulations requiring environmental impact assessments prior to any coastal or marine activities (GCC 2004), and adopted integrated coastal zone management plans. West Asia has more than 30 marine reserves (IUCN 2003), and has signed 18 regional and international agreements related to coastal and marine environments. Accordingly, various conservation measures and regional programmes were undertaken over the last two decades (ROPME 2004).

In the last five years, many activities have taken place in the Red Sea to protect mangroves as part of habitat and biodiversity conservation programmes and regional action plans (PERSGA 2004, ROPME 2004). In 2006, ROPME countries agreed to establish a regional environmental information centre (QEIC) in Oman, which will collect information on mangroves. A regional survey on the globally threatened sea cow began in 1986, and continues through cooperation among Saudi Arabia, Bahrain and the UAE (Preen 1989, ERWDA 2003).

There has been enormous pressure on marine and coastal areas, for example from oil production, although increased efficiency measures have reduced oil spillage significantly. The signing of the International Convention for the Prevention of Pollution from Ships (MARPOL), and the introduction of oil tanker reception facilities will improve the situation, but not all GCC countries have signed the protocol (GCC 2004). A 2000–2001 survey of coastal water contamination in the RSA revealed that petroleum hydrocarbon levels were lower than those reported in the Gulf War of 1990–1991, but sediments near some industrial facilities and harbours had higher levels of trace metals (De Mora and others 2005, ROPME 2004).

URBAN ENVIRONMENT

Urbanization

There has been intense urbanization in West Asia over the past two decades (see Figure 6.51), which has overstretched urban infrastructure, and had significant but varying impacts on the region's environment and natural resources. Natural population growth, rural migration and displacement in the Mashriq, economic

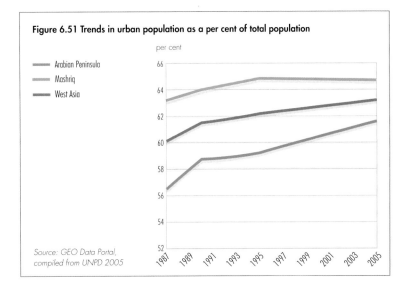

Figure 6.50 Trends in annual per capita fish catch in West Asia

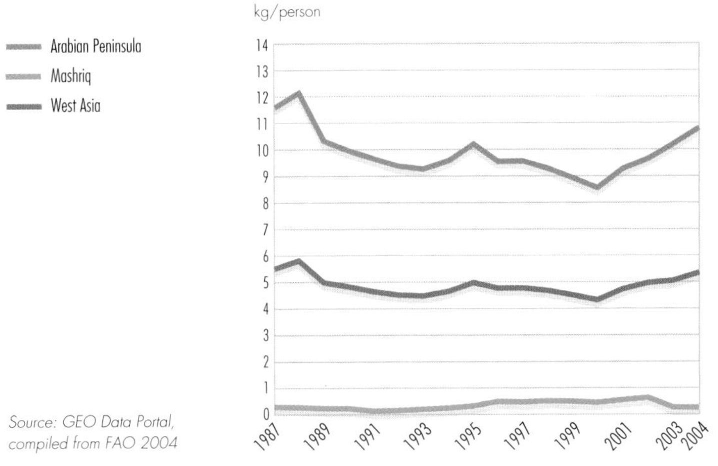

kg/person

- Arabian Peninsula
- Mashriq
- West Asia

Source: GEO Data Portal, compiled from FAO 2004

Figure 6.51 Trends in urban population as a per cent of total population

per cent

- Arabian Peninsula
- Mashriq
- West Asia

Source: GEO Data Portal, compiled from UNPD 2005

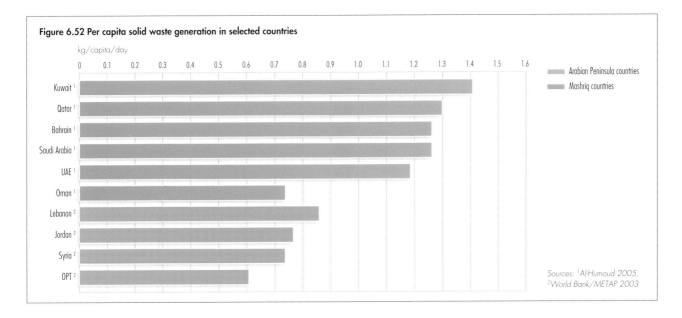

Figure 6.52 Per capita solid waste generation in selected countries

kg/capita/day

Arabian Peninsula countries
Mashriq countries

Sources: [1]Al-Humoud 2005,
[2]World Bank/METAP 2003

transformations, and increased migrant labour in GCC have resulted in higher demand for water and energy, waste management challenges and the deterioration of urban air quality.

Slums and urban poverty

Slum areas have expanded, especially around the Mashriq's major cities. Over the last decade, the number of people living in slum conditions almost doubled in Yemen, and increased by about 15, 25 and 30 per cent in Jordan, Syria and Lebanon, respectively (UN-HABITAT 2003a).

In the OPT and Iraq, military conflicts contributed to the growth in slum and refugee camp populations. By 2005, there were about 400 600 Palestinian refugees in Lebanon, some 424 700 refugees in Syria and about 1.78 million in Jordan. In the West Bank and Gaza there were about 687 500 and 961 650 registered refugees respectively, totalling more than one-third of the total Palestinian population in these two areas (UNRWA 2005).

During Iraq's three Gulf wars, stringent economic sanctions and the continued conflict devastated the environment, and resulted in a severe housing shortage. The shortage was estimated at 1.4 million units in the centre and south, while in the north an estimated one in every three persons lived in grossly substandard housing or neighbourhoods (UN-HABITAT 2003b). In 2003, 32 per cent of people in Iraqi cities lived under or near the poverty line, while a large number lived in refugee camps on the borders

with Syria and Iran (UNPD 2003). Similarly, in the conflict-stricken countries of the OPT and Lebanon, these conditions led to a corresponding rise in levels of urban poverty. In 1997, 27 per cent lived below the poverty line in Lebanon, 67 per cent fell below that level in the OPT in 2004.

Urban waste management

Rapid urbanization, inadequate waste management and lifestyle changes have resulted in increased waste generation. Per capita solid waste in the GCC ranges between 0.73 and 1.4 kg/person/day, compared to 0.61 and 0.86 kg/person/day in the Mashriq (see Figure 6.52). The inability of existing waste management systems to cope has led to significant health and environmental problems. The presence of landfills, burning waste, rodents and odours has also depressed real estate values in surrounding residential areas. Reduce, re-use and recycle initiatives have recently been established in some urban centres.

Energy sector, transport and air pollution

The energy sector, dominated by huge oil and gas facilities and thermal electric power plants, is a primary driver of both economic development and environmental degradation. A balance between the two has yet to be achieved in West Asia. The region holds about 52 per cent of the world's oil and 25.4 per cent of its gas reserves. It is responsible for nearly 23 per cent of global oil and about 8.7 per cent of global gas production (OAPEC 2005), a contribution that is expected to increase. Per capita energy consumption in the region varies greatly between oil

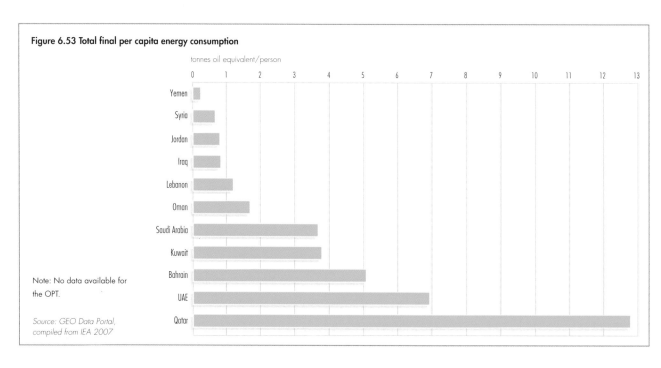

Figure 6.53 Total final per capita energy consumption

tonnes oil equivalent/person

Yemen
Syria
Jordan
Iraq
Lebanon
Oman
Saudi Arabia
Kuwait
Bahrain
UAE
Qatar

Note: No data available for the OPT.

Source: GEO Data Portal, compiled from IEA 2007

producing countries and non-oil producers (see Figure 6.53). The energy sector has adverse impacts on air, water, land and marine resources, and contributes to global climate change (see Chapter 2). Average per capita CO_2 emissions increased from 6 to 7.2 tonnes between 1990 and 2003, compared to a world average of 3.9 tonnes (GEO Data Portal, compiled from UNFCCC-CDIAC 2006).

The CO_2 emission increases result not only from industrial expansion and use of fossil fuel, but also from a growth in the number of vehicles, poor traffic management, energy subsidies, inefficient public transportation, ageing cars and congested roads, especially in the Mashriq sub-region. In GCC countries, power, petrochemical, aluminium and fertilizer plants, as well as motor vehicles are the primary sources of CO_2 and other air pollutants, such as SO_2 and NO_2. Fine particles from seasonal sand and dust storms add to the air pollution burden in the whole region. Air pollution has a considerable impact on human health. In Jordan, for example, it is estimated that over 600 people die prematurely each year due to urban air pollution, while another 10 000 disability adjusted life years (DALYs) are lost annually due to related illnesses (World Bank 2004a).

Addressing the urban challenges

Government responses to these challenges have been varied and thus far inadequate. To curb growth in slum areas, the GCC guarantees housing

for all citizens. Some countries have developed energy efficiency codes and standards for buildings and home appliances. Efforts to create integrated waste management programmes, and to monitor and legislate air pollution are underway. In their efforts to alleviate urban air pollution, all GCC countries, as well as Lebanon, Syria and the OPT, have introduced unleaded gasoline (see Box 6.38). To meet global market specifications, refineries in Kuwait, Saudi Arabia, Bahrain and the UAE have pledged to reduce sulphur content in petroleum products. Gas flaring and other hydrocarbon releases are decreasing.

Box 6.38 Phasing out leaded gasoline in Lebanon

The introduction of unleaded gasoline and the use of catalytic converters in Lebanon decreased lead emissions from some 700 tonnes/year in 1993 to almost 400 tonnes in 1999. However, lead concentrations at monitored urban and suburban locations still average 1.86 µ/m³ and 0.147 µ/m³ respectively. These levels are much higher than in countries where leaded gasoline has been completely phased out. The cost of lead-related pollution in Lebanon is estimated at US$28–40 million/year, or 0.17 to 0.24 per cent of GDP, associated mainly with impaired neurological development in children. This is a reason to continue vigorously with measures to reduce lead emissions.

Sources: Republic of Lebanon/MOE 2001, World Bank 2004a

Policies aimed at switching to natural gas are another response to mitigate air pollutants and GHG emissions. Planned regional integration of gas projects, such as the natural gas pipeline or Dolphin Project, which is expected to deliver 82 million m³ of Qatar gas to the UAE in 2005, will improve energy availability, economic efficiency and environmental quality (UNESCWA, UNEP, LAS and OAPEC 2005). Some countries have been developing and promoting renewable energy resources, such as wind and solar energy (see Box 6.39).

Assessing urbanization's full impact on West Asia's environment continues to be an elusive goal. Major efforts are needed in multisectoral planning, monitoring, legislation and public awareness campaigns in the sectors concerned. Harmonized approaches and data collection need to be applied across countries to enable better comparative and regional assessments. The recent creation of a Beirut-based regional monitoring organization to coordinate efforts in the different countries is a step towards this goal. Without the legislative power to enforce changes, however, these efforts will remain ineffective.

PEACE, SECURITY AND THE ENVIRONMENT
War and conflict

Armed conflict in West Asia has harmed human well-being, and resulted in the degradation of natural resources and ecological habitats. While the impacts are severe, reliable data remain scarce, except for a few selected locations (Butayban 2005, Brauer 2000), making long-term assessment difficult.

The Gulf War of 1990–1991 resulted in serious environmental damage, especially in Iraq, Kuwait and Saudi Arabia, and is fully documented in previous GEO editions and other reports (Al-Ghunaim 1997, Husain 1995, UNEP 1993). Fifteen years later, ecosystems are still showing clear symptoms of damage (Omar and others 2005, Misak and Omar 2004). The situation further deteriorated during the 2003 invasion of Iraq. The construction of military fortifications, laying and clearance of mines, and movement of military vehicles and personnel severely disrupted Kuwait's and Iraq's ecosystems and protected areas (Omar and others 2005). In the desert, these activities accelerated soil erosion, and increased sand movement, along with dust and sandstorms.

In Iraq, concern has risen over the use of depleted uranium munitions in the wars of 1991 and 2003 (Iraq Ministry of Environment 2004, UNEP 2005c). In addition, even several years after the end of major wars, unexploded ordnance (UXO) and land mines are still killing civilians and hampering reconstruction (UNAMI 2005). A detailed assessment of five priority industrial areas revealed serious threats to human health and the environment, and called for urgent action to contain hazardous materials (UNEP 2005c).

In the armed conflict in Lebanon in 2006, extensive oil pollution occurred along the country's coast after Israeli bombers hit the Jiyeh power station oil depots, south of Beirut. Environmentalists have described it as the worst environmental disaster in Lebanon's history (UNEP 2006g), with additional threats to human health from air and water pollution.

Removal of contaminated sludge from electroplating tanks in Al-Qadissiya.

Credit: UNEP/Post Conflict Branch 2006

The cumulative impact of decades of occupation and neglect in the OPT has resulted in serious environmental problems, including the degradation of scarce water resources, and pollution by solid and liquid waste (UNEP 2003c).

The consequences of these wars include the disruption of health services, deepened poverty, destroyed institutions and the inability to enforce environmental legislation (Kisirwani and Parle 1987). Among children under five in Baghdad, for example, 7 out of 10 suffered from diarrhoea between 1996 and 2000, due to the lack of clean water, poor sanitation and large amounts of uncollected garbage (UNICEF 2003). The non-violent death rates in Iraq increased in 2005 and 2006, which may reflect deterioration in health services, and environmental health (Burnham and others 2006).

The issue of refugees and internally-displaced people in West Asia cannot be overemphasized. Successive wars have increased their numbers to about 4 million (UNHCR 2005, UNRWA 2005). They live in poor socio-economic conditions, with high population densities and inadequate basic environmental infrastructure, adding pressure to fragile environments. Dense populations in Gaza's refugee camps contributed to aquifer depletion, which resulted in saltwater intrusion and saline water unsuitable for irrigation (Weinthal and others 2005, Homer-Dixon and Kelly 1995). During the 2006 hostilities along the borders of Israel and Lebanon, about 1 million people were temporarily displaced in Lebanon alone, in addition to those displaced in northern Israel, raising serious concerns over their well-being (UNEP 2007b).

War results in heavy infrastructural damage. Bombardments of military and civilian targets resulted in altered Iraqi and Lebanese rural and cityscapes. In the OPT, occupation forces demolished a major section of the Jenin refugee camp (UNEP 2003c). The economic infrastructure in the Gaza Strip was damaged in the May 2004 hostilities, aggravating existing environmental problems (World Bank 2004b).

About 150 000 land mines were placed indiscriminately in Lebanon between 1975 and 1990 (Wie 2005). In Iraq, the total number of individual UXOs may range from 10 000 to 40 000 (UNEP 2005c). UNEP's preliminary post-conflict assessment of the recent conflict in Lebanon indicated that approximately 100 000 unexploded cluster "bomblets" were identified, a figure that is expected to rise (UNEP 2006h). The detonation of UXOs potentially releases contaminants into the air and soil.

Addressing the impacts of war

The hidden and long-term environmental cost of war in the region is enormous, and cannot be easily estimated. Since the 1990 Gulf War, a mechanism to address environmental claims resulting from wars and conflicts was introduced. Countries neighbouring Iraq have submitted environmental claims for compensation from Iraq through the United Nations Compensation Commission (UNCC 2004). This mechanism could help prevent policies that threaten human and environmental well-being. On-site response to war-related environmental damage in affected countries has included monitoring and assessment of damage, mine clearance, and cleaning and restoration measures. Internationally, some conflict resolution techniques have been implemented, including agreements, mutual understandings, the promotion of peace, cultural exchanges and other reconciliation measures.

POLAR REGIONS
DRIVERS OF CHANGE
Governance

There have been major political events in the Arctic since the 1987 report of the World Commission on Environment and Development, *Our Common Future*. The dissolution of the Soviet Union was followed by a drop of one-quarter in the population in the Russian Arctic (AHDR 2004), and the withdrawal of government support to indigenous economies (Chapin and others 2005). There has been some political restructuring in the Arctic, in part inspired by international human rights development. It includes some delegation of management authority to local people in Finland and Scandinavia, and increases in indigenous self-government in Canada and Greenland (AHDR 2004). The settlement of land claims, and associated changes in resource management and ownership, starting with the Alaska Native Claims Settlement Act 1971, continue to be important political trends in the North American Arctic.

Antarctica is subject to sovereignty claims that are not universally recognized. The continent was without a governance regime until the Antarctic Treaty in 1959. The Antarctic is now governed by an international multilateral regime, under which measures are implemented through domestic legislation. Today, 46 nations, including all but one of the Arctic nations, are parties to the Antarctic Treaty. The system is centred on principles of peaceful use, international scientific cooperation and environmental protection. The current treaty signatories, along with invited expert and observer groups, meet on an annual basis to effectively govern the region, and provide a forum for discussion and resolution of issues. The most significant legal development since 1987 was the 1991 Protocol on Environmental Protection, which designated Antarctica as "a natural reserve devoted to peace and science." In 2005, Annex VI to the Protocol was adopted, addressing liability for environmental emergencies in the Antarctic region.

By contrast, large parts of the Arctic have governance regimes based on state sovereignty. The Arctic includes all or part of eight nations: Canada, Denmark (Greenland), Finland, Iceland, Norway, the Russian Federation, Sweden and the United States. National domestic laws remain the primary legal controls in the Arctic. Since 1987, a series of "soft law" agreements and cooperative arrangements have been created (Nowlan 2001) at both regional and circumarctic levels. The Arctic Environmental Protection Strategy (1991) was absorbed into the work of the newly-formed Arctic Council in 1996. The council develops assessments, recommendations and action plans on a broad range of environmental and socio-economic issues. It is composed of the eight Arctic nations, six indigenous peoples' organizations, who sit on the council as permanent participants, and additional nations and international organizations holding observer status.

Multilateral environmental agreements

Multilateral environmental agreements (MEAs), and international policies and guidelines play an increasing role in both polar legal systems. The concept of sustainable development and the MEAs that embrace it have had a profound resonance in the Polar Regions, particularly in the vulnerable Arctic.

Integrating sustainability of Arctic communities and the natural environment are key components of circumarctic agreements and programmes (AC 1996). This is only possible with reference to the aspirations,

traditional lifestyles and values of indigenous and local people, and with their involvement in decision making. Monitoring and projections from scientific models demonstrate that these MEAs can be effective. However, the current suite of MEAs is inadequate to meet the challenges from climate change and the many harmful substances that remain unregulated internationally. In common with other parts of the world, progress in implementing these actions has been relatively slow, despite the institutionalization of sustainability principles (Harding 2006).

SELECTED ISSUES

The Polar Regions are among the world's last great wilderness areas, but they are undergoing rapid and accelerating change, stressing ecosystems in both the Arctic and Antarctic, and affecting the well-being of Arctic residents. These regions are of vital importance to the health of the planet (see Box 6.40), and the changes are of global significance.

Box 6.40 Global-scale ecosystem services provided by the Polar Regions

Regulate climate
Without the global thermohaline marine current exchanging waters between the Polar Regions and the tropics (see Box 6.42 and Chapter 4), the tropics would become much warmer (or too hot), while polar and temperate regions would become much colder.

Store freshwater
They account for about 70 per cent of the world's total water stored as ice.

Provide resources
The Arctic holds 28 per cent of the global marine commercial fish catch. Antarctic fisheries add 2 per cent. The Arctic has rich mineral resources, and undeveloped stores of oil and gas, including an estimated 25 per cent of the world's undiscovered petroleum reserves.

Store carbon
The Arctic stores one-third of the global carbon pool, an important sink for greenhouse gases (GHGs).

Support migratory species
About 300 species of fish, marine mammals and birds migrate between the Polar Regions and mid-latitudes each year. Between 500 million and 1 billion birds annually connect with almost every part of the globe. More than 20 whale species migrate between polar and tropical waters.

Form an essential part of our global heritage
Antarctica is by far the largest wilderness area on Earth, while 7 of the 11 other largest wilderness areas are in the Arctic. These are crucial not only for tangible ecosystem services, such as the preservation of biological diversity, but also for intrinsic values related to aesthetics and culture.

Sources: ACIA 2005, CAFF 2001, FAO 2004, Lysenko and Zöckler 2001, Scott 1998, Shiklomanov and Rodda 2003, USGS 2000

There are important geographical and political
distinctions between the Arctic and Antarctic.
About 4 million people live in the Arctic, of whom
approximately 10 per cent are indigenous (AHDR
2004). The Antarctic has no indigenous population;
the only residents are transient scientists and staff in
research stations. The Arctic is a partially-frozen ocean,
surrounded by a diversity of landscapes, including
sparsely-vegetated barren lands, tundra, wetlands
and forests, influenced by ice, seasonal snow cover
and permafrost. There are low numbers of known
terrestrial species compared to mid-latitudes, but large
and widespread populations of key species, several
of which are of major importance to indigenous and
local cultures and economies. Agriculture in the Arctic
is severely limited, and subsistence economic activities
mainly involve hunting and fishing, reindeer herding,
trapping and gathering.

Antarctica, a continent surrounded by ocean, is
99 per cent ice covered (Chapin and others 2005),
with no native terrestrial vertebrates, but large
populations of marine birds and seals go there to
breed. The small crustacean, krill, is the basis of
the Southern Ocean food web, which supports fish,
marine mammals and birds.

Climate change, the accumulation of persistent
toxic substances and pollutants, damage to the
stratospheric ozone layer, and increasing development
and commercial activity are examples of globally-
driven issues that have particularly affected Polar
Regions. Over the past 20 years, polar research and
assessment, along with the direct involvement of Arctic
residents, particularly indigenous peoples, have been
instrumental in understanding the impacts, and in
bringing these issues to the world's attention.

CLIMATE CHANGE
Melting ice: a local and a global threat
Impacts occur faster in the Polar Regions
Global increases in population, industrialization,
expanding agriculture and deforestation, and
the burning of fossil fuels have resulted in rising
atmospheric concentrations of GHGs, and dramatic
changes in land cover. Scientists agree that it is very
likely that most of the observed increase in global
temperatures in the past half-century is due to human
additions of GHGs to the atmosphere (IPCC 2007a).
It is a major issue for the Polar Regions because these
regions are experiencing impacts faster and to a
greater extent than the global average, and because
climate change in the Polar Regions has major
implications for the Earth.

Climate change is accentuated in the Polar Regions,
mainly because of feedback mechanisms related
to shrinking ice and snow cover (see Chapters 2

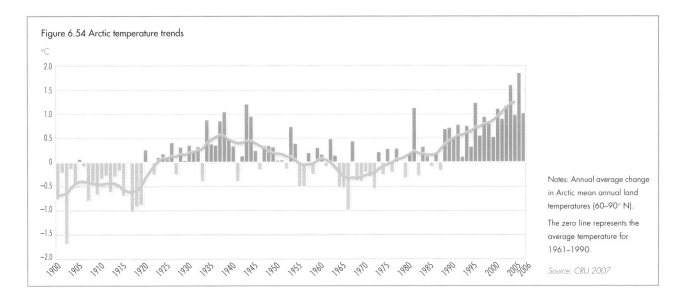

Figure 6.54 Arctic temperature trends

°C

Notes: Annual average change in Arctic mean annual land temperatures (60–90° N).

The zero line represents the average temperature for 1961–1990.

Source: CRU 2007

and 7). The Antarctic shows complex temporal and spatial patterns of both warming and cooling, with the most pronounced warming along the Antarctic Peninsula (UNEP 2007c). The Arctic is warming almost twice as fast as the world average (IPCC 2007a), and most increases occurred in the past 20 years (see Figure 6.54), causing shrinking and thinning of sea ice (see Figure 6.55), melting glaciers and changing vegetation. The land and the sea absorb more heat when there is less ice and snow, resulting in the melting of more ice and snow. Thawing of frozen peat bogs is releasing methane (a potent GHG) at some sites, but it is not known if the circumpolar tundra will be a carbon source or sink in the long-term (Holland and Bitz 2003, ACIA 2005).

In 2005, the world's first major regional multistakeholder climate change assessment, the *Arctic Climate Impact Assessment* (ACIA) was released. It included comprehensive reviews of the state of knowledge on climate variability and change, and on current and projected impacts and vulnerabilities, and it incorporated perspectives based on Arctic indigenous peoples' knowledge. Some major observed trends identified in the ACIA 2005 include:

- sharply rising temperatures, especially in winter, and particularly in Alaska, northwestern Canada and Siberia;
- rainfall increasing but snow cover decreasing;
- glaciers melting and summer sea ice shrinking;
- river flows increasing;
- North Atlantic salinity reducing; and
- thawing of permafrost, and decreasing periods of ice cover on lakes and rivers, in some areas.

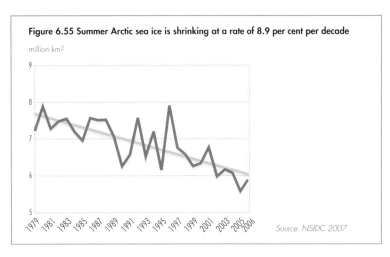

Figure 6.55 Summer Arctic sea ice is shrinking at a rate of 8.9 per cent per decade

million km²

Source: NSIDC 2007

These observed changes have wide-ranging impacts on plants, animals and the well-being of Arctic residents (see Box 6.41 and Box 7.8 in Chapter 7). Impacts affecting people range from those related to melting permafrost and shorter ice seasons (damage to buildings, and shorter seasons for winter roads), to warmer and less predictable weather (more forest fires in some regions, and problems travelling on frozen rivers and through snow for reindeer herders and hunters). Sea ice changes cause increased coastal erosion, necessitating the relocation of coastal communities, such as Shishmaref, Alaska (NOAA 2006), and affecting marine hunters and fishers. Many impacts are indirect; for example, more melting and freezing of snow makes food less accessible to caribou and reindeer, affecting herders and hunters, their economies and cultural integrity. Future impacts are expected to be widespread, and include

The Arctic tundra is a land of meltwater lakes, meandering rivers and wetlands. Analysis of algae in sediment cores from 55 circumpolar Arctic lakes revealed dramatic regime shifts in many lakes over the past 150 years. Lakes have become more productive, and there are more species of algae in the shallow lakes. These ecosystem changes are triggered by climate warming; they are more marked at higher latitudes, and they correspond in timing to climate warming inferred through records such as sediment cores and tree rings. Changes at the base of the aquatic food chain can be expected to have far-reaching effects on other life in and around the lakes.

Polar bears depend on sea ice for hunting, and use ice corridors to move from one area to another. Pregnant females build winter dens

in areas with thick snow cover, and need good spring ice conditions to find food. The mothers emerge with their cubs in the spring, and have not eaten for 5–7 months. Late sea ice formation in the Arctic autumn and earlier breakup of ice in the Arctic spring means a longer period of fasting. During the past two decades, the condition of adult polar bears in western Hudson Bay in Canada has declined. There was a reduction of 15 per cent both in average adult body weight and in the number of cubs born between 1981 and 1998. Some climate models project that if GHG emissions are not curbed drastically there will be an almost complete loss of summer sea ice in the Arctic before the end of this century. Polar bears, along with other marine mammals, such as seals, are unlikely to survive in such a changed environment.

Sources: ACIA 2005, Smol and others 2005

Credits: J. Smol (Lake) and Jon Aaars/Norwegian Polâr Institute (Polar bear)

positive and negative changes, both in economic opportunities and in risks to the environment. One of the big factors is the potential for changes in access due to more open Arctic marine shipping routes (ACIA 2005, UNEP 2007b).

Increased attention is being paid to climate change impacts on Antarctic ecosystems, including new research through the International Polar Year (2007–2008). Seasonal and regional variations in the extent of sea ice have large impacts on ecosystem processes (Chapin and others 2005). Krill, the food source for many birds, fish and marine mammals, rely on the algae that live in sea ice, and cannot survive without ice cover (Siegel and Loeb 1995). Many seabirds are significantly influenced by rising temperatures (Jenouvrier and others 2005), and changes in winter sea ice conditions affect the populations of three of the most ice-dependent species: Adélie penguins, Emperor penguins and Snow petrels

(Croxall and others 2002). Even a small increase in temperature may allow the introduction of non-native plant and animal species that affect native biodiversity.

Global impacts of polar climate change

There are many ways in which the major changes observed and projected for the Polar Regions influence the environment, the economy and human well-being around the globe. Two of the most fundamental of these are ocean circulation and sea-level rise.

The role of the Polar Regions as a driver of ocean circulation (see Box 6.42) is of enormous significance, because of their influence on global climate regimes. For instance, part of this ocean circulation warms Europe by 5–10°C, compared to what would be expected at this latitude. Breakdown of thermohaline circulation could precipitate an abrupt change in global climate regimes (Alley and others 2003).

The global sea level has been rising at a rate of about 3 millimetres/year since 1993, compared to less than 2 mm/year over the previous century (WCRP 2006). This increased rate is very likely due to human-induced climate change, primarily through thermal expansion of warming oceans, and freshwater from melting glaciers and ice sheets (IPCC 2007a, UNEP 2007c, Alley and others 2005). The Greenland and Antarctic ice sheets have the potential to be the largest contributors, because they store so much ice. The rate at which polar ice sheets are contributing to sea-level rise is faster than previously predicted, and there is a lot of uncertainty around the future of the ice sheets. Until a few years ago, most scientists studying the ice sheets believed that the biggest immediate impact of global warming was that it would lead to mass loss from increased surface melting. While increased melting is certainly a concern, it appears that other mechanisms may be at least as important. For example, meltwater reaching the base of the ice causes the ice to flow faster. This accelerated flow is a far more efficient way of rapidly losing large amounts of ice mass than surface melting (Rignot and Kanagaratnam 2006). These dynamic processes of mass loss are not well understood and current models that project future sea-level rise are not able to take them fully into account (UNEP 2007c). This means that there is a lot of uncertainty around projections of future sea-level rise.

Studies of the Greenland Ice Sheet show that ice melt and calving of icebergs is occurring at a greater rate than new ice is being formed (Hanna and others 2005, Luthcke and others 2006). A rise of 3°C in the average annual temperature in Greenland is likely to cause the ice sheet to slowly melt away, leaving only glaciers in the mountains. If GHG emissions rise at currently projected rates, it is expected that by the end of this century the average temperature will be above this tipping point. The meltwater could raise the sea level 7 metres over a period of 1 000 years or more (Gregory and others 2004).

In Antarctica, there are two giant ice sheets: the West and the East Antarctic Ice Sheets. Together they account for about 90 per cent of the world's freshwater ice (Shiklomanov and Rodda 2003), and changes to them would have huge global repercussions. The West Antarctic Ice Sheet is particularly vulnerable, and recent evidence points to instability (Alley and others 2005). Three large

Box 6.42 Polar Regions and ocean circulation

The circulation of water through the oceans is partly driven by differences in density of seawater, determined by temperature and salt content (see Chapter 4). The formation of deep, dense seawater in the Arctic and Antarctic drives this "ocean conveyor belt." This process is disrupted by warming and freshening of surface water, reduction in sea ice and melting of glaciers and ice sheets. Evidence shows that circulation of the deep, cold water of the North Atlantic conveyor belt may have slowed by as much as 30 per cent over the past 50 years. In the Antarctic, recent increases in precipitation have reduced the salinity of surface layers, weakening deep-water formation that drives the southern conveyor belt.

Sources: Bryden and others 2005, Chapin and others 2005

parts of ice shelves in the Antarctic Peninsula have collapsed over the past 11 years, followed by a marked acceleration and thinning of glaciers that previously fed the shelves (Rignot and others 2004, Scambos and others 2004). Over the last decade, the grounded ice shelf in the Amundsen Sea, and ice shelves in Pine Island Bay have thinned significantly; in the latter area there has been a tenfold reduction in ice mass in the past decade (Shepherd and others 2004.) Some experts think that a full collapse of the West Antarctic Ice Sheet is conceivable in this century (New Scientist 2005). Were this to happen, the sea level would rise by about 6 m (USGS 2005) (see Figure 6.56).

Figure 6.56 The potential impact of a 5-metre sea level rise in Florida (above) and Southeast Asia (below)

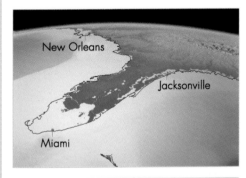

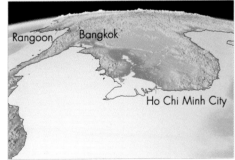

Note: The black lines show the current coast lines. The reconstruction shows that with a 5-metre sea-level rise the coastlines would recede drastically, and cities such as Bangkok, Ho Chi Minh City, Jacksonville, Miami, New Orleans and Rangoon would disappear from the land map.

Credit: W. Haxby/Lamont-Doherty Earth Observatory

The East Antarctic Ice Sheet is more stable, and increased snowfall has resulted in local mass gains, which partially compensate for the contributions to the oceans of water from the West Antarctic and Greenland ice sheets, and mountain glaciers (Davis and others 2005). However, a 2006 satellite-based estimate of overall losses and gains of the Antarctic ice sheets concluded that there was a net loss of 152 ±80 cubic kilometres of ice annually between 2002 and 2005 (Velicogna and Wahr 2006).

Responding to climate change

There are two policy response categories to climate change from the polar perspective: accelerate efforts to reduce GHG emissions and, at the same time, adapt to changing conditions. The policy document issued by the Arctic ministers through the ACIA (ACIA 2005) recognizes that action must be taken both on mitigation and adaptation, and establishes broad guidelines for action. Recommended mitigation actions include meeting commitments under the Kyoto Protocol in reducing GHG emissions.

Adaptation measures include identifying vulnerable regions and sectors, assessing risks and opportunities associated with climate change, and developing and implementing strategies to increase the capacity of Arctic residents to adapt to change (see Box 6.43).

As Arctic nations are responsible for 40 per cent of global carbon dioxide emissions (see Chapter 2) (Chapin and others 2005), implementing these recommendations would have a significant positive impact globally. However, the world's response has been slow, and emissions continue to rise, while the scale of this issue and the lag time between action and ecosystem response require immediate action, both on mitigation and on adaptation. To protect environmental quality, biodiversity and human well-being, policy responses must take cumulative impacts into consideration, and all polar policies now need to be evaluated in the context of climate change.

PERSISTENT POLLUTANTS

Contamination

Many toxic chemicals released into the environment from industry and agriculture at lower latitudes are transported to the Polar Regions by wind, ocean currents and migratory wildlife (Chapin and others 2005). Persistent organic pollutants (POPs), such as

---

Box 6.43 Hunters adapting to climate change

An example of climate change adaptation by Arctic residents is Inuit use of modern technology in hunting at the ice edge. Because of rapid change, ice conditions are becoming increasingly difficult to predict based only on traditional knowledge. Satellite imagery is now also used routinely by indigenous hunters in the Canadian Arctic as a tool for safe and efficient navigation in the icescape.

Sources: Ford and others 2006, Polar View 2006

Credit: Roger Debreu/CIS

DDT and PCBs, are long-lived, fat-soluble chemicals that build up to higher levels through the food chain. Arctic animals are especially vulnerable, since they store fat to survive when food is not available. Metals differ from POPs in that they occur naturally in the environment, but levels are elevated as a result of industrial activities around the world, including transport (lead), coal burning (mercury) and waste disposal. There are also local sources of industrial metals in the Arctic, especially the smelters on the Kola Peninsula and at Norilsk in Russia. Emissions of metals transported through the air from industry in Europe and Asia, however, are the largest sources (AMAP 2002a). The Arctic Monitoring and Assessment Programme of the Arctic Council (AMAP) and national programmes research and report on toxics in the Arctic (AMAP 2002a, INAC 2003). Some results of this work are shown in Figure 6.57. The graph shows declining levels of regulated POPs, and rising levels of mercury in the eggs of Thick-billed murres on Prince Leopold Island, Nunavut, Canada. In the past 20–30 years, DDT and PCBs have generally declined in Arctic animals, while mercury has risen in some species and regions, and remained unchanged in others. Rising mercury levels may be from anthropogenic sources, from ecosystem changes related to climate warming, or a combination of these factors.

The levels of POPs that are banned or being phased out tend to be lower in Antarctic animals than in Arctic animals, though high concentrations of PCBs have been found in south Polar skuas (Corsolini and others 2002). In Antarctica, limited work on mercury indicates that the increases seen in some Arctic seabirds are not occurring in the Antarctic. Feathers of King penguins collected in 2000–2001 showed a reduction in mercury concentrations of 34 per cent compared with levels in feathers from a 1970s collection (Scheifler and others 2005). Types of POPs that are still in use and are not adequately regulated continue to build up in both Polar Regions in birds, seals and whales, and, in Antarctica, in ice and krill (Chiuchiolo and others 2004, Braune and others 2005).

POPs and mercury pose a threat to the integrity of traditional food systems and the health of indigenous peoples (see Chapters 1 and 5). The highest exposures – to Inuit populations in Greenland and northeastern Canada – are linked to consumption of marine species as part of traditional diets. Unborn and young children are most susceptible (AMAP 2003). There are also potentially widespread impacts on polar animals. Effects that have been demonstrated include reduced immunological response in polar bears, leading to increased susceptibility to infection, multiple health effects in glaucous gulls and reproductive failure from eggshell thinning in peregrine falcons (AMAP 2004a, AMAP 2004b).

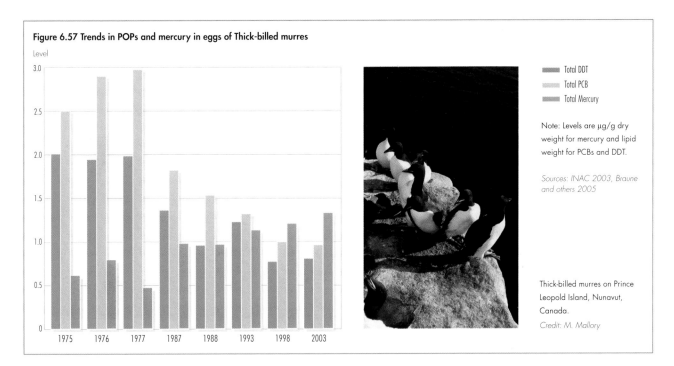

Figure 6.57 Trends in POPs and mercury in eggs of Thick-billed murres

Level

■ Total DDT
■ Total PCB
■ Total Mercury

Note: Levels are μg/g dry weight for mercury and lipid weight for PCBs and DDT.

Sources: INAC 2003, Braune and others 2005

Thick-billed murres on Prince Leopold Island, Nunavut, Canada.
Credit: M. Mallory

Response measures

Balancing and communicating the risks of contaminants in relation to other health risks and against the known benefits of breastfeeding and eating a traditional diet continues to be a challenge (Furgal and others 2005). Indigenous organizations, concerned about the safety of traditional foods, have taken a lead in directing and collaborating on studies, and on providing balanced information about risks and benefits of traditional foods (AMAP 2004c, Ballew and others 2004, ITK 2005).

Arctic indigenous peoples' organizations, working with the Arctic scientific community and AMAP, have pushed for international action on POPs, and directly participated in developing the global Stockholm Convention on POPs, which entered into force in 2004, and commits governments to reducing and eliminating the use of specific POPs. This success story of circumpolar cooperation between indigenous peoples and scientists (Downie and Fenge 2003) is now looked upon as a model for taking global action on climate change.

There is still work to be done on the issue of POPs in Polar Regions. POPs now in use, such as brominated flame retardants, are accumulating in polar ecosystems (Braune and others 2005), and are not yet included in the POPs convention. Although there are moves to find alternative products, many of these chemicals remain in widespread and increasing use (AMAP 2002a). In the Arctic, there are also local sources of POPs from past industrial and military activities, and from electric installations in Russia. As a response to this, the Arctic Council initiated a project to assist Russia in phasing out PCBs and managing PCB-contaminated waste (AMAP 2002b). In addition, the 1998 Protocol on Heavy Metals of the Convention on Long-Range Transboundary Air Pollution (the Protocol came into force in 2003), calls for a reduction in emissions of mercury, lead and cadmium to below 1990 levels (UNECE 2006b).

Ongoing monitoring and assessment of trends is needed to determine whether these international control measures are reducing toxic substances in polar environments, and to assess emerging issues. This includes identifying problem toxic substances currently in use, and assessing how climate change interacts with the accumulation of toxic substances in plants and animals.

DAMAGE TO THE OZONE LAYER

Ozone-depleting substances

The use of ozone-depleting substances leads to destruction of stratospheric ozone. This destruction has been most dramatic in the Antarctic, but the stratosphere over the Arctic is also affected. In September 2006, the Antarctic ozone hole was the largest recorded (NASA 2006). The ozone layer over the Arctic is not pierced by a hole as in the Antarctic, but in the winter of 2004–2005 it was the thinnest layer on record (University of Cambridge 2005) (see Chapter 2).

When the Antarctic ozone hole occurs, most of the coast is covered with 2–3 m of seasonal sea ice, which acts as a protective barrier for marine organisms. Microalgae in the sea ice are potentially adversely affected by increases in ultraviolet (UV-B) radiation resulting from ozone depletion (Frederick and Lubin 1994), and a reduction in sea ice may affect primary production in the entire region. Even with the ozone barrier, enough UV-B is transmitted through the annual ice to damage or kill embryos of the sea urchin *Sterechinus neumayeri* (Lesser and others 2004).

In the Arctic, young people today are likely to receive a lifetime dose of UV-B that is about 30 per cent higher than any prior generation, with increased risk of skin cancer. Studies show that increased UV-B is causing changes to Arctic lakes (see Chapter 2), forests and marine ecosystems (ACIA 2005). Despite the Montreal Protocol's success in markedly reducing ozone-depleting substances, the ozone layer's recovery is expected to take more than another half century (WMO and UNEP 2006).

INCREASING DEVELOPMENT AND COMMERCIAL ACTIVITY

Multiple development pressures – cumulative impacts

The biggest and fastest-growing development in the past 20 years in the Arctic is the expansion of oil and gas activity to meet growing global energy needs. Arctic oil and gas activity has been focused on onshore oil development in Siberia, the Russian Far East and Alaska. There has been offshore activity in the Barents and Beaufort Seas. Expanded, new and proposed petroleum developments, including access corridors and pipelines, are at various stages of preparation and implementation around the Arctic, especially in

Siberia, Alaska, the Canadian western Arctic and the Barents Sea.

Mineral exploitation in the Antarctic is prohibited under the 1991 Protocol on Environmental Protection to the Antarctic Treaty. The effect of the protocol on rights relating to the exploitation of Antarctic seabed resources arising from the United Nations Convention on the Law of the Sea has not yet been tested. Mining is widespread in the Arctic, and has decreased in some areas and expanded in others. At the same time, logging has decreased overall in northern Russia, but expanded in some regions of Siberia, and remains an important economic activity in northern Scandinavia and Finland (Forbes and others 2004).

There are many pressures associated with these activities, including emissions, leaks, spills and other releases of contamination from operating and decommissioned mines and oil facilities. As well, there are impacts that build up slowly with piecemeal development, such as habitat fragmentation and disturbance to wildlife (see Box 6.44). Spills in Arctic marine and coastal regions could have disastrous consequences for the livelihoods of residents who hunt and fish in these regions.

Box 6.44 Habitat loss and fragmentation

The destruction and breakup of large areas of habitat into patches has negative impacts on many species. Some examples of observed trends and impacts related to *Rangifer* (caribou and reindeer) include:

■ **North America.** Woodland caribou habitat is being taken over by roads and logging. By 1990, caribou in Ontario, Canada, were found only in the northern half of the lands that they had occupied in 1880, coinciding with the gradual northward shift in logging.

■ **Alaska's North Slope.** Petroleum drilling has resulted in infrastructure growth far beyond the initial Prudhoe Bay development, leading to avoidance of former calving grounds by barren ground caribou.

■ **Scandinavia.** Piecemeal development associated with recreational cabins, hydropower dams, bomb-testing ranges, power lines and road construction in particular, have led to an estimated 25–35 per cent loss of the central summer ranges for the reindeer of Saami herders. Projections indicate that up to 78 per cent may be lost in the coming decades.

■ **Yamal Peninsula, Western Siberia.** Destruction of vegetation by oil facilities, pipelines and vehicle use resulted in reindeer herds concentrating into a smaller area. This led to overgrazing, with impacts on ecosystems, local economies and human well-being.

Sources: Cameron and others 2005, Forbes 1999, Joly and others 2006, Schaefer 2003, Vistnes and Nellemann 2001

Caribou, *Rangifer tarandus* – bull in fall colours in the Denali National Park, Alaska.

Credit: Steven Kazlowski/Still Pictures

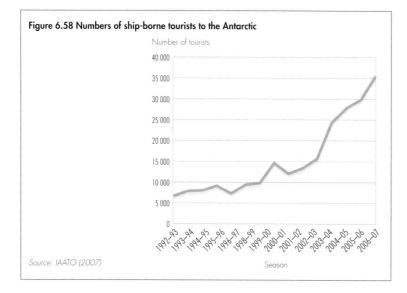

Figure 6.58 Numbers of ship-borne tourists to the Antarctic

Number of tourists

Season

Source: IAATO (2007)

chemicals that may have commercial applications currently takes place in the Antarctic without dedicated management.

Furthermore, there is a diversifying and expanding tourism industry in the Antarctic (see Figure 6.58) that has seen a great increase in ship-borne passengers (ASOC and UNEP 2005). The Antarctic Treaty Consultative Meeting (ATCM) is examining the regulation of tourism (ACTM 2005). The increase in visitors, combined with changing conditions related to global warming, risks the introduction of non-native species to this isolated part of the world (Frenot and others 2004) (see Chapter 5).

Economy, environment and culture: striking a balance
Long-term planning and effective environmental policies are needed to balance economic development with environmental and cultural considerations. Increasingly, cumulative effects are considered when impacts of large-scale industrial development in some parts of the Arctic are assessed (see, for example, Johnson and others 2005). However, smaller projects and infrastructure are rarely assessed in terms of their cumulative effects and in terms of how they interact with impacts from other developments and climate change (see Box 6.45). Countering the combined actions of many

Development pressures, such as global energy demands, combine and interact with climate change, persistent toxics and other pressures on polar ecosystems. In the marine environment, commercial fishing (see Chapter 4) is a significant pressure for both polar regions, including the ongoing problem of illegal, unregulated and unreported (IUU) fishing. In Arctic waters, increased shipping brings increased risk of spills, contamination and disturbance to wildlife. In the Antarctic, even the growth of scientific activities adds new pressures, as does bioprospecting (Hemmings 2005). The search for naturally occurring

Box 6.45 The importance of monitoring and assessing species distribution and abundance

Climate change is the big unknown factor in assessing vulnerability, and projecting cumulative impacts from multiple pressures.

Polar bears are threatened by the accumulation of POPs at the same time as their primary habitat, coastal ice, is shrinking due to climate change (see Box 6.41). An assessment of the interactions between contaminants and climate change concluded that it is difficult to predict whether climate change will lead to decreased or increased contaminant levels in Arctic ecosystems in the long-term, because there are so many factors to consider. There may be changes in winds, ocean currents and temperatures, and even changes in migration patterns of birds and fish that carry contaminants from lower latitudes.

The Canadian population of the Ivory gull, which lives along the ice edge year-round, has declined by 80 per cent since the early 1980s, with a total count in 2005 of only 210 birds, and there are indications that the species may be declining in the rest of its range. There are several factors that singly or in some combination could be implicated in this decline, including changes in sea ice in the winter range, hunting during migration through northwest Greenland, disturbance from diamond exploration and high levels of mercury in their eggs.

These examples highlight the importance of monitoring and assessing species distribution and abundance to detect and respond to changes in biodiversity. Recent initiatives have identified gaps, and recommended improvements in Arctic monitoring and assessment (NRC 2006). The Circumpolar Biodiversity Monitoring Programme was launched by the Arctic Council to improve monitoring and assessment of biodiversity and ecosystems to help meet the Arctic target of the Convention on Biological Diversity.

Sources: ACIA 2005, AMAP 2002b, Braune and others 2006, Gilchrist and Mallory 2005, Muir and others 2006, NRM 2005, Petersen and others 2004, Stenhouse and others 2006

pressures represents one of the most serious gaps in the Arctic management regime (EEA 2004). Effective measures include integrated planning that incorporates protection of representative ecosystems, key habitats and vulnerable areas, especially along the Arctic coastline.

Throughout the Arctic, governments and industry face great challenges in minimizing environmental and social impacts, and in including local residents in decision making for new and expanding developments. Priorities in responding to these issues include ensuring that local residents share in the opportunities and benefits from petroleum development, and that adequate technology, policy, planning and systems are in place to protect vulnerable regions, and to prevent and respond to accidents.

In the Antarctic, monitoring of cumulative impacts and consideration of management measures based on the precautionary approach are under discussion (Bastmeijer and Roura 2004). Site-specific guidelines have been adopted, but the question remains whether this will be sufficient for comprehensive protection.

REGIONAL ENVIRONMENTAL CHALLENGES
Progress has been made, challenges remain
Countries in developed regions have progressively invested in solving "conventional" or easy-to-manage environmental problems, and have achieved relative success, but such issues are still daunting to developing nations. Since the mid-1980s, many global conferences on the environment have been convened, a diversity of multilateral environmental agreements have been adopted (see Figure 1.1), and governments and other stakeholders continue to pursue sustainable development. But challenges remain as environmental issues become more complex and onerous. They are often cumulative, diffuse, indirect and/or persistent. For example, as Europe and North America addressed discrete and obvious pollution sources (point sources), they found they needed to deal with diffuse and scattered non-point sources. Non-point pollution is often hard to control, and its impacts difficult to measure. The Polar Regions identified cumulative and interacting pressures as a key priority. The causes, consequences and solutions for such complex problems cut across economic sectors and political

portfolios. All regions are now aware of the health and economic costs associated with air pollution, including weather-related hazards. They are also aware of the savings to be had through prevention and mitigation.

The world's most crucial environmental challenges, such as climate change, start with many actions at the local level that accumulate to have global effects. The reach and magnitude of transboundary issues is seen in the impacts of persistent organic pollutants on the Polar Regions, and the distances travelled by dust storms. New environmental issues arise quickly, and can have important human health impacts before existing policies can be used, or new policies put in place to address them. Examples of such new issues include: electronic waste, pharmaceuticals, hormones and other organic contaminants, and commercial exploitation of the Antarctic. As pointed out by the Polar Regions, a very important lesson learned is that there is a long lag time between dealing with complex global environmental issues and seeing improvements, as is the case with climate change.

In addition to complexity, progress in addressing regional environmental issues is challenged by counteracting forces and diminishing returns. For example, gains in potable water provision in many urban areas are being offset by the rising numbers of urban residents, for example, in North East Asia. In some regions, improved energy efficiency is being offset by an increase in the number of cars and other energy uses. Increased consumption and production, coupled with a lack of prevention often outstrips efficiency gains in waste management. Another limitation expressed in several regions is that despite progress in introducing environmental policies, there is inadequate monitoring to inform new environmental policies, regulations and other measures. Some report a lack of coordination among different decision making agencies, insufficient public participation or a lack of transboundary collaboration. This is, for example, a challenge in the Mediterranean Basin with its long, common history and geography, but large differences in culture and economic development. Box 6.46 describes inter-regional efforts to establish international implementation programmes in the Mediterranean.

Box 6.46 The Mediterranean Sea: taking a holistic approach

The Mediterranean Sea is bordered by 21 countries. More than 130 million people live permanently along its coastline, a figure that doubles during the summer tourist season. The sea and its shores are the biggest tourist destination on Earth. Because of its geographical and historical characteristics, and its distinctive natural and cultural heritage, the Mediterranean is a unique ecoregion. Although the Mediterranean countries fall within three different GEO regions, the sea and the surrounding land mass must be dealt with as one ecosystem, with common issues and problems.

Local, regional and national authorities, international organizations and financing institutions have devoted a great deal of effort to protecting the Mediterranean region's environment, but many environmental problems continue to plague it. In recent decades, environmental degradation has accelerated. Valuable agricultural land is being lost to urbanization and salinization (80 per cent of arid and semi-arid areas in the southern Mediterranean countries are affected by desertification, as well as 63 per cent of the semi-arid land in the northern bordering countries). Scarce, overused water resources are threatened with depletion or degradation. Traffic congestion, noise, poor air quality and the rapid growth of waste generation are compromising urban standards of living and health. Coastal areas and the sea are affected by pollution and coastlines are being built

up and/or eroded, while fish resources are being depleted. In short, overexploitation is disrupting the Mediterranean's unique landscapes and biodiversity.

In addition, the region is increasingly vulnerable to flooding, landslides, earthquakes, tsunamis, droughts, fires and other ecological disturbances, which have a direct and immediate impact on the livelihood and welfare of a large proportion of the population. Although it is difficult and risky to assign specific values, the costs of environmental degradation are clearly very significant. In addition, environmental pressures are likely to increase considerably over the coming 20 years, especially in the tourism, transport, urban development and energy sectors.

There are two major current initiatives to improve the state of the environment in the Mediterranean region. The Mediterranean Strategy for Sustainable Development, developed by UNEP's Mediterranean Action Plan and adopted in 2005, focuses on seven priority fields of action: water resources management, energy, transport, tourism, agriculture, urban development, and the marine and coastal environments. Complementary to this is the Horizon 2020 initiative under the Euro-Mediterranean Partnership. The aim of this initiative is to "de-pollute the Mediterranean by 2020" through tackling all the major sources, including industrial emissions, and municipal waste, particularly urban wastewater.

Sources: EEA 2006e, Plan Bleu 2005

Inequities prevail

The 1987 Brundtland Commission report, *Our Common Future*, and subsequent global, regional and national processes have highlighted the need for sustainable development, which integrates improvements in economic, social and environmental well-being. Sustainable development calls for increased intra- and intergenerational equity, so that environmental goods-and-services are shared fairly among people today, and are passed on to future generations. As shown in this chapter, however, environmental inequities continue to grow. They exist in many of the world's cities, where the poor are generally less well served by municipal water and waste systems, and are more exposed to pollution. The poor are the main victims of environmental degradation (Henninger and Hammond 2002). Indeed, poor people suffer more than the wealthy when water, land and the air are degraded and polluted. Not only are they deprived of livelihood options but their health is also impaired. In

developing countries, environmental risk factors are a major source of health problems for the poor in particular (DFID and others 2002).

The poor are also disproportionately affected by natural hazards. Before the devastating death tolls of the 2004 Indian Ocean Tsunami and the 2005 earthquake in Pakistan, from 1970 to 2002, some 3 million people, mostly in low-income countries, died as a result of natural disasters (UNEP 2002). The majority of the rural poor live in ecologically fragile areas. The environments in which the urban poor live and work are often fraught with hazards. Faced with a disaster, they suffer more from the loss of income and assets, and have greater difficulty in coping with the aftermath. Climate change and environmental degradation increase the frequency and impact of natural hazards, such as droughts, floods, landslides and forest fires, which often lead to the loss of land, food insecurity and migration (Brocklesby and Hinshelwood 2001, World Bank 2002c).

Another message in Chapter 6 is that gender inequities with environmental links continue in many regions. In Africa and South East Asia, for example, women often have limited access to land, water and other resources, and they are exposed to the health risks of indoor air pollution from the burning of biomass fuels. In many cases, indigenous peoples also continue to face inequities related to land rights, access to resources, and provision of potable water and wastewater services, even in some developed countries.

Disparities in ecological impacts prevail
Although the regions have made significant progress in reducing some environmental threats since the 1980s, those with growing economies are suffering from increased traffic, waste and greenhouse gas emissions. Asia and the Pacific, for example, reports that its economic growth surpassed the 5 per cent suggested by *Our Common Future*, (the Brundtland Commission report) but ecosystems and human health continue to deteriorate. Biodiversity loss and global climate change have irreversible consequences that income growth cannot restore (UNDP 2005c).

This chapter suggests that some progress on the environmental front in developed regions has been achieved at the expense of developing countries. This imbalance is expressed by the notion of "ecological debt." Experts agree that this term describes the ecological damage that production and consumption patterns in some countries cause in others or in ecosystems beyond their borders, at the expense of the equitable rights to the ecosystem goods-and-services by those other countries or peoples (Paredis and others 2004). For example, the outsourcing of energy, food and industrial production can increase efficiency in one region at the expense of others through the displacement of impacts (see Figure 6.59). The European regional perspective points out that the relocation of highly-polluting industries to Eastern European countries is contributing to higher energy use per unit of industrial output there, while improving energy efficiency and decreasing polluting emissions in Western Europe. Other examples include the export of electronic waste to Southeast Asia, where those who recycle it are exposed to hazardous materials, and the fact that Arctic peoples suffer the consequences of POPs that originate outside the region.

A prime example of the disproportionate effect of developed regions on the global environment is the former's generally higher per capita emission of greenhouse gases, which contribute to climate change, while impacts are and will be greater among the poor and other vulnerable people, nations and regions (Simms 2005). Poor people in tropical countries will be particularly vulnerable to climate change impacts, such as water shortages, declining crop yields and disease (Wunder 2001), while indigenous peoples in the Arctic suffer from the accelerated impact of climate change. Continued environmental degradation in all regions is unfairly shifting burdens onto future generations, and contradicts the principle of intergenerational equity.

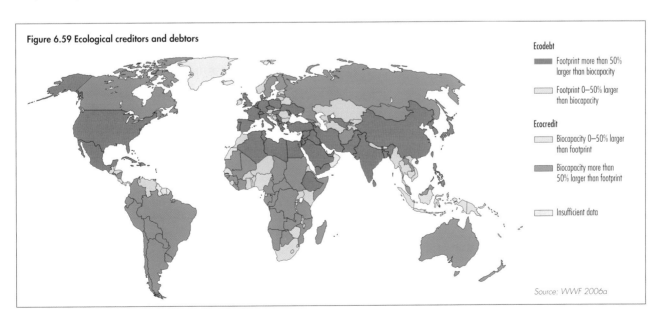

Figure 6.59 Ecological creditors and debtors

Ecodebt
- Footprint more than 50% larger than biocapacity
- Footprint 0–50% larger than biocapacity

Ecocredit
- Biocapacity 0–50% larger than footprint
- Biocapacity more than 50% larger than footprint

- Insufficient data

Source: WWF 2006a

One sign of a "Northern" development model is the accelerating growth of urban development based on car dependency.

Credit: Ngoma Photos

One of *Our Common Future*'s recommendations was to remove subsidies for intensive agriculture, which are discussed in the introduction to this chapter. Since environmental assets, such as fish, forests and crops, make up a larger share of national wealth in developing countries than in high-income countries, subsidy reform can improve rural livelihoods, and increase the equity between developed and developing regions. The regional perspectives reveal that though there has been some recent progress in debt relief and subsidy reform, developing countries still face unfavourable trade policies and external debt burdens, while a number of developed countries continue to enjoy subsidies.

Economy and environment not mutually exclusive

Although there are signs that environmental issues are being addressed more holistically than they were 20 years ago, the environment is generally still treated "apart" from social and economic considerations. Asia and the Pacific, for example, notes the lack of integration between environmental and economic policies as the major constraint on effective environmental management in the region. As can be seen throughout this chapter, a "Northern" development model still prevails (one sign is the accelerating growth of urban development based on car dependency), and despite progress on some fronts there is too much evidence of development to the detriment of environment, and too few signs of environment for development.

Economic growth and environmental protection are not mutually exclusive; efforts towards poverty alleviation and environmental protection can be mutually reinforcing. Improving the productivity of environmental resources (soils and fish stocks, for example), and investing in protecting and rehabilitating land and water resources can secure poverty reduction (UNDP 2005c). When the ecosystems on which the developing world's rural poor depend are kept healthy enough to provide them with food and income-generating opportunities, they are less likely to migrate to already overcrowded cities and emigrate to other countries. The economic value of ecosystem goods-and-services needs to be fully recognized, and countries need to strengthen their national policies to fully incorporate these values. Given the observed ecological impacts, and the projected consequences for human well-being that all regions point to, climate change needs to be addressed in a more concerted and aggressive fashion in all regions, and by the international community.

Reducing extreme poverty and hunger, the first Millennium Development Goal (MDG), requires work towards achieving MDG 7, which refers to the sustainable management of land, water and biodiversity resources, and the adequate provision of urban sanitation, potable water and waste management (World Bank 2002d). Both poverty and consumption are factors in environmental degradation. All people – rich and poor, urban and rural, and in all the world's regions – depend on environmental goods-and-services. The challenge is to foster "environment for development" in the developing world, while simultaneously slowing consumption in the developed world.

References

AC (1996). *Declaration on the Establishment of the AC.* Arctic Council Archive. http://www.arctic-council.org (last accessed 16 May 2007)

ACIA (2004). *Impacts of a warming Arctic.* Arctic Climate Impact Assessment. Cambridge University Press, Cambridge

ACIA (2005). *Arctic Climate Impact Assessment.* Cambridge University Press, Cambridge

ACSAD (2003). *Selected satellite images.* RS and GIS Unit Archive, Arab Center for the Studies in Arid Zones and Drylands, Damascus

ACSAD (2005). *Hydrogeological Study of Northern Palmyride Area,* Syria. Arab Center for the Studies in Arid Zones and Drylands, Damascus

ACSAD, CAMRE and UNEP (2004). *State of Desertification in the Arab World* (Updated Study)(In Arabic). Arab Center for the Studies in Arid Zones and Drylands, Damascus

ADB (2005). *Asia Development Outlook 2005.* Asian Development Bank, Manila (http://www.adb.org/Documents/Books/ADO/2005/default.asp (last accessed 5 May 2007)

ADB and GEF (2005). *The Master Plan for the Prevention and Control of Dust and Sandstorms in North-East Asia.* Asian Development Bank, Manila and Global Environment Facility, Washington, DC

AfDB (2004). *African Development Report 2004: Africa in the Global Trading System.* African Development Bank and Oxford University Press, Oxford

AfDB (2005). *African Development Bank Report 2005: Africa in the World Economy – Public Sector Management in Africa: Economic and Social Statistics on Africa.* African Development Bank and Oxford University Press, Oxford

AfDB (2006b). *Gender, Poverty and Environmental Indicators on African Countries.* Vol VII. Statistics Division, Development Research Department, African Development Bank, Tunis

AFRODAD (2005). *The Illegitimacy of External Debts: The Case of the Democratic Republic of Congo.* African Forum and Network on Debt and Development, Harare http://www.afrodad.org/downloads/publications/Illegitimate%20Debts%20-%20DRC.pdf (last accessed 5 May 2007)

AHDR (2004). *Arctic Human Development Report.* Stefansson Arctic Institute, Akureyi

Al-Dhabi, H., Koch, M., Al-Sarawi, M., and El-Baz, F. (1997). Evolution of sand dune patterns in space and time in north-western Kuwait using Landsat images. In *Journal of Arid Environments* 36:15-24

Al-Ghunaim, A. Y. (1997). *Devastating oil wells as revealed by Iraqi Documents.* Kuwait Institute for Scientific Research, Kuwait

Al-Humoud, J. M. (2005). Municipal solid waste recycling in the Gulf Cooperation Council States. In *Resources, Conservation and Recycling* 44:142-158

Al-Kassas, M. A. (1999). *Desertification; Land degradation in Dry Areas.* Alam Almarifah series No. 242 (In Arabic). The National Council for Culture, Art and Literature of Kuwait, Kuwait

Allen, C. R. (2006). Sprawl and the resilience of humans and nature: an introduction to the special feature. In *Ecology and Society* 11(1):36

Alley, R. B, Marotzke, J., Nordhaus, W. D., Overpeck, J. T., Peteet, D. M., Pielke Jr., R. A., Pierrehumbert, R. T., Rhines, P. B., Stocker, T. F., Talley, L. D. and Wallace, J. M. (2003). Abrupt climate change. In *Science* 299:2005-2010

Alley, R. B., Clark, P. U., Huybrechts, P. and Joughin, I. (2005). Ice-sheet and sealevel changes. In *Science* 310:456-460

Al-Rewaee, H. M. H. (2003). Water use efficiency to cultivate vegetable crops using soil less culture. MS.c. Thesis. Desert and Arid Zones Sciences Programme, Arabian Gulf University, Bahrain

Altamirano, T. (2003). From country to city: internal migration — focus on Peru. In *Revista: Harvard Review of Latin America* 2:58-61 (In Spanish) http://drclas.fas.harvard.edu/revista/articles/view_spanish/206 (last accessed 21 April 2007)

Al-Zubari, W. K. (2005). Groundwater Resources Management in the GCC Countries: Evaluation, Challenges, and Suggested Framework. Presented at *Water Middle East 2005 Conference,* Bahrain

Amann, M., Bertok, I., Cofala, J., Gyarfas, F., Heyes, C., Klimont, Z., Schöpp, W., and Winiwarter, W. (2005). *Baseline Scenarios for the Clean Air for Europe (CAFE) Programme.* International Institute for Applied Systems Analysis, Laxenburg

AMAP (2002a). *Arctic Pollution 2002 (Persistent Organic Pollutants, Heavy Metals, Radioactivity, Human Health, Changing Pathways).* Arctic Monitoring and Assessment Programme, Oslo

AMAP (2002b). *The Influence of Global Climate Change on Contaminant Pathways to, within, and from the Arctic.* Arctic Monitoring and Assessment Programme, Oslo

AMAP (2003). *AMAP Assessment 2002: Human Health in the Arctic.* Arctic Monitoring and Assessment Programme, Oslo

AMAP (2004a). *AMAP Assessment 2002: Heavy Metals in the Arctic.* Arctic Monitoring and Assessment Programme, Oslo

AMAP (2004b). *AMAP Assessment 2002: Persistent Organic Pollutants in the Arctic.* Arctic Monitoring and Assessment Programme, Oslo

AMAP (2004c). *Persistent Toxic Substances, Food Security and Indigenous Peoples of the Russian North.* Arctic Monitoring and Assessment Programme, Oslo

American Rivers (2005). *America's Most Endangered Rivers of 2005.* Washington, DC http://www.americanrivers.org/site/PageServer?pagename=AMR_endangeredrivers (last accessed 17 May 2007)

ASEAN (2006). Press Statement First Meeting of the Sub-Regional Ministerial Steering Committee (MSC) on Transboundary Haze Pollution http://www.aseansec.org/18807.htm (last accessed on 5 May 2007)

ASOC and UNEP (2005). *Antarctic Tourism Graphics, An overview of tourism activities in the Antarctic Treaty Area.* XXVIII ATCM Information Paper, Agenda Item 12. Submitted by the Antarctic and Southern Ocean Coalition and the United Nations Environment Programme to the XXVIII ATCM, Stockholm http://www.asoc.org/pdfs/2005%20XXVIII%20ATCM%20ASOC%20IP%20119%20Antarctic%20Tourism%20Graphics.pdf (last accessed 21 April 2007)

Asperen, P.C.M. van, and Zevenbergen, J.A. (2006). Towards effective pro-poor tools for land administration in Sub-Saharan Africa. In Gollwitzer, T., Hillinger, K. and Villikka, M. (eds.) *Shaping the Change; XXIII international FIG congress.* International Federation of Surveyors, Copenhagen

ATCM (2005). *Final Report XXVIII Antarctic Treaty Consultative Meeting.* Antarctic Treaty Secretariat, Buenos Aires

AWEA (2006). Annual industry rankings demonstrate continued growth of wind energy in the United States. *American Wind Energy Association News Releases, 15 March* http://www.awea.org/news/Annual_Industry_Rankings_Continued_Growth_031506.html (last accessed 5 May 2007)

Azmier, J. J. and Dobson, S. (2003). *The Burgeoning Fringe:* Western Canada's Rural Metro-Adjacent Areas. Canada West Foundation http://www.cwf.ca/V2/files/BurgeoningFringe.pdf (last accessed 5 May 2007)

Bails, J., Beeton, A., Bulkley, J., DePhilip, M., Gannon, J., Murray, M., Regier, H. and Scavia, D. (2005). *Prescription for Great Lakes Ecosystem Protection and Restoration Avoiding the Tipping Point of Irreversible Changes.* Healing Our Waters-Great Lakes Coalition http://restorethelakes.org/PrescriptionforGreatLakes.pdf (last accessed 5 May 2007)

Baldock, D., Beaufoy, G. and Clark, J. (eds.) (1994). *The Nature of Farming. Low Intensity Farming Systems in Nine European Countries.* Joint Nature Conservation Committee, Peterborough

Ballance, R. and Pant, B. D. (2003). *Environmental statistics in Central Asia – Progress and prospects,* ERD (Economics and Research Department), Working paper series No. 36, Asian Development Bank, Manila

Ballew, C., Ross, A., Wells, R. S. and Hiratsuka, V. (2004). *Final Report on the Alaska Traditional Diet Survey.* Alaska Native Health Board and Alaska Native Epidemiology Center http://www.anthc.org/cs/chs/epi/upload/traditional_diet.pdf (last accessed 21 April 2007)

Baker, L. (2000). Growing pains/malling America: the fast-moving fight to stop urban sprawl. In *EMagazine* 11: 3

Barbier, Edward B. (1997). Introduction to the Environmental Kuznets Curve Special Issue. In *Environment and Development Economics* 2(4): 369-81

Barth, H. J. (1999). Desertification in the Eastern Province of Saudi Arabia. In *Journal of Arid Environments* (1999)43:399-410. http://www.uni-regensburg.de/Fakultaeten/phil_Fak_III/Geographie/phygeo/downloads/bartharid43.pdf (last accessed 5 May 2007)

Basheer, C., Obbard, J. P. and Lee, H. K. (2003). Persistent organic pollutants in Singapore's coastal marine environment: Part I, seawater and Part II, sediments. In *Water Air and Soil Pollution* 149(1-4):295-313; 315-325

Bass, F. and Beamish, R. (2006). Development Inches Toward National Parks. Discovery News http://dsc.discovery.com/news/2006/06/19/nationalpark_pla.html?category=earth&guid=20060619120030&dcitc=w19-502-ak-0000 (last accessed 5 May 2007)

Bastmeijer, K. and Roura, R. (2004). Regulating Antarctic tourism and the precautionary principle. In *American Journal of International Law* 98:763-781

Beach, D. (2002). *Coastal Sprawl: The Effects of Urban Design on Aquatic Ecosystems in the United States.* Pew Oceans Commission http://www.pewtrusts.org/pdf/env_pew_oceans_sprawl.pdf (last accessed 5 May 2007)

Beaulieu, M. S. (2004). Manure Management in Canada. In *Farm Environmental Management in Canada* 1(2) http://www.statcan.ca/english/research/21-021-MIE/21-021-MIE2004001.pdf (last accessed 5 May 2007)

Bell, G., Blake, E., Landsea, C., Mo, K., Pasch, R., Chelliah, M. and Goldenberg, S. (2005). *The 2005 North Atlantic Hurricane Season: A Climate Perspective.* NOAA Climate Prediction Center, National Hurricane Center, and the Hurricane Research Division http://www.cpc.noaa.gov/products/expert_assessment/hurrsummary_2005.pdf (last accessed 5 May 2007)

Bengston, D. N., Fletcher, J. O. and Nelson, K. C. (2004). Public policies for managing urban growth and protecting open space: policy instruments and lessons learned in the United States. In *Landscape and Urban Planning* 69:271-286 http://www.ncrs.fs.fed.us/pubs/jrnl/2003/nc_2003_bengston_001.pdf (last accessed 1 June 2007)

Boko, M., I. Niang, A. Nyong, C. Vogel, A. Githeko, M. Medany, B. Osman-Elasha, R. Tabo and P. Yanda, 2007: Africa. *Climate Change 2007: Impacts, Adaptation and Vulnerability. Contribution of Working Group II to the Fourth Assessment Report of the Intergovernmental Panel on Climate Change,* M.L. Parry, O.F. Canziani, J.P. Palutikof, P.J. van der Linden and C.E. Hanson, Eds., Cambridge University Press, Cambridge UK, 433-467.

Bosworth, D. (2003). We need a new national debate. In *Izaak Walton League, 81st Annual Convention, 17 July,* Pierre, SD http://www.fs.fed.us/news/2003/speeches/07/bosworth.shtml (last accessed 17 May 2007)

Boyd, D. R. (2006). *The Water We Drink: An International Comparison of Drinking Water Quality Standards and Guidelines.* David Suzuki Foundation, Vancouver, BC http://www.davidsuzuki.org/WOL/Publications.asp (last accessed 17 May 2007)

Braga, M.C.B. and Bonetto, E.R. (1993). Solid Waste Management in Curitiba, Brazil - Alternative Solutions. In *The Journal of Solid Waste Technology and Management* 21(1)

Brauer, J. (2000). The Effect of War on the Natural Environment. In *Arms, Conflict, Security and Development Conference, 16-17 June, Middlesex University Business School, London* http://www.aug.edu/~sbajmb/paper-london3.PDF (last accessed 17 May 2007)

Braune, B. M., Outridge, P. M., Fisk, A. T., Muir, D. C. G., Helm, P. A., Hobbs, K., Hoekstra, P. F., Kuzyk, Z. A., Kwan, M., Letcher, R. J., Lockhart, W. L., Norstrom, R. J., Stern, G. A. and Stirling, I. (2005). Persistent organic pollutants and mercury in marine biota of the Canadian Arctic: an overview of spatial and temporal trends. In *Science of the Total Environment* 4(56):351-352

Braune, B. M., Mallory, M. L. and Gilchrist, H. G. (2006). Elevated mercury levels in a declining population of ivory gulls in the Canadian Arctic. In *Marine Pollution Bulleting* PMID: 16765993 in process

Bravo, H., Roy-Ocotla, G., Sanchez, P., and Torres, R. (1992). La contaminación atmosférica por ozono en la zona Metropolitana de la Ciudad de México. En I. Restrepo (coord.) In *La contaminación del aire en México: Sus causas y efectos en la salud.* Comisión Nacional de los Derechos Humanos, Mexico, DF

Brigden, K., Labunska, I., Santillo, D. and Allsopp, M. (2005). *Recycling of Electronic Wastes in China and India: Workplace and Environmental Contamination.* Greenpeace Research Laboratories, Department of Biological Sciences, University of Exeter, Exeter http://www.greenpeace.org/raw/content/china/en/press/reports/recycling-of-electronic-wastes.pdf (last accessed 23 April 2007)

Brocklesby, M. A. and Hinshelwood, E. (2001). *Poverty and the Environment: What the Poor Say: An Assessment of Poverty-Environment Linkages in Participatory Poverty Assessments.* Department for International Development, London http://www.dfid.gov.uk/Pubs/files/whatthepoorsay.pdf (last accessed 21 April 2007)

Browder, J. D., and Godfrey, B. J. (1997). *Rainforest Cities: Urbanization, Development and Globalization of the Brazilian Amazon.* Columbia, New York, NY

Bryant, D., Rodenburg, E. Cox, T. and Nielsen, D. (1996). *Coastlines at Risk: an Index of Potential Development-Related Threats to Coastal Ecosystems.* World Resources Institute, Washington, DC

Bryant, D., Burke, L. McManus, J. and Spalding, M. (1998). *Reefs at Risk. A Map-Based Indicator of Threats to the World's Coral Reefs.* World Resources Institute, Washington, DC

Bryceson, I., De Souza, T. F., Jehangeer, I., Ngoile, M. A. K. and Wynter, P. (1990). *State of the Marine Environment in the East African Region.* UNEP Regional Seas Reports and Studies no. 113. United Nations Environment Programme, Nairobi

Bryden, H. L., Longworth, H. R. and Cunningham, S. A. (2005). Slowing of the Atlantic meridional overturning circulation at 25°N. In *Nature* 438:655-657

Burnham, G., Doocy, S., Dzeng, E., Lafta, R. and Robert, L. (2006). *The Human Cost of the War in Iraq A Mortality Study, 2002-2006.* Bloomberg School of Public Health, Johns Hopkins University, Baltimore, Maryland; School of Medicine, Al Mustansiriya University, Baghdad, Iraq ; and the Center for International Studies, Massachusetts Institute of Technology, Cambridge, Massachusetts. http://web.mit.edu/CIS/pdf/Human_Cost_of_War.pdf (last accessed 16 May 2007)

Burke, L., Kura, Y., Kassem, K., Revenga, C., Spalding, M., and McAllister, D. (2001). *Pilot Analysis of Global Ecosystems: Coastal Ecosystems.* World Resources Institute, Washington, DC

Butayban, N. (2005). *An Overview of Land Based Sources of Marine Pollution in ROPME Sea Area.* Environment Public Authority, Kuwait

CAFF (2001). *Arctic Flora and Fauna: Status and Conservation.* Edita, Helsinki

CAI-Asia and APMA (2004). Air Quality in Asian Cities. Clean Air Initiative – Asia and Air Pollution in the Mega-cities Project http://www.cleanairnet.org/caiasia/1412/articles-59689_AIR.pdf (last accessed 21 April 2007)

Cambers, G. (1997). Beach changes in the Eastern Caribbean Islands: Hurricane impacts and implications for climate change. In *Journal of Coastal Research* Special Issue 24:29-47

Cameron, R. D., Smith, W. T., White, R. G. and Griffith, B. (2005). Central Arctic caribou and petroleum development: distributional, nutritional, and reproductive implications. In *Arcti* 58:1-9

Carabias, J. (2002). Conservación de los Ecosistemas y el Desarrollo Rural sustentable en América Latina: Condiciones, limitantes y retos. In Leff, E., Ezcurra, E., Pisanty, I., Romero-Lankau, P. (coords). *La transición haca el desarrollo sustentable. Perspectivas desde América Latina y El Caribe*. Instituto Nacional de Ecología, México, DF

Carius, A., Dabelko, G. D. and Wolf, A. T. (2006). Water, conflict, and cooperation. United Nations and Global Security Initiative http://www.un-globalsecurity.org/pdf/Carius_Dabelko_Wolf.pdf (last accessed 21 April 2007)

Carrington, W. J. and Detragiache, E. (1999). How extensive is the brain drain? In *Finance and Development* 36(2) http://www.imf.org/external/pubs/ft/fandd/1999/06/index.htm (last accessed 17 May 2007)

CBP (2007). Chesapeake Bay 2006 Health and Restoration Assessment: Part One, Ecosystem Health. Chesapeake Bay Program http://www.chesapeakebay.net/press.htm (last accessed 24 April 2007)

CBO (2002 Draft). Future Investment in Drinking Water and Wastewater Infrastructure. Congressional Budget Office http://www.cbo.gov/ftpdocs/39xx/doc3983/11-18-WaterSystems.pdf (last accessed 21 April 2007)

CCME (2005). Canada-Wide Standards for Mercury Emissions from Coal-Fired Electric Power Generation Plants. Canadian Council of Ministers of the Environment (Draft report) http://www.ccme.ca/assets/pdf/canada_wide_standards_hgepg.pdf (last accessed 17 May 2007)

CEC (2004). *North American Air Quality and Climate Change Standards, Regulations, Planning and Enforcement at the National, State/Provincial and Local Levels*. Commission for Environmental Cooperation of North America, Montreal

CEC (2006). *Children's Health and the Environment in North America. A First Report on Available Indicators and Measures*. Commission for Environmental Cooperation, Montreal http://www.cec.org/files/pdf/POLLUTANTS/CEH-Indicators-fin_en.pdf (last accessed 17 March 2007)

CEPAL (2005). *Objetivos de Desarrollo del Milenio: una mirada desde América Latina y el Caribe*. Comisión Económica de las Naciones Unidas para América Latina, LC/G.2331, Junio, Santiago de Chile

CEPAL (2006). Energía y desarrollo sustentable en América Latina: Enfoques para la política energética. Presentación de Hugo Altomonte en *Regional Implementation Forum on Sustainable Development, 19-20 enero*, Santiago de Chile

CFS (2004). Wildland–Urban Interface. Canadian Forest Service, Natural Resources Canada http://fire.cfs.nrcan.gc.ca/research/management/wui_e.htm (last accessed 17 May 2007)

CGLG (2005). *Governors and Premiers sign agreements to protect Great Lakes Water*. Council of Great Lakes Governors http://www.cglg.org/projects/water/docs/12-13-05/Annex_2001_Press_Release_12-13-05.pdf (last accessed 17 May 2007)

Chapin, F. S., III, Berman, M., Callaghan, T. V., Crepin, A.-S., Danell, K., Forbes, B. C., Kofinas, G., McGuire, D., Nuttall, M., Pungowiyi, C., Young, O. and Zimov, S. (2005). Polar systems. In R. Scholes (ed). *Millennium Ecosystem Assessment*. Island Press, Washington, DC

Charkasi, D. (2000). Balancing the use of old and new agricultural varieties to sustain agrobiodiversity. In *Dryland Agrobio* No. 3, October-December http://www.icarda.org/gef/newsLetter34.html (last accessed 17 May 2007)

Chen, B., Hong, C. and Kan, H. (2001). *Integrated Assessment of Energy Options and Health Benefits in Shanghai*. Final report to USEPA and USNREL. (in English & Chinese). http://www.epa.gov/ies/documents/shanghai/full_report_chapters/ch9pdf (last accessed 20 June 2007)

Chiuchiolo, A. L., Dickhut, R. M., Cochran, M. A. and Ducklow, H. W. (2004). Persistent organic pollutants at the base of the Antarctic marine food web. In *Environ. Sci. Technology* 38:3551

CHMI (2003). *Air pollution in the Czech Republic in 2003*. Czech Hydrometeorological Insitute, Air Quality Protection Division http://www.chmi.cz/uoco/isko/groce/gr03e/akap3.html (last accessed 21 April 2007)

Christ, C., Hillel, O., Matus, S. and Sweeting, J. (2003). *Tourism and Biodiversity: Mapping Tourism's Global Footprint*. Conservation International, Washington, DC http://www.unep.org/PDF/Tourism_and_biodiversity_report.pdf (last accessed 17 May 2007)

Christian Reformed Church (2005). *Global Debt. An OSJHA Fact Sheet*. Office of Social Justice and Hunger Action http://www.crcna.org/site_uploads/uploads/factsheet_globaldebt.doc (last accessed 21 April 2007)

Ciais, Ph., Reichstein, M., Viovy, N., Granier, A., Ogée, J., Allard, V., Aubinet, M., Buchmann, N., Bernhofer, Chr., Carrara, A., Chevallier, F., De Noblet, N., Friend, A. D., Friedlingstein, P., Grünwald, T., Heinesch, B., Keronen, P., Knohl, A., Krinner, G., Loustau, D., Manca, G., Matteucci, G., Miglietta, F., Ourcival, J. M., Papale, D., Pilegaard, K., Rambal, S., Seufert, G., Soussana, J. F., Sanz, M. J., Schulze, E. D., Vesala, T. and Valentini, R. (2005). Europe-wide reduction in primary productivity caused by the heat and drought in 2003. In *Nature* 437(7058):529-533

CIFOR (2007). *Nature, wealth and power to defeat poverty in Africa* http://www.cifor.cgiar.org/Publications/Corporate/NewsOnline/NewsOnline35/defeat_poverty.htm (last accessed on 28 April 2007)

CLAES (2003). *Ambiente En América Latina: Los seis hechos ambientales más importantes en América Latina. La tendencia sobresaliente en la gestión ambiental*. Centro Latino Americano de Ecología Social, Montevideo http://www.ambiental.net/noticias/ClaesAmbienteAmericaLatina.pdf (last accessed 5 May 2007)

Cohen, A. J., Anderson, H. R., Ostra B., Pandey, K. D., Krzyzanowski, M., Künzli, N., Gutschmidt, K., Pope, A., Romieu, I., Samet, J. M. and Smith, K. (2005). The global burden of disease due to outdoor air pollution. In *Journal of Toxicology and Environmental Health* 68 (1):1-7

Cohn, J. P. (2004). Colorado River Delta. In *BioScience* 54(4):386-91

Columbia Encyclopedia (2003). Sahara. In *The Columbia Encyclopedia* Sixth Edition, 2001-05. Columbia University Press, New York, NY http://www.bartleby.com/65/sa/Sahara.html (last accessed 17 May 2007)

CONABIO (2006). *Capital Natural y Bienestar Social*. Comisión nacional para el conocimiento y uso de la biodiversidad, Mexico, DF

CONAPO (2004). Informe de Ejecución 2003-2004 del programa nacional de la poblacion 2001-2006. Consejo Nacional de la Población, Mexico, DF

Conner, R., Seidl, A., VanTassel, L. and Wilkins, N. (2001). *United States Grasslands and Related Resources: An Economic and Biological Trends Assessment*. Land Information Systems, Texas A&M Institute of Renewable Natural Resources, Tamu http://landinfo.tamu.edu/presentations/grasslands.cfm (last accessed 21 April 2007)

Conservation International (2006). World's Largest Tropical Forest Reserve Created in Amazon. http://www.conservation.org/xp/news/press_releases/2006/120406.xml (last accessed 26 June 2007)

Corsolini, S., Kannan, K., Imagawa, T., Focardi, S. and Giesy, J. P. (2002). Polychloronaphthalenes and other dioxin-like compounds in Arctic and Antarctic marine food webs. In *Environ. Sci. Technol*, 36(16):3490-3496

Council of Europe (2003a). 3rd International Symposium of the Pan-European Ecological Network - Fragmentation of habitats and ecological corridors - Proceedings, Riga, October 2002. In Environmental Encounters No. 54. Council of Europe Publishing, Strasbourg

Court, T. de la (1990). *Beyond Brundtland: Green Development in the 1990s*. (Translated by Bayens, E. and Harle, N.) New Horizons Press, New York, Zed Books Ltd, London and New Jersey

CPCB (1996). *Annual Report 1995-1996*. Central Pollution Control Board, New Delhi

Croxall, J. P., Trathan, P. N. and Murphy, E. J. (2002). Environmental change and Antarctic seabird populations. In *Science* 297:1510-1514

CRU (2007). *CRUTEM3v dataset*. Climate Research Unit, University of East Anglia. http://www.cru.uea.ac.uk/cru/data/temperature (last accessed 6 April 2007)

CSB (1987-2004). *China Statistical Yearbook 1987-2004* (in Chinese). China Statistical Bureau, China Statistics Press, Beijing

Cunningham, A. (2001). *Applied Ethnobotany: People, Wild Plant Use and Conservation*. Earthscan Publications Ltd, London

Darkoh, M. B. (1993). Desertification: the scourge of Africa. In *Tiempo (Tiempo Climate Cyberlibrary)* 8 http://www.cru.uea.ac.uk/cru/tiempo/issue08/desert.htm (last accessed 1 June 2007)

Davis, C. H., Yonghong, L., McConnell, J. R., Frey, M. M. and Hanna, E. (2005). Snowfall-driven growth in East Antarctic ice sheet mitigates recent sea-level rise. In *Science* 308:1898-1901

DeCoster, L. A. (2000). Summary of the Forest Fragmentation 2000 Conference. In Decoster, L. A. (ed.) *Fragmentation 2000 – A Conference on Sustaining Private Forests in the 21st Century* Annapolis, MA http://www.sampsongroup.com/acrobat/fragsum.pdf (last accessed 17 May 2007)

Defenders of Wildlife (2006). Issues in Multilateral Trade Agreements with Environmental Impacts. http://www.defenders.org/international/trade/issues.html (last accessed 21 April 2007)

De Mora, S., Scott, W., Imma, T., Jean-Pierre, V. and Chantal, C. (2005). Chlorinated hydrocarbons in marine biota and coastal sediments from the Gulf and Gulf of Oman. In *Marine Pollution Bulletin* 50

DFID, EC, UNDP and World Bank (2002). *Linking Poverty Reduction and Environmental Management: Policy Challenges and Opportunities*. Department for International Development, European Commission, United Nations Development Programme, and The World Bank, Washington, DC http://www.undp.org/pei/pdfs/LPREM.pdf (last accessed 6 May 2007)

Dinerstein, E., Olson, D. M., Graham, D. J., Webster, A. L., Primm, S. A., Bookbinder, M. P. and Ledec, G. (1995). In *Una Evaluación del Estado de Conservación de las Eco-regiones Terrestres de América Latina y el Caribe*. Banco Mundial en colaboración con el Fondo Mundial para la Naturaleza, Washington, DC

Donahue, W. F. and Schindler, D.W. (2006). Whiskey's for drinkin' and water's for fightin': Climate change and water supply in the Western Canadian Prairies. In *59th Canadian Conference for Fisheries Research, 5-7 January*. Calgary, Alberta

Downie, D. L. and Fenge, T. (eds.) (2003). *Northern Lights Against POPs: Combatting Toxic Threats in the Arctic*. McGill-Queen's University Press, Montreal and Kingston

DPI (2005). Dubai Property Investment. Palm Islands http://dubai.property-investment.com (last accessed 17 May 2007)

Dufour D. L. and Piperata, B. A. (2004). Rural-to-Urban Migration in Latin America: An Update and Thoughts on the Model. In *American Journal of Human Biology* 16:395-404

Dybas, C. L. (2005). Dead zones spreading in world oceans. In *BioScience* 55(7):552-557

EAP Task Force (2006). *Regional Meeting on Progress in Achieving the Objectives of the EECCA Environment Strategy*, Kiev, 18-19 May 2006

Easterling, W., Hurd, B. and Smith, J. (2004). *Coping with Global Climate Change: The Role of Adaptation in the United States*. Pew Center on Global Climate Change, Arlington, VA http://www.pewclimate.org/global-warming-in-depth/all_reports/adaptation/index.cfm (last accessed 5 May 2007)

EC (2004). *Forest fires in Europe 2003 fire campaign. European Commission*. Official Publication of the European Communities, SPI.04.124 EN. Luxembourg

EC (2005a). *Proposal for a directive of the European Parliament and of the Council on ambient air quality and cleaner air for Europe*. COM(2005) 447. European Commission, Brussels

EC (2005b). *Thematic Strategy on Air Pollution*. COM(2005) 446 final. Commission the European Communities, Brussels http://eur-lex.europa.eu/LexUriServ/site/en/com/2005/com2005_0446en01.pdf (last accessed 17 April 2007)

EC (2006a). *Ministerial Declaration on Enhanced energy co-operation between the EU, the Littoral States of the Black and Caspian Seas and their neighbouring countries*. 30 November 2006, Astana http://www.inogate.org/en/news/30-november-2006/ (last accessed 17 May 2007)

EC (2006b). *Environmental Impact of Products (EIPRO), Analysis of the life cycle environmental impacts related to the final consumption of the EU-25*. Main report, European Commission, Brussels

EC (2006c). *Halting the Loss of Biodiversity by 2010 and Beyond – Sustaining Ecosystem Services for Human Well-Being*. COM(2006) 216 final. European Commission, Brussels

EC (2007a). Presidency Conclusions of the Brussels European Council (8/9 March 2007)

EC (2007b). *Limiting Global Climate Change to 2 degrees Celsius: The way ahead for 2020 and beyond*. COM(2007) 2 final. Communication from the Commission to the Council, the European Parliament, the European Economic and Social Committee and the Committee of the Regions, Brussels

ECHAVARRÍA, M. (2002). Water user associations in the Cauca Valley, Colombia. A voluntary mechanism to promote upstream -downstream cooperation in the protection of rural watersheds. In: *FAO Land-Water Linkages in Rural Watersheds Case Study Serie*. Food and Agriculture Organization of the United Nations, Rome

ECI (2006). Climate Change: An Evangelical Call to Action. Evangelical Climate Initiative http://www.christiansandclimate.org/statement (last accessed 21 April 2007)

EEA (1999). *Sustainable water use in Europe - Part :1 Sectoral use of water*. Environmental assessment report No. 1. European Environment Agency, Copenhagen http://reports.eea.europa.eu/binaryeenviasses01pdf/en (last accessed 9 May 2007)

EEA (2001). *Sustainable water use in Europe - Part 2: Demand management*. Environmental issue report No. 19. European Environment Agency, Copenhagen http://reports.eea.europa.eu/Environmental_Issues_No_19/en (last accessed 9 May 2007)

EEA (2003). *Europe's water: An indicator-based assessment*. EEA topic report 1/2003. European Environment Agency, Copenhagen http://reports.eea.europa.eu/topic_report_2003_1/en (last accessed 9 May 2007)

EEA (2004a). *High nature value farmland. Characteristics, trends and policy challenges*. EEA report No 1/2004. European Environment Agency, Copenhagen http://reports.eea.europa.eu/report_2004_1/en (last accessed 9 May 2007)

EEA (2004b). *Agriculture and the environment in the EU accession countries. Implications of applying the EU common agricultural policy*. Environmental issue report No 37. European Environment Agency, Copenhagen http://reports.eea.europa.eu/environmental_issue_report_2004_37/en (last accessed 9 May 2007)

EEA (2004c). *Air pollution by ozone in Europe in summer 2003 - Overview of exceedances of EC ozone threshold values during the summer season April-August 2003 and comparisons with previous years*. Topic report No 3/2003. European Environment Agency, Copenhagen http://reports.eea.europa.eu/topic_report_2003_3/en (last accessed 9 May 2007)

EEA (2004d). *Arctic environment: European perspectives*. Environmental issue report No 38/2004. European Environment Agency, Copenhagen http://reports.eea.europa.eu/environmental_issue_report_2004_38/en (last accessed 9 May 2007)

EEA (2005a). *Agriculture and environment in the EU 15 – the IRENA indicator report*. EEA report No 6/2005. European Environment Agency, Copenhagen http://reports.eea.europa.eu/eea_report_2005_6/en (last accessed 9 May 2007)

EEA (2005b). *Household Consumption and the Environment*. EEA report No 11/2005. European Environment Agency, Copenhagen http://reports.eea.europa.eu/eea_report_2005_11/en (last accessed 9 May 2007)

EEA (2006a). *Greenhouse gas emission trends and projections in Europe 2006*. EEA Report No. 9/2006. European Environment Agency, Copenhagen http://reports.eea.europa.eu/eea_report_2006_9/en (last accessed 9 May 2007)

EEA (2006b). *Energy and environment in the European Union - Tracking progress towards integration*. EEA Report No 8/2006 European Environment Agency, Copenhagen http://reports.eea.europa.eu/eea_report_2006_8/en (last accessed 9 May 2007)

EEA (2006c). *Transport and environment: facing a dilemma*. EEA report No. 3/2006. European Environment Agency, Copenhagen http://reports.eea.europa.eu/eea_report_2006_3/en (last accessed 9 May 2007)

EEA (2006d). *Exceedance of air quality limit values in urban areas (CSI 004)*. EEA Core Set of Indicators. http://themes.eea.europa.eu/IMS/ISpecs/ISpecification20041001123040/IAssessment1153220262064/view_content (last accessed 9 May 2007)

EEA (2006e). *Priority issues in the Mediterranean environment (revised edition)*. EEA Report No 4/2006. European Environment Agency, Copenhagen http://reports.eea.europa.eu/eea_report_2006_4/en

EEA (2007). *Europe's Environment: the Fourth Assessment*. European Environment Agency, Copenhagen

Eilers, W. and Lefebvre, A. (2005). National and regional summary. In *Environmental Sustainability of Canadian Agriculture: Agri-Environmental Indicator Report Series - Report #2*, Lefebvre, A., Eilers, W. and Chunn, B. (eds.). Agriculture and Agri-Food Canada http://www.agr.gc.ca/env/naharp-pnarsa/pdf/2005_AEI_report_e.pdf (last accessed 21 April 2007)

EIP (2005). *Backed Up: Cleaning Up Combined Sewer Systems in the Great Lakes*. Environmental Integrity Project, Washington, DC http://www.environmentalintegrity.org/pubs/EIP_BackedUp_fnl.pdf (last accessed 5 May 2007)

ElHadj, E. (2004). *The household water crisis in Syria's Greater Damascus Region*. SOAS Water Research Group Occasional paper 47. School of Oriental and African Studies and King's College, London http://www.soas.ac.uk/waterissues/occasionalpapers/OCC47.pdf (last accessed 5 May 2007)

El Nasser, H. and Overberg, P. (2001). A comprehensive look at sprawl in America: the USA Today sprawl index. In *USA Today* 22 February http://www.usatoday.com/news/sprawl/main.htm (last accessed 5 May 2007)

EM-DAT (undated). *Emergency Events Database: The OFDA/CRED International Disaster Database* (in GEO Data Portal). Université Catholique de Louvain, Brussels

Environment Canada (2001). *Urban Water Indicators: Municipal Water Use and Wastewater Treatment*. Environment Canada http://www.ec.gc.ca/soer-ree/English/Indicators/Issues/Urb_H2O/default.cfm (last accessed 5 May 2007)

Environment Canada (2003). *Environmental Signals: Canada's National Environmental Indicator Series 2003*. Environment Canada http://www.ec.gc.ca/soer-ree/English/Indicator_series (last accessed 17 May 2007)

Environment Canada (2007). Canada's new government announces mandatory industrial targets to tackle climate change and reduce air pollution. *Environment Canada News Releases*, 26 April http://www.ec.gc.ca/default.asp?lang=En&n=714D9AAE-1&news=4F2292E9-3EFF-48D3-A7E4-CEFA05D70C21 (last accessed 27 April 2007)

Environment Canada and USEPA (2005). *State of the Great Lakes 2005: Highlights*. Environment Canada and the U.S. Environmental Protection Agency http://www.epa.gov/glnpo/solec/solec_2004/highlights/SOGL05_e.pdf (last accessed 21 April 2007)

ERWDA (2003). *Report on Conservation of Dugong in the UAE*. Environmental Research and Wildlife Development Agency, UAE

ESA (2004). Artificial island arises off Dubai. In *ESA News: Protecting the Environment (European Space Agency)* http://www.esa.int/esaCP/SEMKRXZO4HD_Protecting_0.html (last accessed 21 April 2007)

Ewel, K. C., Twilley, R. R. and Ong, J. E. (1998). Different kinds of mangrove forests provide different goods and services. In *Global Ecology and Biogeography Letters* 7(1)83-94

EWG (2005). *Farm Subsidy Database: New EWG farm subsidy database re-ignites reform efforts*. Environmental Working Group's Farm Subsidy Database http://www.ewg.org/farm/region.php?fips=00000 (last accessed 17 May 2007)

Ewing, R., Kostyack, J., Chen, D., Stein, B. and Ernst, M. (2005). *Endangered by Sprawl: How Runaway Development Threatens America's Wildlife*. National Wildlife Federation, Smart Growth America, Nature Serve, Washington, DC http://www.nwf.org/nwfwebadmin/binaryVault/EndangeredBySprawlFinal.pdf (last accessed 5 May 2007)

Export Processing Zones Authority (2005). *Tea and Coffee Industry in Kenya*. Export Processing Zones Authority, Nairobi

Ezcurra, E., Mazari, M., Pisanty, I. and Guillermo, A. (2006). *La Cuenca de México: Aspectos Ambientales Críticos Y Sustentabilidad*. Fondo de cultura económica, México, DF

Ezzati, M., Rodgers, A. D., Lopez, A. D. and Murray, C. J. L., (eds) (2004a). *Comparative Quantification of Health Risks: Global and Regional Burden of Disease Due to Selected Major Risk Factors*. 3 vols. World Health Organization, Geneva

Ezzati M., Bailis, R., Kammen, D. M., Holloway, T., Price, L., Cifuentes, L. A., Barnes, B., Chaurey, A. and Dhanapala, K. N. (2004b). Energy management and global health. In *Annual Review of Environment and Resources* 29:383-419

FAO (1997). *Irrigation Potential in Africa: A Basin Approach*. FAO Land and Water Bulletin 4, Land and Water Development Division, Food and Agriculture Organization of the United Nations, Rome http://www.fao.org/docrep/W4347E/w4347e0o.htm (last accessed 25 September 2006)

FAO (2002). *Comprehensive Africa Agriculture Development Programme, New Partnership for Africa's Development (NEPAD)*. Food and Agriculture Organization of the United Nations, Rome http://www.fao.org/documents/show_cdr.asp?url_file=/docrep/005/Y6831E/y6831e00.htm (last accessed 3 June 2007)

FAO (2003a). *Forestry Outlook Study for Africa - African Forests: A View to 2020*. European Commission, African Development Bank and the Food and Agriculture Organization of the United Nations, Rome

FAO (2003b). *Status and Trends in Mangrove Area Extent Worldwide*. Food and Agriculture Organization of the United Nations, Rome http://www.fao.org/docrep/007/j1533e/j1533e00.htm (last accessed 4 June 2007)

FAO (2003c). FAO Gender and Development Plan of Action (2002-2007). http://www.fao.org/WAICENT/FAOINFO/SUSTDEV/2002/PE0103_en.htm (last accessed 27 September 2006)

FAO (2004). *The State of World Fisheries and Aquaculture 2004*. Food and Agricultural Organization of the United Nations, Rome

FAO (2005). *Global Forest Resources Assessment 2005*. Food and Agriculture Organization of the United Nations, Rome (in GEO Data Portal)

FAO (2006a). *Global Forest Resources Assessment 2005. Report on fires in the Central Asian Region and adjacent countries*. Fire Management Working Paper 16. FAO-Forestry Department, Food and Agriculture Organization of the United Nations, Rome http://www.fire.uni-freiburg.de/programmes/un/fao/FAO-Final-12-Regional-Reports-FRA-2005/WP%20FM16E%20Central%20Asia.pdf (last accessed 17 April 2007)

FAO (2006b). *Global Forest Resources Assessment 2005. Report on fires in the Balkan Region and adjacent countries*. Fire Management Working Paper 11. FAO-Forestry Department, Food and Agriculture Organization of the United Nations, Rome http://www.fire.uni-freiburg.de/programmes/un/fao/FAO-Final-12-Regional-Reports-FRA-2005/WP%20FM11E%20Balkan.pdf (last accessed 17 April 2007)

FAO (2007a). *State of the World's Forests 2007*. Food and Agriculture Organization of the United Nations, Rome http://www.fao.org/docrep/009/a0773e/a0773e00.htm (last accessed 3 June 2007)

FAO (2007b). *The State of Food and Agriculture 2006. Food Aid or Food Security*. Food and Agriculture Organization of the United Nations, Rome http://www.fao.org/docrep/009/a0800e/a0800e00.htm (last accessed 4 June 2007)

FAOSTAT (2004). Food and Agriculture Organization Statistical Database (in GEO Data Portal)

FAOSTAT (2005). FAO Statistical Databases. Food and Agriculture Organization of the United Nations, Rome (in GEO Data Portal)

FAOSTAT (2006). FAO Statistical Databases. Food and Agriculture Organization of the United Nations, Rome (in GEO Data Portal)

FAOSTAT (2007). FAO Statistical Databases. Food and Agriculture Organization of the United Nations, Rome (in GEO Data Portal)

FAO AQUASTAT (2007). FAO's Information System on Water in Agriculture. Food and Agriculture Organization of the United Nations, Rome (in GEO Data Portal)

FAO TERRASTAT (2003). Land resource potential and constraints statistics at country and regional level. http://www.fao.org/ag/agl/agll/terrastat (last accessed 9 May 2007)

FCM (2005). *Partners for Climate Protection*. Federation of Canadian Municipalities http://kn.fcm.ca/ev.php?URL_ID=2805andURL_DO=DO_TOPICandURL_SECTION=201andreload=1122483013 (last accessed 27 July 2005)

FEWSNET (2005). FEWS Somalia food security emergency 25 August 2005: poor harvest and civil insecurity hit South. *Relief Web. Famine Early Warning System Network (FEWS NET)*, 25 August http://www.reliefweb.int/rw/RWB.NSF/db900SID/RMOI-6FM7CV?OpenDocument (last accessed 10 May 2007)

Fischlowitz-Roberts, B. (2002). Air pollution fatalities now exceed traffic fatalities by 3 to 1. In *Earth Policy Institute Eco-Economy Updates*, 17 September http://www.earth-policy.org/Updates/Update17.htm (last accessed 1 June 2007)

Forbes, B. C. (1999). Land use and climate change in the Yamal-Nenets region of northwest Siberia: Some ecological and socio-economic implications. In *Polar Research* 18:1-7

Forbes, B. C., Fresco, N., Shvidenko, A., Danell, K. and Chapin III, F. C. (2004). Geographic variations in anthropogenic drivers that influence the vulnerability and resilience of social-ecological systems. In *Ambio* 33:377-382

Ford, J. D., Smit, B. and Wandel, J. (2006). Vulnerability to climate change in the Arctic: a case study from Arctic Bay, Canada. In *Global Environmental Change* 16:145-160

Frederick, J. E. and Lubin, D. (1994). Solar ultraviolet irradiance at Palmer Station, Antarctica. In *Ultraviolet Radiation in Antarctica: Measurement and Biological Effects*, Weiler, C. S. and Penhale, P. A. (eds). Antarctic Research Series 62, American Geophysical Union

Frenot, Y., Chown S. L., Whinam, J., Selkirk P. M., Convey P., Skotnicki, M. and Bergstrom D. M. (2004). Biological invasions in the Antarctic: extent, impacts and implications. In *Biological Review* 79:1-28

Frumkin, H., Frank, L. and Jackson, R. (2004). *Urban Sprawl and Public Health: Designing, Planning, and Building for Healthy Communities*. Island Press, Washington, DC

Furgal, C. M., Powell, S. and Myers, H. (2005). Digesting the message about contaminants and country foods in the Canadian North: a review and recommendations for future research and action. In *Arctic* 58:103-114

Gabaldón, A. J. and Rodríguez Becerra, M. (2002). Evolución de las políticas e instituciones ambientales: ¿Hay motivo para estar satisfechos? In *La Transición Hacia el Desarrollo Sustentable: Perspectivas de América Latina y El Caribe*. Leff, E., Ezcurra, E., Pisanty, I. and Romero-Lankau, P. (2002). Instituto Nacional de Ecología, Universidad Autónoma Metropolitana and Programa de Naciones unidas para el Medio Ambiente, México, DF

Galafassi, G. P. (2002). Ecological crisis, poverty and urban development in Latin America. In *Democracy and Nature* 8(1):17-131

GAO (2005). Increased Permitting Activity Has Lessened BLM's Ability to Meet Its Environmental Protection Responsibilities. *US Government Accountability Office Highlights*, June http://www.gao.gov/highlights/d05418high.pdf (last accessed 21 April 2007)

Garza, G. (2002). Evolución de las ciudades Mexicanas en el siglo XX. In *Revista de información y análisis* 19:7-16

Gauthier, D. A., Lafon, A., Toombs, T., Hoth, J. and E.Wiken (2003). *Grasslands: Toward a North American Conservation Strategy*. Canadian Plains Research Center, Regina, SK, and Commission for Environmental Cooperation, Montreal, QC

GCC (2004). *Role of GCC States in Protecting the Environment and Conserving Natural Resources*. General Secretariat of the Gulf Cooperation Council, Riyadh

GEO Data Portal. *UNEP's online core database with national, sub-regional, regional and global statistics and maps, covering environmental and socio-economic data and indicators*. United Nations Environment Programme, Geneva http://www.unep.org/geo/data or http://geodata.grid.unep.ch (last accessed 1 June 2007)

GeoHive (2006). Global Statistics http://www.geohive.com/ (last accessed 21 April 2007)

GFN (2004). *Ecological Creditors and Debtors*. Global Footprint Network. http://www.footprintnetwork.org/gfn_sub.php?content=creditor_debtor (last accessed 21 April 2007)

Gilchrist, H. G. and Mallory, M. L. (2005). Declines in abundance and distribution of the Ivory Gull (*Pagophila eburnea*) in Arctic Canada. In *Biological Conservation* 121:303-309

Githeko, A.K, Lindsay, S.W., Confalonier, U.E. and Patz, J.A. (2000). Climate change and vector-borne diseases: a regional analysis. In *Bulletin of the World Health Organization* 79(8):1-20

Global Ballast Water Management Programme (2006). International Maritime Organization, London http://globallast.imo.org/index.asp (last accessed 10 May 2007)

GOI (2003). Auto Fuel Policy. Ministry of petroleum and Natural Gas, Government of India, New Dehli http://petroleum.nic.in/autoeng.pdf (last accessed 1 June 2007)

Goldammer, J.G., Sukhinin, A. and Csiszar, I. (2003). The Current Fire Situation in the Russian Federation: Implications for Enhancing International and Regional Cooperation in the UN framework and the Global Programs on Fire Monitoring and Assessment. In *International Forest Fire News* 29:89-111 http://www.fire.uni-freiburg.de/iffn/iffn_29/Russian-Federation-2003.pdf (last accessed 17 April 2007)

Goulder, L. H. and Kennedy, D. (1997). Valuing ecosystem services: philosophical bases and empirical methods. In G. Daily (ed.) In *Nature's Services Societal Dependence on Natural Ecosystems* 23-47. Islands Press, Washington, DC

Government of Canada (2005). *Project Green: Moving Forward on Climate Change: A Plan for Honouring our Kyoto Commitment*. Government of Canada, Ottawa http://collaboration.cin-ric.ca/file_download.php/GOC+Climate+Change+Plan.pdf?URL_ID=1839&filename=11211880921GOC_Climate_Change_Plan.pdf&filetype=application%2Fpdf&filesize=2013181&name=GOC+Climate+Change+Plan.pdf&location=user-S/ (last accessed 20 June 2007)

Government of Canada (2006). Government Notices: Department of the Environment, Canadian Environmental Protection Act, 1999: Notice of intent to develop and implement regulations and other measures to reduce air emissions. In *Canada Gazette* 140:42 http://canadagazette.gc.ca/partI/2006/20061021/pdf/g1-14042.pdf (last accessed 1 June 2007)

Greenpeace (2007). Greenpeace Southeast Asia Photos. http://www.greenpeace.org/seasia/en/photosvideos/photos/boy-runs-to-catch-the-school-b (last accessed 21 June 2007)

Gregory, J. M., Huybrechts, P. and Raper, S. C. B. (2004). Threatened loss of the Greenland ice-sheet. In *Nature* 428:616

Guimaraes, R. and Bárcena, A. (2002). El desarrollo sustentable en América Latina y el Caribe desde Río 1992 y los nuevos imperativos de la institutcionalidad. In Leff, E., Ezcurra, E., Pisanty, I. and Romero-Lankau, P. (2002). *La Transición Hacia el Desarrollo Sustentable. Perspectivas de América Latina y El Caribe*. Instituto Nacional de Ecología, Universidad Autónoma Metropolitana and Programa de Naciones Unidas para el Medio Ambiente, México, DF

Gupta, S. K. (2001). Rethinking waste management in India. In *Humanscape Magazine* 9:4 http://www.humanscape.org/Humanscape/new/april04/rethinking.htm (last accessed 1 June 2007)

Hales, S., de Wet, N., Maindonald, J., and Woodward, A. (2002). Potential effect of population and climate changes on global distribution of dengue fever: an empirical model. In *Lancet* 360:830-34

Halweil, B. (2002). Farming in the public interest. In L. Starke (ed.) *State of the World 2002: A Worldwatch Institute Report on Progress Toward a Sustainable Society*. The Worldwatch Institute, Washington, DC

Hammond, S. V. (2002). *Can City and Farm Coexist? The Agricultural Buffer Experience in California*. Great Valley Center, Agricultural Transactions Program, Modesto, CA http://www.greatvalley.org/publications/pub_detail.aspx?pId=132 (last accessed 1 June 2007)

Hanna, E., Huybrechts, P., Cappelen, J., Steffen, K. and Stephens, A. (2005). Runoff and mass balance of the Greenland ice sheet: 1958-2003. In *Journal of Geophysical Research* 110

Harding, R. (2006). Ecologically sustainable development: origins, implementation and challenges. In *Desalination* 187:229-239

Harlow, T. (2005). *Water 2025: preventing crises and conflict in the West*. US Department of Interior http://www.usbr.gov/newsroom/presskit/factsheet/factsheetdetail.cfm?recordid=3 (last accessed 1 June 2007)

Health Canada (2001). *Climate Change and Health and Well-Being: A Policy Primer*. http://www.hc-sc.gc.ca/ewh-semt/pubs/climat/policy_primer_north-nord_abecedaire_en_matiere/index_e.html (last accessed 1 June 2007)

Heimlich, R. E. and Anderson, W. D. (2001). *Development at the Urban Fringe and Beyond: Impacts on Agriculture and Rural Land*. US Department of Agriculture, Economic Research Service, http://www.ers.usda.gov/publications/aer803/ (last accessed 1 June 2007)

Hemmings, A. D. (2005). A question of politics: bioprospecting and the Antarctic Treaty System. In Hemmings, A. and Rogan-Finnemore, M. (eds), *Antarctic Bioprospecting*. University of Canterbury, Christchurch

Henninger, N. and Hammond, A. (2000). *Environmental Indicators Relevant to Poverty Reduction: A Strategy for the World Bank*. World Resources Institute, Washington, DC

Hermansen, L. A. (2003). The Wildland-Urban Interface: An Introduction. In *APA National Planning Conference Proceedings, 2 April*, Denver, CO http://www.design.asu.edu/apa/proceedings03/HERMAN/herman.htm (last accessed 1 June 2007)

Hoffmann, N. (2001). Urban consumption of agricultural land. In *Rural and Small Town Canada Analysis Bulletin* 3:2 http://www.statcan.ca/english/freepub/21-006-XIE/21-006-XIE2001002.pdf (last accessed 1 June 2007)

Hoguane, A. M. (1997). *Marine Science Country Profiles: Mozambique*. Intergovernmental Oceanographic Commission Western Indian Ocean Marine Science Association, Zanzibar

Holland, M. M. and Bitz, C. M. (2003). Polar amplification of climate change in coupled models. In *Clim. Dyn.* 21:221-232

Homer-Dixon, T. and Kelly, K. (1995). *Environmental Scarcity and Violent Conflict: The Case of Gaza*. Project on Environment, Population and Security. American Association for the Advancement of Science and the University of Toronto, Toronto

Holmgren, M., M. Scheffer, E. Ezcurra, J.R. Gutiérrez, y G.M.J. Mohren. (2001). El Niño effects on the dynamics of terrestrial ecosystems. In *Trends in Ecology & Evolution* 16(2):59-112

Huang, Shaopeng (2006). 1851–2004 annual heat budget of the continental landmasses. In *Geophysical Research Letters* 33

Huang, Zhenli, Wu, Bingfang and Ao, Liang-gui (2006). *Studies on Ecological and Environmental Monitoring Systems for Three Gorges Dam* (in Chinese). Science Press, Beijing

HUD (2000). *The State of the Cities 2000: Megaforces Shaping the Future of the Nation's Cities*. US Department of Housing and Urban Development, Washington, DC

Huggins, C. (2004). Communal conflicts in Darfur Region, Western Sudan, In *Africa Environment Outlook: Case Studies*, UNEP, Earthprint, Hertfordshire

Husain, T. (1995). *Kuwait Oil Fires: Regional Environmental Perspectives*. Elsevier Science Ltd., Dhahran

Hutson, S. S., Barber, N. L., Kenny, J. F., Linsey, K. S., Lumia, D. S. and Maupin, M. A. (2004). *Estimated Use of Water in the United States in 2000*. US Geological Survey http://pubs.usgs.gov/circ/2004/circ1268/ (last accessed 1 June 2007)

IAATO (2007). *IAATO Overview of Antarctic Tourism - 2006-2007 Antarctic Season*. Information Paper 121. XXX Antarctic Treaty Consultative Meeting. International Association of Antarctic Tour Operators http://www.iaato.org (last accessed 1 June 2007)

ICARDA (2002). Conservation and Sustainable Use of Dryland Agrobiodiversity, International Center for Agricultural Research in the Dry Areas. http://www.icarda.cgiar.org/Gef/Agro10_11.pdf (last accessed 22 April 2007)

IEA (2006). *World Energy Outlook 2006*. International Energy Agency, Paris

IEA (2007). *Energy Balances of OECD Countries and Non-OECD Countries: 2006 edition*. International Energy Agency, Paris (in GEO Data Portal).

IFAD (2000). *The Land Poor: Essential Partners for the Sustainable Management of Land Resources*. International Fund for Agricultural Development, Rome http://www.ifad.org/pub/dryland/e/eng1.pdf (last accessed 1 June 2007)

IFAD and GEF (2002). *Tackling Land Degradation and Desertification*. International Fund for Agricultural Development and Global Environment Facility, Rome http://www.ifad.org/events/wssd/gef/GEF_eng.pdf (last accessed 1 June 2007)

ILO (2006). *Global Employment Trends Model 2006*. Employment Trends Team, International Labour Office, Geneva http://www.ilo.org/public/english/employment/strat/global.htm (last accessed 20 May 2007)

INAC (2003). *Canadian Arctic Contaminants Assessment Report II*. Indian and Northern Affairs Canada, Ottawa

INAC (2006). Government announces immediate action on First Nations drinking water. *INAC News Releases*, 21 March http://www.ainc-inac.gc.ca/nr/prs/j-a2006/2-02757_e.html (last accessed 1 June 2007)

INPE (2006). http://www.amazonia.org.br/english (last accessed 10 May 2007)

INSMET (2004). *El proceso de sequía del 2003-2004: antecedentes, actualidad y futuro*. Declaración Oficial del Instituto de Meteorología relacionada con el actual proceso de sequía que afecta a Cuba, Instituto de Meteorología de Cuba, Havana http://www.insmet.cu (last accessed 10 May 2007)

International Crane Foundation (2003). Africa: Water, Wetlands and Wattled Cranes http://www.savingcranes.org/conservation/our_projects/article.cfm?cid=3&aid=74&pid=1 (last accessed 22 April 2007)

International Year of Freshwater (2003). *Virtual Water*. http://www.wateryear2003.org/en/ev.php-URL_ID=5868&URL_DO=DO_TOPIC&URL_SECTION=201.html (last accessed 17 May 2007)

IPCC (2001a). *Climate Change 2001 – Impacts, Adaptation and Vulnerability*. Contribution of Working Group II to the Third Assessment Report of the Intergovernmental Panel on Climate Change. McCarthy, J.J., Canziani, O.F., Leary, N. A., Dokken, D.J. and White, K.S. (eds). Cambridge University Press, Cambridge and New York, NY

IPCC (2001b). *Climate Change 2001: Synthesis Report*. A Contribution of Working Groups I, II and III to the Third Assessment Report of the Intergovernmental Panel on Climate Change. Cambridge University Press, Cambridge and New York, NY

IPCC (2001c). Climate Change 2001: The Scientific Basis. Contribution of Working Group I to the third Assessment Report of the Intergovernmental Panel on Climate Change. Cambridge University Press, Cambridge and New York, NY

IPCC (2007a). *Climate Change 2007: The Physical Science Basis*. Contribution of Working Group I to the Fourth Assessment Report of the Intergovernmental Panel on Climate Change, Geneva http://www.ipcc.ch/WG1_SPM_17Apr07.pdf (last accessed 5 April 2007)

IPCC (2007b). *Climate Change 2007: Climate Change Impacts, Adaptation and Vulnerability*. Contribution of Working Group II to the Fourth Assessment Report of the Intergovernmental Panel on Climate Change, Geneva http:// www.ipcc.ch/SPM6avr07.pdf (last accessed 27 April 2007)

Iraq Ministry of Environment (2004). *The Iraqi Environment: Problems and Horizons*. Ministry of Environment, Baghdad

ITK (2005). *Effects on Human Health*. Inuit Tapariit Kanatami. http://www.itk.ca/environment/contaminants-health-risks.php (last accessed 1 June 2007)

IUCN (1987). *Saudi Arabia: Assessment of biotopes and coastal zone management requirements for the Arabian Gulf Coast*. MEPA Coastal and Marine Management Series, Report 5. World Conservation Union (International Union for the Conservation of Nature and Natural Resources), Gland

IUCN (2003). 2003 *UN List of Protected Areas*. Chape, S., S. Blyth, L. Fish, P. Fox and M. Spalding (compilers). World Conservation Union (International Union for the Conservation of Nature and Natural Resources), Gland and UNEP-World Conservation Monitoring Centre, Cambridge http://www.unep-wcmc.org/wdpa/unlist/2003_UN_LIST.pdf (last accessed 22 April 2007)

IUCN (2006). *The IUCN Red List of Threatened Species: Summary Statistics, Table 5*. World Conservation Union (International Union for the Conservation of Nature and Natural Resources) http://www.redlist.org/info/tables/table5 (last accessed 22 April 2007)

James, C. (2004). Preview: Global status of commercialized Biotech/GM crops 2004. In *ISAA Briefs* 32. International Service for the Acquisition of Agri-Biotech Applications, Ithaca, NY

Japan Environmental Council (2005). *The State of the Environment in Asia 2005/2006*. Japan Environmental Council, Toyoshinsya

Jenouvrier, S., Barbraud, C., Cazelles, B. and Weimerskirch, H. (2005). Modelling population dynamics of seabirds: importance of the effects of climate fluctuations on breeding proportions. In *Oikos* 108:511-522

Johnson, C. J., Boyce, M. S., Case, R. L., Cluff, H. D., Gau, R. J., Gunn, A. and Mulders, R. (2005). Cumulative effects of human developments on arctic wildlife. In *Wildlife Monographs* 160:1-36

Joly, K., Nellemann, C. and Vistnes, I. (2006). A re-evaluation of caribou distribution near an oilfield road on Alaska's North Slope. In *Wildlife Society Bulletin* 34(3):866–869

Judek, S., Jessiman, B., Stieb, D. and Vet, R. (2005). Estimated number of excess deaths in Canada due to air pollution. In *Health Canada News Releases* 3 November http://www.hc-sc.gc.ca/ahc-asc/media/nr-cp/2005/2005_32bk2_e.html (last accessed 1 June 2007)

Kalkstein, L. S., Greene, J. S., Mills, D. M. and Perrin, A. D. (2005). Extreme weather events. In Epstein, P. R. and Mills, E. (eds.) *Climate Change Futures: Health, Ecological and Economic Dimensions*. The Center for Health and the Global Environment, Harvard Medical School, 53-9 http://www.climatechangefutures.org/pdf/CCF_Report_Final_10.27.pdf (last accessed 1 June 2007)

Kennett, M. and Steenblik, R. (2005). *Environmental Goods and Services: A Synthesis of Country Studies*. OECD Trade and Environment Working Paper No. 2005-03. Organisation for Economic Co-operation and Development, Paris http://www.oecd.org/dataoecd/43/63/35837583.pdf (last accessed 1 June 2007)

Kisirwani, M. and Parle, W. M. (1987). Assessing the impact of the post civil war period on the Lebanese bureaucracy: a view from inside. In *Journal of Asian and African Studies* XXII:1-2

Kolowski, M. L. and Laquintinie, M. L. (2006). Heavy metals in recent sediments and bottom-fish under the influence of tanneries in south Brazil. In *Water, air, and soil pollution* 176:307-327

Kolpin, D. W., Furlong, E. T., Meyer, M. T., Thurman, E. M., Zuagg, S. D., Barber, L. B. and Buxton, H. T. (2002). Pharmaceuticals, hormones, and other organic wastewater contaminants in U.S. steams, 1999-2000: A national reconnaissance. In *Environmental Science and Technology* 36(6):1202-1211

Kouyoumjian, H. H. and Nouayhed, M. (2003). *Proceedings of the International Workshop on Mediterranean Vermeted Terraces and Migratory/Invasive Organisms*, 19-21 December, Beirut. INOC Publications

Krauskopf, R. B. and Retamales Saavedra, R. (2004). *Guidelines for Vulnerability Reduction in the Design of New Health Facilities*. Pan American Health Organization and World Health Organization, Washington, DC

Kuznets, Simon (1995). Economic Growth and Income Inequality. In *American Economic Review* 45(1):1-28

Lal, M. (2005). Climate change – implications for India's water resources. In Mirza, M. M. Q. and Ahmad, Q. K. (eds) *Climate Change and Water Resources in South Asia*. A. A. Balkema Publishers, Leiden

Landsat.org (2006). *FREE Global Orthorectified Landsat Data via FTP* http://www.landsat.org/ortho/index.htm (last accessed 26 June 2007)

Larsen, J. (2004). *Dead Zones Increasing in World's Coastal Waters*. Earth Policy Institute, Washington, DC http://www.earth-policy.org/Updates/Update41.htm (last accessed 1 June 2007)

Lebel, L., Fuchs, D., Garden, P., Giap, D. H., Hobson, K., Lorek, S., Shamshub, H. (2006). Linking *knowledge and action for sustainable production and consumption systems.* USER Working Paper WP-2006-09. Unit for Social and Environmental Research, Chiang Mai

Lesser, M. P. Lamare, M. L. and Barker, M. F. (2004). Transmission of ultraviolet radiation through the Antarctic annual sea ice and its biological effects on sea urchin embryos. In *Limnology & Oceanography* 49:1957–1963

Lightburn, S. (2004). Hybrids pick up speed in the race to go green. In *The Galt Global Review* 18 February: http://www.galtglobalreview.com/business/hybrid_race.html (last accessed 1 June 2007)

Linacre, E. and Geerts, B. (1998). The climate of the Kalahari Desert. In *Resources in Atmospheric Sciences,* University of Wyoming, Laramie, WY http://www-das.uwyo.edu/~geerts/cwx/notes/chap10/sec3.html (last accessed 1 June 2007)

LMMA (2006). *The Locally-Managed Marine Areas Network-Improving the Practice of Marine Conservation: 2005 Annual Report – A Focus on Lessons Learned* http://www.lmmanetwork.org (last accessed 1 June 2007).

Lubowski, R., Vesterby, M. and Bucholtz, S. (2006). Land use. In Wiebe, K. and Gollehon, N. (eds.) *Agricultural Resources and Environmental Indicators,* 2006 Edition. US Department of Agriculture, Economic Research Service http://www.ers.usda.gov/publications/arei/eib16/eib16_1-1.pdf (last accessed 1 June 2007)

Luthcke, S. B., Zwally, H. J., Abdalati, W., Rowlands, D. D., Ray, R. D., Nerem, R. S., Lemoine, F. G., McCarthy, J. J. and Chinn, D. S. (2006). Recent Greenland ice mass loss by drainage system from satellite gravity observations. In American Association for the Advancement of Science *Express Reports* 1130776v1

Lysenko, I. and Zöckler, C. (2001). The 25 largest unfragmented areas in the Arctic. UNEP-WCMC and UNEP Grid Arendal for the WWF Arctic Programme, unpublished

MA (2005). *Ecosystems and Human Well-being: Synthesis.* Millennium Ecosystem Assessment, World Resources Institute, Washington, DC

Maffi, L. (ed.) (2001). *On Biocultural Diversity: Linking Language, Knowledge, and the Environment.* Smithsonian Institution Press, Washington, DC

Marsalek, J., Diamond, M., Kok, S. and Watt, W. E. (2001). Urban runoff. In Environment Canada (ed.) *Threats to Sources of Drinking Water and Aquatic Ecosystem Health in Canada.* National Water Research Institute, Burlington, ON http://www.nwri.ca/threats/threats-eprint.pdf (last accessed 1 June 2007)

Marsalek, J., Watt, W. E., Lefrancois, L., Boots, B. F. and Woods, S. (2002). Municipal water supply and urban developments. *Threats to Water Availability in Canada.* National Water Research Institute, Environmental Conservation Service of Environment Canada, Burlington, ON

Martinelli, L. A., Howarth, R. W., Cuevas, E., Filoso, S., Austin, A. T., Donoso, L., Huszar, V., Keeney, D., Lara, L. L., Llerena, C., McIssac, G., Medina, E., Ortiz-Zayas, J., Scavia, D., Schindler, D.W., Soto D. and Towsend, A. (2006). Sources of reactive nitrogen affecting ecosystems in Latin America and the Caribbean: current trends and future perspectives. In *Biogeochemestry* 79:3-24

Martínez, J. and Fernández, A. (comps) (2004). *Cambio climático: una vision desde México.* Instituto Nacional de Ecología, Semarnat, México, DF

Maya, S. 2003. African Savanna. In *Blue Planet Biomes* http://www.blueplanetbiomes.org/african_savanna.htm (last accessed 1 June 2007)

Mayaux, P., Bartholomé, E., Fritz, S. and Belward A. (2004). A new land-cover map of Africa for the year 2000. Institute for Environment and Sustainability, Joint Research Centre of the European Commission. In *Journal of Biogeography* 31:861-877

McMichael, A.J., Campbell-Lendrum, D.H., Corvalán, C.F., Ebi, K.L., Githeko, A.K., Scheraga, J.D. and Woodward A. (eds.) (2003). *Climate change and human health: Risks and responses.* World Health Organization, Geneva

Middleton, N. (1999). *The Global Casino: An Introduction to Environmental Issues.* 2nd. Edition. Arnold, London

Ministry of the Environment of Japan (2005). Aichi Statement. *Regional EST Forum, 1-2 August 2005, Nagoya* http://www.env.go.jp/press/file_view.php3?serial=7047&hou_id=6242 In Japanese (last accessed 1 June 2007)

Ministry of Environment, Republic of Korea (1996). National waste generation and treatment. http://eng.me.go.kr/docs (in Korean) (last accessed 20 June 2007)

Mirza, M. Q. (2002). Global warming and changes in the probability of occurrence of floods in Bangladesh and implications. In *Global Environmental Change* 12:127-138

Mirza, M. M. Q., Warrick, R. A., Erickson, N. J. and Kenny, G. J. (2005). Are floods getting worse in the GBM Basins? In Mirza, M. M. Q. and Ahmad, Q. K. (eds.) *Climate Change and Water Resources in South Asia* A. A. Balkema Publishers, Leiden

Misak, R. F. and Omar, S. A. (2004). Military operations as a major cause of soil degradation and sand encroachment in Arid Regions (the case of Kuwait). In *Journal of Arid Land Studies* 14S:25-28

Mitchell, A. T. and Grau, H. (2004). Globalization, migration, and Latin American Ecosystems. In *Science* 305:5692

Molina, T. L. and Molina, M. J. (2002). *Air quality in the Mexico megacity: an integrated assessment.* Kluwer Academic Publishers, London

Montenegro, R. A. and Stephens, C. (2006). Indigenous health in Latin America and the Caribbean. In *Indigenous Health* 367(3):1859-1869

Moser, M., Crawford, P. and Scott, F. (1996). *A Global Overview of Wetland Loss and Degradation.* Wetlands International, http://www.ramsar.org/about/about_wetland_loss.htm (last accessed 10 May 2007)

Mountain Partnership (2001). Did You Know? http://www.mountainpartnership.org/issues/resources/didyouknow.html (last accessed 22 April 2007)

Muir, D. C. G., Backus, S., Derocher, A. E., Dietz, R., Evans, T. J., Gabrielsen, G. W., Nagy, J., Norstrom, R. J., Sonne, C., Stirling, I., Taylor, M. K. and Letcher, J. J. (2006). Brominated flame retardants in Polar bears (*Ursus maritimus*) from Alaska, the Canadian Arctic, East Greenland, and Svalbard. In *Env. Sci. Tech.* 40:449-455

NASA (2006). *Ozone hole watch.* National Aeronautics and Space Administration. http://ozonewatch.gsfc.nasa.gov/index.html (last accessed 10 May 2007)

Naylor, R. L., Steinfeld, H., Falcon, W. P., Galloway, J., Smil, V., Bradford, E., Alder, J. and Mooney, H. A. (2005). Losing the links between livestock and land. In *Science* 310(5754):1621-1622

Nefedova, T. G. (2003). Selskaya Rossiya na pereput'e: geographicheslie ocherki (Rural Russia at a Cross-Roads: Geographical Essays). Novoe izdatelstvo, Moscow (in Russian)

New Agriculturalist (2005). Crisis What crisis? In *New Agriculturalist Online,* 1 November, http://www.new-agri.co.uk/05-6/focuson/focuson1.html (last accessed 1 June 2007)

New Scientist (2005). Antarctic ice sheet is an 'awakened giant'. In *NewScientist.com news service.* http://www.newscientist.com/article.ns?id=dn6962 (last accessed 1 June 2007)

NMA (2006). *Coal industry poised for national, global growth.* National Mining Association Press Releases, 14 July, http://www.nma.org/newsroom/press_releases.asp (last accessed 8 September)

NOAA (2006). *Human and economic indicators- Shishmaref.* National Oceanic and Atmospheric Administration. http://www.arctic.noaa.gov/detect/human-shishmaref.shtml (last accessed 1 June 2007)

Nowlan, L. (2001). *Arctic Legal Regime for Environmental Protection.* World Conservation Union (International Union for the Conservation of Nature and Natural Resources), Gland and International Council of Environmental Law, Bonn

Nowlan, L. (2005). *Buried Treasure: Groundwater Permitting and Pricing in Canada.* West Coast Environmental Law and Sierra Legal Defence Fund http://www.sierralegal.org/reports/Buried_Treasure.pdf (last accessed 1 June 2007)

NRC (2006). *Toward an Integrated Arctic Observing Network.* Committee on Designing an Arctic Observing Network: National Research Council, Washington, DC

NRCan (2004). *Climate Change Impacts and Adaptation: A Canadian Perspective.* Climate Change Impacts and Adaptation, Natural Resources Canada, Ottawa http://adaptation.nrcan.gc.ca/perspective/index_e.php (last accessed 22 April 2007)

NRCan (2005). *Improving Energy Performance in Canada – Report to Parliament Under the Energy Efficiency Act For the Fiscal Year 2004-2005.* Natural Resources Canada, Office of Energy Efficiency, Gatineau, QC http://oee.nrcan.gc.ca/Publications/statistics/parliament04-05/summary.cfm?attr=0 (last accessed 22 April 2007)

NRCS (1999). *Summary Report, 1997 National Resources Inventory, Revised December 2000.* US Department of Agriculture, Natural Resources Conservation Service. http://www.nrcs.usda.gov/technical/NRI/1997/summary_report/table5.html (last accessed 1 June 2007)

NRCS (2003). *National Resources Inventory 2001 Annual NRI: Urbanization and Development of Rural Land.* US Department of Agriculture, Washington, DC http://www.nrcs.usda.gov/Technical/land/nri01/nri01dev.html (last accessed 1 June 2007)

NRCS (2005). *Conservation Innovation Grants.* US Department of Agriculture, Washington, DC http://www.nrcs.usda.gov/programs/cig/ (last accessed 1 June 2007)

NRM (2005). *Factsheet: Russian-Norwegian Seabird Collaboration.* Norwegian Polar Institute, Tromsø http://doksenter.svanhovd.no/faktaark/2005/Faktaark_Sjofugl_2005_ENG.pdf (last accessed 22 April 2007)

NSIDC (2006). *Arctic sea-ice extent.* National Snow and Ice Data Center News Release, 28 September 2005 ftp://sidads.colorado.edu/DATASETS/NOAA/G02135/Sep/N_09_area.txt (last accessed 15 May 2007)

OAPEC (2005). Arab Energy Data. Organization of Arab Petroleum Exporting Countries. http://www.oapecorg.org/images/DATA/ (last accessed 22 April 2007)

OECD (2004). *OECD Environmental Performance Reviews: Canada.* Organisation for Economic Co-operation and Development, Paris

OECD (2005). *OECD Environmental Performance Reviews: United States.* Organisation for Economic Co-operation and Development, Paris

OECD (2006a). Improving water management – Recent OECD Experience. In *OECD Policy Brief* February http://www.oecd.org/dataoecd/31/41/36216565.pdf (last accessed 1 June 2007)

OECD (2006b). *Environment Performance Reviews – Water: The Experience of OECD Countries.* Organisation for Economic Co-operation and Development, Paris http://www.oecd.org/dataoecd/18/47/36225960.pdf (last accessed 1 June 2007)

OECD and Republic of Armenia (2004). *Financing Strategy for Urban Wastewater Collection and Treatment infrastructure in Armenia.* Final Report prepared by State Committee of Water Economy and Ministry of Finance and Economy of the Republic of Armenia in cooperation with the EAP Task Force, Joint edition of OECD and Republic of Armenia, Yerevan

Office of the Governor (2006). Statement by Governer Schwarzenegger on historic agreement with legislature to combat global warming. *Press Release,* 30 August http://gov.ca.gov/index.php?/press-release/3722/ (last accessed 1 June 2007)

OLADE (2005). *Prospectiva energética de América Latina y el Caribe 2005.* Organización Latinoamericana de Energía, Quito

Oldeman, L.R., Hakkeling, R.T.A. and Sombroek, W.G. (1991). *World Map of the Status of Human-Induced Soil Degradation: A Brief Explanatory Note.* International Soil Reference Information Centre-ISRIC (currently called World Soil Information), Wageningen

Olson, D. M, Dinerstein, E., Wikramanayake, E.D., Burgess, N.D., Powell, G.V.N., Underwood, E.C., D'amico, J.A., Itoua, I., Strand, H.E., Morrison, J.C., Loucks, C.J., Allnutt, T.F., Ricketts, T.H., Kura, Y., Lamoreux, J.F., Wettengel, W.W., Hedao, P. and Kassem, K.R. (2001). Terrestrial Ecoregions of the World: A New Map of Life on Earth. BioScience 51:933-938Omar, S. A., Bhat, N. R., Shahid, S. A. and Asem, A. (2005). Land and vegetation degradation in war affected areas in the Sabah Al Ahmad Nature Reserve of Kuwait. A Case Study of Umm Ar Rimam. In *Journal of Arid Environments* 62:475-490

OPS (2005). Informe Regional sobre la Evaluación de los Servicios de Manejo de Residuos Sólidos Municipales en la Región de América Latina y el Caribe. Washington, DC

OPS (2006). Datos básicos de cobertura en agua potable y saneamiento para la región de las Américas. http://www.bvsde.paho.org/AyS2004/AguayS2004.html (last accessed 1 June 2007)

Otto, B., Ransel, K., Todd, J., Lovaas, D., Stutzman, H. and Bailey, J. (2002). *Paving Our Way to Water Shortages: How Sprawl Aggravates the Effects of Drought.* American Rivers, the Natural Resources Defense Council and Smart Growth America http://www.smartgrowthamerica.org/DroughtSprawlReport09.pdf (last accessed 1 June 2007)

Paredis, E., Lambrecht, J., Goeminne, G. and Vanhove, W. (2004). *Elaboration of the concept of 'ecological debt': VLIR-BVO project 2003.* Centre for Sustainable Development (CDO) – Ghent University, Ghent http://cdonet.ugent.be/noordzuid/onderzoek/ecological_debt/ (last accessed 1 June 2007)

Pendall, R., Martin, J. and Fulton, W. (2002). *Holding the Line: Urban Containment in the United States.* The Brookings Institution Center on Urban and Metropolitan Policy, Washington, DC

Peopleandplanet.net (2003). *People and Food and Agriculture: Production trends.* FactFile, 8 August, http://www.peopleandplanet.net/doc.php?id=344 (last accessed 22 April 2007)

Peters, C. (1996). Observation of sustainable exploitation of non-timber forest products. An ecologist's perspective. In *Current Issues in Non-Timber Forest Products Research.* Centre for International Forestry Research, Bogor

Peters, C. (1997). Sustainable use of biodiversity: myths, realities and potential. In Grifo, F. and Rosenthal, J. (eds.). In *Biodiversity and human health.* Island Press, Washington, DC

Petersen, A., Zöckler , C. and Gunnarsdottir, M. V. (2004). *Circumpolar Biodiversity Monitoring Programme – Framework Document. CAFF CBMP Report No.1.* Conservation of Arctic Flora and Fauna International Secretariat, Akureyri

Pew Center on Global Climate Change (2006). Learning from state action on climate change. June 2006 update. In *Brief: Innovative Policy Solutions to Global Climate Change* http://www.pewclimate.org/docUploads (last accessed 1 June 2007)

Philippi, A., Soares Tenório, J.A. and Calderoni, S. (2002). In Leff, E., Ezcurra, E., Pisanty, I. and Romero-Lankau, P. (2002). *La Transición Hacia el Desarrollo Sustentable. Perspectivas de América Latina y El Caribe.* Instituto Nacional de Ecología, Universidad Autónoma Metropolitana and Programa de Naciones Unidas para el Medio Ambiente, México, DF

Pimentel, D. and Pimentel, M. (2004). Land, water and energy versus the ideal U.S. population. NPG (Negative Population Growth) Internet Forum Series, http://www.npg.org/forum_series/forum0205.html (last accessed 1 June 2007)

Plan Bleu (2005). *A Sustainable Future for the Mediterranean. The Blue Plan's Environment and Development Outlook.* Plan Bleu – Regional Activity Centre of UNEP/Mediterranean Action Plan, Valbonne and Earthscan, London

Polar View (2006). *Ice Edge Monitoring.* Global Monitoring for Environment and Security http://www.polarview.org/services/iem.htm (last accessed 11 May 2007)

Postel, S. (2005). Liquid Assets: *The Critical Need to Safeguard Freshwater Ecosystems.* Worldwatch Institute, Washington, DC

PNUMA (1999). "Evaluación sobre las fuentes terrestres y actividades que afectan al medio marino, costero y de aguas dulces asociadas en la región del Gran Caribe", in Informes y Estudios del Programa de Mares Regionales del PNUMA Nº 172, PNUMA/ Oficina de Coordinación del PAM/ Programa Ambiental del Caribe, Mexico, DF

PNUMA (2004). *Perspectivas del Medio Ambiente Urbano en América Latina y el Caribe.* Programa de las Naciones Unidas para el Medio Ambiente, México, DF

PERSGA (2003). *Regional Action Plan for the Conservation of Coral Reefs in the Red Sea and Gulf of Aden.* Technical Report Series No. 3. Protection of the Environment of the Red Sea and Gulf of Aden, Jeddah

PERSGA (2004). *Regional Action Plan for the Conservation of Mangroves in the Red Sea and Gulf of Aden.* Technical Report Series No.12. Protection of the Environment of the Red Sea and Gulf of Aden, Jeddah

PERSGA and GEF (2003). *Status of Mangroves in the Red Sea and Gulf of Aden.* Technical Report Series No. 11. Protection of the Environment of the Red Sea and Gulf of Aden, Jeddah

PPG7 (2004). *The Sustainable BR-163 Plan within the Framework of Government Policies for the Amazon Brasilia.* Pilot Program to Conserve the Brazilian Rain Forest, International Advisory Group (IAG), Report on the 21st Meeting, 26 July-6 August

Preen, A. (1989). *Dugongs.* Technical Report. Meteorological and Environmental Protection Administration, Jiddah

Price, A., and Robinson, H. (1993). The 1991 Gulf War: Coastal and marine environmental consequences. In *Marine Pollution Bulletin* 27

Prishchepov, A. V., Alcantara, P. C. and Radeloff, V. C. (2006). Monitoring agricultural land abandonment in Eastern Europe with multitemporal MODIS data products. Presented at *The 2006 Meeting of the Association of American Geographers,* 7-11 March 2006, Chicago, Illinois

Raad, T. and Kenworthy, J. (1998). The U.S. and Us: Canadian Cities are going the way of their U.S. Counterparts into car-dependent sprawl. In *Alternatives* 24(1):14-22

Reich, P. F., Numbem, S. T., Almaraz, R. A. and Eswaran, H. (2001). Land Resources Stresses and Desertification in Africa. In Bridges, E. M., Hannam, I. D., Oldeman, L. R., Pening de Vries, F. W. T., Scerr, S J. and Sompatpanit, S. (eds.) *Responses to Land Degradation. Proc. 2nd International Conference on Land Degradation and Desertification,* Khon Kaen. Oxford Press, New Dehli

Republic of Lebanon (2001). *Lebanon State of the Environment Report.* Ministry of the Environment, Beirut

RFA (2005). *Homegrown for the Homeland: Ethanol Industry Outlook 2005.* Renewable Fuels Association http://www.ethanolrfa.org/objects/pdf/outlook/outlook_2005.pdf (last accessed 17 May 2007)

Ribaudo, M. and Johansson, R. (2006). Water quality: impacts of agriculture. In Wiebe, K. and Gollehon, N. (eds.) *Agricultural Resources and Environmental Indicators, 2006 Edition.* US Department of Agriculture, Economic Research Service, Washington, DC

Ricketts, T. and Imhoff, M. (2003). Biodiversity, urban areas, and agriculture: locating priority ecoregions for conservation. In *Conservation Ecology* 8(2):1 http://www.consecol.org/vol8/iss2/art1/ (last accessed 1 June 2007)

RICYT (2003). Indicadores de ciencia y tecnología en Iberoamérica http://www.ricyt.org/interior/interior.asp?Nivel1=1&Nivel2=2&Idioma= (last accessed 26 June 2007)

Riegel, B. (2003). Climate change and coral reefs: different effects in two high latitude areas (Arabian Gulf, South Africa). In *Coral Reefs* 22:433-446

Rignot, E. and Kanagaratnam, P. (2006). Changes in the velocity structure of the Greenland Ice Sheet. In *Science* 311:986-990

Rignot, E., Cassasa, G., Gogineni, P., Krabill, W., Rivera, A. and Thomas, R. (2004). Accelerated discharge from the Antarctic Peninsula following the collapse of the Larsen B ice shelf. In *Geophysical Research Letters* 31

Robinson, W. D. (2005). Biodiversity and its health in urbanizing landscapes. In *Emerging Issues Along Urban/Rural Interfaces: Linking Science And Society, 13-16 March 2005,* Atlanta, GA

Rodriguez, J. P., Tatiana, G. and Dirzo, R. (2005). Diversitas y el reto de la conservación de la biodiversidad latinoamericana. In *Inci* 30(8):449-449

ROPME (2004). *State of the Marine Environment Report, 2003.* Regional Organization for the Protection of the Marine Environment of the sea area surrounded by Bahrain, I.R. Iran, Iraq, Kuwait, Oman, Qatar, Saudi Arabia and the United Arab Emirates, Kuwait

Rosa, H., Kandel, S., and Dimas, L. (2003). Compensation for environmental services and rural communities. PRISMA, San Salvador

Ruiz Marrero, C. (2005). *Water Privatization in Latin America.* Global Policy Forum. http://www.globalpolicy.org/socecon/gpg/2005/1018carmelo.htm (last accessed 10 May 2007)

Russian 3rd Nat. Comm. (2002). ТРЕТЬЕ НАЦИОНАЛЬНОЕ СООБЩЕНИЕ РОССИЙСКОЙ ФЕДЕРАЦИИ, Moscow http://unfccc.int/resource/docs/natc/rusncr3.pdf (last accessed 17 April 2007)

Saunders, S. and Maxwell, M. (2005). *Less Snow, Less Water: Climate Disruption in the West.* Rocky Mountain Climate Organization, Louisville, CO http://www.rockymountainclimate.org/website%20pictures/Less%20Snow%20Less%20Water.pdf (last accessed 1 June 2007)

Sawahel, W. (2004). Gulf's first wind power plant is opened. *SciDev.Net News,* 2 November, http://www.scidev.net/content/news/eng/gulfs-first-wind-power-plant-is-opened.cfm (last accessed 1 June 2007)

Scambos, T. A., Bohlander, J. A., Shuman, C. A. and Skvarca, P. (2004). Glacier acceleration and thinning after ice shelf collapse in the Larsen B embayment, Antarctica. In *Geophysical Research Letters* 31

Schaefer, J. A. (2003). Long-term range recession and the persistence of caribou in the taiga. In *Conservation Biology* 17:1435-1439

Scheifler, R., Gauthier-Clerc, M., Le Bohec, C. and Crini, N. (2005). Mercury concentrations in King Penguin (*Aptenodytes patagonicus*) feather at Crozet Island (Sub-Antarctic): temporal trend between 1996-1974 and 2000-2001. In *Environmental Toxicology and Chemistry* 24:125

Scherr, S. J. and Yadav, S. (2001). Land degradation in the developing world: issues and policy options for 2020. In IFPRI (ed.) *The Unfinished Agenda: Perspectives on Overcoming Hunger, Poverty and Environmental Degradation.* International Food Policy Research Institute, Washington, DC

Schmidt, C. W. (2004). Sprawl: the new manifest destiny? In *Environmental Health Perspectives* 112(11):A620-A627

Schneider, C. G. and Hill, L. B. (2005). *Diesel and Health in America: The Lingering Threat.* Clean Air Task Force, Boston, MA http://www.catf.us/publications/reports/Diesel_Health_in_America.pdf (last accessed 1 June 2007)

Scholes, R. J. and Biggs, R. (eds.) (2004). *Ecosystems Services in Southern Africa: A Regional Assessment.* Council for Scientific and Industrial Reseach, Pretoria

Scott, D. A. (1998). *Global Overview of the Conservation of Migratory Arctic Breeding Birds outside the Arctic.* CAFF Technical Report No.4. Conservation of Arctic Flora and Fauna, Akureyri

Sea Around Us, 2006. A global database on marine fisheries and ecosystems. The Fisheries Centre, University British Columbia, Vancouver, BC http://www.seaaroundus.org (last accessed 10 May 2007)

Sedell, J. R., Bennett, K., Steedman, R., Foster, N., Ortuno, V., Campbell, S. and Achouri, M. (2002). Integrated Watershed Management Issues in North America. In *21st Session of the North American Forestry Commission, Food and Agriculture Organization of the United Nations, 22-26 October,* Kona, Hawaii www.fs.fed.us/global/nafc/2002/meeting_info/technical_papers/watershed.doc (last accessed 1 June 2007)

SEMARNAT (2002). ACUERDO por el que se establece como área de refugio para proteger a las especies de grandes ballenas de los subórdenes Mysticeti y Odontoceti, las zonas marinas que forman parte del territorio nacional y aquellas sobre las que la nación ejerce su soberanía y jurisdicción. Diario Oficial de la Federación. Viernes 24 de mayo de 2002. México, DF

SEPA (2004). *Report on the State of the Environment in China 2003.* State Environmental Protection Administration, Beijing http://www.sepa.gov.cn/plan/zkgb/2003 (in Chinese) (last accessed 22 April 2007)

SGN (2005). Smart Growth Network http://www.smartgrowth.org/sgn/default.asp (last accessed 22 April 2007)

Sgro, J. (2002). *Canada's Urban Strategy: A Blueprint for Action, Final Report, Prime Minister's Caucus Task Force on Urban Issues.* http://www.udiontario.com/reports/pdfs/UrbanTaskForce_0211.pdf (last accessed 24 April 2007)

Shepherd, A., Wingham, D. and Rignot, E. (2004). Warm ocean is eroding West Antarctic Ice Sheet. In *Geophysical Research Letters* 31

Sheppard, C. (2003). Predicted recurrence of mass coral mortality in the Indian Ocean. In *Nature* 425:294-297

Sheppard, C., Price, A. and Roberts, C. (1992). *Marine Ecology of the Arabian Region: Patterns and Processes in Extreme Tropical Environments.* Academic Press, London

Shiklomanov, I. A. (2004). Summary of the Monograph "World Water Resources at the Beginning of The 21st Century" Prepared in the Framework of IHP UNESCO. International Hydrological Programme, UNESCO, Paris

Shiklomanov, I. A. and Rodda, J. C. (2003). *World Water Resources at the Beginning of the 21st Century.* Cambridge University Press, Cambridge

Shorbagy, M. A. (1986). Desertification of rangeland in the Arab world: causes, indications, impacts and ways to combat. In *Journal of Agriculture and Water* 4:68-83 (ACSAD publication in Arabic)

Siddeek, M., Fouda, M. and Hermosa, G. (1999). Demersal fisheries of the Arabian Sea, the Gulf of Oman and the Arabian Gulf. In *Estuarine Coastal and Shelf Science* 49:87-97

Siegel, V. and Loeb, V. (1995). Recruitment of Antarctic krill *Euphasia Superba* and possible causes for its variability. In *Marine Ecology Progress Series* 123:45-56

Simms, A. (2005). Ecological Debt – the Health of the Planet and the Wealth of Nations. Pluto Books, London

Smith, G. (2005). Present day drought conditions in the Colorado River Basin. In *2005 Colorado River Symposium: Sharing the Risks: Shortage, Surplus, and Beyond, 28-30 September,* Santa Fe, NM http://www.cbrfc.noaa.gov/present/2005/GSmith_SantaFe.pdf (last accessed 22 April 2007)

Smith, R. (2003). Canada's freshwater resources: toward a national strategy for freshwater management. In *Water and the Future of Life on Earth: Workshop and Think Tank,* Simon Fraser University, Vancouver, BC http://www.sfu.ca/cstudies/science/water/pdf/Appendix_3.pdf (last accessed 22 April 2007)

Smol, J. P., Wolfe, A. P., Birks, H. J. B., Douglas, M. S. V., Jones, V. J., Korhola, A., Pienitz, R., Ruhlanda, K., Sorvarii, S., Antoniades, D., Brooks, S. J., Fallu, M-A., Hughes, M., Keatley, B. E., Laing, T. E., Michelutti, T., Nazarova, L., Nymani, M., Paterson, A. M., Perren, B., Quinlan, R., Rautio, R., Saulnier-Talbot, J. E., Siitonen, S., Solovieva, N. and Weckstrom, J. (2005). Climate-driven regime shifts in the biological communities of arctic lakes. *Proceedings of the National Academy of Sciences of the United States of America* 102:4397-4402

Sperling, D. and Kurani, K. (2003). Sustainable urban transport in the 21st Century: A new agenda. In *Transportation, Energy, and Environmental Policy: VIII Biennial Asilomar Conference Proceedings.* Transportation Research Board, Keck Center of the National Academies, Washington, DC

Stafford, L. (2005). Drought in the Horn of Africa. In *Geotime,* April 29, 2005 http://www.geotimes.org/apr05/WebExtra042905.html (last accessed 22 April 2007)

Stenhouse, I. J., Gilchrist, H. G., Mallory, M. L. and Robertson, G. J. (2006). Unsolicited Status report on Ivory Gull (Pagophila eburnea). Committee on the Status of Endangered Wildlife in Canada, Ottawa

Suresh, V. (2000). Sustainable development of water resources in urban areas. Proceedings 10th National Symposium on Hydrology, July 18-19, Central Soil and Materials Research Station, New Delhi

Surfrider Foundation (2005). Coastal A-Z. http://www.surfrider.org/a-z/index.asp (last accessed 11 May 2007)

Taylor, A., Bramley, M. and Winfield, M. (2005). *Government Spending on Canada's Oil and Gas Industry: Undermining Canada's Kyoto Commitment.* The Pembina Institute http://www.pembina.org/publications_item.asp?id=181 (last accessed 22 April 2007)

TEI (2006). *Thai Green Label Scheme.* Thailand Environment Institute, Nonthaburi http://www.tei.or.th/greenlabel/GL_home_main.htm (last accessed 11 May 2007)

Terazono, A., Moriguchi, Y., Yamamoto, Y. S., Sakai, S., Inanc, B., Yang, J., Siu, S., Shekdar, A. V., Lee, D-H., Idris, A. B., Magalang, A. A., Peralta, G. L., Lin, C-C., Vanapruk, P. and Mungcharoen, T. (2005). Waste management and recycling in Asia. In *International Review for Environmental Strategies* 5(2)

TerrAfrica (2004). *Terrafrica: Halting Land Degradation.* http://web.worldbank.org/WBSITE/EXTERNAL/COUNTRIES/AFRICAEXT/0,,contentMDK:20221507~menuPK:258659~pagePK:146736~piPK:146830~theSitePK:258644,00.html (last accessed 11 May 2007)

The White House (2002). *Global Climate Change Policy Book.* http://www.whitehouse.gov/news/releases/2002/02/climatechange.html (last accessed 1 June 2007)

The White House (2005). *President Bush Signs into Law a National Energy Plan.* Office of the Press Secretary http://www.whitehouse.gov/news/releases/2005/08/20050808-4.html (last accessed 1 June 2007)

The White House (2006). President Bush delivers State of the Union Address. In *News and Policies,* 31 January

Theobald, D. (2005). Landscape patterns of exurban growth in the USA from 1980 to 2020. In *Ecology and Society* 10(1):32

Thomas, C. D., Cameron, A., Green, R. E., Bakkenes, M., Beaumont, L. J., Collingham, Y. C., Erasmus, B. F. N., Ferreira de Siqueira, M., Grainger, A., Hannah, L., Hughes, L., Huntley, B., van Jaarsveld, A.S., Midgley, G.F., Miles, L., Ortega-Huerta, M.A., Townsend Peterson, A., Phillips, O. L. and Williams, S.E. (2004). Extinction risk from climate change. In *Nature* 427(6970):145-148

Toledo, V.M. (2002). Ethnoecology: A conceptual framework for the study of indigenous knowledge of nature. In Stepp, J.R., Wyndham, F.S. and Zarger, R.S. (eds.) *Ethnobiology and Biocultural Diversity: Proceedings of the Seventh International Congress of Ethnobiology.* International Society of Ethnobiology, Athens, GA

Toledo, A. (2005). Marco conceptual: Caracterización ambiental del Golfo de México. In Botello, A.V., Rendón von Osten, J., Gold-Bouchot, G. and Agraz-Hernández C. (eds.). *Golfo de México. Contaminación e impacto ambiental: diagnóstico y tendencias. 2ª edición.* Universidad Autónoma de Campeche, Universidad Nacional Autónoma de México, Instituto Nacional de Ecología, México, DF

Toledo, V. M. and Castillo, A. (1999). La ecología en Latinoamérica: siete tesis para una ciencia pertinente en una región en crisis. In *Interciencia* 24(3):157-168

Tompkins, E.L. and Adger, W.N. (2003). *Building resilence to climate change through adaptive management of natural resources.* Working paper 27. Tyndall Center for Climate Change Research, Norwich

Torras, M. (2003). An ecological footprint approach to external debt relief. In *World Development* 31(12):2161-71

Tucker, C. J., Dregnc, H. F., and Newcomb, W. W. (1991). Expansion and contraction of the Sahara Desert from 1980 to 1990. In *Science* 253:299-301

UN (2005a). *The Millennium Development Goals Report 2005.* United Nations, New York, NY

UN (2005b). *The Millennium Development Goals in the Arab Regions*: 2005 Summary. United Nations, New York, NY

UNAMI (2005). *UN-Iraq Reconstruction and Development Update - August 2005.* United Nations Assistance Mission for Iraq

UNCC (2004). *Exhibits to the oral submissions of the State of Kuwait to the F4 Panel of Commissioners (Procedural Order No 3).* United Nations Compensation Commission, Governing Council http://www2.unog.ch/uncc/ (last accessed 1 June 2007)

UNCCD (2001). *Global Alarm: Dust and Sandstorms from the World's Drylands.* UN Convention to Combat Desertification, Bonn http://www.unccd.int/publicinfo/duststorms/part0-eng.pdf (last accessed 22 April 2007)

UNCCD Secretariat (2004). *The Secretary-General: Message on the World Day to Combat Desertification.* UNCCD Newsroom, 17 June http://www.unccd.int/publicinfo/statement/annan2004.php (last accessed 22 April 2007)

UNCTAD (2005). *El transporte maritime en 2005.* United Nations, New York and Geneva

UNDP (2001). *Human Development Report 2001.* United Nations Development Programme, New York, NY

UNDP (2004). *Arab Human Development Report 2004: Towards Freedom in the Arab World.* United Nations Development Programme, New York, NY

UNDP (2005a). *Sub-Saharan Africa – The Human Costs of the 2015 "Business-as-usual" scenario,* Human Development Report Office, United Nations Development Programme

UNDP (2005b). *The Waste Business.* United Nations Development Programme, New York, NY

UNDP (2005c). *Assessing Environment's Contribution to Poverty Reduction: Environment for the MDGs.* United Nations Development Programme, Poverty-Environment Partnership http://www.povertyenvironment.net/pep/?q=assessing_environment_s_contribution_to_poverty_reduction (last accessed 1 June 2007)

UNDP (2006). *Human Development Report 2006: Beyond Scarcity: Power, Poverty and the Global Water Crisis.* United Nations Development Programme, New York, NY http://hdr.undp.org/hdr2006/pdfs/report/HDR06-complete.pdf (last accessed 22 April 2007)

UNECA(2000). *Transboundary River/Lake Basin Water Development in Africa: Prospects, Problems and Achievements.* ECA/RCID/052/00, United Nations Economic Commission for Africa, Addis Ababa http://www.uneca.org/publications/RCID/Transboundary_v2.PDF (last accessed 5 May 2007)

UNECA(2004). *Assessing Regional Integration in Africa.* Economic Commission for Africa, Addis Ababa

UNECE (2003a). *Kyiv Resolution on Biodiversity.* ECE/CEP/108, Fifth Ministerial Conference, Environment For Europe, Kiev, Ukraine, 21-23 May 2003

UNECE (2003b). *National report on the State of the Environment in Armenia in 2002.* United Nations Economic Commission for Europe, Yerevan http://www.unece.org/env/europe/monitoring/Armenia/ http://www.countdown2010.net/documents/biodiv_resolution_Kiev.pdf (last accessed 17 April 2007)

UNECE (2006a). Annual Bulletin of transport statistics for Europe and North America. UN Economic Commission for Europe, Geneva

UNECE (2006b). Convention on Long-range Transboundary Air Pollution: Protocol on Heavy Metals. United Nations Economic Commission for Europe, Geneva http://www.unece.org/env/lrtap/hm_h1.htm (last accessed 1 June 2007)

UNECE-EMEP (n.d.). Official country reports to the Cooperative Programme for Monitoring and Evaluation of the Long-range Transmission of Air Pollutants in Europe (EMEP) of the United Nations Economic Commission for Europe http://www.emep.int/ (last accessed 20 June 2007)

UNCLAC (2002). *The sustainability of development in Latin America and the Caribbean: challenges and opportunities.* Economic Commission for Latin America and the Caribbean, Regional Office for the Latin America and the Caribbean, Santiago de Chile

UNEP (1992). *World Atlas of Desertification.* Edward Arnold, London

UNEP (1993). *Updated Scientific Report on the Environmental Effects of the Conflict Between Iraq and Kuwait.* United Nations Environment Programme, Nairobi

UNEP (1999). *Overview of Land-based Sources and Activities Affecting the Marine Environment in the ROPME Sea Area.* UNEP Regional Seas Reports and Studies No. 168. UNEP/GPA & ROPME, Nairobi

UNEP (2002a). *Global Environment Outlook 3: Past, Present and Future Perspectives.* Earthprint, Hertfordshire, England

UNEP (2002b). *Africa Environment Outlook: Past, Present and Future Perspectives.* EarthScan, London

UNEP (2002c). *Vital Waste Graphics.* United Nations Environment Programme. Basel Convention, GRID-Arendal, UNEP Division of Early Warning and Assessment-Europe), Arendal

UNEP (2003a). *Global Environment Outlook (GEO) – 3. Fact Sheet – Africa.* United Nations Environment Programme, http://www.unep.org/GEO/pdfs/GEO-3FactSheet-Africa.pdf (last accessed 1 June 2007)

UNEP (2003b). UNEP support to NEPAD: Period of Support 2004-2005. United Nations Environment Programme, Nairobi (unpublished report) http://www.un.org/africa/osaa/2005%20UN%20System%20support%20for%20NEPAD/UNEP.pdf (last accessed 10 May 2007)

UNEP (2003c). *Desk study on the Environment in the Occupied Palestinian Territories.* United Nations Environment Programme, Nairobi http://postconflict.unep.ch/publications/INF-31-WebOPT.pdf (last accessed 22 April 2007)

UNEP (2004a). *Freshwater in Europe. Facts, Figures and Maps.* UNEP-Division of Early Warning and Assessment, Office for Europe, Geneva http://www.grid.unep.ch/product/publication/freshwater_europe.php (last accessed 17 April 2007)

UNEP (2004b). *GEO Latin America and the Caribbean Environment Outlook 2003.* United Nations Environment Programme, Nairobi

UNEP (2005a). *After the Tsunami, Rapid Environmental Assessment.* United Nations Environment Programme, Nairobi http://www.unep.org/tsunami/tsunami_rpt.asp (last accessed 22 April 2007)

UNEP (2005b). *E-waste: the hidden side of IT equipment's manufacturing and use.* Early Warning of Emerging Environmental Threats, Issue 5. UNEP Division of Early Warning and Assessment GRID-Europe, Geneva

UNEP (2005c). *Assessment of Environmental "Hot Spots" in Iraq.* United Nations Environment Programme, Nairobi

UNEP (2006a). *Africa Environment Outlook 2: Our Environment, Our Wealth.* United Nations Environment Programme, Nairobi

UNEP (2006b). World Environment Day Factsheet. UNEP, Nairobi Africa

UNEP (2006c). *Asia-Pacific Lead Matrix.* Partnership for Clean Fuels and Vehicles, United Nations Environment Programme, Geneva http://www.unep.org/pcfv/PDF/LeadMatrix-Asia-Pacific-Jan07.pdf (last accessed 1 June 2007)

UNEP (2006d). The Regional Seas Programme, 2006. United Nations Environment Programme http://www.unep.org/regionalseas (last accessed 22 April 2007)

UNEP (2006e). *Assessment Reports on Priority Ecological Issues in Central Asia.* United Nations Environment Programme, Ashgabat

UNEP (2006f). *GEO Year Book 2006: An Overview of Our Changing Environment.* United Nations Environment Programme, Nairobi

UNEP (2006g). *The Crisis in Lebanon: Environmental Impact.* United Nations Environment Programme, Nairobi www.unep.org/Lebanon/ (last accessed 26 September 2006)

UNEP (2006h). *Situation Report #8, Environmental Issues Associated with the Conflict in Lebanon.* United Nations Environment Programme, Post-Conflict Branch, Nairobi

UNEP (2006i). *Marine and coastal ecosystems and human well-being: A synthesis report based on the findings of the Millennium Ecosystem Assessment.* DEW/0785/NA. United Nations Environment Programme, Nairobi

UNEP (2007a). *Central and Eastern Europe + Central Asia lead matrix.* Partnership for Clean Fuels and Vehicles, United Nations Environment Programme, Geneva http://www.unep.org/pcfv/PDF/MatrixCEELeadMarch07.pdf (last accessed 1 June 2007)

UNEP (2007b). *GEO Year Book 2007.* United Nations Environment Programme, Nairobi

UNEP (2007c). *Global Outlook for Ice and Snow.* United Nations Environment Programme, Nairobi

UNEP/DEWA/GRID-Europe (2006). *Gridded Population of the World version 3.* UNEP Division of Early Warning and Assessment GRID–Europe, Geneva

UNEP/GPA (2006a). *The State of the Marine Environment: Trends and Processes.* UNEP Global Programme of Action for the Protection of the Marine Environment from Land-based Activities, The Hague

UNEP/GPA (2006b). *Implementation of the GPA at regional level: The role of regional seas conventions and their protocols.* UNEP-Global Programme of Action for the Protection of the Marine Environment from Land-based Activities, The Hague

UNEP, UNESCWA and CAMRE (2001). *World Summit on Sustainable Development. Progress Assessment Report for the Arab Region.* United Nations, New York, NY

UNEP-WCMC (2007). *World Conservation Monitoring Centre database* (in GEO Data Portal). Cambridge http://www.unep-wcmc.org/ (last accessed 4 June 2007)

UNEP/MAP (2005). *Mediterranean Strategy for Sustainable Development.* UNEP-Mediterranean Action Plan, Athens

UNEP/PCAU (2004). UNEP-Post Conflict Assessment Unit http://Postconflict.Unep.Ch/ (last accessed 1 June 2007)

UNESCAP (2005a). *Asia-Pacific in Figures 2004.* Statistics Division, UN Economic and Social Commission for Asia and the Pacific, Bangkok

UNESCAP (2005b). *Review of the State of the Environment in Asia and the Pacific 2005.* UN Economic and Social Commission for Asia and the Pacific http://www.unescap.org/mced/documents/presession/english/SOMCED5_1E_SOE.pdf (last accessed 22 April 2007)

UNESCO (2003). Water quality indicator values in selected countries. In *The 1st UN World Water Development Report: Water for People, Water for Life.* United Nations Educational, Scientific and Cultural Organization, Paris http://www.unesco.org/bpi/wwdr/WWDR_chart2_eng.pdf (last accessed 24 April 2007)

UNESCWA (2001). *Strengthening Institutional Arrangements for the Implementation of Water Legislation and Improvement of Institutional Capacity.* Report No. E/UNESCWA/ENR/2001/11, United Nations Economic and Social Commissions for Western Asia, New York, NY

UNESCWA (2003a). *Governance for Sustainable Development in the Arab Region: Institutions and Instruments for Moving Beyond an Environmental Management Culture.* United Nations Economic and Social Commissions for Western Asia, New York, NY

UNESCWA (2003b). *Sectoral Water Allocation Policies in Selected UNESCWA Member Countries: An Evaluation of the Economic, Social and Drought Related Impact.* Report No. E/UNESCWA/SDPD/2003/13. United Nations Economic and Social Commissions for Western Asia, New York, NY http://www.escwa.org.lb/information/publications/edit/upload/sdpd-03-13.pdf (last accessed 1 June 2007)

UNESCWA (2003c). *Updating the Assessment of Water Resources in UNESCWA Member Countries.* Report No. E/UNESCWA/ENR/1999/13. United Nations Economic and Social Commissions for Western Asia, New York, NY

UNESCWA (2004). *Survey of Economic and Social Developments in the UNESCWA Region 2002-2003.* United Nations Economic and Social Commissions for Western Asia, New York, NY

UNESCWA (2005). Integrated Water Resources Management in UNESCWA Member Countries (Draft Report in Arabic). United Nations Economic and Social Commissions for Western Asia, New York, NY

UNESCWA and API (2002). *Economic Diversification in the Arab World.* United Nations Economic and Social Commissions for Western Asia, New York, NY

UNESCWA, UNEP, LAS and OAPEC (2005). *Energy for sustainable development for the Arab region, a framework for action.* United Nations Economic and Social Commissions for Western Asia, New York, NY

UNFCCC-CDIAC (2006). Greenhouse Gases Database. United Nations Framework Convention on Climate Change, Carbon Dioxide Information Analysis Centre (in GEO Data Portal) http://unfccc.int/ghg_emissions_data/items/3800.php (last accessed 16 May 2007)

UN-HABITAT (2001). *The State of the World's Cities 2001.* United Nations Centre for Human Settlements (Habitat), Nairobi

UN-HABITAT (2003a). *Guide to Monitoring Target 11: Improving the Lives of 100 Million Slum Dwellers.* United Nations-Habitat, Nairobi http://www.povertyenvironment.net/?q=guide_to_monitoring_target_11_improving_the_lives_of_100_million_slum_dwellers_2003 (last accessed 1 June 2007)

UN-HABITAT (2003b). *Observation of the Housing Sector.* United Nations-Habitat, Nairobi http://www.unhabitat.org/content.asp?cid=688&catid=203&typeid=13&subMenuId=0 (last accessed 1 June 2007)

UN-Habitat (2006). *State of the World's Cities 2006/7.* United Nations-Habitat, Nairobi

UNHCR (2005). *2004 Global Refugee Trends.* Population and Geographical Data Section, Division of Operational Support, United Nations High Commission for Refugees, Geneva

Unisys Corp. (2005). *Atlantic Hurricane Database.* Atlantic Oceanographic and Meteorological Laboratory, US National Oceanic and Atmospheric Administration http://weather.unisys.com/hurricane/atlantic/1987/index.html (last accessed 10 May 2007)

Universidad de Chile (2006). *Estado del Medio Ambiente en Chile 2005.* Informe país. GEO Chile. Universidad de Chile, Centro de Análisis de Políticas Públicas, Santiago

University of Cambridge (2005). *Large Ozone Losses over the Arctic.* University of Cambridge, Cambridge http://www.admin.cam.ac.uk/news/press/dpp/2005042601 (last accessed 22 April 2007)

UNPD (2003). *World Urbanization Prospects: The 2003 Revision Population Database.* United Nations Population Division, Economics and Social Affairs, New York, NY

UNPD (2005). *World Urbanisation Prospects: The 2005 Revision* (in GEO Data Portal). UN Population Division, New York, NY http://www.un.org/esa/population/unpop.htm (last accessed 4 June 2007)

UNPD (2007). *World Population Prospects: The 2006 Revision* (in GEO Data Portal). UN Population Division, New York, NY http://www.un.org/esa/population/unpop.htm (last accessed 4 June 2007)

UNRWA (2005). *Total Registered Refugees per Country and Area*. United Nations Relief and Works Agency for Palestine Refugees in the Near East http://www.un.org/unrwa/publications/pdf/rr_countryandarea.pdf (last accessed 16 May 2007)

UNU (2002). INWEH leads project to reduce blue baby syndrome in Syria. *UNU Update: The newsletter of United Nations University and its network of research and training centres and programmes* 14 http://update.unu.edu/archive/issue14_6.htm (last accessed 22 April 2007)

UNSD (2005). *UN Statistics Division Transport Statistics Database, UN Statistical Yearbook*. United Nations, New York, NY (in GEO Data Portal)

USEIA (2005a). *Annual Energy Review 2004*. US Department of Energy, Energy Information Administration, Washington, DC http://www.eia.doe.gov/emeu/aer/contents.html (last accessed 21 April 2007)

USEIA (2005b). *Country Analysis Briefs: Canada*. US Department of Energy, Energy Information Administration http://www.eia.doe.gov/emeu/cabs/canada.html (last accessed 21 April 2007)

USEIA (2006a). *International Energy Outlook, 2006*. Energy Information Administration, Office of Integrated Analysis and Forecasting, US Department of Energy, Washington, DC http://www.eia.doe.gov/oiaf/ieo/pdf/0484(2006).pdf (last accessed 10 May 2007)

USEIA (2006b). *Emissions of Greenhouse Gases in the United States 2005*. US Department of Energy, Energy Information Administration ftp://ftp.eia.doe.gov/pub/oiaf/1605/cdrom/pdf/ggrpt/057305.pdf (last accessed 21 April 2007)

USEPA (2000). *National Water Quality Inventory*. http://www.epa.gov/305b/2000report/ (last accessed 5 May 2007)

USEPA (2002). *Index of Watershed Indicators: An Overview*. US Environmental Protection Agency http://www.epa.gov/iwi/iwi-overview.pdf (last accessed 17 April 2007)

USEPA (2003). *Draft Report on the Environment*. US Environmental Protection Agency, Environmental Indicators Initiative http://www.epa.gov/indicators/roe/html/roePDF.htm (last accessed 5 May 2007)

USEPA (2004). *Smart Growth*. US Environmental Protection Agency http://www.epa.gov/smartgrowth/index.htm (last accessed 5 May 2007)

USEPA (2005a). *Acid Rain Program 2004 Progress Report*. US Environmental Protection Agency, Clean Air Markets Division, Office of Air and Radiation http://www.epa.gov/airmarkets/progress/docs/2004report.pdf (last accessed 17 May 2007)

USEPA (2005b). USEPA Announces New Rules that Will Further Improve and Protect Drinking Water. US Environmental Protection Agency http://yosemite.epa.gov/opa/admpress.nsf/d9bf8d9315e942578525701c005e573c/7b40d09f9a90e02f852570d80066e978!OpenDocument (last accessed 5 March 2007)

USEPA (2006a). *Energy Star Overview of 2005 Achievments*. US Environmental Protection Agency, http://www.epa.gov/appdstar/pdf/CPPD2005.pdf (last accessed 24 April 2007)

USEPA (2006b). *Draft Wadeable Streams Assessment: A Collaborative Survey of the Nation's Streams*. US Environmental Protection Agency, Office of Water, Washington, DC http://www.cpcb.ku.edu/datalibrary/assets/library/projectreports/WSAEPAreport.pdf (last accessed 20 June 2007)

USGS (2000). *US Geological Survey World Petroleum Assessment 2000 — Description and Results*. United States Geological Survey, Washington, DC

USGS (2005a). *Distance to Nearest Road in the Conterminous United States*. Fact Sheet 2005-3011. US Geological Survey, Washington, DC http://www.fort.usgs.gov/products/publications/21426/21426.pdf (last accessed 1 June 2007)

USGS (2005b). *Coastal-Change and Glaciological Maps of Antarctica*. Fact Sheet FS 2005-3055. US Geological Survey, Washington, DC http://pubs.usgs.gov/fs/2005/3055/ (last accessed 1 June 2007)

US Mayors (2005). Adopted resolution reported out of the Standing Committees. In *73rd Annual US Conference of Mayors, 10-14 June, Chicago, IL* http://www.usmayors.org/uscm/resolutions/73rd_conference/resolutions_adopted_2005.pdf (last accessed 1 June 2007)

Vázquez-Botello, A., Rendón von Osten, J., Gold-Bouchot, G. and Agraz-Hernández, C. (2005). *Golfo de México. Contaminación e impacto ambiental: diagnóstico y tendencias*. 2ª edición. Universidad Autónoma de Campeche, Universidad Nacional Autónoma de México, Instituto Nacional de Ecología

Velicogna, I. and Wahr, J. (2006). Measurements of time-variable gravity show mass loss in Antarctica. In *Science* 311:1754-1756

Venema, H. D. (2006). *From Cumulative Threats to Integrated Responses: A Review of Ag-Water Policy Issues in Prairie Canada*. International Institute for Sustainable Development, Winnipeg, MB

Venetoulis, J. and Talberth, J. (2005). *Ecological Footprint of Nations: 2005 Update*. Redefining Progress, Oakland CA http://www.ecologicalfootprint.org/pdf/Footprint%20of%20Nations%202005.pdf (last accessed 10 May 2007)

Vestreng, V., Breivik, K., Adams, M., Wagener, A., Goodwin, J., Rozovskkaya, O. and Pacyna, J. M. (2005). Inventory Review 2005, Emission Data reported to LRTAP Convention and NEC Directive, Initial review of HMs and POPs. UNECE-EMEP Technical report MSC-W 1/2005. Meteorological Synthesising Centre-West, Norwegian Meteorological Institute, Oslo http://www.emep.int/ (last accessed 20 June 2007)

Vistnes, I. and Nellemann, C. (2001). Avoidance of cabins, roads and powerlines by reindeer during calving. In *Journal of Wildlife Management* 65:915-925

Wang Shu-cheng (2005). Report on water saving, protecting and reasonable utilization. Address at: The 13th *Session of the Standing Committee of the 10th National People's Congress of the People's Republic of China* http://www.mwr.gov.cn/bzzs/20050124/50676.asp 2005-7-16 (in Chinese) (last accessed 10 May 2007)

WBCSD (2005). *Facts and Trends: Water*. World Business Council for Sustainable Development, Geneva http://www.wbcsd.org/web/publications/Water_facts_and_trends.pdf (last accessed 1 June 2007)

WCED (1987). *Our Common Future: The World Commission on Environment and Development*. Oxford University Press, Oxford

WCRP (2006). *Summary Statement from the World Climate Research Programme Workshop: Understanding Sea-level Rise and Variability*. World Climate Research Programme, 6-9 June 2006. IOC/United Nations Educational, Scientific and Cultural Organization, Paris http://copes.ipsl.jussieu.fr/Workshops/SeaLevel/Reports/Summary_Statement_2006_1004.pdf (last accessed 22 April 2007)

Wear, D. N. (2005). Forest Sustainability along Rural Urban Interfaces. In *Emerging Issues Along Urban/Rural Interfaces: Linking Science And Society*, 13-16 March 2005, Atlanta, GA http://www.urbanforestrysouth.org/Resources/Library/copy5_of_Citation.2005-04-28.5241/ (last accessed 24 April 2007)

Webster, P. J., Holland, G. J., Curry, J. A. and Chang, H.-R. (2005). Changes in tropical cyclone number, duration, and intensity in a warming environment. In *Science* 309(5742):1844-1846

Weinthal, E., Vengosh A., Gutierrez A. and Kloppmann, W. (2005). The water crises in Gaza Strip: prospects for resolution. In *Ground Water* 43:653-660

WFP (2005). 2004 Food Aid Flows, International Food Aid Information System (INTERFAIS). In *Food Aid Monitor* May 2005 http://www.wfp.org/interfais/index.htm (last accessed 22 April 2007)

WHO (2000a). *Guidelines for Air Quality*. WHO/SDE/OEH/00.02. World Health Organization, Geneva http://whqlibdoc.who.int/hq/2000/WHO_SDE_OEH_00.02_pp1-104.pdf (last accessed 1 June 2007)

WHO (2000b). World Health Organization: Preparation of WHO water quality guidelines for desalination - A Preliminary Note (Unpublished Report)

WHO (2006). *CDC Dengue Map: Distribution of Aedes aegypti in the Americas*. CDC Division of Vector-Borne Infectious Diseases (DVBID), World Health Organization, Geneva

WHO (2007). Children's health and the environment in Europe: a baseline assessment. World Health Organization Regional Office for Europe, Copenhagen (in press)

WHO and UNICEF (2006). *MDG Drinking Water and Sanitation Target: Assessment Report 2006*. World Health Organization, Geneva and United Nations Children's Fund, New York, NY (in GEO Data Portal)

Wie, H. (2005). Landmines in Lebanon: An historic overview and the current situation. In *Journal of Mine Action* October

Wilkinson, C. (ed.) (2000). *Status of Coral Reefs of the World: 2000*. Australian Institute of Marine Science, http://www.aims.gov.au/pages/research/coral-bleaching/scr2000/scr-00.html (last accessed 1 June 2007)

Wilkinson, C. (ed.) (2004). *Status of Coral Reefs of the World: 2004*. Australian Institute of Marine Science, Townsville http://www.aims.gov.au/pages/research/coral-bleaching/scr2004/index.html (last accessed 1 May 2007)

Winchester, L. (2005). Sustainable human settlements development in Latin America and the Caribbean. In *Medio Ambiente y Desarrollo No. 99*, February. Sustainable Development and Human Settlements Division, UNCLAC, Santiago de Chile

WMO and UNEP (2006). *Executive Summary- Scientific Assessment of Ozone Depletion: 2006*. World Meteorological Organization/United Nations Environment Programme, Geneva/Nairobi http://www.wmo.ch/web/arep/reports/ozone_2006/exec_sum_18aug.pdf (last accessed 22 April 2007)

Wood, S., Sebastian, K. and Scherr, S. (2000). *Pilot Analysis of Global Ecosystems (PAGE)*. Agroecosystems Technical Report. World Resources Institute, Washington, DC

Wood, D. B. (2005). More tests, more closed shores. *Christian Science Monitor*, 9 August http://www.csmonitor.com/2005/0809/p01s01-usgn.html (last accessed 1 June 2007)

World Bank (1999). *What a Waste: Solid Waste Management in Asia*. Urban Development Sector Unit, East Asia and Pacific Region, The World Bank, Washington, DC http://siteresources.worldbank.org/INTEAPREGTOPURBDEV/Resources/whatawaste.pdf (last accessed 1 June 2007)

World Bank (2002a). *Philippines Environment Monitor 2002*. The World Bank, Washington, DC

World Bank (2002b). *Thailand Environment Monitor 2002*. The World Bank, Washington, DC

World Bank (2002c). *Environment Matters at the World Bank: Annual Review*. The World Bank, Washington, DC

World Bank (2002d). *The Environment and the Millennium Development Goals*. The World Bank, Washington, DC

World Bank (2003a). *the science of health impacts of particulate matter*. South Asia Urban Air Quality Management Briefing Note: Note no. 9. The World Bank, Washington, DC

World Bank (2003b). *Jobs, Growth and Governance in the Countries of West Asia: Unlocking the Potential for Prosperity*. The World Bank, Washington, DC http://lnweb18.worldbank.org/mna/mena.nsf/Attachments/Integrative+Report+-+English/$File/intergrativepaper.pdf (last accessed 1 June 2007)

World Bank (2004a). *Cost of Environmental Degradation — The Case of Lebanon and Tunisia*. The World Bank, Washington, DC

World Bank (2004b). *Four Years — Intifada, Closures and Palestinian Economic Crisis: An Assessment*. The World Bank, Washington, DC

World Bank (2005a). *World Development Indicators 2005*. The World Bank, Washington, DC

World Bank (2005b). *Middle East and North Africa Economic Developments and Prospects 2005, Oil Booms and Revenue Management*. The World Bank, Washington, DC http://web.worldbank.org/WBSITE/EXTERNAL/COUNTRIES/MENAEXT/0,,contentMDK:20449345~pagePK:146736~piPK:226340~theSitePK:256299,00.html (last accessed 1 June 2007)

World Bank (2005c). *Poverty in MENA, Sector Brief*. The World Bank, Washington, DC http://siteresources.worldbank.org/INTMNAREGTOPPOVRED/Resources/POVERTY-ENG2006AM.pdf (last accessed 1 June 2007)

World Bank (2006). *World Development Indicators 2006*. The World Bank, Washington, DC (in GEO Data Portal)

World Bank/METAP (2003). Regional Solid Waste Management Project in METAP Mashreq and Maghreb Countries, Inception Report (Final Version). Mediterranean Environmental Technical Assistance Program http://lnweb18.worldbank.org/mna/mena.nsf/METAP+Documents/062332B943F17C8685256CD90013EF8A?OpenDocument (last accessed 1 June 2007)

Woynillowicz, D., Severson-Baker, C. and Raynolds, M. (2005). *Oil Sands Fever: The Environmental Implications of Canada's Oil Sands Rush*. The Pembina Institute, Drayton Valley, AB http://www.oilsandswatch.org/pub/203 (last accessed 1 June 2007)

WSSCC (2006). *Partnerships in Action*. Water Supply and Sanitation Collaborative Council http://www.wash-cc.org/ (last accessed 1 June 2007)

WRI (1995). *World Resources 1994-1995: People and the environment, resource consumption, population growth and women*. World Resources Institute, in collaboration with United Nations Environment Programme and United Nations Development Programme. Oxford University Press, New York, NY

WRI (2000). *World Resources 2000-2001: People and Ecosystems, The Fraying Web of Life*. World Resources Institute, in collaboration with United Nations Development Programme, United Nations Environment Programme and The World Bank, Washington, DC

Wunder, S. (2001). Poverty alleviation and tropical forests — what scope for synergies? In *World Development* 29(11):1817-33

WWF (2005). *An Overview of Glaciers, Glacier Retreat, and Subsequent Impacts in Nepal, India and China*. World Wide Fund for Nature Nepal Program http://assets.panda.org/downloads/himalayaglaciersreport2005.pdf (last accessed 22 April 2007)

WWF (2006a). *Living Planet Report*. World Wide Fund for Nature, Gland http://assets.panda.org/downloads/living_planet_report.pdf (last accessed 4 June 2007)

WWF (2006b). *Paraguay: Zero Deforestation Law contributes significantly to the conservation of the Upper Parana Atlantic Forest*. World Wide Fund for Nature, Gland http://assets.panda.org/downloads/final_fact_sheet_eng_llp_event_asuncion_aug_30__2006.pdf (last accessed 4 June 2007)

Yefremov, D.F. and Shvidenko, A.Z. (2004). Long-terms Environmental Impact of Catastrophic Forest Fires in Russia's Far East and their Contribution to Global Processes. In *International Forest Fire News*. No. 32. 2004:43-39 http://www.fire.uni-freiburg.de/iffn/iffn_32/06-Yefremov.pdf (last accessed 17 April 2007)

Human Dimensions of Environmental Change

Chapter 7 **Vulnerability of People and the Environment: Challenges and Opportunities**

Chapter 8 **Interlinkages: Governance for Sustainability**

Many people, individually and collectively, contribute, often inadvertently, to the suffering of others while improving their own well-being. This can result from environmental changes which are linked across scales and between geographical regions through both biophysical and social processes.

Chapter 7

Vulnerability of People and the Environment: Challenges and Opportunities

Coordinating lead authors: Jill Jäger and Marcel T.J. Kok

Lead authors: Jennifer Clare Mohamed-Katerere, Sylvia I. Karlsson, Matthias K.B. Lüdeke, Geoffrey D. Dabelko, Frank Thomalla, Indra de Soysa, Munyaradzi Chenje, Richard Filcak, Liza Koshy, Marybeth Long Martello, Vikrom Mathur, Ana Rosa Moreno, Vishal Narain, and Diana Sietz

Contributing authors: Dhari Naser Al-Ajmi, Katrina Callister, Thierry De Oliveira, Norberto Fernandez, Des Gasper, Silvia Giada, Alexander Gorobets, Henk Hilderink, Rekha Krishnan, Alexander Lopez, Annet Nakyeyune, Alvaro Ponce, Sophie Strasser, and Steven Wonink

In memory of Gerhard Petschel-Held

Chapter review editor: Katharina Thywissen

Chapter coordinator: Munyaradzi Chenje

Credit: Munyaradzi Chenje

Main messages

Vulnerability depends on exposure, sensitivity to impacts and the ability or inability to cope or adapt. It needs to be seen within a global context of demographic change, patterns of poverty, health, globalization, conflict and governance. Broad representative patterns of vulnerability to environmental and socio-economic changes are identified in this chapter. This provides a basis for an analysis of the interacting pressures. It shows opportunities for reducing vulnerability and increasing human well-being, while protecting the environment. The following are the main messages:

Significant improvements in human well-being have been achieved over the last 20 years. However, there are more than 1 billion poor people. They are found in all regions. They lack essential services, making them vulnerable to environmental and socio-economic changes. Many countries will not meet the Millennium Development Goals' 2015 targets. But, dealing with vulnerability provides opportunities to meet these goals.

Analysis of patterns of vulnerability shows the unequal distribution of risks for specific groups of people. The most vulnerable groups include the poor, indigenous populations, women and children in both developing and developed countries.

Improving human well-being – the extent to which individuals have the ability to live the kind of lives they value, and the opportunities to achieve their potential – is at the heart of development. This is not just a moral imperative, but is also a critical aspect of human rights. It is essential for reducing vulnerability and achieving sustainable use of the environment.

Gains in life expectancy and per capita health expenditures, as well as declines in child mortality have been systematically greater in those countries with more equitable income distribution and access to medical treatment. It is, however, paradoxical that opulence and consumerism, as well as relative poverty, contribute to ill health in many wealthier societies.

International trade has helped increase income, and has helped millions of people out of poverty, but it is also sustaining unequal patterns of consumption. Outsourcing the extraction of natural resources, as well as production and manufacturing to developing countries, means they must struggle to deal with their resulting hazardous wastes and other environmental impacts.

Conflicts, violence and persecution regularly displace large civilian populations, forcing millions of people into marginal ecological and economic areas within countries and across international boundaries. This undermines, sometimes for decades, sustainable livelihoods and economic development as well as the capacities of societies and nations. The resulting poverty, often tied to shortages or degradation of natural resources, contributes directly to lower levels of human well-being and higher levels of vulnerability.

Exposure to natural hazards has increased as a result of climate change and such actions as the destruction of mangroves that protect coasts from tidal surges. Risks are also increasing as a result of the continuing concentration of people in highly-exposed areas. Over the past 20 years, natural disasters have claimed more than 1.5 million lives, and affected more than 200 million people annually. More than 90 per cent of the people exposed to disasters live in the developing world, and more than half of disaster deaths occur in countries with a low human development index. Capacity to adapt is being eroded through, for example, reduced state social protection schemes, undermining of informal safety nets, poorly built or maintained infrastructure, chronic illness and conflict.

Poverty must be addressed in all countries if vulnerability to both environmental and socio-economic changes is to be reduced. Relative poverty is increasing in many countries despite general affluence. Improved access to material assets at the household level (income, food, drinking water, shelter, clothing, energy, natural and financial resources) and at the societal level (physical and service infrastructure) can help break the cycle of impoverishment, vulnerability and environmental degradation. This means that being poor need not mean staying poor.

To achieve sustainable development, governance must be integrated from the local to the global levels, across a range of sectors, and over a longer time frame for policy making. Over the past 20 years, governance has become increasingly multi-level, with more interaction and interdependence. Local governments, community-based groups and other non-governmental actors now engage more widely in international cooperation, contributing to a better grounding of global policy in experiences of local vulnerability.

Integrating development, health and environment policies provides an opportunity, since health and education are the cornerstones of human capital. Continued investment remains critical for increasing the capacity to adapt to environmental and other changes. While under-five mortality rates have improved considerably, large regional differences still exist.

Empowering women not only contributes to the widely-shared objective of equity and justice, but also makes good economic, environmental and social sense. Practice shows that finance schemes that especially target women can have higher than usual payoffs. Better access to education increases maternal health, creating a better starting point for the next generation. Gender-sensitive poverty alleviation in both rural and urban

settings is a central component of strategies to address environment and health issues.

Environmental cooperation creates an effective path to peace by promoting sustainable resource use and equity within and between countries. Investing in cooperation is an investment in the future, because both scarcity and abundance of environmental resources can exacerbate existing tensions, and contribute to conflict between groups, especially in societies that lack the capacity to effectively and equitably manage competition for control over resources.

Official development assistance must be stepped up to meet the agreed global target of 0.7 per cent of GNI. The decline in support for agriculture and infrastructure investment must be reversed if developing countries are to build their economies and increase their capacity to adapt to environmental and socio-economic change. Making international trade fairer, and including environmental concerns will also increase such adaptive capacity.

The potential for science and technology to reduce vulnerability is still very unevenly distributed worldwide. Partnerships that deliver, and increased investments could improve this situation. However, science and technology have also undoubtedly added to the risks faced by people and the environment, particularly by driving environmental change.

There are strong synergies between improving human well-being and reducing vulnerability from environment, development and human rights perspectives. The call for action to protect the environment needs to be strongly focused on human well-being. It also underlines the importance of implementing existing obligations made by governments at the national and international levels.

INTRODUCTION

There are strong causal relationships among the state of the environment, human well-being and vulnerability. Understanding how environmental and non-environmental changes affect human well-being and vulnerability is the critical foundation for addressing challenges to and the opportunities for improving human well-being while also protecting the environment.

Vulnerabilities are often driven by actions taken at a great distance, highlighting worldwide interdependencies. Within the context of vulnerability, the chapter illustrates how current policies on mitigation, coping and adaptive capacity support the contribution of environmental policies to internationally agreed development goals, particularly the Millennium Development Goals (MDGs). This analysis also evaluates whether environmental governance adequately links with other relevant policy domains, such as poverty alleviation, health, science, and technology and trade. It underlines the need for mainstreaming environment into these domains to reduce vulnerability. This provides strategic directions for policy making to reduce vulnerability and enhance human well-being (see Chapter 10).

As the World Commission on Environment and Development (Brundtland Commission) stated in *Our Common Future*, "A more careful and sensitive consideration of their (vulnerable groups) interests is a touchstone of sustainable development policy" (WCED 1987). The vulnerability approach applied here (see Box 7.1) shows the potential for strong negative consequences for well-being of, for example, reduced access to resources, such as food and drinking water, and the existence of thresholds beyond which health and survival are severely threatened. Patterns of vulnerability to environmental and socio-economic changes, here referred to as "archetypes," describe the impacts of these changes on human well-being.

GLOBAL CONTEXT OF VULNERABILITY

A number of factors shape the vulnerability of people and the environment, including poverty, health, globalization, trade and aid, conflict, changing levels of governance, and science and technology.

Poverty

Poverty (see Chapter 1) reduces the ability of individuals to respond and adapt to environmental change. Although the multidimensional nature of poverty is widely recognized, income and consumption remain the most common measures. Most regions have made progress in meeting the first Millennium Development Goal (MDG 1) on reducing extreme poverty and hunger (see Figure 7.1), although many will not achieve the 2015 targets. In developing countries, extreme poverty (those living on less than US$1/day) fell from 28 per cent in 1990 to 19 per cent in 2002. Actual numbers decreased from 1.2 billion to just over 1 billion in 2002 (World Bank 2006). The percentage of people in the world with insufficient food to meet their daily needs has declined, but actual numbers

---

Box 7.1 The concept of vulnerability

Vulnerability is an intrinsic feature of people at risk. It is multidimensional, multidisciplinary, multisectoral and dynamic. It is defined here as a function of exposure, sensitivity to impacts and the ability or lack of ability to cope or adapt. The exposure can be to hazards such as drought, conflict or extreme price fluctuations, and also to underlying socio-economic, institutional and environmental conditions. The impacts not only depend on the exposure, but also on the sensitivity of the specific unit exposed (such as a watershed, island, household, village, city or country) and the ability to cope or adapt.

Vulnerability analysis is widely used in the work of many international organizations and research programmes concerned with poverty reduction, sustainable development and humanitarian aid organizations. These include FAO, the Red Cross and Red Crescent Societies, UNDP, UNEP and the World Bank. This kind of work helps to identify the places, people and ecosystems that may suffer most from environmental and/or human-induced variability and change, and identifies the underlying causes. It is used to develop policy relevant recommendations for decision-makers on how to reduce vulnerability and adapt to change.

The concept of vulnerability is an important extension of traditional risk analysis, which focused primarily on natural hazards. Vulnerability has become a central aspect of studies of food insecurity, poverty and livelihoods and climate change. While earlier research tended to regard vulnerable people and communities as victims in the face of environmental and socio-economic risks, more recent work increasingly emphasizes the capacities of different affected groups to anticipate and cope with risks, and the capacities of institutions to build resilience and adapt to change.

The complementary concept of resilience has been used to characterize a system's ability to bounce back to a reference state after a disturbance, and the capacity of a system to maintain certain structures and functions despite disturbance. If the resilience is exceeded, collapse can occur.

Sources: Bankoff 2001, Birkmann 2006, Blaikie and others 1994, Bohle, Downing and Watts 1994, Chambers 1989, Chambers and Conway 1992, Clark and others 1998, Diamond 2004, Downing 2000, Downing and Patwardhan 2003, Hewitt 1983, Holling 1973, Kasperson and others 2005, Klein and Nicholls 1999, Pimm 1984, Prowse 2003, Watts and Bohle 1993, Wisner and others 2004

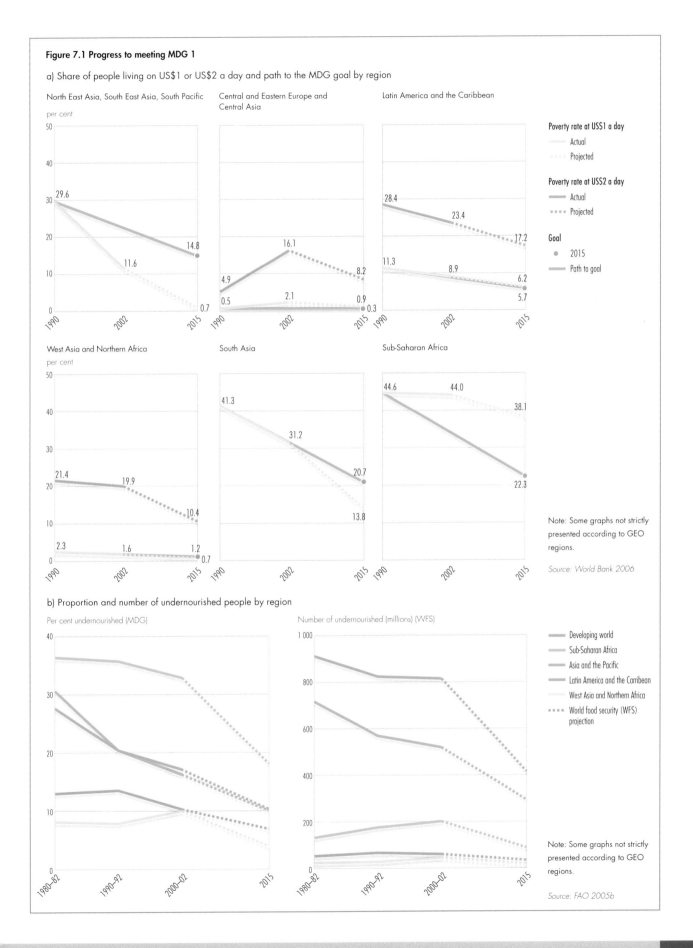

Figure 7.1 Progress to meeting MDG 1

a) Share of people living on US$1 or US$2 a day and path to the MDG goal by region

North East Asia, South East Asia, South Pacific

Central and Eastern Europe and Central Asia

Latin America and the Caribbean

Poverty rate at US$1 a day
Actual
Projected

Poverty rate at US$2 a day
Actual
Projected

Goal
● 2015
Path to goal

West Asia and Northern Africa

South Asia

Sub-Saharan Africa

Note: Some graphs not strictly presented according to GEO regions.

Source: World Bank 2006

b) Proportion and number of undernourished people by region

Per cent undernourished (MDG)

Number of undernourished (millions) (WFS)

Developing world
Sub-Saharan Africa
Asia and the Pacific
Latin America and the Carribean
West Asia and Northern Africa
World food security (WFS) projection

Note: Some graphs not strictly presented according to GEO regions.

Source: FAO 2005b

increased between 1995 and 2003 (UN 2006), when about 824 million people suffered chronic hunger. Sustained growth in China and India has contributed to sharp decreases in extreme poverty in Asia (Dollar 2004, Chen and Ravallion 2004). Where inequity is high, including in some of the transition countries of Europe and Central Asia, economic growth does not necessarily translate into less poverty (WRI 2005, World Bank 2005). In many countries, relative poverty is increasing despite general affluence. In the United States, for example, the number of people living below the national poverty line has risen since 2000, reaching almost 36 million in 2003 (WRI 2005). Structural economic adjustment, ill health, and poor governance affected progress in some regions, including sub-Saharan Africa (Kulindwa and others 2006).

Health

Health is central to the achievement of the MDGs because it is the basis for job productivity, the capacity to learn, and the capability to grow intellectually, physically and emotionally (CMH 2001). Health and education are the cornerstones of human capital (Dreze and Sen 1989, Sen 1999). Ill health reduces the capacity to adapt to environmental and other changes. Under-five mortality rates have improved considerably, though there are still large regional differences (see Figure 7.2), and more than 10 million children under five still die every year – 98 per cent of

them in developing countries. Some 3 million die due to unhealthy environments (Gordon and others 2004).

WHO identified the major health risks for developing and developed countries, as shown in Table 7.1. They include traditional risks associated with underdevelopment (such as underweight, unsafe water and lack of sanitation), and those associated with consumptive lifestyles (such as obesity and physical inactivity).

Health gains are unequal across regions and within population groups. In the least favourable health situations, people suffer persistent communicable diseases associated with deficient living conditions, including poverty and progressive environmental degradation. AIDS has become a leading cause of premature deaths in sub-Saharan Africa, and the fourth largest killer worldwide (UN 2006). By the end of 2004, an estimated 39 million people were living with HIV/AIDS. The epidemic has reversed decades of development progress in the worst-affected countries, contributing to strong increases in vulnerability.

Globalization, trade and aid

The rapid growth of trade and financial flows is creating more global interdependence. The trade and development agendas have so far not been reconciled, and the gulf between the rich and the poor is still growing. Poor countries are moving to market solutions and pragmatic arrangements for increasing

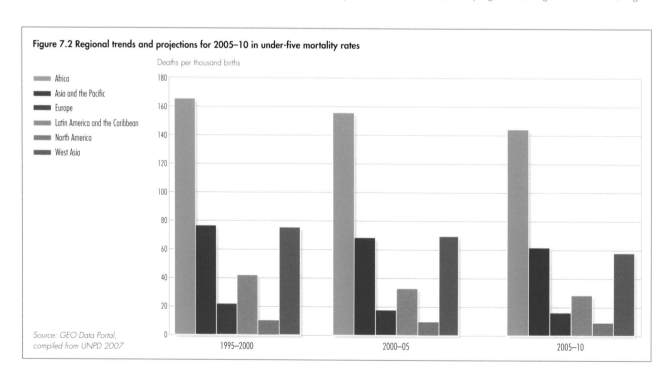

Figure 7.2 Regional trends and projections for 2005–10 in under-five mortality rates

Deaths per thousand births

Legend:
- Africa
- Asia and the Pacific
- Europe
- Latin America and the Caribbean
- North America
- West Asia

(x-axis: 1995–2000, 2000–05, 2005–10)

Source: GEO Data Portal, compiled from UNPD 2007

Table 7.1 Estimated attributable and avoidable burdens of 10 leading selected risk factors

Developing countries high mortality (per cent)		Developing countries low mortality (per cent)		Developed countries (per cent)	
Underweight	14.9	Alcohol	6.2	Tobacco	12.2
Unsafe sex	10.2	Blood pressure	5.0	Blood pressure	10.9
Unsafe water, sanitation and hygiene	5.5	Tobacco	4.0	Alcohol	9.2
Indoor smoke from solid fuel	3.6	Underweight	3.1	Cholesterol	7.6
Zinc deficiency	3.2	Overweight	2.4	Overweight	7.4
Iron deficiency	3.1	Cholesterol	2.1	Low fruit and vegetable intake	3.9
Vitamin A deficiency	3.0	Low fruit and vegetable intake	1.9	Physical inactivity	3.3
Blood pressure	2.5	Indoor smoke from solid fuel	1.9	Illicit drugs	1.8
Tobacco	2.0	Iron deficiency	1.8	Unsafe sex	0.8
Cholesterol	1.9	Unsafe water, sanitation and hygiene	1.8	Iron deficiency	0.7

Note: percentage causes of disease burden expressed in Disability Adjusted Life Years.

Source: WHO 2002

trade and foreign direct investment (FDI) to create more jobs and alleviate poverty (Dollar and Kraay 2000, UNCTAD 2004). The outcomes are highly uneven (see Figure 7.3). The failure of the Doha round of the WTO talks continues to hurt the poorest of the poor who often depend on agricultural markets.

With the growing interest in markets, the aid agenda has also changed. Most of the recent increases in aid have been used to cancel debt, and meet humanitarian and reconstruction needs following

emergencies (UN 2006). The share of total official development assistance (ODA) going to basic human needs has doubled since the mid-1990s, but the share going to agriculture and physical infrastructure has diminished. These two sectors need support if countries are to feed their own people, build their economies (UN 2006), and increase their adaptive capacity. Africa remains the most aid dependent region by far, while West Asia's dependence on aid has varied considerably over the past 20 years (see Figure 7.3). Together the figures suggest a bleak

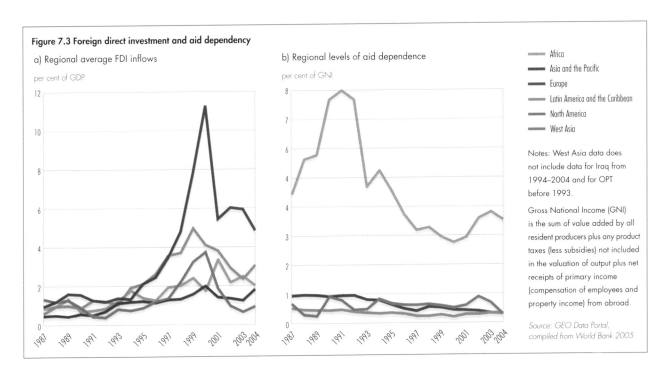

Figure 7.3 Foreign direct investment and aid dependency

a) Regional average FDI inflows

per cent of GDP

b) Regional levels of aid dependence

per cent of GNI

— Africa
— Asia and the Pacific
— Europe
— Latin America and the Caribbean
— North America
— West Asia

Notes: West Asia data does not include data for Iraq from 1994–2004 and for OPT before 1993.

Gross National Income (GNI) is the sum of value added by all resident producers plus any product taxes (less subsidies) not included in the valuation of output plus net receipts of primary income (compensation of employees and property income) from abroad.

Source: GEO Data Portal, compiled from World Bank 2005

reality. FDI, which is productive capital, is a great deal lower than aid in many regions. In 2005, the 191 million migrants worldwide (up from 176 million in 2000) contributed in excess of US$233 billion to productive capital of which US$167 billion went to developing countries (IOM 2005).

Conflict

The end of the Cold War in the late 1980s has reduced the threat of nuclear war from great power rivalry, although fears of continued nuclear proliferation among states and non-state actors remain (Mueller 1996). Civil conflicts continue to be the biggest threat, although incidences have decreased dramatically in recent years (see Box

7.2 and Figure 7.4). International involvement in civil wars, primarily in peacemaking and peacekeeping capacities, is at an all-time high due to humanitarian pressures. The increase in the number of formal democracies is unprecedented; this trend may contribute to the decreasing incidence of civil wars, although the transition to democracy is often a highly unstable period (Vanhanen 2000). All regions of the world have seen a decrease in armed violence except for sub-Saharan Africa and West Asia (Strand and others 2005).

Despite the positive global trends in armed violence, persistent conflicts have very negative impacts on well-being. More than 8 million people

---

Box 7.2 A less violent world

Since World War II, the number of interstate armed conflicts (conflicts between states) has remained relatively low, and no such conflict has been recorded since 2003. Extrasystemic armed conflicts (colonial conflicts and other conflicts between an independent state and non-state groups outside its own territory) had disappeared by the mid-1970s. Intrastate armed conflicts (civil conflicts or conflicts between a government and an organized internal rebel group) rose steadily until 1992, after which they declined steeply. Internationalized intrastate conflicts (intrastate conflicts with armed intervention from other

governments) have been frequent since the early 1960s. The lower threshold for conflicts recorded here is 25 battle-related deaths in a given year. The graph does not include state violence against unorganized people ('one-sided violence' or genocide and politicide) or violence between groups where the government is not a party to the fighting ('non-state violence' or communal violence). It is a stacked graph, meaning that the number of conflicts in each category in a given year is indicated by the height in that year of the area of a particular colour.

Source: Harbom and Wallensteen 2007

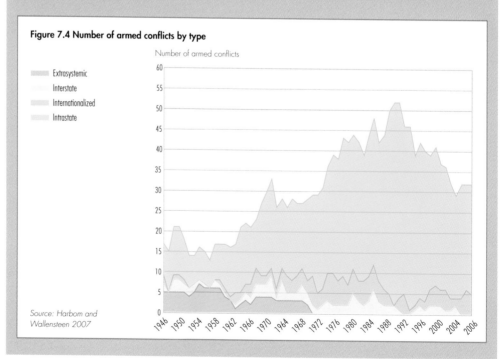

Figure 7.4 Number of armed conflicts by type

Extrasystemic
Interstate
Internationalized
Intrastate

Source: Harbom and Wallensteen 2007

have died directly or indirectly as a result of war in Africa since 1960 (Huggins and others 2006). Conflicts, violence and fear of persecution regularly displace large civilian populations, forcing millions of people into marginal ecological and economic areas within countries and across international boundaries. The UNHCR estimates that there were 11.5 million refugees, asylum seekers and stateless persons, and 6.6 million internally displaced persons globally in 2005 (UNHCR 2006). The forced movement of people into marginal areas undercuts, sometimes for decades, sustainable livelihoods, economic development, and societal and state capacities. The resulting poverty, often tied to shortages or degradation of natural resources, contributes directly to lower levels of well-being and higher levels of vulnerability.

Changing levels of governance

Over the past 20 years, governance has become increasingly multi-level, with more interaction and interdependence between different levels. The effectiveness of national policies (see Figure 7.5) remains mixed, but the capacity and political will of governments to take action has increased. In combination, these trends increase opportunities to reduce vulnerability. The early years after the end of the Cold War witnessed a renewed optimism in multilateralism and global governance. In parallel, regional cooperation made significant progress around the world, even if its forms and intensity differ.

There has also been a trend towards political and fiscal decentralization from national to sub-national levels, including in countries of the Organisation for Economic Cooperation and Development (OECD) (Stegarescu 2004) and in Africa and Latin America (Stein 1999, Brosio 2000). This may not necessarily mean that local authorities have been empowered, as decentralization without devolution of power can be a way to strengthen the presence of the central authority (Stohr 2001). Local governments, community-based groups and other non-governmental actors now engage more widely in international cooperation, contributing to a better grounding of global policy in experiences of local vulnerability. Global corporations' influence has extended beyond the economic arena (De Grauwe and Camerman 2003, Graham 2000, Wolf 2004), and many choose to develop voluntary environmental codes, and to increase self regulation (Prakash 2000).

Science and technology

Developments in science and technology have helped reduce human vulnerability to environmental and non-environmental change, although the pace and levels at which different regions achieve progress vary widely (UNDP 2001). Expenditures on research and development in OECD countries between 1997 and 2002 were 2.5 per cent of GDP compared to 0.9 per cent of GDP in developing countries (UNDP 2005). While the number of researchers was 3 046 per million people in OECD countries between 1990 and

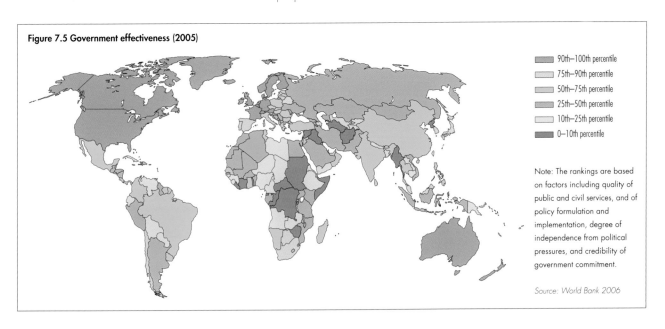

Figure 7.5 Government effectiveness (2005)

- 90th–100th percentile
- 75th–90th percentile
- 50th–75th percentile
- 25th–50th percentile
- 10th–25th percentile
- 0–10th percentile

Note: The rankings are based on factors including quality of public and civil services, and of policy formulation and implementation, degree of independence from political pressures, and credibility of government commitment.

Source: World Bank 2006

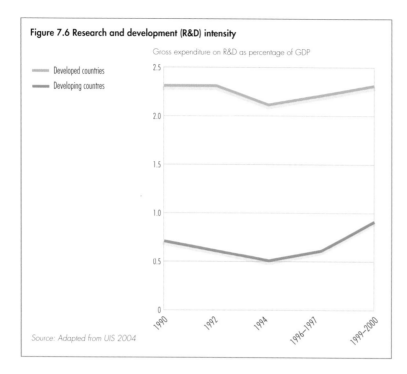

Figure 7.6 Research and development (R&D) intensity

Gross expenditure on R&D as percentage of GDP

— Developed countries
— Developing countres

Source: Adapted from UIS 2004

2003, it was 400 per million in developing countries (UNDP 2005). The potential for science and technology to reduce vulnerability remains very unevenly distributed worldwide (see Figure 7.6). This illustrates the need to improve technology transfers between regions.

For example, new farming technologies and practices since 1960 increased food production, and decreased food prices, addressing undernutrition and chronic famine in many regions, but access to these technologies remains unevenly distributed. In the 1980s, oral rehydration therapies and vaccines suitable for use in developing countries were critical in reducing under-five mortality. New information and communication technologies give unprecedented opportunities for early warning systems, and for generating local entrepreneurship. However, science and technology have undoubtedly also added to the risks faced by people and the environment, particularly by driving environmental change.

HUMAN WELL-BEING, ENVIRONMENT AND VULNERABILITY

Development challenges

Improving human well-being – the extent to which individuals have the ability to live the kind of lives they value, and the opportunities they have to achieve their potential – is at the heart of development. This is not just a moral imperative, but also a critical aspect of

fundamental human rights (UN 1966, UN 1986, UN 2003), and is essential for reducing vulnerability and achieving sustainable use of the environment.

Since the 1987 Brundtland Commission report emphasized the environment-development link, different policy statements and multilateral environmental agreements, including the 1992 Rio Declaration (Principle 1) and the conventions on biological diversity and climate change, have highlighted the opportunities the environment holds for development (see Chapter 1). Increased convergence between these international approaches and those at national level is evident from the highest-level recognition of environmental rights as human rights (Ncube and others 1996, Mollo and others 2005). Importantly, environmental rights approaches have moved from a focus on environmental quality to incorporating basic needs, development, and intergenerational and governance concerns (UN 2003, Gleick 1999, Mollo and others 2005). However, progress in meeting development objectives has been uneven.

Improvements in well-being – for some
Despite significant improvements in well-being over the last 20 years, with gains in income, nutrition, health, governance and peace, there are many on-going challenges (see global context section and Chapters 1–6) (UNDP 2006). Millions of people across all regions are poor, and lacking essential services that are now common among the wealthy. Many countries will not meet the MDGs' 2015 targets (UN 2006, World Bank 2006). But the environment provides opportunities to meet these goals, and to enhance well-being through the various goods-and-services it provides.

The link between environment and well-being is complex, non-linear and influenced by multiple factors, including poverty, trade, technology, gender and other social relations, governance, and the different aspects of vulnerability. Global interconnectedness – through a shared natural environment and globalization – means that achieving human well-being in one place may be affected by practices elsewhere.

How people actually live and the opportunities they have are closely connected to the environment (Prescott-Allen 2001, MA 2003) (see Chapters

1–6). As the Brundtland Commission warned, environmental degradation contributes to "the downward spiral of poverty" and amounts to "a waste of opportunities and of resources" (WCED 1987). Good health, for example, is directly dependent on good environmental quality (see Chapters 1–6) (MA 2003). Many national constitutions now recognize a healthy environment as a fundamental human right. Despite some improvements, pollution continues to be a problem, sometimes spurred on by factors outside the control of its victims (see global commons and contaminated sites archetypes). Associated risks and costs are unevenly distributed across society (see Figure 7.7). Although the incidence of ill health has been reduced globally, the costs remain monumental.

Notwithstanding improvements in access to water and sanitation (see Figure 4.3), the poorest people suffer the greatest water deficit as a result of location, poor infrastructure and lack of financial resources (see Figure 7.8). Consequently, they experience ill health and indignity (UNDP 2006). In many developing countries, poor people in cities pay more for water than wealthier inhabitants.

Poor access to material assets at the household level (income, food, water, shelter, clothing, energy, natural and financial resources) and at the societal level (physical and service infrastructure) is part of a cycle of impoverishment, vulnerability and environmental change. It is part of a sequence of becoming poor and staying poor (Brock 1999, Chronic Poverty Centre 2005). In developed countries too, relative poverty, age and gender are critical factors in the distribution of benefits. The energy archetype illustrates the vulnerabilities that arise through lack of access to energy, as well as those related to dependency on energy imports. Investing in physical and service infrastructural development can improve well-being by increasing marketing opportunities, security, and access to energy, clean water and technologies for efficient and sustainable natural resources use.

In countries with a low human development index, people also live shorter lives (see Figure 7.9), because they have reduced health, due to hunger, unsafe water, sanitation and hygiene (lack of water), and suffer from other environmental problems, such

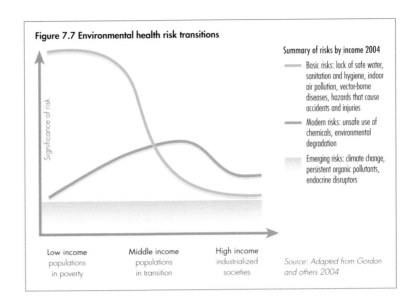

Figure 7.7 Environmental health risk transitions

Summary of risks by income 2004

— Basic risks: lack of safe water, sanitation and hygiene, indoor air pollution, vector-borne diseases, hazards that cause accidents and injuries

— Modern risks: unsafe use of chemicals, environmental degradation

Emerging risks: climate change, persistent organic pollutants, endocrine disruptors

Significance of risk

Low income populations in poverty   Middle income populations in transition   High income industrialized societies

Source: Adapted from Gordon and others 2004

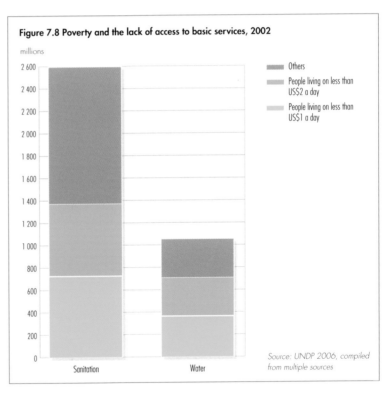

Figure 7.8 Poverty and the lack of access to basic services, 2002

millions

Others
People living on less than US$2 a day
People living on less than US$1 a day

Sanitation   Water

Source: UNDP 2006, compiled from multiple sources

as indoor and outdoor air pollution (see Figure 2.12 in Chapter 2), lead exposure, and climate change. Gains in life expectancy, child mortality and per capita health expenditures have been systematically greater in those countries with more equitable income distribution and access to medical treatment (PAHO 2002). Costa Rica, for example, has a higher average life expectancy than the United States. In many wealthier societies, opulence and consumerism, as well as relative poverty, contribute to ill health.

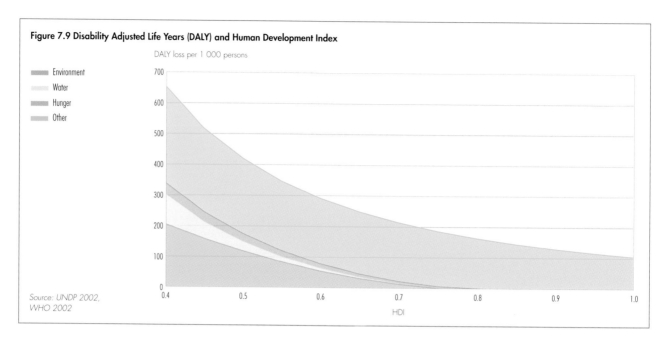

Figure 7.9 Disability Adjusted Life Years (DALY) and Human Development Index

DALY loss per 1 000 persons

Environment
Water
Hunger
Other

Source: UNDP 2002, WHO 2002

HDI

Investing in human and social capital reduces vulnerability

Environmental assets can provide important opportunities for improving well-being but, as shown in the archetypes, too often the benefits from these resources do not reach the most vulnerable. The distribution of environmental benefits is affected by access to networks (for example NGOs, governments and the private sector) and relations of trust, reciprocity and exchange (Igoe 2006). Development processes that arbitrarily extinguish local rights (see technological approaches archetype) and degrade the environment, as well as global trade regimes are also important factors influencing distribution.

Several policy interventions respond to these challenges, but slow progress in achieving the MDGs in many countries suggests that not enough has been done. The Convention on Biological Diversity (CBD), for example, emphasizes the importance of more equitable sharing of conservation benefits. Agenda 21, the Rio Declaration, and the CBD all prioritize public participation as essential for sustainable development. Increasing income from benefit sharing may strengthen efforts to meet MDG 1, and as household resources increase, the education and health-related MDGs may be more achievable. Countries with low access to improved drinking water have lower equity in access to education. Worldwide, girls and women spend about 40 billion hours collecting water – equivalent to a

year's labour for the entire workforce in France (UNDP 2006). In many developing countries, women and girls spend more than 2 hours a day collecting water (UNICEF 2004b). There are strong positive linkages between progress on the different MDGs, with, for example, improved access to water (MDG 7), resulting in girls spending less time collecting water, and increasing their opportunities to attend school (MDG 3) (UNICEF 2004b, UNDP 2006). For many countries, effectively implementing an interlinkages approach is challenging (see Chapter 8).

Meeting basic needs, such as education and health, provides the basis for valued choices, and enhances the day-to-day capacity of individuals, including that for environmental management (Matthew and others 2002). Education and access to technology are particularly important in poor communities, where they provide a potential route to a better situation and reduced vulnerability (Brock 1999).

Basic capabilities and rights to be treated with dignity, to have access to information, to be consulted and to be able to give prior informed consent where one's livelihood or assets are affected, are increasingly recognized as social and economic rights (UN 1966, UN 1986). The 1986 UN Declaration on the Right to Development represents a global consensus, but for many, these rights are inaccessible as a result of weak national and regional governance systems, undercutting capacity and opportunities. Women remain particularly disadvantaged. Notwithstanding improvements in maternal health (MDG 5) resulting, for example, from improved access to technologies and energy in rural hospitals, and access to education (MDG 3) in all regions since 1990, women continue to be among the most disadvantaged. They are under-represented in the economy and decision making (UN 2006).

Women are under-represented in important parts of society, due to a combination of factors. Socio-cultural attitudes, education, employment policies, and a lack of options for balancing work and family responsibilities and for family planning affect opportunities for employment and participation in community affairs (UN 2006).

Personal security – being protected from or not exposed to danger, and the ability to live a life one values (Barnett 2003) – may be threatened by declining social cohesion, poor living standards, inequity, unfair distribution of benefits and

environmental change (Narayan and others 2000). In some circumstances, environmental change creates a security challenge for entire cultures, communities, countries or regions (Barnett 2003). Where (cultural) identities are closely associated with natural resources, as in the Arctic and many Small Island Developing States (SIDS), social conflict and breakdown may be directly linked to habitat destruction or decreasing availability of environmental services. Other contributing factors include low levels of rural growth, high income inequity, ill health (especially HIV prevalence), climatic factors, such as drought, and environmental degradation (see Chapters 3 and 6, and Box 7.11).

Conflict also affects food security because of its long-lasting disruption of the productive base, and its impact on overall human well-being (Weisman 2006). In many cases, countries involved in conflict, and those with high levels of inequity, experience higher than expected levels of food emergencies (FAO 2003b) (see Figure 7.10).

Personal security is threatened by poor living standards. Below, makeshift houses such as these grow and spread along flooded estuaries exposing residents to grave risks.

Credit: Mark Edwards/Still Pictures

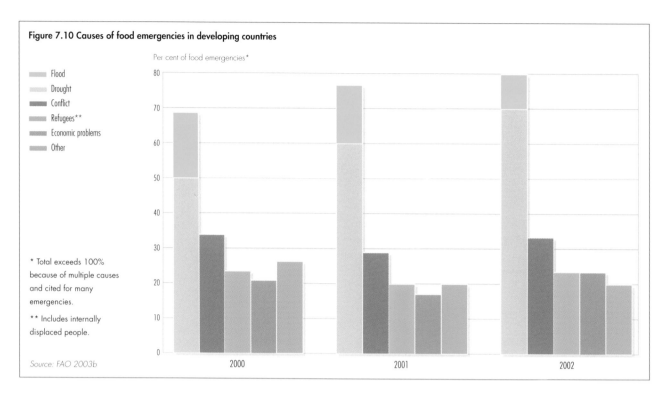

Figure 7.10 Causes of food emergencies in developing countries

Per cent of food emergencies*

Legend:
- Flood
- Drought
- Conflict
- Refugees**
- Economic problems
- Other

* Total exceeds 100% because of multiple causes and cited for many emergencies.

** Includes internally displaced people.

Source: FAO 2003b

2000   2001   2002

Investing in good social relations, building social capital through better governance, improving cooperation, and empowering women not only supports conservation efforts, but builds opportunities for peace, development and improving well-being. Developed countries' experiences suggest a number of factors that help hedge the impact of disasters: a well-financed government, an insurance industry, transport and communication infrastructure, democratic participation and personal affluence (Barnett 2003) (see Boxes 7.3 and 7.11). Improving capacity and access to technology, as envisaged under the Johannesburg Plan of Implementation (JPOI) and the Bali Strategic Plan for Technology Support and Capacity Building (BSP), can improve coping capacity. However, progress towards developing the global partnership to support this access remains slow (see Figure 7.27). More far-sighted and equitable approaches to the movement of resources, goods and people are critical to address the new levels of stress the most vulnerable communities will face as a result of environmental change (see the archetypes on drylands, SIDS and global commons).

Aspects of vulnerability

Although vulnerability is context and site specific, certain common elements can be observed across various regions, scales and contexts. Overarching vulnerability issues, such as equity, the export and import of vulnerability from one place or generation to another, and the causal relationships with conflict, hazards and the environment, deserve special attention, since they represent strategic entry points for effective vulnerability reduction and policy making.

Box 7.3 Environmental justice

Over the last three decades, a substantial environmental justice movement has emerged, although not always under that name. It was propelled by community struggles against unequal treatment and discrimination in the distribution of adverse environmental effects. The demand for environmental justice is closely linked to environmental rights: the right of every individual to an environment adequate for his/her well-being. A just system requires policies that protect people from harm, counter the tendency to maximize profits at the environment's expense, and distribute opportunities, risks and costs in a fairer way. It requires accessible institutions (courts), and fair processes. Governments have responded to this need by broadening laws and policies to include the polluter-pays-principle, environmental impact assessments, principles of good neighbourliness, environmental taxes, redistributive mechanisms, participatory and inclusive processes, access to information and right to know provisions, and compensation (see Chapter 10).

Inequalities, equity and vulnerable groups

Vulnerability varies across categories, including among men and women, poor and rich, and rural and urban, as can be observed in all archetypes. Refugees, migrants, displaced groups, the poor, the very young and old, women and children are often the groups most vulnerable to multiple stresses. Factors such as ethnicity, caste, gender, financial status or geographical location underlie processes of marginalization and disempowerment, which all lower the capacity to respond to changes. For example, the access of women and children to health care is often inequitably distributed, resulting in unfair and unjust outcomes that entrench disadvantage. Gender inequalities, reflected, for example, in male and female differences in wages, nutrition and participation in social choice, are illustrated in the contaminated sites archetype. Addressing MDG 3, to promote gender equality, empower women and eliminate gender disparity in primary and secondary education, is essential for increasing women's opportunities, reducing their vulnerability, and improving their ability to create sustainable and sufficient livelihoods.

One response by communities and governments to the unequal distribution of vulnerability and the impacts of multiple stressors on human well-being has been to focus on issues of environmental justice (see Box 7.3).

Export and import of vulnerability

Vulnerability is created or increased remotely, in many cases, through cause-and-effect relationships that persist over long distances in space or time. Many vulnerability archetypes demonstrate the phenomenon of "vulnerability export." Decreasing the vulnerability of some, for example through provision of shelter, increases the vulnerability of others far away, for example through land degradation and contamination around mining areas for building materials (Martinez-Alier 2002). At the same time, many people in industrialized nations, and the new consumers in the developing countries do not feel most of the impacts on the environment that result from their behaviour. These negative effects on the environment and well-being (especially health, security and material assets) are felt most strongly by those, especially the poor, living where the resources are extracted or the waste is dumped. This is illustrated in Figure 7.11, which shows the declining mineral extraction in

the European Union, and the increasing import of minerals. The emissions and land degradation associated with extraction and processing of the materials are increasing in developing countries, while the high-value end products are consumed in industrialized countries. Similarly, food imports often mean that environmental degradation and social impacts occur in the producing land, rather than where the goods are consumed (see, for example, Lebel and others 2002).

Vulnerability is imported where, for example, there is agreement to import waste and hazardous materials to locations where it cannot be safely disposed of or managed (see Chapters 3 and 6). The vulnerability of local populations is created or reinforced by poor governance and a lack of capacity to deal with the hazardous materials. Inadequate storage and poor stock management often result from insufficient storage capacity for pesticides, inappropriate storage conditions, insufficient training of responsible staff in stock management, poor distribution systems, inappropriate handling during transport, and unavailability of analytical facilities (FAO 2001).

While international trade can lead to increases in income, and has helped millions of people out of poverty, it is also sustaining unequal patterns of consumption, and in outsourcing the extraction of natural resources, much of the production and manufacturing, and also the generation and disposal of their hazardous wastes (Grether and de Melo 2003, Schütz and others 2004).

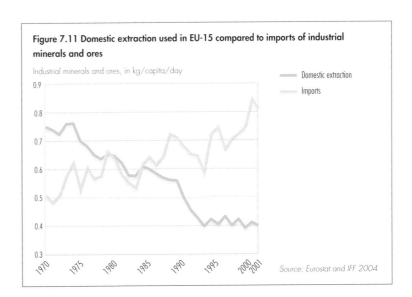

Figure 7.11 Domestic extraction used in EU-15 compared to imports of industrial minerals and ores

Industrial minerals and ores, in kg/capita/day

— Domestic extraction
— Imports

Source: Eurostat and IFF 2004

Recently, however, there have been some attempts to include the external impacts of trade policies into decision making processes, for example, through sustainability impact assessments in the European Union.

Vulnerability, environment and conflict

Many of the patterns of vulnerability represent a potential for or have already led to conflict. The relationship between environmental problems and international and civil conflict has been the subject of a great deal of academic research in the post-Cold War period (Diehl and Gleditsch 2001, Homer-Dixon 1999, Baechler 1999, Gleditsch 1999). Both scarcity and abundance of environmental resources can exacerbate existing tensions, and contribute to conflict between groups, especially in societies that lack the capacity to effectively and equitably manage competition for control over resources (Homer-Dixon 1999, Kahl 2006). These dynamics tend to be most common in the developing world. However, the export of vulnerability (see above) from developed to developing countries, can mean that even conflicts that appear localized have critical external connections.

A combination of environmental change, resource capture and population growth decreases the per capita availability of natural resources, and can threaten well-being for large segments of societies, particularly the poorest who depend on these natural resources for survival. The resulting social effects – migration, intensified unsustainable behaviour and social sub-grouping – strain the state's ability to meet its citizens' demands, and can contribute to violent outcomes (Homer-Dixon 1999, Kahl 2006). In the dryland archetype, conflict potential is related to unequal access to scarce water, forest and land resources, exacerbated by desertification and climate variability. Migration, a traditional coping strategy, sometimes heightens conflict when migrants create new competition for resources, or upset tenuous cultural, economic or political balances in the receiving area (Dietz and others 2004). In other cases, the scarcities heighten tensions between nomadic and pastoralist communities. Where this migration occurs across international boundaries, it can contribute to inter-state tension and new civil strife. Even when a state's natural resource base is high, conflict can

erupt over control of these valuable resources, if the potential cost of waging war is lower than the potential gains associated with securing access to the resources for export.

In the archetype on technological approaches to water problems, conflicts and tensions surrounding the distribution, access and quality of water resources arise. Megaprojects, such as dam construction, often carry considerable costs, including forced displacement for riparian dwellers, who may receive few of the resulting benefits (WCD 2000). These costs may include tensions between the state and riparian users, as well as between upstream and downstream riparian groups. The overexploitation of global commons, such as fisheries, the focus of another archetype, brings smaller-scale fisher groups and their governments into conflict with transnational or foreign-flagged ships that venture into exclusive economic zones from the depleted commons. Future energy generation and climate change directly link to security concerns for both oil-importing and oil-exporting countries. In rapidly urbanizing coastal zones and SIDS, conflicts emerge over competition for the environment for tourism-related activities, or for its environmental services associated with marine ecosystems and local livelihoods. Greater attention to proper management of ecosystems and valuable resources promises lower vulnerability to violence and greater overall well-being.

Vulnerability, well-being and natural hazards

Over the past 20 years, natural disasters have claimed more than 1.5 million lives, and affected more than 200 million people annually (Munich Re 2004b). One of the main drivers of increased vulnerability to hazards is global environmental change. Natural hazards, such as earthquakes, floods, droughts, storms, tropical cyclones and hurricanes, wildfires, tsunamis, volcanic eruptions and landslides threaten everyone. Proportionally, however, they hurt the poor most of all. Global datasets on extreme events indicate that the number of natural hazards is increasing (EM-DAT, Munich Re 2004b, Munich Re 2006). Two-thirds of all disasters are hydrometeorological events, such as floods, windstorms and extreme temperatures. Between 1992 and 2001, floods were the most frequent natural disaster, killing nearly 100 000 and affecting more than 1.2 billion people worldwide (Munich Re 2004b). More than 90 per cent of the people

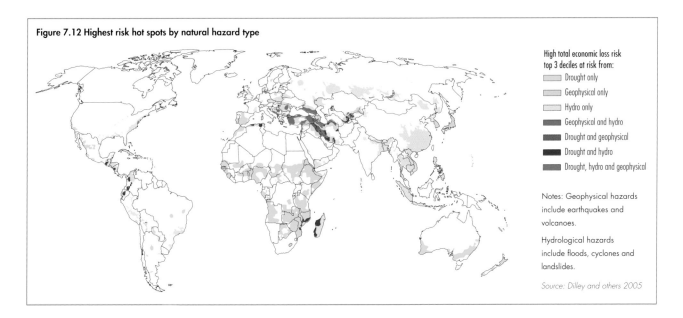

Figure 7.12 Highest risk hot spots by natural hazard type

High total economic loss risk
top 3 deciles at risk from:

▢ Drought only
▢ Geophysical only
▢ Hydro only
▢ Geophysical and hydro
▢ Drought and geophysical
▢ Drought and hydro
▢ Drought, hydro and geophysical

Notes: Geophysical hazards
include earthquakes and
volcanoes.

Hydrological hazards
include floods, cyclones and
landslides.

Source: Dilley and others 2005

exposed to disasters live in the developing world (ISDR 2004), and more than half of disaster deaths occur in countries with a low human development index (UNDP 2004a). Figure 7.12 shows the global distribution of highest-risk hot spots.

The consequences of disasters can have a lasting impact, threaten achievements in development and undermine resilience. Natural hazards affect food security, water supply, health, income and shelter (Brock 1999). These impacts are illustrated in several of the archetypes. Insecurity is driven by a multiplicity of environmental, political, social and economic factors, and is also closely related to issues of material access and social relations. Inefficient and poor governance, as well as inadequate or inefficient early warning and response systems, exacerbate vulnerability and the risks associated with environmental change and natural disasters. In some cases, short-term disaster relief even contributes to increasing long-term vulnerability.

Exposure to hazards has increased as a result of climate change and, for example, destruction of mangroves that protect coasts from tidal surges, but also through the continuing concentration of population in highly exposed areas. Adaptive capacity is also being eroded through, for example, reduced state social protection schemes, undermining of informal safety nets, poorly built or maintained infrastructure, chronic illness and conflict (UNDP 2004a).

PATTERNS OF VULNERABILITY

Recurring patterns of vulnerability can be found in numerous places around the world, including industrialized and developing regions, and urban and rural areas. With the recognition of the relevance of multiple pressures, and the close interlinkages among local, regional and global scales, vulnerability analyses become increasingly complex. For detailed local vulnerability case studies, there is the question of their relevance for other parts of the world, but it is possible to recognize some similarities between cases and to draw policy-relevant lessons from them.

A limited number of typical patterns or so-called "archetypes of vulnerability" are distinguished in this chapter (see Table 7.2 for an overview). An archetype of vulnerability is defined as a specific, representative pattern of the interactions between environmental change and human well-being. They do not describe one specific situation, but rather focus on the most important common properties of a multitude of cases that are "archetypical." The approach is inspired by the syndrome approach, which looks at non-sustainable patterns of interaction between people and the environment, and unveils the dynamics behind them (Petschel-Held and others 1999, Haupt and Müller-Boker 2005, Lüdeke and others 2004). The archetype approach is broader, as it includes opportunities offered by the environment to reduce vulnerability and improve human well-being (Wonink and others 2005) (see Table 7.4).

The archetypes presented here are simplifications of real cases, to show the basic processes whereby vulnerability is produced within a context of multiple pressures. This may allow policy-makers to recognize their particular situations in a broader context, providing regional perspectives and important connections between regions and the global context, and insights into possible solutions. The patterns of vulnerability are not mutually exclusive. In some ecosystems, countries, sub-regions, regions and globally, a mosaic of these and other patterns of vulnerability may exist. This makes policy response a complex challenge.

The archetypes of vulnerability have been identified through the GEO-4 assessment, ensuring regional

Table 7.2 Overview of archetypes analysed for GEO-4			
Archetype	Linkages to other chapters	Key human well-being issues	Key policy messages
Contaminated sites	Chapter 3 Chapter 6 - Asia Pacific – waste management - Polar – persistent toxics - Polar – industry and related development activities	Health hazards – main impacts on the marginalized in terms of people (forced into contaminated sites) and nations (hazardous waste imports)	- Better laws and better enforcement against special interests - Increase participation of the most vulnerable in decision-making
Drylands	Chapter 3 Chapter 6 - Africa – land degradation - West Asia – land degradation and desertification	Worsening supply of potable water, loss of productive land, conflict due to environmental migration	- Improve security of tenure (for example through cooperatives) - Provide more equal access to global markets
Global commons	Chapters 2 and 5 Chapter 6 - LAC * – degraded coasts and polluted seas - LAC – shrinking forests - Polar – climate change - West Asia – degraded coasts	Decline or collapse of fisheries, with partly gender-specific poverty consequences Health consequences of air pollution and social deterioration	- Integrated regulations for fisheries and marine mammal conservation and oil exploration - Use the promising persistent organic pollutants policies for heavy metals
Securing energy	Chapter 2 Chapter 6 - Europe – energy and climate change - LAC – energy supply and consumption patterns - North America – energy and climate change	Affects material well-being, marginalized mostly endangered by rising energy prices	- Secure energy for the most vulnerable, let them participate - Foster decentralized and sustainable technology - Invest in the diversification of the energy-systems
Small Island Developing States	Chapter 4 Chapter 6 - LAC – degraded coasts and polluted seas - Asia Pacific – alleviating pressures on valuable ecosystems	Livelihoods of users of climate-dependent natural resources most endangered, migration and conflict	- Adapt to climate change by improving early warning - Make economy more climate independent - Shift from "controlling of" to "working with nature" paradigm
Technology-centred approaches to water problems	Chapter 4 Chapter 6 - Asia Pacific – balancing water resources and demands - North America – freshwater quantity and quality - West Asia – water scarcity and quality	Forced resettlement, uneven distribution of benefits from dam building, health hazards from water-borne vectors	- The World Commission on Dams (WCD) framework, and the UNEP Dams and Development Project (WCD and UNEP-DDP) path of stakeholder participation should be further followed - Dam alternatives, such as small-scale solutions and green engineering, should play an important role
Urbanization of the coastal fringe	Chapter 6 - North America – urban sprawl - LAC – growing cities - LAC – degraded coasts - West Asia – degradation of coastal and marine environments - West Asia – management of the urban environment	Lives and material assets endangered by floods and landslides, health endangered by poor sanitary conditions due to rapid and unplanned coastal urbanization, strong distributional aspects	- Implementation of the Hyogo Framework of action - Bring forward green engineering solutions that integrate coastal protection and livelihood opportunities

* LAC = Latin America and the Caribbean.

relevance and balance. The seven archetypes presented here are not meant to provide an exhaustive overview of all possible patterns of vulnerability. However, they provide a good basis for identifying challenges and exploring opportunities for reducing vulnerability while protecting the environment.

Exposing people and the environment to contaminants

The archetype concerns sites at which harmful and toxic substances occur at concentrations:

- above background levels and pose or are likely to pose an immediate or long-term hazard to human health or the environment; or
- exceed levels specified in policies and/or regulations (CSMWG 1995).

As shown in Chapters 3 and 6, people and ecosystems are exposed to widespread contamination due to persistent organic pollutants and heavy metals, urban and industrial sites, military activity, agro-chemical stockpiles, leaking oil pipelines and waste dumps.

Global relevance

Much work is still needed to quantify the extent of contamination due to toxic and hazardous substances, and to make governments and civil society aware of the problems. However, a considerable amount of contamination has been documented.

In addition to contamination generated in particular locations, transport and deposition of waste is a major threat. More than 300 million tonnes of waste, including hazardous and other wastes, were generated worldwide in 2000, of which less than 2 per cent was exported. About 90 per cent of the exported waste was classified as hazardous, with about 30 per cent believed to be persistent organic pollutants (POPs) (FAO 2002). The principal waste export (see Figure 7.13) by volume was lead and lead compounds, bound for recycling (UNEP 2004).

Contaminated sites are also legacies of past industrial and economic development, and a heritage of present production and consumption patterns that affect both current and future generations. Abandoned industrial sites can present a serious risk to people and the environment. Governments face problems of holding polluters accountable for site clean-ups. Therefore, clean-up costs are imposed on state budgets, or on people from surrounding areas exposed to health risks and environmental deterioration.

Sometimes, abandoned industrial sites are in relatively isolated areas around former factories or mines, and, sometimes, whole regions are affected by the problem (see Box 7.4). Short-term profit interests, lack of regulations or corruption, and weak law enforcement

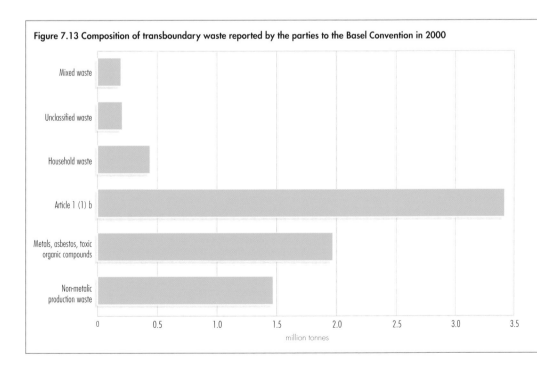

Figure 7.13 Composition of transboundary waste reported by the parties to the Basel Convention in 2000

million tonnes

Note: Article 1(1) b concerns wastes that are defined as, or are considered to be, hazardous wastes by the domestic legislation of the Party of export, import or transit, such as contaminated soil or sewage sludge.

Source: UNEP 2004, compiled by UNEP/GRID-Arendal, based on Basel Convention reports

are among the factors that have led and may still lead to the creation of present and future environmental hazards from contaminated sites (UNEP 2000).

Vulnerability and human well-being

In developing countries, chemical mixtures in the vicinity of small-scale enterprises, such as smelters, mines, agricultural areas and toxic waste disposal sites, are often a human health hazard (Yanez and others 2002). For example, about 60 per cent of the smelters of the world are located in developing countries, while developed countries import the metals (Eurostat and IFF 2004). Health effects, such as cancer and neuropsychological disorders, have also been reported around smelters (Benedetti and others 2001, Calderon and others 2001). For example, in Torreon, Mexico, 77 per cent of the children living closest to a lead smelter had lead levels twice as high as the reference level (Yanez and others 2002).

Mercury contamination associated with small-scale gold mining and processing presents a

---

Box 7.4 Contamination in Central Asia's Ferghana-Osh-Khudjand area

The Ferghana-Osh-Khudjand area in Central Asia (also referred to as the Ferghana Valley) is shared by Uzbekistan, Kyrgyzstan and Tajikistan (see Figure 7.14). The region is a typical example of former centrally planned economies, where development plans paid little attention to local conditions (especially environmental), and social progress was planned to be achieved through large-scale industrial projects. In the Ferghana Valley, the construction of enormous irrigation schemes made the region a major cotton producer. It also became a heavy industrial area, based on mining and oil, gas and chemical production. Discoveries of uranium ore led to extensive mining, and it became an important source of uranium for the former Soviet Union's civilian and military nuclear projects.

Several factors – population density in disaster-prone areas, high overall population growth, poverty, land and water use, failure to comply with building codes, and global climate change – make the region particularly vulnerable to natural as well as human-made hazards. Cumulative risks from different industrial facilities, deteriorating

infrastructure and contaminated sites threaten not only the inhabitants living directly in the polluted zones, but also have transboundary impacts in the three countries that share the valley. Even though past spills and accidents have created tensions among the countries, officials do not consistently regard environmental pollution by existing facilities as a security problem.

In the immediate wake of the breakup of the Soviet Union, pollution and, particularly, shared water resources in this newly internationalized river basin, created tensions among the new states. Officials point to the potential this area has to serve as an example of international cooperation in addressing legacies of the past. However, without extensive international aid, this task is impossible for the local governments. Also, in the absence of alternative development plans and access to environmentally-friendly technologies and management practices, some of the abandoned facilities may be reopened.

Source: UNEP and others 2005

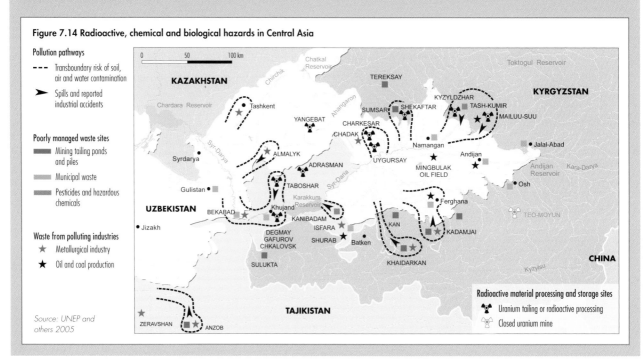

Figure 7.14 Radioactive, chemical and biological hazards in Central Asia

Source: UNEP and others 2005

major hazard for environment and human health in at least 25 countries in Africa, Asia and the Pacific, and Latin America and the Caribbean, (Malm 1998, Appleton and others 1999, van Straaten 2000). Harmful health effects have been reported for individuals exposed to mercury in gold mining areas (Lebel and others 1998, Amorin and others 2000).

Pesticides can contribute to water pollution, and seriously threaten the health of both rural and urban residents, especially the poorest people. Organochlorine compounds, such as DDT, dieldrin and HCH, which have been withdrawn or banned for human health and/or environmental reasons (FAO 1995), are still found in dumps, particularly in developing countries. Long-term exposure to pesticides can increase the risk of developmental and reproductive disorders, disruption of the immune and endocrine systems, and can impair the function of the nervous system, and is associated with the development of certain cancers. Children are at higher risk from exposure than are adults (FAO and others 2004).

The international traffic in hazardous wastes exposes local populations to health risks. For example, in 1998, about 2 700 tonnes of industrial waste, containing high levels of toxic compounds such as mercury and other heavy metals, were shipped illegally to Sihanoukville, Cambodia. An estimated 2 000 residents were exposed to the waste and at least six deaths and hundreds of injuries were associated with the incident (Hess and Frumkin 2000).

An emerging issue is the great volume of electronic waste exported to developing countries, where it is recycled by workers who often lack protection. They are exposed to mercury, lead, cadmium and other toxic chemicals (see Chapter 6). In one Chinese city where electronic waste is recycled, sediment samples had heavy metal concentrations far above the guidelines of the US Environmental Protection Agency (Basel Action Network 2002). Similarly, workers are exposed to contaminants that pose serious risks to their health in locations where ships are broken up for recycling (Basel Action Network 2006).

Abandoned factories and industrial sites are most likely to be found in poor communities,

which can be home to marginalized newcomers. Contamination of air, water and land decreases land productivity, making agricultural products unsuitable for markets. Children are particularly at risk from contaminated sites (as places of play and work), and women are especially at risk for physiological reasons. A survey conducted in the United Kingdom (Walker and others 2003) about the social status of people living close to integrated pollution control sites (IPC), confirmed that in England there is strong evidence of a socially unequal distribution of IPC sites and their associated potential impacts. Out of about 3.6 million people living in one-kilometre radius of an IPC site, there were six times more people from the most deprived groups than from the least deprived groups.

Responses

Over the years, a series of measures have been adopted to deal with the risks that hazardous materials and chemicals pose for both people and the environment. Principle 14 of the Rio Declaration, calls on countries to "effectively cooperate to discourage or prevent the relocation and transfer to other States of any activities and substances that cause severe environmental degradation or are found to be harmful to human health." The UN Commission on Human Rights has appointed a special rapporteur on adverse effects of the illicit movement and dumping of toxic and dangerous products and wastes on the enjoyment of human rights (UN).

Responses to the problem of contaminants now include 17 multilateral agreements (see Chapter 3), together with numerous intergovernmental organizations and coordination mechanisms. They include the 1989 Basel Convention on the Control of Transboundary Movements of Hazardous Waste and Their Disposal, the 1998 Rotterdam Convention on Prior Informed Consent Procedure for Certain Hazardous Chemicals, the 2001 Stockholm Convention on Persistent Organic Pollutants, as well as the 2006 Strategic Approach to International Chemicals Management.

Other responses to contamination have created opportunities for building trust in post-conflict societies. For example, joint scientific assessment of threats from radioactive contamination in the Russian northwest provided an opportunity for Russian, Norwegian, and American exchange as the Cold

War ended and the superpowers began to develop links for confidence building among scientists and military personnel. The low politicization of environmental issues actually facilitated face-to-face dialogue among military foes in a highly militarized and sensitive region.

The success of the existing instruments for dealing with contamination depends strongly on institutional capacity and political will (see Chapter 3). Important areas for future action include:

- strengthening the ability of international organizations to monitor and enforce multilateral agreements, such as the Basel and Rotterdam conventions;
- promoting global environmental and social standards to avoid dumping;
- investing in technology and technology transfer for improved risk assessment, monitoring, information and communication, and clean-up;
- increasing corporate social and environmental responsibility;
- investing in assets, especially skills and knowledge, to avoid exposure or to mitigate health effects from exposure to hazardous material;
- improving state capacity to monitor and enforce laws, as this may reduce risk, and improve local coping capacity;
- providing opportunities for participation, and addressing the social situation of people affected by contaminated sites;
- better incorporation of established international legal principles – including the precautionary approach, producer liability, polluter pays, prior informed consent and right to know – into national, regional and global frameworks;
- increasing support for research on causes and effects (especially cumulative effects) of industrial production and chemicals; and
- increasing support for life cycle analyses and environmental impact assessments.

In situations of contaminated sites, formal institutions, better laws at national and international levels, and better enforcement of existing laws are crucial for reducing vulnerability. This requires strong and well-functioning states, with law-making, implementation and enforcement branches working towards the same goals (Friedmann 1992). Measures that strengthen the capacity of states can also help strengthen coping capacity at local levels, if this is supported by higher levels of governance.

Increasing the participation of the most vulnerable groups in planning and governance, and giving both local and higher levels of governance opportunities to articulate their challenges is a major factor in strengthening their coping capacity. Giving the vulnerable a voice requires that they be actively empowered to raise their voices, for example by having access to relevant environmental information – as enshrined in Principle 10 of the Rio Declaration – and capacity building for taking part in the governance process. The 1992 UN Conference on Environment and Development (UNCED) provided the basic institutional change for increasing participation in environment-related decision making. This has been reinforced, for example, in the Aarhus Convention (UNECE 2005). The Basel and Rotterdam conventions are important for giving countries a voice in the context of vulnerability to contamination.

Disturbing the fragile equilibrium in drylands

In this archetype, current production and consumption patterns (from global to local levels) disturb the fragile equilibrium of human-environment interactions that have developed in drylands, involving sensitivity to variable water supplies and resilience to aridity. The result is new levels of vulnerability. For thousands of years, drylands populations have been dependent on the proper functioning of these ecosystems for their livelihoods (Thomas 2006). These resilient ecosystems have considerable productive potential – supporting, for example, 50 per cent of the world's livestock (Allen-Diaz and others 1996) – but are increasingly at risk. Moreover, governance and trade patterns mean that much dryland wealth remains hidden or poorly used, constituting missed opportunities for improving well-being.

Global relevance

Drylands are widespread, occur in developed and developing countries, and support significant populations (see Chapter 3). Worldwide, 10–20 per cent of drylands are degraded, directly affecting well-being of drylands populations, and indirectly affecting people elsewhere through biophysical (see Chapter 3) and socio-economic impacts. Globally-driven processes, including climate change, have direct impacts on well-being in drylands (Patz and others 2005).

Vulnerability and human well-being

There are a number of factors that influence the vulnerability of dryland communities, including:

■ biophysical features, especially water availability;

■ access to natural and economic resources, levels of development, and conflict and social instability;

■ interlinkages between dryland and non-dryland areas through migration, remittances and trade; and

■ global governance regimes (Safriel and others 2005, Dobie 2001, Griffin and others 2001, Mayrand and others 2005, Dietz and others 2004).

People in the drylands of industrialized countries – such as in Australia and the United States – typically have a diversity of livelihood options, and can adapt more to land degradation and water scarcity more easily than can rural people in drylands in developing countries who directly depend on environmental resources for their livelihoods. They are most vulnerable. Although high land productivity and a strong manufacturing sector, such as in North

---

Box 7.5 Analysing different types of vulnerability in drylands

Systematic analysis of the diverse socio-economic and natural conditions in drylands enhances understanding of the specific patterns of vulnerability. The global distribution of vulnerability is investigated here using a cluster analysis.

The following indicators were used to characterize the main underlying processes of vulnerability:

■ water stress, to show the relationship between water demand and availability;

■ soil degradation;

■ human well-being as indicated by infant mortality;

■ availability of infrastructure, indicated by road density; and

■ the climatic and soil potential for agriculture.

The table legend to the map shows the qualitative values of the indicators that are typical for the eight clusters:

+ = high value for the specific indicator

– = low value for the specific indicator

0 = intermediate value for the specific indicator

Together these indicators cluster into eight constellations, or "clusters of socio-economic and natural conditions" in drylands, depicted by colours ranging from bright red for the most vulnerable, to neutral grey for the least vulnerable cluster (see Figure 7.15). Humid regions are shown in white.

The analysis shows a need for the wise and efficient use of resources, based on best available knowledge and technological options:
Clusters 1 to 6 are all vulnerable (with low to medium levels of well-being).
Clusters 1 and 2 are most problematic, with high water stress, soil degradation and infant mortality, low agricultural potential, and intermediate infrastructure.
Clusters 3 and 4 are large areas, which exhibit a better level of human well-being compared to clusters 1 and 2 under very similar levels of exploitation of the water and, in some places, even more severe overuse of soil resources. This shows that the worst expressions of vulnerability are not a necessary fate.
Clusters 5 and 6 illustrate that improved water use on its own does not guarantee improved well-being.
Clusters 7 and 8, in contrast, are the least vulnerable regions, with only intermediate infrastructure restrictions and infant mortality.

Sources: Alcamo and others 2003, ArcWorld ESRI 2002, CIESIN 2006, GAEZ 2000, Kulshreshta 1993, Murtagh 1985, Oldeman and others 1991

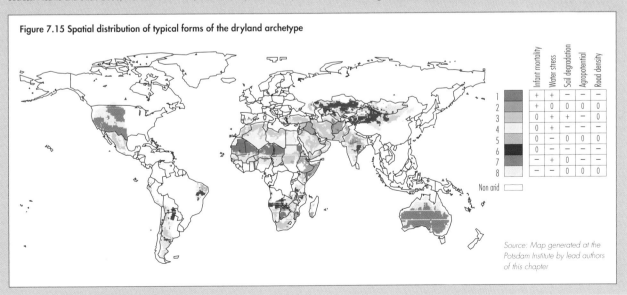

Figure 7.15 Spatial distribution of typical forms of the dryland archetype

	Infant mortality	Water stress	Soil degradation	Agropotential	Road density
1	+	+	–	–	–
2	+	0	0	0	0
3	0	+	+	–	0
4	0	+	–	–	–
5	0	–	0	0	0
6	0	–	–	–	–
7	–	+	0	–	–
8	–	–	0	0	0
Non arid					

Source: Map generated at the Potsdam Institute by lead authors of this chapter

America, can decrease vulnerability, the distribution of access to natural and economic resources, and participation in decision making trigger the vulnerability pattern (see Box 7.5).

Desertification (see Chapter 3) is a challenge for development and improving well-being. Globally, some 60 000 square kilometres of productive land and about US$42 billion in income are lost annually, due to declining agricultural productivity (UNDP and GEF 2004). Since 1975, the incidence of drought has increased fourfold from 12 to 48 episodes (UNDP and GEF 2004). Where there is high agricultural dependency, droughts may undercut food security and economic performance, lessening the opportunity to meet MDG 1 (see Figure 7.16). In Pakistan, for example, drylands are increasingly threatened by declining soil fertility and flash floods – early warnings of a looming crisis (UNDP and GEF 2004).

The seemingly low production potential of drylands has made them less favoured for the systematic investments (in water and land) needed to offset negative effects of land use and sustain their productive capacity (see Chapter 3). Freshwater availability in drylands is projected to be further reduced from an average of 1 300 cubic metres/person/year in 2000, which is already below the threshold of 2 000 m3 required for minimum human well-being and sustainable development (Safriel and others 2005). In arid and semi-arid areas, water shortages are predicted to be the most significant constraint for socio-economic development (Safriel and others 2005, GIWA 2006) (see Chapter 4). In some countries, the reduced supply of potable water will mean women and girls will be forced to travel longer distances to collect water.

The high number of transboundary aquifers under stress (GIWA 2006) may, in some instances, add a regional dimension to the risk of tensions related to water scarcity. In some situations, adaptation strategies, such as irrigation of water-intensive crops, lead to clashes between rural and urban users, as well as between agriculturalists and pastoralists. In the US southwest for example, multistakeholder dispute resolution mechanisms, including judicial systems and significant technological and financial resources, keep most of these conflicts from turning violent. In areas with higher vulnerability, such as the Sahel, shortages of arable land and water, particularly in drought periods, have sometimes led to violent conflicts along a number of

lines of division: rural-urban, pastoralist-agriculturalist and ethnic group-ethnic group (Kahl 2006, Lind and Sturman 2002, Huggins and others 2006).

Movement of "dryland refugees" to new areas, including cities, has the potential to create local and regional ethnic, social and political conflict (Dietz and others 2004). Seasonal and cyclic migrations are important coping strategies for pastoral dryland peoples. Pastoral societies (found in all regions) are critically exposed to ecosystem change, which can increase their vulnerability, affect their capital stocks, hinder coping strategies, decrease the productive performance of livestock and generate tensions with other herder and host farmer communities (Nori and others undated).

Responses

Given the extent of drylands, the roughly 2 billion people they support and the biological diversity they hold, the maintenance and restoration of their ecosystem functions is essential for achieving the CBD 2010 biodiversity targets and the MDGs. The UN Convention to Combat Desertification (UNCCD) provides the overall framework for addressing land degradation (see Chapter 3). It is complemented by the CBD, UNFCCC, Agenda 21, WSSD and other multilateral agreements.

The UNCCD supports national action to combat desertification and improve opportunities from land management. This includes the development of national (NAP), sub-regional (SRAP) and regional (RAP) action programmes. By 2006, a significant number of countries had developed NAPS, with 34 in Africa, 24 in Asia, 21 in Latin America and the Caribbean, and eight in Europe. The CBD provides for management based on equitable benefit-sharing, which helps to increase local resource-based income. Successful applications in drylands include co-management initiatives for wildlife (Hulme and Murphree 2001), and the development of markets for non-timber forest products (NTFPs) (Kusters and Belcher 2004). Intergovernmental initiatives, including the WSSD, UNCCD and the UNEP-led BSP, that focus on capacity building and transfer of technology to enhance management, production and marketing, offer opportunities for building on these successes.

Early warning systems (EWS) are widely used to improve the ability to respond to environmental

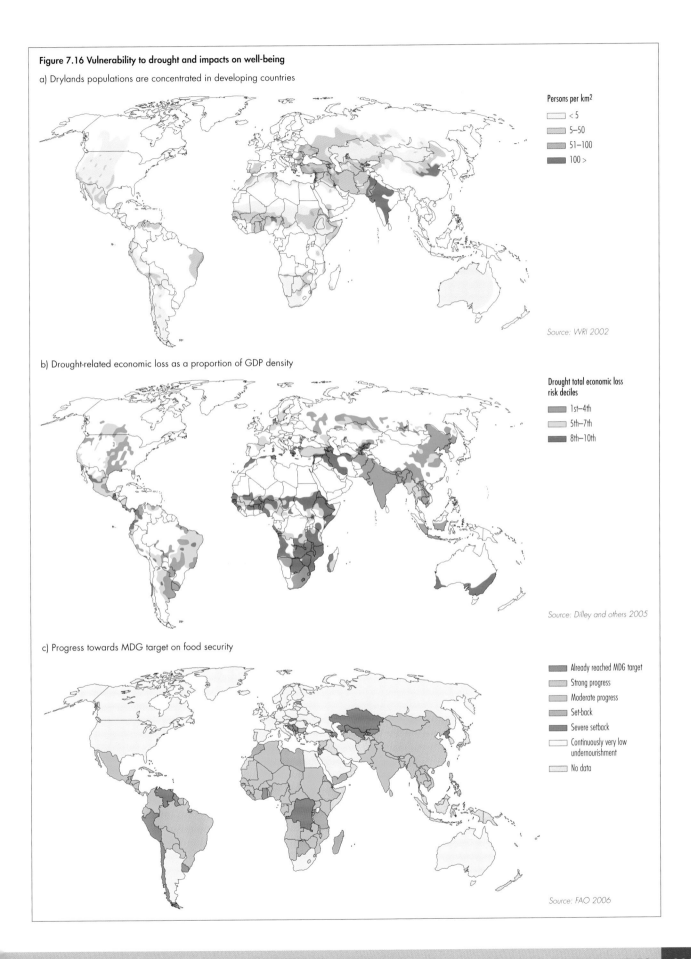

Figure 7.16 Vulnerability to drought and impacts on well-being

a) Drylands populations are concentrated in developing countries

Persons per km²
- < 5
- 5–50
- 51–100
- 100 >

Source: WRI 2002

b) Drought-related economic loss as a proportion of GDP density

Drought total economic loss risk deciles
- 1st–4th
- 5th–7th
- 8th–10th

Source: Dilley and others 2005

c) Progress towards MDG target on food security

- Already reached MDG target
- Strong progress
- Moderate progress
- Set-back
- Severe setback
- Continuously very low undernourishment
- No data

Source: FAO 2006

pressures. The UNEP/FAO Land Degradation Assessment in Drylands (LADA) systematically observes land degradation to increase understanding of drought and desertification processes and their effects. In addition, national, sub-regional and global EWS enhance capacity to respond to potential food insecurity. In Eastern Africa, for example, the Intergovernmental Authority on Development (IGAD) links conflict monitoring (through its Conflict Early Warning and Response Mechanism) to environmental EWS (through its Drought Monitoring Centre), because drought and other environmental pressures may trigger pastoral conflict.

Effective responses to the multiple and complex drivers of land degradation demands interlinked approaches, adequate funding and sufficient capacity (see Box 7. 6). For example, attempts to reverse water degradation trends are constrained by a number of factors. They include: poverty, slow economic development, deficiencies in the technical, administrative and managerial capacity of water management institutions, weak national and regional legal frameworks, and a lack of international cooperation (GIWA 2006) (see Chapter 4). Developing systems for managing water scarcity, which deal with rainwater and run-off, and mediating between competing water claims, including environmental claims, has proved difficult. The failure to harness different kinds of knowledge, including

traditional farming knowledge, in management and policy means that the full range of options for improving dryland farming is not taken up (Scoones 2001, Mortimore 2006). Insufficient funding, including for NAPs (White and others 2002), and failure to respond to early warnings (FAO 2004a), are constraints.

Experience shows that financial investments and loans to dryland farmers can produce significant returns, but this approach continues to be underused (Mortimore 2006). Although women play a pivotal role in environmental and agricultural management, they have limited support. Institutional and governance factors, coupled with insufficient capacity, limit the financial benefits that producers reap from drylands products, such as crops and NTFPs (Marshall and others 2003, Katerere and Mohamed-Katerere 2005). In 2005, UNCCD COP 7 acknowledged that insufficient decentralization and insecure tenure undermine management and reduce opportunities. Potential income is lost to intermediaries: in Namibia, devil's claw (*Harpagophytum* species) producers receive just a fraction of the retail price, ranging from 0.36 per cent, when dealing with intermediaries, to 0.85 per cent when selling directly to exporters (Wynberg 2004).

Global trade regimes, particularly protectionist tariffs and agricultural subsidies in developed country markets (Mayrand and others 2005), affect income of drylands producers in developing countries. These tariffs and subsidies have, for example, reduced the competitiveness of developing country cotton, even though developing countries are among the lowest-cost producers (Goreux and Macrae 2003). Conflict can also be an important factor inhibiting product and market development in drylands (UNDP 2004b).

Addressing these constraints can improve opportunities for increasing well-being. Options include (see Chapter 3):

- improving tenure, and recognizing the value of traditional knowledge to encourage farmer investments in soil and water conservation, which lead to more profitable agriculture;

- addressing resource-related conflicts through multi-level environmental and development cooperation, including bringing all stakeholders together to negotiate sharing benefits from interdependent resources, such as transboundary water. This helps to build trust through cooperative environmental management; and

---

Box 7.6 Institutional reform for poverty alleviation in drylands

Long-term social and ecological transformation in Machakos District in Kenya is widely cited as a success story of how a combination of efforts can lead to improved well-being in dryland areas. This involved dealing with a series of interconnected domains:

- ecosystem management (conservation of biodiversity, soil and water management);
- increasing land productivity (increased market access to agricultural products, improved crop yields, increased value and price of products);
- land investments; and
- social welfare (investments in education, diversification of employment and income opportunities, and stronger linkages to urban centres).

Between the 1930s and the 1990s, despite a sixfold increase in the population, erosion had been largely brought under control on private farmlands through small investments and extension support. During the same period, the value of agricultural production per capita increased sixfold. This was due to developments in agricultural technology, increased emphasis on livestock production, intensive farming, integration of crops with livestock production, and improved production and marketing of higher-value commodities, such as fruit, vegetables and coffee. This was done in tandem with investments in education, and the provision of employment opportunities outside the district.

Source: Mortimore 2005

- ensuring more equitable access to global markets, to improve opportunities for agriculture and livelihood diversification.

Misusing the global commons

Another archetype is a pattern of vulnerability resulting from misuse of the global commons, which include the deep oceans and seabed beyond national jurisdiction and the atmosphere. In some contexts biodiversity (where species concerned are found in the global commons) and Antarctica are also included in the list of global commons, but the focus here is on the oceans and the atmosphere. The misuse of these global commons leads to the exposure of people and the environment to pollution (such as heavy metals and persistent organic pollutants in the Arctic), to resource depletion (such as in fisheries) and to environmental changes (in particular as a result of climate change). Very often those that are extremely vulnerable to the changes resulting from misuse of the commons are not responsible for the misuse itself.

Global relevance

Resources that cannot be governed under the normal governance framework of national sovereignty are usually referred to as 'global commons.' The global commons physically envelop the globe and humanity. The oceans have the character of both a common

(re)source – for example, providing large amounts of fish – and a common 'sink' – receiving large amounts of pollution from ships, land and the atmosphere (see Chapter 4). The atmosphere is a decisive (re)source for life on this planet, both because it protects people from the harmful rays of the sun and provides the climate system, and because the oxygen in its lower parts is also the source of the air most organisms need to sustain life. The atmosphere is heavily misused as a sink for pollution from a wide range of human activities (see Chapter 2).

Vulnerability and human well-being

Marine living resources provide a significant proportion of protein in the human diet (see Chapter 4). Two-thirds of the total food fish supply is from capture fisheries in marine and inland waters (WHO 2006b). However, fisheries are declining, formerly abundant species are now rare, food webs are being altered, and coastal ecosystems are being polluted and degraded (Crowder and others 2006). In some cases, fisheries have collapsed, and the livelihoods of entire communities have been destroyed. A well-known example is the collapse of much of the Canadian cod fishery. In the early 1980s, the Canadian catches of Atlantic groundfish peaked, and then declined rapidly. This is illustrated in Figure 7.17 in Box 7.7, which also shows the sharp decline in the number of fishers (Higashimura 2004).

Box 7.7 Conflicts over marine resources

At the international level, conflict can occur between states acting on behalf of vulnerable local users and the states of large industrial users of the global commons. One example occurred in 1995 between Canada and Spain on the Grand Banks, a rich fishing zone just off Canada's east coast. Industrial foreign trawlers were fishing for turbot, a resource also used by local fishermen in Newfoundland, a Canadian province. The Canadian government was under great domestic political pressure from the local

fishers, who claimed their way of life was threatened because fishers from countries fishing the Grand Banks, including Spain, did not respect catch quotas. Canada forcibly boarded a Spanish fishing trawler in international waters and arrested its crew after the Canadians alleged repeated incursions into Canada's 200-mile Exclusive Economic Zone. The Spanish referred to this incident as an act of piracy, touching off a series of high seas encounters and diplomatic clashes referred to as the "Turbot War."

Sources: McDonald and Gaulin 2002, Soroos 1997)

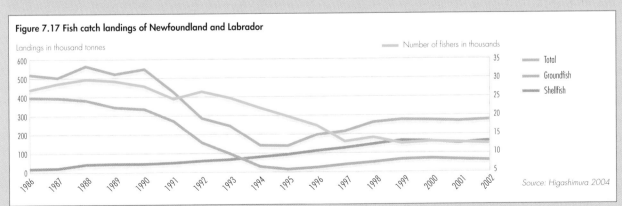

Figure 7.17 Fish catch landings of Newfoundland and Labrador

Source: Higashimura 2004

The Mediterranean Sea is currently part of the global commons, since many surrounding countries have not exercised their right to establish 200-mile exclusive economic zones. As a result of overfishing and pollution in the Mediterranean, catches of the high-value bluefin tuna reached a high of 39 000 tonnes in 1994, but had dropped by nearly half that amount by 2002 (FAO 2005a).

More recently, after the decline of traditional stocks, such as cod, attention has turned to deep-sea fishing (deeper than about 400 m), where fish are particularly vulnerable to overfishing because of their slow ability to reproduce (see Chapter 4). Several deep-sea stocks are now heavily exploited, and, in some cases, severely depleted (ICES 2006). A very small number of countries land most of the fish catch from the high seas (see Figure 7.18).

Many coastal communities have no capacity to fish in the global commons of the high seas, and are thus deprived of the food and revenue the resource provides. The disruption of small-scale fisheries by high-technology competition often leads to a vicious cycle of fisheries depletion, poverty, and loss of cultural identity. It can also lead to conflict (see Box 7. 7).

Figure 7.18 Landings in high seas by major fishing countries

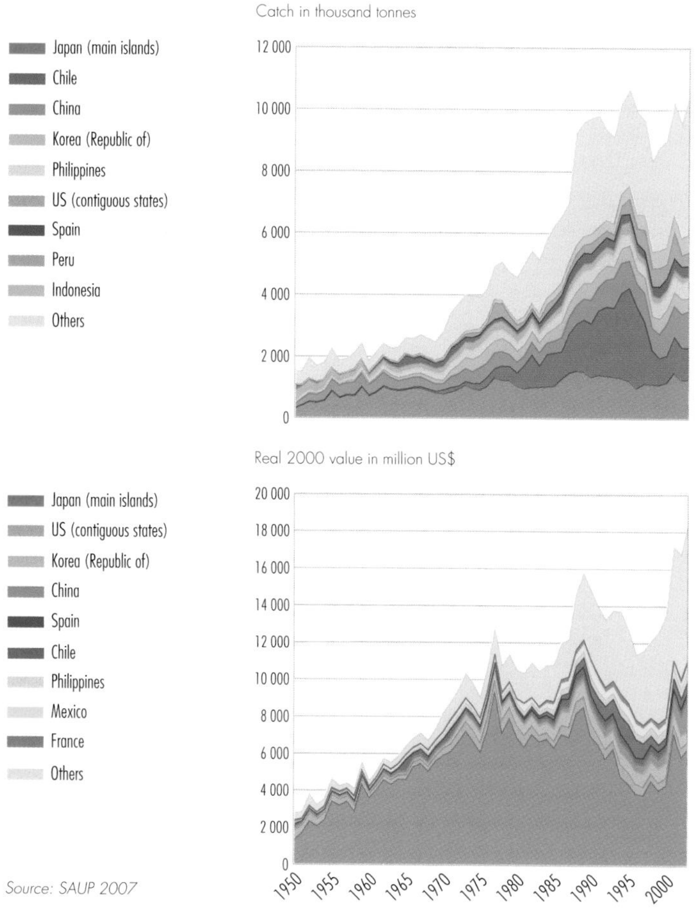
Catch in thousand tonnes
Japan (main islands)
Chile
China
Korea (Republic of)
Philippines
US (contiguous states)
Spain
Peru
Indonesia
Others
12 000
10 000
8 000
6 000
4 000
2 000
0
Real 2000 value in million US$
Japan (main islands)
US (contiguous states)
Korea (Republic of)
China
Spain
Chile
Philippines
Mexico
France
Others
20 000
18 000
16 000
14 000
12 000
10 000
8 000
6 000
4 000
2 000
0
1950
1955
1960
1965
1970
1975
1980
1985
1990
1995
2000

Source: SAUP 2007

An example of the impacts on human well-being from air pollution is the long-range transport (via air and oceans) of persistent organic pollutants (POPs) and heavy metals, which disproportionately affect indigenous people of the Arctic (see Box 7.8, Figure 7.19, and Polar Regions section in Chapter 6). These same communities are also vulnerable to the adverse impacts of climate change.

Responses

People from more than 190 countries use the global commons, but no global authority exists to enforce a management regime. Agreements built around consensus are often very weak. In some cases, countries do not sign or accede to the agreements, leading to the "free rider" problem. The multilateral agreements covering the atmosphere are listed in Table 2.4, Chapter 2, and agreements on the oceans are discussed in Chapter 4.

The wide range of agreements now covering the use of ocean resources beyond national jurisdiction includes the UN Convention on the Law of the Sea (UNCLOS), the UN Agreement on Straddling Fish Stocks and Highly Migratory Fish Stocks, the Convention on Biological Diversity, the International Plan of Action on Illegal, Unreported and Unregulated Fishing, and a range of regional fisheries agreements. However, management responses have been unable to keep pace with the repeated pattern in deep-sea fishing of exploration, discovery, exploitation and depletion. Gaps in the high seas governance regime contribute to the depletion of deepwater fish stocks (IUCN 2005). There is a strong need for integrated approaches instead of separate regimes for fisheries, aquaculture, marine mammal conservation, shipping, oil and gas, and mining. A multiplicity of sectoral agreements cannot deal with conflicts across sectors, or with cumulative effects (Crowder and others 2006).

Over the past decade, multilateral agreements have been adopted to deal with persistent organic pollutants (Eckley and Selin 2002). The global Stockholm Convention on POPs (2001) and the regional UNECE/CLRTAP POPs protocol (1998) both seek to phase out the production and use of a number of harmful substances. POPs are also subject to strong policy actions under the European Union, the Convention on the Protection of the Marine Environment of the Baltic Sea Area, the Convention for the Protection of the Marine Environment of the North-East Atlantic (OSPAR), and the North American Agreement on Environmental Cooperation (NAAEC). These overlapping international agreements, together with increasing domestic regulations have, in many cases, resulted in declining pollution levels and reduced threats to human health.

There is no global heavy metals (HM) agreement. The HM agreement with the largest geographical coverage is the 1998 UNECE/CLRTAP Heavy Metals Protocol. HMs are also subject to regulations under the European Union, HELCOM, and OSPAR. Mercury is also targeted under the NAAEC. Global efforts to address mercury led to a mercury assessment (UNEP 2002a), and the UNEP Mercury Programme. HM emission reductions measures, such as limiting allowed emissions from major stationary sources and bans on lead in gasoline, have helped to reduce emissions. Despite these actions, environmental levels of some HMs do not seem to be declining, and in some cases, are even increasing, raising concerns for human health (Kuhnlein and Chan 2000).

It has been possible to misuse the oceans and the atmosphere for long periods of time with only slowly emerging visible repercussions. Their volumes are very large, their composition very complex, the lag times between cause-and-effect are long, and their physical "location" can be distant from people. Furthermore, the response capacity of the international community has been predominantly low, with exception of protecting the stratospheric ozone layer. It has been difficult to overcome the challenges, and to manage these global commons as collective resources of humankind, because of the weak institutional architecture at the global level.

Despite these challenges, international treaty regimes to protect global commons signal an unprecedented level of international cooperation, and are giving rise to a number of policy innovations in global environmental governance, such as emissions trading schemes (the Kyoto Protocol) and shared revenues from using resources (UNCLOS). But reducing vulnerability related to the degradation of global commons requires

a number of responses beyond international treaties alone. Some of the opportunities that deserve closer attention are:

- integrating governance from the local to the global level by supporting governance measures at all levels, and going beyond providing resources and capacity building for

---

Box 7.8 Indigenous Arctic Peoples

While many Arctic residents would not receive a high human development index score, they do not consider their quality of life as inferior to that of other societies. About 400 000 indigenous peoples living in the Arctic contribute very little to climate change, yet they are already experiencing its effects. Countries emitting large amounts of greenhouse gases essentially export climate change to the Arctic where, according to the *Arctic Climate Impact Assessment*, climate change is occurring sooner and more rapidly than in other regions, with many large changes projected for the future. Indigenous peoples make up a small percentage of the region's nearly 4 million residents, but they form the main group in many parts of the region. They are the Arctic inhabitants most directly affected by current and future effects of climate change (see Figure 7.19) (see Chapters 6 and 8).

The exposure of the Arctic population to POPs and heavy metals (HMs) is likely to have a severe impact on human well-being, indigenous cultures and food security. POPs and HMs have been associated with a number of human health risks, which include negative effects on the development and maintenance of female characteristics of the body (oestrogenic effects), disruption of endocrine functions, impairing the way the immune system works and affecting reproduction capabilities. Evidence suggests that exposure of people to levels of POPs and HMs found in traditional foods may adversely affect human health, particularly during early development (see Chapter 1).

Sources: ACIA 2004, ACIA 2005, AHDR 2004, Ayotte and others 1995, Colborn and others 1996, Hild 1995, Kuhnlein and Chan 2000

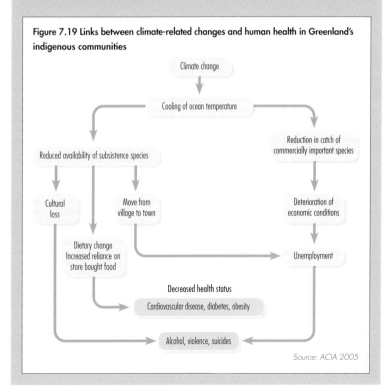

Figure 7.19 Links between climate-related changes and human health in Greenland's indigenous communities

Source: ACIA 2005

national agencies in charge of implementing global agreements;

- strengthening the voices of vulnerable communities in global processes, helping to bridge different types of knowledge, and to build a culture of responsibility for action;

- institutionalizing longer time horizons and intergenerational equity in research efforts, impact assessments, decision making and law, which is essential for reversing the pattern of misuse of the global commons, and which will need consistent incremental decisions and policies over years and decades to effect change;

- paying attention to mitigation and adaptation to help the communities most vulnerable to degradation of the global commons in ways that are sensitive to their local cultures, for example in the global treaties that until now have their strongest focus on reducing the degradation of the commons; and

- resolving conflicts with stronger multilateral fish stock management.

Securing energy for development

This archetype is about vulnerabilities as a consequence of efforts to secure energy for development, particularly in countries that depend on energy imports. The dramatic increase in energy use in the last 150 years (Smil 2001) has been a key factor in economic and social development. In those countries and sectors of population that do not yet benefit from modern energy, development is hindered and energy security and increasing energy access are therefore high on the national agendas. Vital societal functions depend on reliable energy supply. The dominating energy production patterns (centralized production systems, fossil fuel dominance and lack of diversification) have created increased technical and political risks for disrupted supplies as well as a host of negative health and environmental effects.

Global relevance

Since the 1970s, each 1 per cent increase in GDP in industrialized countries has been accompanied by a 0.6 per cent increase in primary energy consumption (IEA 2004). A further increase of over 50 per cent in energy use – mostly in developing countries – is expected in 2–3 decades (IEA 2004, IEA 2005). In 2000, about 1.6 billion people had no access to electricity while 2.4 billion people still relied on traditional uses of biomass, a burden that falls mainly on women (IEA 2002). Although there are no MDGs for energy access, the WSSD warned that without access to modern energy supplies, and fundamental changes in energy use, poverty reduction and sustainable human development would be difficult to realize (UN 2002).

Oil and gas are expected to remain the dominant sources of energy over the next 2–3 decades, if current trends continue (IEA 2006). Energy supply security is becoming a problem, due to increasing competition for oil and natural gas among Europe, the United States and the rapidly-growing economies in Asia. Among the factors affecting supply security are (IEA 2007):

- oil exports are from a smaller number of countries;
- geo-political tensions;
- uncertainty over when the global resource base for oil and gas may become critical, with mainstream energy analysis suggesting this is unlikely in the next 2–3 decades, while others believe that oil production already is peaking; and
- the impacts of extreme weather events on energy production, such as the heat wave in Europe in 2003, and hurricanes in the Gulf of Mexico in 2005.

About 90 per cent of the global anthropogenic greenhouse gas emissions are energy related, and dramatic shifts towards low greenhouse gas emitting production and consumption systems are necessary to address climate change problems, especially in developed and rapidly developing countries (Van Vuuren and others 2007).

Oil has become increasingly important in total energy consumption of low-income regions (see Figure 7.20a). In contrast, in high-income countries, the share of oil in energy use has declined, although absolute consumption of oil still increases. The share of oil that is imported is increasing in both high- and low-income countries, following a decline in the 1970s and 1980s due to the oil crises (see Figure 7.20b). Since the early 1970s, oil intensity has almost halved in high-income regions. Although oil intensity is declining in low-income regions, the ratio is significantly higher, indicating that oil price shocks are having a far greater impact on their economies (see Figure 7.20c).

Vulnerability and human well-being

Impacts of energy use on human well-being due to air pollution and climate change, as well as the importance of energy for realizing the MDGs are analysed in Chapter 2. For energy-importing countries, securing the supply of affordable energy is directly linked with human well-being. There could be a "vulnerability paradox" regarding energy: the less vulnerable a country's energy sector becomes, the greater the impacts could be from energy problems (see Box 7.9). Since society has become very dependent on energy, there could even be a "double vulnerability paradox." Both the decreased vulnerability of the energy supply as well as the increased dependency on a reliable energy supply contribute to an increasing vulnerability of society to disturbances in the energy supply (Steetskamp and van Wijk 1994). For households, energy becomes an issue of concern with increasing energy prices. This especially affects lower-income groups in industrialized and developing countries. For instance, the United Kingdom has had a fuel poverty strategy since 2001 (DTI 2001) that recognizes that fuel poverty is caused by a combination of low income, lack of energy efficiency measures and unaffordable energy, especially for the elderly (Burholt and Windle 2006).

For developing countries without fossil fuel reserves, the security of supply is an even more pressing problem. Again, this affects the poorer population groups, because transport and food prices are affected most. Rural areas are especially vulnerable, as are small

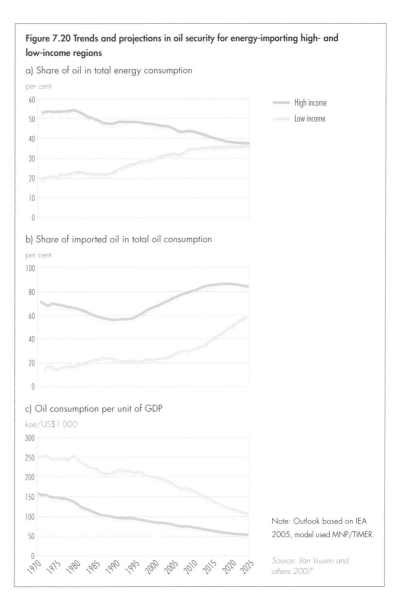

Figure 7.20 Trends and projections in oil security for energy-importing high- and low-income regions

a) Share of oil in total energy consumption

per cent

High income
Low income

b) Share of imported oil in total oil consumption

per cent

c) Oil consumption per unit of GDP

koe/US$1 000

Note: Outlook based on IEA 2005, model used MNP/TIMER.

Source: Van Vuuren and others 2007

Box 7.9 The resource paradox: vulnerabilities of natural resource rich, exporting countries

Oil-exporting countries have a different set of human well-being and vulnerability challenges connected to fossil fuels. Populations living near points of extraction often suffer direct health effects, or indirectly, as a result of degraded ecosystems. On a national scale, the lucrative single commodity often lowers incentives to diversify the economies, while offering considerable financial incentives for poor governance and corruption.

The "natural resource curse" describes the large number of resource-abundant economies that exhibit high levels of corruption in the public and private sectors. This overdependence on natural resource abundance in a weak or corrupt political system lowers economic growth. It can underlie the generation of human vulnerability and ill-being, and even result in violent conflict.

Taking the "problem" of resource wealth out of the political sphere is thought to be a healthy if difficult approach. For oil-exporting countries, diversification of their economies would reduce their dependency on import revenues. Countries such as Norway managed the problem of large resource rents by creating a fund for health and education, managed by an independent central bank. Botswana introduced social transparency policies to effectively and equitably manage its mineral wealth. The World Bank's transparency and social investment conditions put onto the Chad-Cameroon oil pipeline illustrates how more equitable sharing of resource rents is pursued. Not spending resource wealth for a poor country makes little sense, but it is generally argued that equitable and transparent spending of the revenue is possible without deindustrializing a nation's economy through an increase in the exchange rate.

Sources: Auty 2001, Bulte, Damania, and Deacon 2005, Collier and others 2003, De Soysa 2002a, De Soysa 2002b, De Soysa 2005, Lal and Mynt 1996, Leite and Weidmann 1999, Papyrakis and Gerlagh 2004, Ross 2001, Sachs and Warner 2001, Sala-I-Martin 1997

and medium enterprizes that often cannot cope with the volatility of oil prices (ESMAP 2005). Rises in energy prices also result in macro-economic losses, indirectly affecting human well-being. In OECD countries, although oil intensity is already decreasing, an increase of US$10/barrel is estimated to result in 0.4 per cent in lost GDP in the short-term (IEA 2004). For the poorest countries, IEA (2004) estimates are about a 1.47 per cent GDP loss per US$10 rise per barrel. Some of the lowest-income countries suffer losses of up to 4 per cent of GDP (ESMAP 2005).

Responses

Countries have pursued a variety of options to improve their energy security, including diversifying energy supply, improving regional energy trade arrangements, reducing dependence on imports by promoting energy efficiency, using domestic sources and alternative options, including renewable energy (see Box 7.10). In most countries, the buildup of energy infrastructure is extensively regulated by governments. With liberalization in many industrialized and developing countries over the last decade, this situation has changed. The internal market in Europe had two opposing effects with respect to energy security and the environment. It improved the overall efficiency of the energy system, and created a market for more energy-saving technologies. However, it also made investments that require large capital input, or have long payback times. R&D became more short-term oriented, and budgets were reduced and often not aligned with sustainable development objectives.

Public support remains necessary to stimulate new technologies (European Commission 2001). Many

development strategies treat energy only in the context of large-scale infrastructure projects, where energy access issues are usually ignored, and the focus is on electricity, neglecting fuel availability and rural energy development. Out of 80 MDG country reports, only 10 mention energy outside discussions in relation to environmental sustainability (MDG 7). Only one-third of Poverty Reduction Strategy Papers allocate financial resources to national energy priorities (UNDP 2005). Implementation of sustainable energy systems is hindered by a number of issues, including a finance gap, subsidies biased towards fossil fuels, lack of stakeholder involvement, and regulatory and sector management problems (IEA 2003, Modi and others 2005).

Energy has long been considered the exclusive prerogative of national governance, and with the exception of nuclear energy, has lacked both an organizational home and a coherent normative framework in the UN system. This has begun to change in recent years with energy for sustainable development being discussed as a theme by the Commission on Sustainable Development in 2001 and 2006–07. At the World Summit on Sustainable Development (WSSD), energy received high priority in the action plan. Converging agendas seem to be pushing for strengthened global governance of energy through its links to climate change, poverty (especially MDG 1), health and security (CSD 2006). Following the WSSD, a number of multistakeholder partnerships were established to implement various elements of the international energy agenda. As a follow-up to the 2005 G8 Gleneagles energy initiative, the World Bank completed in 2006 an investment framework for clean energy and sustainable development. There has also been some efforts to create mechanisms for coordinating energy work, most recently through UN-Energy, an interagency mechanism established to support the implementation of WSSD energy-related decisions (UN-Energy).

The policies to move away from oil dependence have had some impacts in industrialized countries (see Figure 7.20). One of the reasons for the limited impact of policies is the long lifespan (40–50 years and longer) of energy infrastructure. This means that technology and investment decisions from decades ago have created a path dependency for today's production and consumption patterns. It also means that the decisions made today will have major impacts for decades

Box 7.10 The ethanol programme in Brazil

Pró-Alcool, the Brazilian ethanol programme, was launched in 1975 to respond to the declining trend in sugar prices and the increasing cost of oil. Brazil has since developed a large ethanol market, and widely uses ethanol produced from sugar cane as a transport fuel. With higher oil prices, ethanol became a cost-effective substitute for gasoline, and the official alcohol programme was phased out. The programme helped to reduce dependency on imported oil, saved about US$52 billion (January 2003 US$) between 1975 and 2002 in foreign exchange, created 900 000 relatively well-paid jobs, considerably reduced local air pollution in the cities, and cut greenhouse gas emissions. With the possible increase of ethanol exports from countries such as Brazil to Europe, the United States and Japan, concerns are increasing about the sustainability of large-scale biomass production, especially in terms of competition for land for food production, biodiversity and energy crops.

Source: La Rovere and Romeiro 2003

to come, and there are few incentives in place for considering the well-being of future generations.

Given the large scope for synergies among policies related to energy security, health and air pollution and climate change (see Chapter 2), there are many opportunities to reduce vulnerability of people and communities, including:

- focusing energy policies on improving access to appropriate energy services for the most vulnerable, such as women, the elderly and children, as part of broad development planning;
- improving the opportunities for the most vulnerable to have voice in energy issues, for instance, in designing new energy systems;
- investing in the diversification of both centralized and decentralized technologies, with technology transfer playing an important role; and
- strengthening the capacity for sustainable energy technology innovation and production in cooperation with vulnerable communities, as a way to create jobs and increase coping capacity.

Coping with multiple threats in Small Island Developing States

Small Island Developing States (SIDS) are vulnerable to climate change impacts in the context of external shocks, isolation and limited resources, creating another archetype of vulnerability. SIDS are highly prone to natural disasters, such as tropical storms and storm surges (IPCC 2007, UNEP 2005a, UNEP 2005b, UNEP 2005c). Limited institutional, human and technical capacities highly constrain their ability to adapt and respond to climate change, variability and extremes (IPCC 2007). Current vulnerabilities are further exacerbated by growing populations. For example, the total fertility rate of most Pacific islands is greater than four. The international trading regime and WTO compliance are increasingly demanding for SIDS. With eroding access to protected markets for their export commodities, such as sugar, bananas and tuna, and with declines in commodity prices triggering economic volatility, they are highly sensitive to globalization and trade liberalization (Campling and Rosalie 2006, FAO 1999, Josking 1998).

Global relevance

SIDS are located in the Pacific, Indian and Atlantic Oceans, and the Wider Caribbean and South China Seas. In UNEP regional terms, 6 SIDS are in Africa, 23 in Latin America and the Caribbean,

and 22 in Asia and the Pacific. The Environmental Vulnerability Index (EVI) scores for 47 SIDS illustrate that none are ranked resilient and almost three-quarters are highly (36 per cent) or extremely (36 per cent) vulnerable (Figure 7.21). EVI was prepared by various organizations, including UNEP.

Vulnerability and human well-being

Natural hazards have severe adverse impacts on lives and socio-economic development in SIDS. A high proportion of the total population of 56 million (UNEP 2005d) is frequently exposed to natural hazards. For example, in 2001 nearly 6 million people were affected by natural disasters in the Caribbean (see Figure 1.2 in Chapter 1). In 1988, the cumulative economic damage attributed to disasters was as high as 43 per cent of GDP in Latin America and the Caribbean (Charveriat 2000).

Sea-level rise, and the increasing frequency and severity of extreme events threaten livelihoods and limit adaptation options. These pressures have forced some people to abandon their homes and assets, and to migrate to other countries. New Zealand, for example, amended its Government Residence Policy in March 2006 to allow a small number of citizens from Tonga, Tuvalu, Kiribati and Fiji to immigrate each year (NZIS 2006). Sea-level rise is likely to induce large-scale migration in the longer term, and large migrations have at times led to conflict (Barnett 2003, Barnett and Adger 2003). Abandoning islands would also result in

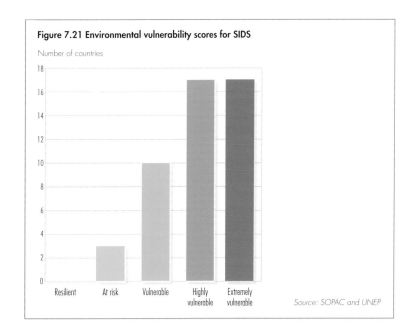

Figure 7.21 Environmental vulnerability scores for SIDS

Number of countries

Source: SOPAC and UNEP

Box 7.11 Disaster preparedness and well-being

The graph below illustrates linkages between vulnerability to natural disasters and poverty (Figure 7.22). With more money to spend, a country can better prepare its people against disaster. Looking at more detailed statistics, in 2004, Hurricane Jeanne claimed more than 2 700 victims in Haiti, while in the Dominican Republic fewer than 20 lost their lives. This was no coincidence. Dominicans are, on average, four times richer, are better prepared in terms of education and training, and benefit from improved infrastructure and housing.

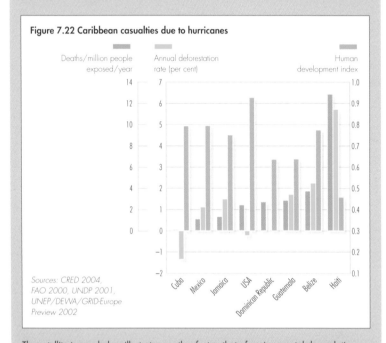

Figure 7.22 Caribbean casualties due to hurricanes

Deaths/million people exposed/year — Annual deforestation rate (per cent) — Human development index

Sources: CRED 2004, FAO 2000, UNDP 2001, UNEP/DEWA/GRID-Europe Preview 2002

The satellite image below illustrates another factor, that of environmental degradation. The Dominican Republic has over 28 per cent forest cover, while Haiti had reduced its forest cover from 25 per cent in 1950 to 1 per cent by 2004. In the image, deforested Haiti is to the left, while the Dominican Republic is the greener area to the right. This environmental aspect is significant, because many victims drowned or died in mudflows, phenomena strongly influenced by land cover change.

Credit: NASA 2002

the loss of sovereignty, and highlights the need to reconsider traditional development issues as matters of national and regional security (Markovich and Annandale 2000), as well as issues of equity and human rights (Barnett and Adger 2003).

Climate-related hazards cause socially differentiated impacts, and tend to affect the poor and disadvantaged groups disproportionately. Most exposed to hazards are people living on atolls and low-lying islands, and in high-risk coastal settlements with substandard housing and infrastructure. The livelihoods most affected include those depending on climate-sensitive natural resources, such as subsistence and commercial farming, and on coastal tourism (Douglas 2006, FAO 2004b and 2005b, UNICEF 2004a, Nurse and Rawleston 2005, Pelling and Uitto 2001).

The most severe impacts on human well-being include the loss of livelihood assets, displacement, increased water- and vector-borne diseases, and the loss of life in natural disasters. The loss of livelihood assets is predominantly caused by reduced or lost ecosystem services due to recurring natural hazard impacts, loss of productive land due to coastal erosion, salinization of land and irrigation water, estuaries and freshwater systems (IPCC 2007), and other forms of environmental degradation, such as deforestation (see Box 7.11 and Figure 7.22). In addition, degradation and overexploitation harm resources such as coral reefs, seagrass beds and mangroves that provide a natural coastal protection as well as the basis for subsistence and commercial activities (see Chapter 5). Hoegh-Guldberg and others (2000) estimate that coral bleaching will reduce future GDP by 40–50 per cent by 2020 in smaller Pacific islands. Furthermore, SIDS are faced with biodiversity loss and impacts on agriculture, due to invasive alien species.

Deteriorating resource access has led to growing competition at community, national and regional levels, though pressures are spatially variable (IPCC 2007, Hay and others 2004, UNEP 2005a, UNEP 2005b, UNEP 2005c). Further stresses, including social pressures from eroding customary resource tenure and security of land titles, have been highlighted as key issues for the management of some marine ecosystems (Cinner and others 2005, Graham and Idechong 1998, Lam 1998).

Higher exposure to natural hazards can have negative impacts on tourist infrastructure and investments, and can reduce tourism income. At the same time, tourism adds to increased pressures on ecosystems (Georges 2006, McElroy 2003). In some coastal locations, inappropriate development in risk-prone areas, due to inadequate consideration of impacts of natural hazards and climate change effects, demonstrates a failure to adapt.

Responses

Recognizing the vulnerabilities of SIDS, the Barbados Programme of Action for the Sustainable Development of Small Island Developing States was adopted internationally in 1994. The Commission on Sustainable Development reviewed the implementation of the Barbados Programme of Action in 1996 and 1998. In 2005, the programme was reviewed at a UN Conference in Mauritius, at which the opening statement pointed out that a decline of international support and resources had hindered implementation. The Mauritius Strategy was adopted at the 2005 conference, laying out a comprehensive multilateral agenda for the sustainable development of SIDS.

The Cooperative Initiative on Invasive Alien Species (IAS) on Islands deals with invasive species that threaten biodiversity, as well as agriculture and human well-being. Innovative initiatives also link ecotourism with eradication of IAS (see Box 7.12).

While some adaptation options are already being implemented in SIDS, specific adaptation strategies offer opportunities for more efficient adaptation, including the use of traditional knowledge based on typical regional or cultural conditions. For example, traditional food preservation techniques, such as burying and smoking food for use in drought periods, can improve food security in rural areas. Box 7.13 illustrates an example of community-based marine resource management that improves both coastal resources and human-well-being. Traditional building materials and designs help reduce infrastructure damage and loss from natural hazards. Renewable resources, such as biofuels (such as bagasse), wind and solar power, show a great potential for energy diversification, and for improving the energy resource potential and energy supply for SIDS. This can also increase resilience in the face of recurring extreme events.

Box 7.12 Ecotourism: paying the costs of invasive alien species control

In many SIDS, tourism is the main economic activity. Seychelles has created a win-win situation for development and environment by linking ecotourism and indigenous species restoration.

Two invasive species, *Rattus rattus* and *R. norvegicus*, have a significant impact on Seychelles' endemic biodiversity. In central Seychelles (41 islands) six species and one subspecies of land birds are endangered and threatened by rats. Rat eradication is essential for re-establishing indigenous bird populations that support ecotourism.

Protected area status is sought after by the ecotourism sector. By linking the awarding of protected area status to the ability to maintain predator-free islands, the government has successfully brought the private sector into IAS management. With the lure of potential future ecotourism revenue, operators in three islands participated in an eradication programme, funding their own costs of nearly US$250 000.

Source: Nevill 2001

Box 7.13 Twinning marine protection and resource replenishment in community-based conservation in Fiji

Coastal marine resources in many parts of Fiji are being overfished by both commercial fishing and subsistence harvesting. These practices have largely affected rural communities – about half of Fiji's population of 900 000 – that rely on communal marine resources for their traditional subsistence-based livelihoods. Food security and accessibility have been reduced. Women gleaning off mudflats, for instance, expend more fishing effort for subsistence species such as clams. Some 30–35 per cent of rural households in Fiji live below the national poverty line.

In response to these concerns, Fijians have established Locally Managed Marine Areas (LMMA), and strengthened traditional marine resource management to replenish marine stock. Communities work with *Qoliqoli* (officially recognized customary fishing rights areas), imposing temporary closures of these fishing zones, and *tabu* (no-take for certain species). Communities typically set aside 10–15 per cent of the village's fishing waters to protect spawning and overexploited areas for resource recovery. While the communities receive external technical expertise, they make the decisions, making an LMMA significantly different from a marine reserve or marine protected area. Prized local species, such as mangrove lobster, have increased up to 250 per cent annually, with a spillover effect of up to 120 per cent outside the *tabu* area in the village of Ucunivanua. The establishment of LMMAs has increased household income and improved nutrition.

As a result of the success of Fiji's LMMAs, villagers have been increasing the pressure on the government to return legal ownership of the country's 410 qoliqolis to their traditional owners.

Source: WRI 2005

To achieve this overarching goal of successfully improving human well-being in SIDS, vulnerability and adaptation assessments need to be further mainstreamed into national policies and development activities at all levels and scales. A number of options are available to reduce vulnerability, and to build capacities in SIDS:

- enhancing early warning systems to support disaster preparedness and risk management systems (IFRCRCS 2005) helps adaptation to short-term variability (Yokohama Strategy and Plan of Action for a Safer World 1994 and the Hyogo framework) (see Box 7.14);

- improving integrated planning for climate-robust, long-term development, especially that of livelihood assets, improves access to resources for local people. Water resource and Integrated Coastal Zone Management (ICZM) can contribute to improving the long-term adaptive capacity of vulnerable communities (UNEP 2005a, UNEP 2005b, UNEP 2005c). This requires governance systems that take possible long-term changes into account;

- using participatory approaches to integrate traditional ecological knowledge in conservation and resource management empowers communities for disaster preparedness and resource management;

- developing technologies for reducing vulnerability can shift from a "controlling nature" to a "working with nature" paradigm. This includes the technology and capacity to assess impacts and adaptation options, document traditional coping mechanisms and develop alternative energy solutions;

- investing in improved regional cooperation can better address environmental challenges and improve coping capacity. An example would be development and strengthening of global and regional bodies, such as Alliance of Small Island States (AOSIS) and the Indian Ocean Commission, to build early warning systems for environmental stresses;

- strengthening of cooperation and partnerships at the national, regional and international levels, including pooling of resources for the implementation of activities and Multilateral Environmental Agreements (MEAs) (Hay and others 2003, IPCC 2001, Tompkins and others 2005, Smith and others 2000, Reilly and Schimmelpfennig 2000, IFRC 2005); and

- recognizing in international negotiations that basic rights laid down in the Universal Declaration of Human Rights are at risk in the case of climate change effects on atoll countries (Barnett and Adger 2003).

Taking technology-centred approaches to water problems

Poorly planned or managed large-scale water projects that commonly involve massive reshaping of the natural environment can create another archetype of vulnerability. Examples include certain irrigation and drainage schemes, the canalization and diversion of rivers, large desalinization plants and dams. Dam projects are prominent and important examples, although many of the conclusions often apply to other vulnerability-inducing water management schemes. Dams

Box 7.14 The Hyogo Framework for action

Disaster reduction strategies have the potential to save lives and protect livelihoods by even the simplest of measures. Acknowledging this and recognizing that much more needs to be done to reduce disasters, governments adopted in January 2005, the Hyogo Framework for Action 2005–2015, Building the Resilience of Nations and Communities to Disasters. This framework defines strategic goals and five priorities for disaster reduction. Priority Four deals with environmental and natural resource management to reduce risk and vulnerability. It encourages the sustainable use and management of ecosystems, and the integration of climate change concerns into the design of specific risk reduction measures.

For the MDGs to be realized, the burden of natural disasters needs to be reduced. Disaster risk reduction policies should be incorporated into development plans and programmes, and into multilateral and bilateral development assistance, particularly that related to poverty alleviation, natural resource management and urban development. The implementation of disaster risk reduction is promoted through the International Strategy for Disaster Reduction (ISDR), a partnership between governments, non-governmental organizations (NGOs), UN agencies, funding institutions, the scientific community and other relevant stakeholders in the disaster reduction community.

Source: UNISDR

have both positive and negative impacts: they satisfy human needs (water for food security and renewable energy), and protect existing resources by providing flood control. However, they may have severe impacts on the environment through river fragmentation (see Chapters 4 and 5), and on social structure. Some dams provide benefits without major negative effects. But many do not due to the inadequate consideration given to social and ecological impacts from poor dam planning and management. This is a result of the prevailing technology-centred development paradigm (WBGU 1997). Reducing vulnerability here means either to reduce the negative consequences of these projects, or to find alternative means to fulfil the demand for energy, water and flood protection (see Box 1.13 in Chapter 1 on restoration of ecosystems through decommissioning of dams).

Global relevance

The dynamics described here occur worldwide. Important examples are the planned Ebro water scheme in Spain, large-scale water management schemes in the US southwest, the Narmada in India, the Nile in Africa and the Three Gorges Dam in China. Major irrigation schemes built in the 20th century and new, multifunctional mega-dams (over 60 m in height) have had significant impacts on water resources. There are more than 45 000 large dams in 140 countries, about two-thirds of these in the developing world (WCD 2000). The actual trend is characterized by a decline in the annual number of new large dams, while no decline is observed for the mega-dams. The geographical distribution of new dam construction continues to shift from the industrialized countries to the newly industrialized and developing countries (ICOLD 2006). The effects of these large-scale installations are rarely confined to the local area, but can assume far-reaching and even international proportions (see Chapter 4).

Vulnerability and human well-being

Currently, large dams are typically built in remote areas of developing countries. The integration of such peripheral regions into the world market through dam projects leads to an extensive transformation of social conditions for the indigenous population. Consideration must be given to the social consequences, which may range from resettlement of the local population, to

intensification of economic disparities, and domestic and international conflicts (McCully 1996, Pearce 1992, Goldsmith and Hildyard 1984). According to estimates (WCD 2000), 40–80 million people have been forced to leave their homes since 1950 because of large dam projects. Forced resettlement, lack of stakeholder participation in planning and decision making, and lack of sharing in the benefits of the projects may marginalize and victimize the local people in development (see for example, Akindele and Senyane 2004). The distribution of the benefits gained from dam construction (power generation and irrigated agriculture) can be very uneven, reinforcing the widening of social and economic disparities and poverty.

Tensions may build up, and can escalate into national and international conflicts (Bächler and others 1996). Although widespread organized violence is rare, local protests against large water projects are common. Despite high levels of political attention to future "water wars" between states, cooperation between states has been more common than conflict over the last half of the 20th century. A comprehensive analysis of bilateral and multilateral state-to-state interactions over water between 1948 and 1999 found that of more than 1 830 events, 28 per cent were conflicts, 67 per cent were cooperative, and the remaining 5 per cent were neutral or not significant (Yoffe and others 2004). International water cooperation institutions, such as basin commissions, have fostered international cooperation, for example in the cases involving the Itaipu and Corpus Christi dams in Argentina, Brazil and Paraguay. In some cases, a key to fostering cooperation appears to be moving parties, often through external facilitation, from asserting competing rights to water to identifying needs for water, and finally to negotiating the sharing of the benefits of water (Sadoff and Grey 2002). Further examples of cooperation involve the Zambezi, Niger, Nile and Rhine rivers.

Other negative impacts on human well-being are health hazards in the form of water-based vectors (for example, mosquitoes and snails), which occur due to the changes in the run-off regimes. This exacerbates the risk of malaria and other diseases in many subtropical and tropical regions. Figure 7.23 shows the relationship between the distance from a dam and the occurrence of water-related

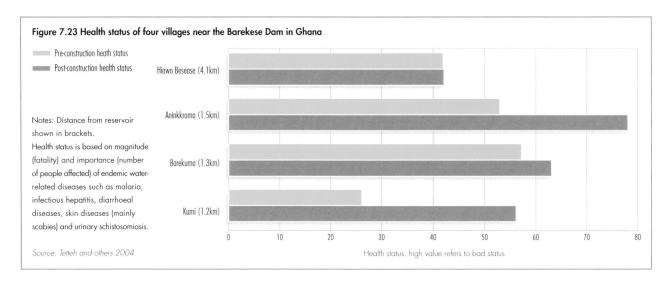

Figure 7.23 Health status of four villages near the Barekese Dam in Ghana

Pre-construction heath status

Post-construction health status

Hiawo Besease (4.1km)

Aninkkroma (1.5km)

Barekuma (1.3km)

Kumi (1.2km)

Notes: Distance from reservoir shown in brackets.

Health status is based on magnitude (fatality) and importance (number of people affected) of endemic water-related diseases such as malaria, infectious hepatitis, diarrhoeal diseases, skin diseases (mainly scabies) and urinary schistosomiasis.

Source: Tetteh and others 2004

Health status: high value refers to bad status

diseases in four villages near the Barekese Dam in Ghana (Tetteh and others 2004). In Hiawo Besease Village, more than 4 km from the dam, there was hardly any change in health status after the dam was built. In the other three villages, located only 1.2–1.5 km from the dam, the health status declined after the dam was built.

Responses

In 2000, the international multi-stakeholder World Commission on Dams (WCD) evaluated the development effectiveness of large dams, and developed international guidelines for dam building. Their final report (WCD 2000) identified five core values, and formulated seven strategic priorities (see Table 7.3).

Building synergies between biodiversity concerns (as in the CBD, RAMSAR Convention on Wetlands and the Convention on Migratory Species), and development is an important concern. As a follow-up to the WCD framework, the UNEP Dams and Development Project (UNEP-DDP) was launched in 2001. Recognizing that for many developing countries hydropower and

irrigation remain priorities to meet energy and food security needs, UNEP-DDP focuses on how to support building and management of dams sustainably. At national and sub-regional levels, countries have responded by increasingly accepting social and environmental impact assessments (EIAs) of large dam projects prior to construction (Calcagno 2004). The trend towards shared river management, acknowledged in the 1997 UN Convention on the Non-Navigational Uses of International Watercourses, has created new opportunities for addressing such concerns.

Nevertheless, the effectiveness of these measures is mixed. In some places, it is evident that stakeholder expectations regarding participation, transparency and accountability in dam planning and development is changing. The WCD recommendations provided a new, authoritative reference point for NGOs trying to influence government decisions, but has had different levels of success. The value of cooperation between states is increasingly recognized, but in practice this has played out in different ways. For example, the controversial Ilisu Dam project in Turkey came to a halt in 2001 when the European construction

Table 7.3 Some findings of the World Commission on Dams	
Five core values were identified	**Seven strategic priorities were formulated**
■ Equity ■ Efficiency ■ Participatory decision making ■ Sustainability ■ Accountability	■ Gaining public acceptance ■ Comprehensive options assessment ■ Addressing existing dams ■ Sustaining rivers and livelihoods ■ Recognizing entitlements, and sharing benefits ■ Ensuring compliance ■ Sharing rivers for peace, development and security

Source: WCD 2000

firms withdrew from the project, citing outstanding economic and social issues, and the difficulty of meeting conditions imposed on their effort to procure a US$200 million export credit guarantee from the British government. In contrast, the World Bank and the African Development Bank forged ahead with the controversial US$520 million Bujagali Dam project in Uganda despite strong transnational NGO opposition, and the earlier withdrawal from the project by bilateral funding agencies in the United Kingdom, France, Germany, Sweden and the United States (IRN 2006).

Several relevant international policy initiatives deal with problems of insufficiently and inequitably fulfilled water demand (see Chapter 6). One important aspect of MDG 7 on ensuring environmental sustainability is to "reduce by half the proportion of people without sustainable access to safe drinking water." The implementation plan calls for an approach to "promote affordable and socially and culturally acceptable technologies and practices." These needs can be met, as suggested by the World Water Vision (World Water Council 2000), through a mix of large and small dams, groundwater recharge, traditional, small-scale water storage techniques and rainwater harvesting, as well as water storage in wetlands (see Box 7.15).

It is clear that maladapted and mainly supply-oriented technological approaches will, at least in the medium-term, fail to realize the desired development benefits.

Well-planned water management can reduce vulnerability, and contribute to development. There are a number of options (see Chapter 4):

- improving access to water as an essential asset for household needs and agricultural production. Distributional aspects should be given much more attention;
- increasing opportunities for more effective local participation in basin and catchment management, as local rights and values may be in conflict with those held by the state. This requires supportive and inclusive institutions, and governance processes;
- trading, including the import of "virtual water" via food imports, may substitute for irrigation water consumption in arid regions;

Box 7.15 Substituting micro-catchment for large-scale water projects

A promising alternative to large reservoirs for irrigation is micro-catchment management, which uses natural run-off directly, and in a decentralized way. A good example is the water-harvesting technique used in Tunisia, consisting of ancient terraces and recharge "jessour" wells. These decentralized techniques allow for the cultivation of olive trees in arid zones while conserving and even ameliorating the soil. Furthermore, the efficient control of sediment flows reduces the danger of floods downstream.

Source: Schiettecatte 2005

Traditional terracing to harvest water and control overland flow near Tataouine in Southern Tunisia.
Credit: Mirjam Schomaker

- improving cooperative water basin management can increase development opportunities, and reduce potential for conflict. Developing transboundary river basin institutions offers important opportunities for building on environmental interdependence to foster collaboration and contribute to conflict prevention. The SADC Water Protocol of 2000, the Nile Basin Initiative (NBI) and the Niger Basin Authority (NBA) are good examples in Africa of riparian dwellers and stakeholders developing shared visions for water and development, while integrating international legal norms, such as prior notification and causing no significant harm; and
- investing in local capacities and employing alternative technologies can improve water access and use. This strategy is an important way to enhance coping capacities, and ensure consideration of a broader range of alternatives to conventional, large-scale solutions (see Box 7.15).

Rapidly urbanizing the coastal fringe

Rapid and poorly planned urbanization in often ecologically sensitive coastal areas increases vulnerabilities to coastal hazards and climate change impacts. In recent decades, many of the world's coastal areas have experienced significant and sometimes extremely rapid socio-economic and environmental changes. Limited institutional, human and technical capacities have led to severe hazard impacts, and constrain the ability of many coastal communities, particularly those in the developing world, to adapt to changing conditions.

Global relevance

Many of the world's coastal areas have been experiencing rapidly-growing concentrations of people and socio-economic activities (Bijlsma and others 1996, WCC'93 1994, Sachs and others 2001, Small and Nicholls 2003). The average population density in coastal areas is now twice as high as the global average (UNEP 2005d). Worldwide, more than 100 million people live in areas no more than 1 m above sea level (Douglas and Peltier 2002). Of the world's 33 mega-cities, 26 are located in developing countries, and 21 are in coastal areas (Klein and others 2003). Figure 7.24 shows coastal population and shoreline degradation.

Much of this development has been occurring in low-lying floodplains, river deltas and estuaries that are highly exposed to coastal hazards, such as storms, hurricanes, tidal surges, tsunamis and floods. In many cities, major rezoning of former industrial waterfront areas is being undertaken in flood-prone locations to accommodate the tremendous requirement for housing. Examples include Brooklyn and Queens in New York (Solecki and Leichenko 2006), and the Thames Gateway, a 60-km-long corridor along the Thames River between London and the Thames Estuary that is currently undergoing considerable urban regeneration.

Poor urban planning and inappropriate development in highly exposed coastal locations, in combination with rapid population growth, sea-level rise and other climate change impacts, have led to a considerable increase in socio-economic impacts

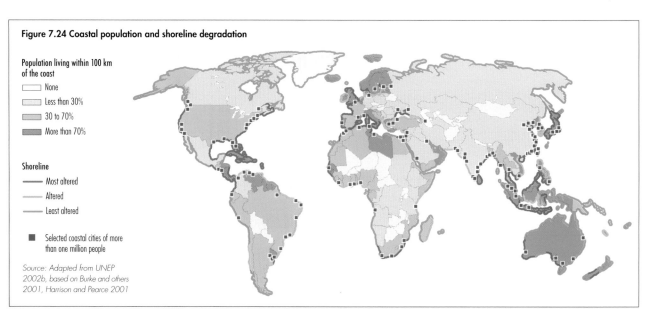

Figure 7.24 Coastal population and shoreline degradation

Population living within 100 km of the coast
- None
- Less than 30%
- 30 to 70%
- More than 70%

Shoreline
— Most altered
— Altered
— Least altered

■ Selected coastal cities of more than one million people

Source: Adapted from UNEP 2002b, based on Burke and others 2001, Harrison and Pearce 2001

from coastal hazards. The EM-DAT global datasets on extreme events indicate (see Figure 7.25) that annual economic losses from extreme events have increased tenfold from the 1950s to the 1990s. In the decade between 1992 and 2001, floods were the most frequent natural disaster, killing nearly 100 000 people and affecting more than 1.2 billion people. Munich Re (2004a) documented an increasing concentration of the loss potential from natural hazards in mega-cities. Only a small proportion of these losses were insured.

Environmental change is expected to exacerbate the exposure of many coastal urban areas to natural hazards from rising sea levels, increased erosion and salinity, and the degradation of wetlands and coastal lowlands (Bijlsma and others 1996, Nicholls 2002, IPCC 2007). There is also a concern that climate change might, in some areas, increase the intensity and frequency of coastal storms and hurricanes (Emanuel 1988), but there is no scientific consensus (Henderson-Sellers and others 1998, Knutson and others 1998). In a recent global assessment of storm surges, Nicholls

(2006) estimated that in 1990 some 200 million people were living in areas vulnerable to storm surge flooding. The North Sea, the Bay of Bengal and East Asia are considered as notable hot spots, but other regions, such as the Caribbean, and parts of North America, Eastern Africa, Southeast Asia and Pacific states are also vulnerable to storm surges (Nicholls 2006).

Increasing development in coastal areas causes fragmentation of coastal ecosystems and conversion to other uses, including infrastructure and aquaculture development, and rice and salt production (see Chapter 4). This negatively affects the condition and functioning of ecosystems, and their ability to provide ecosystem services. An assessment of the status of the world's mangroves FAO (2003a) found that their extent has been reduced by 25 per cent since 1980 (see Chapters 4 and 5).

Vulnerability and human well-being
The relationship between increasing urbanization and growing vulnerability to natural hazards is most

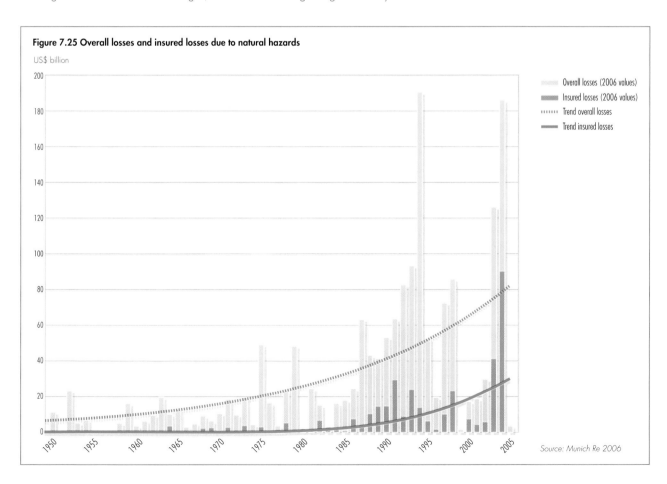

Figure 7.25 Overall losses and insured losses due to natural hazards

US$ billion

Overall losses (2006 values)
Insured losses (2006 values)
Trend overall losses
Trend insured losses

Source: Munich Re 2006

pronounced, but not exclusive (see Box 7.16) in developing countries due, to a large extent, to rural-urban migration (Bulatao-Jayme and others 1982, Cuny 1983, Mitchell 1988, Mitchell 1999, Smith 1992, Alexander 1993, Bakhit 1994, Zoleta-Nantes 2002). This often affects cities facing severe constraints on their institutional, human, financial and technical capacities to develop integrated approaches to urban planning. As a result of the lack of affordable housing options, poor migrants tend to inhabit informal settlements, which are often located in the most undesirable and hazardous areas of the city. According to UN Habitat (2004), more than 750 million of the world's more than 1 billion poor people live in urban areas, without adequate shelter and basic services. Unsafe living conditions, the lack of secure livelihoods and access to resources and social networks, and exclusion from decision making processes limit the capacity of poor urban people to cope with a range of hazards.

Estimates of the number of additional people at risk from coastal flooding in the future vary widely, but all indicate a considerable increase. For example, Nicholls (2006) estimates that the number of people living in areas vulnerable to storm surge flooding to increase by nearly 50 per cent (or 290 million) by the 2020s compared to 1990, while Parry and others (2001) estimate about 30 million more people at risk from coastal flooding due to climate change by the 2050s and 85 million more by the 2080s.

Response

In recent decades, particularly since the 1990s, the dramatic increase of losses and suffering due to natural disasters has brought the issue of disaster risk reduction increasingly onto the political agenda. From the International Decade for Natural Disaster Reduction (IDNDR) to the International Strategy for Disaster Reduction (ISDR), to the Hyogo Framework of Action (see Box 7.14), the disaster risk reduction community has been calling for renewed commitment, and the integration of disaster risk reduction, with the pursuit of sustainable development as a strategic goal.

The Hyogo Framework calls for the incorporation of disaster risk assessments into the urban planning and management of disaster-prone human settlements. It prioritizes the issues of informal or non-permanent housing, and the location of housing in high-risk areas. This reflects the ISDR (2002) estimate that 60–70 per cent of urbanization in the 1990s was unplanned. One consequence of this framework is that international organizations, like UNESCO, reviewed their present activities with respect to the suggested actions for disaster risk reduction.

Most of the urbanization challenges are still the result of a lack of integrated environmental and urban planning. Policies for more sustainable patterns of urbanization are frequently not implemented. Short-sighted concessions for economic gain, weak institutions and corruption are major factors in the proliferation of planning "oversights," "exceptions" and other forms of inappropriate development in urban areas.

Networks such as the African Urban Risk Analysis Network (AURAN) aim at mainstreaming disaster risk reduction in the management of urban planning and governance in Africa. Here, community-based action research is supported. Projects like "Engaging in awareness-raising activities and household surveys on local perceptions of flood risk in flood-prone districts of

---

Box 7.16 Increasing disaster vulnerability in urban areas: the New Orleans flood of 2005

The scenario of a major hurricane striking the US Gulf Coast had been extensively researched and rehearsed among scientists and emergency managers. Many called New Orleans "a disaster waiting to happen." There had been tremendous loss of coastal salt marshes in the Gulf Coast, particularly in the Mississippi Delta, with estimates of up to 100 square kilometres/year. This meant that many urban areas were increasingly exposed to high winds, water levels and waves. There was an increased flood risk from a combination of land subsidence caused by drainage and soil compaction, and about 80 per cent of the city lies below sea level. There was also the accelerated loss of sediment from salt marshes and barrier islands, the modification of waterways and a rising sea level.

While all of these factors had gradually increased the city's hurricane risk over recent decades, it was the catastrophic failure of the engineered flood protection infrastructure in the face of Hurricane Katrina in 2005, in combination with high social and institutional vulnerability that contributed to the largest natural disaster in recent US history. Cutter and colleagues (2006) demonstrate that there are clear patterns of losses related to the lack of access to resources and to social disadvantage.

An estimated 21.4 per cent of the city's residents did not heed evacuation messages, because they did not have the means to escape. As the hurricane struck before welfare payments at the end of the month, many poor people were short of money. Despite a growing awareness of the increasing physical exposure to coastal hazards, the socially created vulnerabilities had been largely ignored. This was true for New Orleans, and is the case for many other places. It arises at least in part from the difficulty of measuring and quantifying the factors that contribute to social vulnerability.

Sources: America's Wetland 2005, Blumenthal 2005, Cutter 2005, Cutter and others 2006, Fischetti 2001, Travis 2005

Saint Louis, Senegal" reduce vulnerabilities locally, and generate transferable knowledge for other cases.

Science increasingly recognizes the importance of sustainable resource management and biodiversity for ecological resilience and livelihood security in the face of extreme environmental shocks (Adger and others 2005). For example, the detrimental consequences of the loss of coastal ecosystems and their buffering capacity against natural hazards has recently been documented in relation to the 2004 Indian Ocean Tsunami (Liu and others 2005, Miller and others 2006, Solecki and Leichenko 2006) and the impacts of Hurricane Katrina on New Orleans (see Box 7.16).

Environmental actions that reduce vulnerability are seldom promoted in disaster reduction strategies, and many opportunities to protect the environment and reduce disaster risk are missed. Integrated coastal zone management (ICZM) and even further integrated coastal area and river basin management (ICARM) are important tools in reconciling multiple uses of coastal resources and promoting ecological resilience. They provide an institutional framework to implement, enforce, monitor and evaluate policies for the protection and restoration of coastal ecosystems, and to place more value on the goods-and-services (cultural values, natural protection of coastal zones, recreation and tourism and fisheries) they provide. There are significant opportunities to reduce hazard vulnerability:

- integrating of risk reduction and adaptation strategies with existing sectoral development policies in areas such as integrated coastal zone management, urban planning, health care planning, poverty reduction, environmental impact assessment and natural resource management (Sperling and Szekely 2005, IATF Working Group on Climate Change and Disaster Reduction 2004: Task Force on Climate Change, Vulnerable Communities and Adaptation 2003, Thomalla and others 2006);

- strengthening education and awareness raising to deal with the multiple risks associated with rapid coastal urbanization, and with possible response options;

- providing more opportunities for local participation in urban development. The challenge for institutional development is to be responsive to change. One approach is to focus on processes in which local users become active "makers and shapers" of the rights, management and use regimes upon which their livelihoods are based (Cornwall and Gaventa 2001). Participation of women is a critical component in such approaches (Jones 2006); and

- green engineering can help to protect coastlines using mangroves and reefs. It can help to maintain forests and protect soils to avoid the risk posed by landslides, floods, drought and tsunamis.

Green engineering can help to protect coastlines using mangroves.

Credit: BIOS- Auteurs Gunther Michel/Still Pictures

CHALLENGES POSED BY PATTERNS OF VULNERABILITY

The seven patterns of vulnerability show how environmental and non-environmental changes affect human well-being. Some of the different human-environment systems throughout the world share certain vulnerability-creating conditions. The different patterns reflect vulnerabilities across the full range of geographic and economic contexts: developing and industrialized countries, and countries with economies in transition. This allows putting particular situations within a broader context, providing regional perspectives, and showing important connections between regions and globally, as well as possible opportunities to address the challenges in a more strategic manner. Furthermore, the analysis of the archetypes underlines findings of other vulnerability research:

- Research on the underlying causal structures of human vulnerability to environmental change increasingly recognizes that vulnerability arises through complex interactions of multiple socio-political, ecological and geophysical processes that operate in different areas and at different times, resulting in highly differentiated impacts in and across regions (Hewitt 1997), social groups (Flynn and others 1994, Cutter 1995, Fordham 1999) and individuals.

- Environmental risks affect a wide range of natural, economic, political and social activities and processes. Therefore, vulnerability reduction should be integrated as a strategic goal into overall development planning across many sectors, including education, health, economic development and governance. Reducing vulnerability in one area often results in increasing vulnerabilities far away, or moving it into the future, which needs to be taken into account.

- Environmental change has the potential to spur conflict. However, managed environmental change (for example conservation and cooperation) can also make tangible contributions to conflict prevention, de-escalation and post-conflict reconstruction (Conca and Dabelko 2002, Haavisto 2005).

- Human vulnerability and livelihood security are closely linked to biodiversity and ecosystem resilience (Holling 2001, Folke and others 2002, MA 2005). Sustainable environmental and resource management is important in poverty and vulnerability reduction. Extreme events, such as the Indian Ocean Tsunami, show that environmental degradation and poorly planned development activities increase communities' vulnerability to shocks (Miller and others 2005).

- Vulnerability is determined, to a large extent, by a lack of options, due to the unequal distribution of power and resources in society, including the most vulnerable population groups throughout the world, such as indigenous people, and the urban or rural poor. Economic sectors heavily dependent on environmental services are also vulnerable. Resilience increases with diversification of livelihood strategies, and with access to social support networks and other resources.

- For successful use of vulnerability research findings, the policy arena should recognize that vulnerability arises from multiple stressors, which are dynamic over space and time. If vulnerability is reduced to a static indicator, the richness and complexity of the processes that create and maintain vulnerability over time are lost.

- The analysis of the patterns of vulnerability also helps identify a number of opportunities to reduce vulnerability and improve human well-being. Taking these opportunities would also support reaching the MDGs and examples of this are given in Table 7.4, which also illustrates how vulnerability works against the achievement of the goals.

OPPORTUNITIES FOR REDUCING VULNERABILITY

Policy-makers can use vulnerability analysis to target policies for groups that most need them. Vulnerability analysis helps to examine the sensitivity of a human-environment system (such as a watershed or coastal town) to various social and environmental changes, and its ability to adapt or accommodate such changes. Therefore, evaluations of vulnerability include attention to exposure, sensitivity and resilience to multiple pressures. The evaluations consider the degree to which a system is affected by particular pressures (exposure), the degree to which a set of pressures affect the system (sensitivity), and the ability of the system to resist or recover from the damage (resilience). Policies can address each of these components of vulnerability. The analysis, which is most often at the sub-national level is, however, frequently hampered by lack of and/or unreliable data, as well as the challenges of showing the links between environmental degradation and human well-being.

Table 7.4 Links between vulnerability and the achievement of the MDGs, and opportunities for reducing vulnerability and meeting the MDGs		
MDGs and selected targets	**Vulnerability affects potential to achieve the MDGs**	**Adopting strategies to reduce vulnerability contributes to reaching the MDGs**
Goal 1 Eradicate extreme poverty and hunger Targets: Halve the proportion of people living on less than US$1/day. Halve the number of people who suffer from hunger.	■ Contaminated sites damage health and thus the ability to work; this undercuts opportunities to eradicate extreme poverty and hunger. ■ In drylands land degradation, insufficient investments and conflict contribute to low agricultural productivity, threatening food security and nutrition.	■ Improving environmental management and restoring threatened environments will help protect natural capital, and increase opportunities for livelihoods and food security. ■ Improving governance systems – through wider inclusion, transparency and accountability – can increase livelihood opportunities as policies and investments become more responsive to the needs of poor people.
Goal 2 Achieve universal primary education Target: Ensure that all boys and girls complete a full course of primary school	■ Children are particularly at risk when they play, live or attend school near contaminated sites. Lead and mercury contamination presents specific risks for child development. ■ The time-consuming activity of fetching water and fuelwood reduces school attendance, particularly for girls.	■ Sustainable resource management can decrease the environmental health risks children face, and thus increase school attendance. ■ Improved and secure access to energy supports learning at home and at school. It is essential for access to IT-based information, and opportunities to engage in scientific and other experimentation.
Goal 3 Promote gender equality and empower women Target: Eliminate gender disparity in primary and secondary education	■ Women with poor access to education are at greater risk of ill health than men. For example, in many SIDS, more women than men have HIV. ■ Women play a pivotal role as resource managers, but are marginalized in decision making, often have insecure tenure rights and lack access to credit.	■ Redressing inequities – in access to health care and education – is critical in improving coping capacity. ■ Strategies that link health and housing, nutrition, education, information and means increase opportunities for women, including in decision making.
Goal 4 Reduce child mortality Target: Reduce by two-thirds the under-five child mortality	■ Contaminated sites affect mortality of all, but children are particularly vulnerable to pollution-related diseases. ■ Some 26 000 children die annually from air pollution-related diseases.	■ Interlinked environment-development-health strategies, improved environmental management and ensuring access to environmentally derived services can contribute to reducing child mortality and reducing vulnerability.
Goal 5 Improve maternal health Target: Reduce by three-quarters the maternal mortality ratio	■ The accumulation of POPs in food sources affects maternal health. ■ Dams may increase the risk of malaria, which, in turn, threatens maternal health. Malaria increases maternal anaemia, threatening healthy foetal development.	■ Improved environmental management can improve maternal well-being by improving nutrition, reducing risks from pollutants and providing essential services. ■ Integrated environment-health strategies can contribute to achieving this goal by reducing vulnerability.
Goal 6 Combat HIV/AIDS, malaria and other diseases Targets: Halt and begin to reverse the spread of HIV/AIDS Halt and begin to reverse the incidence of malaria and other major diseases	■ Contaminated sites are a huge risk for individuals already exposed to HIV/AIDS, potentially further compromising their health. ■ Climate change is likely to increase the disease burden of poor people, including the incidence of malaria.	■ Integrated environment-health planning and management is critical. ■ Acknowledging and acting on the shared responsibility of developed and developing countries for the adverse impacts of climate change on the most vulnerable is essential.
Goal 7 Ensure environmental sustainability Targets: Integrate the principles of sustainable development into planning and programmes Reduce by half the proportion of people without access to safe drinking water Achieve significant improvement in the lives of at least 100 million slum dwellers	■ Water contamination from dumps, industry and agriculture, water-borne diseases, and growing water scarcity threaten well-being at all levels. ■ The lack of access to energy limits opportunities for investment in technologies, including those for water provisioning and treatment.	■ Improving governance systems, including strengthening institutions and laws and policies, and adopting interlinked strategies, are critical to contributing to environmental sustainability and reducing vulnerability. ■ Securing energy is critical to improving the living conditions of the growing number of slum dwellers.

MDGs and selected targets	Vulnerability affects potential to achieve the MDGs	Adopting strategies to reduce vulnerability contributes to reaching the MDGs
Goal 8 Develop a global partnership for development Targets: An open trading and financial system Cancellation of official bilateral debt, and more generous ODA In cooperation with the private sector, ensure developing counties have access to the benefits of new technologies Address the special needs of landlocked developing countries and SIDs	■ Unfair trade regimes reduce earnings from agricultural products in developing countries. Low-income countries rely on agriculture for close to 25 per cent of GDP. ■ Poor access to energy undermines the investments and technologies that can be used in productive land and natural resource management. ■ Sea-level rise is threatening the security and socio-economic development of SIDS and low-lying coastal areas. More than 60 per cent of the global population lives within 100 km of the coastline, and 21 of the world's 33 mega-cities are located in coastal zones in developing countries.	■ Transparent and fair global processes, especially in trade, are essential to increasing opportunities in developing countries, and can help increase local investments in environmental capital. ■ Massive investments, and technology-sharing in clean energy and transport systems can reduce poverty, increase security and stabilize greenhouse gas emissions. It has been estimated that about US$16 trillion will be required for global infrastructure investment in the energy sector in less than 25 years. ■ Building partnerships for addressing climate change, and honouring technology transfer promises are essential for increasing adaptive and coping capacity in low-lying areas.

The archetypes of vulnerability described above highlight responses that have been taken, primarily at the global or regional level, to address patterns of vulnerability. They also point to opportunities for addressing vulnerability by reducing exposure and sensitivity, and through enhancing adaptive capacity. Many of these opportunities are not directly related to environmental policy processes but to poverty reduction, health, trade, science and technology, as well as to general governance for sustainable development. This section pulls together the opportunities to provide strategic directions for policy making to reduce vulnerability and improve human well-being.

Given the localized nature of vulnerability to multiple stresses, opportunities exist for national decision-makers to target the most vulnerable groups. Decision-makers should clearly identify provisions in their own policies that create and reinforce vulnerability in their countries, and deal with them. At the same time, collaboration at regional and international levels plays a supportive and important role. The opportunities underline the importance of increasing awareness worldwide about the consequences of policy choices for people and the environment in other countries.

Integrating governance across levels and sectors
Increasing the coping and adaptive capacity of the most vulnerable people and communities requires integration of policies across governance levels and sectors, and over time to address the coping and adaptive capacities of future generations.

A consistent focus on increasing the well-being of the most vulnerable can involve costs for other actors, but it helps promote equity and justice. For some issues, there are clearly win-win situations between short- and long-term goals and priorities, but for many there are considerable trade-offs, not necessarily on a societal level, but clearly for certain groups or sectors in society, and even for individuals. Opportunities include the integration of knowledge and values to underpin and support institutional design and compliance. This involves integrating local and global knowledge, for example on impact and adaptation, and integrating concern for neighbours with concern for all humanity and future generations.

Strengthening coping and adaptive capacity, and reducing the export of vulnerability require much more cooperation among different governance levels and sectors. Such integrated governance requires mutually supportive policies and institutions at all governance levels, from the local to the global (Karlsson 2000). This can be a considerable challenge, as illustrated repeatedly in the implementation of MEAs. In many cases, it requires higher governance levels to provide the resources, knowledge and capacity at lower levels to implement plans and policies. This is in line with the Bali Strategic Plan on Technology Support and Capacity Building (BSP), as well as other capacity-building initiatives. For example, adaptation to climate change among vulnerable communities in the Arctic needs support from national governments and regional organizations. To facilitate successful adaptation, stakeholders must promote and enable adaptive measures. In addition, governments should consider revising policies that hinder adaptation. Self-determination and self-government, through ownership and management of land and natural resources, are important for empowering indigenous Arctic peoples to maintain their self-reliance, and to face climate change on their own terms (see Chapter 6, Polar Regions) (ACIA 2005). Another related strategy of integrating governance across levels is special organizational forms that facilitate cross-level interaction, such as co-management of natural resources (Berkes 2002).

Different sectoral priorities should be reconciled and integrated through cooperation and partnership, especially when there are trade-offs between them and these affect vulnerability. One strategy is to integrate, in organizational terms, a focus on strengthening coping capacity and reducing export of vulnerability. For example, when councils, task forces, even ministries are set up, their mandates should cover inter-related sectors, and their staff should have the appropriate training and attitudes to implement broader mandates. Another strategy has been to "mainstream" attention to vulnerability through policy. Mainstreaming of the environment has been tried at various governance levels, including in the UN system, with varying degrees of success (Sohn and others 2005, UNEP 2005e). A third strategy is to ensure that planning and governance processes include all relevant stakeholders from various sectors, as in successful integrated coastal zone management (see Chapter 4). A fourth strategy is to address integration between environment and other sectors, using economic valuation, which raises the parity of natural capital in comparison with other types of capital (see Chapter 1).

The integration of longer time horizons in governance is an even larger challenge given that decision making in governments and other sectors of society tends to be biased towards much shorter time horizons than sustainable development and the well-being of future generations require (Meadows and others 2004). Strategies that change the time horizon of decision-makers should be further explored. Such strategies can include: setting clear long-term goals and intermittent targets, extending the time horizons considered in formal planning, developing indicators and

accounting measures that illustrate intergenerational impacts, and the institutionalization of long-term liability from harmful activities. These strategies are unlikely to be implemented, however, unless people across societies expand their time horizons for development.

Improving health

The well-being of present and future generations is threatened by environmental change and social problems, such as poverty and inequity, which are contributing to environmental degradation. Preventive or proactive solutions for many contemporary health problems need to address the links among environment, health and other factors that determine well-being. Opportunities include better integration of environment and health strategies, economic valuation, targeting the most vulnerable, education and awareness, and the integration of environment and health into economic and development sectors.

Measures to ensure ecological sustainability to safeguard ecosystem services will benefit health, so these are important in the long-term. The emphasis on environmental factors has been a central part of the public health tradition. In recent years, several international policies made provisions for improved consideration of health in development. Global initiatives include the World Health Organization's 2005 recommendations for health impact assessment. At the regional level, the Strategic Environmental Assessment Protocol (1991) to the UNECE Convention on Environmental Impact Assessment emphasizes consideration of human health. More effective impact assessment procedures are needed in both developed and developing countries.

Economic valuation can help ensure that environment and health impacts are given adequate consideration in policy. An integrated economic analysis of such impacts can capture the hidden costs and benefits of policy options, as well as the synergies and institutional economies of scale that may be achieved through complementary policies that support sustainable development.

In most countries, mainstreaming of environment and health considerations into all government sectors and economic endeavours remains a challenge (Schütz and others (in press)). Policies and practices regarding health, environment, infrastructure and

Millions of people continue to be displaced and to be negatively affected by conflict, which reduces societal capacity to adapt to environmental change, while making sustained environmental management difficult.

Credit: UN Photo Library

economic development should be considered in an integrated manner (UNEP and others 2004). As environmental pollutants affect health through a variety of pathways, environmental monitoring and epidemiological surveillance systems should be strengthened. Health indicators and strategies are needed for specific groups at risk, such as women and children, the elderly, the disabled and the poor (WHO and UNEP 2004).

It is important to raise awareness not only in the health sector, but also in sectors such as energy, transport, land-use development, industry and agriculture, through information on the likely health consequences of decisions. Not only health professionals, but also all other stakeholders need the means to evaluate and influence policies that have impacts on health. A better understanding of the dynamic linkages between ecosystems and public health is leading to new and diverse opportunities for interventions early in processes that could become direct threats to public health (Aron and others 2001). Building awareness about environment and health problems, tools and policy options requires sustained and comprehensive communication strategies.

Resolving conflict through environmental cooperation

Despite the decrease in civil wars globally in recent years, millions of people continue to be displaced and negatively affected by violent conflict. Armed conflict often, but not always, causes heavy damage to the environment. It reduces societal capacity to adapt to global change, while making sustained environmental management difficult. Reducing violent conflict, whether related to natural resources or not, would reduce a major source of vulnerability, and

would better support human well-being in many parts of the world. Environmental cooperation offers several opportunities for achieving these ends.

Policy tools aimed at identifying the contribution of the environment to violent conflict and breaking those links would help redress key stresses. Developing and deploying such tools requires collaboration across a range of environment, development, economic, and foreign policy institutions, including the UN agencies. Such collaboration recognizes interlinkages across the biophysical components of the environment as well governance regimes (also see Chapter 8). Environmental assessment and early warning activities by UNEP and other stakeholders can play an active role in collecting and disseminating lessons learned. This may support the implementation of the UN Secretary-General's call during the UN General Assembly in 2006 for integrating environmental considerations in conflict prevention strategies.

Environmental cooperation has historically had two main areas of focus. At the international level, emphasis has been on multilateral treaties aimed at mitigating the effects of global change. At the sub-regional level, cooperation has focused on equitable sharing of natural resources, such as regional seas (Blum 2002, VanDeveer 2002) and shared water resources (Lopez 2005, Swain 2002, Weinthal 2002), as well as on improving conservation, through transboundary conservation areas (also known as transfrontier parks), to support integration and development-related activities such as tourism (Ali 2005, Sandwith and Besançon 2005, Swatuk 2002). Environmental cooperation – for conflict avoidance and peacemaking – could be employed across all levels of political organization.

In the rush to pursue policy interventions to sever the links between environment and conflict, analysts and practitioners alike have neglected the prospect for building upon environmental interdependencies to achieve confidence building, cooperation and, perhaps, peace (Conca and Dabelko 2002, Conca and others 2005). Environmental peacemaking is a strategy for using environmental cooperation to reduce tensions by building trust and confidence between parties in disputes. Environmental peacemaking opportunities will remain untested and underdeveloped until more systematic policy attempts are made to achieve these windfalls in a larger number of cases across resource types and across political levels.

Pursuing environmental pathways to confidence building would capitalize on environmental interdependence, and the need for long-term, iterated environmental cooperation to reduce conflict-induced vulnerability and improve human-well-being. Such policy interventions could:

- help prevent conflict among states and parties;
- provide an environmental lifeline for dialogue during times of conflict;
- help end conflicts with environmental dimensions; and
- help restart economic, agricultural and environmental activities in post-conflict settings.

Not all environmental cooperation lowers vulnerability and increases equity. Systematic assessment of experiences can increase opportunities. Comparing lessons learned across environmental peacemaking cases helps identify environmental management approaches that instigate, rather than ameliorate conflict, such as early examples of transboundary peace parks that neglected wide consultation with local peoples (Swatuk 2002). The ultimate goal of pursuing environmental peacemaking opportunities is reduction of vulnerability and assaults on human well-being created by the still-numerous local, national, and regional conflicts.

Pursuing environmental peacemaking opportunities will require focusing on local, national and regional institutional settings, rather than the historical emphasis on multilateral environmental agreements. Trying to capture these environmental and conflict prevention benefits requires considerable capacity building among stakeholders, including public and private interests in the conflict, as well as facilitators, such as bilateral donors or UN entities.

Strengthening local rights

The fast-paced changes of social and political values create challenges for developing effective responses that address human vulnerability and well-being, while ensuring complementarity among priorities. Strengthening local rights can offer opportunities for ensuring that local and national conservation and development priorities are recognized at higher levels of decision making.

Globalization has resulted in a growing emphasis on free exchange of commodities and ideas, and individual ownership and rights. In some circumstances, this may not support national or regional development goals (Round and Whalley 2004, Newell and Mackenzie 2004). Changing values associated with gender, traditional institutions, and democracy and accountability make the management of environmental resources extremely complex, and present challenges for institutional development. For example, the authority and right of both the state and traditional institutions to manage are increasingly contested. This is evident in conflicts around conservation areas (Hulme and Murphree 2001), water (Bruns and Meizin-Dick 2000, Wolf and others 2003) and forests (Edmonds and Wollenberg 2003). Such conflicts often have negative consequences for conservation and livelihoods, and may also have regional implications, where resources are shared.

Mediating these different interests and perspectives requires responses at the national, regional and global levels. Developing more inclusive institutions that recognize the rights and values of local natural resource users can be an effective response, and can facilitate the inclusion of local concerns into governance processes at a higher level (Cornwall and Gaventa 2001). This can also lead to better information sharing, and more equitable distribution of financial and other resources (Edmonds and Wollenberg 2003, Leach and others 2002). Inclusive processes can reduce the tension between local values and rights, and those held by state institutions (Paré and others 2002). Making these approaches effective requires investing in capacity building. Scaling these approaches up to the national or regional level can be appropriate, particularly where resource use has implications for users elsewhere, as in the case of water management (Mohamed-Katerere and van der Zaag 2003). Recognizing existing local institutions, including common property institutions, instead of creating new institutions, may be environmentally and socially beneficial, especially where they have a high degree of local legitimacy.

Building better links between local aspirations and the strategies and policies adopted at the global level is more challenging. It is constrained by international law and governance, but is not impossible (Mehta and la Cour Madsen 2004). Building negotiating capacity can be an important strategy for increasing the development focus of international governance systems (Page 2004). In some sectors, regional cooperation has proved effective in creating synergies between global governance and development objectives.

Promoting freer and fairer trade

Trade has far-reaching effects on livelihoods, well-being and conservation. Freer and fairer trade can be a useful tool for promoting growth and reducing poverty (Anderson 2004, Hertel and Winters 2006), improving resilience through transfer of food and technology (Barnett 2003), and improving governance.

Environmental and equity issues should be at the centre of global trade systems (DfID 2002). This is particularly important to ensure that poor people are not taken advantage of when it comes to trade, especially in products, such as hazardous materials that threaten well-being. The trade regime, particularly in agriculture and textiles, is characterized largely by preferential trade agreements (PTAs), bilateral agreements and quotas. High-income countries negotiate bilateral PTAs with poor countries, but such agreements cause more harm than good (Krugman 2003, Hertel and Winters 2006).

Poor countries, which have abundant labour, are expected to gain from access to larger markets' elsewhere, and high-income countries should ensure them such market access. Since small countries have smaller internal markets, lowering trade barriers would provide them with opportunities to exploit economies of scale, so that the poor can garner employment and better wages. Most models show the liberalization of trade under the current Doha round in the World Trade Organization is expected to reduce poverty, particularly if developing countries adjust their policies accordingly (Bhagwati 2004).

Trade facilitates learning-by-doing, which can drive higher productivity and industrialization (Leamer and others 1999). Contact between industrialized and developing countries can be an effective vehicle for diffusion of best practices, through the transfer of capital and knowledge. As poor countries, particularly primary commodity exporters, are vulnerable to price shocks and other market failures, diversification is a good option for reducing vulnerability (UNCTAD 2004), and may contribute to sustainable natural resource use.

Higher levels of income, sophisticated markets and the increased power of non-state actors may enhance the prospects for democracy and liberty (Wei 2000, Anderson 2004). Since trade requires large amounts of arm's-length transactions, better institutions are required for it to work smoothly (Greif 1992). Trade may not only raise incomes, but also indirectly and directly promotes better international governance, societal welfare (Birdsall and Lawrence 1999), and international and civil peace, which reinforces and is reinforced by prosperity (Barbieri and Reuveny 2005, De Soysa 2002a, De Soysa 2002b, Russett and Oneal 2000, Schneider and others 2003, Weede 2004).

Trade, like almost all other economic activity, creates winners and losers, and carries externalities. For some, adjustment costs of increased competition can be high (see section on export and import of vulnerability). These problems could be addressed by compensating losers, and encouraging increased adaptation through better public investment in education and infrastructure (Garrett 1998, Rodrik 1996). Trade contributes most to increasing incomes when combined with good governance (Borrmann and others 2006). Good governance, local capacities to regulate trade, and the regulation of industry in ways that encourage the adoption of best practices all help mitigate externalities, including those stemming from disposal of hazardous waste and pollution from increased consumption.

Securing access to and maintaining natural resource assets

For many people in developing countries as well as indigenous peoples, farmers and fishers in developed countries, secure entitlements to productive assets, such as land and water, are central to ensuring sustainable livelihoods (WRI 2005, Dobie 2001). Continued natural resource availability and quality, involving good conservation practices, is essential for the livelihoods of many in developing countries. Existing policies often compromise this. Strengthening access regimes can offer opportunities for poverty

Continued natural resource availability and quality, involving good conservation practices, is essential for the livelihoods of many in developing countries.

Credit: Audrey Ringler

eradication, as well as improving conservation and long-term sustainability. This national level action can be important for attaining globally agreed objectives, such as those in the MDGs, the CBD and UNCCD.

Secure entitlement refers to conditions under which users are able to plan and manage effectively. Secure access to natural resources can be an important stepping stone out of poverty, as it provides additional household wealth, which may support investments in health and education (WRI 2005, Pearce 2005, Chambers 1995). Moreover, it may contribute to better natural resource management by supporting long-term vision that keep future generations and options in mind, and may encourage investment (Hulme and Murphree 2001, Dobie 2001, UNCCD 2005). Specifically addressing women's tenure rights is vital, as they play key roles in managing natural resources, and are particularly affected by environmental degradation (Brown and Lapuyade 2001). Intergovernmental initiatives, such as dam development, should not undercut local resource rights by shifting responsibility from the local to the national or regional level (Mohamed-Katerere 2001, WCD 2000). To be effective, secure access rights may need to be complemented by addressing other barriers to sustainable and productive use, such as global trade regimes, insufficient access to capital and information, inadequate capacity and lack of technology. Valuation strategies, including payment for environmental services, can help ensure greater returns for local resource managers. Ensuring access to credit for small farmers and those directly reliant on ecosystem services is extremely important. Practice shows that finance schemes that especially target women can have higher than usual payoffs. Credit schemes, such as the Grameen banks in Bangladesh, can be designed to compensate those who ensure that environmental services are maintained.

Improved local authority over natural resources can help diversify livelihood options, reducing pressures on resources that are under threat (Hulme and Murphree 2001, Edmonds and Wollenberg 2003). Devolution of authority is one such mechanism (Sarin 2003). Despite a growing trend towards decentralization and devolution since the 1980s, and a broad policy commitment to give users greater authority, the institutional reform required to improve livelihoods is often lacking (Jeffrey and Sunder 2000). Devolution needs to be complemented by capacity building and empowerment initiatives, improved tenure, and better trade and value-adding options.

Building and bridging knowledge to enhance coping capacity

The roles of knowledge, information and education in reducing vulnerability converge around the learning process. The strengthening of learning processes for three specific objectives emerges as a key strategy to increasing coping capacity in a rapidly changing and complex environment.

Building knowledge about the environmental risk that threatens well-being, both among vulnerable communities themselves and among decision-makers at higher levels, is important. This involves both improved monitoring and assessment of the environmental, social and health-related aspects of pollution. It also involves mechanisms such as early warning systems (EWS) and indicators (for example, the Environmental Vulnerability Index) (see Gowrie 2003) for communicating and disseminating information on environmental change. These systems should be integrated into mainstream development. One tool that has proved useful in this regard is poverty mapping (see Figure 7.26). Poverty maps are spatial representations of poverty assessments. Poverty maps also allow easy comparison of indicators of poverty or well-being with data from other assessments, such as availability and condition of natural resources. This can assist decision-makers in the targeting and implementation of development projects, and the communication of information to a wide range of stakeholders (Poverty Mapping 2007). The map in Figure 7.26 shows the amount of resources needed to raise the population in each area to lift the poor out of poverty. It shows the uneven distribution of poverty density in Kenya. Most of the administrative areas in Kenya's arid and semi-arid lands require less than 4 000 Kenya Shillings (US$ 57 at US$1 = Ksh70) per square kilometre per month as a result of the low density of people. In contrast, at least 15 times that amount is needed in the densely populated areas west of Lake Victoria.

Bridging knowledge for better decision making is also key. This includes vulnerable communities learning about and from the national and global science advisory and decision making processes, and learning to raise their voices in these arenas, as illustrated in Box 7.17. At the same time, the scientists and decision-makers should learn to listen to and to talk with these communities, and consider their unique, specialized knowledge that centres on human-environment relationships and the use of natural resources (see for example Dahl 1989), even if it is not cast in the language of science.

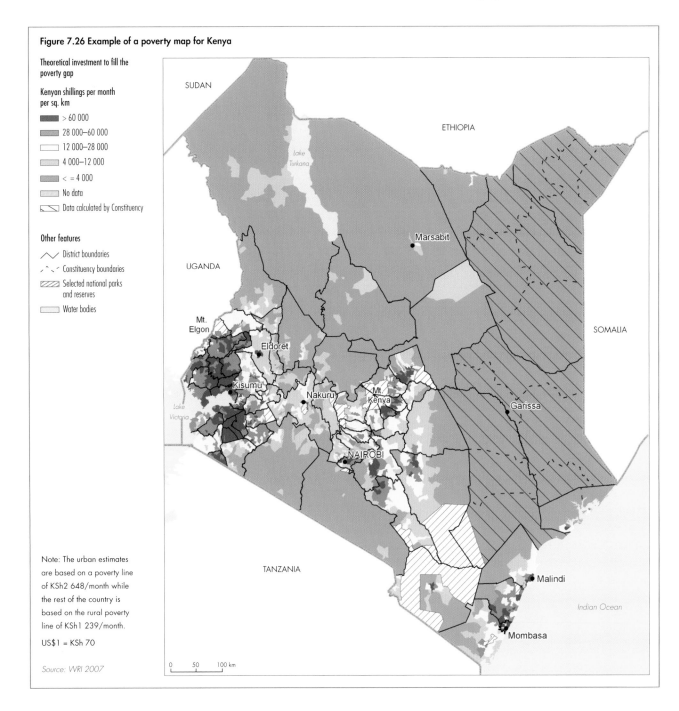

Figure 7.26 Example of a poverty map for Kenya

Theoretical investment to fill the poverty gap

Kenyan shillings per month per sq. km

- > 60 000
- 28 000–60 000
- 12 000–28 000
- 4 000–12 000
- < = 4 000
- No data
- Data calculated by Constituency

Other features

- District boundaries
- Constituency boundaries
- Selected national parks and reserves
- Water bodies

Note: The urban estimates are based on a poverty line of KSh2 648/month while the rest of the country is based on the rural poverty line of KSh1 239/month.

US$1 = KSh 70

Source: WRI 2007

The most vulnerable should learn competencies and skills that enable them to adapt and cope with risks. The foundation for this and the learning processes above lies in a good basic education, as set out in MDG 2. This increases the ability both to understand information from public awareness and early warning campaigns about specific sources of vulnerability and to develop coping and adaptation strategies. For example, it was the poorest and least educated groups who did not heed the evacuation warnings for Hurricane Katrina in 2005 (Cutter and others 2006). Educating the most vulnerable groups improves their coping capacities, and is also important for equity reasons. For example, the education of girls is one of the key means to break the intergenerational cycle of poverty. It is strongly associated with healthier children and families (UN Millennium Project 2005), and more sustainable environmental management.

Investing in technology for adaptation

Science, technology, and traditional and indigenous knowledge are important resources for reducing vulnerability. Policies that facilitate the development, application and transfer of technology to vulnerable communities and areas can improve access to basic materials, enhance risk assessment practices and EWS, and foster communication and participation. Policies should support technologies that ensure equitable access and the safety of water, air and energy, and that provide transportation, housing and infrastructure. They should be socially acceptable in the local context. The opportunity lies in investing in a diversity of technologies, including small-scale technologies that allow decentralized solutions. Some types of technology can also be important resources for promoting social connectedness, stability and equality through democratization. Policies that facilitate communication, education and governance via information technologies, and that improve the status of underprivileged groups, are particularly valuable.

Developing countries stand to derive many benefits from technologies developed elsewhere, but they also face the greatest challenges in accessing these technologies and managing their risks. Commitments made in the Johannesburg Plan of Implementation (JPOI) remain largely unfulfilled. Computer and information and communications technologies, biotechnology, genetics and nanotechnology (UNDP 2001) remain unavailable to vast numbers of people in the developing world. Past experience has shown the importance of attending to the appropriateness of technology's multiple connections with broader society,

its fit (or lack thereof) in particular social, cultural and economic contexts, and its implications for gender. An important strategy to ensure this is to invest much more in capacity building in the countries for technology innovation and production. See Figure 7.6, which illustrates in global context how big a leap many countries need to make. The UN Task Force on Science, Technology and Innovation made a number of recommendations including: focusing on platform technologies, existing technologies with broad economic impacts (for example biotechnology, nanotechnology, and information and communication technology); providing adequate infrastructure services as a foundation for technology; investing in science and technology education; and promoting technology-based business activities (UNMP 2005).

Building a culture of responsibility

The export and import of vulnerability is a recurring feature of the seven archetypes, meaning that many people – individually and collectively – contribute, often inadvertently, to the suffering of others while improving their own well-being. In this context, vulnerable communities need support to cope and adapt, so there is a need to build a stronger culture of "responsibility to act." Educating people about how their production and consumption patterns export vulnerability to other areas, continents and generations, and how this affects the prospects for living together at local scales, can contribute to a culture of responsibility. UNESCO's Education for All emphasizes the need to expand the view of education to include learning "life skills," such as learning "to live together" and learning "to be" (UNESCO 2005).

However, the chain of interactive drivers is far too complex to allow individual and collective actors to be aware of their own contributing roles and to feel more responsible to act (Karlsson 2007). In addition, the institutional frameworks for addressing legal responsibilities to protect the global commons are often weak, particularly when issues cross international borders and happen over different time frames. A response strategy is needed where a culture of responsibility is based more on global solidarity for present and future generations as a way of integrating neighbourhood values with global solidarity (Mertens 2005). Such solidarity can be actively nurtured through, for example, education (Dubois and Trabelsi 2007), processes of cooperative interaction (Tasioulas 2005), or the design of institutions that strengthen cosmopolitan aspirations and commitments (Tan 2005).

Education for the purpose of learning to care for and feel empathy for neighbours, and through this build a culture of responsibility to act can be readily integrated into the overall strategy for both formal and informal education. Enabling learners to participate directly in environmental problem solving is one effective way to enhance conservation behaviour (Monroe 2003). Examples of teaching environmentally-relevant life skills include the education initiatives related to the Earth Charter and various programmes on global and world citizenship and human rights (Earth Charter Initiative Secretariat 2005).

Building institutions for equity

There is very little equity or justice in who is vulnerable to environmental change. The poor and marginalized are almost always hit hardest by the degrading environment (Harper and Rajan 2004, Stephens 1996).

Poor governance, social exclusion and powerlessness limit the opportunities poor people have to participate in the decision making related to a country's resources and environment and how these have an impact on their well-being (Cornwall and Gaventa 2001). Improved governance and tenure regimes may not work for the poorest people if the opportunities for their participation are not specifically strengthened. Improving opportunities for participation in governance and planning processes at local and higher levels of governance can help strengthen their coping capacity. Box 7.17 gives an example of a recent initiative of Arctic indigenous communities and SIDS to combine their voices in the face of climate change.

The UN Conference on Environment and Development (Rio conference) provided the basic institutional change for increasing participation in environment-related decision making. However, having a voice without being listened to and having an impact on outcome can lead to greater estrangement. Weaknesses in this aspect is a recurring complaint, for example, in the multistakeholder dialogues at the global level (IISD 2002, Hiblin and others 2002, Consensus Building Institute 2002). Existing responses need to be strengthened, and active strategies to empower the most vulnerable could be developed by, for example, improving access to relevant environmental information, as provided for in Principle 10 of the Rio Declaration. This has already been implemented in many countries (Petkova and others 2002, UNECE 2005). Capacity building is also essential.

Putting a strong focus on the equity aspects of the outcome of governance is another essential aspect of enhancing coping capacity and the legitimacy of governance. Equity-centred strategies involve identifying the most vulnerable groups and communities, assessing the impacts of suggested policies first and foremost on these groups, and taking measures to improve equity in access to resources, capital and knowledge.

Building capacity for implementation

"Implementation failure" is common. There are many elaborate regional and global level multilateral agreements and action plans that have not been successfully implemented at the national level. The reasons behind the implementation failure are complex, and there are no simple solutions. Addressing this requires a multilevel approach. Three important opportunities can be identified: improving funding, investing in capacity, and developing effective monitoring and evaluation of existing plans and policies. International partnership is critical to success.

Increased financial commitment is essential to promote adaptation activities, increase human capability, support the implementation of MEAs and stimulate development. In developing countries, where financial resources are often constrained, creating better synergies between environment and development objectives is important. For example, there could be more interlinked health-environment strategies or poverty-environment initiatives (Kulindwa and others 2006). The incorporation of environment into Poverty Reduction Strategy Papers

---

Box 7.17 Many Strong Voices – building bonds

Many Strong Voices is a project, launched at the 2005 Conference of the Parties to the UN Framework Convention on Climate Change, which aims to build strategies for climate change awareness raising and adaptation among the vulnerable in the Arctic and Small Island Developing States (SIDS).

The purpose of this project is to link the vulnerable in the Arctic and SIDS to stimulate a dialogue that will:
- support regional initiatives in education, training and public awareness raising;
- develop partnerships that will allow people in these areas to exchange information about efforts underway to raise awareness about, and to develop adaptation strategies for climate change;
- support efforts of local inhabitants so they will be able to influence the debate on, and participate in, decisions on adaptation; and
- facilitate regional efforts to influence global efforts on adaptation and mitigation.

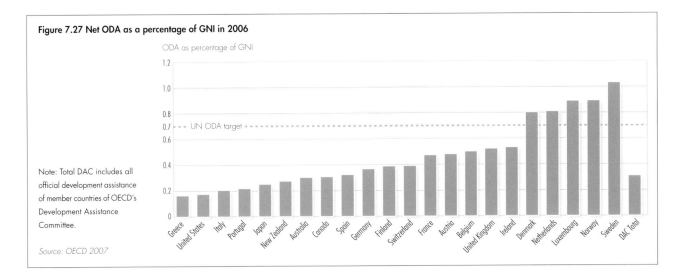

Figure 7.27 Net ODA as a percentage of GNI in 2006

ODA as percentage of GNI

Note: Total DAC includes all official development assistance of member countries of OECD's Development Assistance Committee.

Source: OECD 2007

is one opportunity that can be more effectively used (Bojö and Reddy 2003, WRI 2005).

Official development assistance (ODA) continues to lag behind agreed targets. At the 1992 Rio conference, most countries pledged to increase ODA towards the UN target of 0.7 per cent of GNI (Parish and Looi 1999). In 1993, the average level of ODA was 0.3 per cent of GNI (Brundtland 1995). Describing the international redistributive system as is "in shameful condition," Brundtland emphasized that "the cost of poverty, in human suffering, in the wasteful use of human resources, and in environmental degradation, has been grossly neglected" (Brundtland 1995). The 2002 Monterrey Consensus recommitted developed countries to meeting the UN target. Since then, there has been a steady increase in aid, and by 2004 average ODA was 0.42 per cent of GNI. However, only five countries have met the UN target and by 2006 the average was down to 0.3 per cent again (see Figure 7.27). The IMF's 15 richest member states have agreed to allocate at least 0.51 per cent of GNI by 2010, increasing this share to 0.7 per cent by 2015 (Gupta and others 2006).

Investing in capacity building and necessary technology support, as envisaged in the JPOI and the BSP, can enhance ability to develop and implement required measures. Targeting capacity building at the right level is essential. Improved land management might require local capacity building, whereas addressing illegal movement of hazardous waste will require capacity building of the relevant agencies. In some areas, such as biodiversity management, capacity of some developed and developing countries to develop and implement interlinked strategies is

lacking (CBD 2006). Pooling resources, sharing best practices and collaborating in joint capacity building at the regional level have been successful.

Improving monitoring and evaluation capacity also hinges on increased investment in capacity building, and appropriate institutional and governance development. In some situations, there is a need for stronger government institutions, as well as national and international laws to ensure that standards are abided by. Better institutional and governance mechanisms, including measures to ensure access to relevant information and the courts, are necessary to support people in safeguarding their interests.

CONCLUSION

The patterns of vulnerability to environmental and socio-economic changes that have been highlighted are not mutually exclusive, nor are they the only ones that exist within countries, in and across regions, and globally. They present an environment and development paradox for decision-makers at different levels: millions of people remain vulnerable to multiple and interacting pressures in a world of unprecedented wealth and technological breakthroughs. Addressing the challenges presented by the patterns of vulnerability will, however, contribute to overall human well-being and to meeting the MDGs. There is a range of strategic approaches, many of them not in the environmental policy domain that could be taken. At the same time, implementation of obligations already made in a wide range of policy domains, ranging from basic human rights to development aid, trade and to environment, would reduce vulnerability and increase human well-being.

References

ACIA (2004). *Impacts of a Warming Arctic: Arctic Climate Impact Assessment.* Cambridge University Press, Cambridge

ACIA (2005). *Arctic Climate Impact Assessment.* Arctic Council and the International Arctic Science Committee. Cambridge University Press, Cambridge http://www.acia.uaf.edu/pages/scientific.html (last accessed 27 June 2007)

Adger, N., Hughes, T.P., Folke, C., Carpenter, S.R. and Rockström, J. (2005). *Social-Ecological Resilience to Coastal Disasters.* In *Science* 309:1036-1039

Akindele, F. and Senaye, R. (eds.) (2004). *The Irony of "White Gold".* Transformation Resource Centre, Moraja http://www.trc.org.ls/publications/ (last accessed 14 June 2007)

Alcamo, J., Döll, P., Henrichs, T., Kaspar, F., Lehner, B., Rösch, T. and Siebert, S. (2003). Global estimation of water withdrawals and availability under current and "business as usual" conditions. In *Hydrological Science* 48 (3):339-348

Alexander, D. (1993). *Natural disasters.* Chapman and Hall, New York, NY

Ali, Saleem H. 2005. "Conservation and Conflict Resolution: Crossing the Policy Frontier." *Environmental Change and Security Program Report* (11):59-60

Allen-Diaz, B. (1996). *Rangelands in a changing climate: Impacts, adaptations, and mitigation.* Cambridge University Press, Cambridge

America's Wetland (2005). *Wetland Issues Exposed in Wake of Hurricane Katrina.* America's Wetland, Press Releases http://www.americaswetland.com/article.cfm?id=292&cateid=2& pageid=3&cid=16 (last accessed 27 April 2007)

Anderson, K. (2004). Subsidies and Trade Barriers. In *Global Crises, Global Solutions* (ed. B. Lomborg). Cambridge University Press, Cambridge

Appleton, J.D., Williams, T.M., Breward, N., Apostol, A., Miguel, J. and Miranda, C. (1999). Mercury contamination associated with artisinal gold-mining on the island of Mindanao, The Philippines. In *Sci. Total Environment* 228:95-109

ArcWorld ESRI (2002) *ESRI Data & Maps 2002.* CD-ROM

Aron J.L., Ellis J.H. and Hobbs B.F. (2001). Integrated Assessment. In *Ecosystem Change and Public Health. A Global Perspective,* Aron, J.L. and Patz, J.A. (eds.). The Johns Hopkins University Press, Baltimore and London

Amorin, M.I., Mergier, D., Bahia, M.O., Dubeau, H., Miranda, D., Lebel, J., Burbano, R.R. and Lucotte, M. (2000). Cytogenic damage related to low levels of methyl mercury contamination in the Brazilian Amazon. In *Ann. Acad. Bras. Cienc.* 72:497-507

Ayotte, P., Dewailly, E., Bruneau, S., Careau, H. and A. Vezina (1995). Arctic Air-Pollution and Human Health - What Effects Should Be Expected. In *Science of the Total Environment* 160/161:529-537

AHDR (2004). *Arctic Human Development Report.* Stefansson Arctic Institute, Akureyri

Auty, R. M., ed. (2001). *Resource Abundance and Economic Development.* UNU/WIDER studies in development economics. Oxford University Press, Oxford

Bächler, G., Böge, V., Klötzli, S., Libiszewski, S. and Spillmann, K. R. (1996). *Kriegsursache Umweltzerstörung: Ökologische Konflikte in der Dritten Welt und Wege ihrer friedlichen Bearbeitung. Volume 1.* Rüegger, Chur, Zürich

Baechler, G. (1999), Internationale und binnenstaatliche Konflikte um Wasser. *Zeitschrift für Friedenspolitik* 3:1-8

Bakhit, A.H. (1994). Mubrooka: a study in the food system of a squatter settlement in Omdurman, Sudan. In *Geojournal* 34 (3):263-268

Bankoff, G. (2001). Rendering the world unsafe: 'vulnerability' as Western discourse. In *Disasters* 21(1):19-35

Barbieri, K. and Reuveny, R. (2005). Economic Globalization and Civil War. In *Journal of Politics* 67 (4)

Barnett, J. (2003). Security and Climate Change. In *Global Environmental Change* 13(1):7-17

Barnett, J. and Adger, N. (2003). Climate Dangers and Atoll Countries. In *Climatic Change* 61(3):321-337

Basel Action Network (2002). *Exporting harm: the high tech trashing of Asia.* Basel Action Network, Seattle http://www.ban.org/E-waste/technotrashfinalcomp.pdf (last accessed 13 June 2007)

Basel Action Network (2006). *12 Human Rights/Green Groups Call for an Immediate Halt on Scrapping of 'Toxic Ships' Following Recent Findings of Death and Disease in India.* Toxic Trade News. http://www.ban.org/ban_news/2006/060914_immediate_halt.html (last accessed on 24 August 2007)

Benedetti, M., Lavarone, I., Combe, P. (2001). Cancer risk associated with residential proximity to industrial sites: A review. In *Arch. Environ. Health* 56:342-349

Berkes, F. (2002). Cross-Scale Institutional Linkages: Perspectives from the Bottom Up. In *The Drama of the Commons,* Ostrom, E., Dietz, T., Dolsak, N., Stern, P. C., Stonich S. and Weber, E.U. (eds.) National Academy Press, Washington, DC

Bhagwati, J. (2004). *In Defense of Globalization.* Oxford University Press, Oxford

Bijlsma, L., Ehler, C.N., Klein, R.J.T., Kulshrestha, S.M., McLean, R.F., Mimura, N., Nicholls, R.J., Nurse, L.A., Perez Nieto, H., Stakhiv, E.Z., Turner, R.K. and Warrick, R.A. (1996). Coastal Zones and Small islands. In *Impacts, Adaptations and Mitigation of Climate Change: Scientific Technical Analyses,* Watson, R.T., Zinyowera, M.C. and Moss, R.H. (eds). Cambridge University Press, Cambridge

Birdsall, N. and Lawrence, R. Z. (1999). Deep Integration and Trade Agreements: Good For Developing Countries? In *Global Public Goods: International Cooperation in the 21st Century,* Kaul, I. Grunberg, I. and Stern, M. (eds). Oxford University Press, Oxford

Birkmann, J. (ed.) (2006). *Measuring Vulnerability to Natural Hazards: towards disaster resilient societies.* United Nations University Press, Tokyo

Blacksmith Institute (2006). *The World's Worst Polluted Places. The Top Ten.* Blacksmith Institute, New York, NY http://www.blacksmithinstitute.org/get10.php (last accessed 27 April 2007)

Blaikie, P., Cannon T., Davis, I. and Wisner, B. (1994). *At Risk: Natural Hazards, People's Vulnerability and Disasters.* Routledge, London

Blum, Douglas W. (2002). Beyond Reciprocity: Governance and Cooperation around the Caspian Sea. In *Environmental Peacemaking,* Conca, K. and Dabelko, G.D. (eds). The Woodrow Wilson Center Press and the Johns Hopkins University Press, Washington, DC and Baltimore

Bojo, J. and Reddy, R.C. (2003). *Status and Evolution of Environmental Priorities in the Poverty Reduction Strategies. An Assessment of Fifty Poverty Reduction Strategy Papers.* Environment Economic Series Paper No. 93. The World Bank, Washington, DC

Borrmann, A., Busse M., Neuhaus S. (2006). Institutional Quality and the Gains from Trade. In *Kyklos* 59(3):345 - 368

Blumenthal, S. (2005). No-one Can Say They Didn't See It Coming. In *Spiegel International* 31 August 2005

Bohle, H. G., Downing, T. E. and Watts, M. (1994). Climate Change and Social Vulnerability: The Sociology and Geography of Food Insecurity. In *Global Environmental Change:* 4 (1):37-48

Brock, K. (1999). *It's not only wealth that matters – it's peace of mind too: a review of participatory work on poverty and ill-being.* Voices of the Poor Study paper. The World Bank, Washington, DC

Brosio, G. (2000). *Decentralization in Africa.* Africa Department of the International Monetary Fund, Washington, DC

Brown, K. and Lapuyade S. (2001). A livelihood from the forest: Gendered visions of social, economic and environmental change in southern Cameroon. In *Journal of International Development* 13:1131-1149.

Brundtland, G.H. (1995). *A Shameful condition. Progress of Nations.* United Nations Children's Fund, New York, NY http://www.unicef.org/pon95/aid-0002.html (last accessed 27 April 2007)

Bruns, B.R. and Meizin-Dick, R. (2000). Negotiating Water Rights in Contexts of Legal Pluralism: Priorities for Research and Action. In *Negotiating Water Rights,* Bruns, B.R. and Meizin-Dick, R. (eds.). Intermediate Technologies Publications, London

Bulatao-Jayme, Fr.J., Villavieja, G.M., Domdom, A.C and Jimenez, D.C. (1982). Poor urban diets: causes and feasible changes. In *Geoj. Suppl. Iss.* 4:3-82

Bulte, E.H., Damania, R. and Deacon, R.T. (2005). Resource Intensity, Institutions, and Development. In *World Development* 33 (7):1029–1044

Burholt, V. and Windle, G. (2006). Keeping warm? Self-reported housing and home energy efficiency factors impacting on older people heating homes in North Wales. In *Energy Policy* (34):1198-1208

Calderon, J., Navarro, M.E., Jiminez-Capdeville, M.E., Santos-Diaz, MA., Golden, A., Rodriguez-Leyva, I., Bonja-Aburto, V.H. and Diaz-Barriga, F. (2001). Exposure to arsenic and lead and neuripsychological development in Mexican children. In *Environ. Res.* 85:69-76

Calcagno, A.T (2004). *Effective environmental assessment: Best practice in the planning cycle.* Comprehensive options assessment http://www.un.org/esa/sustdev/sdissues/energy/op/hydro_calcagno_environ_assessment.pdf (last accessed 27 April 2007)

Campling, L. and Rosalie, M. (2006). Sustaining social development in a small island developing state? The case of the Seychelles. In *Sustainable Development* 14(2):115-125

CBD (2006). *Report of the Eighth Meeting of the Parties to the Convention on Biological Diversity.* UNEP/CBD/COP/8/31. Convention on Biological Diversity http://www.cbd.int/doc/meeting.aspx?mtg=cop-08 (last accessed 15 June 2007)

Chambers, R. (1989). Vulnerability, coping and policy. Institute of Development Studies, University of Sussex. In *IDS Bulletin* 20:1-7

Chambers, R. (1995). Poverty and livelihoods: Whose Reality counts. In *Environment and Urbanization* 1(7):173-204

Chambers, R. and G. R. Conway (1992). *Sustainable rural livelihoods: Practical concepts for the 21st century.* Discussion Paper 296, Institute of Development Studies, Sussex

Charveriat, C. (2000). *Natural Disasters in Latin America and the Caribbean: An Overview of Risk. Inter-American Development Bank* (IADB), Washington, DC

Chen, S. and Ravallion,M. (2004) How have the World's Poorest Fared since the early 1980s? In *World Bank Research Observer,* v. 19 (2):141-70

Chronic Poverty Research Centre (2005). *Chronic Poverty Report 2004-05.* Chronic Poverty Research Centre, Oxford

CIESIN (2006). *Global distribution of Poverty. Infant Mortality Rates.* http://sedac.ciesin.org/povmap/ds_global.jsp (last accessed 10 May 2007)

Cinner, J.E., Marnane, M.J., McClanahan, T.R., Clark, T.H. and Ben, J. (2005). Trade, Tenure, and Tradition: Influence of Socio-cultural Factors on Resource Use in Melanesia. In *Conservation Biology* 19(5):1469-1477

CMH (2001). *Macroeconomics and Health: Investing in Health for Economic Development.* Commission on Macroeconomics and Health, World Health Organization, Geneva

Colborn, T., Dumanoski, D. and Myers, J. P. (1996). *Our Stolen Future.* Dutton, New York, NY

Collier, P., Elliot, L., Hegre, H., Hoeffler, A., Reynal-Querol M. and Sambanis, N. (2003). *Breaking the Conflict Trap: Civil War and Development Policy.* Oxford University Press, Oxford

Conca, K. and Dabelko, G.D. (eds.) (2002). *Environmental Peacemaking.* The Woodrow Wilson Center Press and the Johns Hopkins University Press, Washington, DC and Baltimore

Conca, K., Carius, A. and Dabelko, G.D. (2005). Building peace through environmental cooperation. In *State of the World 2005: Redefining Global Security.* Worldwatch Institute, Norton, New York, NY

Consensus Building Institute (2002). *Multi-stakeholder Dialogues: Learning From the UNCSD Experience. Background Paper No.4.* United Nations Department of Economic and Social Affairs. Commission on Sustainable Development Acting as the Preparatory Committee for the World Summit on Sustainable Development. Third Preparatory Session. 25 March - 5 April 2002, New York, NY

CSMWG (1995). *Definition of a Contaminated Site.* Contaminated Sites Management Working Group of the Canadian Government. http://www.ec.gc.ca/etad/csmwg/en/index_e.htm (last accessed 27 April 2007)

Cornwall, A. and Gaventa, J. (2001) *From Users and Choosers to Makers and Shapers: Repositioning Participation in Social Policy,* IDS Working Paper 127, Institute for Development Studies, Brighton

Crowder, L.B., Osherenko, G., Young, O., Airame, S., Norse, E.A., Baron, N., Day, J.C., Douvere, F., Ehler, C.N., Halpern, B.S., Langdon, S.J., McLeod, K.L., Ogden, J. C., Peach, R.E., Rosenberg, A.A. and Wilson, J.A. (2006). Resolving mismatches in U.S. Ocean Governance. In *Science* 313:617-618

CSD (2006). *14th Session of the Commission on Sustainable Development.* Chairman's Summary. http://www.un.org/esa/sustdev/csd/csd14/documents/chairSummaryPartI.pdf (last accessed 27 April 2007)

Cuny, 1983 and Cuny, F.C. (1983). *Disasters and development.* Oxford University Press, New York, NY

Cutter, S.L. (1995). The forgotten casualties: women, children and environmental change. In *Global Environmental Change* 5(3):181-194

Cutter, S.L (2005). The Geography of Social Vulnerability: Race, Class, and Catastrophe. In Understanding Katrina: Perspectives from the Social Sciences. Social Science Research Council http://understandingkatrina.ssrc.org/Cutter/ (last accessed 27 April 2007)

Cutter, S.L., Emrich, C.T., Mitchell, J.T., Boruff, B.J., Gall, M., Schmidtlein, M.C., Burton, C.G. and Melton, G. (2006). The Long Road Home: Race, Class, and Recovery from Hurricane Katrina. In *Environment* 48(2):8-20

Dahl, A. (1989). Traditional environmental knowledge and resource management in New Caledonia. In *Traditional Ecological Knowledge: a Collection of Essays* R.E. Johannes (ed.). IUCN, Gland and Cambridge http://islands.unep.ch/dtradknc.htm (last accessed 13 June 2007)

De Grauwe, P. and Camerman, F. (2003). How Big Are the Big Multinational Companies? In *World Economics* 4(2):23–37

DfID (2002). *Trade and Poverty.* Background Briefing Trade Matters Series. UK Depratment for International Development. http://www.dfid.gov.uk/pubs/files/bg-briefing-tradeandpoverty.pdf (last accessed 13 June 2007)

De Soysa, I. (2002a). Ecoviolence: Shrinking Pie or Honey-Pot? In *Global Environmental Politics* 2(4):1-34

De Soysa, I. (2002b). Paradise is a Bazaar? Greed, Creed, and Governance in Civil War, 1989–1999. *Journal of Peace Research* 39(4):395–416

De Soysa, I. (2005). Filthy Rich, Not Dirt Poor! How Nature Nurtures Civil Violence. In *Handbook of Global Environmental Politics,* P. Dauvergne (ed). Edward Elgar, Cheltenham

Diamond, J. (2004). *Collapse: How Societies Choose to Fail or Succeed*. Penguin Books, London

Diehl, P.F. and Gleditsch, N.P. (eds.) (2001). *Environmental Conflict*. Westview Press, Boulder, CO

Dietz, A.J., Ruben, R. and Verhagen, A. (2004). *The Impact of Climate Change on Drylands*. Kluwer Academic Publishers, Dordrecht

Dilley, M., Chen, R., Deichmann, U., Lerner-Lam, A.L. and Arnold, M. (with Agwe, J., Buy, P., Kjekstad, O., Lyon, B. and Yetman, G.) (2005). *Natural Disaster Hotspots: A Global Risk Analysis. Synthesis Report*. The World Bank, Washington, DC and Columbia University, New York, NY

Dobie P. (2001). *Poverty and the Drylands. The Global Drylands Development Partnership*. United Nations Development Programme, Nairobi

Dollar, D. (2004). *Globalization, poverty, and inequality since 1980*. Policy Research Working Paper Series 3333. The World Bank, Washington, DC

Dollar, D. and Kraay, A. (2000). *Trade, Growth, and Poverty*. Development Research Group, The World Bank, Washington, DC

Douglas, B.C. and Peltier, W.R. (2002). The puzzle of global sea-level rise. In *Physics Today* 55:35–41

Douglas, C.H. (2006). Small island states and territories: sustainable development issues and strategies - challenges for changing islands in a changing world. In *Sustainable Development* 14(2):75-80

Downing, T.E. (Ed.) (2000). *Climate, Change and Risk*. Routledge, London

Downing, T. E. and Patwardhan, A. (2003). Technical Paper 3: Assessing Vulnerability for Climate Adaptation. In UNDP and GEF *Practitioner Guide, Adaptation Policy Frameworks for Climate Change: Developing Strategies, Policies and Measures*. Cambridge University Press, Cambridge

Dreze, J. and Sen, A. (1989). *Hunger and Public Action*. Clarendon Press, Oxford

DTI (2001). *Fuel Poverty*. UK Department of Trade and Industry, London http://www.dti.gov.uk/energy/fuel-poverty/index.html (last accessed 27 April 2007)

Dubois, J.-L. and Trabelsi, M. (2007). Social Sustainability in Pre- and Post-Conflict Situations: Capability Development of Appropriate Life-Skills. In *International Journal of Social Economics* 34

Earth Charter Initiative Secretariat (2005). *Bringing Sustainability into the Classroom. An Earth Charter Guidebook for Teachers*. The Earth Charter Initiative International Secretariat, Stockholm and San José http://www.earthcharter.org/resources/ (last accessed 27 April 2007)

European Commission (2001). *Towards a European strategy for the security of energy supply*. Green Paper. European Commission, Brussels

Eckley, N. and Selin, H. (2002). The Arctic Vulnerability Study and Environmental Pollutants: A Strategy for Future Research and Analysis. Paper presented at the *Second AMAP International Symposium on Environmental Pollution of the Arctic, Rovaniemi, Finland, 1-4 October, 2002*

Edmonds, D. and Wollenberg, E. (2003). Whose Devolution is it Anyway? Divergent Constructs, Interests and Capacities between the Poorest Forest Users and the States. In *Local Forest Management. The Impacts of Devolution Policies*, Edmonds, D. and Wollenberg, E. (eds). Earthscan, London

Emanuel, K.A. (1988). The Dependency of Hurricane Intensity on Climate. In *Nature* 326:483 – 485

EM-DAT. *The International Disaster Database*. http://www.em-dat.net/ (last accessed 13 June 2007)

ESMAP (2005). *ESMAP Annual Report 2005*. Energy Sector Management Assistance Program. International Bank for Reconstruction and Development, Washington, DC

Eurostat and IFF (2004). *Economy-wide Material Flow Accounts and Indicators of Resource Use for the EU-15: 1970-2001, Series B*. Prepared by Weisz, H., Amann, Ch., Eisenmenger, N. and Krausmann, F. Eurostat, Luxembourg

FAO (1995). *Prevention of accumulation of obsolete pesticide stocks*. Food and Agriculture Organization of the United Nations, Rome http://www.fao.org/docrep/v7460e/V7460e00.htm (last accessed 27 April 2007)

FAO (1999). *Trade issues facing small island developing states*. Food and Agriculture Organization of the United Nations, Rome http://www.fao.org/docrep/meeting/X1009E.htm (last accessed 27 April 2007)

FAO (2001). *Baseline study on the problem of obsolete pesticide stocks*. Food and Agriculture Organization of the United Nations, Rome

FAO (2002). *Stockpiles of obsolete pesticides in Africa higher than expected*. Food and Agriculture Organization, Rome http://www.fao.org/english/newsroom/news/2002/9109-en.html (last accessed 13 June 2007)

FAO (2003a). *Status and trends in mangrove area extent worldwide*. By Wilkie, M.L. and Fortuna, S. Forest Resources Assessment Working Paper No. 63. Forest Resources Division. Food and Agricultural Organization of the United Nations, Rome (Unpublished)

FAO (2003b). *The State of Food Insecurity in the World; monitoring progress towards the World Food Summit and Millennium Development Goals*. Food and Agriculture Organization of the United Nations, Rome

FAO (2004a). *Advance Funding for Emergency and Rehabilitation Activities*. 127th Session Council. Food and Agricultural Organization of the United Nations, Rome http://www.fao.org/docrep/meeting/008/j3631e.htm (last accessed 27 April 2007)

FAO (2004b). *FAO and SIDS: Challenges and Emerging Issues in Agriculture, Forestry and Fisheries*. Food and Agriculture Organization, Rome

FAO (2005a). *Mediterranean fisheries: as stocks decline, management improves*. Food and Agriculture Organization, Rome http://www.fao.org/newsroom/en/news/2005/105722/index.html (last accessed 21 June 2007)

FAO (2005b). *The state of food security in the world 2005; eradicating world hunger - key to achieving the Millennium Development Goals*. Food and Agricultural Organization of the United Nations, Rome

FAO (2006). *Progress towards the MDG target*. Food security statistics. Food and Agriculture Organization of the United Nations, Rome http://www.fao.org/es/ess/faostat/foodsecurity/FSMap/mdgmap_en.htm (last accessed 27 April 2007)

FAO, UNEP and WHO (2004). *Pesticide Poisoning: Information for Advocacy and Action*. United Nations Environment Programme, Geneva

Fischetti, M. (2001). Drowning New Orleans. In *Scientific American* 285(4):76-85

Flynn, J., Slovik, R. and Mertz, C.K. (1994). Gender, race and perception of environmental health risks, Oregon. In *Decision Research* March 16

Folke, C., Carpenter, S., Elmqvist, Th., Gunderson, L., Holling, C. S. and Walker, B. (2002). Resilience and Sustainable Development: Building Adaptive Capacity in a World of Transformations. In *Ambio* 31(5):437-440

Fordham, M.H. (1999). The intersection of gender and social class in disaster: balancing resilience and vulnerability. In *International Journal of Mass Emergencies and Disasters* 17 (1):15-37

Friedmann, J. (1992). *Empowerment: The politics of alternative development*. Blackwell Publishers, Cambridge, MA

GAEZ (2000). *Global Agro-Ecological Zones*. Food and Agricultural Organization of the United Nations and International Institute for Applied Systems Analysis, Rome http://www.fao.org/ag/agl/agll/gaez/index.htm (last accessed 13 June 2007)

Garrett G. (1998). Global Markets and National Politics: Collision Course or Virtuous Circle? In *International Organization* 52 (4):787–824

Georges, N.M. (2006). Solid Waste as an Indicator of Sustainable Development in Tortola, British Virgin Islands. In *Sustainable Development* 14:126-138

GEO Data Portal. *UNEP's online core database with national, sub-regional, regional and global statistics and maps, covering environmental and socio-economic data and indicators*. United Nations Environment Programme, Geneva http://www.unep.org/geo/data or http://geodata.grid.unep.ch (last accessed 1 June 2007)

GIWA (2006). *Challenges to International Waters; Regional Assessments in a Global Perspective*. United Nations Environment Programme, Nairobi http://www.giwa.net/publications/finalreport/ (last accessed 13 June 2007)

Gleditsch, N.P. (ed.) (1999). *Conflict and the Environment*. Kluwer, Dordrecht, Boston, London

Gleick, P. (1999). The Human Right to Water. In *Water Policy* 1(5):487-503

Goldsmith, E. and Hildyard, N. (1984). *The Social and Environmental Effects of Large Dams*. Sierra Club Books, San Francisco

Gordon, B., Mackay, R. and Rehfuess, E. (2004). *Inheriting the World. The Atlas of Children's Health and the Environment*. World Health Organization, Geneva

Goreux, L. and Macrae, J. (2003). *Reforming the Cotton Sector in Sub-Saharan Africa*. Africa Regional Working Paper Series 47. The World Bank, Washington, DC. http://www.worldbank.org/afr/wps/wp47.pdf (last accessed 27 April 2007)

Gowrie, M. N. (2003). Environmental vulnerability index for the island of Tobago, West Indies. In *Conservation Ecology* 7(2:11 http://www.consecol.org/vol7/iss2/art11/ (last accessed 27 April 2007)

Graham, Edward, M. (2000). *Fighting the Wrong Enemy: Antiglobal Activists and Multinational Enterprises*. Institute for International Economics, Washington, DC

Graham, T. and Idechong, N. (1998). Reconciling Customary and Constitutional Law - Managing Marine Resources in Palau, Micronesia. In *Ocean and Coastal Management* 40(2):143-164

Greif, A. (1992). Institutions and International Trade: Lessons from the Commercial Revolution (Historical Perspectives on the Economics of Trade). In *AEA Papers and Proceedings* 82(2):128-133

Grether, J.M. and De Melo, J. (2003). *Globalization and Dirty Industries: Do Pollution Havens Matter?* NBER Working Papers 9776, National Bureau of Economic Research, Cambridge, MA

Griffin, D.W., Kellogg, C.A and Shinn, E.A. (2001). Dust in the wind:Long range transport of dust in the atmosphere and its implications for global public and ecosystem health. In *Global Change & Human Health* 2(1)

Gupta, S.; Patillo, C. and Wagh, S. (2006). *Are Donor Countries Giving More or Less Aid?*. Working Paper WP/06/1. International Monetary Fund, Washington, DC http://www.imf.org/external/pubso/ft/wp/2006/wp0601.pdf (last accessed 27 April 2007)

Haavisto, P. (2005). Environmental impacts of war. In *State of the World 2005: Redefining Global Security* Worldwatch Institute, Norton, New York, NY

Harbom, L. and Wallensteen, P. (2007). Armed Conflict, 1989-2006. In *Journal of Peace Research* 44(5) http://www.ucdp.uu.se (last accessed 29 June 2007)

Harper, K. and Rajan, S.R. (2004). *International Environmental Justice: Building Natural Assets for the Poor*. Working Paper Series, 87. Political Economy Research Institute, http://www.peri.umass.edu/Publication.236+M53cb8b79b72.0.html (last accessed 13 June 2007)

Harrison, P. and Pearce, F. (2001). AAAS Atlas of Population and Environment.American Association for the Advancement of Science. University of California Press, California http://www.ourplanet.com/aaas/pages/about.html (last accessed 27 April 2007)

Haupt, F., Muller-Boker, U. (2005). Grounded research and practice - PAMS - A transdisciplinary program component of the NCCR North-South. In *Mountain Research and Development* 25(2):101-103

Hay, J.E., Mimura, N., Campbell, J., Fifita, S., Koshy, K., McLean, R.F., Nakalevu, T., Nunn, P. and de Wet, N. (2003). *Climate Variability and change and sea-level rise in the Pacific Islands Region: A resource book for policy and decision makers, educators and other stakeholders*. SPREP, Apia, Samoa

Henderson-Sellers, A., Zhang, H., Berz, G., Emanuel, K., Gray, W., Landsea, C., Holland, G., Lighthill, J., Shieh, S.-L., Webster, P. and McGuffie, K.(1998). Tropical Cyclones and Global Climate Change: A Post-IPCC Assessment. In *Bulletin of the American Meteorological Society* 79:19–38

Hertel, T.W. and Winters, A.L. (eds.) (2006). *Poverty and the WTO: Impacts of the Doha Development Agenda*. The World Bank, Washington, DC

Hess, J. and Frumkin, H. (2000). The International trade in toxic waste: The case of Sihanoukville, Cambodia. In *Int J Occup Environ Health* 6:263-76

Hewitt, K. (1983). *Interpretations of Calamity: from the viewpoint of human ecology*. Allen and Unwin, St Leonards, NSW

Hewitt, K. (1997). *Regions of Risk. A geographical introduction to disasters*. Addison Wesley Longman, Harlow, Essex

Hiblin, B., Dodds, F. and Middleton, T. (2002). Reflections on the First Week – Prep. Comm. II Progress Report. Outreach 2002, 4th February, 1-2

Higashimura, R. (2004). Fisheries in Atlantic Canada after the collapse of cod. *Proceedings Twelfth Biennial Conference of the International Institute of Fisheries Economics & Trade (IIFET), July 20-30, 2004, Tokyo*

Hild, C.M. (1995). The next step in assessing Arctic human health. In *The Science of the Total Environment* 160/161:559-569

Holling, C.S. (1973). Resilience and stability of ecological systems. In *Annual Review of Ecology and Systematics* 4:1-23

Holling, C.S. (2001). Understanding the Complexity of Economic, Ecological and Social Systems. In *Ecosystems* 4:390-405

Homer-Dixon, T.F. (1999). *Environment, Scarcity, and Violence*. Princeton University Press, Princeton, NJ

Hoegh-Guldberg, O., Hoegh-Guldberg, H., Stout, D., Cesar, H. and Timmerman, A. (2000). *Pacific in Peril: Biological, Economic and Social Impacts of Climate Change on Pacific Coral Reefs*. Greenpeace, Amsterdam

Huggins, C., Chenje, M. and Mohamed-Katerere, J.C. (2006). Environment for Peace and Regional Cooperation. In UNEP (2006). *Africa Environment Outlook 2. Our Environment, Our Wealth*. United Nations Environment Programme, Nairobi

Hulme, D. and Murphree, M. (eds.) (2001) *African Wildlife and Livelihoods: The promise and performance of community conservation*. James Curry, Oxford

IATF (2004). Report of the Tenth Session of the Working Group on Climate Change and Disaster Reduction, 7–8 October 2004. Inter-Agency Task Force on Disaster Reduction, Geneva

ICES (2006). *Is time running out for deepsea fish?* http://www.ices.dk/marineworld/deepseafish.asp

ICOLD (2006). *Proceedings of the 22nd ICOLD Congress of the International Commission on Large Dams, 18-23 June 2006, Barcelona*

IEA (2002). *World Energy Outlook 2002*. International Energy Agency, Paris

IEA (2003). *World Energy Outlook 2003*. International Energy Agency, Paris

IEA (2004). *World Energy Outlook 2004*. International Energy Agency, Paris

IEA (2005). *World Energy Outlook 2005.* International Energy Agency, Paris

IEA (2006). *World Energy Outlook 2006.* International Energy Agency, Paris

IEA (2007). *Energy Security and Climate Change Policy.* International Energy Agency, Paris http://www.iea.org/Textbase/publications/free_new_Desc.asp?PUBS_ID=1883 (last accessed 15 June 2007)

IFPRI (2006). *2006 Global Hunger Index. A Basis for Cross-Country Comparisons.* International Food Policy Research Institute, Washington, DC

IFRCRCS (2005). *World Disasters Report.* International Federation for the Red Cross and Red Crescent Societies, Geneva

Igoe, J. (2006). Measuring the Costs and Benefits of Conservation to Local Communities. In *Journal of Ecological Anthropology* 10:72-77

IISD (2002). WSSD PrepCom II Highlights: Monday, 28 January 2002. Earth Negotiations Bulletin. International Institute for Sustainable Development, Winnepeg

IOM (2005). *World Migration 2005: The Costs and Benefits of International Migration.* International Organization for Migration, Geneva

IPCC (2001). *Climate Change 2001 – Impacts, Adaptation and Vulnerability.* Contribution of Working Group II to the Third Assessment Report of the Intergovernmental Panel on Climate Change. McCarthy, J.J., Canziani, O.F., Leary, N. A., Dokken, D.J. and White, K.S. (eds). Cambridge University Press, Cambridge and New York, NY

IPCC (2007). *Climate Change 2007: Climate Change Impacts, Adaptation and Vulnerability, Summary for Policymakers.* Contribution of Working Group II to the Fourth Assessment Report of the Intergovernmental Panel on Climate Change, Geneva http://www.ipcc.ch/SPM6avr07.pdf (last accessed 27 April 2007)

IRN (2006). *IRN's Bujagali Campaign.* International River Network http://www.irn.org/programs/bujagali/ (last accessed 14 June 2007)

ISDR (2002). *Natural disasters and sustainable development: understanding the links between development, environment and natural disasters.* United Nations International Strategy for Disaster Reduction (ISDR), Geneva

ISDR (2004). *Living with Risk: A global review of disaster reduction initiatives.* International Secretariat for Disaster Reduction, Geneva

IUCN (2005). Constraints to the sustainability of deep sea fisheries beyond national jurisdiction. *IUCN Committee on Fisheries. Twenty-sixth Session, Rome, Italy, 7–11 March 2005*

Jeffrey, R. and Sunder, N. (2000). *A New Moral Economy for India's Forests? Discourses of Community and Participation.* Sage Publications, New Delhi

Jones, R. (2006). *Slum politics: how self-government strategies are improving futures for slum-dwellers.* Association for Women's Rights in Development. http://www.awid.org/go.php?stid=1584 (last accessed 27 April 2007)

Josking, T. (1998). Trade in Small Island Economies: Agricultural Trade Dilemmas for the OECS. Paper prepared for *IICA/NCFAP Workshop on Small Economies in the Global Economy, Grenada*

Kahl, C. (2006). *States, Scarcity, and Civil Strife in the Developing World.* Princeton University Press Princeton, NJ

Karlsson, S. (2000). Multilayered Governance. Pesticides in the South - Environmental Concerns in a Globalised World. PhD Dissertation, Linköping University, Linköping

Karlsson, S. (2002). The North-South Knowledge Divide: Consequences for Global Environmental Governance. In *Global Environmental Governance: Options and Opportunities*, Esty, D.C. and Ivanova, M.H. (eds). Yale School of Forestry and Environmental Studies, New Haven

Karlsson, S.I. (2007). Allocating Responsibilities in Multi-level Governance for Sustainable Development. *International Journal of Social Economics* 34

Kasperson, J.X., Kasperson, R.E., Turner II, B.L., Hsieh, W. and Schiller, A. (2005). Vulnerability to Global Environmental Change. In *The Human Dimensions of Global Environmental Change*, Rosa, E. A., Diekmann, A., Dietz, T., Jaeger, C.C. (eds.) MIT Press, Cambridge MA

Katerere, Y. and Mohamed-Katerere, J.C. (2005). From Poverty to Prosperity: Harnessing the Wealth of Africa's Forests. In *Forests in the Global Balance – Changing Paradigms*, Mery, G., Alfaro, R., Kanninen, M. and Lobovikov, M. (eds.). IUFRO World Series Vol. 17. International Union of Forest Research Organizations, Helsinki

Klein, R. J. T. and Nicholls, R. J. (1999). Assessment of Coastal Vulnerability to Climate Change. In *Ambio* 28:182-187

Klein, R.J.T., Nicholls, R.J. and Thomalla, F. (2003). Resilience to Weather-Related Hazards: How Useful is this Concept? In *Global Environmental Change Part B: Environmental Hazards* 5:35-45

Knutson, T.R., Tuleya, R.E. and Kurihara, Y.. (1998). Simulated Increase in Hurricane Intensities in a CO2-Warmed Climate. In *Science* 279:1018–1020

Krugman, P. (2003). *The Great Unraveling: Losing Our Way in the New Century.* Norton, New York, NY

Kuhnlein, H. V. and H. M. Chan (2000). Environment and contaminants in traditional food systems of northern indigenous peoples. In *Annual Review of Nutrition* 20:595-626

Kulindwa, K., Kameri-Mbote, P., Mohamed-Katerere and J.C., Chenje, M. with Sebukeera, C. (2006). The Human Dimension. In *Africa Environment Outlook 2. Our Environment, Our Wealth.* United Nations Environment Programme, Nairobi

Kulshreshta, S.N. (1993). *World water resources and regional vulnerability. Impact of future changes.* RR-93-10. International Institute for Applied Systems Analysis, Laxenburg.

Kusters, K. and Belcher, B. (Eds) (2004). *Forest products, livelihoods and conservation: case studies of non-timber forest product systems. Vol. 1, Asia.* Center for International Forestry Research, Bogor

Lal, Deepak and Mynt, Hla (1996). *The Political Economy of Poverty, Equity, and Growth.* Clarendon, Oxford

Lam, M. (1998). Consideration of Customary Marine Tenure System in the Establishment of Marine Protected Areas in the South Pacific. In *Ocean and Coastal Management* 39(1):97-104

Leach, M., Scoones, I. and Thompson, L. (2002). Citizenship, science and risk: conceptualising relationships across issues and settings. In *IDS Bulletin* 33(2):83 -91. Institute of Developing Studies, University of Sussex, Brighton

Leamer, E.E., Maul, H., Rodriguez, S. and Schott, P.K. (1999). Does Natural Resource Abundance Increase Latin American Income Inequality. In *Journal of Development Economics* 59:3–42

Lebel, J., Mergier,D., Branches, F., Lucotte, M., Amorim, M., Larribe, F., Dolbec, J. (1998). Neurotoxic effects of low-level methyl mercury contamination in the Amazonian Basin. In *Environ. Res.* 79:20-32

Lebel, L., Tri, N.H., Saengnoree, A., Pasong, S., Buatama, U. and Thoa, L.K. (2002). Industrial transformation and shrimp aquaculture in Thailand and Vietnam: Pathways to ecological, social and economic sustainability? In *Ambio* 31(4):311-323

La Rovere, E.L. and Romeiro, A.R. (2003). *Country study Development and Climate: Brazil*, COPPE/UFRJ and UNICAMP/EMBRAPA, Rio de Janeiro, http://www.developmentfirst.org/Publications/BrazilCountryStudy.pdf.

Leite, C. and J. Weidmann (1999). *Does Mother Nature Corrupt? Natural Resources, Corruption, and Economic Growth.*, International Monetary Fund, Washington, DC

Lind, J. and Sturman, K. (eds.). (2002). *Scarcity and Surfeit - The ecology of Africa's conflicts*, African Centre for Technology Studies and Institute for Security Studies, South Africa

Liu, P. F., Lynett, P, Fernando, H., Jaffe, B. E., Fritz, H., Higman, B., Morton, R., Goff, J. and Synolakis C. (2005). Observations by the International Tsunami Survey Team in Sri Lanka. In *Science* 308(5728):1595

Lopez, P. D. (2005). *International Environmental Regimes: Environmental Protection as a Means of State Making?* No. 242. Oficina do CES, Centro de Estudos Sociais. Coimbra http://www.ces.uc.pt/publicacoes/oficina/242/242.php (last accessed 15 June 2007)

Lüdeke, M. K. B., Petschel-Held, G. and Schellnhuber, J. (2004). Syndromes of global change: The first panoramic view. In *GAIA* 13(1)

Malm, O. (1998). Gold mining as a source of mercury exposure in the Brazilian Amazon. In *Environ. Res.* 77:73-78

Markovich, V. and Annandale, D. (2000). Sinking without a life-jacket? Sea Level Rise and the Position of Small Island States in International Law. In *Asia-Pacific Journal of Environmental Law* 5(2):135-154

Marshall, E., Newtron, A. C. and Schrekenberg, K. (2003). Commercialization of non-timber forest products: first steps in the factors influencing success. In *International Forestry Review* 5(2):128-137

Martinez-Alier, J. (2002). *Environmentalism of the poor.* Edward Elgar, Cheltenham

Matthew, R., Halle, M. and Switzer, J (2002). *Conserving the Peace: Resources, Livelihoods, and Security.* International Institute for Sustainable Development and IUCN – The World Conservation Union, Winnipeg

Mayrand, K., Paquin, M. and Dionne, S. (2005). *From Boom to Dust?* Agricultural Trade Liberalization, Poverty and Desertification in Rural Drylands: The Role of UNCCD. http://www.unisfera.org/?id_article=216&pu=1&ln=1 (last accessed 27 April 2007)

McCully, P. (1996). *Silenced Rivers. The Ecology and Politics of Large Dams.* Zed Books, London, New Jersey

McDonald, B. and Gaulin, T. (2002). Environmental Change, Conflict, and Adaptation: Evidence from Cases. Presented at the *Annual Meeting of the International Studies Association, March 24-27, 2002*

McElroy, J.L. (2003). Tourism Development in Small Islands Across the World. In *Geografiska Annaler* (86):231-242

Meadows, D., Randers, J. And Meadows, D. (2004). *Limits to Growth. The 30-Year Update.* Green Publishing Company, White River Junction, Vermont, Chelsea

Mertens, T. (2005). International or Global Justice? Evaluating the Cosmopolitan Approach. In *Real World Justice*, Follesdal, A. and Pogge, T. (eds.). Springer, Dordrecht

Metha, L. and La Cour Madsen, B. (2004). Is the WTO after your water? The General Agreement on Trade in Services (GATS) and poor people's right to water. In *Natural resources forum: a United Nations Journal* 2 (2):154-164

Miller, F., F. Thomalla and J. Rockström (2005). Paths to a Sustainable Recovery after the Tsunami. In *Sustainable Development Update* 5(1)

Miller, F., Thomalla, F., Downing T. E. and Chadwick, M. (2006). Resilient Ecosystems, Healthy Communities: Human Health and Sustainable Ecosystems after the Tsunami. In *Oceanography* 19(2):50-51

MA (2003). *Ecosystems and Human Well-being; a framework for assessment.* Island Press, Washington, DC

MA (2005). *Ecosystems and Human Well-being: Synthesis.* Island Press, Washington, DC

Mitchell, J.K. (1988). Confronting natural disasters: an international decade for natural hazard reduction. In *Environment* 30 (2):25-29

Mitchell, J.K. (1999). *Crucibles of hazard: mega-cities and disasters in transition.* United Nations University Press, New York, NY

Modi, V., McDade, S., Lallement, D. and Saghir, J. (2005). *Energy and the Millennium Development Goals.* Energy Sector Management Assistance Programme, United Nations Development Program, UN Millenium Project, and The World Bank, New York, NY

Mohamed-Katerere, J.C. (2001). *Review of the Legal and Policy Framework for Transboundary Natural Resource Management in Southern Africa.* Paper No 3, IUCN-ROSA Series on Transboundary Natural Resource Management. IUCN – The World Conservation Union, Harare

Mohamed-Katerere, J.C. and Van der Zaag, P. (2003). Untying the knot of silence – making water law and policy responsive to local normative systems. Hassan, F.A. Reuss, M. Trottier, J. Bernhardt, C. Wolf, A.T. Mohamed-Katerere, J. C. and Van der Zaag, P. (eds.). In *History and Future of Shared Water Resources.* UNESCO-International Hydrological Porgramme, Paris

Mollo, M., Johl, A., Wagner, M., Popovic, N., Lador, Y., Hoenninger, J., Seybert, E. and Walters, M. (2005). Environmental Rights Report. Human Rights and the Environment. Materials for the *61st Session of the United Nations Commission on Human Rights, Geneva, March 14 – April 22, 2005.* Earthjustice, Oakland

Monroe, M. C. (2003). Two Avenues for Encouraging Conservation Behaviours. In *Human Ecology Review* 10(2):113-125

Mortimore, M. (2005). Dryland development: success stories from West Africa. In *Environment* 47:8-21

Mortimore, M. (2006). Why invest in drylands? Synergies and strategy for developing ecosystem services. In *Drylands' hidden wealth. Integrating Dryland Ecosystem Services into National Development Planning. Conference Report. Amman, Jordan, 26 – 27 June 2006* http://www.iucn.org/themes/CEM/documents/drylands/amman_drylands_wreport_noppt_sept2006.pdf (last accessed 27 April 2007)

Mueller, H. (1996). *Nuclear Non-Proliferation Policy 1993-1995.* Peter Lang Publishing

Munich Re (2004a). *Megacities – Megarisks: Trends and Challenges for Insurance and Risk Management.* Munich Re Group, Munich

Munich Re (2004b). *Topics 2/2004. IFRS – New Accounting Standards. Flood Risks. Rising Costs of Bodily Injury Claims.* Munich Re Group, Munich

Munich Re (2006). *Topics Geo Annual Review: Natural Catastrophes 2005.* Munich Re Group, Munich

Murtagh, F. (1985). *Multidimensional Clustering Algorithms.* Physica-Verlag

Narayan, D., Chambers, R., Shah, M. and Petesch P. (2000). *Voices of the Poor – Crying Out for Change.* Oxford University Press, New York, NY

NASA (2002). Haitian Deforestation. Goddard Space Flight Center. http://svs.gsfc.nasa.gov/vis/a000000/a002600/a002640/ (last accessed 14 June 2007)

Ncube, W., Mohamed-Katerere, J. C. and Chenje, M. (1996). Towards the Constitutional Protection of Environmental Rights in Zimbabwe. In *Zimbabwe Law Review*

Nevill, J. (2001). Ecotourism as a source of funding to control invasive species. In *Invasive Alien Species: A Toolkit of Best Prevention and Management Practices*, Wittenberg, R. and Cock, M.J.W. (eds.). CAB International, Wallingford

Newell, P. and Mackenzie, R. (2004). Whose rules rule? Development and the global governance of biotechnology. Centre for the Study of Globalisation and Regionalisation, University of Warwick. In *IDS Bulletin* 35(1):82-92

Nicholls, R. J. (2002). Analysis of global impacts of sea-level rise: A case study of flooding. In *Physics and Chemistry of the Earth* 27:1455-1466

Nicholls, R.J. (2006). Storm Surges in Coastal Areas. In *Natural Disaster Hotspots - Case Studies*, Arnold, M., Chen, R.S., Deichmann, U., Dilley, M., Lerner-Lam, A.L., Pullen, R.E. and Trohanis, Z. (eds.). The World Bank, Washington, DC

Nori, M., Switzer, J. and Crawford, A (no date). *Herding on the Brink. Towards a Global Survey of Pastoral Communities and Conflict.* Occasional Working Paper. IUCN Commission on Environmental, Economic and Social Policy. IUCN – The World Conservation Union and International Institute for Sustainable Development, Gland

Nurse, L. and Rawleston, M. (2005). Adaptation to Global Climate Change: An Urgent Requirement for Small Island Developing States. In *RECIEL* 14(2):100-107

NZIS (2006). *Immigration New Zealand Online Operations Manual, April 2006 Update.* New Zealand Immigration Service, Wellington www.immigration.govt.nz/migrant/general/generalinformation/operationsmanual (last accessed 27 April 2007)

OECD (2007). *Reference DAC Statistical Tables. Net ODA in 2006 (updated April 2007).* Organisation for Economic Cooperation and Development, Paris http://www.oecd.org/dataoecd/12/8/38346276.xls (last accessed 15 June 2007)

Oldeman, L.R., Hakkeling, R.T. A. and Sombroek, W.G. (1991) *World map of human-induced soil degradation: A brief explanatory note.* ISRIC and United Nations Environment Programme, Wageningen

Page, S. (2004). Developing Countries in International Negotiations: how they influence Trade and Climate change Negotiations.. University of Sussex, Institute of Development Studies, Brighton. In *IDS Bulletin 35(1) Globalization and Poverty*

PAHO (2002). *Health in the Americas, 2002 Edition.* Pan-American Health Organisation http://www.paho.org/English/DD/PUB/HIA_2002.htm (last accessed 27 April 2007)

Papyrakis, El. and Gerlagh, R. (2004). The Resource Curse Hypothesis and its Transmission Channels. In *Journal of Comparative Economics* 32:181–193

Paré, L., Robles, C. and Cortéz, C. (2002). Participation of Indigenous and Rural People in the Construction of Development and Environmental Public Policies in Mexico. Development Studies Institute, University of Sussex, Brighton. In *IDS Bulletin* 33(2) Making Rights Real: Exploring Citizenship, Participation and Accountability

Parish, F. and Looi, C.C. (1999). *Mobilising financial support from bilateral and multilateral donors for the implementation of the Convention.* Ramsar COP7 DOC. 20.4. The Ramsar Convention on Wetlands, Gland http://www.ramsar.org/cop7/cop7_doc_20.4_e.htm (last accessed 15 June 2007)

Parry, M. L., Arnell, N., McMichael, T., Nicholls, R., Martens, P., Kovats, S., Livermore, M., Rosenzweig, C., Iglesias, A. and Fischer, G. (2001). Millions at Risk: Defining Critical Climate Change Threats and Targets. In *Global Environmental Change* 11(3):181-83

Patz, J.A., Campbell-Lendrum, D., Holloway, T. and Foley, J.A. (2005). Impact of regional climate change on human health. In *Nature* 438(7066):310-317

Pearce, D. (2005). *The Critical Role of Environmental Improvement in Poverty Reduction.* Report prepared for the Poverty Environment Partnership Pep Mdg7 Initiative of the United Nations Development Programme and United Nations Environment Programme, Washington, DC and Nairobi

Pearce, F. (1992). *The Dammed – Rivers, Dams, and the Coming World Water Crises.* The Bodley Head, London

Pelling, M. and Uitto, J.I. (2001). Small Island Developing States: Natural Disaster Vulnerability and Global Change. In *Environmental Hazards* 3:49-62

Petkova, E., Maurer, C., Henninger, N. and Irwin, F. (2002). *Closing the Gap: Information, Participation and Justice in Decision-making for the Environment.* World Resources Institute, Washington, DC

Petschel-Held, G., Block, A., Cassel-Gintz, M., Kropp, J., Lüdeke, M.K.B., Moldenhauer, O. and Reusswig, F. (1999). Syndromes of global change, a qualitative approach to assist global environmental management. In *Environmental Modelling and Assessment* 4:315-326

Pimm, S.L. (1984). The complexity and stability of ecosystems. In *Nature* 307:321-326

Poverty Mapping (2007). http://www.povertymap.net (last accessed 14 June 2007)

Prakash, A. (2000). Responsible Care: An Assessment. In *Business & Society* 39(2):183-209

Prescott-Allen, R. (2001). *The Well-being of Nations. A Country-by-Country Index of Quality of Life and the Environment.* Island Press, Washington DC

Prowse, M. (2003). *Towards a clearer understanding of "vulnerability" in relation to chronic poverty.* University of Manchester, Chronic Poverty Research Centre, WP24, Manchester

Reilly, J. and Schimmelpfennig, D. (2000). Irreversibility, Uncertainty, and Learning: Portraits of Adaptation to Long-Term Climate Change. In *Climate Change* 45(1), pp. 253-278(26)

Rodrik, D. (1996). *Why Do More Open Countries Have Bigger Governments?* National Bureau of Economic Research, Cambridge, MA

Round, J. and Whalley, J. (2004). *Globalisation and Poverty: Implications of South Asian Experience for the Wider Debate.* Centre for the Study of Globalisation and regionalisation, University of Warwick, *IDS Bulletin* 35(1):11-19

Ross, M.L. (2001). Does Oil Hinder Democracy? In *World Politics* 53:325–361

Russett, B. and Oneal, J. (2000). *Triangulating Peace: Democracy, Interdependence, and International Organizations.* The Norton Series in World Politics. W.W. Norton and Company, London

Sachs, J. D. and Warner, A. (2001). The Curse of Natural Resources. In *European Economic Review* 45(4-6):827–838

Sadoff, C.W. and Grey, D. (2002). Beyond the river: the benefits of cooperation on international rivers. In *Water Policy*, 4, 5:389-403

Safriel, U., Adeel, Z., Niemeijer, D., Puigdefabres, J., White, R., Lal, R., Winslow, M., Ziedler, J., Prince, S., Archer, E and King, C. (2005). Drylands Systems. In MA (2005). State and Trends. Volume 1.

Sala-I-Martin, X.X. (1997). I Just Ran Two Million Regressions (What Have We Learned From Recent Empirical Growth Research?). In *AEA Papers and Proceedings* 87(2):178-183

Sandwith, T. and Besançon, C. (2005). Trade-offs among multiple goals for transboundary conservation. In *Environmental Change and Security Program Report* 11:61-62

Sarin, M. (2003). *Devolution as a threat to democratic decision-making in forestry? Findings from three states in India.* Working Paper 197. Overseas Development Institute, London

SAUP (2007). *Landings in High Seas.* Web Products: High Seas Areas. http://www.seaaroundus.org/eez/SummaryHighseas.aspx?EEZ=0 (last accessed 26 April 2007)

Schiettecatte, W., Ouessarb, M., Gabrielsa, D., Tanghea, S., Heirmana, S. and Abdellib, F. (2005). Impact of water harvesting techniques on soil and water conservation: a case study on a micro catchment in southeastern Tunisia. In *Journal of Arid Environments* 6:297–313

Schneider, G., Barbieri, K. and Gleditsch, N. P. (eds.) (2003). *Globalization and Armed Conflict.* Rowman and Littlefield, Oxford

Schütz, H., Bringezu, S. and Moll, S. (2004). *Globalisation and the shifting environmental burden. Material trade flows of the European Union.* Wuppertal Institute, Wuppertal

Schütz G. Hacon, S. Moreno AR and Nagatani K. Principales marcos conceptuales para indicadores de salud ambiental aplicados en América Latina y Caribe. *Revista de la Organización Panamericana de la Salud.* (In press)

Scoones, I (ed.) (2001). *Dynamics and Diversity. Soil fertility and farming livelihoods in Africa.* Earthscan, London

Sen, A. (1999). *Development as Freedom.* Alfred A. Knopf, New York, NY

Small, C. and Nicholls, R.J. (2003). A Global Analysis of Human Settlement in Coastal Zones. In *Journal of Coastal Research* 19(3):584 – 599

Smil, V. (2001). *Enriching the Earth.* MIT Press, Cambridge MA

Smith, K. (1992). *Environmental hazards: assessing risk and reducing disaster.* Routledge, New York, NY

Smith, B., Burton, I., Klein, R.J.T., Wandel, J., (2000). An Anatomy of Adaptation to Climate Change and Variability. In *Climatic Change* 45(1):223-251

Sohn, J., Nakhooda, S. and Baumert, K. (2005). *Mainstreaming Climate Change Considerations at the Multilateral Development Banks.* World Resources Institute, Washington, DC

Solecki, W.D. and Leichenko, R.M. (2006). Urbanisation and the Metropolitan Environment: Lessons from New York and Shanghai. In *Environment* 48(4):6 – 23

SOPAC and UNEP. *Environmental Vulnerability Index . EVI Results.* South Pacific Applied Geoscience Commission and United Nations Development Porgramme, Suva http://www.vulnerabilityindex.net/EVI_Results.htm (last accessed 14 June 2007)

Soroos, M.S. (1997). The Canadian-Spanish 'Turbot War' of 1995: A Case Study in the Resolution of International Fishery Disputes. In *Conflict and the Environment* Gleditsch, N.P. (ed). Kluwer Publishers, Dordrecht

Sperling, F. and Szekely, F. (2005). Disaster Risk Management in a Changing Climate. Informal discussion paper prepared on behalf of the Vulnerability and Adaptation Resource Group (VARG) for the *World Conference on Disaster Reduction in Kobe, Japan, 18–22 January 2005*

Stegarescu, D. (2004). *Public Sector Decentralization: Measurement Concepts and Recent International Trends.* Discussion Paper 04-74, Zentrum für Europäische Wirtschaftsforschung ftp://ftp.zew.de/pub/zew-docs/dp/dp0474.pdf (last accessed 27 April 2007)

Steetskamp, I. and Van Wijk A. (1994). *Stroomloos. Kwetsbaarheid van de samenleving; gevolgen van verstoringen van de electriciteitsvoorziening* (in Dutch). Rathenau Insituut, The Hague

Stephens, C. (1996). Review Article: Healthy cities or Unhealthy Islands? The health and social implications of urban inequalities. In *Environment and Urbanization* 8(2):9-30

Stein, E. (1999). Fiscal Decentralization and Government Size in Latin America. In *Journal of Applied Economics* II (2):357-91

Stohr, W. (2001). Introduction. In *Decentralization, Governance and the New Planning for Local-level Development,* Stohr, W., Edralin, J. and Mani, D. (eds.). Greenwood Press, Westport

Strand, H., Carlsen, J., Gleditsch, N.P., Hegre, H., Ormhaug, C. and Wilhelmsen, L. (2005). *Armed Conflict Dataset Codebook.* Version 3-2005 http://www.prio.no/cscw/armedconflict (last accessed 27 April 2007)

Swain, A. (2002). Environmental Cooperation in South Asia. In *Environmental Peacemaking,* Conca K. and Dabelko, G.D. (eds.). The Woodrow Wilson Center Press and the Johns Hopkins University Press, Washington, DC and Baltimore

Swatuk, L. (2002). Environmental cooperation for regional peace and security in Southern Africa. In *Environmental Peacemaking,* Conca K. and Dabelko, G.D. (eds.). The Woodrow Wilson Center Press and the Johns Hopkins University Press, Washington, DC and Baltimore

Tan, K.-C. (2005). Boundary Making and Equal Concern. In *Global Institutions and Responsibilities: Achieving Global Justice,* Barry, C. and Pogge, T.W. (eds.). Blackwell Publishing, Oxford

Tasioulas, J. (2005). Global Justice Without End? In *Global Institutions and Responsibilities: Achieving Global Justice* Barry, C. and Pogge, T.W. (ed.). Blackwell Publishing, Oxford

Tetteh, I.K., Fremponga, E. and Awuahb, E. (2004). An analysis of the environmental health impact of the Barekese Dam in Kumasi, Ghana. In *Journal of Environmental Management* 72:189–194

Thomalla, F., Downing, T.E., Spanger-Siegfried, E., Han, G. and Rockström, J. (2006). Reducing Hazard Vulnerability: Towards a Common Approach Between Disaster Risk Reduction and Climate Adaptation. In *Disasters* 30(1):39-48

Thomas, D. (2006). *People, deserts and drylands in the developing world. Policy Briefs.* Science and Development Network http://www.scidev.net/dossiers/index.cfm? (last accessed 27 April 2007)

Tompkins, E.L., Nicholson-Cole, S.A., Hurlston, L., Boyd, E., Hodge, G.B., Clarke, J., Gray, G., Trotz, N. and Varlack, L. (2005). *Surviving Climate Change in Small Islands – A Guidebook.* Tyndall Centre for Climate Change Research, University of East Anglia, Norwich

Travis, J. (2005). Hurricane Katrina: Scientists' Fears Come True as Hurricane Floods New Orleans. In *Science* 309:1656-1659

UIS (2004). A Decade of Investment in Research and Development (R&D): 1990-2000. In *UIS Bulletin on Science and Technology Statistics 1.* UNESCO Institute for Statistics, Paris http://www.uis.unesco.org/template/pdf/S&T/BulletinNo1EN.pdf (last accessed 26 June 2007)

UN. *Terms of reference for the special rapporteur on the effects of illicit movement and dumping of toxic and dangerous products and waste on the enjoyment of human rights.* UN Office of the High Commissioner on Human Rights, New York, NY http://www.unhchr.ch/html/menu2/7/b/toxtr.htm (last accessed 14 June 2007)

UN (1966). *International Covenant on Economic, Social and Cultural Rights.* Office of the High Commissioner for Human Rights. United Nations, New York and Geneva http://www.unhchr.ch/html/menu3/b/a_cescr.htm (last accessed 27 April 2007)

UN (1986). *Declaration on the Right to Development.* Office of the High Commissioner for Human Rights. United Nations, New York and Geneva http://www.unhchr.ch/html/menu3/b/74.htm (last accessed 15 June 2007)

UN (2002). Plan of Implementation of the World Summit on Sustainable Development. In *Report of the World Summit on Sustainable Development.* Johannesburg, South Africa, 26 August - 4 September. A/CONF.199/20. United Nations, New York, NY

UN (2003). *Substantive Issues Arising in the Implementation of the International Covenant on Economic, Social and Cultural Rights. General Comment No. 15 (2002). The Right to Water (arts. 11 and 12).* E/C.12/2002/11. Committee on Economic, Social and Cultural Rights Twenty-ninth session, Geneva, 11-29 November 2002. Economic and Social Council, United Nations, Geneva http://www.unhchr.ch/tbs/doc.nsf/0/a5458d1d1bbd713fc1256cc400389e94/$FILE/G0340229.pdf (last accessed 27 April 2007)

UN (2005). *The Millennium Development Goals Report.* United Nations, New York, NY

UN (2006). *Millennium Development Goals Report 2006.* United Nations, New York, NY

UNCCD (2005). *The consequences of desertification.* Fact Sheet 3. United Nations Convention to Combat Desertification, Berlin http://www.unccd.int/publicinfo/factsheets/showFS.php?number=3 (last accessed 27 April 2007)

UNCTAD (2004). *Trade Performance and Commodity Dependence: Economic Development in Africa.* United Nations Conference on Trade and Development, Geneva

UNDP (2001). *Human Development Report 2001: Making New Technologies Work for Human Development.* United Nations Development Programme, New York, NY

UNDP (2004a). *Reducing disaster risk: A challenge for development.* United Nations Development Programme, New York, NY http://www.undp.org/bcpr/whats_new/rdr_english.pdf (last accessed 19 June 2007)

UNDP (2004b). *Analysis of conflict as it relates to the production and marketing of drylands products. The case of Turkana (Kenya) and Karamoja (Uganda) cross-border sites. Baseline Survey Results.* http://www.undp.org/drylands/docs/marketaccess/ Baselines-Conflict_and_Markets_Report.doc (last accessed 27 April 2007)

UNDP (2005). *International cooperation at a crossroads: Aid, trade and security in an unequal world. Human Development Report.* United Nations Development Programme, New York, NY

UNDP (2006) *Human Development Report 2006. Beyond scarcity: Power, poverty and the global water crisis.* United Nations Development Programme, New York, NY

UNDP and GEF (2004). *Reclaiming the Land Sustaining Livelihoods.* United Nations Development Programme and Global Environment Facility, New York, NY

UNECE (2005). *Aarhus Convention. Synthesis Report on the Status of Implementation of the Convention.* ECE/MP.PP/2005/18. United Nations Economic Commission for Europe, Geneva http://www.unece.org/env/documents/2005/pp/ece/ece. mp.pp.2005.18.e.pdf (last accessed 15 June 2007)

UN-Energy. *Welcome to UN-Energy, the interagency mechanism on energy.* http://esa. un.org/un-energy/ (last accessed 14 June 2007)

UNEP (2000). *Post-Conflict Environmental Assessment—Albania.* United Nations Environment Programme , Nairobi

UNEP (2002a). *Global Mercury Assessment.* United Nations Environment Programme, Geneva http://www.chem.unep.ch/MERCURY/Report/GMA-report-TOC.htm (last accessed 15 June 2007)

UNEP (2002b). *Vital Water Graphics. Coastal population and shoreline degradation.* UNEP/ GRID-Arendal Maps and Graphics Library http://maps.grida.no/go/collection/CollectionID/ 70ED5480-E824-413F-9B63-A5914EA7CCA1 (last accessed 27 April 2007)

UNEP (2004). *Vital Waste Graphics. Composition of transboundary waste reported by the Parties in 2000.* The Basel Convention, United Nations Environment Porgramme, and UNEP/GRID-Arendal http://maps.grida.no/go/collection/CollectionID/ 17F46277-1AFD-4090-A6BB-86C7D31FD7E7 (last accessed 15 June 2007)

UNEP (2005a). *Atlantic and Indian Oceans Environment Outlook.* Special Edition for the Mauritius International Meeting for the 10-year Review of the Barbados Programme of Action for the Sustainable Development of Small Island Developing States. United Nations Environment Programme, Nairobi

UNEP (2005b). *Pacific Environment Outlook.* Special Edition for the Mauritius International Meeting for the 10-year Review of the Barbados Programme of Action for the Sustainable Development of Small Island Developing States. United Nations Environment Programme, Nairobi

UNEP (2005c). *Caribbean Environment Outlook.* Special Edition for the Mauritius International Meeting for the 10-year Review of the Barbados Programme of Action for the Sustainable Development of Small Island Developing States. United Nations Environment Programme, Nairobi

UNEP (2005d). *GEO Year Book 2004/5.* An Overview of Our Changing Environment. United Nations Environment Programme, Nairobi

UNEP (2005e). *Report of the High-Level Brainstorming Workshop for Multilateral Environmental Agreements on Mainstreaming Environment Beyond Millennium Development Goal 7.* United Nations Environment Programme, Nairobi

UNEP, UNDP, OSCE and NATO (2005). *Environment and Security: Transforming risks into cooperation – Central Asia Ferghana/ Osh/ Khujand area.* United Nations Environment Programme, United Nations Development Programme, Organization for Security and Co-operation in Europe and the North Atlantic Treaty Organization, Geneva http://www.osce.org/publications/eea/2005/10/16671_461_en.pdf (last accessed 15 June 2007)

UNESCO (2005). *Contributing to a More Sustainable Future: Quality Education, Life Skills and Education for Sustainable Development.* http://unesdoc.unesco.org/ images/0014/001410/141019e.pdf

UNHCR (2006). *The State of the World's Refugees.* UN High Commission on Refugees, Geneva

UNICEF (2004a). *Children's Well-being in Small Island Developing States and Territories.* United Nations Children's Fund, New York, NY

UNICEF (2004b). *State of the World's Children 200. Childhood under threat.* United Nations Children's Fund, New York, NY http://www.unicef.org/publications/index_ 24433.html (last accessed 27 April 2007)

UNISDR. *Hyogo Framework for Action 2005-2015: Building the resilience of nations and communities to disasters (HFA).* http://www.unisdr.org/eng/hfa/hfa.htm (last accessed 14 June 2007)

UNMP (2005). *Environment and human well-being: a practical strategy.* Report of the task force on environmental sustainability. UN Millennium Project. Earthscan, London

UNPD (2007). *World Population Prospects: The 2006 Revision.* UN Population Division, New York, NY (in GEO Data Portal)

VanDeveer, S.D. (2002). Environmental Cooperation and Regional Peace: Baltic Politics, Programs, and Prospects. In K. Conca and G.D. Dabelko (eds.). *Environmental Peacemaking,* The Woodrow Wilson Center Press and the Johns Hopkins University Press, Washington, DC and Baltimore

Vanhanen, T. (2000). A New Dataset for Measuring Democracy, 1810–1998. In *Journal of Peace Research* 37(2):251–265

Van Straaten, P. (2000). Mercury contamination associated with small-scale gold mining in Tanzania and Zimbabwe. In *Sci. Total Environment* 259:95-109

Van Vuuren, D., M. den Elzen, P. Lucas, B. Eickhout, B. Strengers, B. van Ruijven, S. Wonink, R. van Houdt (2007). Stabilizing Greenhouse Gas Concentrations at Low Levels: An Assessment of Reduction Strategies and Costs, Climatic Change (accepted for publication)

Walker, G., Fairburn, J., Smith, G. and Michell, G. (2003). *Environmental Quality and Social Deprivation.* Environment Agency, Bristol

Watts M. J. and Bohle H. G. (1993). The space of vulnerability:The causal structure of hunger and famine. In *Progress in Human Geography* 17(1):43-67

WBGU (1997). *World in Transition: Ways Towards Sustainable Management of Freshwater Resources.* German Advisory Council on Global Change. Springer Verlag, Heidelberg

WCC'93 (1994). Preparing to Meet the Coastal Challenges of the 21st Century. *Report of the World Coast Conference, NoordwijkNovember 1–5, 1993. Ministry of Transport, Public Works and Water Management,* The Hague

WCD (2000). *Dams and Development. A New Framework for Decision Making.* Report of the World Commission on Dams. Earthscan, London

WCED (1987). *Our Common Future.* World Commision on Environment and Development. Oxford University Press, Oxford and New York, NY

Weede, E. (2004). On Political Violence and Its Avoidance. In *Acta Politica* 39:152-178

Wei, S. (2000). *Natural Openness and Good Government.* National Bureau of Economic Research, Cambridge, MA

Weinthal, E. (2002). The Promises and Pitfalls of Environmental Peacemaking in the Aral Sea Basin. In *Environmental Peacemaking,* Conca, K. and Dabelko, G.D. (eds.). The Woodrow Wilson Center Press and the Johns Hopkins University Press, Washington and Baltimore

Weisman, D. (2006). *Global Hunger Index. A basis for cross-country comparisons.* International Food Policy Research Institute, Washington, DC

White, R.P., Tunstall, D. and Henninger, N. (2002). *An Ecosystem Approach to Drylands: Building Support for New Development Policies.* Information Policy Brief 1. World Resources Institute, Washington, DC

WHO (2002). *The world health report 2002, reducing risks, promoting healthy life.* World Health Organization, Geneva.

WHO (2006b). Global and regional food consumption patterns and trends. Chapter 3 In *Diet, Nutrition and the Prevention of Chronic Diseases.* Report of the Joint WHO/FAO Expert Consultation. WHO Technical Report Series, No. 916 (TRS 916). World Health Organization, Geneva http://www.who.int/dietphysicalactivity/publications/trs916/ download/en/index.html (last accessed 15 June 2007)

WHO and UNEP (2004). *The health and the environment linkages initiative.* World Health Organization, Geneva

Wisner, B., Blaikie, P., Cannon, T. and Davis, I. (2004). *At Risk: Natural Hazards, Peoples Vulnerability and Disasters,* 2nd edition. Routledge, London

Wolf, M. (2004). *Why Globalization Works: The Case for the Global Market Economy.* Yale University Press, New Haven

Wolf, A.T., Yoffe, S.B. and Giordano M. (2003). International waters: Identifying basins at risk. In *Water Policy* 5:29-60

Wonink, S.J., Kok, M.T.J. and Hilderink, H.B.M. (2005). *Vulnerability and Human Well-being.* Report 500019003/2005. Netherlands Environmental Assessment Agency, Bilthoven

World Bank (2005). *The World Development Report 2006. Equity and Development.* Oxford University Press, Oxford and The World Bank, Washington, DC

World Bank (2006). *World Development Indicators 2006.* The World Bank, Washington, DC (in GEO Data Portal)

World Water Council (2000). *World Water Vision: Making Water Everybody's Business.* http://www.worldwatercouncil.org/index.php?id=961&L=0 (last accessed 27 April 2007)

WRI (2002). *Drylands, People, and Ecosystem Goods and Services: A Web-based Geospatial Analysis.* World Resources Institute. http://www.wri.org (last accessed 27 April 2007)

WRI (2005). *World Resources. The Wealth of the Poor – Managing Ecosystems to Fight Poverty.* World Resources Institute in collaboration with United Nations Development Programme, United Nations Environment Programme and The World Bank, Washington, DC

WRI (2007). *Nature's Benefits in Kenya. An Atlas of Ecosystems and Human Well-Being.* World Resources Institute, Department of Resource Surveys and Remote Sensing, Ministry of Environment and Natural Resources, Kenya, Central Bureau of Statistics, Ministry of Planning and National Development, Kenya, and International Livestock Research Institute. World Resources Institute, Washington, DC and Nairobi

Wynberg, R. (2004). Achieving a fair and sustainable trade in devil's claw (*Harpagophytum spp*). In *Forest Products, Livelihoods and Conservation.* Case Studies of Non-Timber Forest Products. Vol. 2 – Africa. Sunderland, T. and Ndoye, O. (eds.). Centre for International Forestry Research, Bogor

Yanez, L., Ortiz, D., Calderon, J., Batres, L., Carrizales, L., Mejia, J., Martinez, L., Garcia-Nieto, E. and Diaz-Barriga, F. (2002). Overview of Human Health and Chemical Mixtures: Problems facing developing countries. In *Environmental Health Perspectives* 110 (6):901 – 909

Yoffe, S.B., Fiske, G., Giordano, M., Giordano, M.A., Larson, K., Stahl K. and Wolf, A.T. (2004). Geography of international water conflict and cooperation: Data sets and applications. In *Water Resources Research* 40(5):1-12

Yokohama Strategy and Plan of Action for a Safer World (1994). *International Strategy for Disaster Reduction* http://www.unisdr.org/eng/about_isdr/bd-yokohama- strat-eng.htm (last accessed 15 June 2007)

Zoleta-Nantes, D. (2002). Differential Impacts of Flood Hazards among the Street children, the Urban Poor and Residents of Wealthy Neighborhood in Metro Manila, Philippines. In *Journal of Mitigation and Adaptation Strategies for Global Change* 7(3):239-266

Chapter 8

Interlinkages: Governance for Sustainability

Coordinating lead authors: Habiba Gitay, W. Bradnee Chambers, and Ivar Baste

Lead authors: Edward R. Carr, Claudia ten Have, Anna Stabrawa, Nalini Sharma, Thierry De Oliveira, and Clarice Wilson

Contributing authors: Brook Boyer, Carl Bruch, Max Finlayson, Julius Najah Fobil, Keisha Garcia, Elsa Patricia Galarza, Joy A. Kim, Joan Eamer, Robert Watson, Steffen Bauer, Alexander Gorobets, Ge Chazhong, Renat A. Perelet, Maria Socorro Z. Manguiat, Barbara Idalmis Garea Moreda, Sabrina McCormick, Catherine Namutebi, Neeyati Patel, and Arie de Jong

Chapter review editors: Richard Norgaard and Virginia Garrison

Chapter coordinators: Anna Stabrawa and Nalini Sharma

Credit: McPhoto/Still Pictures

Main messages

The Earth functions as a system: atmosphere, land, water, biodiversity and human society are all linked in a complex web of interactions and feedbacks. Environment and development challenges are interlinked across thematic, institutional and geographic boundaries through social and environmental processes. The state of knowledge on these interlinkages and implications for human well-being are highlighted in the following messages:

Environmental change and development challenges are caused by the same sets of drivers. They include population change, economic processes, scientific and technological innovations, distribution patterns, and cultural, social, political and institutional processes. Since the report of the World Commission on Environment and Development (Brundtland Commission), these drivers have become more dominant. For instance, the world population has increased by almost 30 per cent and world trade has increased almost three times. During the past two decades it has resulted in a situation where:
- *human societies have become more interconnected* through globalization driven by increasing flows of goods, services, capital, people, technologies, information, ideas and labour;
- *development challenges have become more demanding* as evident in the efforts to meet the Millennium Development Goals (MDGs); and
- *pressures on the environment, and consequently the rate, extent, interconnectedness and magnitude of environmental change, have increased*, as have their impacts on human well-being.

The responsibility for the drivers that create the pressures on the environment is not equally distributed throughout the world. Economic processes are a good example. In 2004, the total annual income of the nearly 1 billion people in the richest countries was nearly 15 times greater than that of the 2.3 billion in the poorest countries. Also that year, the Annex 1 countries of the UN Framework Convention on Climate Change, contained 20 per cent share of the world population, produced 57 per cent of world GDP, based on purchasing power parity, and accounted for 46 per cent of greenhouse gas (GHG) emissions. Africa's share of the GHG emissions was 7.8 per cent.

One form of human activity can cause several reinforcing environmental effects and affect human well-being in many ways. Emissions of carbon dioxide, for example, contribute both to climate change and to acidification of oceans. In addition, land, water and atmosphere are linked in many ways, particularly through the carbon, nutrient and water cycles, so that one form of change leads to another. For example, changes in the structure and functioning of ecosystems caused in part by climate change will, in turn, affect the climate system, particularly through the carbon and nitrogen cycles. Human activities, such as agriculture, forestry, fisheries and industrial production, have increasingly altered ecosystems, and the ways in which they provide services in support of human well-being.

Social and biophysical systems are dynamic, and characterized by thresholds, time-lags and feedback loops. Thresholds – sometimes also referred to as tipping points – are common in the Earth system, and represent the point of sudden, abrupt, or accelerating and potentially irreversible change triggered by natural events or human activities. Examples of thresholds being crossed due to sustained human activities include: collapse of fisheries, eutrophication and deprivation of oxygen (hypoxia) in aquatic systems, emergence of diseases and pests, introduction and loss of species, and regional climate change. Biophysical and social systems also have the tendency to continue to change, even if the forces that caused the initial change are removed. For example, even if atmospheric concentrations of greenhouse gases were to be stabilized today, increases in land and ocean temperatures due to these emissions would continue for decades, and sea levels would continue to rise for centuries, due to the time-lags associated with climate processes and feedbacks.

The complexity of human-ecological systems, and the limitations in our current state of knowledge of the dynamics of these systems, make it hard to predict precisely where critical thresholds lie. These are the points where an activity results in an unacceptable level of harm, for example in terms of ecological change, and requires a response. This uncertainty also makes it difficult to identify measures for pre-empting the crossing of critical thresholds. This is of significant concern for human well-being, as past examples such as in Mesopotamia and Easter Island show how crossing some thresholds can contribute to the catastrophic disruption of societies.

The complexity, magnitude and the interconnectedness of environmental change do not mean that decision-makers are faced with the stark choice of "doing everything at once in the name of integrated approaches or doing nothing in the face of complexity." Identifying interlinkages offers opportunities for more effective responses at the national, regional and global levels. It may facilitate the transition towards a more sustainable society. It provides the basis for applying measures where they are most effective, based on trade-offs among different interests in society, and in a complementary manner.

Consideration of interlinkages among environmental challenges can facilitate more effective treaty compliance, while respecting the legal autonomy of the **treaties.** This would highlight areas for cooperation and joint programming among the treaties, and for more effective enforcement and compliance at the national level, as well as for related capacity building and technology support. Considerations of the overall normative basis for environmental governance may help identify new opportunities for more effective institutional structures for international environmental cooperation.

Collaboration across existing governance regimes can strengthen the integration of environmental concerns into the wider development agenda. Significant opportunities in this respect are offered by the UN reform process, due to its particular focus on system-wide coherence in the area of environment, and the "One UN" approach at the country level. Approaches such as mitigation, including carbon storage, and adaptation to climate change that consider links with other environment and development challenges, may potentially address multiple environmental and development challenges simultaneously.

Governance approaches that are flexible, collaborative and learning-based may be *responsive* and *adaptive*, and better able to cope with the challenges of integrating environment and development. Such adaptive governance approaches are well placed to address complex interlinkages, and to manage uncertainty and periods of change. They are likely to result in incremental and cost-effective evolution of institutional structures, and reduce the need for more fundamental institutional restructuring. Tools for dealing with interlinkages, such as assessments, valuation techniques and integrated management approaches that link environment to development, provide a critical foundation for adaptive governance.

INTRODUCTION

The World Commission on Environment and Development (Brundtland Commission) referred to the environmental, development and energy crises as "the interlocking crises" (WCED 1987). The interconnectedness of the environment and human society is emphasized throughout the Brundtland Commission report, and it is central to the concept of sustainable development (WCED 1987). It is also fundamental to the GEO conceptual framework, which focuses on the interaction between environment and society. Preceding chapters have assessed the linkages among and between drivers, pressures, environmental change, ecosystem services, human well-being and policy responses to the environmental challenges. They have also demonstrated how the patterns of the human-society interactions change with scale and time, how the environmental changes vary from one geographic region to another, and how different groups are vulnerable to various forms of environmental change.

Twenty years after the Brundtland Commission report was published, its findings are more pertinent than ever. The global pattern of the human-society interactions is changing. From a human perspective, the world is becoming smaller. For example, the amount of land per capita has been reduced to about one-quarter of what it was a century ago due to population growth (see Figure 8.1), and is expected to be further reduced to about one-fifth of the 1900 level by 2050 (GEO Data Portal,

from UNPD 2007 and FAOSTAT 2006). Social change processes, in terms of population growth, scientific and technological innovation, economic growth, and consumption and production patterns, are increasingly seen as the major drivers of environmental change (Young 2006, Schellnhuber 1999, Vitousek and others 1997). Trends for some of these major drivers of change are also illustrated in Figure 8.1.

The world is witnessing a pattern of globalization characterized by increasing flows of goods, services, capital, technologies, information, ideas and labour at global level, driven by liberalization policies and technological change (Annan 2002). In particular, the rapid development of the Internet (see Figure 1.9 in Chapter 1) is revolutionizing the communication abilities and interconnectedness of people, and can be harnessed to level the playing field for nations and individuals (Friedman 2006).

With an increasingly interconnected global society ever more potently driving environmental change, there is a need to understand how and by whom the environmental challenges best can be addressed. The report, *"Protecting Our Planet – Securing Our Future,"* (Watson and others 1998) and the *Millennium Ecosystem Assessment* (2005), demonstrated how environmental problems are often linked to one another. In drawing on the findings of previous chapters, this chapter further pursues

Gro Harlem Brundtland, then Prime Minister of Norway, addressing the UN General Assembly in 1987. The interconnectedness of the environment and human society is a common thread that runs throughout the Brundtland Commission report and the *GEO-4* assessment.

Credit: UN Photo

the current understanding of human-environment interlinkages. It examines how the different drivers, human activities and environmental changes are interlinked through complex cause-and-effect relationships embedded in both biophysical and social processes. This part of the chapter also examines to what extent the increasingly complex set of human pressures on the environment may exceed critical thresholds, and result in potentially sudden, unexpected effects and irreversible changes.

Environmental governance regimes have evolved in response to the environmental changes, but these mechanisms have often lagged behind the problems they address. These mechanisms have thus faced major challenges in being effective (Schmidt 2004, Najam and others 2006). As previous chapters have shown, some environmental challenges, such as point source pollution, are characterized by linear cause-effect interactions, and are relatively easy to deal with. Others are characterized by complex, often nested sets of linkages that are more persistent and difficult to address. These linkages need to be addressed in a systematic, sustained, integrated and coherent manner across administrative borders at various scales. Sustainable development is contingent upon an environmental governance regime that adapts to the evolving environmental challenges of the Earth system.

This chapter discusses how understanding these interlinkages and applying a systems approach can strengthen the effectiveness and complementarity of the environmental governance regimes at national,

regional and international levels. It considers how interventions within and among response regimes can be aligned through adaptive governance, supported by enhanced knowledge and information infrastructure. These considerations include the implications of such approaches for the enforcement and compliance regimes under the various multilateral environmental agreements.

HUMAN-ENVIRONMENT INTERLINKAGES

Previous chapters have assessed the state of knowledge with respect to key environmental challenges. They have demonstrated that there are interlinkages within and between changes such as climate change, ozone depletion, air pollution, biodiversity loss, land degradation, water degradation and chemical pollution. Environmental changes are linked across scales and between geographical regions through both biophysical and social processes. This section uses the GEO conceptual framework as a basis for an overarching and integrated analysis of these human-environment linkages (see the Reader's Guide). More specifically, this section provides an overview of how:

■ human drivers of environmental change cause and link various forms of environmental change, and how the social and economic sectors shape the human-environment linkages;

■ human activities and pressures create multiple environmental changes, and how various forms of environmental changes are connected through complex systems involving feedback loops and biophysical thresholds; and

Understanding and addressing the human-environment interlinkages will strengthen the effectiveness of governance regimes at all levels.

Credit: Shehab Uddin/Still Pictures

- an increasingly complex set of environmental changes and potential system-wide changes can exceed biophysical thresholds, leading to sudden and unexpected effects on human well-being.

Drivers of change

Environmental change and human development are all driven by the same factors, such as demographics, economic processes, scientific and technological innovations, distribution patterns, and cultural, social, political and institutional processes. These processes are complex and vary, depending on social and ecological circumstances. The pressure on the environment and consequently the rate, extent and magnitude of environmental changes have grown larger. The development challenges have also become more demanding as evidenced, for example, in the efforts to meet the MDGs.

Population growth is creating an increasing pressure on the planet, as illustrated by the shrinking size of land per capita since 1900 as the population increased (see Figure 8.1). According to estimates used in this report, the world population is expected to rise to 9.2 billion by 2050 from about 6.7 billion in 2007. The population in less developed regions is expected to rise from 5.5 billion in 2007 to 8 billion in 2050. In contrast, the population of the more developed regions is expected to remain largely unchanged at 1.2 billion, and would have declined were it not for the expected migration from developing to developed countries (GEO Data Portal, from UNPD 2007). Programmes to address population issues need to be closely related to other policies, such as those for economic development, migration, maternal and reproductive health, and gender equality and empowerment of women (UN 1994).

The impacts of population growth on the environment are inextricably related to people's consumption patterns. Consumption, particularly in the richer nations, has been increasing at a faster rate than that of population growth. Technological innovation has been a critical driver of this trend (Watson and others 1998). Since 1987, the world population has increased by almost 30 per cent (GEO Data Portal, from UNPD 2007), and world trade has increased 2.6 times. As shown in Figure 8.1, global economic output has increased by 67 per cent, also increasing the average per capita income in the same period. However, changes in

per capita income vary greatly among regions, from a decrease of more than 2 per cent in a few African countries to a doubling in some countries in Asia and the Pacific since 1987 (World Bank 2006a). The graphs in Figure 8.1 give an indication of such pressures and environmental changes from human activities.

Resources are not equitably distributed around the world. The world's poorest countries – mainly in Africa, Asia and the Pacific and Latin America and the Caribbean – had, in 2004, an average annual per capita income of US$2 100. The richest regions and countries – Europe, North America, Australia and Japan respective – had an average annual per capita income of US$30 000. On average, the total annual income of the nearly 1.2 billion people in the richest countries, is nearly 15 times greater than that of the 2.3 billion people in the poorest countries (Dasgupta 2006). Also in 2004, the Annex 1 countries of the UN Framework Convention on Climate Change, had 20 per cent of the world population, produced 57 per cent of world GDP, based on purchasing power parity, and accounted for 46 per cent of greenhouse gas (GHG) emissions. Africa's share of the GHG emissions was 7.8 per cent, while it had 13 per cent of world population (IPCC 2007a).

Increased consumption of raw materials and the related production of waste place tremendous pressure on the environment. Sixty per cent of the ecosystem services studied by the Millennium Ecosystem Assessment (MA) are being degraded or used unsustainably. Their degradation could grow significantly worse before 2050 due to rapidly growing demands for food, freshwater, timber, fibre and fuel, as well as from increasing pollution and climate change (MA 2005a).

Changes in the biosphere over the last few decades have contributed to substantial net gains in human well-being and economic development (MA 2005a). Formal and informal social and economic sectors have transformed natural resources (equated to natural capital) into forms that support development and human well-being.

In the poorest countries, natural resources are estimated to make up 26 per cent of the total wealth, forming the basis for subsistence and a

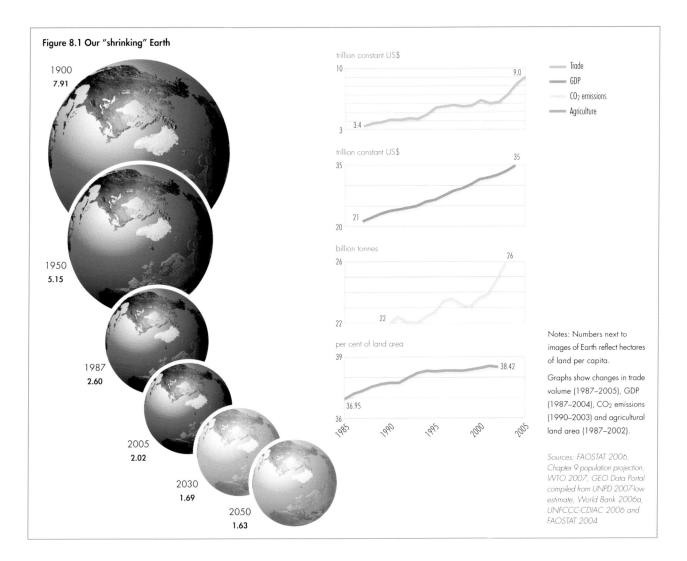

Figure 8.1 Our "shrinking" Earth

1900
7.91

1950
5.15

1987
2.60

2005
2.02

2030
1.69

2050
1.63

trillion constant US$

- Trade
- GDP
- CO₂ emissions
- Agriculture

10
9.0
3
3.4

trillion constant US$
35
35
21
20

billion tonnes
26
26
22
22

per cent of land area
39
38.42
36.95
36

1985 1990 1995 2000 2005

Notes: Numbers next to images of Earth reflect hectares of land per capita.

Graphs show changes in trade volume (1987–2005), GDP (1987–2004), CO₂ emissions (1990–2003) and agricultural land area (1987–2002).

Sources: FAOSTAT 2006, Chapter 9 population projection, WTO 2007, GEO Data Portal compiled from UNPD 2007-low estimate, World Bank 2006a, UNFCCC-CDIAC 2006 and FAOSTAT 2004

source of development finance (World Bank 2006b). Agriculture is the most important sector in low-income countries', responsible for 25–50 per cent of their gross domestic product (GDP) (CGIAR and GEF 2002). Agricultural growth is directly correlated to well-being, notably in terms of income and livelihood of farmers. For every dollar earned by farmers in low-income countries, there is a US$2.60 increment in incomes in the economy as a whole (CGIAR and GEF 2002). Therefore, an increase in crop yields has a significant impact on the upward mobility of those living on less than a dollar a day. The World Bank estimates that a 1 per cent increase in crop yields reduces the number of people living under US$1/day by 6.25 million. Natural capital can be transformed into forms of material capital, such as infrastructure and machines, as well as human capital, for example, knowledge and social capital, such as governance structures. These capitals determine the ability of individuals to exercise their

freedoms of choice and to take actions to achieve their material needs.

The observed net gains in human well-being facilitated by the social and economic sectors have, however, been at the cost of growing environmental changes, and the exacerbation of poverty for some groups of people (MA 2005a). Sustainable development relies on an effective integration of environmental concerns into development policies. A critical component of a strengthened international environmental governance regime is that it is able to support such integration (Berruga and Maurer 2006). Environmental impacts are, however, often not factored into operations of the social and economic sectors as a cost, and hence these impacts are referred to as externalities. The externalization of such costs does not allow for a true trade-off in terms of costs and benefits when development decisions are taken. These sectors are instrumental in utilizing ecosystem services and natural resources. They also affect ecosystem services, and are affected by ecosystem change (see Figure 8.2).

The agricultural sector, for example, interlinks a number of environmental changes, including climate change, biodiversity loss, land degradation, and water degradation. Chemicals are also a factor in envirnmental change. Agriculture is, however, also highly dependent on ecosystem services, such as predictable climatic conditions, genetic resources, water regulation, soil formation, pest regulation, and primary productivity of land and water. These services must be secured if the sector is to meet the demand for food. Chapter 3 concludes that a doubling of global food production will be required to meet the MDG on hunger, given projections that the world's population will increase to more than 9.2 billion by 2050. In the four GEO-4 scenarios, the human population is projected to between 8 billion and 9.7 billion in 2050 (see Chapter 9).

Measures for responding to environmental changes will often be implemented by government authorities, the private sector, civil society, communities and individuals associated with social and economic sectors. Responses will, as outlined in Figure 8.2, be in the form of either mitigation of or adaptation to environmental change. Both mitigation and adaptation can take the form of informal and formal approaches to altering human behaviour as they relate not only to drivers, but also to pressures and impacts. Response strategies need to take into account that roles, rights and responsibilities of women and men are socially defined, culturally based, and are reflected in formal and informal power structures that influence how management decisions are taken (Faures and others 2007). Management of common resources and complex systems are particularly challenging, and may require a broad set of multi-scaled governance tools, and an adaptive approach (Dietz and others 2003). Responses are an integral part of the

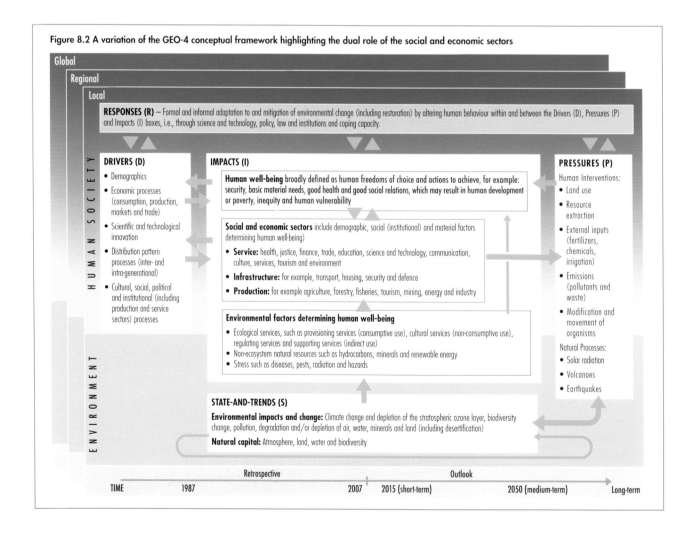

Figure 8.2 A variation of the GEO-4 conceptual framework highlighting the dual role of the social and economic sectors

Global

Regional

Local

RESPONSES (R) – Formal and informal adaptation to and mitigation of environmental change (including restoration) by altering human behaviour within and between the Drivers (D), Pressures (P) and Impacts (I) boxes, i.e., through science and technology, policy, law and institutions and coping capacity.

HUMAN SOCIETY

DRIVERS (D)
- Demographics
- Economic processes (consumption, production, markets and trade)
- Scientific and technological innovation
- Distribution pattern processes (inter- and intra-generational)
- Cultural, social, political and institutional (including production and service sectors) processes

IMPACTS (I)

Human well-being broadly defined as human freedoms of choice and actions to achieve, for example: security, basic material needs, good health and good social relations, which may result in human development or poverty, inequity and human vulnerability

Social and economic sectors include demographic, social (institutional) and material factors determining human well-being)
- **Service:** health, justice, finance, trade, education, science and technology, communication, culture, services, tourism and environment
- **Infrastructure:** for example, transport, housing, security and defence
- **Production:** for example agriculture, forestry, fisheries, tourism, mining, energy and industry

Environmental factors determining human well-being
- Ecological services, such as provisioning services (consumptive use), cultural services (non-consumptive use), regulating services and supporting services (indirect use)
- Non-ecosystem natural resources such as hydrocarbons, minerals and renewable energy
- Stress such as diseases, pests, radiation and hazards

ENVIRONMENT

STATE-AND-TRENDS (S)

Environmental impacts and change: Climate change and depletion of the stratospheric ozone layer, biodiversity change, pollution, degradation and/or depletion of air, water, minerals and land (including desertification)

Natural capital: Atmosphere, land, water and biodiversity

PRESSURES (P)

Human Interventions:
- Land use
- Resource extraction
- External inputs (fertilizers, chemicals, irrigation)
- Emissions (pollutants and waste)
- Modification and movement of organisms

Natural Processes:
- Solar radiation
- Volcanoes
- Earthquakes

Retrospective		Outlook	
TIME	1987	2007	2015 (short-term)	2050 (medium-term)	Long-term

human-environment interlinkages. A response to one environmental change may, therefore, directly or indirectly affect other environmental changes, and in itself contribute to the interlinkages among them.

Impacts and consequences of human activities on biophysical processes

Efforts to integrate environmental concerns into development and to promote sustainable consumption and production patterns need to factor in the ways in which environmental challenges are linked through human activities (pressures) and biophysical processes. Human activities have multiple direct impacts on the environment, and thus on ecosystem services and human well-being. Emissions of carbon dioxide, for example, contribute both to climate change (see Chapter 2) and to acidification of oceans (see Chapter 4). Human activities, such as agriculture, forestry and fisheries, meet human needs, especially in the short-term and thus have a positive impact on human well-being (see next subsection). However, if such activities are not managed sustainably, they can have a negative impact on the environment.

Human activities result in multiple impacts on the environment because of biophysical interlinkages.

Land, water and atmosphere are linked in many ways, but particularly through the carbon, nitrogen (see Chapter 3) and water cycles, which are fundamental to maintaining life on Earth. Feedbacks and thresholds affect the boundaries, composition and functioning of ecological systems. A classic case of feedback loops is seen in the interactions that influence the Arctic (see Box 8.1) (see Chapters 2 and 6).

Examining the interlinkages among multiple environmental challenges is similar to applying a systems approach by looking at the interlinkages within and between the wider global system or a sub-system. The biophysical interlinkages constitute an important characteristic of the environmental challenges themselves. System properties such as non-linear changes, thresholds, inertia and switches (see Box 8.2) are important characteristics. When developing management options, there is a need to consider the cause-effect chains, as these system properties (Camill and Clark 2000) are often cumulative in time and space.

A key example of how a human activity has resulted in multiple environmental impacts is the release of reactive nitrogen (Nr) from the burning of fossil fuels and use of fertilizers, discussed in more detail

---

Box 8.1 Feedback loops in the Arctic

Feedback
This describes a process by which the output of a system is used or allowed to modify its input, leading to either positive or negative results. In the climate system, a "feedback loop" has been described as a pattern of interaction where a change in one variable, through interaction with other variables in the system, either reinforces the original process (positive feedback) or suppresses the process (negative feedback). It is becoming apparent that there are major feedbacks in the Arctic systems associated with the rapid changes in the regional climate (see Chapters 2 and 6). It is clear that the Arctic system is very dynamic, and different sets of variables form feedbacks at different times, highlighting the complexity of feedbacks and interlinkages.

Temperature-albedo feedback
Rising temperatures increase melting of snow and sea ice, not only reducing surface reflectance, but also increasing solar absorption, raising temperatures further, and changing vegetation cover. The feedback loop can also work in reverse. For example, if temperatures were to cool, less snow and ice would melt in summer, raising the albedo and causing further cooling as more solar radiation is reflected rather than absorbed. The temperature-albedo feedback is positive because the initial temperature change is amplified.

Temperature-cloud cover-radiation feedbacks
Feedbacks among temperature, cloud cover, cloud types, cloud albedo and radiation play an important role in the regional climate. There is some indication that, except in summer, Arctic clouds seem to have a warming effect, because the blanket effect of clouds tends to dominate over reduction in shortwave radiation to the surface caused by the high cloud albedo. This appears to be different when compared with other regions of the world. The temperature-cloud cover-radiation feedback is negative as the initial temperature change is dampened. However, cloud cover also acts as a blanket to inhibit loss of long wave radiation from the Earth's atmosphere. By this process, an increase in temperature leading to an increase in cloud cover could lead to a further increase in temperature – a positive feedback.

Melting of permafrost and methane emissions
Permafrost areas of the Arctic, in particular tundra bogs, contain methane trapped since the last glaciation, about 10 000–11 000 years ago. Climate change is resulting in melting of the permafrost, and the gradual release of methane, a gas with warming potential more than 20 times as great as CO_2 (see Chapter 2 and 3). This is a positive feedback, which could lead to significant acceleration of climate change.

Sources: ACIA 2004, Stern and others 2006, UNEP 2007a

in Chapter 3. Nr creation has increased tenfold since 1860 (UNEP 2004). The benefits from use of fertilizers have been increased food production to support a growing population and increasing per capita food consumption. Many factors influence how much nitrogen is applied and used, including soil moisture, timing of fertilizer application, labour availability, inherent soil quality and type, farming systems, and major macro-nutrient availability (N-P-K) (see Chapter 3). It is recognized that to increase food production in Africa, there is a need for improved soil quality and fertility, with some improvements coming from the addition of inorganic fertilizers (Poluton and others 2006). However, in other regions, excess nitrogen is being lost to the environment, partly due to inefficient farming practices related to the quantity and timing of fertilizer application. Reactive nitrogen adversely affects many components of terrestrial and aquatic ecosystems and the atmosphere, as illustrated in Figure 8.3. For example, nitrogen released to the atmosphere from fossil fuel combustion and fertilizer use can, in sequence, increase tropospheric ozone concentration, decrease atmospheric visibility and increase precipitation acidity. Following deposition it can increase soil acidity, decrease biodiversity, pollute groundwater and cause coastal eutrophication. Once emitted back to the atmosphere it can contribute to climate change and decreased stratospheric ozone (UNEP 2004). The impacts continue as long as the nitrogen remains active in the environment, and it ceases only when Nr is stored for a very long time, or is converted back to non-reactive forms. Policy options aimed at addressing only a single impact and thus only one substance can lead to pollutant swapping. This illustrates the need for an approach that considers the multiple and linked impacts, and prevents the creation of reactive nitrogen.

Another example of multiple impacts from human activity is climate change. The links between climate

Box 8.2 System properties: thresholds, switches, tipping points and inertia

Identification and assessment of key human-environment interlinkages needs to take into account that most social and biophysical systems are characterized by dynamic system properties. These properties include thresholds, switches, inertia and time-lags, as well as feedback loops, illustrated in Box 8.1

Thresholds are sometimes referred to as tipping points. They are common in the Earth system, and represent the point of sudden, abrupt, or accelerating and potentially irreversible change switched on by natural events or human activities. For example, there is evidence to show that a decrease in vegetation cover in the Sahara several thousand years ago was linked to a decrease in rainfall, promoting further loss of vegetation cover, leading to the current dry Sahara. Examples of thresholds being crossed due to sustained human activities include the collapse of fisheries, eutrophication and deprivation of oxygen (hypoxia) in aquatic systems, emergence of diseases and pests, the introduction and loss of species, and regional climatic change.

Another example of switches or thresholds and interlinkages in environmental change is illustrated by the change from grass dominance to shrubland. Changes in the grazing and fire regime associated with land management practices during the past century are thought to have increased the woody plant density over significant areas of Australia and Southern Africa. Large-scale ecosystem changes (such as savannah to grassland, forest to savannah, shrubland to grassland) clearly occurred in the past (such as during the climatic changes associated with glacial and interglacial periods in Africa). Because these changes took place over thousands of years, diversity losses were ameliorated, since species and ecosystems had time to undergo geographical shifts. Changes in disturbance regimes

and climate over the coming decades are likely to produce equivalent threshold effects in some areas, but over a much shorter time frame.

Biogeochemical and social systems have time lags and inertia, the tendency to continue to change even if the forces that cause the change are relieved. For example, even if greenhouse gas concentrations in the atmosphere were to be stabilized today, increases in land and ocean temperatures due to these emissions will continue for decades, and sea levels will rise for centuries, due to time scales associated with climate processes and feedbacks (see Chapter 2). Time lags associated with human societies include the time between development of technologies, their adoption and behavioural changes needed, for example, for climate change mitigation.

Critical thresholds are the points where activities result in unacceptable levels of harm, for example, in terms of ecological change, and require responses. The complexity of the coupled human-ecological systems and our current state of knowledge of the dynamics of the system makes it hard to predict precisely where such thresholds lie. It also makes it challenging to identify measures to pre-empt the crossing of such thresholds. Consequently, society is often left coping with harmful environmental changes through mitigation, and if mitigation proves difficult, through adaptation to the change. With the unprecedented and increasing socio-economic impacts of humanity on ecological systems, there is concern that these systems may be nearing or have exceeded some critical thresholds, and as a result, it is increasingly likely that they will experience large, rapid and non-linear changes. The crossing of such thresholds is of significant concern for human well-being, as in the past they have led to the catastrophic disruption of societies.

Sources: Australian Government 2003, Diamond 2005, IPCC 2001a, IPCC 2001b, IPCC 2007b, Linden 2006, MA 2005a

Figure 8.3 The nitrogen cascade and associated environmental impacts

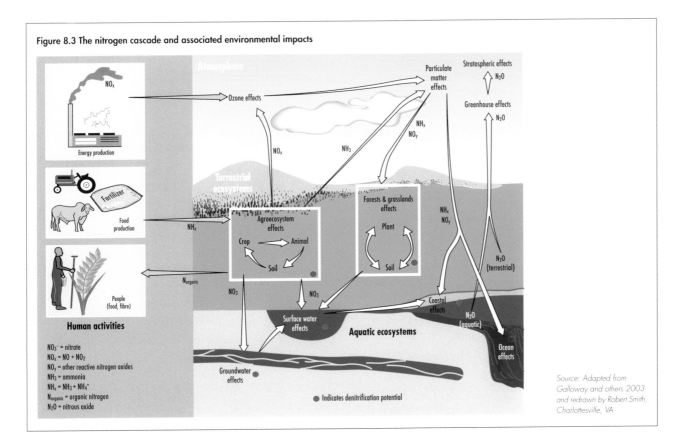

Source: Adapted from Galloway and others 2003 and redrawn by Robert Smith, Charlottesville, VA

change and biodiversity – both aquatic and terrestrial – are illustrative of the links between land, water and atmosphere (see Figure 8.4). Biodiversity is, in many instances, under multiple pressures. These can include land degradation, land and water pollution, and invasive alien species. Changes in climate exert additional pressures, which have affected biodiversity (see Chapter 5). These include the timing of reproduction of animals and plants and/or migration of animals, the length of the growing season, species distribution and population size, especially the poleward and upward shifts in ranges in plant and animal species, and the frequency of pest and disease outbreaks. Bleaching of coral reefs in many parts of the world has been associated with increased seasonal sea surface temperatures. Changes in regional temperatures have contributed to changes in stream-flow, and the frequency and intensity of extreme climatic events, such as floods, droughts and heat waves. These changes have affected biodiversity and ecosystem services (IPCC 2002, IPCC 2007b, CBD 2003, Root and others 2003, Parmesan and Yohe 2003). In high-latitude ecosystems in the northern hemisphere, there have been changes in species composition and even ecosystem types. For example, some boreal forests in central Alaska have been transformed into extensive wetlands

during the last few decades of the 20th century. The area of boreal forest burned annually in western North America has doubled in the last 20 years, in parallel with the warming trend in the region. Large fluctuations in the abundance of marine birds and mammals across parts of the Pacific and western Arctic may be related to climate variability and extreme events (CBD 2006). Species and ecosystems appear to be changing and/ or adapting at differing rates, which may also disrupt species relationships and ecosystem services.

The case of ongoing environmental change in the Arctic, discussed in detail in Chapter 6, also illustrates the land-water-climate change links. Some of the feedbacks and linkages are highlighted in Box 8.1. Ongoing changes in the Arctic include the effect of regional climate change on land cover, permafrost, biodiversity, sea ice formation and thickness, and meltwater intrusion into ice sheets, which increases the speed of their disintegration on the seaward edge. Feedbacks can result in further changes, with adverse impacts on human well-being, both in the Arctic and around the world.

A major interlinkage that occurs is due to changes in land use, particularly land cover. Changes in

Figure 8.4 Linkages and feedback loops among desertification, global climate change and biodiversity loss

Notes: Green text: major components of biodiversity involved in the linkages.

Bold text: major services impacted by biodiversity losses.

The major components of biodiversity loss (in green) directly affect major dryland services (in bold). The inner loops connect desertification to biodiversity loss and climate change through soil erosion. The outer loop interrelates biodiversity loss and climate change. On the top section of the outer loop, reduced primary production and microbial activity reduce carbon sequestration and contribute to global warning. On the bottom section of the outer loop, global warming increases evapotranspiration, adversely affecting biodiversity; changes in community structure and diversity are also expected because different species will react differently to the elevated CO_2 concentrations.

Source: MA 2005a

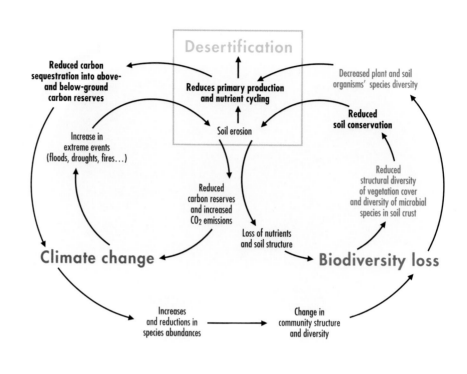

land use and/or land cover, such as deforestation and conversion to agriculture, affect biodiversity and waterbodies and contribute to land degradation (see Chapters 2–5). These activities not only change the biodiversity at the species level, but also result in habitat loss, fragmentation and alteration of ecosystems, as well as contribute to climate change by altering the local energy balance, reducing plant cover and loss of soil carbon. However, some changes in land use, such as afforestation and reforestation, can also result in an increase in biodiversity and increased local energy balance.

Land degradation can lead to the loss of genetic and species diversity, including the ancestors of many cultivated and domesticated species. This means losing potential sources of medicinal, commercial and industrial products. In addition, change from forest to agricultural or degraded lands affects biophysical and biogeochemical processes, particularly the hydrological cycle. The reduced water holding capacity of cleared land results in increased flooding, erosion and loss of the more fertile topsoil, resulting in less water and organic matter retained in the soil.

Consequently the siltation results in the degradation of waterbodies, such as rivers and lakes, by soil. In freshwater and coastal systems, land degradation affects sediment mobilization and transport. This, in turn, can affect biodiversity (Taylor and others 2007), such as that of coral reefs, mangroves and sea grasses, in adjacent coastal and shelf environments. In some cases, these effects are exacerbated by particle-reactive contaminants, including persistent organic pollutants (POPs), which are adsorbed onto soil particles.

Water resource management affects terrestrial, freshwater, coastal and nearshore (marine) systems. For example, water withdrawals and the rerouting of inflows, affect biodiversity, terrestrial and aquatic ecosystem functioning, and land cover. Chapters 3, 4 and 5 provide details on how pollution, siltation, canalization and water withdrawals adversely affect biodiversity (terrestrial, near coastal and aquatic), and change ecosystem functioning and composition upstream and downstream. They can also result in land degradation, especially salinization, and an increase in invasive alien species.

Increased levels of UV-B radiation are reaching the Earth's surface due to the depletion of the ozone layer by ozone-depleting substances. This has had a number of impacts on the biosphere. UV-B radiation affects the physiology and development of plants, influencing plant growth, form and biomass, although the actual responses vary significantly among species and cultivars. Increased UV-B radiation will probably affect biodiversity through changes in species composition, as well as affecting ecosystems through changes in competitive balance, herbivore composition, plant pathogens and biogeochemical cycles. Increased UV-B radiation reduces the production of marine phytoplankton, which is the foundation for aquatic food webs, and a major sink for atmospheric CO_2. It has also been found to cause damage to fish, shrimp, crabs, amphibians and other marine fauna during early development (see Chapters 2 and 6).

Environmental changes and human well-being

Environmental changes are not only interlinked through various human activities and biophysical processes, but also through how they affect human well-being. The different constituents of human well-being, including basic material needs (food, clean air and water), health and security, can all be influenced by single or multiple environmental changes through the alteration of ecosystem services (MA 2005a). Well-being exists on a continuum with poverty, which has been defined as "pronounced deprivation in well-being." Linked with

these are concepts of natural, human, social, financial and physical capital and the issue of substitution among these capitals (MA 2003).

Socio-economic sectors that are highly dependent on ecosystem services, such as agriculture, forestry and fisheries, have contributed to substantial net gains in human well-being, especially through provisioning services (such as food and timber) (MA 2005a). However, this has been at the cost of increased poverty for some groups, and environmental changes, such as land degradation and climate change. It is therefore important to consider the trade-offs and synergies that can arise between and among ecosystem services and human well-being when developing management options. More detailed analysis of the numerous impacts of environmental changes on human well-being is found in Chapters 2–5.

As seen in Chapter 7, the degree to which some groups are vulnerable to such changes depends on both their coping capacity and the state of land and water. For example, environmental changes, such as land degradation, have enhanced the destructive potential of extreme climatic events, such as floods, droughts, heat waves and storm surges. The increase in the frequency and intensity of extreme climate-related disasters in the last four decades provides evidence of this trend (Munich Re Group 2006). About 2 billion people were affected by such disasters in the 1990s:

Poor land-use policies contribute to land degradation which adversely affects human health, security and limits livelihood options.

Credit: Ngoma Photos

40 per cent of the population in developing countries, compared to a few per cent in developed countries (see Figure 8.5). A combination of the observed and projected figures for the first decade of the 21st century shows more than 3.5 billion people or 80 per cent of the population in developing countries affected by such disasters, while still only a few per cent are affected in developed countries (see Figure 8.5). The variation between developing and developed countries is a reflection of the multiple environmental changes that the different populations face, the socio-economic status of the countries, and the fact they are located in areas that are sensitive to climate variability and change, water scarcity, and, in some cases, conflict. Some of the increase is due to more people living on marginal (such as semi-arid and arid) land, and in coastal zones prone to disasters, such as storm surges (IPCC 2001b). Part of this increase in the number affected is attributed to the accelerated rate and magnitude of climate change and variability, land degradation and the scarcity of clean water in many parts of the world (UN 2004).

Environmental changes may affect human well-being in more than one way (see Figure 8.6). For example, land degradation not only threatens food production and contributes to water shortages, but may also have impacts across spatial and temporal scales and boundaries which means that human well-being in one locality may be influenced by drivers, pressures and changes caused outside the locality. Human well-being may also be affected by drivers and human impacts stemming from many different sectors.

There are increasing and cumulative human pressures on the Earth system, creating a variety of interacting forms of environmental change. The amount of change taking place begs the question as to whether there are biophysical thresholds and limits within which humanity must stay to avoid significant disruption to the planet's life support systems (Upton and Vitalis 2002). The history of past societies may provide insight into such thresholds and limits. Environmental degradation has been deduced to have played a key role in the decline and even collapse of whole societies. This includes societies in Mesopotamia 7 000 years ago (Watson and others 1998), as well as the Easter Island society and the Norse society in Greenland within the last millennium. For the Maya in Central America, there are multiple hypotheses, including one of periodic droughts acting as added stress on top of other environmental changes, especially deforestation and overgrazing (Diamond 2005, Linden 2006, Gallet and Genevey 2007). The studies of those societal declines suggest that the environment-society interaction may have gone beyond a point of no return, whereby society did not have the capacity to reverse the ecological degradation that eventually undermined its existence (Diamond 2005). However, it must be understood that the scale of contemporary environmental changes is far greater than that which led to the localized collapse of the spatially limited societies mentioned here.

A key challenge in sustainable development is to avoid a development path that could lead society to such points of no return (Diamond 2005). Such efforts could be facilitated by enhancing the understanding of how environmental changes interact within the coupled human-environment system. A strengthened knowledge base should include information on the risk of exceeding thresholds and undermining life-supporting processes, how crossing thresholds may lead to degradation of ecosystem services, and how this would have impacts on development paths in terms of expanding or limiting people's capabilities to be and achieve what they value. Such knowledge would underpin the choices and trade-offs with respect to distribution of access to environmental services and exposure to environmental stress among different groups of people. The knowledge

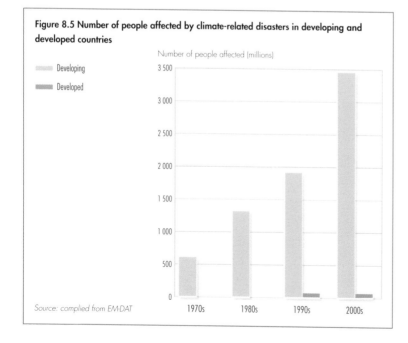

Figure 8.5 Number of people affected by climate-related disasters in developing and developed countries

Developing
Developed

Number of people affected (millions)

Source: complied from EM-DAT

1970s   1980s   1990s   2000s

base would be part of the continued evolution of adaptive environmental governance, which incorporates ideas of environmental management, and the integration of environment into development policies (see last section of this chapter).

INTERLINKAGES AND ENVIRONMENTAL GOVERNANCE

Governance systems can be considered as institutional filters, mediating between human actions and biophysical processes (Kotchen and Young 2006). Interlinked environment-development challenges require effective, linked and coherent governance and policy responses within the framework of sustainable development. Governance for sustainable development requires effective administrative executive bodies, and enabling legal and regulatory frameworks. Progress in this area over the last 20 years is mixed, with limited success. However, there are encouraging developments at international, regional and national levels, including the private sector and civil society, which provide valuable lessons and directions for managing interlinked environment-development

challenges. This includes the emergence of flexible, more adaptive governance entities.

Governance regimes have undergone a significant evolution in response to different environment and development challenges since the Brundtland Commission. Milestones include the UN Conference on Environment and Development and its achievements, including Agenda 21; the Millennium Summit and Declaration; and the 2002 World Summit on Sustainable Development (WSSD) in Johannesburg and the Johannesburg Plan of Implementation (UNEP 2002a, Najam and others 2006). An examination of the landscape of environmental governance over the last 20 years shows that states have created a growing number of institutions, authorities, treaties, laws and action plans to conserve and safeguard the environment, and more recently, to respond to new understanding of the extent and implications of global environmental change. Through summits, states have set common goals and outlined key definitions. Many of the responses that have been

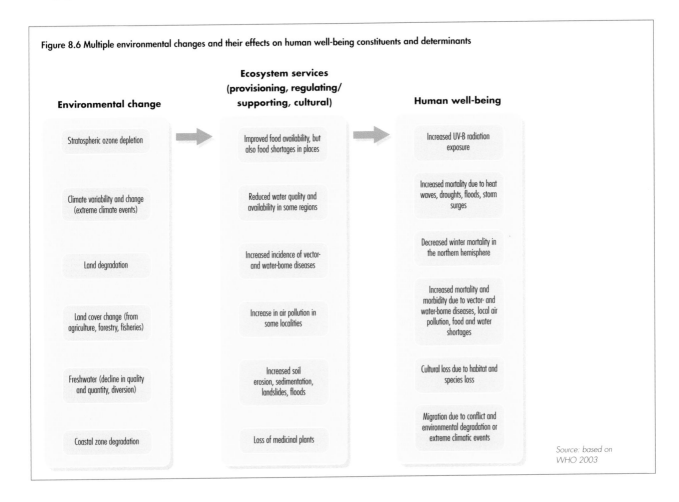

Figure 8.6 Multiple environmental changes and their effects on human well-being constituents and determinants

Environmental change

- Stratospheric ozone depletion
- Climate variability and change (extreme climate events)
- Land degradation
- Land cover change (from agriculture, forestry, fisheries)
- Freshwater (decline in quality and quantity, diversion)
- Coastal zone degradation

Ecosystem services (provisioning, regulating/ supporting, cultural)

- Improved food availability, but also food shortages in places
- Reduced water quality and availability in some regions
- Increased incidence of vector- and water-borne diseases
- Increase in air pollution in some localities
- Increased soil erosion, sedimentation, landslides, floods
- Loss of medicinal plants

Human well-being

- Increased UV-B radiation exposure
- Increased mortality due to heat waves, droughts, floods, storm surges
- Decreased winter mortality in the northern hemisphere
- Increased mortality and morbidity due to vector- and water-borne diseases, local air pollution, food and water shortages
- Cultural loss due to habitat and species loss
- Migration due to conflict and environmental degradation or extreme climatic events

Source: based on WHO 2003

put in place nationally, regionally and internationally
are not necessarily well matched, and there is often
a "problem of fit" between the institutions created,
and the ecological and development concerns being
addressed (Young 2002, Cash and others 2006).

Commonly cited areas of concern regarding
international environmental governance (IEG) include
(Najam and others 2007):

- proliferation of multilateral environmental
  agreements (MEAs), and fragmentation of IEG;
- lack of cooperation and coordination among
  international organizations;
- lack of implementation, enforcement and
  effectiveness of IEG;
- inefficient use of resources;
- the challenge of extending IEG outside the
  traditional environmental arena; and
- involvement of non-state actors in a state-centric
  system.

Informal consultations by the UN General
Assembly on the institutional framework for the
United Nations' environmental activities identified
similar areas of concern among governments.

While the large number of bodies involved with
environmental work has allowed specific issues
to be addressed effectively and successfully, it
has also increased fragmentation, and resulted
in uncoordinated approaches in both policy
development and implementation. It has further
placed a heavy burden on countries in terms of
participation in multilateral environmental processes,
compliance with and effective implementation
of legal instruments, reporting requirements and
national-level coordination. Whereas a large body
of policy work has been developed and continues
to expand, a growing gap remains between
normative and analytical work and the operational
level. The focus of attention and action is shifting
from the development of norms and policies to
their implementation in all countries. In that respect,
capacity building at all levels, especially in
developing countries, is of key importance (Berruga
and Maurer 2006).

This section summarizes developments in
environmental governance at national, regional
and international levels, in the context of how
institutions respond to a situation characterized by

environmental changes that are interacting across themes, as well as across spatial and temporal scales and boundaries. The following section looks at some of the opportunities to change, adapt or reorient this current governance regime towards a system that could more effectively address the human and biophysical interlinkages.

National level

The national environmental governance landscape evolved in a largely linear, sectoral fashion to provide specific services over a short- or medium-time scale, often related to electoral cycles. Such arrangements are not always well suited to respond to more complex, cross-sectoral challenges posed by sustainable development, which has a longer-term intergenerational time horizon, requiring sustained commitment going beyond the typical 4–5 year electoral cycles. With its need for a "triple bottom line" focus on environment, economy and society, sustainable development contradicts the way policies have traditionally been formulated and developed (OECD 2002).

Effective environmental governance depends on a well-functioning executive, legislature and judiciary, as well as participation by all stakeholders, including the electorate, civil society and the private sector. This can result in conflicting interests, and there is a need for well-defined mechanisms and processes to involve the various groups in collective decision making and in finding solutions (OECD 2002). The electorate has become a key stakeholder in the management of the environment, supporting legislative changes, and protecting environmental resources and the rights of communities (Earthjustice 2005). Business and industry are increasingly engaging in responsible corporate citizenship, making efforts to improve and report on their environmental and social performance, particularly related to climate change, and in high-impact industries that face criticism from stakeholders and public institutions (UNEP 2006a).

The effective implementation of environmental policies, particularly in the case of binding international commitments, such as MEAs, involves a simultaneous and interconnected process at the

---

Box 8.3 Examples of national-level mechanisms that bridge environmental governance challenges

Coordinating mechanisms in the prime minister's or president's office including inter-cabinet or inter-ministerial committees, such as the National Environmental Board in Thailand, chaired by the prime minister. **Sustainable development committees**, often established after the UN Conference on Environment and Development, coordinate national and/or international policy related to sustainable development at interdepartmental and interagency levels.

Judicial institutions and mechanisms are central to promoting the goals of sustainable development, interpreting and ensuring effective implementation of legislation, integrating emerging principles of law, handling diverse sectoral laws, and providing an opportunity for society to ensure protection of fundamental rights such as the right to a clean and healthy environment. An important area of activity dealing with interlinked environmental challenges has been the strengthening of national laws and institutional frameworks, both through the development of framework environmental legislation, and the development of integrated sectoral legislation. This seeks to improve the implementation of several MEAs related to one issue, such as biodiversity or chemicals .

National Focal Points (NFPs) or lead agencies are designated for the coordination of the implementation of binding international commitments such as MEAs and for national reporting to CSD, sometimes supported by **national committees**.

National Sustainable Development Strategies (NSDS) that "should build upon and harmonize the various sectoral economic, social and environmental

policies and plans that are operating in the country" were called for in Agenda 21. The WSSD urged states to not only formulate NSDSs, but also to begin implementation by 2005, while integrating the principles of sustainable development into country policies and programmes. This is one of the targets of the Millennium Declaration. There have been mixed results regarding governance structures for NSDSs. Nevertheless NSDSs and associated planning processes provide unique opportunities to address interlinkages, such as those involving local and national development, environmental issues and global environmental threats, through links to the MEAs.

Planning and development bodies and mechanisms, such as commissions and authorities, are crucial macroeconomic institutions that take a long-term view of development issues, and can promote a cross-sectoral, integrated and interlinked approach between economic, social and environmental issues. In developing and middle-income countries, initiatives such as UN Development Assistance Frameworks (UNDAFs), and national planning processes, such as poverty reduction strategies (PRS), include the environment as a key factor to be considered in the context of development, poverty reduction and achieving other aspects of human well-being, such as health, food and security.

Other innovative mechanisms include the creation of a **Commissioner of the Environment and Sustainable Development (CESD)** within the Office of the Auditor-General of Canada to monitor and report on the federal government's performance in environmental and sustainable development areas. Fact-based, independent reports from the commissioner help Parliament to hold the government accountable for its performance in these areas.

Sources: OAG 2007, UNEP 2005, UNEP 2006b, UNESCAP 2000

---

domestic and intergovernmental levels of policy making to follow up on agreements. A number of obstacles to coordination of interlinkages arise at the national level. They may be horizontal in nature, surfacing across government ministries and agencies, such as between MEAs and national focal points for negotiation and policy implementation, or between the environment ministries or agencies and development planning authorities. Institutional constraints may also arise vertically, across different levels of governmental administration, for example, where initiatives at the provincial, district or village level may not support, or may even be contradictory to national policies or programmes (DANCED 2000).

A major impediment faced by many countries is the lack of capacity at national and sub-national (federal, provincial, state and local government) levels. In addition, there may be inadequate financial resources to implement policies and agreements (UNDP 1999, UNESCAP 2000). The proliferation of MEAs, sometimes cited as an indicator of the increased recognition of and response to environmental challenges at the international level, has shown a trend towards greater complexity over time, and placed a huge demand on national-level capacity to implement their requirements (Raustiala 2001). For example, in Thailand the National Environmental Board (NEB) has 42 sub-committees created to oversee the implementation of MEAs and other environmental

policies (UNU 2002). With increasing recognition of this burden, there are efforts to streamline and harmonize implementation among the MEAs in order to reduce the burden at national level, as well as to maximize the synergies and interlinkages (UNU 1999, UNEP 2002b). This has included developing coordinating mechanisms, such as national committees, streamlining legislation and reporting, and capacity building (see Box 8.3).

Regional level

The regional level presents an important middle ground for environmental governance. Regions (bioregions or institutional entities) provide a bounded context within which policies and programmes can be devised and implemented, that are relevant and responsive to local and interlinked conditions and priorities. Though rule making for better environmental governance is primarily a function of the national, international and global levels, the regional level has emerged as an important intermediate link for action and implementation. The pressures of environmental changes come to bear on particular localities, and more often than not cross national boundaries and intersect with development concerns. Responses to environmental challenges are encapsulated by a number of regional institutions and mechanisms that are important for addressing and coordinating such environment-development challenges and interlinkages (see Box 8.4).

Box 8.4 Regional institutions and mechanisms

Regional integration agreements can harmonize standards among member countries (such as the European Union's new Sustainable Development Strategy 2007), and implement programmes that foster regional cooperation in, for example, fisheries, chemicals and hazardous waste management (such as NEPAD's Action Plan of the Environment Initiative).

Regional MEAs or implementation mechanisms can bridge international and national levels (such as Africa's Bamako Convention in response to the Basel Convention on the Control of Transboundary Movements of Hazardous Wastes and their Disposal). They can reinforce and translate international commitments (such as the Andean Community's Regional Biodiversity Strategy to implement the Convention on Biological Diversity).

Regional ministerial arrangements, such as the African Ministerial Conference on the Environment (AMCEN) and the Tripartite Environment Ministers' Meetings (TEMM) between China, Korea and Japan, are high-level political fora that can set regional priorities and agendas, and raise awareness of regional concerns.

Mechanisms attached to regional trade agreements, such as NAFTA's Commission for Environmental Cooperation (CEC) and the ASEAN Agreement on Transboundary Haze Pollution, can address cross-border environmental issues through intergovernmental cooperation.

Regional or sub-regional environment and development organizations, such as the UN regional economic commissions, regional development banks, and the Central American Commission on Environment and Development (CCAD), can play an important role in data collection and analysis, capacity building, and resource allocation and management.

Transboundary or bioregion-based plans and programmes, such as the Mekong River Commission, the Pacific Regional Environment Programme (SPREP) and UNEP's Regional Seas Programme, are important for data collection, analysis and dissemination, sectoral and resource assessment, policy development, capacity development and monitoring.

Regional approaches tend to work partly because of established mechanisms for collective experimenting, and the learning and sharing of experiences. Geographical proximity provides a basis for the rapid diffusion of practices, and reduces the time needed to adapt to new conditions. In addition, actions implemented at the regional level can benefit from the continuous emergence of implementation opportunities provided by other complementary initiatives (Juma 2002). Nevertheless, there are still many challenges to making regional mechanisms work and fulfil their functions or mandates, particularly for developing regions. There are challenges in terms of financial resources, and the human capacity for implementation and institutional interplay for coherence and effectiveness.

International level environmental governance

At the international level, the key actors with respect to governance and management regimes relevant to environment, development and their interlinkages are the United Nations, the MEAs, and regimes dealing with development, trade, finance and related fields. The private sector, research and scientific bodies, civil society, trade unions and other stakeholders are also key players, and their individual and collective actions have been central to mainstreaming the environment into development. The need for institutional coordination and cooperation has become an increasing imperative, due to the heavily fragmented structure of international environmental governance, and similar issues in development governance (UNEP 2002c, Gehring and Oberthur 2006, Najam and others 2007, UN 2006).

The international governance landscape has multiple organizations that were established to address environment and human interactions. Within this landscape there are several distinguishable regimes for environment, development, trade and sustainable development (the latter is the most loosely connected, as it brings the environment and socio-economic components together). Cooperation and coordination under each of the regimes generally takes place through lead organizations (such as UNEP for environment, WTO for trade, UNDP and the World Bank for development, and CSD for sustainable development).

The development of multilateral environmental agreements (MEAs) over the last decades has been remarkable (see Figure 1.1 in Chapter 1). There are now more than 500 international treaties and other agreements related to the environment, of which 323 are regional and 302 date from the period between 1972 and the early 2000s (UNEP 2001a).

The largest cluster of MEAs is related to the marine environment, accounting for over 40 per cent of the total. Biodiversity-related conventions form a second important but smaller cluster, including most of the key global conventions, such as the 1973 Convention on International Trade in Endangered Species of Wild Fauna and Flora (CITES) and the 1992 Convention on Biological Diversity. CITES and the Basel Convention on the Control of Transboundary Movement of Hazardous Wastes and their Disposal are two of a few MEAs that regulate trade. They also highlight some of the interlinkages between environment and trade. One of the challenges faced in enforcement is the growth of illegal trade in both wildlife and hazardous waste. Box 8.5 and Figure 8.7 highlight some of the issues.

Most of these institutions and treaties have independent governing bodies with independent mandates and objectives. The interlinkages among these bodies are complex (see Figure 8.8), and the systems have been described as fragmented and overlapping (UN 1999). With the growth of the number and diversity of actors and organizations, interagency mechanisms, such as the Environmental Management Group (EMG), UN Development Group and the liaison groups between MEA secretariats, have been created to bridge independent agencies and promote greater cooperation. The UN Economic and Social Council and the UN General Assembly play major roles in coordination, and they have created fora for promoting cooperation with other institutions, such as the WTO and Bretton Woods institution that are outside of the UN system.

At the international level, business and industry have played increasingly important roles in connecting the environment, development and trade regimes through direct interaction with global institutions. For example, organizations such the World Business Council for Sustainable Development and processes such as the Global Compact are bridging international action with that of business actions (WBCSD 2007, UN Global Compact 2006). The power of markets has equally played an important role in bridging the interlinkages between environmental change (such as climate change and the carbon markets) and development (such as through the Clean Development Mechanism). The international system of investment and finance fuels global development, and investment decisions – from where to build a dam to which type of automobile to develop – and all have direct impacts on the environment. However, investors are beginning to understand the powerful implications of global environmental change, particularly climate change, on portfolio performances across sectors, and are seeking out various business models to manage environmental risk. The Principles for Responsible Investment (PRI) are a major commitment by signatory institutional investors and asset managers to integrate environmental and social issues into their decision making processes, and provide a significant platform for their inclusion in mainstream investment practices (UNEP 2006d and UNEP 2006e).

Box 8.5 Eco-crime exploits loopholes of legal regimes

Few of the MEAs actually regulate trade. Two exceptions are CITES and the Basel Convention. While enforcement to regulate trade is a key element for the implementation of both conventions, effectiveness of both MEAs is being undermined by illegal trade, highlighting the interlinked challenges of trade and environment, particularly in relation to thriving black markets across the globe.

The basic criteria required to fulfil the mandate of the Basel Convention (Secretariat of the Basel Convention undated) include the existence of a regulatory infrastructure that ensures compliance with applicable regulations, as well as enforcement personnel (competent authorities, police, customs officers, port and airport authorities, and coast guards) trained in technical areas, including procedures and identification of hazardous wastes. However, the lack of human resources, training and equipment are some of the barriers to effective implementation. Others include inadequate industry response to treat, recycle, re-use and dispose of wastes at source and an inadequate information network and alert systems to assist with detection of illegal traffic in hazardous wastes. In an effort to try and address some of these gaps, the Basel Convention parties have developed an illegal traffic guidance manual, while a guide for legal officials is under development and training is provided to developing countries through the Basel Convention Regional Centres.

UNEP estimated the annual revenue from the international illegal wildlife trade to be US$5–8 billion. While enforcement in the trade of wildlife (especially through the use of permits, licences and quotas) has proved effective in many cases, illegal trade (and the subsequent creation of "black markets") will continue as long as consumer demand is high, profits remain enormous and risks remain low. As with many environmental concerns, the characterization of the wildlife trade as a mere "environmental" consequence tends to reduce its importance on national policy making agendas, vis-à-vis security and economic issues, resulting in fewer resources and less attention being committed to it. Another major problem is that CITES itself contains several loopholes which are extensively exploited by black marketers. Such loopholes include trading with non-parties, and exemptions for sports hunting of the captive-breeding programme.

Other MEAs also relate to trade and the environment, but have been undermined by "eco-crimes." Stronger international regulations, effective governance structures for enforcement at all levels and a national commitment to sustainable development can help align developmental and environmental needs.

Sources: Lin 2005, Secretariat of the Basel Convention 1994, Secretariat of the Basel Convention undated, UNEP 1998, UNEP 2006c, YCELP undated

---

Figure 8.7 Waste trafficking

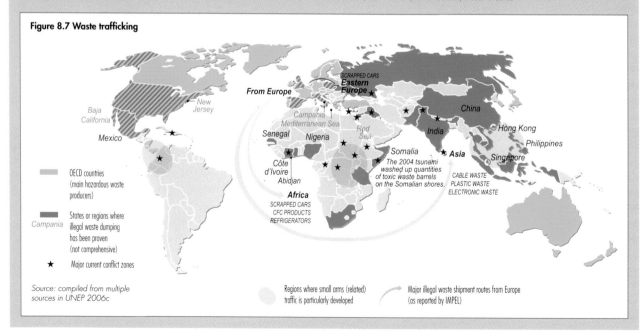

Source: compiled from multiple sources in UNEP 2006c

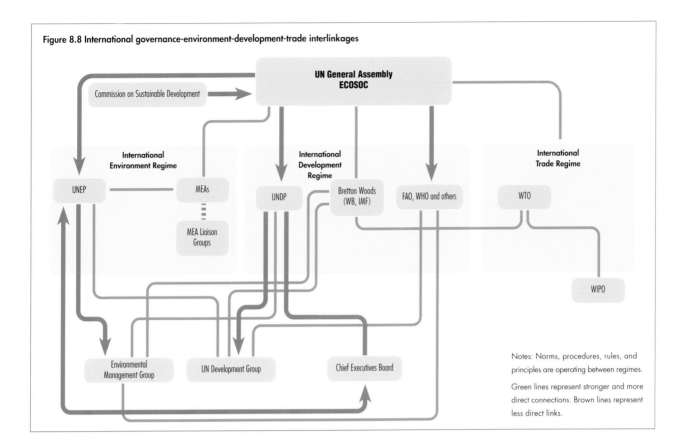

Figure 8.8 International governance-environment-development-trade interlinkages

UN General Assembly ECOSOC

Commission on Sustainable Development

International Environment Regime

UNEP   MEAs

MEA Liaison Groups

International Development Regime

UNDP   Bretton Woods (WB, IMF)   FAO, WHO and others

International Trade Regime

WTO

WIPO

Environmental Management Group   UN Development Group   Chief Executives Board

Notes: Norms, procedures, rules, and principles are operating between regimes.

Green lines represent stronger and more direct connections. Brown lines represent less direct links.

In the last 20 years, there has been a significant rise of international plurality. Civil society has a major role under the international environmental, development and trade regimes, and plays an essential role in providing analysis, advocacy, and awareness raising to these regimes. The vertical interlinkages between national and international levels have been especially well developed in this period, and now many national and local civil society actors (such as NGOs and indigenous groups) play major roles in international decision making, either as observers or as members of national delegations, by providing commentary and analysis, or through protest and civil action. Horizontally, the interlinkages between civil societies are developing, and many have formed umbrella groups (such as the Climate Action Network), and cooperate on common and overlapping issues and interests. Civil society has not, however, adequately developed the issue of interlinkages (among drivers, environmental changes and impacts) as a subject area for its attention. Most civil society groups remain focused on single-issue areas, such as climate change, wildlife conservation, poverty reduction or human rights, and have not recognized the need to address the interlinkages among these issues.

OPPORTUNITIES FOR MORE EFFECTIVE ENVIRONMENTAL GOVERNANCE

The previous section has demonstrated that the environmental governance system is multi-scaled, diverse and extends into development governance regimes. The boundaries separating institutional systems, like those of ecosystems, are often indistinct. Consideration of the interactions between these international arrangements are important in understanding and strengthening their effectiveness in addressing interlinkages between environmental changes, which are interacting across spatial and temporal scales and boundaries (Young 2002). Not only does environmental governance involve many institutional regimes, but it also involves trade-offs and transaction costs that are critical to adaptation to and mitigation of environmental changes, and the improvement of human well-being.

The magnitude of the interconnectedness of environmental changes does not mean, however, that policy-makers are only faced with the choice of "doing everything at once in the name of integrated approaches or doing nothing in the face of complexity" (OECD 1995). Interlinkages offer opportunities for more effective responses at the national, regional and global levels. Sometimes, responses need to be integrated, and occur as a chain of actions to match the complexity of the situation; sometimes more restricted and targeted responses are called for (Malayang and others 2005). Understanding the nature of interlinkages, their interplay, and identifying which linkages need to be acted on at which scale, offers opportunities for more effective responses at the national, regional and global levels.

The complexity and the magnitude of the interlinkages among the environmental changes requires that policy-makers prioritize which interlinkages require immediate attention. Appropriate policies and measures can then be adopted nationally to mitigate the negative impacts, and to maximize the effectiveness of existing policies. Such understanding can also guide parties to MEAs to decide which types of collaboration and which types of joint work programmes could be prioritized and strengthened. A scientific understanding of the key interlinkages among the environmental changes (and between environmental and socio-economic changes) is still not fully developed nor widely understood, and will require future assessments and research in order to guide such policy making. However, it is clear that one of the major interlinkages is driven by climate change, seen in its roles in land and water degradation.

An adaptive approach to environmental governance (see later sections) may address the call for enhanced coordination, and improved policy advice and guidance. Development of a long-term strategic approach for enhancing the infrastructure and capacities for keeping the environmental situation under review may help in identifying key interlinkages at and between both the national and international levels. There is broad agreement on the need for better treaty compliance, while respecting the legal autonomy of the treaties. A process that considers interlinkages may help identify areas for cooperation among the treaties, and for more effective enforcement and compliance at national level as well as for related capacity building and technology transfer.

Considerations on the overall normative basis for environmental governance may help identify more effective institutional structures. Better integration of environmental activities in the broader sustainable development framework at the operational level, including through capacity building, requires an in-depth understanding of interlinkages. Current gaps and needs relating to existing national and international infrastructure and capacities for integrating environment into development could be identified, and a long-term approach for addressing such needs could be explored. The subsequent section assesses the opportunities in the context of interlinkages.

UN reform and system-wide coherence on the environment

Efforts to enhance governance and system-wide coherence have been a recurrent feature of the United Nations (Najam and others 2007). Recent processes within the United Nations itself have acknowledged that it has not been as effective as it could be. The UN Secretary-General's High-Level Panel on UN System-wide Coherence in the Areas of Development, Humanitarian Assistance and the Environment (the Coherence Panel) states for instance that: "The UN has outgrown its original structure. We have seen how weak and disjointed governance and inadequate and unpredictable funding have contributed to policy incoherence, duplicating functions and operational ineffectiveness across the system" (UN 2006).

The importance of UN system-wide coherence in order to address environmental change has also been a recurring theme, particularly over the last decade (Najam and others 2007). Table 8.1 provides a summary of the recommendations of three recent processes. One was a review of the requirements for a greatly strengthened institutional structure for international environmental governance (IEG) in 2000, and adoption of an IEG package (UNEP 2002b). The second was the outcome of the 2005 World Summit, which called for stronger system-wide coherence within and between the policy and operational activities of the United Nations, in particular in the areas of humanitarian affairs, development and environment. The third was the Coherence Panel. The panel's mission has been to explore how the United Nations can be better structured to help countries achieve the MDGs and other internationally agreed development goals, and how the United Nations can better respond to major global challenges such as environmental degradation (UN 2006).

	The International Environmental Governance (IEG) Initiative (UNEP 2002c)	The 2005 World Summit Outcome (UN 2005)	Selected recommendations of the Secretary-General's High-level Panel on UN System-wide Coherence (UN 2006)
UNEP and the environment in the UN	A Strengthened UNEP through: ■ improved coherence in international environmental policy making – the role and structure of the Governing Council/ Global Ministerial Environment Forum; ■ strengthening the role and financing of UNEP; and ■ strengthening the scientific capacity of UNEP.	More efficient UN environmental activities through: ■ enhanced coordination and improved policy advice and guidance; and ■ strengthened scientific knowledge, assessment and cooperation.	■ strengthen and improve IEG coherence by upgrading UNEP with a renewed mandate and improved funding; and ■ UNEP's technical and scientific capacity should be strengthened for monitoring, assessing and reporting on the state of the global environment.
UN system-wide coherence	■ enhanced coordination across the UN system – the role of the Environmental Management Group.	■ stronger system-wide coherence within and between the policy and operational activities of the United Nations, in particular in the areas of humanitarian affairs, development and environment; and ■ agreement to explore the possibility of a more coherent institutional framework. including a more integrated structure.	■ UN Development Policy Operations Group within the Chief Executives Board for Coordination framework bringing together heads of all UN organizations working on development; ■ more effective cooperation among UN agencies, programmes and funds working in different thematic areas of the environment; and ■ an independent assessment of the current UN system of IEG should be commissioned.
MEAs	■ improved coordination among and effectiveness of multilateral environmental agreements (MEAs).	■ better treaty compliance, while respecting the legal autonomy of the treaties.	■ more efficient and substantive coordination to support effective implementation of the major MEAs.
Country-level operations	■ capacity-building, technology transfer and country-level coordination for the environmental pillar of sustainable development.	■ better integration of environmental activities in the broader sustainable development framework at the operational level, including through capacity building.	■ One UN Country Programme to deliver as one at the country level; ■ UNEP to provide substantive leadership and guidance at the country level, including building capacity and mainstreaming environmental costs and benefits into policy making; and ■ UN Sustainable Development Board, reporting to ECOSOC, to oversee the performance of the One UN at country level.

There are clear commonalities in the outcomes and recommendations of these three processes, which relate to UNEP and environment in the UN system, UN system-wide coherence, implementation of the MEAs, and country-level operations.

Calls for a UN or World Environment Organization (UNEO or WEO) have been made since the early 1970s (Charnovitz 2005). There is still much debate about whether there is a need for such an organization, and what form it might take in order to address the shortcomings of the present international environmental governance system (Charnovitz 2005, Speth and Haas 2006). Suggested functions include planning, data gathering and assessment, information dissemination, scientific research, standards and policy setting, market facilitation, crisis response, compliance review, dispute settlement and evaluation (Speth and Haas 2006, Charnovitz 2005).

A number of studies have observed that, despite significant achievements, the current governance regimes are inadequate and unable to deal effectively with the complexity of the interlinked human-biophysical or the social-ecological systems (Najam and others 2007, Kotchen and Young 2006, Olsson and others 2006). The current reform processes and debates offer a significant opportunity for addressing many of the interlinkages within and between environmental change and environmental governance at all scales, because much of what occurs or is agreed at the global level has to be addressed or implemented at the national and sub-national levels.

Better treaty compliance and implementation

The informal consultations by the UN General Assembly on the institutional framework for the United Nations' environment-related activities identified a range of views among member states on how to ensure better treaty compliance. Despite some value in specificity, there was widespread support for a much more coherent system dealing with the multitude of environmental issues currently under discussion. Issues raised included the material limitations to attend and participate meaningfully in a multitude of meetings, as well as the administrative costs and heavy reporting burden. This burden also extended to capacities required to implement legal agreements, affecting the legitimacy of such instruments and thus reinforcing the argument that enhanced capacity building is essential, especially for developing countries. On compliance, there were different perspectives. Some were in favour of improved monitoring and compliance mechanisms, like the establishment of a voluntary peer-review mechanism on compliance,

while others supported capacity building (Berruga and Maurer 2006).

One challenge is that thematic responsibilities often fall under several different MEAs, such as biodiversity which falls under the CBD, CITES, Ramsar, CCD, CMS and the World Heritage Convention. Also, one MEA can contribute to the objectives of other MEAs. For instance, ozone-depleting substances (ODS), which are also greenhouse gases, are regulated under the Montreal Protocol. By 2004, emissions of these gasses were about 20 per cent of their 1990 levels (IPCC 2007a). The fact that the major environmental changes are interlinked offers opportunities for cooperation among the MEAs at many levels.

Some voluntary cooperative mechanisms now act as bridges among secretariats of conventions. There is the Joint Liaison Group on the conventions on climate, biodiversity and desertification, and the Biodiversity Liaison Group, which involves five biodiversity-related conventions. Potential avenues

Biodiversity issues, at all levels – genes, species and ecosystems – are covered by several MEAs such as CBD, CITES, RAMSAR, CCD, CMS and WHC.

Credit: Ferrero J.P./Labat J.M./ Still Pictures

for improved cooperation among MEAs and between MEAs and UNEP have been explored through informal consultations.

While compliance with and enforcement of a treaty is first and foremost the responsibility of the parties to the conventions, the parties frequently call on support from other institutions, individually and collectively. The Global Environment Facility (GEF) is the funding mechanism for multiple MEAs, and therefore has a major influence on the operational activities and priorities of the participants, namely the implementing and executing agencies, and the national or regional institutions involved in implementation. The GEF is therefore well placed to focus activities on interlinkages and exploiting synergies between the focal areas (biodiversity, climate change, international waters, land degradation and persistent organic pollutants (POPs)), and between the respective MEAs. In addition, the GEF finances multifocal area projects to promote sustainable transport, conservation and sustainable use of biodiversity. These are important to agriculture, sustainable land management, adaptation to climate change, and national capacity assessment and development. Other initiatives in support of better treaty compliance include the third Montevideo Programme for the Development and Periodic Review of Environmental Law for the first Decade of the Twenty-first Century (UNEP 2001b), and guidelines on compliance with and enforcement of multilateral environmental agreements, which are complemented by a manual on compliance with and enforcement of MEAs (UNEP 2002c, UNEP 2006b).

Future opportunities for strengthening compliance with and implementation of MEAs at the national level may include greater focus on the creation of integrated or umbrella legislation for MEAs that are related or which overlap. With the growing number of MEAs, and the shift from negotiations to implementation (Bruch 2006), this option is increasingly attractive for countries that have passed the relevant legislation but do not implement it. Benefits of such an umbrella approach could include more coherent national legal frameworks, promotion of institutional coordination, or even cost effectiveness (Bruch and Mrema 2006). Umbrella approaches are relatively new, but there are some good examples of national legislation implementing biodiversity-related and chemical-related MEAs (Bruch and Mrema 2006).

An umbrella format at the international level was already proposed by the Brundtland Commission in 1987. It recommended that "the General Assembly commit itself to preparing a universal Declaration and later a Convention on environmental protection and sustainable development." It stressed the need, in building on existing declarations, conventions and resolutions, to consolidate and extend relevant legal principles on environmental protection and sustainable development (WCED 1987). While the first element of the recommendation from the Brundtland Commission was implemented through the Rio Declaration on Environment and Development, the idea of a universal convention has so far not been pursued by UN member states. The idea was, however, visited by stakeholders, led by the World Conservation Union (IUCN), in the form of a Draft Covenant on Environment and Development. This was launched in 1995 at the United Nations' Congress on Public International Law (IUCN 2004).

The interlinked nature of the environment and development challenges, and the diverse landscape of environmental governance may warrant regular reviews of the overall normative basis for international environmental cooperation. Ideally, the multilateral governance structures would flow from an agreed normative basis relating to the overarching purpose and scope of environmental cooperation and its contribution to development. They would deal with key principles for such cooperation, general rights and obligations of states, and key structures needed to support such intergovernmental cooperation, including capacity building. Considerations on the overall normative basis for environmental governance at both national and international level may help identify more effective institutional structures.

Integrating environment into development

The integration of environmental activities into the broader development framework is at the heart of MDG 7 on achieving environmental sustainability (UN 2000). Recognition of the need for integration of environmental concerns into public and private social and economic sector institutions, which was greatly enhanced by the vision put forward by the Brundtland Commission, has increased tremendously over the last decade at both national and international levels.

A key approach to integration of environment into development is achieving more sustainable patterns of consumption and production (SCP), as facilitated through the Marrakech Process (see Box 8.6). The overarching objective is to decouple economic growth from environmental damage, in both developed and developing countries, through the active engagement of both the public and private sectors. This relates to all stages in the life cycle of goods-and-services, and requires a range of tools and strategies, including awareness raising, capacity building, design of policy frameworks, market-based and voluntary instruments, and consumer information tools.

SCP is becoming a priority for countries worldwide, and there are many initiatives and programmes in addition to the Marrakech Process. Unsustainable patterns of consumption and lifestyles in developed countries have so far proved a particularly intractable problem. These forms of consumption result in, by far, the majority of negative environmental impacts associated with production and consumption of goods-and-services. It is necessary to look at innovative measures to meet (material) needs, and develop new innovative product and service systems. This is especially important when considering the new emerging "global consumer class," with large groups of middle-class consumers showing increasingly similar consumption patterns in rapidly-developing countries, such as Brazil, China and India (Sonnemann and others 2006).

One of the main messages in developing policies for sustainable consumption and production is that one single instrument will not fix the problem; it is necessary to design a package of different instruments, including regulatory frameworks, voluntary measures and economic instruments. Likewise, it is important to actively involve all stakeholders: government,

---

Box 8.6 Sustainable consumption and production: the Marrakech Process

Sustainable consumption involves the choices consumers make, and the design, development and use of products and services that are safe, and energy and resource-efficient. It considers the full life-cycle impacts, including the recycling of waste and use of recycled products. It is the responsibility of all members of society, and includes informed consumers, government, business, labour, consumer and environmental organizations. Instruments to promote sustainable consumption include sustainable or green procurement, economic and fiscal instruments to internalize environmental costs, and use of environmentally sound products, services and technologies.

Sustainable and cleaner production is "the continuous application of an integrated preventive environmental strategy to processes, products, and services to increase overall efficiency, and reduce risks to humans and the environment. Cleaner production can be applied to the processes used in any industry, to products themselves and to various services provided in society." This broad term encompasses such concepts as eco-efficiency, waste minimization, pollution prevention, green productivity and industrial ecology. Cleaner production is not anti-economic growth, but is pro-ecologically sustainable growth. It is also a "win-win" strategy that aims to protect the environment, the consumer and the worker while improving industrial efficiency, profitability and competitiveness.

Central to such efforts is the global, multistakeholder Marrakech Process, which supports regional and national initiatives to promote the shift towards sustainable consumption and production (SCP) patterns. The process responds to the call of the WSSD Johannesburg Plan of Implementation to develop a 10-Year Framework of Programmes on Sustainable Consumption and Production (10YFP). UNEP and UNDESA are the leading agencies of this global process,

with the active participation of national governments, development agencies, the private sector, civil society and other stakeholders. The Commission on Sustainable Development (CSD) will review the theme of SCP during its 2010–2011 two-year cycle.

Activities under the Marrakech Process are undertaken through voluntary task forces led by governments, with the participation of experts from developing and developed countries. Through a Cooperation Dialogue with other partners, they commit themselves to carrying out a set of concrete activities at national or regional level that promote a shift to SCP patterns. The task forces are carrying out activities such as:

- an eco-labelling project in Africa;
- national action plans on SCP;
- developing tools and supporting capacity building to promote sustainable public procurement;
- projects and networks on product policy to encourage more innovation on product eco-design and performance;
- projects on sustainable buildings focusing on energy efficiency;
- the promotion of sustainable lifestyles and education through demonstration projects; and
- developing policy tools and strategies for sustainable tourism.

Another important mechanism for implementing SCP is collaboration with development agencies and regional banks. The Cooperation Dialogue aims to highlight the contribution of SCP policies and tools to poverty reduction and sustainable development, including the MDGs, and better integration of SCP objectives in development plans. A key priority is to contribute to poverty reduction through the promotion of sustainable consumption and production, which is especially relevant for developing countries.

Sources: UNEP 2006f, UNEP 2007b, UNEP 2007c

industry, business, advertising, academia, consumer associations, environmental NGOs, trade unions and the general public. In addition, there is a need for sectoral approaches in order to modify the unsustainable systems of consumption and production (Sonnemann and others 2006).

Integration of environment into development also needs to be addressed at a macro-economic level. Wealth as an index of well-being (Dasgupta 2001), and the idea that an economy's wealth should not decline over time, or should ideally increase, have recently been put forth as powerful concepts serving the cause of sustainable development (Dasgupta 2001, World Bank 2006b). This is based on the idea that a decline in wealth (or assets) signifies an unsustainable path. In accounting terms, it means that depreciation or loss of assets should be recorded as negative. Furthermore, the idea of wealth creation brings with it the twin notions of investment and saving.

A portfolio approach assumes that assets are managed in a way that minimizes risks through, for instance, distribution of assets across a broad range of investment schemes, that profit (rent) is realized, and that there is sustained growth of the various portfolios, which will permit saving and reinvestment (see Box 8.7).

In previous sections, the importance of natural capital, including ecosystem services, was highlighted as being critical in the development of nations. Yet depletion of energy resources, forests, agricultural lands and watersheds, and damage from air and water pollutants are not recorded in the national accounts as depreciation. However, all these sectors through their respective activities create unwanted negative impacts (externalities). An impact analysis and evaluation calls for an assessment of the trade-offs (the pluses and minuses) caused by economic activities and development projects that are necessary for development. In the case of these sectors, the productive base is the natural capital, which provides great sources of well-being.

Evaluation of activities related to these sectors involves assessing the benefits versus the costs that development projects will have on the individual and society in general. The social worth (Dasgupta 2001) of such projects not only looks at the monetary return, but

---

Box 8.7 Portfolio management: analysis of impacts

A portfolio approach to sustainable development takes into account not only the value (both tangible and intangible) of the assets at hand, but also the necessary institutions that go hand in hand with the development process. This ultimately leads to an environmental and social optimum between and across generations.

A portfolio approach to sustainable development presupposes the optimal and long-term management of natural resources. The socially optimal allocation of these stocks, and how to mainstream these resources into the main economy and development process is where the challenges lie. This is also where policies that emerged as a response to the recommendations made in the 1987 Brundtland Commission report have for the most part failed.

Additionally, governmental institutions, mostly those responsible for the management of natural resources, have been for the most part unable to sensitize finance and treasury ministries to the importance of natural resources, both for the development process as well as for human well-being. At the same time, ministries of finance have mostly ignored the analysis of natural resource issues.

Exploring the interlinkages between environment and development, and more specifically the roles and impacts of sectors on the environment and human well-being calls for an impact analysis and evaluation of policies and projects. It requires close scrutiny of the important role played by institutions and governance, and of the instruments and tools available in order to provide the required information for decision making.

Sources: Dasgupta 2001, Dasgupta and Maler 1999, World Bank 2006b

---

also assesses how the quality of life of communities is affected. If the projects or portfolio has negative externalities on the productive base (in this case, natural resources), its social worth might be negative and therefore should be rejected.

It is important for policy and decision making to move accounting of natural resources from satellite accounts to the main accounts, as they provide critical information in the planning and budgetary processes. Use of instruments such as genuine savings is an effort in this direction. Indeed, genuine savings measures the true level of saving in a country after recording depreciation of produced capital (goods), investment in human capital (expenditures on education) and depletion of natural resources (World Bank 2006b). These types of assets accounts are helpful in measuring and monitoring how sustainable or unsustainable countries' activities are.

Accounting for the depletion in stocks provides countries with a picture of how balanced or unbalanced their portfolio of stocks is. For instance, countries and regions, such as Malaysia, Canada, Chile, the European Union and Indonesia, have constructed accounts for forests. Work by Norway

(1998), the Philippines (1999) and Botswana (2000) (see Box 8.8) in resource rent to calculate the value of assets, has illuminated policy decisions with regard to economic efficiency in the management of resources, as well as to the sustainability of the decisions.

In terms of accounting for natural resources, some of the challenges are (World Bank 2006b):

- lack of data in some countries;
- no market for many of these resources;
- some of the intangible services provided by these resources (such as cultural and spiritual services) are difficult or impossible to value;
- few countries have comprehensive environmental accounts; and
- there are difficulties in undertaking international comparisons, because of differences in approaches, coverage and methodologies.

Efforts are needed by a broad range of partners to address these challenges in a coherent and systematic manner.

Coping with interlinkages among environmental changes, which are increasing in rate and magnitude, will become a major challenge for development. The case of climate change is an example of where this is becoming evident. As the impacts of climate change are becoming more obvious, the importance of adaptation to climate change is gaining attention on international and national agendas. It is also clear that climate variability and change do not act in isolation (IPCC 2002, CBD 2003) (see earlier

sections). The status of the natural resources, the other environmental changes (such as land degradation and water stress), and human, social, financial and physical capital can determine the coping capacity of the people and the adaptive capacity of ecosystems (IPCC 2001). In addition, many developing countries cannot cope with the present climatic extremes, and climate change is seen to be a risk to development (Stern and others 2006, World Bank 2007). Thus, adaptation is a necessity (IPCC 2001). A climate risk management approach is being adopted by funding agencies (such as the World Bank and the UK Department for International Development), which takes account of the threats and opportunities arising from both current and future climate variability and change, and the interlinkages among the environmental changes. This approach also necessitates the consideration of interlinkages between and among the environmental changes, ecosystem services and human well-being.

The recent focus on these interlinkages, and not just climate change alone, represents an opportunity for addressing current environment-development challenges more coherently. Mitigation of climate change in terms of carbon storage measures may potentially also address multiple environment and development challenges simultaneously. Such measures need to be supported in the context of development assistance frameworks, and take account of the fact that those groups of people most vulnerable to environmental changes are often different from those causing such changes.

Although achievements have been made in the area of integrating environment into development and internalizing the human-environment interlinkages into social and economic sectors, they have not kept pace with accelerating environmental degradation. Integration of environmental concerns into the wider development agenda requires collaborative efforts across existing governance regimes. Significant opportunities are offered by the UN reform process, due to its particular focus on strengthening system-wide coherence in the area of environment and the "One UN" approach at country level.

Environmental integration remains a formidable challenge for all sectors, but in particular for the environmental institutions, both at national and international levels. It requires a systematic and sustained effort by these institutions, comparable

Box 8.8 Reinvesting resource rent: the case of Botswana

Since its independence in 1966, Botswana, originally one of the world's poorest countries, has shown remarkable economic progress. Botswana has used its mineral wealth to transform the economy, joining the World Bank's category of upper-middle-income countries in the 1990s. The country came up with its own rule of thumb for reinvestment of mineral revenues to account for and offset natural resource depletion. The use of the Sustainable Budget Index in its accounting system requires that all mineral revenues be reinvested. Some of Botswana's achievements include improvements in infrastructure, human capital, and the basic services supplied to its population, for example:

- paved roads: 23 km in 1970, increased to 2 311 km by 1990;
- improved drinking water: 29 per cent of the population in 1970, increased to 90 per cent by 1990;
- telephones: 5 000 connections in 1970, increased to 136 000 by 2001; and
- female literacy: 77 per cent by 1997.

Sources: World Bank 2006b

to those of more established coordinating sectors, such as finance and planning. Current gaps and needs relating to existing national and international infrastructure and capacities for integrating environment into development could be identified. A long-term approach for addressing such needs could also be explored. It could draw on lessons learned from integration of environment into development at the macro-economic level. This could be done through portfolio management, promotion of sustainable production and consumption patterns to decouple economic growth from environmental damage, and approaches for reviews of environmental effectiveness in sectors based on, for example, agreed targets and indicators of achievements.

Strengthened scientific knowledge, assessment and cooperation

The Brundtland Commission report and subsequent environmental policy documents continue to emphasize reliable data and sound scientific information as being key components of sustainable development. Development efforts, including poverty reduction, and humanitarian assistance, need to take full account of knowledge about the contribution of the environment and ecosystem services to the enhancement of human well-being. Investing in infrastructure and capacities for environmental knowledge and information is, therefore, also an investment in sustainable development.

There is a wide range of collaborative processes for monitoring, observing, networking, managing data, developing indicators, carrying out assessments and providing early warnings of emerging environmental threats at international, regional and national levels. Notable achievements include the ozone and climate assessments. Many national and international institutions, including scientific and UN bodies, are active in the field of environmental assessments, monitoring and observing systems, information networks, and research programmes. At the global level, these include the global observing systems and the newly established Group on Earth Observations, with its implementation plan for a Global Earth Observation System of Systems (GEOSS). Efforts also include international scientific programmes, such as those operated by academic institutions around the world and under the International Council for Science (ICSU).

Most MEAs have their own subsidiary scientific advisory bodies, which to varying degrees, analyse scientific information. The UN Framework Convention on Climate Change is, in addition to its

Environmental integration requires bridging gaps, to strengthen scientific knowledge, assessment, and cooperation and improve decision making for sustainable development.

Credit: ullstein-Hiss/Mueller/ Still Pictures

subsidiary scientific advisory body, also supported by a corresponding assessment mechanism, the Intergovernmental Panel on Climate Change (IPCC), for which WMO and UNEP jointly provide the secretariat. Calls have been made for a similar assessment mechanism based on the achievements of the Millennium Ecosystem Assessment to support the ecosystem-related MEAs. The usefulness of such a mechanism is still being debated among governments and experts. In addition, the GEF has its own Scientific and Technical Advisory Panel (STAP).

Many countries in different regions have either national legislative or other provisions for undertaking state of the environment assessments, environmental impact assessments and strategic environmental assessments (SEA). Such assessments offer opportunities for identifying and addressing interlinkages, and promoting coherence, integration of environment into development, and improved management of national environmental endowments. European Union member states, for example, adopted the European Directive (2001/42/EC) on the Assessment of the Effects of Certain Plans and Programmes on the Environment (the SEA Directive), which became effective in 2004 (European Commission 2007). On a pan-European

level, countries have agreed on a Protocol on Strategic Environmental Assessment to the Convention on Environmental Impact Assessment in a Transboundary Context, which opened for signature in 2003. In Canada, a cabinet-level directive provides for an administrative requirement to conduct a SEA on all policies, plans and programmes. In South Africa, some sectoral and planning regulations identify SEA as an approach for integrated environmental management. In the Dominican Republic, legislation refers to SEA or strategic environmental evaluation. Existing environmental impact assessment legislation in other countries requires a SEA-type approach to be applied either to plans (for example, in China), programmes (Belize) or to both policies and programmes (Ethiopia) (OECD 2006).

Adaptive governance as an opportunity for addressing interlinkages

Ideal conditions for governance of human-environment systems are rare. As the preceding pages have shown, more often than not decision-makers are faced with challenges:

- **Problems of complexity.** These include the intricate nature of ecosystems, the differing spatial reach and temporal implications of biophysical processes, thresholds and feedback loops, and the human dimensions shaping ecosystem dynamics.
- **Problems of uncertainty and change.** Science is incomplete on aspects of environmental change, some understanding of biophysical processes and ecosystem dynamics are likely to be wrong, some changes are not predicted and provided for, and existing knowledge is not fully integrated.
- **Problems of fragmentation.** Much of the governance regime is not sufficiently linked or coordinated, resulting in inconsistent or conflicting policy proposals, authorities and mandates of institutions. Administrative structures overlap, decision making is divided, important users and constituents are outside the process, and centralization and decentralization of governance is often not appropriately balanced.

From a governance perspective, the problems of complexity, uncertainty and change, and fragmentation easily result in governance disjunctures (see Box 8.9) (Galaz and others 2006). Moreover, opportunities to shift underperforming existing governance processes and structures to more responsive interlinked ones are rare. Policy-makers

---

Box 8.9 Types of governance disjunctures

Spatial disjuncture
Governance does not match the spatial scales of ecosystem processes. For example, local institutions for management of sea urchins are unable to cope with the development of global markets and highly mobile "roving bandits."

Temporal disjuncture
Governance does not match the temporal scales of ecosystem processes. For example, in the 1950s and 1960s, governments in the West African Sahel promoted agricultural and population development in areas with only temporary productivity due to above-average rainfall. As the areas returned to a low-productivity state, erosion, migration and livelihood collapse resulted.

Threshold behaviour
Governance does not recognize or is unable to avoid, abrupt shifts in social-ecological systems. Application of "maximum sustainable yields" trigger fish stock collapse, due to overharvesting of key functional species.

Cascading effects
Governance is unable to buffer, or amplifies cascading effects between domains. For example, in Western Australia abrupt shifts from sufficient soil humidity to saline soil, and from freshwater to saline ecosystems, might make agriculture a non-viable activity at a regional scale, and trigger migration, unemployment and the weakening of social capital.

Sources: Adapted from Galaz and others 2006

and implementers hardly ever have the luxury of starting from a clean slate; rather they have to work with and within existing interests and structures.

To address complex interactions and interlinkages, and to manage uncertainty and periods of change, adaptive governance approaches have much to offer (Gunderson and Holling 2002, Folke and others 2005, Olsson and others 2006). Adaptive governance emerges from many actors in the state-society complex, and can be institutionalized, though usually in a structure more akin to network governance. Adaptive governance relies on polycentric institutional arrangements that are nested and quasi-autonomous decision making units operating at multiple scales (Olsson and others 2006). The emphasis in adaptive governance is on management and responsibility sharing; it is governance through networks that link individuals, organizations and agencies at multiple levels. A core characteristic of this type of governance is collaborative, flexible and learning-based issue management (Olsson and others 2006).

Adaptive approaches are advocated as more realistic and promising ways to deal with human-ecosystem complexity than, for example, management for optimal use and control of resources (Folke and others 2005). A key strength of adaptive governance approaches is that they start with existing organizations, and seek to link with other relevant entities and stakeholders. Besides the democratic appeal of including all stakeholders, this type of inclusive governance also broadens the knowledge base significantly, and so brings together a range of different experiences and expertise (MA 2005a). With its emphasis on social coordination through networks, rather than the formation of new (often self-contained) institutions, adaptive governance inherently promotes more flexible management arrangements, and is likely to be more responsive to changes in the given human-environment system. It also allows decision-makers to more easily take on board new insights and knowledge to evoke change where necessary, survive change where needed and/or nurture sources of reorganization following change.

Given its diffuse and multi-actor nature, two elements critical for effective adaptive governance are leadership and bridging organizations (see Box 8.10). Leaders are imperative for trust building, managing conflicts, linking key individuals, initiating partnerships among relevant actors, compiling and generating knowledge, developing and communicating vision, recognizing and creating windows of opportunity, mobilizing broad support for change across levels, and gaining and maintaining momentum needed to institutionalize new approaches. Bridging organizations facilitate

---

Box 8.10 Leadership and bridging organizations: bottom-up and top-down collaboration

A response executed by the public sector may be based on ideas and initiatives from any stakeholder. For instance, in Sweden's Kristianstad Wetlands, the vision of one individual sparked a municipal response, and developed into a proposal for collaboration with a few stakeholders across sectors (environment, agriculture, tourism and university). This proposal was adopted by the municipal executive board, and turned into a policy for ecosystem management. The number of stakeholders involved increased during the trust-building and learning process of implementation, resulting in horizontal (multi-sector) and vertical (multi-level) networks. The latter have been important for attracting funds from the national and European Union levels. Thus a bottom-up initiative has resulted in a flexible, cost-effective project organization that succeeded in applying the ecosystem approach and adaptive co-management to water resources without changing the legal framework.

Laguna Lake Basin, Philippines, illustrates successful collaboration through a top-down initiative. The Laguna Lake Development Authority (LLDA) responded to declining water quality by forming River Rehabilitation Councils (RRCs) to address pollution coming from the lake's 22 tributaries. Until then, governance of the basin had been compartmentalized and was non-participatory. The RRCs on the other hand are composed of people's organizations, environmental groups, industry representatives and local government units, with the LLDA acting as the facilitating institution. The involvement of civil society has proven to be crucial to resolving major conflicts (for example, industry versus community, fishery versus industry, agriculture versus conversion of land to other uses). The multisectoral nature of the RRCs has resulted in a sustained clean-up of some tributaries, reducing pollution in the lake. In this way, the RRCs became crucial bridging organizations to build agreement around a new approach, and to include relevant stakeholders.

Sources: MA 2005b, MA 2005c, Malayang and others 2005

collaboration among different actors and entities. They are often at the interface of scientific knowledge and policy, or of local experience and research and policy. They reduce the cost of collaboration significantly, and often perform important conflict resolution functions (Folke and others 2005).

Adaptive governance approaches are a promising avenue for future efforts to address key interlinkages in a way that complements ongoing processes. Key to building adaptive capacity into governance responses is to prioritize the following three principles in the governance structures (Dietz and others 2003):

- **Analytical deliberation:** involves dialogue among interested parties, officials and scientists.
- **Nesting:** involves complex, layered and connected institutions. Nesting refers to solution-oriented processes that are embedded in several layers of governance, so that accountability exists from the local up to the national or even the international level, and includes the temporal and spatial scales of the environmental changes.
- **Institutional variety:** a mix of institutional types that facilitate experimentation, learning and change.

A range of tools and approaches are available to help in developing and implementing more adaptive policies and actions to address interlinkages, especially at national, sub-national and local levels. These are at project or programme level, and can be applied at several stages of project and programme development. These include, but are not limited to, environmental impact assessments (EIAs), strategic environmental assessments (SEAs), decision analytical frameworks, valuation techniques, criteria and indicators and integrated management approaches. At the national level, many of the approaches can be put into a national policy framework and thus covered by legislation. There are other tools and approaches that can help in the trade-offs between environment and development, including economic valuation of ecosystem services (MA 2003). Green accounting can help in the inclusion of ecosystem services and natural capital in national accounts. There is still a clear need for testing these tools and approaches in specific regions and where there are different combinations of environmental changes and development challenges. Lessons from these can help in further development of these tools and approaches.

CONCLUSION

This chapter has illustrated how human-environment interactions and the resulting environmental challenges are interlinked through complex, dynamic biophysical and social processes. Recognizing and addressing these interlinkages offers an opportunity for more effective responses at all levels of decision making. It may facilitate a transition towards a more sustainable society with a low-carbon economy. Such an approach requires collaboration across the existing governance regimes, which, in turn, have to become more flexible and adaptive.

References

ACIA (2004). *Impacts of a Warming Arctic.* Arctic Climate Impact Assessment. Cambridge University Press, Cambridge

Annan, K. (2002). *In Yale University Address, Secretary-General pleads cause of "Inclusive" Globalization.* United Nations News Centre, press release SG/SM/8412, 2 October 2002 http://www.un.org/News/Press/docs/2002/SGSM8412.doc.html (last accessed 18 May 2007)

Australian Government (2003). *Climate Change – An Australian Guide to the Science and Potential Impacts.* Pittock, B. (ed.) Department of the Environment and Water Resources, Australian Greenhouse Office, Canberra http://www.greenhouse.gov.au/science/guide/ (last accessed 10 July 2007)

Berruga, E. and Maurer, P. (2006). *Co-Chairmen's Summary of the Informal Consultative Process on the Institutional Framework for the UN's Environmental Activities.* New York, NY

Bruch, C. (2006). Growing Up. In *Environmental Forum* Volume 23, issue 3/4:28-33

Bruch, C. and Mrema, E. (2006). More than the Sum of its Parts: Improving Compliance with the Enforcement of International Environmental Agreements through Synergistic Implementation. Conference Paper at *4th IUCN International Academic Colloquium on Compliance and Enforcement Towards More Effective Implementations of Environmental Law*, White Plains, New York, NY

Camill, P. and Clark, J. S. (2000). Long-term perspectives on lagged ecosystem responses to climate change: Permafrost in boreal peatlands and the grassland/woodland boundary: Fast Slow Variable in Ecosystems. In *Ecosystems* 3:534-544

Cash, D.W., Adger, W.N., Berkes, F., Garden, P., Lebel, L., Olsson, P., Pritchard, L. and Young, O. (2006). Scale and Cross-Scale Dynamics: Governance and Information in a Multilevel World. In *Ecology and Society* 11(2):8

CBD (2003). *Interlinkages between Biological Diversity and Climate Change; Advice on the integration of biodiversity considerations into the implementation of the United Nations Framework Convention on Climate Change and its Kyoto Protocol.* Watson, R.T. and Berghall, O. (eds) CBD Technical Series No. 10. Secretariat of the Convention on Biological Diversity, Montreal http://www.biodiv.org/doc/publications/cbd-ts-10.pdf (last accessed 10 July 2007)

CBD (2006). *Guidance for promoting synergy among activities addressing biological diversity, desertification, land degradation and climate change.* CBD Technical Series No. 25. Secretariat of the Convention on Biological Diversity, Montreal http://www.biodiv.org/doc/publications/cbd-ts-25.pdf (last accessed 10 July 2007)

CGIAR and GEF (2002). *Agriculture and the Environment. Partnership for a Sustainable Future.* Consultative Group on International Agricultural Research and Global Environment Facility, Washington, DC http://www.worldbank.org/html/cgiar/publications/gef/CGIARGEF2002final.pdf (last accessed 10 July 2007)

Charnovitz, S. (2005). A World Environment Organization. In Chambers, W.B. and Green, J. F. (eds.) *Reforming International Environmental Governance: From Institutional Limits to Innovative Reforms.* United Nations University Press, Tokyo, New York, Paris

DANCED (2000). *Thailand-Danish Country Programme for Environmental Assistance 1998-2001.* Ministry of Environment and Energy, Danish Environmental Protection Agency, Copenhagen

Dasgupta, P. (2001). *Human Well-Being And the Natural Environment.* Oxford University Press, New York, NY

Dasgupta, P. (2006). Nature and the Economy. Text of the *British Ecological Society Lecture delivered at the annual conference of the British Ecological Society*, Oxford, 7 September 2006

Dasgupta, P. and Mäler, K.-G. (1999). Net National Product, Wealth, and Social well-Being. In *Environment and Development Economics* 5:69-93

Diamond, J. (2005). *Collapse: How Societies Choose to Fail or Survive.* Allen and Lane, an imprint of Penguin Books Ltd., London

Dietz, T., Ostrom, E. and Stern, P.C. (2003). The Struggle to Govern the Commons. In *Science* 302(5652):1907-1912

Earthjustice (2005). Environmental Rights Report: Human Rights and the Environment. Materials for the *61st Session of the United Nations Commission on Human Rights.* Earthjustice, Oakland, CA

EM-DAT (undated). The International Disaster Database http://www.em-dat.net/ (last accessed 10 July 2007)

European Commission (2007). *Environmental Assessment.* http://ec.europa.eu/environment/eia/home.htm (last accessed 13 May 2007)

FAOSTAT (2004). *FAO Statistical Databases.* Food and Agriculture Organization of the United Nations, Rome (in GEO Data Portal) http://faostat.fao.org/faostat/ (last accessed 10 July 2007)

FAOSTAT (2006). *FAO Statistical Databases.* Food and Agriculture Organization of the United Nations, Rome http://faostat.fao.org/faostat/ (last accessed 10 July 2007)

Faures, J., Finlayson, C.M., Gitay, H., Molden, D., Schipper, L. and Vallee, D. (2007). Setting the scene. In *Water for food, Water for life. A comprehensive Assessment of Water Management in Agriculture. Synthesis Report.* International Water Management Institute and Earthscan, London

Folke, C., Hahn,T., Olsson, P., Norberg, J. (2005). Adaptive Governance of Social-Ecological Systems. In *Annual Review of Environmental Resources* 30:441-473

Friedman, T.L. (2006). *The World is Flat, A Brief History of the Twenty-first Century.* First updated and expanded edition. Farrar, Straus and Giroux, New York, NY

Galaz, V., Olsson, P., Hahn, T., Folke, C. and Svedin, U. (2006). The Problem of Fit between Ecosystems and Governance Systems: Insights and Emerging Challenges. Paper presented at *Institutional Dimensions of Global Environmental Change Synthesis Conference*, 6-9 December 2006, Bali http://www2.bren.ucsb.edu/~idgec/responses/Victor%20Galaz%20et%20al%20-%20Fit.pdf (last accessed 10 July 2007)

Gallet, Y. and Geneve, A. (2007). The Mayans: climate determinism or geomagnetic determinism. In *EOS, Transactions, American Geophysical Union* 88(11):129-130

Galloway, J.N., Aber, J.D., Erisman, J.W., Seitzinger, S.P., Howarth, R.W., Cowling E.B. and Cosby, B.J. (2003). The nitrogen cascade. In *Bioscience* 53(4):341-356

Gehring, T. and Oberthür, S. (2006). *Introduction to Institutional Interaction in Global Environmental Governance.* The MIT Press, Cambridge, Massachusetts and London

GEO Data Portal. *UNEP's online core database with national, sub-regional, regional and global statistics and maps, covering environmental and socio-economic data and indicators.* United Nations Environment Programme, Geneva http://www.unep.org/geo/data or http://geodata.grid.unep.ch (last accessed 10 July 2007)

Gunderson, L. H. and Holling, C.S. (eds.) (2002). *Panarchy: Understanding Transformations in Human and Natural Systems.* Island Press, Washington, DC

IPCC (2001a). *Climate Change 2001: Synthesis Report. A Contribution of Working Groups I, II, and III to the Third Assessment Report of the Intergovernmental Panel on Climate Change.* Watson, R.T. and others (eds.). Cambridge University Press, New York, NY

IPCC (2001b). Climate Change 2001: Summary for the Policymakers. In *Climate Change 2001: Impacts, Adaptations, and Vulnerability.* Contribution of Working Group II to the Third Assessment Report of the Interngovernmental Panel on Climate Change. McCarthy, J.J., Canziani, O.F., Leary, N.A., Dokken, D.J. and White, K.S. (eds). Cambridge University Press Cambridge, and New York, NY

IPCC (2002). *Climate Change and Biodiversity.* Gitay, H., Suarez, A., Watson, R. T. and Dokken, D. (eds.). IPCC Technical Paper V. World Meteorological Organization and United Nations Environment Programme, IPCC Secretariat, Geneva http://www.ipcc.ch/pub/tpbiodiv.pdf (last accessed 10 July 2007)

IPCC (2007a). *Climate Change 2007: Mitigation of Climate Change: Summary for Policymakers.* Contribution of Working Group III to the Fourth Assessment Report of the Intergovernmental Panel on Climate Change. World Meteorological Organization and United Nations Environment Programme, IPCC Secretariat, Geneva

IPCC (2007b). *Climate Change 2007: The Physical Science Basis: Summary for Policymakers.* Contribution of Working Group I to the Fourth Assessment Report of the Intergovernmental Panel on Climate Change. World Meteorological Organization and United Nations Environment Programme, IPCC Secretariat, Geneva

IUCN (2004). *Draft International Covenant on Environment and Development. Third Edition: Updated Text.* Prepared by the IUCN Commission on Environmental Law in cooperation with the International Council of Environmental Law. World Conservation Union (International Union for the Conservation of Nature and Natural Resources), Gland and Cambridge

Juma, C. (2002). The Global Sustainability Challenge: From Agreement to Action. In *International Journal Global Environment Issues* 2(1/2):1-14

Kotchen, M.J. and Young, O.R. (2006). Meeting the Challenges of the Anthropocene: Toward a Science of Human-Biophysical Systems. In *Global Environmental Change* (forthcoming), Norwich

Lin, J. (2005). *Tackling Southeast Asia's Illegal Wildlife Trade.* (2005) 9 SYBIL: 191-208. Singapore Year Book International Law and Contributors, Singapore www.traffic.org/25/network9/ASEAN/articles/index.html (last accessed 9 July 2007)

Linden, E. (2006). *The Winds of Change: Climate, Weather, and the Destruction of Civilizations.* Simon and Schuster, New York, NY

MA (2003). *Ecosystems and Human Well-being.* Millennium Ecosystem Assessment. Island Press, Washington, DC

MA (2005a). *Ecosytems and Human Well-being. Synthesis.* Island Press, Washington, DC

MA (2005b). *Ecosystems and Human Well-being. Multiscale Assessments, Volume 4. Findings of the Sub-global Assessments Working Group.* Millennium Ecosystem Assessment. Island Press, Washington, DC.

MA (2005c). *Ecosystems and People. The Philippine Millennium Ecosystem Assessment (MA) Sub-global Assessment Synthesis Report.* Millennium Ecosystem Assessment. University of the Philippines, Los Baños

Malayang, B.S., Hahn, T., Kumar, P. and others (2005). Chapter 9: Responses to ecosystem change and their impacts on human well-being. In Capistramo, D., Samper, C., Lee, M. and Raudsepp-Hearne, C. (eds.) *Ecosystems and Human well-being: Multiscale Assessments, Volume 4. Findings of the Sub-Global Assessments Working Group.* Island Press, Washington, DC

Munich Re Group (2006). *Natural Catastrophes 2006: Analyses, assessments, positions.* Munich Re Group, Munich http://www.munichre.com/publications/302-05217_en.pdf (last accessed 10 July 2007)

Najam, A., Papa, M. and Taiyab, N. (2006). *Global Environmental Governance, A reform Agenda.* International Institute for Sustainable Development, Winnipeg, MB

Najam, A., Runnalls, D. and Halle, M. (2007). *Environment and Globalization: Five Propositions.* International Institute for Sustainable Development, Winnipeg, MB http://www.iisd.org/publications/pub.aspx?pno=836 (last accessed 11 July 2007)

OAG (2007). *Office of the Auditor General of Canada* http://www.oag-bvg.gc.ca/domino/oag-bvg.nsf/html/menue.html (last accessed 11 July 2007)

OECD (1995). *Developing Environmental Capacity: A Framework for Donor Involvement.* Organisation for Economic Co-operation and Development, Paris

OECD (2002). *Governance for Sustainable Development: Five OECD Case Studies.* Organisation for Economic Co-operation and Development, Paris

OECD (2006). *Environmental Performance Review of China – Conclusions and Recommendations (Final).* Organisation for Economic Co-operation and Development, Paris http://www.oecd.org/dataoecd/58/23/37657409.pdf (last accessed 11 July 2007)

Olsson, P., Gunderson, L.H., Carpenter, S.R., Ryan, P., Lebel, L., Folke, C. and Holling, C.S. (2006). Shooting the rapids: navigating transitions to adaptive governance of social-ecological systems. In *Ecology and Society* 11(1):18

Parmesan, C. and Yohe, G. (2003). A globally coherent fingerprint of climate change impacts across natural systems. In *Nature* 421:37–42

Poluton, C., Kydd, J. and Dorward, A. (2006). *Increasing fertilizer use in Africa: What have we learned?* Agriculture and Rural Development Discussion Paper 25. The World Bank, Washington, DC

Raustiala, K. (2001). *Reporting and Review Institutions in 10 Selected Multilateral Environmental Agreements.* United Nations Environment Programme, Nairobi

Root, T.L., Price, J.T., Hall, K.R., Schneider, S.H., Rosenzweig, C. and Punds, J.A. (2003). Fingerprints of global warming on wild animals and plants. In *Nature* 421:57-60

Schellnhuber, H. J. (1999). "Earth system" analysis and the second Copernican revolution. In *Nature* 402(6761supp): C19

Schmidt, W. (2004). Environmental Crimes: Profiting at the Earth's Expense. *Environmental Health Perspectives* Volume 112 (2). Secretariat of the Basel Convention (1994). *The Basel Convention Ban Amendment.* Secretariat of the Basel Convention, Geneva http://www.basel.int/pub/baselban.html (last accessed 11 July 2007)

Secretariat of the Basel Convention (undated). National Enforcement Requirements. Secretariat of the Basel Convention, Geneva http://www.basel.int/pub/enforcementreqs.pdf (last accessed 11 July 2007)

Sonnemann, G., Zacarias, A. and De Leeuw, B. (2006). Promoting SCP at the International Arena. In Sheer, D. and Rubik, F. (eds.) *Governance of Integrated Product Policy.* Greenleaf Publishing, Sheffield

Speth, J.G. and Haas, P. M. (2006). *Global Environmental Governance.* Island Press, Washington, DC

Stern, N. and others (2006). *Stern Review: the Economics of Climate Change.* Final Report.The Office of the Chancellor of the Exchequer, London

Taylor, D.S., Reyier, E.A., Davis, W.P. and McIvor, C.C. (2007). Mangrove removal in the Belize cays: effects on mangrove-associated fish assemblages in the intertidal and subtidal. In press

UN (1994). *Programme of Action of the United Nations International Conference on Population and Development in Cairo 1994.* http://www.iisd.ca/Cairo/program/p00000.html (last accessed 11 July 2007)

UN (1999). UN General Assembly Resolution UNGA/53/242. United Nations, New York, NY http://www.un.org/Depts/dhl/resguide/r53.htm (last accessed 12 July 2007)

UN (2000). Resolution Adopted by the General Assembly. 55/2 *United Nations Millennium Declaration.* Document A/RES/55/2. United Nations, New York, NY http://www.un.org/millennium/declaration/ares552e.pdf (last accessed 11 July 2007)

UN (2004). *A more secure world: Our shared responsibility.* Report of the Secretary-General's High-level Panel on Threats, Challenges and Change. United Nations, New York, NY

UN (2005). Resolution Adopted by the General Assembly. 60/1 2005 *World Summit Outcome.* Document A/RES/60/1 of 24 October 2005. United Nations, New York, NY http://daccessdds.un.org/doc/UNDOC/GEN/N05/487/60/PDF/N0548760.pdf?OpenElement (last accessed 11 July 2007)

UN (2006). Final Draft to Co-Chairs "Delivering as One" Report of the Secretary-General's High-Level Panel. 17 October 2006. Secretary-General's High-level Panel on UN System-wide Coherence in the Areas of Development, Humanitarian Assistance, and the Environment. United Nations, New York, NY

UN Global Compact (2006). *What is the Global Compact?* http://www.unglobalcompact.org/AboutTheGC/index.html (last accessed on 12 July 2007)

UNDP (1999). Synergy in National Implementation: The Rio Agreements. Paper submitted by UNDP to the *International Conferences on Interlinkages: Synergies and Coordination between Multilateral Environmental Agreements,* Tokyo, July 1999

UNEP (1998). *Policy Effectiveness and Multilateral Environmental Agreements.* Environment and Trade Series. United Nations Environment Programme, Geneva

UNEP (2001a). *Open-ended Intergovernmental Group of Ministers or their Representatives on International Environmental Governance.* International Environmental Governance Report of the Executive Director, UNEP/IGM/4/3. United Nations Environment Programme, Nairobi

UNEP (2001b). *The third Montevideo Programme for development and periodic review of environmental law for the first decade of the twenty-first century.* Decision UNEP/GC.21/23 of 9 February 2001. United Nations Environment Programme, Nairobi

UNEP (2002a). *Global Enviornment Outlook 3: Past, present and future perspectives.* United Nations Environment Programme and Earthscan, London

UNEP(2002b). *UNEP's guidelines on compliance with and enforcement of multilateral environmental agreements.* Decision UNEP/GCSS.VII/4 of 15 February 2002. United Nations Environment Programme, Nairobi

UNEP (2002c). *Report of the Governing Council on the Work of its Seventh Special Session/Global Ministerial Environment Forum, Annex I: Decision SS.VII/1.* Adopted by the Governing Council Decision UNEP/SS.VII/1 (2002). United Nations Environment Programme, Nairobi

UNEP (2004). Emerging Challenges – New Findings: The Nitrogen Cascade: Impacts of Food and Energy Production on the Global Nitrogen Cycle. In *GEO Year Book 2003.* United Nations Environment Programme, Nairobi

UNEP (2005). Divided, yet United. Discussion Paper for the *UNEP High Level Workshop Mainstreaming Environment beyond MDG 7,* 13-14 July 2005, Nairobi http://www.unep.org/dec/docs/Discussion_paper.doc (last accessed 11 July 2007)

UNEP (2006a). *Class of 2006: Industry Report Cards on environment and social responsibility.* UNEP Division of Technology, Industry and Economic, Paris http://www.unep.fr/outreach/csd14/docs/Classof2006_press_release.pdf (last accessed 11 July 2007)

UNEP (2006b). *Training Manual on International Environmental Law.* United Nations Environment Programme, Nairobi http://www.unep.org/DPDL/law/Publications_multimedia/index.asp (last accessed 11 July 2007)

UNEP (2006c). *Vital Waste Graphics Update.* United Nations Environment Programme, Nairobi and Arendal http://www.vitalgraphics.net/waste2/download/pdf/VWG2_p36and37.pdf (last accessed 10 July 2007)

UNEP (2006d). *Principles for Responsible Investment: An investor initiative in partnership with UNEP Finance Initiative and the UN Global Compact* http://www.unpri.org/files/pri.pdf (last accessed 11 July 2007)

UNEP (2006e). *Show Me the Money: Linking Environmental, Social and Governance Issues to Company Value.* UNEP Finance Initiative Asset Management Working Group http://www.unepfi.org/fileadmin/documents/show_me_the_money.pdf (last accessed 11 July 2007)

UNEP (2006f). Sustainable Consumption and Production. How Development Agencies make a difference. Review of Development Agencies and SCP-related projects. DRAFT. UNEP Division of Technology, Industry and Economics, Paris http://www.uneptie.org/pc/sustain/reports/general/Review_Development_Agencies.pdf (last accessed 11 July 2007)

UNEP (2007a). *Global Outlook for Ice and Snow.* United Nations Environment Programme, Nairobi

UNEP (2007b). *Cleaner Production – Key Elements.* UNEP Division of Technology, Industry and Economics, Paris http://www.unep.fr/pc/cp/understanding_cp/home.htm (last accessed on 11 July 2007)

UNEP (2007c). *United Nations Guidelines for Consumer Protection: Section G. Promotion of Sustainable Consumption.* UNEP Division of Technology, Industry and Economics, Paris http://www.uneptie.org/pc/sustain/policies/consumser-protection.htm (last accessed 11 July 2007)

UNESCAP (2000). Integrating Environmental Considerations into Economic Policy Making: Institutional Issues. In *Development Paper 21.* United Nations, New York, NY

UNFCCC-CDIAC (2006). *Greenhouse Gases Database.* United Nations Framework Convention on Climate Change, Carbon Dioxide Information Analysis Centre (in GEO Data Portal)

UNPD (2007). *World Population Prospects: the 2006 Revision Highlights.* United Nations Department of Social and Economic Affairs, Population Division, New York, NY (in GEO Data Portal)

UNU (1999). Inter-linkages between the Ozone and Climate Change Conventions. In *Interlinkages: Synergies and Coordination between Multilateral Environmental Agreements.* United Nations University, Tokyo

UNU (2002). National and Regional Approaches in Asia and the Pacific. In *Interlinkages: Synergies and Coordination between Multilateral Environmental Agreements.* United Nations University, Tokyo

Upton, S. and Vitalis, V. (2002). Poverty, Demography, Economics and Sustainable Development: Perspectives from the Developed and Developing Worlds: What are the Realistic Prospects for Sustainable Development in the first decade of the new Millennium? Speech delivered by the Rt. Hon. Simon Upton at *the Annual Meeting of the Alliance for Global Sustainability,* ETH-MIT-UT-Chalmers in cooperation with the Instituto Centro Americano de Administracion de Empresas (INCAE) 21-23 March 2002 in San Jose, Costa Rica

Vitousek, P.M, Mooney, H.A., Lubchenko, J. and Melillo, J. M. (1997). Human Domination of Earth's Ecosystems. In *Science* 277(5325):494-9

Watson, R.T., Dixon, J.A., Hamburg, S.P., Janetos, A.C. and Moss, R.H. (1998). *Protecting Our Planet Securing Our Future: Linkages Among Global Environmental Issues and Human Needs.* United Nations Environment Programme, US National Aeronautics and Space Administration and The World Bank, Washington, DC

WBCSD (2007). *Then & Now: Celebrating the 20th anniversary of the "Brundtland Report".* 2006 WBCDS Annual Review. World Business Council for Sustainable Development, Geneva http://www.wbcsd.org/DocRoot/BfNGWxUk4gSKBfZfbYV7/annual-review2006.pdf (last accessed 11 July 2007)

WCED (1987). *Our Common Future.* World Commission on Environment and Development. Oxford University Press, Oxford and New York, NY

WHO (2003). *Climate change and human health: risks and responses.* Summary. World Health Organization, Geneva http://www.who.int/globalchange/climate/en/ccSCREEN.pdf (last accessed 11 July 2007)

World Bank (2006a). *World Development Indicators 2006.* The World Bank, Washington, DC (in GEO Data Portal)

World Bank (2006b). *Where is the Wealth of Nations? Measuring Capital for the 21st Century.* The World Bank, Washington, DC

World Bank (2007). *An Investment Framework for Clean Energy and Development. A platform for convergence of public and private investments.* The World Bank, Washington, DC http://siteresources.worldbank.org/EXTSDNETWORK/Resources/AnInvestmentFrameworkforCleanEnergyandDevelopment.pdf?resourceurlname=AnInvestmentFrameworkforCleanEnergyandDevelopment.pdf (last accessed 11 July 2007)

WTO (2007). *Statistics Database.* World Trade Organization, Geneva http://www.wto.org/english/res_e/statis_e/statis_e.htm (last accessed 9 July 2007)

YCELP (undated). *Improving Enforcement and Compliance with the Convention on International Trade in Endangered Species.* Yale Centre for Environmental Law and Policy, New Haven , CT http://www.yale.edu/envirocenter/clinic/cities.html (last accessed 11 July 2007)

Young, O. R. (2002). *The Institutional Dimensions of Environmental Change: Fit, Interplay and Scale.* MIT Press, London

Young, O. R. (2006). Governance for Sustainable Development in a World of Rising Interdependencies. Background Paper for the *Workshop on Governance for Sustainable Development,* at the Donald Bren School of Environmental Science and Management, University of California, Santa Barbara, 12-14 October 2006

Section **E**

The Outlook – Towards 2015 and Beyond

Chapter 9 **The Future Today**

The scenarios point to both risks and opportunities in the future. Of particular significance are the risks of crossing thresholds, the potential of reaching turning points in the relationship between people and the environment, and the need to account for interlinkages in pursuing a more sustainable path.

The Future Today

Coordinating lead authors: Dale S. Rothman, John Agard, and Joseph Alcamo

Lead authors: Jacqueline Alder, Waleed K. Al-Zubari, Tim aus der Beek, Munyaradzi Chenje, Bas Eickhout, Martina Flörke, Miriam Galt, Nilanjan Ghosh, Alan Hemmings, Gladys Hernandez-Pedresa, Yasuaki Hijioka, Barry Hughes, Carol Hunsberger, Mikiko Kainuma, Sivan Kartha, Lera Miles, Siwa Msangi, Washington Odongo Ochola, Ramón Pichs Madruga, Anita Pirc-Velkarvh, Teresa Ribeiro, Claudia Ringler, Michelle Rogan-Finnemore, Alioune Sall, Rüdiger Schaldach, David Stanners, Marc Sydnor, Bas van Ruijven, Detlef van Vuuren, Peter Verburg, Kerstin Verzano, and Christoph Zöckler

Chapter review editor: Christopher Magadza

Chapter coordinators: Munyaradzi Chenje and Marion Cheatle

Credit: Ron Gilling/Still Pictures

Credit: Munyaradzi Chenje

Main messages

This chapter builds on previous chapters by exploring how current social, economic and environmental trends may unfold along divergent development paths in the future, and what this might mean for the environment, development and human well-being. It presents four scenarios to the year 2050, using narrative storylines and quantitative data to explore different policy approaches and societal choices at both global and regional levels. The main messages of the scenarios – *Markets First*, *Policy First*, *Security First* and *Sustainability First* – are:

There is a need to address interlinkages among numerous environmental issues, such as air and water pollution, land degradation, climate change, and biodiversity loss. There is also a need to link environment with development issues, such as extreme poverty and hunger, implementation of the MDGs, and addressing human vulnerability and well-being. This addresses one of the statements in *Our Common Future*, which says "the ability to choose policy paths that are sustainable requires that the ecological dimensions of policy be considered at the same time as the economic, trade, energy, agricultural, industrial and other dimensions – on the same agendas and in the same national and international institutions."

For a range of indicators, the rate of global environmental change slows or even reverses towards the middle of the century. In all scenarios, the rates of cropland expansion and forest loss steadily decline over the scenario period. The rate of water withdrawals eventually decreases in all scenarios, except *Security First*. Some scenarios also show a slackening in the tempo of species loss, greenhouse gas build-up, and temperature increase. The slowing down of these global indicators is due to

the expected completion of the demographic transition, the saturation of material consumption, and technological advances. This slowing down is important because it gives us hope that the society and nature can more successfully catch up to the pace of change and adjust to it before experiencing many negative consequences.

Despite a possible slowing down of global environmental change, the peak rate and end point of change differs strongly among scenarios. The higher the rate of change, the greater the risk that thresholds in the Earth system will be exceeded in the coming decades, resulting in sudden, abrupt or accelerating changes, which could be irreversible. Differing rates of change lead to very different end points for the scenarios. Under *Markets First*, 13 per cent of all original species are lost between 2000 and 2050 as compared to 8 per cent under *Sustainability First*. The range in 2050 for atmospheric CO_2 concentration is over 560 ppm in *Markets First* as compared to about 475 ppm under *Sustainability First*. It is expected that the risk of exceeding thresholds increases with a higher level of change, and that this change could be sudden rather than gradual. For example, the GEO-4 scenarios showing the fastest rate of increase in fish catches are also accompanied by a significant decline in marine biodiversity, leading to a higher risk of fisheries collapse by mid-century.

Investing in environmental and social sustainability does not impair economic development. Scenarios, including increased investment in health, education, and environmentally benign technologies result in equally large and more equitably distributed economic growth on a per capita basis in most regions as those that do not. The levels of GDP per capita are particularly higher in *Sustainability First* and *Policy First* than *Markets First* and *Security First* in nearly all of the currently less developed regions.

Relying on the market alone is unlikely to achieve key environmental and human well-being goals. The extreme emphasis on markets in *Markets First* results in significant increases in environmental pressures and only slow down advances in achieving social targets. Alternatively, the increased levels of investments in health, education and the environment, along with increased development assistance and new approaches to lending in *Policy First* and *Sustainability First* make for significantly faster progress, without sacrificing economic development in most regions.

Greater integration of policies across levels, sectors and time, strengthening local rights, and building capacity help achieve most environmental and human well-being goals. Additional action under *Sustainability First* – integrating governance across levels, sectors and time, strengthening local rights, and building capacity – lead to greater improvements and slower degradation than in *Policy First*. Much of this is related to the increased ownership of the issues by the broader public, and the greater legitimacy of policies. Interaction between global and regional processes suggests that concentrating environmental governance at one scale is unlikely to result in appropriate responses to environmental problems with their feedbacks.

Both trade-offs and synergies exist in the efforts to achieve key environmental and human well-being goals. Competition for land is likely as a result of competing goals: the production of biofuels to achieve climate goals, the production of food to achieve food security, and designation of areas for biodiversity. Competition can be expected for water use between the provision of adequate supplies for human activities and the maintenance of adequate in-stream flows for the integrity of aquatic ecosystems. Furthermore, achieving these goals may require the acceptance of rates of economic growth, as presently measured, in the currently highly-developed countries that, while still significant, are lower than would be the case otherwise. Key synergies result from policies that address the drivers of many of the problems. These include investments in health and education, particularly of females, which directly achieve key human well-being goals, and help to address current and future environmental goals by improving environmental management and reducing population growth.

The diversity and multiplicity of trade-offs and opportunities for synergy increases complexity for decision-makers, requiring new and adaptive approaches. This complexity should not be ignored. It, however, points to the need for innovative approaches for exploring the options for action to address the intertwined environmental and developmental challenges the world faces.

INTRODUCTION

What lies ahead? Which of the current environmental, social and economic trends will continue, and which will see dramatic shifts? What will this mean for the environment and human well-being, particularly the most vulnerable ecosystems and groups in society? What will this mean for individual sub-regions and regions, and, collectively, at the global level? Finally, what role can society play today in shaping and sustaining our common future?

To imagine what might happen over the next half-year, much less the next half-century is daunting. To imagine the future across national, sub-regional, regional and global levels is even more complex. Given that many processes are already in motion as a result of past decisions, it may be relatively easy to visualize certain trends continuing in the short-term. Still, history shows that much can change, expectedly or unexpectedly, over short periods, and it is unlikely that most trends would continue unabated for decades without changing course. History also shows that some policy decisions take many decades to unfold, for example, sustainable development and mainstreaming the environment. Both have been on the international and national agenda for the 20 years since the report of the World Commission on Environment and Development, *Our Common Future*, was published, but increasing their uptake remains as urgent today as it did then.

Choices made today on issues of environment for development may only begin to reveal their effects after decades. A major challenge is, therefore, to present stories that make sense in both the short- and long-term. This includes keeping one eye on the status of upcoming milestones. For example, the 2010 Convention on Biological Diversity target to significantly reduce the current rate of biodiversity loss at the global, regional and national levels, and the 2015 internationally agreed targets of the Millennium Declaration, such as those on water and sanitation. At the same time, it is necessary to look further ahead in time to more distant goals, such as stabilizing the levels of greenhouse gas concentrations in the atmosphere.

Based on both regional and global consultations and processes involving many stakeholders, including governments and other organizations, this chapter explores these and other issues by considering the future through the lens of environment for development. The four scenarios consider the priority and cross-cutting issues discussed in the preceding chapters.

They focus on the implications of various actions, approaches and societal choices at regional and global levels for the future of the environment and human well-being. Each scenario outlines a pathway into the future up to the year 2050, shaped by divergent assumptions about these actions, approaches, and choices. Each looks at who is making the key decisions (the dominant actors), how these decisions are made (the dominant approaches to governance) and why these decisions are made (the dominant priorities). The nature and the names of the scenarios are characterized by the theme that dominates the particular future envisioned, that is what comes first. Briefly, the scenarios assume the following:

- *Markets First*: the private sector, with active government support, pursues maximum economic growth as the best path to improve the environment and human well-being. Lip service is paid to the ideals of the Brundtland Commission, Agenda 21 and other major policy decisions on sustainable development. There is a narrow focus on the sustainability of markets rather than on the broader human-environment system. Technological fixes to environmental challenges are emphasized at the expense of other policy interventions and some tried-and-tested solutions.

- *Policy First*: government, with active private and civil sector support, initiates and implements

It is unlikely that most trends would continue unabated for decades without changing course.

Credit: Munyaradzi Chenje

strong policies to improve the environment
and human well-being, while still emphasizing
economic development. *Policy First* introduces
some measures aimed at promoting sustainable
development, but the tensions between
environment and economic policies are biased
towards social and economic considerations.
Still, it brings the idealism of the Brundtland
Commission to overhauling the environmental
policy process at different levels, including
efforts to implement the recommendations and
agreements of the Rio Earth Summit, the World
Summit on Sustainable Development (WSSD),
and the Millennium Summit. The emphasis is
on more top-down approaches, due in part to
desires to make rapid progress on key targets.

- *Security First*: government and private sector
compete for control in efforts to improve, or at least
maintain, human well-being for mainly the rich and
powerful in society. *Security First*, which could
also be described as *Me First*, has as its focus a
minority: rich, national and regional. It emphasizes
sustainable development only in the context of
maximizing access to and use of the environment
by the powerful. Contrary to the Brundtland doctrine
of interconnected crises, responses under *Security
First* reinforce the silos of management, and the UN
role is viewed with suspicion, particularly by some
rich and powerful segments of society.

- *Sustainability First*: government, civil society
and the private sector work collaboratively to
improve the environment and human well-being,
with a strong emphasis on equity. Equal weight
is given to environmental and socio-economic
policies, and accountability, transparency and
legitimacy are stressed across all actors. As in
Policy First, it brings the idealism of the Brundtland
Commission to overhauling the environmental
policy process at different levels, including strong
efforts to implement the recommendations and
agreements of the Rio Earth Summit, WSSD,
and the Millennium Summit. Emphasis is placed
on developing effective public-private sector
partnerships not only in the context of projects but
also that of governance, ensuring that stakeholders
across the spectrum of the environment-
development discourse provide strategic input to
policy making and implementation. There is an
acknowledgement that these processes take time,
and that their impacts are likely to be more long-
term than short-term.

As is the case for most scenarios, these four are
caricatures in that the real future will include elements
of all four and many others. Furthermore, scenarios are
not predictions, nor should they be taken as the most
likely of the myriad of possible futures. At most, they
paint pictures of a limited number of plausible futures,
based upon a coherent and internally consistent set
of assumptions about choices by key actors, the
progression of other social processes, and underlying
system relationships (Robinson 2003). Finally, in any
scenario exercise, there are inherent uncertainties
related to both the current state and the behaviour
of human and ecological systems. Thus, individual
scenarios represent conditional projections based
upon assumptions about the underlying human
and ecological systems, as well as the actions,
approaches, and choices noted above (Yohe and
others 2005).

Despite these challenges, the scenarios presented
here provide valuable insights for decision making
today. The narratives and numerical elements
complement each other, reflecting the approach
of most recent scenario exercises (Cosgrove and
Rijsberman 2000, IPCC 2000, MA 2005, Raskin
and others 2002, Alcamo and others 2005, Swart
and others 2004). The Technical Annex at the end
of this chapter briefly reviews how the scenarios
were developed.

FUNDAMENTAL ASSUMPTIONS BEHIND
THE SCENARIOS
Scenario development is traditionally characterized
by identifying key drivers and critical uncertainties
surrounding their future evolution, making assumptions
about how these critical uncertainties will evolve,
and exploring the broader implications of these

Each scenario outlines a
pathway into the future up to the
year 2050, shaped by divergent
assumptions about actions,
approaches, and choices.

Credit: Munyaradzi Chenje

developments. In the *GEO-4* conceptual framework, the key drivers of environmental change include: institutional and socio-political frameworks, demographics, economic demand, markets and trade, scientific and technological innovation, and value systems. This list is much the same as used in *GEO-3*, as well as in the Millennium Ecosystem Assessment (Nelson 2005) and other recent scenario activities.

Behind these different drivers are the decisions by key actors, such as whether to act reactively or proactively with respect to environmental change. In addition, assumptions are made about key system relationships, such as the precise sensitivity of the climate system to increased concentrations of greenhouse gases (GHGs), or the exact effect of a reduction of crop yields on the health of some groups. From this perspective, the evolution of many of the drivers, as well as the pressures, state and impacts, are themselves part of the unfolding of the scenarios and not *a priori* assumptions. As such, this presentation of the assumptions underpinning the *GEO-4* scenarios differs somewhat from similar exercises.

Figure 9.1 and Table 9.1 summarize the assumptions underpinning and distinguishing the four scenarios. Table 9.1 considers a series of questions grouped by the key drivers in the *GEO-4* conceptual framework. Using the set of opportunities for reducing vulnerability in human-environment systems and improving human well-being presented in Chapter 7, Figure 9.1 illustrates the strength of the investments targeted to

these opportunities. Together, these provide more specific information, building on the assumptions provided in the Introduction. They highlight the general character of the scenarios; differences will certainly exist across regions and over time in any given future, just as they do today.

Other than for trade, technology and resource access, investments are assumed to be lower in *Markets First* than in either *Policy First* or *Sustainability First*. *Sustainability First* is distinguished from *Policy First* by the added emphasis placed on equity and shared governance, particularly at the local level. Not surprisingly, the overall level of investments in these opportunities is assumed to be the lowest in *Security First*, although this does not rule out significant efforts by particular groups. Each scenario provides challenges and opportunities in the way society addresses environmental issues.

With respect to assumptions about other aspects of the current state and the behaviour of human and ecological systems, key system relationships, such as the levels of environmental robustness and the physical availability of natural resources, are held constant across the scenarios. While it is clear that there is significant uncertainty about many of these factors, varying them across the scenarios would complicate efforts to understand the impacts of the different assumptions about individual and societal choices, which is the primary focus of this exercise.

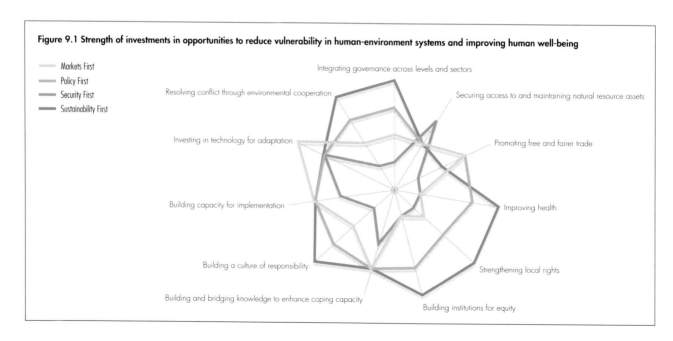

Figure 9.1 Strength of investments in opportunities to reduce vulnerability in human-environment systems and improving human well-being

Markets First
Policy First
Security First
Sustainability First

Integrating governance across levels and sectors
Resolving conflict through environmental cooperation
Securing access to and maintaining natural resource assets
Investing in technology for adaptation
Promoting free and fairer trade
Building capacity for implementation
Improving health
Building a culture of responsibility
Strengthening local rights
Building and bridging knowledge to enhance coping capacity
Building institutions for equity

Table 9.1 Key questions related to scenario assumptions

Driver category	Critical uncertainty	Fundamental assumption			
		Markets First	Policy First	Security First	Sustainability First
Institutional and socio-political frameworks	What is the dominant scale of decision making?	International	International	National	None
	What is the general nature and level of international cooperation?	High, but with focus on economic issues (trade)	High	Low	High
	What is the general nature and level of public participation in governance?	Low	Medium	Lowest	High
	What is the power balance between government, private and civil sector actors?	More private	More government	Government and certain private	Balanced
	What is the overall level and distribution of government investment across areas (e.g., health, education, military and R&D)?	Medium, fairly evenly distributed	Higher, more emphasis on health and education	Low, focus on military	Highest, more emphasis on health and education
	What is the general nature and level of official development assistance?	Low	Higher, increasingly as grants not loans	Lowest	Highest, increasingly as grants not loans
	To what degree is there mainstreaming of social and environmental policies?	Low, for example, little or no specific climate policy, reactive policies with respect to local air pollutants	High, or example, aims at stabilization of CO_2-equivalent concentration at 650 ppmv, proactive policies on local air pollutants	Lowest, or example, little or no specific climate policy, reactive policies with respect to local air pollutants	Highest, or example, aims at stabilization of CO_2-equivalent concentration at 550 ppmv, proactive policies on local air pollutants
Demographics	What actions are taken related to international migration?	Open borders	Fairly open borders	Closed borders	Open borders
	How many children do women want to have when the choice is theirs to make?	Continued trend towards fewer births as income rises	Accelerated trend	Slowing trend	Accelerated trend
Economic demand, markets and trade	What actions are taken related to the openness of international markets?	Move to increased openness, with few controls	Increasingly open, with some embodiment of fair trade principles	Moves towards protectionism	Increasing open, with strong embodiment of fair trade principles
	To what degree is there an emphasis on sectoral specialization vs. diversification in the economy?	Specialized	Balanced	Diverse, but with emphasis on sectors of interest to governments and powerful private sector actors	Diverse
	How much do people choose to work in the formal economy?	Most work in formal economy	Most work in formal economy	Larger underground economies	Variable by region and societal groups
	What is the general level and emphasis of government intervention in the economy?	Low, efficient markets	High, efficient but also fair markets	Variable by region and sector	Medium, greater emphasis on fairness of markets

Table 9.1 Key questions related to scenario assumptions *continued*

Driver category	Critical uncertainty	Fundamental assumption			
		Markets First	**Policy First**	**Security First**	**Sustainability First**
Scientific and technological innovation	What are the levels, sources, and emphases of R&D investment?	High, primarily private or by government at behest of private sector, for profit	High, primarily government Benign, but still with eye on profit	Variable, government and certain private sector actors Military/security	High, from range of sources Benign, appropriate
	What is the emphasis in terms of energy technologies?	Focus on economic efficiency	Focus on general efficiency and environmental impact	Emphasis on security of supply	Focus on general efficiency, environmental impact
	What is done with respect to the access and availability of new technologies?	What you can pay for, primarily through trade	Promotion of technology transfer and diffusion	Closely guarded	Promotion of technology transfer and diffusion, and encouragement of open source technologies
Value systems	What actions are taken related to cultural homogenization vis-à-vis diversity?	Little overt action	Little overt action	Diverse, tending towards xenophobia	Efforts to maintain diversity and tolerance
	What is the emphasis on individualism vis-à-vis the community?	Individual	More community	Individual	Community
	What is the relative rank of conflicting priorities in fisheries?	Profits	Balance between profits, total catch and jobs	Total catch	Focus on ecosystem restoration, but also emphasis on jobs and landings
	What are the key priorities with regard to protected areas?	"Sustainable use," emphasizing tourism development and some genetic resource protection	Species conservation and ecosystem services Maintenance, then sustainable use, including benefit sharing	Tourism development, and some genetic resource protection	Sustainable use, including benefit sharing, then ecosystem services maintenance and species conservation
	How do resource demands shift, independent of changing prices and income?	Follow traditional patterns	Follow traditional patterns for most resources, but some relative reduction in water use	Follow traditional patterns	Slower uptake of meat consumption, energy use, water use and other resource use with rising income

SNAPSHOTS OF FOUR FUTURES

Looking back to 1987, it is clear that many dramatic changes have occurred in the world. Not surprisingly, it is possible to see developments and trends during this period that support each of the four pathways to the future, as well as other possible futures.

To some, increased international cooperation on climate change issues is an example of the benefits that high-level policy actions can offer for environmental protection. The entry into force of the Kyoto Protocol, the development of global regulations enabling carbon capture technologies and emissions trading, the implementation of national strategies to reduce GHG emissions, and the adoption of various multilateral environmental agreements to address a diversity of challenges, all point to the success of negotiated agreements. The establishment of biodiversity targets for 2010 under the Convention on Biological Diversity provides another example of international agreement on common goals. Recent policy reforms at the regional level have also seen greater integration of policies, sectors and standards across groups of countries, for example, with respect to water management and agricultural practices in the enlarged European Union.

Others are encouraged by what they see as a continued shift in favour of a stronger social and environmental agenda among both governments and citizens. Concerted efforts to promote universal primary and secondary education, and mainstreaming of environmental and social adjustments to GDP figures represent two movements in this direction. The adoption of internationally agreed targets of the Millennium Declaration reflects commitment across the world to address sustainable development challenges. At the local level, a growing level of grassroots and civil society engagement directs energy and attention towards livelihood issues with both local and global relevance, including fair trade.

Less encouragingly, some see an unsettling pattern of conflict and entrenched interests playing out both in and between nations in today's world, characterized by increasing inequality and social isolation. Heightened security measures that restrict human movement and increase military expenditure lend weight to this view of the world. Instability and conflict have a critical effect on quality of life for millions. Certain international trade policies protect the existing balance of power through increased tariffs and protectionism, while local enclaves can be seen in the form of highly-secure housing developments in cities.

The market economy is seen as the dominant paradigm for fostering growth and human well-being, with diverging opinions about its success. Proponents see the continued rise in oil consumption and prices as fuel for considerable growth, while sceptics focus on its negative social and environmental consequences. Some argue that the role of governments is tilted in favour of economic objectives, even while it may be shrinking overall in the face of increasing corporate influence in policy decisions and trade agreements.

These varied aspects of the recent and today's world exert very different pressures on human decisions and actions, with implications for the environment and human well-being. A continuation or change in any of these patterns could have a pivotal influence on major issues at local, regional and global levels. Government leadership, market incentives, protectionist motives or unconventional approaches could mean the difference between marked improvement and steady decline on such prevailing environmental concerns as freshwater quality and availability, land degradation, biodiversity conservation, and energy use with its associated GHG emissions and climate change effects. Socially, these different approaches could translate into radically different situations regarding equity and distribution of wealth, peace and conflict, access to resources and health services, and opportunities for political and economic engagement.

Which of these trends will be the most dominant over the next decades? This is open to debate. In the end, the answer will likely differ across regions and over time. This section presents snapshots of the four futures considered in this chapter.

Markets First

The dominant characteristic of this scenario is the tremendous faith placed in the market to deliver not only economic advances, but also social and environmental improvements. This takes several forms: an increased role of the private sector in areas that were previously dominated by governments, a continued movement towards freer trade, and the commoditization of nature. A key question it poses is: how risky is it to put the markets first?

Most regions see a significant increase in the privatization of education, health and other social services, extending even to the military, as governments seek to achieve economic efficiency and reduce their financial burden. Research and development becomes increasingly dominated by private organizations. Assistance to developing countries moves even further in the direction of direct investment and private donations, with little change in official development assistance.

International trade accelerates as the World Trade Organization grows. Although no global free trade zone is achieved, pre-existing regional free trade agreements are strengthened and new ones emerge, for example, in South Asia (SAFTA). In addition international economic cooperation grows, both within and among regions. Growing South-South cooperation, such as between Asia and the Pacific, Africa, and Latin America and the Caribbean, stands out in this regard.

Efforts to increase privatization and trade are accompanied by an increase in measures to put prices on ecosystem services and turn them into commodities. Although this forces people to better recognize the value of these services, it is not the primary intent of these efforts, which are driven more by ideological aims. The commoditization and economic exchange of goods such as water, genetic material, and traditional knowledge and culture, dramatically increases. With these changes, the size of the "commons," both globally and locally, shrinks significantly.

Formal environmental protection progresses slowly, as it competes against efforts to increase economic investment and expand trade. The Kyoto Protocol is only ineffectively enforced and there is no significant international follow-up after its expiry in 2012. Multilateral environmental agreements generally defer to trade and other economic agreements when they come into conflict.

The effects of these choices are seen in many aspects of society and the environment. The growing economy, with its seemingly insatiable demand for energy, the continued dominance of fossil fuels, and the limited efforts to reduce emissions result in continued rapid growth in equivalent CO_2 emissions for the world as a whole.

In terms of regional air pollutants, the pattern varies by region as increasing incomes bring calls for greater controls. In regions such as North America and Western Europe, reductions continue, although these slow somewhat over time. Regions where economic growth reaches sufficient levels see increases followed by declines, particularly for those pollutants most detrimental to human health, such as particulates and SO_2. Other regions, such as parts of Latin America and the Caribbean, Africa and Central Asia continue to see rises in pollutant levels.

A number of forces, most notably the increased demand for food, freer trade, the phasing out of agricultural subsidies, technological advances, the growth of cities, and increased demand for biofuels affect land use in quite different ways across the world. Globally, there is actually a slight decline in land devoted to food crops, but a rise in grazing land. The total forest area declines, but starts to recover later in the period, albeit with a continued slight decline in mature forests. All regions see an intensification of agriculture, bringing increased worries about soil degradation. In Latin America and the Caribbean, and Africa, where the intensification is not accompanied by a net reduction in cropland, these concerns are severe.

The privatization of water and improvements in technology lead to increases in water use efficiency in most regions, but the emphasis is primarily on the augmentation of supply. At the same time, the decline in subsidies in most regions affects those less able to pay, be they agricultural, industrial or domestic users. Still, with growing populations, particularly in regions where demand reaches a point of saturation or where climate change results in reduced precipitation, the number of people living in river basins with severe water stress grows significantly. Even though the percentage of wastewater treated grows, the total volume of untreated wastewater continues to increase rapidly.

Terrestrial and marine biodiversity pay a high price. There is a continued decrease in mean species abundance globally, with the largest losses in sub-Saharan Africa, parts of South America, and some areas in Asia and the Pacific. The poor quality of the management of some protected areas, the opening up of others, and the introduction of alien invasives and genetically modified species all contribute to this decline. Although agriculture, through its effect on land use, historically played the dominant role in the reductions in terrestrial biodiversity, its share of changes is eclipsed by climate change and the growth of infrastructure. In fact, except for Africa, and Latin America and the Caribbean, shifts in land-use patterns reduce the pressure that agriculture puts on terrestrial biodiversity. The continued growth in landings from marine fisheries in many regions belies the increasing loss of marine biodiversity.

Policy First

The dominant characteristic of this scenario is the highly centralized approach to balancing strong economic growth with a lessening of the potential environmental and social impacts. A key question is whether the slow and incremental nature of this approach will be adequate.

The first decades of the 21st century see concerted efforts by governments to solve the pressing problems facing the world as it entered the new millennium. Many of these, for example the HIV/AIDS crisis and the lack of access to safe water in many parts of the world, were already evident. Others, such as climate change, make their presence felt, portending much more serious consequences in the future if action is not taken.

The pattern of responses to the environmental challenges is characterized by a move towards a more "holistic" approach to governance, particularly in the management of the economy. Economic growth, while seen to be necessary, is no longer pursued without significant consideration of its social and environmental impacts. More specifically, uncontrolled markets are recognized as being limited in their ability to provide many of the public goods-and-services societies hold dear, including the maintenance of key ecosystem services and the stewardship of non-renewable resources. New theories point to the importance of these goods-and-services to longer-term economic sustainability, nationally and internationally. These help to lend support to the increased public investments in, among others, health, education (particularly of women), R&D and environmental protection, even when this requires increased government expenditures. It is also reflected in the richer nations ultimately meeting targets for foreign aid to poorer countries set in the previous century.

National governments and international institutions, including the United Nations and regional organizations, lead in these efforts. In fact, the increasing economic and political integration in the regions is one of the hallmarks of the changes. Previously existing institutions, such as the European Union, expand, while new ones such the Asia Pacific Community for Environment and Development, are formed. For the most part, the private and civil sectors follow and support the efforts of governments.

Although specific actions taken related to environmental governance vary across and within regions, there are common elements, due in large part to the increased coupling of national institutional arrangements to international agreements. "Perverse" subsidies, which encourage the overexploitation of resources, ranging from fossil fuels to water to agricultural land to marine fisheries, are gradually reduced, if not eliminated. Public investments in science and technology grow, and increasingly emphasize environmental concerns, particularly those of the most vulnerable groups. The designation of both terrestrial and marine protected areas increases, and efforts are broadly, albeit not uniformly, effective in terms of preventing land-use change in these areas.

The effects of these choices are seen in many aspects of society and the environment. Climate change and its associated impacts remain a primary concern. A series of international agreements, the removal of subsidies and investments in R&D motivate concerted efforts to increase energy efficiency and move to more low-carbon and renewable sources, including biofuels. Still, total energy consumption continues to increase. Furthermore, in spite of significant growth in renewable sources of energy, oil and gas continue to dominate fuel supplies.

The increased demand for biofuels and food, even in the presence of technological advances and the phasing out of most agricultural subsidies, results in a significant increase in land devoted to pasture, even as land for crops falls slightly after reaching a peak. Much of this increase comes at the expense of forest land.

Strong investments to increase supply and reduce demand, particularly through efficiency improvements, help to alleviate concerns over freshwater availability in much of the world. Still, growing populations and economic activity continue to strain resources, particularly in the developing regions. Globally, the population living under severe water stress continues to rise, with almost all of this increase occurring in those regions exhibiting continued population growth. Social and political institutions, with efforts to better manage shared resources, help to limit the impacts of this stress in most regions.

The increased demand also places a strain on the quality of water resources. While treatment of wastewater expands in all regions, it trails the requirements. The total global volume of untreated wastewater continues to grow, even as the percentage treated increases.

Climate change has a dramatic effect on terrestrial biodiversity. Agriculture is the other significant contributor to these losses. The most severely affected areas are in Central Africa, parts of Latin America and the Caribbean, and parts of Central Asia, because these regions see the greatest changes in land use as biodiversity protection has to compete with food production and the harvesting of biofuels.

The demand for food extends to the world's oceans, with most parts experiencing an increase in landings. However, in most cases this also involves fishing further down the food chain. The two areas experiencing the most improvement in diversity of the catch – the northwest Atlantic and the south Pacific around Antarctica – do so in part by reducing their landings.

Security First

The dominant characteristic of this scenario is the emphasis on security, which consistently overshadows other values. It is a fairly narrow notion of security, implying increased limits on how people live, both physically and psychologically. Whether people reside behind actual walls or outside, their movements are not nearly as free as might have been imagined looking from the start of the century. Where increased restrictions on migration have reduced the movement of people, the continuation and extension of trade barriers limit the movement of goods across borders. Much of this is driven by continued conflict in many parts of the world, government mandate and lack of resources for many individuals. Thus, at the same time that the world becomes more crowded as population grows, it feels even smaller by many of the choices made by society. A key question is: what might be the broader implications of security first?

Expenditures on security, both public and private, grow at the expense of investments on other priorities, including in R&D in science and technology. Many governments hand over the provision of public services to private interests to improve efficiency and save costs. Both official development assistance (ODA) and foreign direct investment (FDI) contract overall, or become more focused and subject to greater conditionality. International trade follows similar patterns. Internationally, the more unsavoury aspects of the ideas championed by many anti-globalization campaigners in the past prevail. Domestically, broad-based social safety nets either do not develop or deteriorate.

Governments, particularly those that retain strong control at the national level, continue to play a strong role in decision making, but they are increasingly influenced by multinational corporations and other private interests. Very little progress is achieved in reducing corruption in official circles. International institutions, both at the regional and global levels, see their authority decline. Public participation and the role of the civil sector, both domestically and internationally, are increasingly marginalized

Not surprisingly, environmental governance suffers as a result of these wider changes; where it is "successful," this is usually to the benefit of particular sectors of society. Most new technologies pay little regard to environmental impacts, and there is a degree of regression in practices, such as the use of inorganic fertilizers. There are various patterns in terms of formal incentives and disincentives related to resource use, but the logic behind these is rarely from an environmental perspective. Globally, there is no expansion of the terrestrial or marine protected area network, and an overall decrease is seen in the level of protection from exploitation of existing protected areas. Also, key environmental services are increasingly the focus of competition and conflict.

The effects of these choices are seen in many aspects of society and the environment. Total energy use increases significantly, reflecting a very slow rate of improvements in energy efficiency. Furthermore, after slow growth in the early decades of the century, there is a dramatic resurgence in the use of coal to the point where it is rapidly approaching the levels of use of natural gas and oil. The net result of these and other forces is a strong rise in the level of atmospheric CO_2, with no sign of a slowing of the rate of increase. The planet continues to warm, with little hint of a slowing down of the rate of increase.

Total SO_X emissions change little. Reductions in Europe, North America and West Asia have been balanced by increases elsewhere. NO_X emissions climb in every region. The health effects of these emissions, particularly in the increasingly crowded urban areas, are felt across the globe.

With the changing climate, the extent of forest in the Arctic increases as species spread north. Europe also sees some increase in its forest area, as does North America, although much of the increase in the latter

is not considered mature forest. These patterns are the exception though, as most regions and the world as a whole witness loss of forests, which are converted to food crops and especially to grazing land. Africa, and Latin America and the Caribbean stand out in this regard. The slow growth in income and continued concentration of land ownership in these regions, to a certain extent, slow these trends. The downside to this is seen in the accompanying slow growth in food availability, which is also reflected in continuing high levels of childhood malnutrition in these regions.

The changing climate also combines with growing populations and greater economic activity to further strain freshwater resources around the globe. The slow advances in water use efficiency are not able to prevent dramatic increases in water stress. Globally, there is a dramatic rise in the number of people living in river basins facing severe water stress. The number in Africa alone is nearly as many people as lived in the entire region at the start of the century. Conflicts are witnessed over shared resources, both within and between countries.

Equally concerning is the quality of water. The volume of wastewater produced vastly exceeds the increases in treatment capacity; the net result is a dramatic rise in the amount of untreated wastewater. Again, the poorer regions of the world face the greatest impact, with the increases being significantly higher in places such as West Asia and Africa. The resulting effects in the form of water-borne diseases are significant.

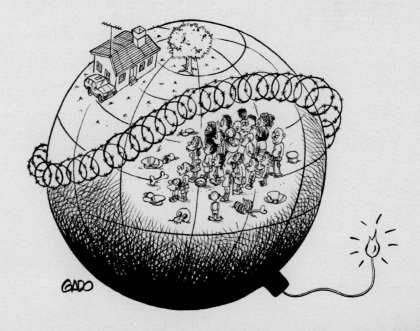

In the absence of concerted efforts, climate change, general population growth, urbanization, and the growth in demand for food and traditional biofuels have significant impacts on terrestrial biodiversity. The expansion of agriculture is eclipsed by increased infrastructure and the changing climate as the primary drivers of biodiversity loss. The loss of species abundance is widespread but certain areas, for example sub-Saharan Africa, parts of Latin America and the Caribbean, and parts of Asia and the Pacific, see greater losses. In addition to these broad patterns of change, some localized areas experience staggering losses as a result of armed conflict.

The pressure on the world's oceans increases dramatically, particularly in the first few decades of the century. Fish catches increase in most areas, but with a decline in the quality of the catch in most cases. There is some decline in the catch in the later years with the changes in quality varying by area. At the same time, efforts to expand aquaculture and mariculture in many regions increase at the expense of critical ecosystems, including mangrove forests and coral reefs.

Sustainability First

The dominant characteristic of this scenario is the assumption that actors at all levels – local, national, regional and international, and from all sectors, including government, private and civil – actually follow through on the pledges made to date to address environmental and social concerns. This implies behaviour that honours not only the letter, but also the spirit of these promises.

The start of the 21st century sees strong calls on governments at all levels to address the myriad of problems facing the world, reflected by national and international responses such as the Millennium Declaration. At the same time, groups from the private and civil sectors under such rubrics as corporate social responsibility, environmental justice, fair trade, socially responsible investment, and organic and slow food, as well as key individuals with significant personal resources, do not wait for governments to act. They gain momentum and increase influence as their numbers of adherents pass key thresholds.

Reforms take place in both national and international institutions, opening these up to more balanced participation. The rules governing international trade are gradually reformed over time to address broader issues than just economic efficiency. The nature and amount of ODA and FDI evolve to make these more beneficial to all parties. The world witnesses a significant increase in the allocation of public resources to social and environmental concerns, and less towards the military. Underpinning much of this is an underlying but not always explicit compact between the richer and the poorer nations to more seriously address the needs of the latter.

Governments play an important role through actions taken to address social and environmental concerns, particularly in integrating these into all aspects of decision making. The biggest impacts, however, result from their willingness to create the space for, and learn from, actions in the private and civil sectors. The more open and partnership-based approaches result in higher levels of cooperation and compliance, stemming from the increased relevance and legitimacy of government actions. The stage is set for different actors to more easily play their appropriate roles in addressing issues of common concern, drawing on the strengths and minimizing the weaknesses of each.

The evolution of environmental governance reflects both the complementarities and competition between social and environmental goals. In areas such as energy and water provision, efforts are made to balance the desire to reduce overall resource use with the need to address issues such as fuel, poverty and water stress. Increased public and private investments in water infrastructure and energy resources and technologies emphasize meeting these and other challenges in more environmentally friendly ways. Choices have to be made with respect to land use in balancing biodiversity protection and food security, not to mention the increased demand for biofuels. There is an increase in the number of terrestrial and marine areas designated for protection; however, the designations emphasize sustainable use and ecosystem services maintenance, rather than simply species conservation.

The effects of these choices are seen in many aspects of society and the environment. Climate change continues to remain a persistent problem. Through significant efforts, the growth in the level of atmospheric CO_2 is limited, but it will still be a few decades before stabilization is reached. After rising, the rate of change in global temperature falls and continues to decline. Still, it is not possible to avoid potentially significant warming and sea-level rise. At the same time, hope is seen in the transformations in the energy sector. Total energy use increases, but the mix of fuels changes significantly. Oil use peaks, and the use of coal declines to the point that more energy is produced by solar and wind. Both modern biofuels and the latter make up a significant fraction of total energy supply, with natural gas as the overall dominant source of energy.

With respect to more local air pollutants, marked declines are seen in NO_X and SO_X emissions. North America and Europe were already seeing reductions at the start of the century, but all regions follow their lead and at a rapid rate.

With climate change, the extent of forest in the Arctic increases as species spread north. Efforts to address climate change also have an effect on land use, with significant amounts of land devoted to the growth of biofuels. Increased area devoted to food crops in Africa, and Latin America and the Caribbean, even in the light of improved yields, is offset by land taken out of production elsewhere. The expansion of grazing land primarily comes at the expense of forests. The increase in food availability is fundamental in reducing hunger, however. Furthermore, the loss of forest land slows significantly over time.

The widespread adoption of integrated water management strategies with a strong emphasis on demand management and conservation helps to reduce the growth in water stress. Still, in part due to the varying patterns of population growth and shifting patterns of precipitation as part of the changing climate, increases are seen in water stress in some regions, notably Africa, Asia and the Pacific, and West Asia. In almost all regions,

though, programmes have been put in place to help people deal with this concern.

The efforts to reduce the growth in water demand also play a role in maintaining and improving water quality around the world. Treatment capacity keeps pace with the increasing amounts of wastewater, such that the total volume of untreated wastewater changes very little. However, the story differs among regions. Some, such as North America almost completely eliminate untreated wastewater while others, such as Latin America and the Caribbean, see small increases in volume even as the percentage treated rises.

Efforts to turn the tide on biodiversity loss are significant, but these face strong challenges due to competing demands for food and fuel, and, most importantly, climate change. The latter becomes, by far, the most important driver of species loss. Parts of Africa, Asia and the Pacific, and Latin America and the Caribbean, also face increased stress due to agricultural expansion, resulting in more significant losses in these areas.

Driven by increased food demand, many parts of the oceans experience increased pressure from fishing, but some experience decreased pressure. Significantly, the mean trophic level of the fish caught stays the same or increases in many parts of the ocean. Designated marine sanctuaries play a key role in these cases. In addition, efforts are made to reduce the potential impacts of aquaculture and mariculture on vulnerable coastal ecosystems.

IMPLICATIONS OF THE SCENARIOS

The previous section has provided glimpses of how the future might play out under the assumptions of each of the four scenarios. What are the implications in each case for the environment and human well-being? Following the structure of this report, this section will look in turn at the atmosphere, land, water and biodiversity, and then human well-being and vulnerability. Since they drive many of the results, it begins with a brief look at some overall demographic and economic developments in the different scenarios.

Demographic and economic change

Global population continues to grow in each of the scenarios (see Figure 9.2). It reaches its highest level, around 9.7 billion, by 2050 in *Security First*. In *Sustainability First*, there are just under 8 billion people at this time, and very little further growth is expected. *Policy First* and *Markets First* see global population reach about 8.6 and 9.2 billion people, respectively. In comparison, the latest UN projections (UN 2007) are 7.79, 9.19, and 10.76 billion people by 2050 for the low, medium and high variants, respectively. These differences

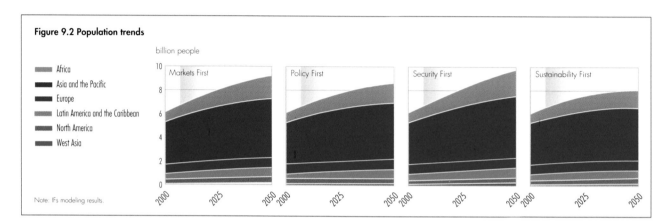

Figure 9.2 Population trends

billion people

Africa
Asia and the Pacific
Europe
Latin America and the Caribbean
North America
West Asia

Markets First   Policy First   Security First   Sustainability First

Note: IFs modeling results.

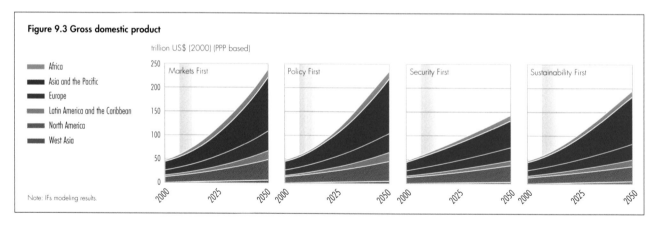

Figure 9.3 Gross domestic product

trillion US$ (2000) (PPP based)

Africa
Asia and the Pacific
Europe
Latin America and the Caribbean
North America
West Asia

Markets First   Policy First   Security First   Sustainability First

Note: IFs modeling results.

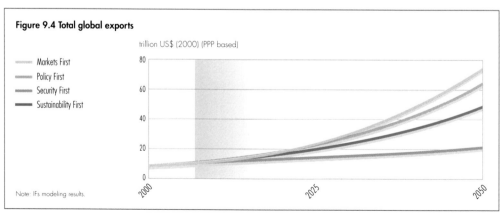

Figure 9.4 Total global exports

trillion US$ (2000) (PPP based)

Markets First
Policy First
Security First
Sustainability First

Note: IFs modeling results.

reflect a number of factors, including differences in female education, population policies and income growth in the different scenarios. The largest absolute growth occurs in Asia and the Pacific, but in terms of percentage growth, it is much larger in Africa and West Asia. Europe is the only region that experiences absolute declines during this period, although these are quite small, particularly in *Sustainability First*.

Global economic activity grows significantly over the scenario period, particularly in *Markets First* and *Policy First*, both of which see an approximate fivefold increase in global GDP (see Figure 9.3). Even in *Security First*, there is nearly a tripling of economic activity. For a comparison, the latest Global Economic Prospects (World Bank 2007) describes three scenarios with average annual growth ranging from 2.8–3.7 per cent between 2005 and 2030 (using market exchange rates); the scenarios presented here have growth rates ranging from 2.6–3.9 per cent over the same period (also using market exchange rates). As shown in Figure 9.4, this growth is accompanied by significant increases in global trade, most notably in *Markets*

First. Due in part to their more rapid population growth, the absolute size of the economies in Africa and West Asia grow at about the same rate as the Asia and the Pacific economy in *Markets First*, *Policy First* and *Sustainability First* and somewhat faster in *Security First*.

Given its somewhat lower population growth and similar economic growth, *Policy First* sees faster growth than *Markets First* in global average GDP per capita, with an increase of nearly 3.5 times over the period of the scenarios (see Figure 9.5). Slightly slower growth occurs in *Sustainability First*, but global average GDP per capita still more than triples; *Security First* sees less than a doubling. The most rapid growth occurs in Asia and the Pacific in all scenarios. As with the other currently less well-off regions, somewhat greater growth is seen in *Policy First* and *Sustainability First* than *Markets First*, with *Security First* having the lowest growth in all regions and the least convergence across regions.

Figures 9.6 and 9.7 shed further light on the convergence of incomes in the different scenarios. *Security First* exhibits growing inequality measured

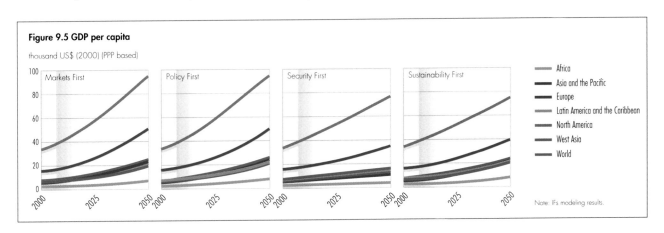

Figure 9.5 GDP per capita

thousand US$ (2000) (PPP based)

Markets First · Policy First · Security First · Sustainability First

Africa · Asia and the Pacific · Europe · Latin America and the Caribbean · North America · West Asia · World

Note: IFs modeling results.

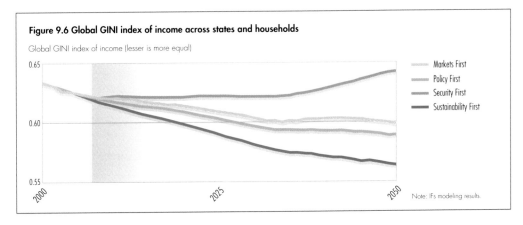

Figure 9.6 Global GINI index of income across states and households

Global GINI index of income (lesser is more equal)

Markets First · Policy First · Security First · Sustainability First

Note: IFs modeling results.

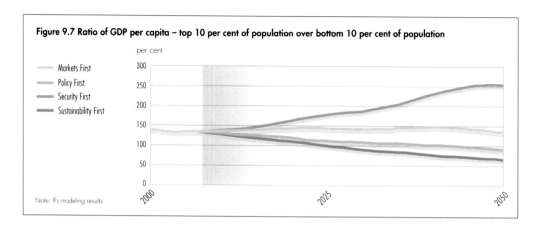

Figure 9.7 Ratio of GDP per capita – top 10 per cent of population over bottom 10 per cent of population

per cent

- Markets First
- Policy First
- Security First
- Sustainability First

Note: IFs modeling results.

by both the GINI index and the ratio of income between the wealthiest and poorest 10 per cent of the global population. A slight improvement is seen in *Markets First* using the former measure, but not the latter. *Sustainability First* shows the most significant improvements in both cases.

Atmosphere

Chapter 2 highlighted the key issues related to the atmosphere. Beginning with energy use, a key pressure, the scenarios illustrate dramatically different possible futures for the atmosphere.

Energy use

Globally, world energy use is expected to increase in all scenarios, driven mostly by increasing energy use in low-income countries (see Figure 9.8). However, per capita energy use in high-income countries remains at a much higher level than in low-income countries (see Figure 9.9). Primary energy use in *Policy First* and *Security First* increases from about 400 EJ in 2000 to 600–700 EJ in 2030 and around 800–900 EJ in 2050. This trajectory is consistent with mid-range scenarios in literature (see for example IEA 2006). In relative terms, population growth is a more important

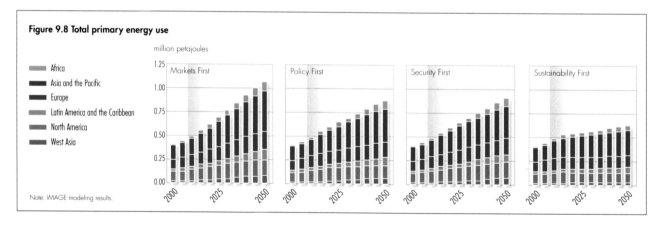

Figure 9.8 Total primary energy use

million petajoules

- Africa
- Asia and the Pacific
- Europe
- Latin America and the Caribbean
- North America
- West Asia

Note: IMAGE modeling results.

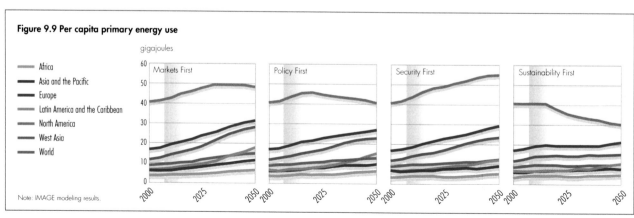

Figure 9.9 Per capita primary energy use

gigajoules

- Africa
- Asia and the Pacific
- Europe
- Latin America and the Caribbean
- North America
- West Asia
- World

Note: IMAGE modeling results.

factor in growth in *Security First*, while income growth plays a more important role in *Policy First*. The trajectory in *Markets First* lies substantially above the other two scenarios, driven by rapid income growth and more material-intensive lifestyles. In contrast, *Sustainability First* follows a lower trajectory. Here, a less material-intensive orientation and considerably higher efficiency – partly induced by global climate policy – contributes to the lower energy use.

In terms of the energy mix, fossil fuels continue to dominate energy supply in all four scenarios (see Figure 9.10). Nevertheless, important differences exist across the scenarios. In *Markets First*, relaxation of current tensions in international energy markets allow for rapid growth of oil and natural gas use worldwide. In *Policy First*, moderate climate policies reduce growth of oil demand, bring down coal use, and stimulate the use of bioenergy and zero-carbon options, such as wind, solar and nuclear power. Some of the remaining fossil fuel use in the power sector is combined with carbon capture and storage. In *Security First*, a totally different picture emerges. Here, growth of oil and natural gas is reduced due to remaining tensions in

international energy markets. This is replaced by an increase in coal use. Finally, in *Sustainability First* a similar picture emerges as in *Policy First*, but trends are much stronger. Here, as a result of stringent climate policy not only is coal use reduced, but so is oil use. Oil is partly replaced by a strong increase in bioenergy use. While natural gas use increases, its consumption in the power sector is, after 2020, mostly combined with carbon capture and storage.

Emissions of regional air pollutants and greenhouse gases

At the global level, energy use dominates the anthropogenic emissions of both regional air pollutants, using SO_x emissions as a general indicator, (see Figure 9.11) and GHGs (see Figure 9.12). The relationship between total energy use and emissions is strongly influenced by a variety of other factors, particularly government policies directed at emission controls.

The total emissions of regional air pollutants decline in all scenarios other than *Security First*. This is a clear consequence of the lack of emissions controls in that scenario. The dramatic declines in *Policy First* and

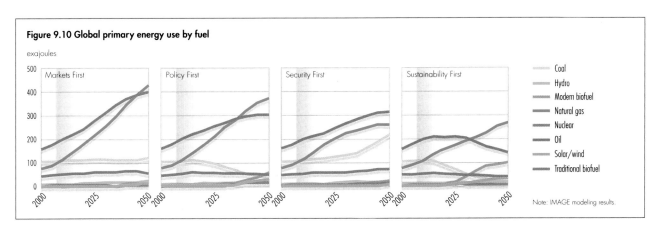

Figure 9.10 Global primary energy use by fuel

exajoules

Markets First | Policy First | Security First | Sustainability First

Coal
Hydro
Modern biofuel
Natural gas
Nuclear
Oil
Solar/wind
Traditional biofuel

Note: IMAGE modeling results.

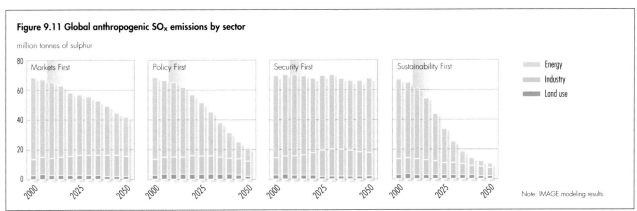

Figure 9.11 Global anthropogenic SO$_x$ emissions by sector

million tonnes of sulphur

Markets First | Policy First | Security First | Sustainability First

Energy
Industry
Land use

Note: IMAGE modeling results.

Sustainability First reflect a combination of strong policy efforts to reduce emissions per unit energy use as well as relatively slower overall growth in energy use and moves towards cleaner fuels. *Markets First* exhibits an overall decline, but the overall increase in economic activity keeps it from matching the reductions in *Policy First* and *Sustainability First*.

The largest increase in GHG emissions over the scenario horizon, more than a doubling, occurs under *Markets First*, reflecting its increase in energy use and the lack of effective mitigation policies, highlighted by

a lack of progress in reducing per capita emissions (see Figure 9.13). For similar reasons, *Security First* also shows a large increase, although this is somewhat smaller, because of more moderate economic growth. In comparison, emissions under *Policy First* and *Sustainability First* peak and subsequently decline during the scenario period. This is mainly driven by the implementation of policies to reduce GHG emissions. In the early years, however, *Policy First* actually has the highest emission of all scenarios due to higher levels of emissions from land-use changes. These emission levels are all within the range of projections

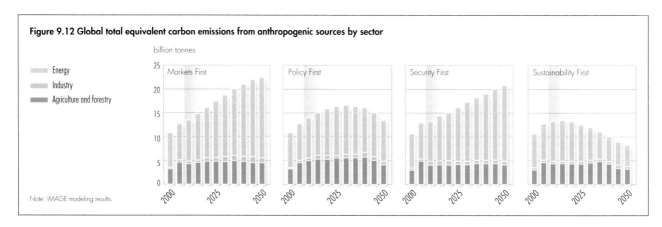

Figure 9.12 Global total equivalent carbon emissions from anthropogenic sources by sector

Note: IMAGE modeling results.

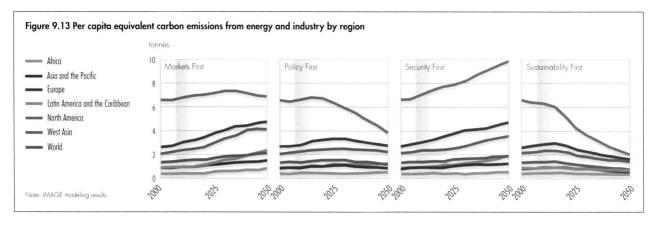

Figure 9.13 Per capita equivalent carbon emissions from energy and industry by region

Note: IMAGE modeling results.

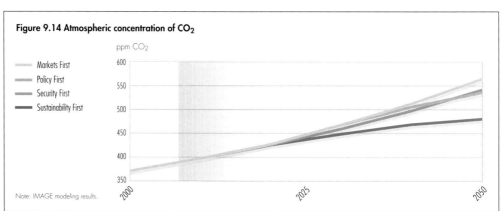

Figure 9.14 Atmospheric concentration of CO$_2$

Note: IMAGE modeling results.

considered in the latest IPCC reports (IPCC 2007a). (For further detail on comparing the climate related projections in the scenarios with those presented in the IPCC, see Box 9.1)

Atmospheric CO_2 concentration and global mean surface temperature

The trend in global CO_2 concentration reflects the trends in emissions and the uptake of atmospheric CO_2 by the ocean and biosphere. The highest CO_2 concentration is reached in *Markets First*, exceeding 550 ppm in 2050 (see Figure 9.14). *Policy First* and *Security First* result in about the same concentration, around 540 ppm in 2050, despite having distinctly different pathways over the period. In case of *Security First*, the increases in CO_2 concentration are lower in the beginning of the scenario period because of the lower emissions, but they continue to increase at an accelerating rate. *Policy First* actually has the highest increases in CO_2 concentration at the start of the scenario period, but the rate of increase slows significantly by the end of the period. *Sustainability First* results in the lowest concentration by a significant margin, about 475 ppm in 2050, and is also the only scenario where the concentration has approached stabilization.

All scenarios show a distinct increase of the global mean temperature, ranging from about 1.7°C above pre-industrial levels in 2050 in *Sustainability First* to about 2.2°C in *Markets First*, with *Policy First* and *Security First* reaching about 2.0°C (see Figure 9.15). These represent the actual changes in temperature in 2050; due to inertia in the climate system, additional warming would be expected in all of the scenarios irrespective of emissions that might occur after 2050.

Box 9.1 Comparing these climate projections with the Fourth IPCC assessment

The models being used in this scenario exercise are also being used by the Intergovernmental Panel on Climate Change (IPCC), ensuring consistency between the projections in this report and in the latest IPCC reports, published in 2007 as its Fourth Assessment Report (see Chapter 2). Due to timing issues, not all model parameters have been updated to the latest IPCC findings. The consequences for the conclusions are marginal, as described here:

- One of the key uncertainties in climate science is the value of the climate sensitivity, such as the expected equilibrium change in global temperature as a result of a doubling of CO_2 in the atmosphere over pre-industrial levels. In the latest IPCC report, the estimated range is 2.0–4.5°C, reflecting an increase in the lower value. The medium value increased from 2.5°C to 3.0°C. The IMAGE 2.4 model used the previous value for this study, reflecting the scientific understanding at the time of the model runs. The results in 2050 are only marginally affected by this difference, since the climate sensitivity is an indication of the final temperature increase at equilibrium, which is only met after 100 years; the temperature in the scenarios would increase by a maximum of 0.2–0.3°C in 2050.

- Another crucial uncertainty is in the estimates of sea-level rise. As with the climate sensitivity, the IMAGE 2.4 model used settings from the previous IPCC report, reflecting the scientific understanding at the time of the model runs. Therefore, the value in 2000 is low compared with the medium estimate of IPCC (17 cm sea-level rise in the 20th century). In the 21st century, the IPCC projects another 20–60 cm, due to expansion of the oceans, a further melting of glaciers, and a constant contribution of Greenland and Antarctica (at a rate of 0.4 mm/year). These values of IPCC are comparable to the values in Figure 9.21 for the scenario period (oceanic expansion contributes 11–13 cm, glaciers 9–10 cm and Greenland and Antarctica 2 cm). The largest uncertainty, an increased ice sheet flow rate from Greenland and Antarctica, is not considered here or in IPCC 2007a. IPCC states that the understanding of these phenomena is too limited to assess their likelihood, or to provide a best estimate of an upper bound for sea-level rise.

Sources: Bouwman and others 2006, IPCC 2000, 2007a, 2007b

Sea-level rise

The processes governing sea-level rise as a result of climate change, for example thermal expansion of oceans and melting of ice, have long response times. Hence, sea-level responds slowly to the changes in temperature. The computed sea-level rise in all of the scenarios is about 30 centimetres in 2050

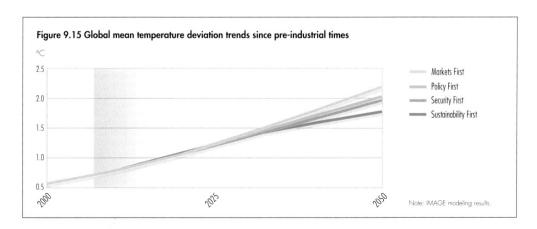

Figure 9.15 Global mean temperature deviation trends since pre-industrial times

°C

Markets First
Policy First
Security First
Sustainability First

Note: IMAGE modeling results.

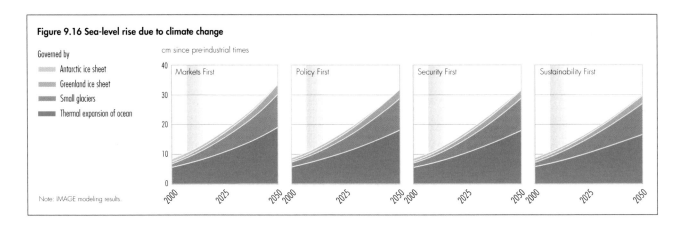

Figure 9.16 Sea-level rise due to climate change

Governed by
- Antarctic ice sheet
- Greenland ice sheet
- Small glaciers
- Thermal expansion of ocean

cm since pre-industrial times

Markets First   Policy First   Security First   Sustainability First

Note: IMAGE modeling results.

relative to the pre-industrial era, with only the smallest of differences between them (see Figure 9.16). This magnitude of sea-level rise implies increasing risk of coastal flooding during storm surge events, accelerating beach erosion, and other changes to the world's coastal zones. As with the global mean surface temperatures, the sea level continues to rise far beyond the time horizon of these scenarios, which is indicated by the steady rate of increase observed at the end of the scenario period.

Land

One of the major environmental challenges is the conservation of land to maintain its ability to supply ecosystem goods-and-services (see Chapter 3). The growth of population, economic wealth and consumption leads to an increase in the overall pressure on land use in each scenario, as well as increased competition between different uses.

Land for agriculture, biofuels and forests

In all scenarios, the use of land for traditional agricultural purposes – food crops and pasture and fodder – increases the most in regions where arable land is still available, notably Africa, and Latin America and the Caribbean (see Figure 9.17). These shifts also imply differences among regions in terms of the reliance on land expansion versus aggressive improvements in yields for agricultural growth. In *Security First*, agricultural land expansion is the smallest, since low economic growth keeps the increase of human demands for land within limits. *Markets First* and *Sustainability First* show comparable results but for different reasons. In *Markets First*, the growth in demand for land is partially compensated by technological developments, whereas in *Sustainability First*, such improvements are counterbalanced by greater concern for food availability. In *Policy First*,

the total area is highest, due to the similar concerns and higher population levels than in *Sustainability First*. In *Policy First* and *Sustainability First*, which include strong targets for the mitigation of GHGs, there is an added demand on land for the production of biofuel crops (see Figure 9.18). The effect of these demands for agriculture and biofuels is reflected in the changes in forest land (see Figure 9.19). Latin America and the Caribbean, and Africa see significant declines in forest land in all scenarios, most notably in *Policy First*, where nearly all of Africa's forests are lost. Meanwhile, Europe and North America see small increases, particularly in *Markets First*.

Land degradation

The continuation of food production from agricultural land is threatened in different ways. First, rainfall erosion increases, due to increases in precipitation because of climate change. Precipitation increase is strongest under *Markets First*, although the differences among scenarios are still small in 2050, because of the inertia in the climate system. Water erosion is greatest in agricultural areas, independent of the soil and climatic conditions.

Combining trends in climate and land-use change and the erodibility index allows a calculation of the water erosion risk index. Compared with the present situation, the area with a high water erosion risk increases by almost 50 per cent in all scenarios (see Figure 9.20). Differences among the scenarios up to 2050 are relatively small. The risks under *Sustainability First* and *Markets First* are somewhat less than under the other scenarios, although there is a period during which they rise in the former as more biofuel crops are introduced. The increases are largest under *Policy First*, mainly due to larger food demand, combined with an increased demand for biofuel crops.

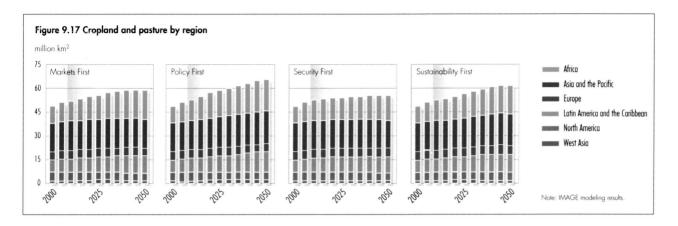

Figure 9.17 Cropland and pasture by region

million km²

Markets First   Policy First   Security First   Sustainability First

Africa
Asia and the Pacific
Europe
Latin America and the Caribbean
North America
West Asia

Note: IMAGE modeling results.

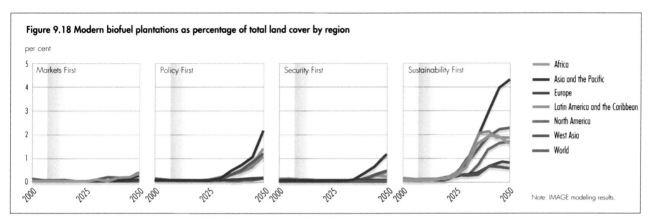

Figure 9.18 Modern biofuel plantations as percentage of total land cover by region

per cent

Markets First   Policy First   Security First   Sustainability First

Africa
Asia and the Pacific
Europe
Latin America and the Caribbean
North America
West Asia
World

Note: IMAGE modeling results.

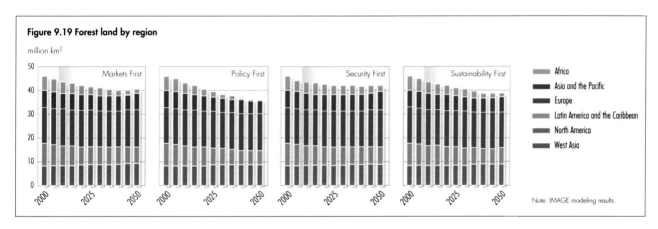

Figure 9.19 Forest land by region

million km²

Markets First   Policy First   Security First   Sustainability First

Africa
Asia and the Pacific
Europe
Latin America and the Caribbean
North America
West Asia

Note: IMAGE modeling results.

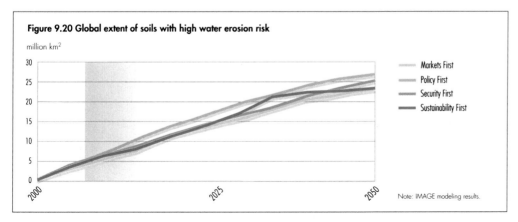

Figure 9.20 Global extent of soils with high water erosion risk

million km²

Markets First
Policy First
Security First
Sustainability First

Note: IMAGE modeling results.

Desertification

Another threat to crop production is desertification. It has been identified as a major social, economic and environmental problem for many countries around the world. Just like land degradation, desertification results from natural factors (such as change in precipitation) and human causes (such as land clearance and excessive land use) or a combination of both.

Changes in arid areas (as a result of climate change) are relatively small. This follows from the fact that climate change results in increasing precipitation, but also increasing evaporation (as a result of temperature increase). For desertification, however, the increase in arid areas is less important than the pressure on these areas. Therefore, the combination of agricultural land expansion in arid areas leads to an increased vulnerability to climatic shocks.

Yields and food availability

The changes in land use and quality, as well as advances in technology and general economic developments, such as trade, are reflected in the changes in agricultural yields and food availability. All regions experience increasing cereal yields per unit area in all scenarios, but these are significantly lower in *Security First*, reflecting slower developments in technology and poorer land management practices (see Figure 9.21). The increasing demands for food, along with greater investments in technology, result in the largest increases in *Markets First* and *Policy First*, with some differences among regions. *Sustainability First* shows slightly lower growth, but this is counterbalanced by a lower overall population.

Figure 9.22 highlights the projected changes in per capita food availability. Overall food production increases in all four scenarios, but per capita food availability is also influenced by the different rates in the growth of the population. Significant increases are seen in *Markets First*, *Policy First* and *Sustainability First*, with the latter achieving global levels 10 per cent and 5 per cent higher than the first two, respectively. In *Security First*, food production barely keeps up with population growth after 2020, and the beginnings of a decline are seen around 2040, with that in Africa happening much sooner. By 2050, there is more than a 30 per cent difference in per capita food availability between *Security First* and *Sustainability First* globally, and a 70 per cent difference in Africa.

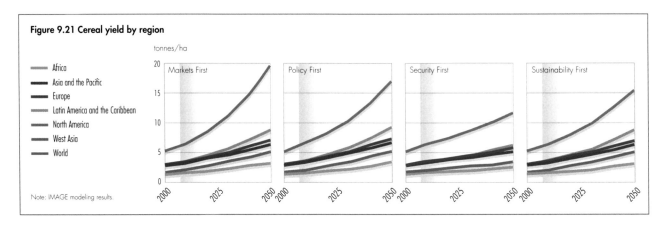

Figure 9.21 Cereal yield by region

- Africa
- Asia and the Pacific
- Europe
- Latin America and the Caribbean
- North America
- West Asia
- World

tonnes/ha

Markets First Policy First Security First Sustainability First

Note: IMAGE modeling results.

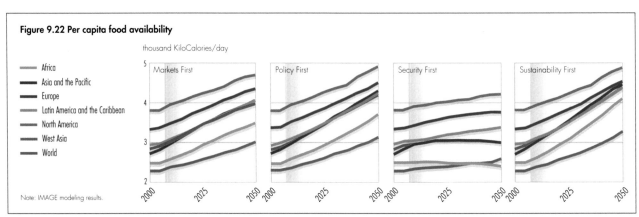

Figure 9.22 Per capita food availability

- Africa
- Asia and the Pacific
- Europe
- Latin America and the Caribbean
- North America
- West Asia
- World

thousand KiloCalories/day

Markets First Policy First Security First Sustainability First

Note: IMAGE modeling results.

Water

As discussed in Chapter 4, water – both its quantity and quality – is fundamental to the environment and human well-being. The scenarios show that very different futures for water are plausible depending on our choices in the near future

Water use

One of the consequences of the rapid push for better material standard of living in *Markets First* is a rapid growth in water use in all socio-economic sectors, resulting in a large increase in withdrawals from surface and groundwaters (see Figure 9.23). Trends differ greatly from country to country; in many industrialized countries water use reaches a saturation point during the scenario period, whereas the growing incomes in developing countries lead to an increasing demand for modern water services. In *Markets First*, the privatization of water services and improvements in technology lead to a moderate but steady increase in the efficiency of water use in most regions. Nevertheless, the water sector emphasizes the expansion of water supply rather than water conservation. Under *Policy First*, a change in water-use behaviour in households and industries, together with rapid improvements in the efficiency of water use in all sectors, leads to a decrease in water withdrawals in many industrialized countries, and a slower growth elsewhere. Under *Security First*, a growing population and neglect of water conservation tends to push water withdrawals upwards. Yet, slower economic growth tends to slow the increase. *Sustainability First* assumes widespread adoption of integrated water management strategies, with strong emphasis on demand management and conservation. These developments, together with slower population growth rates, lead to slower increases in overall water use.

Persons living in areas with severe water stress

The extent of severe water stress will be complicated by the effect of climate change on the future water supply in all scenarios. Increasing precipitation will increase the annual availability of water in most river basins, but warmer temperatures and decreasing precipitation will decrease annual water availability in some arid regions, such as West Asia, southern parts of Europe, and northeastern parts of Latin America and the Caribbean. Changes in climate could also lead to more frequent periods of high and low run-off (not shown). By 2050, the occurrence of droughts could become more common in already arid areas, such as Australia, southern India and Southern Africa. Meanwhile, increasing precipitation could cause more frequent high run-off events in parts of Asia and the Pacific, Latin America and the Caribbean, and North America.

These factors combine with the increases in demand and population growth to determine the number of people living in river basins with severe water stress (see Box 9.2 and Figure 9.24). In *Markets First*, the affected population grows from around 2.5 billion people in 2000 to nearly 4.3 billion people in 2050. In *Policy First*, actions to slow the growth in water use help alleviate concerns about freshwater

---

Box 9.2 Water stress

The concept of "water stress" is used in many water assessments to obtain a first estimate of the extent of society's pressure on water resources. Severe water stress is defined as a situation where withdrawals exceed 40 per cent of renewable resources. It is assumed here that the higher the levels of water stress the more likely that chronic or acute water shortages will occur.

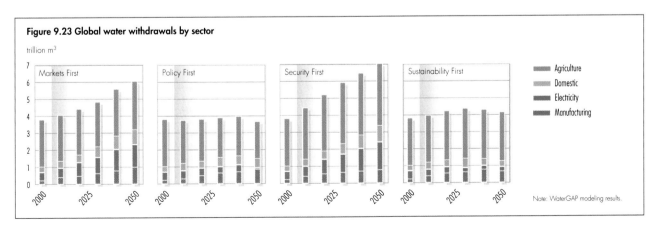

Figure 9.23 Global water withdrawals by sector

trillion m^3

Markets First | Policy First | Security First | Sustainability First

Legend: Agriculture, Domestic, Electricity, Manufacturing

Note: WaterGAP modeling results.

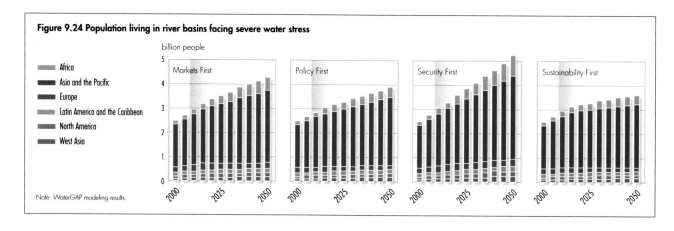

Figure 9.24 Population living in river basins facing severe water stress

billion people

Africa
Asia and the Pacific
Europe
Latin America and the Caribbean
North America
West Asia

Markets First   Policy First   Security First   Sustainability First

Note: WaterGAP modeling results.

availability in much of the world. Nevertheless, growing populations and economic activity continue to put a strain on resources in some places, particularly in the developing regions. The global population living under severe water stress rises by 40 per cent to nearly 3.9 billion people. The net effect of high population and increased demand in *Security First* is that the population living in river basins under severe water stress in 2050 exceeds 5.1 billion people. The number living with water scarce conditions in Africa approaches 800 million, nearly as many people as lived in that region at the start of the century. The developments under *Sustainability First* with respect to water use, together with slower population growth rates, lead to significant reductions in water stress in many river basins. Still, in part due to the varying patterns of population growth and shifting patterns of precipitation as part of the changing climate, there are increases in some regions, notably Africa, Asia and the Pacific, and West Asia. The number of people living in river basins with severe water stress increases by more than 1.1 billion globally. In both *Sustainability First* and *Policy First*, it is expected that many actions will be taken to help

people in river basins facing stress to better cope with water scarcity. These include programmes to reduce water wastage in the distribution of water and highly efficient programmes to manage surface- and groundwater.

Wastewater treatment

A consequence of the rapid increase in water withdrawals in *Markets First* is a similarly rapid growth in the production of wastewater. Although treatment plant capacity expands, it cannot keep up with the increase in volume of wastewater. Hence, the total worldwide volume of untreated wastewater from the domestic and manufacturing sectors doubles between 2000 and 2050 (see Figure 9.25). Since most of this wastewater is discharged into inland surface waters, the world experiences a serious spread of water pollution problems and health risks. In *Policy First*, the level of wastewater treatment increases from around 50 to about 80 per cent between 2000 and 2050, but because of growing population, the total volume of untreated wastewater still increases by about 25 per cent during this time (see Figure 9.25). Yet, the global average hides significant disparities among regions. While the total volume of untreated wastewater shrinks

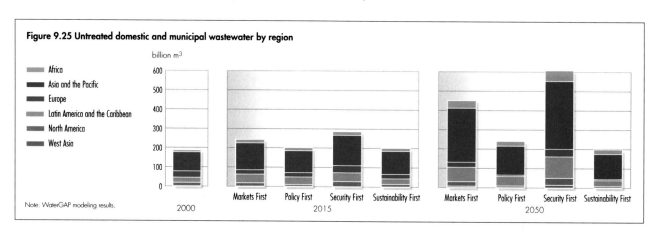

Figure 9.25 Untreated domestic and municipal wastewater by region

billion m³

Africa
Asia and the Pacific
Europe
Latin America and the Caribbean
North America
West Asia

Markets First   Policy First   Security First   Sustainability First
2000   2015   2050
Markets First   Policy First   Security First   Sustainability First

Note: WaterGAP modeling results.

by more than half in Europe, it nearly doubles in LAC. Since the coverage of communities with wastewater treatment plants is also relatively low under *Security First*, the volume of untreated wastewater increases by more than a factor of three between 2000 and 2050. The discharge of these large volumes of untreated wastewater to surface waters causes widespread water contamination, worsening health risks and degrading aquatic ecosystems. Under *Sustainability First*, efforts to reduce the growth in water demand also contribute to maintaining and improving water quality around the world. Treatment capacity manages to keep pace with the increasing amounts of wastewater, such that the total volume of untreated wastewater has changed very little since the turn of the century (see Figure 9.25). This, however, masks large differences among regions. In North America, the volume of wastewater is drastically reduced, while in Latin America and the Caribbean the volume increases by a small amount.

Biodiversity

Across the scenarios and regions, global biodiversity continues to be threatened, with strong implications for ecosystem services and human well-being as described in Chapter 5. This is the case for both terrestrial and marine biodiversity. However, there are clear differences among the scenarios in the magnitude and location of change.

Terrestrial biodiversity

All regions continue to experience declines in terrestrial biodiversity in each of the scenarios. Figure 9.26 shows the levels of Mean Species Abundance as of 2000 and the changes in each of the scenarios from 2000 to 2050; Figures 9.27–9.28 summarize these changes by region and contribution. The greatest losses are seen in *Markets First*, followed by *Security First*, *Policy First* and *Sustainability First* for most regions. Africa, and Latin America and the Caribbean experience the greatest losses of terrestrial biodiversity by 2050 in all four scenarios, followed by Asia and the Pacific. The differences among the regions are largely a result of the broad-scale land-use changes already described, especially increases in pastureland and areas dedicated to biofuel production. The overall changes in terrestrial biodiversity though, are influenced by a number of other factors, including infrastructure development, pollution and climate change, as well as public policy and conflict.

Agriculture, including crop and livestock production, has greater overall biodiversity impacts in *Policy First* and *Sustainability First* than in the other scenarios, both because food security is highly valued in these worlds, and because there is a greater uptake of biofuels based on agricultural products. Tropical forests continue to be particularly vulnerable to conversion.

On a global scale, much more biodiversity loss is seen in *Markets First* than in any other scenario, with infrastructure development playing a major role. In *Markets First*, global population growth is more limited, and road construction and urban development are more regulated than in *Security First*. However, the drivers of development are stronger in *Markets First*: international markets for goods are strengthened, infrastructure is developed to promote access to natural resources, and wealth creation is valued more highly than conservation. In *Markets First* and *Security First*, the biodiversity loss continues to accelerate as the scenario progresses, but within *Policy First* and *Sustainability First*, the global rate of loss stabilizes by 2050.

The impacts of climate change are modelled as being similar in each scenario, but in reality, these will be moderated by the ability of species and ecosystems to adapt and migrate. Resilient, well-connected ecosystems suffer fewer ill effects from climate change than fragmented, overexploited ecosystems, such as those seen in *Security First* and *Markets First*. The rate at which the global temperature continues to rise has a profound influence on the survival chances of many of the worlds' species until 2050 and beyond.

Under *Policy First* and *Sustainability First*, the protected area network is expanded to create ecologically

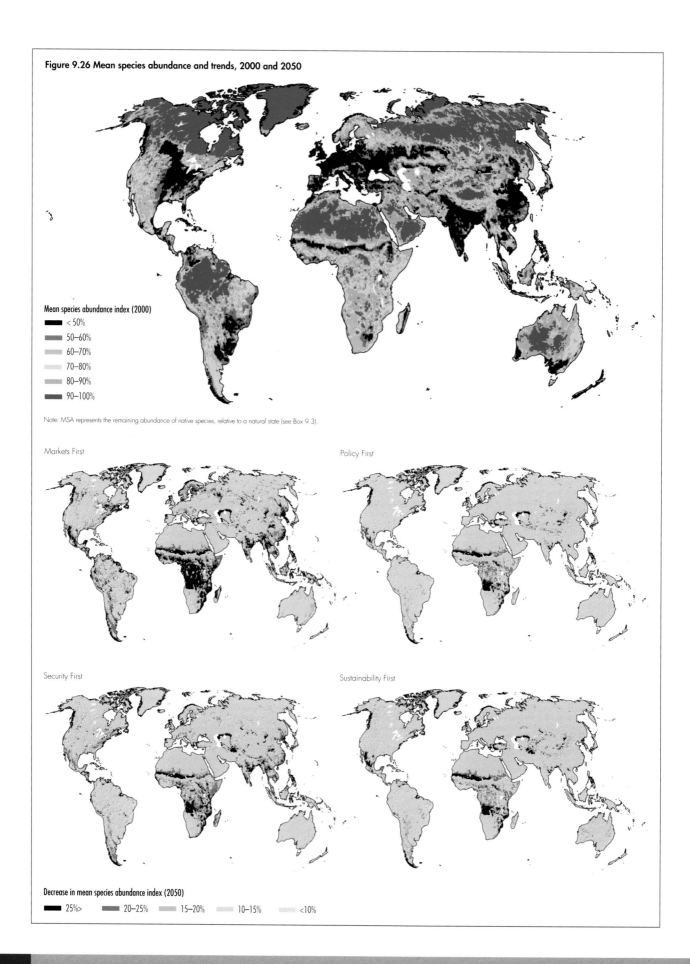

Figure 9.26 Mean species abundance and trends, 2000 and 2050

Mean species abundance index (2000)

- ▬ < 50%
- ▬ 50–60%
- ▬ 60–70%
- ▬ 70–80%
- ▬ 80–90%
- ▬ 90–100%

Note: MSA represents the remaining abundance of native species, relative to a natural state (see Box 9.3).

Markets First

Policy First

Security First

Sustainability First

Decrease in mean species abundance index (2050)

▬ 25%> ▬ 20–25% ▬ 15–20% ▬ 10–15% ▬ <10%

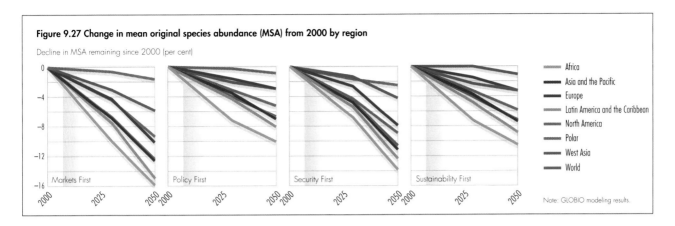

Figure 9.27 Change in mean original species abundance (MSA) from 2000 by region

Decline in MSA remaining since 2000 (per cent)

Markets First | Policy First | Security First | Sustainability First

Africa
Asia and the Pacific
Europe
Latin America and the Caribbean
North America
Polar
West Asia
World

Note: GLOBIO modeling results.

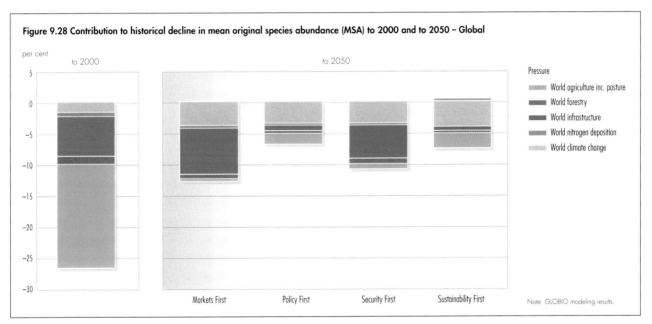

Figure 9.28 Contribution to historical decline in mean original species abundance (MSA) to 2000 and to 2050 – Global

per cent

to 2000   to 2050

Markets First | Policy First | Security First | Sustainability First

Pressure
World agriculture inc. pasture
World forestry
World infrastructure
World nitrogen deposition
World climate change

Note: GLOBIO modeling results.

representative national and regional systems of protected areas. A minor expansion is seen in *Markets First*, and virtually none in *Security First*; the investment dedicated to and effectiveness of protected area management within the different scenarios follows the same pattern. While the new protected areas do not limit the overall amount of wildland converted to agricultural use, they protect some of the most critical remaining habitat, including that inhabited by endangered species with restricted ranges. This effect is not seen in the MSA (mean original species abundance) indicator, as it is insensitive to these specific, rare and unique species and ecosystems. Some agricultural use is possible in some protected areas, but there is a high potential for land-use conflicts by 2050. This is most visible within *Policy First*. In sub-regions such as Meso-America and Southern Africa, there is so much demand for agricultural farmland that wilderness outside protected areas is crowded out,

and the areas themselves are isolated in an agricultural matrix. Sustainable agriculture, with farm design paying explicit attention to biodiversity conservation, is especially important under these circumstances.

Finally, the increased frequency of armed conflict in *Security First* creates unpredictable risks for biodiversity, as well as for people. International funds for conservation action are often frozen as the situation deteriorates. As well as increasing the availability of guns, conflict reduces agricultural production, making illegal and unsustainable hunting more attractive. As rural people struggle to survive, militias seek resources to fund their wars, and unscrupulous companies take advantage of the chaos. Protected areas in the conflict zones are looted for meat, minerals and timber (Draulans and van Krunkelsven 2002, Dudley and others 2002).

Marine biodiversity

Marine biodiversity continues to decline in all scenarios, due to increased pressure on marine fisheries to meet food demand (see Figure 9.29). The declines are smallest in *Sustainability First*, due to the smaller increases in population and shifting diets. Even with its greater population, *Security First* does not see as large an increase as *Markets First* or *Policy First*, due to lower average incomes as well as slower advances in technology that would allow for greater catches.

The scenarios also differ with respect to the types of fish caught. Figure 9.30 shows that in *Sustainability First* there is an attempt to fish lower on the food chain, reflecting the goal of maintaining marine ecosystems. While these differences may seem marginal, in combination with the lower overall catch level, the effect can be important as is shown in Figure 9.31. The total biomass of large demersal fish is seen to grow significantly in *Sustainability First*, and slightly in *Policy First* and *Security First* by 2050, while it decreases in *Markets First*. With respect to large and small pelagic fish, the effect is seen in slower declines and small increases in biomass, respectively.

Human well-being and vulnerability

What do the scenarios indicate with respect to human well-being – the extent to which individuals have the ability and opportunity to achieve their aspirations? How do they compare in terms of personal and environmental security, access to materials for a good life, good health and good social relations, all of which are linked to the freedom to make choices and take action?

To a certain degree, the scenarios exhibit greater or lesser levels of certain aspects of human well-being by design. *Markets First* and *Sustainability First* assume a greater emphasis on individuals' freedom

to make choices and take action than do either *Policy First* or *Security First*. More prominence is given to improving health, strengthening local rights and building capacity in *Policy First* and *Sustainability First* than in either *Markets First* or *Security First*.

Using the Millennium Development Goals as a guide, Table 9.2 (and associated figures) summarizes how the scenarios fare with respect to improvements in human well-being. Here also, some of the results should be seen as assumptions rather than outcomes. In particular, the development of a global partnership for development (MDG 8) and the integration of the principles into country policies and programmes (a key aspect of MDG 7) are fundamental assumptions of *Sustainability First*. These are also assumed, but to a lesser degree, in *Policy First*. In *Markets First*, to the extent that these developments arise, it is assumed this happens only where they fit with the broader goal of increased economic growth. Little or no progress in these areas is assumed in *Security First* (see Box 9.4).

The full picture of human well-being can only be seen by considering the detailed developments within the scenarios. For most regions and sub-regions, there is a fairly consistent pattern of improvements moving from Security to Markets to Policy to *Sustainability First*. The currently wealthier regions and sub-regions experience slower growth in per capita income in *Sustainability First*, but this must be weighed against improvements in other indicators. Even in *Sustainability First*, achieving the MDG targets for example, reducing the percentage of population whose income is less than US$1/day to half their 1990 levels by 2015, is not achieved in all regions.

Looking beyond the MDGs, personal security for most people is significantly lower in *Security First*, but there are also strong tensions and potential conflicts in *Markets First*. Combined with the increasing pressures on the environment in all scenarios, these will significantly affect environmental security, with *Markets First* placing the greatest stress on the global environment and *Security First* on local environments. These changes will be reflected in the vulnerability of people and the environment. This is borne out by considering how the scenarios differ with respect to a few of the archetypes discussed in Chapter 7, specifically those focusing on the commons, Small Island Developing States (SIDS) and water stress.

Box 9.4 Capturing the impact of environmental change on human well-being

As described in the *GEO-4* conceptual framework, the impacts of environmental change on human well-being are strongly mediated by social and institutional factors. Furthermore, the explicit links between environmental change and certain aspects of human well-being, such as food availability and water stress, are better understood than, say, those related to education, personal security, good social relations and overall access to materials for a good life. The scenarios presented here, specifically their quantitative elements, do not fully capture the impacts of environmental change on well-being, particularly for these latter categories. Assuming that these are enhanced by positive environmental change, it is likely that the results presented here underestimate the differences in well-being across the scenarios.

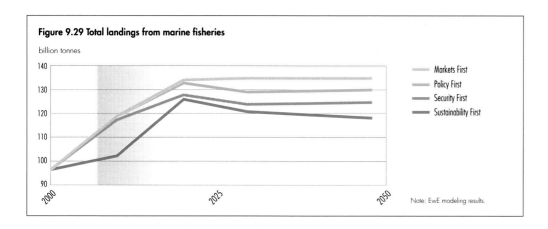

Figure 9.29 Total landings from marine fisheries

billion tonnes

Markets First
Policy First
Security First
Sustainability First

Note: EwE modeling results.

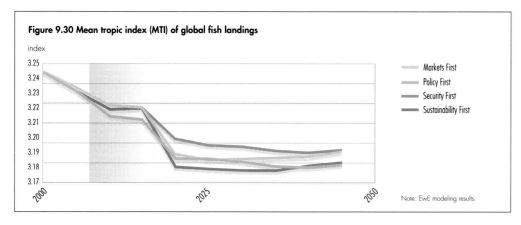

Figure 9.30 Mean tropic index (MTI) of global fish landings

index

Markets First
Policy First
Security First
Sustainability First

Note: EwE modeling results.

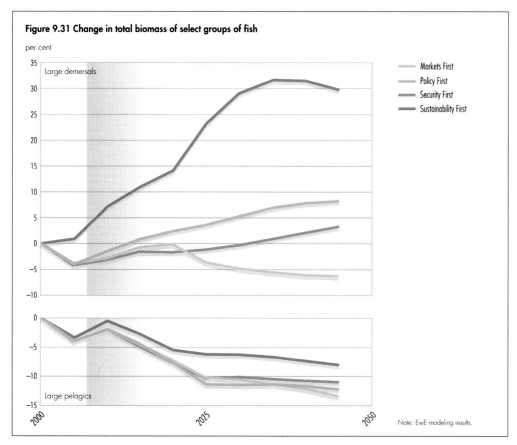

Figure 9.31 Change in total biomass of select groups of fish

per cent

Large demersals

Markets First
Policy First
Security First
Sustainability First

Large pelagics

Note: EwE modeling results.

Table 9.2: Progress on the MDGs across the scenarios*	
MDG and associated targets**	**Progress in the Scenarios**
Goal 1 Eradicate extreme poverty and hunger	Extreme poverty and hunger are influenced by a number of factors, including not only overall economic growth and food production, but also their distribution. At the global level, the income target of halving the share of the population with incomes less than US$1/day is reached in all scenarios by the target date of 2015, led primarily by strong progress in Asia and the Pacific (see Figure 9.32). This is not the case for all regions, however. Latin America and the Caribbean, and Africa lag behind, particularly in *Markets First* and *Security First*; in the latter scenario, Africa never achieves the target, and Latin America and the Caribbean only does so late in the period. In the longer term, the improvement is greatest in *Sustainability First* and *Policy First* in all regions. In *Security First*, there is actually a reversal in the trend around the mid-point of the period, driven largely by slower growth in Africa, but also in West Asia. In the latter case, this reflects in part the dependence of their economies on the oil and gas sector, which is facing a transition as resources become constrained. The same effect is seen to a smaller degree in the other scenarios. Hunger rates show similar declines in all scenarios other than *Security First*, where percentages decline only slightly, implying significant increases in the numbers of malnourished (see Figure 9.33, noting that data are not available for North America or Europe). Africa and Asia and the Pacific continue to have the highest levels of malnourished people in all scenarios.
Goal 2 Achieve universal primary education	All regions reach their highest levels of primary education in *Sustainability First*, followed by *Policy First*, reflecting among other factors the greater emphasis on investments in education (see Figure 9.34). Gradual progress is also made in *Markets First*. Africa and West Asia still lag behind somewhat, but show significant progress in catching up to other regions. In *Security First*, after increases early in the period, there is a slight reversal in efforts to reach this target at the global level. This is due to slower growth in enrolment in Africa and West Asia, and some declines in Latin America and the Caribbean and Asia and the Pacific.
Goal 3 Promote gender equality and empower women	At the global level, the gender disparity in primary and secondary education gradually declines in all scenarios, with the slowest declines seen in *Security First* (see Figure 9.35). Parity in secondary education is achieved earlier than in primary education. The pattern of change is similar in most regions. In Latin America and the Caribbean, North America and Europe, parity is already seen at the start of the century. Asia and the Pacific lags behind the global average and continues to do so, particularly in *Security First*. West Asia and Africa show rapid improvements in all scenarios, particularly the latter in secondary education. Still they generally continue to stay behind the other regions.
Goal 4 Reduce child mortality	Although progress is made in all regions in all scenarios, it is not clear that the 2015 target will be met. Reflecting the more rapid and equitably distributed economic growth, along with greater investments in education and health, the most significant advances are expected in *Sustainability First* and *Policy First*. For the opposite reason, the slowest progress is expected in *Security First*. Combined with the larger population growth, this also implies much higher absolute levels of children dying before reaching their fifth birthday.
Goal 5 Improve maternal health	Similar to child mortality, although progress is made in all regions in all scenarios, it is not clear that the 2015 target will be met. For the same reasons, the most significant advances are expected in *Sustainability First* and *Policy First*, and the slowest in *Security First*.
Goal 6 Combat HIV/AIDS, malaria and other diseases	HIV infection rates peak between 2010 and 2015 in all scenarios, after increases globally, primarily in Asia and the Pacific, Africa, and parts of Eastern Europe early in the century. The peak of AIDS deaths occurs a few years later, with the highest rates in *Security First* and the lowest in *Sustainability First*. The differences in the death rates among the scenarios reflects not only the higher infection rates in *Security First*, but also the less effective public health services, which affect how long and how well people can live with the infection. Similar patterns are also expected across the scenarios for malaria and other major diseases. This is partly reflected in the differences in life expectancies across the scenarios (see Figure 9.36).
Goal 7 Ensure environmental sustainability	In *Markets First* and *Security First*, limited progress is seen in integrating the principles of sustainable development. Strong progress is made in *Policy First* and *Sustainability First*. The larger overall populations in *Markets First* and *Security First*, as well as the more unequal income distributions imply larger numbers of slum dwellers. The relative lack of specific policies to address their concerns also points to less progress in improving the lives of these groups. With respect to physical measures of environmental sustainability, the results point to a general pattern of more positive trends going from *Security* to *Markets* to *Policy* to *Sustainability First*.
Goal 8 Develop a global partnership for development	In *Markets First*, limited progress is achieved on this goal; where it does occur, it is primarily in the context of development defined as economic growth. Very little progress is seen in *Security First*, as groups increasingly focus on more local concerns. In *Policy First* and *Sustainability First*, strong progress is achieved. In the former, this primarily involves the establishment and expansion of fairly centralized institutions. In the latter, more complementary institutions at international, regional, national and local levels are established, and a broad definition of development is adopted.

* The results presented in this table reflect a combination of the narrative and quantitative elements of the scenarios. Certain results, in particular for Goal 8, represent assumptions rather than outcomes of the scenarios.

**UN (2003) describes the specific targets and indicators that are being used to monitor progress towards the achievement of these goals.

All of the scenarios present challenges to the global commons, but in different ways and to different degrees. A scenario such as *Markets First* presents significant challenges; in addition to the growth of population and economic activity, there is relatively less attention paid to social and environmental issues. More fundamentally, the drive towards increased privatization implies that what is now treated as common property will increasingly fall under private control. Although this can have positive or negative implications in terms of environmental protection, it will almost certainly lead to more limited access. In *Security First*, the global commons may actually benefit from several factors: lower levels of economic activity, reduced trade and stricter control in particular areas. Where the commons are accessible, however, it is likely that they will be severely affected. More attention is paid to preserving and sharing the benefits of the global commons in *Policy First* and *Sustainability First*. Still, the relatively more rapid increases in incomes in the poorer regions in these scenarios, and the desire to meet both environmental and human well-being goals may

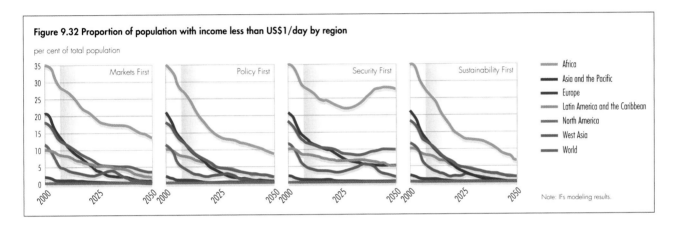

Figure 9.32 Proportion of population with income less than US$1/day by region

per cent of total population

Markets First — Policy First — Security First — Sustainability First

- Africa
- Asia and the Pacific
- Europe
- Latin America and the Caribbean
- North America
- West Asia
- World

Note: IFs modeling results.

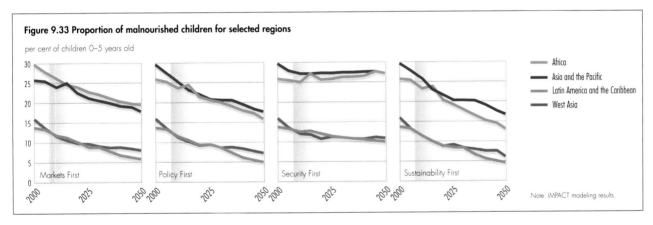

Figure 9.33 Proportion of malnourished children for selected regions

per cent of children 0–5 years old

Markets First — Policy First — Security First — Sustainability First

- Africa
- Asia and the Pacific
- Latin America and the Caribbean
- West Asia

Note: IMPACT modeling results.

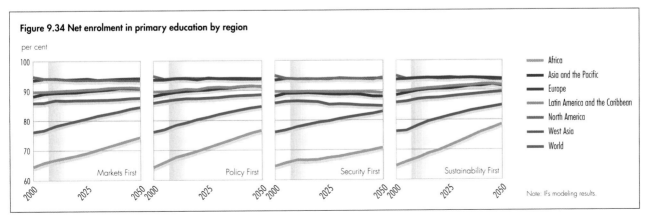

Figure 9.34 Net enrolment in primary education by region

per cent

Markets First — Policy First — Security First — Sustainability First

- Africa
- Asia and the Pacific
- Europe
- Latin America and the Caribbean
- North America
- West Asia
- World

Note: IFs modeling results.

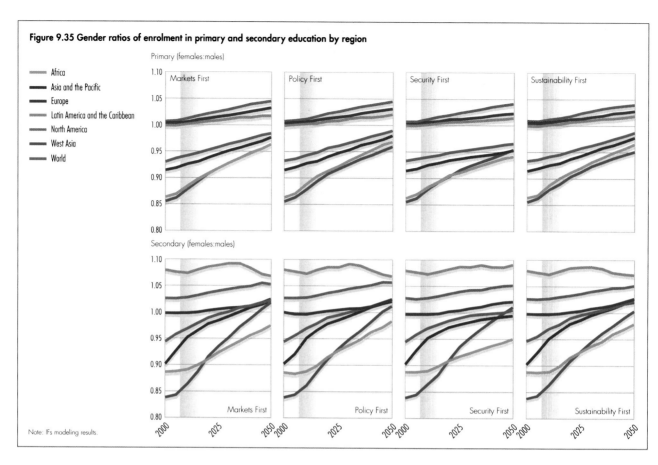

Figure 9.35 Gender ratios of enrolment in primary and secondary education by region

Legend:
- Africa
- Asia and the Pacific
- Europe
- Latin America and the Caribbean
- North America
- West Asia
- World

Primary (females:males)

Markets First | Policy First | Security First | Sustainability First

Secondary (females:males)

Markets First | Policy First | Security First | Sustainability First

Note: IFs modeling results.

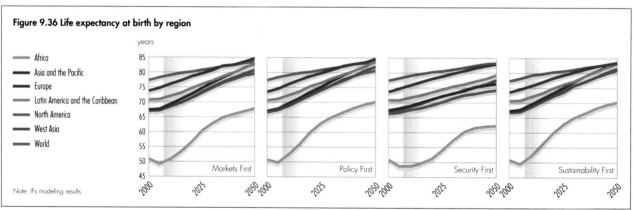

Figure 9.36 Life expectancy at birth by region

Legend:
- Africa
- Asia and the Pacific
- Europe
- Latin America and the Caribbean
- North America
- West Asia
- World

years

Markets First | Policy First | Security First | Sustainability First

Note: IFs modeling results.

lead to conflicts, putting increased pressure on the global commons. In particular, the need to meet increased demands for food and biofuels results in increased pressure on forests and protected areas. This is most likely in *Policy First*, with its larger population.

The fate of many SIDS is closely linked to the impacts of climate change, in particular sea-level rise. Their outlook is not bright in any of the scenarios (see Figure 9.16), with all indicating a further 20-cm increase by mid-century, which will

result in more impacts from tropical storms and storm surges. The scenarios differ, however, in other factors related to the vulnerability of SIDS. *Security First* brings larger populations, relatively lower levels of international trade, lower incomes and increased limits on international migration. Together, these factors imply severe vulnerability for SIDS. Technological developments in *Markets First*, along with increased trade and mobility, may help to temper the vulnerabilities. The lower levels of population growth and relatively larger increases in incomes in the poorer SIDS in *Policy First* and

Sustainability First will increase the adaptive capacity of populations in these locations.

Water stress is an issue that is also present in all of the scenarios. As populations grow, so do their demands for services as the scenarios with greater population growth naturally imply greater demands. This is tempered by the lower economic growth in *Security First* in all regions, and lower growth in the currently wealthy regions in *Sustainability First*. Equally important are the ways the scenarios differ in how these demands are satisfied, including both augmenting supply and improving the efficiency by which services are delivered. In *Markets First*, privatization, the reduction of subsidies and water pricing all work to reduce the effective demand for water. There is still a strong emphasis on supply augmentation, using technology-centred approaches, such as dam building, deeper drilling for groundwater and large desalination plants. Similar approaches to meeting supply are taken in *Security First*, although with less efficient implementation. Furthermore, less attention is paid to the environmental implications of these activities, and vulnerable groups are less equipped to cope with the impacts. *Policy First* and *Sustainability First* see larger efforts to reduce overall demand, although more subsidies remain in place and greater efforts are taken to improve access, particularly for the poor. The net trade-off is a slightly higher exposure to water problems than would otherwise be the case, but a greater capacity to cope with these problems.

KEY MESSAGES FROM THE REGIONS

The regions of the world will not necessarily confront a single future. As discussed elsewhere in this report, particularly in Chapter 6, the challenges faced differ markedly across regions. As such, the key issues of concern and the precise nature of the developments over the scenario period also differ across the regions. This section summarizes the key regional messages coming out of the scenarios.

Africa

Population increases remain an overriding driver in all scenarios. Population distribution, migration, urbanization, age structure, growth and composition are affected by economic and migration policies in Africa and the other regions. Another common factor is that the achievement of the energy goals

set by the New Partnership for Africa's Development (NEPAD) must not ignore environmental considerations. These include developing more clean energy sources, improving access to reliable and affordable commercial energy supply, improving the reliability as well as lowering the cost of energy supply to productive activities, and reversing environmental degradation associated with the use of traditional fuels in rural areas. This will involve integrating the energy economies of the member states of the African Union to ensure the success of NEPAD, particularly by ensuring that poverty reduction strategies incorporate consideration of sustainable energy supplies.

Severe land degradation in *Markets First* and *Security First* results from, respectively, intensified profit-driven agricultural practices and unsustainable practices. This leads to attendant environmental and human well-being impacts. The privatization and amalgamation of sectors in *Markets First* leads to some improvements in human development, but limited environmental stewardship and globalization trade-offs in *Markets First* show significant negative consequences by 2050. In *Security First*, poor economic policies lead to the overexploitation of water, land and mineral resources. In *Policy First*, environmental and social policies assist the attainment of environmental stewardship and social equity. In *Sustainability First*, positive changes in value systems, environmental consciousness and favourable demographic, economic and technological trends lead to environmental conservation, with a marked decline in land degradation. In both *Policy First* and *Sustainability First*, favourable economic policies, regional integration and economic and environmental stewardship guided by the regulatory frameworks of NEPAD and the African Ministerial Conference on the Environment, create an environment conducive to the attainment of environmental and human development goals.

The scenarios indicate that policies affecting the environment require time, and governments should avoid policy reversals through efforts to build institutional capacity for the development, implementation and monitoring of policy. Policy formulation should not be a technocratic exercise, but a process of dialogue and engagement with the citizenry, scientists and implementers. The outcome of the policies formulated will also depend

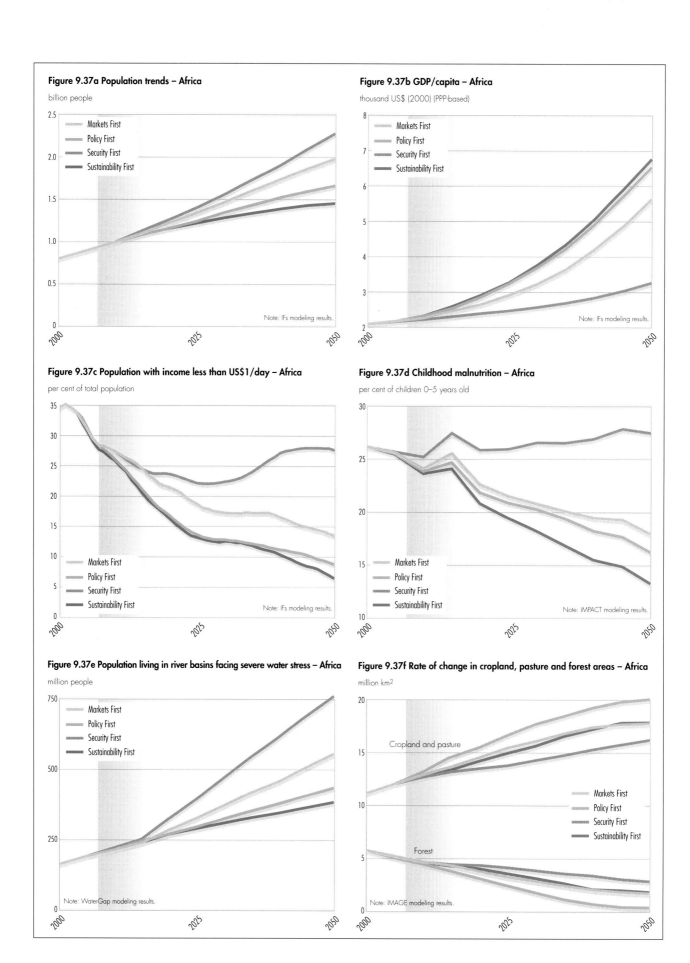

Figure 9.37a Population trends – Africa

billion people

Markets First
Policy First
Security First
Sustainability First

Note: IFs modeling results.

Figure 9.37b GDP/capita – Africa

thousand US$ (2000) (PPP-based)

Markets First
Policy First
Security First
Sustainability First

Note: IFs modeling results.

Figure 9.37c Population with income less than US$1/day – Africa

per cent of total population

Markets First
Policy First
Security First
Sustainability First

Note: IFs modeling results.

Figure 9.37d Childhood malnutrition – Africa

per cent of children 0–5 years old

Markets First
Policy First
Security First
Sustainability First

Note: IMPACT modeling results.

Figure 9.37e Population living in river basins facing severe water stress – Africa

million people

Markets First
Policy First
Security First
Sustainability First

Note: WaterGap modeling results.

Figure 9.37f Rate of change in cropland, pasture and forest areas – Africa

million km²

Cropland and pasture

Forest

Markets First
Policy First
Security First
Sustainability First

Note: IMAGE modeling results.

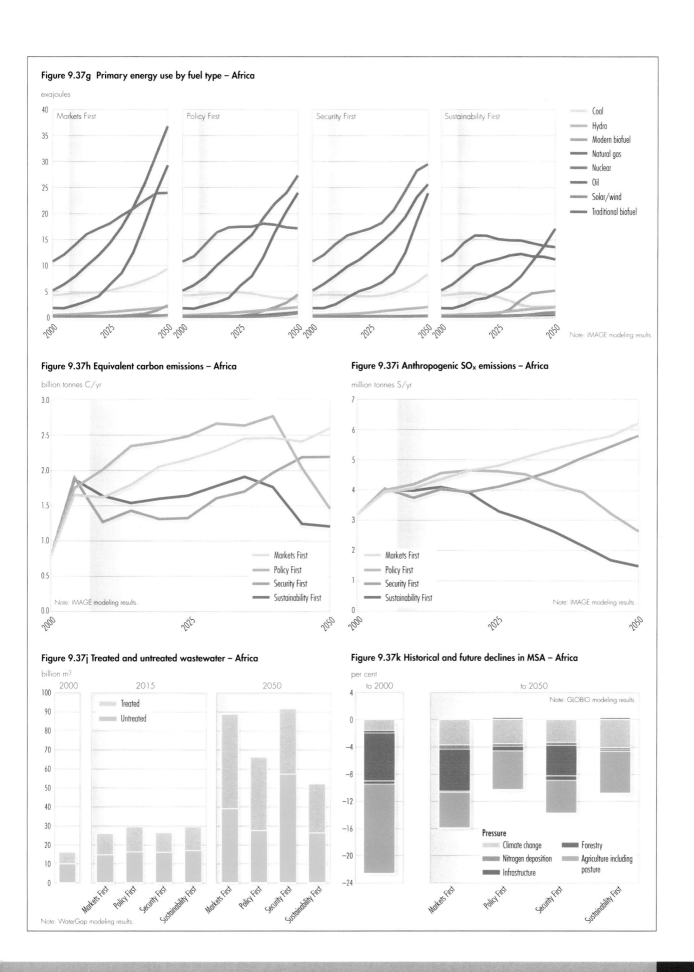

Figure 9.37g Primary energy use by fuel type – Africa

exajoules

Markets First
Policy First
Security First
Sustainability First

Coal
Hydro
Modern biofuel
Natural gas
Nuclear
Oil
Solar/wind
Traditional biofuel

Note: IMAGE modeling results.

Figure 9.37h Equivalent carbon emissions – Africa

billion tonnes C/yr

Markets First
Policy First
Security First
Sustainability First

Note: IMAGE modeling results.

Figure 9.37i Anthropogenic SO$_x$ emissions – Africa

million tonnes S/yr

Markets First
Policy First
Security First
Sustainability First

Note: IMAGE modeling results.

Figure 9.37j Treated and untreated wastewater – Africa

billion m^3

2000   2015   2050

Treated
Untreated

Markets First
Policy First
Security First
Sustainability First

Note: WaterGap modeling results.

Figure 9.37k Historical and future declines in MSA – Africa

per cent

to 2000   to 2050

Note: GLOBIO modeling results.

Pressure
Climate change
Nitrogen deposition
Infrastructure
Forestry
Agriculture including pasture

Markets First
Policy First
Security First
Sustainability First

on the nature of value systems inculcated. Moving environment from its current peripheral situation in the region to the core of development is pivotal to sustainable development. The figures under 9.37 highlight the possible futures of the region.

Asia and the Pacific

There is a danger that increasing the wealth and material well-being of the region's citizens may come at the cost of environmental deterioration and resource depletion, unless countermeasures are also taken. In *Markets First*, the average standard of living improves in the region, but the diversity and stability of marine fisheries are threatened, water scarcity intensifies and pollution control efforts cannot keep up with the increasing pressures. Material well-being also increases under *Policy First*, but in this case, the negative side effects are mitigated by enlightened centralized governmental policies that emphasize conservation and environmental protection. The standard of living also increases for the region's citizens under *Sustainability First*, but here population stabilizes and individuals do not consume as much as in *Markets First* and *Policy First*. As a result, the pressure on the natural environment under *Sustainability First* is less than in the other two scenarios.

Governance will play a key role both in achieving prosperity, and in restoring and maintaining environmental quality. The breakdown of governance in *Security First* contributes to the decline of nearly all indicators of economic well-being, as well as to the degradation of the state of the environment. Conflicts over water scarcity widen, marine fisheries decline, and air and water quality deteriorates. By comparison, new governance structures put into the place under the other scenarios (such as the Asia Pacific Community for Environment and Development) provide a political means for achieving environmental goals. *Sustainability First* suggests that these governance structures are more effective if they are built up from the communities rather than imposed by central governments.

The scenarios also make it clear that investments in technology and research are key for sustainable development in the region. They can lead to improvements in energy efficiency, water use and the consumption of resources, lightening the load on the natural environment. The figures under 9.38 highlight the possible futures of the region.

Europe

All four scenarios illustrate the vulnerability of Europe to environmental change in different ways. Europe is not a leading economic power in any of the scenarios, but it is in position to influence other global regions through its support of environmental and sustainable development technologies, and experience in governance and crisis management in the field of environment. Under unfavourable conditions, however, Europe might be dependent on policy alliances and natural resources from other regions.

A particular uncertainty uncovered by the scenarios is future migration and how this will affect the growth of the European population, especially in interactions with other regions. While ageing of the population is an important issue, equally important are the scale of future programmes in education and research that will reduce possible brain drain from Europe, and will enhance environmentally-related innovation and technological development. The scenarios show that such developments have significant scope to help temper and overcome many socio-economic or environmental crises in the wider region. However, the level of investments in R&D and education programmes required to bring this about can be rather high.

Under two of the four scenarios, environmental changes affecting Europe result in negative effects to both society and nature. In *Markets First*, striving for a higher standard of living in a globalized economy leads to higher production efficiencies in the western part of Europe, but also to higher consumption levels across the region. GHG emissions sharply increase, biodiversity declines, and pressures on water resources increase. Many indicators of the state and trends of the environment also become less favourable in *Security First*, but for different reasons. Europe, in this scenario, experiences a general weakening of its institutions and their control on environmental pollution. High increases in GHG emissions result from the low efficiency of energy use, and from high levels of diffuse emissions from land-based sources. Wastewater discharges and destruction of habitats put increased pressure on aquatic ecosystems in both scenarios.

Policy First and *Sustainability First* scenarios illustrate different pathways that Europe can follow to a more sustainable future. One is to become skilled in managing climate change and other crises, and

Figure 9.38a Population trends – Asia and the Pacific

billion people

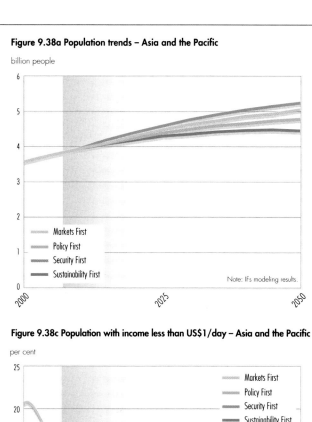

Note: IFs modeling results.

Figure 9.38b GDP/capita – Asia and the Pacific

thousand US$2 000 (PPP-based)

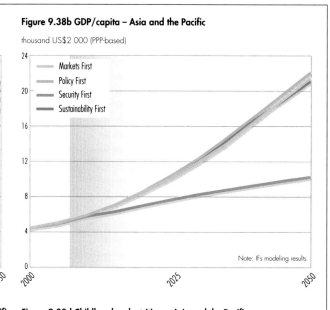

Note: IFs modeling results.

Figure 9.38c Population with income less than US$1/day – Asia and the Pacific

per cent

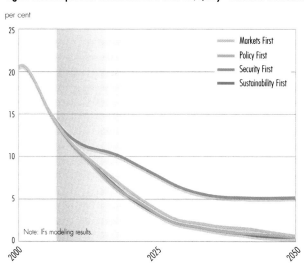

Note: IFs modeling results.

Figure 9.38d Childhood malnutrition – Asia and the Pacific

per cent of children

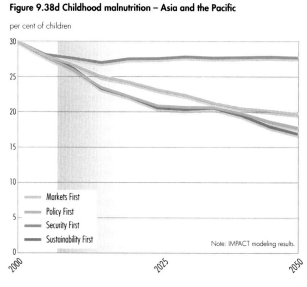

Note: IMPACT modeling results.

Figure 9.38e Population living in river basins facing severe water stress – Asia and the Pacific

billion people

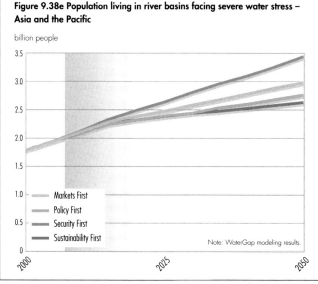

Note: WaterGap modeling results.

Figure 9.38f Rate of change in cropland, pasture and forest areas – Asia and the Pacific

million km²

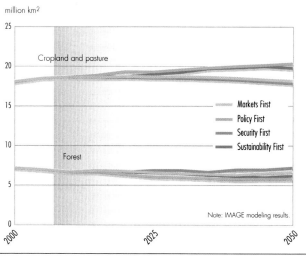

Cropland and pasture

Forest

Note: IMAGE modeling results.

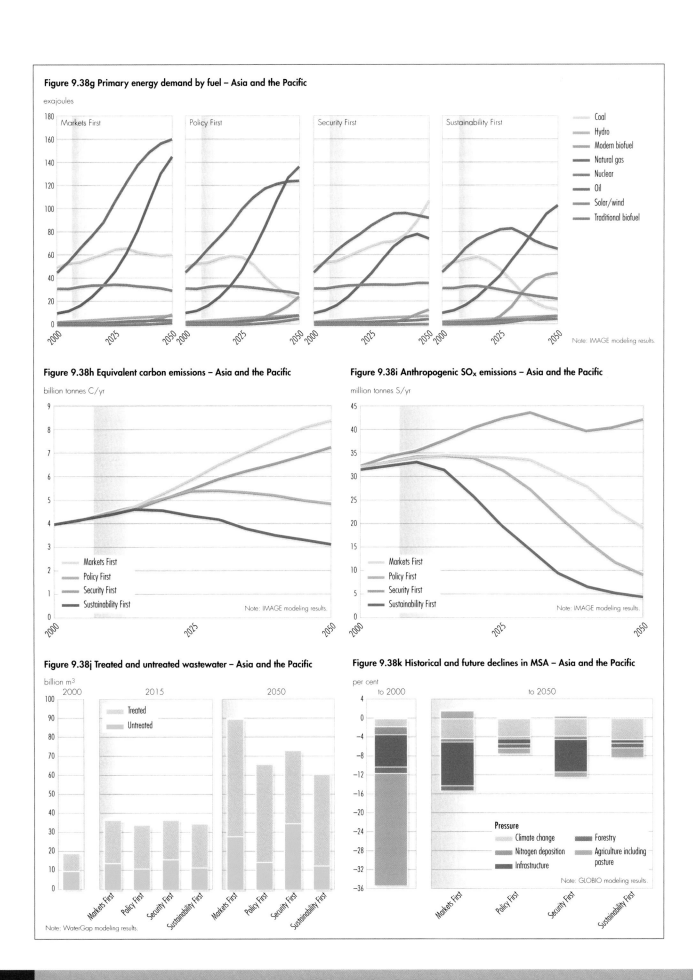

Figure 9.38g Primary energy demand by fuel – Asia and the Pacific

exajoules

Markets First | Policy First | Security First | Sustainability First

Legend:
- Coal
- Hydro
- Modern biofuel
- Natural gas
- Nuclear
- Oil
- Solar/wind
- Traditional biofuel

Note: IMAGE modeling results.

Figure 9.38h Equivalent carbon emissions – Asia and the Pacific

billion tonnes C/yr

- Markets First
- Policy First
- Security First
- Sustainability First

Note: IMAGE modeling results.

Figure 9.38i Anthropogenic SO$_x$ emissions – Asia and the Pacific

million tonnes S/yr

- Markets First
- Policy First
- Security First
- Sustainability First

Note: IMAGE modeling results.

Figure 9.38j Treated and untreated wastewater – Asia and the Pacific

billion m³

2000 | 2015 | 2050

- Treated
- Untreated

Markets First, Policy First, Security First, Sustainability First

Note: WaterGap modeling results.

Figure 9.38k Historical and future declines in MSA – Asia and the Pacific

per cent

to 2000 | to 2050

Pressure
- Climate change
- Nitrogen deposition
- Infrastructure
- Forestry
- Agriculture including pasture

Note: GLOBIO modeling results.

Markets First, Policy First, Security First, Sustainability First

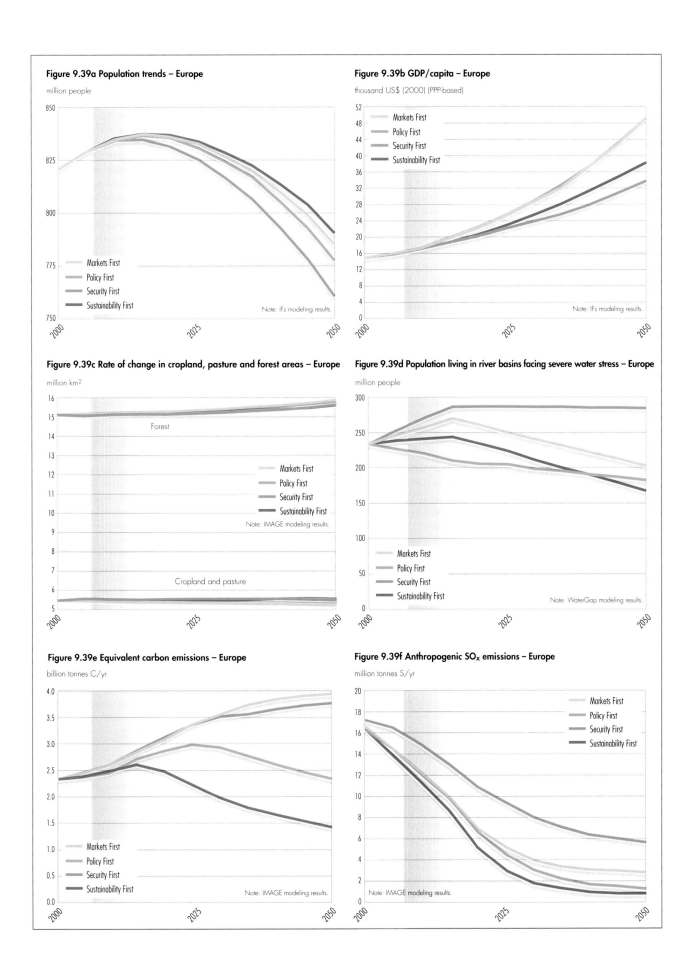

Figure 9.39a Population trends – Europe

million people

Figure 9.39b GDP/capita – Europe

thousand US$ (2000) (PPP-based)

Markets First
Policy First
Security First
Sustainability First

Note: IFs modeling results.

Figure 9.39c Rate of change in cropland, pasture and forest areas – Europe

million km²

Forest

Markets First
Policy First
Security First
Sustainability First

Note: IMAGE modeling results.

Cropland and pasture

Figure 9.39d Population living in river basins facing severe water stress – Europe

million people

Markets First
Policy First
Security First
Sustainability First

Note: WaterGap modeling results.

Figure 9.39e Equivalent carbon emissions – Europe

billion tonnes C/yr

Markets First
Policy First
Security First
Sustainability First

Note: IMAGE modeling results.

Figure 9.39f Anthropogenic SO$_x$ emissions – Europe

million tonnes S/yr

Markets First
Policy First
Security First
Sustainability First

Note: IMAGE modeling results.

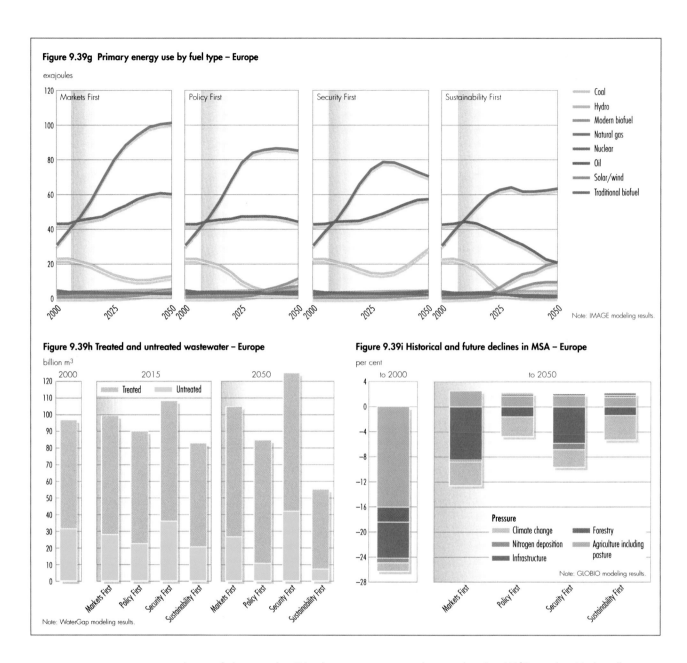

Figure 9.39g Primary energy use by fuel type – Europe

exajoules

Markets First | Policy First | Security First | Sustainability First

Legend:
- Coal
- Hydro
- Modern biofuel
- Natural gas
- Nuclear
- Oil
- Solar/wind
- Traditional biofuel

Note: IMAGE modeling results.

Figure 9.39h Treated and untreated wastewater – Europe

billion m³

2000 | 2015 | 2050

Treated | Untreated

Markets First | Policy First | Security First | Sustainability First

Note: WaterGap modeling results.

Figure 9.39i Historical and future declines in MSA – Europe

per cent

to 2000 | to 2050

Pressure
- Climate change
- Nitrogen deposition
- Infrastructure
- Forestry
- Agriculture including pasture

Markets First | Policy First | Security First | Sustainability First

Note: GLOBIO modeling results.

another is to further strengthen EU policy practices, and spread these further to the eastern part of Europe. Robust strategies include the exchange of technologies, integrated management and stakeholder participation in the decision making processes. The figures under 9.39 highlight the possible futures of the region.

Latin America and the Caribbean
Historically, the implementation of economic policies and programmes in Latin America and the Caribbean has often imposed additional pressures on social conditions as well as on environmental and natural resources. Inequity and poverty increase markedly under *Markets First* and *Security First*, although this is not necessarily captured in measures such as persons

living on less than US$1 per day. Modest alleviation is seen in *Policy First*, and there is a notable contraction under *Sustainability First*. Foreign debt remains as an obstacle for sustainable development under *Markets First* and *Policy First*, with a marked increase under *Security First* and a reduction to manageable levels under *Sustainability First*.

Forests and biodiversity represent crucial components of the region's natural resources, with implications not only for the region but also for the world. Deforestation increases and forest cover falls markedly under *Markets First*, leading to further habitat loss and fragmentation. Key forest areas of interest to the "elites" are preserved under *Security First*, but outside of

Figure 9.40a Population trends – Latin America and the Caribbean

million people

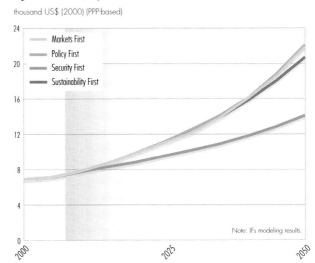

Note: IFs modeling results.

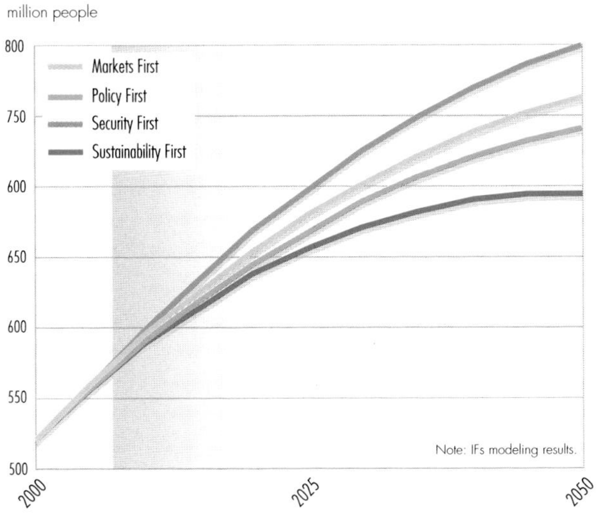

Figure 9.40b GDP/capita – Latin America and the Caribbean

thousand US$ (2000) (PPP-based)

Note: IFs modeling results.

Figure 9.40c Population with income less than US$1/day – Latin America and the Caribbean

per cent of total population

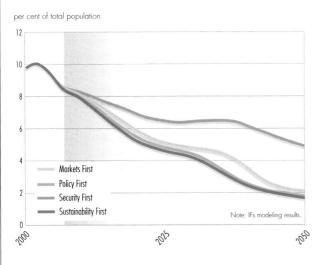

Note: IFs modeling results.

Figure 9.40d Childhood malnutrition – Latin America and the Caribbean

per cent of children 0–5 years old

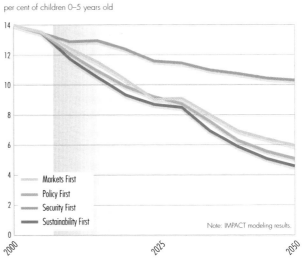

Note: IMPACT modeling results.

Figure 9.40e Population living in river basins facing severe water stress – Latin America and the Caribbean

million people

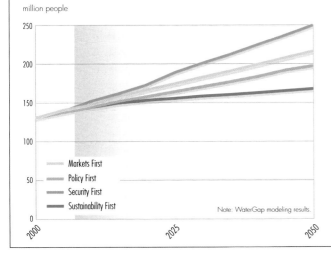

Note: WaterGap modeling results.

Figure 9.40f Rate of change in cropland, pasture and forest areas – Latin America and the Caribbean

million km²

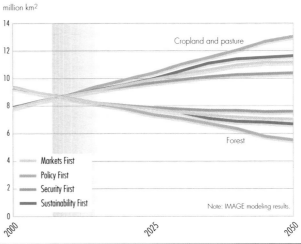

Cropland and pasture

Forest

Note: IMAGE modeling results.

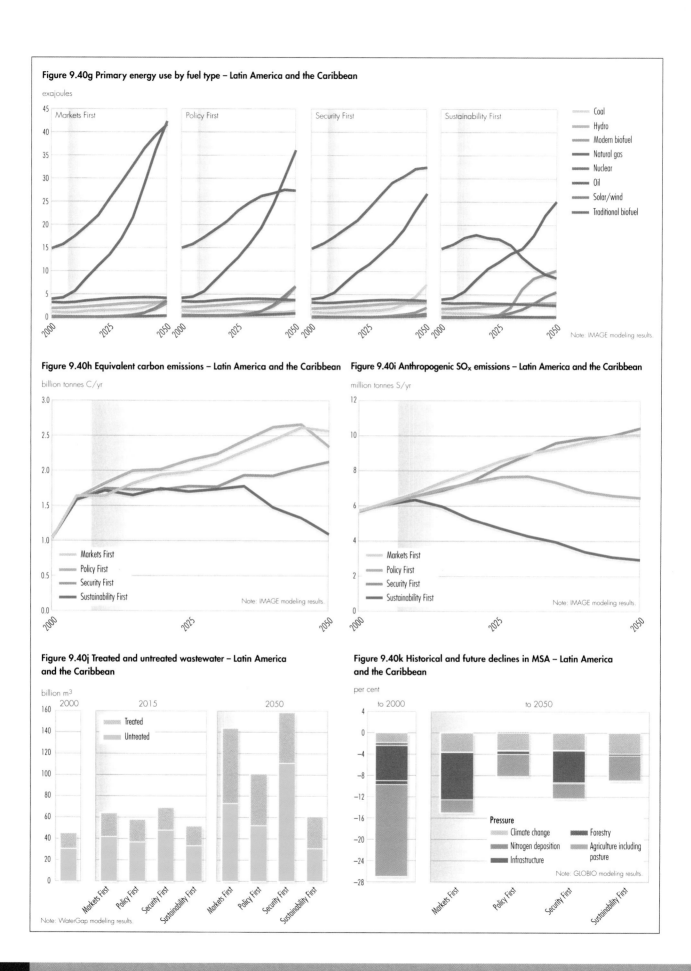

Figure 9.40g Primary energy use by fuel type – Latin America and the Caribbean

exajoules

Markets First   Policy First   Security First   Sustainability First

Coal
Hydro
Modern biofuel
Natural gas
Nuclear
Oil
Solar/wind
Traditional biofuel

Note: IMAGE modeling results.

Figure 9.40h Equivalent carbon emissions – Latin America and the Caribbean

billion tonnes C/yr

Markets First
Policy First
Security First
Sustainability First

Note: IMAGE modeling results.

Figure 9.40i Anthropogenic SO$_x$ emissions – Latin America and the Caribbean

million tonnes S/yr

Markets First
Policy First
Security First
Sustainability First

Note: IMAGE modeling results.

Figure 9.40j Treated and untreated wastewater – Latin America and the Caribbean

billion m^3

2000   2015   2050

Treated
Untreated

Markets First   Policy First   Security First   Sustainability First

Note: WaterGap modeling results.

Figure 9.40k Historical and future declines in MSA – Latin America and the Caribbean

per cent

to 2000   to 2050

Pressure
Climate change
Nitrogen deposition
Infrastructure
Forestry
Agriculture including pasture

Markets First   Policy First   Security First   Sustainability First

Note: GLOBIO modeling results.

protected areas deforestation rapidly increases. *Policy First* shows a moderate reduction in deforestation and habitat fragmentation, due to improved regulations and enforcement mechanisms, while mechanisms to rehabilitate affected forest ecosystems are implemented in *Sustainability First*, stopping the loss and fragmentation of these key habitats.

Increasing pressures on regional water resources persist under the four scenarios by 2050, but qualitative differences can be identified. In *Markets First* and *Security First*, quality and quantity of surface- and groundwater diminish, while in *Policy First*, increase in water withdrawals are tempered by investments in new water saving technologies, which lead to a strong improvement in water use in economic sectors. In *Sustainability First*, special efforts are introduced to manage conflicts in this area, improve efficiency in water use and change the water use behaviour of the population.

Access to and control of energy resources remain a key source of conflict in *Markets First* and, to a greater extent, in *Security First*, with very limited improvement in energy diversification out of fossil fuels and energy efficiency under these two scenarios. In contrast, energy diversification, with greater use of renewable sources, energy efficiency and regional energy cooperation, are promoted in *Policy First* and strongly reinforced in *Sustainability First*.

Urbanization is also a key driver, with Latin America and the Caribbean being the most urbanized region in the developing world. The urbanization process proceeds in all scenarios, but with significant differences. Uncontrollable expansion of urbanization occurs under *Markets First* and *Security First*, and less chaotic urbanization is seen in *Policy First*. In *Sustainability First*, urbanization continues mainly in medium and small cities in a context based on long-term planning for cities development.

A continuous increase in migratory pressures, within the region and to North America, occurs in *Markets First*, due to the deterioration in social conditions for many groups. Under *Security First*, migratory pressures considerably increased in the border areas, but migratory legislation becomes more restrictive. Emigration pressures are reduced in *Policy First* and *Sustainability First*. In the latter, emigration tends to be a matter of choice rather

than of necessity. The figures under 9.40 highlight the possible futures of the region.

North America

A key distinguishing feature among the scenarios is the degree to which this region responds proactively and in a coordinated fashion to environmental problems. As *Markets First* illustrates, markets are phenomenally successful at innovating new products and responding to consumer demand. However, they are not terribly effective in providing solutions to environmental problems if there is no policy guidance, as illustrated by *Policy First*. If, in addition to the market dynamism of *Markets First* and the policy guidance of *Policy First*, there is a further element of cultural awareness and social engagement, as in *Sustainability First*, then civil society can motivate the private sector and policy-makers towards even greater achievements on the environmental front.

A clear distinction is seen in GHG emissions, which in *Policy First* are nearly halved compared to *Markets First*, while in *Sustainability First*, they fall even further. With regard to water resources, *Sustainability First* and *Policy First* also show a much more proactive approach than *Security First* and *Markets First*. In the latter two, the degradation of major aquifers and surface water resources takes its toll, especially within the agricultural sector and in the domestic domain, with the fraction of the population living in water stressed basins rising steadily.

The problems of sprawl, climate and water resources tax the region's policy making capacity. They are diffuse and unfocused problems that worsen slowly but inexorably. They demand action from many different, uncoordinated actors, and require a rethinking of notions of progress and well-being.

Thus, without a more determined and conscious effort, North America could fail to put in place measures that are needed to protect and preserve freshwater resources, to shift to a dramatically lower-carbon economy, and to break the trend towards ever more land-intensive development. The solutions to these problems will ultimately require ambitious policies, such as market-based mechanisms to value natural resources, such as watersheds, support for technological innovation, and forward-thinking "smart growth" strategies. Moreover, the rise in cultural and individual awareness of these problems and sensitivity

Figure 9.41a Population trends – North America

million people

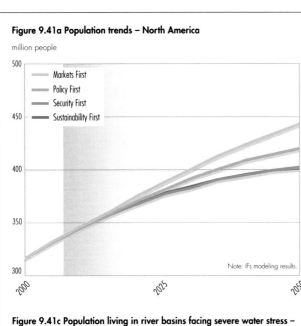

Note: IFs modeling results.

Figure 9.41b GDP/capita – North America

thousand US$ (2000) (PPP-based)

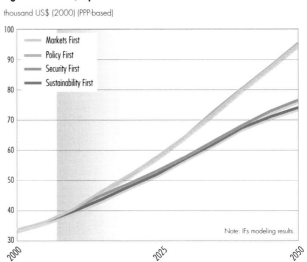

Note: IFs modeling results.

Figure 9.41c Population living in river basins facing severe water stress – North America

million people

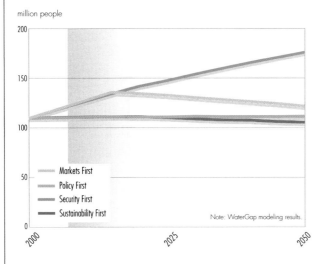

Note: WaterGap modeling results.

Figure 9.41d Rate of change in cropland, pasture and forest areas – North America

million km²

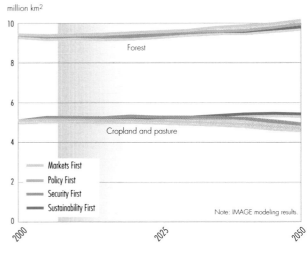

Note: IMAGE modeling results.

Figure 9.41e Equivalent carbon emissions – North America

billion tonnes C/yr

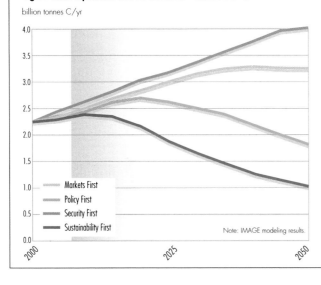

Note: IMAGE modeling results.

Figure 9.41f Anthropogenic SOₓ emissions – North America

million tonnes S/yr

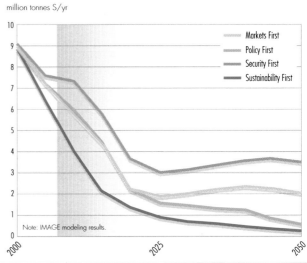

Note: IMAGE modeling results.

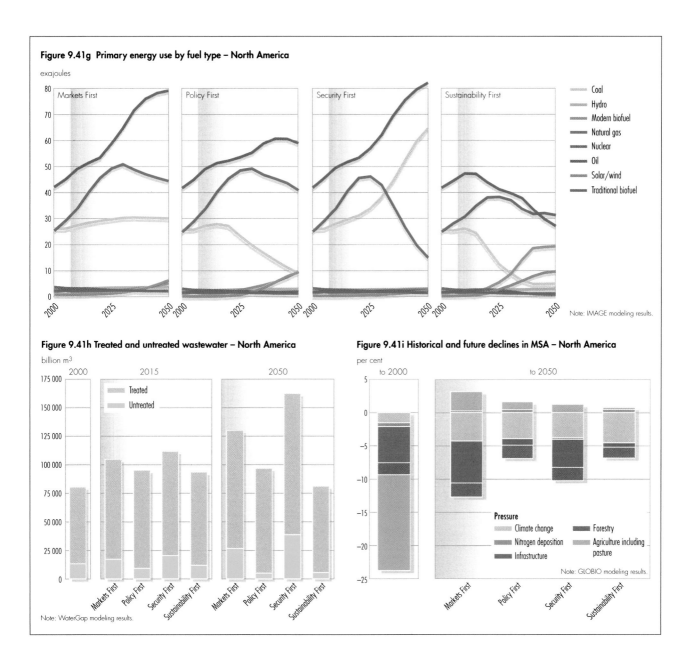

Figure 9.41g Primary energy use by fuel type – North America

exajoules

Markets First | Policy First | Security First | Sustainability First

- Coal
- Hydro
- Modern biofuel
- Natural gas
- Nuclear
- Oil
- Solar/wind
- Traditional biofuel

Note: IMAGE modeling results.

Figure 9.41h Treated and untreated wastewater – North America

billion m³

2000 | 2015 | 2050

- Treated
- Untreated

Note: WaterGap modeling results.

Figure 9.41i Historical and future declines in MSA – North America

per cent

to 2000 | to 2050

Pressure
- Climate change
- Nitrogen deposition
- Infrastructure
- Forestry
- Agriculture including pasture

Note: GLOBIO modeling results.

to their solutions, as illustrated in *Sustainability First*, might be a necessary ingredient to catalyse the needed response in the policy and market realms. A worst case, but not implausible, scenario could see deterioration in environmental and socio-economic conditions to a point that seems to defy repair.

Finally, although income levels are similar in *Sustainability First* and *Security First*, the quality of life is qualitatively better in *Sustainability First*, and arguably better than in *Markets First* and *Policy First* despite their higher income. *Markets First* is highly successful at providing products to consumers; *Policy First* helps ensure that the environmental impacts are mitigated; *Sustainability First*, however, invests also in the non-material aspects

of well-being, such as a healthy environment and a strengthened sense of community, reflecting a reinforced social compact that provides more equitable access to critical resources, such as health care, education and political processes. The figures under 9.41 highlight the possible futures of the region.

West Asia

The scenarios illustrate the different pathways and futures that might be taken by the region's societies, and the relative and complex impacts of the various drivers in shaping its future in terms of human well-being and environmental change. *Markets First* is a depressing scenario for West Asia; although the market stimulates needed improvements in resource efficiency and socio-

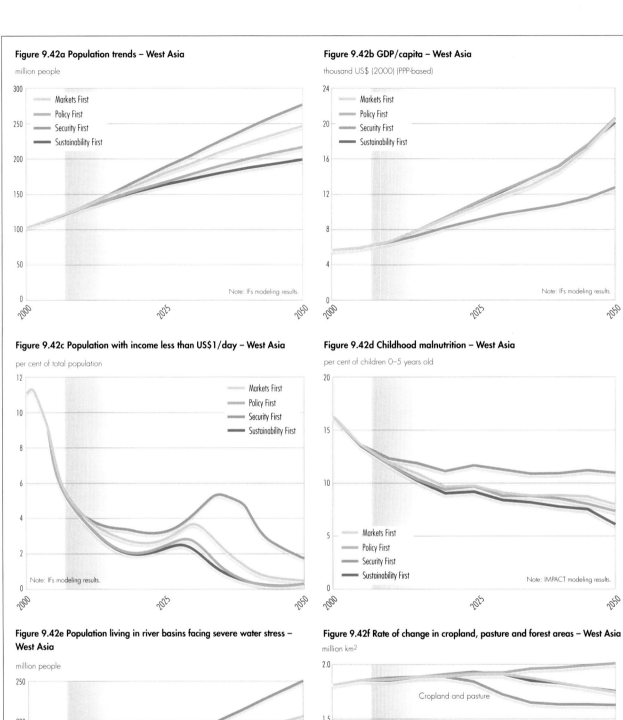

Figure 9.42a Population trends – West Asia

million people

Markets First
Policy First
Security First
Sustainability First

Note: IFs modeling results.

Figure 9.42b GDP/capita – West Asia

thousand US$ (2000) (PPP-based)

Markets First
Policy First
Security First
Sustainability First

Note: IFs modeling results.

Figure 9.42c Population with income less than US$1/day – West Asia

per cent of total population

Markets First
Policy First
Security First
Sustainability First

Note: IFs modeling results.

Figure 9.42d Childhood malnutrition – West Asia

per cent of children 0–5 years old

Markets First
Policy First
Security First
Sustainability First

Note: IMPACT modeling results.

Figure 9.42e Population living in river basins facing severe water stress – West Asia

million people

Markets First
Policy First
Security First
Sustainability First

Note: WaterGap modeling results.

Figure 9.42f Rate of change in cropland, pasture and forest areas – West Asia

million km²

Cropland and pasture

Markets First
Policy First
Security First
Sustainability First

Note: IMAGE modeling results.

Forest

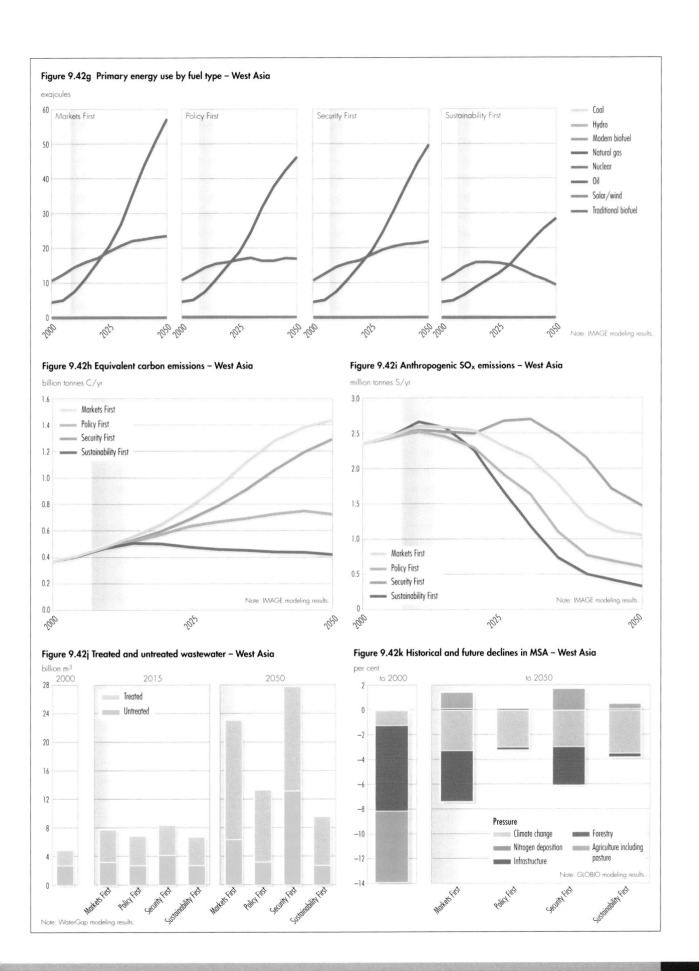

Figure 9.42g Primary energy use by fuel type – West Asia

exajoules

Markets First

Policy First

Security First

Sustainability First

Coal
Hydro
Modern biofuel
Natural gas
Nuclear
Oil
Solar/wind
Traditional biofuel

Note: IMAGE modeling results.

Figure 9.42h Equivalent carbon emissions – West Asia

billion tonnes C/yr

Markets First
Policy First
Security First
Sustainability First

Note: IMAGE modeling results.

Figure 9.42i Anthropogenic SO$_x$ emissions – West Asia

million tonnes S/yr

Markets First
Policy First
Security First
Sustainability First

Note: IMAGE modeling results.

Figure 9.42j Treated and untreated wastewater – West Asia

billion m^3

2000   2015   2050

Treated
Untreated

Markets First Policy First Security First Sustainability First

Markets First Policy First Security First Sustainability First

Note: WaterGap modeling results.

Figure 9.42k Historical and future declines in MSA – West Asia

per cent

to 2000   to 2050

Pressure
Climate change   Forestry
Nitrogen deposition   Agriculture including
Infrastructure   pasture

Note: GLOBIO modeling results.

Markets First Policy First Security First Sustainability First

economic indicators, the region faces considerable environmental, health and social problems, which in the long-term will undermine economic development.

In *Policy First*, strong policy constraints are placed by governments on market forces to minimize their undesirable effects on the environment and human well-being. Environmental and social costs are factored into policy measures, regulatory framework and planning processes to achieve greater social equity and environmental protection, which leads to a decrease in environmental degradation, and an improvement in human well-being. However, pressures from investment policies continue to be high.

In *Security First*, an extreme case of *Markets First* from the perspective of the region, national and regional political tensions and conflicts remain unresolved for a long time. They continue to be major drivers, negatively influencing the region's overall development, and leading eventually to the disintegration of the social and economic fabric of the region. Human well-being, the environment and natural resources are victimized to meet security demands.

In *Sustainability First*, the improvement of governance and a sustained link among social, economic and environmental policies provides a solution to the sustainability challenge in the region. Integration, cooperation, and dialogue at the national, regional and inter-regional levels replace tensions and armed conflicts. Human well-being and the environment are central to planning, and governments adopt long-term integrated strategic planning, with the objective of achieving a superior quality of life and a healthy environment. There is heavy investment in human resources development, aimed at establishing a knowledge-based society. Major funds are allocated to research and development in science and technology to solve the community's social, economic and environmental problems.

A common denominator for the scenarios is that water stress, land degradation, food insecurity and biodiversity loss continue, though occurring at different rates, due to the prevailing natural aridity in the region and its fragile ecosystem, and the pressures exerted by population size and growth rates. Active, adaptive management, with continuous monitoring and evaluation and capacity building, will be required to cope and adapt to future stresses on people and environment.

Perhaps the most important policy lesson that these scenarios offer to the countries of the region is that investment in human resources development and R&D, governance improvement, and regional cooperation and integration are key issues in the long and intricate path to sustainability for the region. The figures under 9.42 highlight the possible futures of the region.

Polar Regions

Climate change is the predominant and overarching issue across both sub-regions and across all scenarios, with long-term and accelerating effects throughout the time period and well beyond 2050. The impact of climate change on the Polar Regions reaches far beyond the immediate sub-regions, and has major global implications during the scenario period and beyond, such as severe interruptions in the marine ecosystems and sea-level rise, jeopardizing the sustainability of millions in coastal communities globally. The profile and consequences of global climate change are essentially the same across all four scenarios for the period up to 2050. This is a consequence of the enormous inertia in the polar and global marine systems, with reaction time lags of several decades. Differences among the scenarios only become evident (and then only marginally) after 2050, because of new targets aiming to significantly cut carbon emissions under *Policy First* and *Sustainability First*.

The Polar Regions store about 70 per cent of the world's freshwater in the form of ice. As a consequence of climate change, the average annual freshwater run-off into the Arctic and North Atlantic Oceans increases, with pronounced differences among the scenarios, ranging from 4 600 km²/year at present to almost 6 000 km²/year by 2050 in the *Markets First* scenario.

The Polar Regions are a global storehouse, with huge potential for exploitation. There are distinct differences between the sub-regions, but also across the scenarios, ranging from extensive and devastating in *Markets First*, to local but intensive in *Security First*, and more controlled and resourceful in *Policy First*. With increasing accessibility of polar ecosystems the last top global pristine wilderness areas and their unique biodiversity are put at risk by an ever-increasing global demand on polar resources in *Markets First* and *Security First*, with distinct areas for conservation preserved in *Policy First*, and a slow recovery in *Sustainability First* (Figure 9.43). The consideration of anything polar as a global resource or commodity

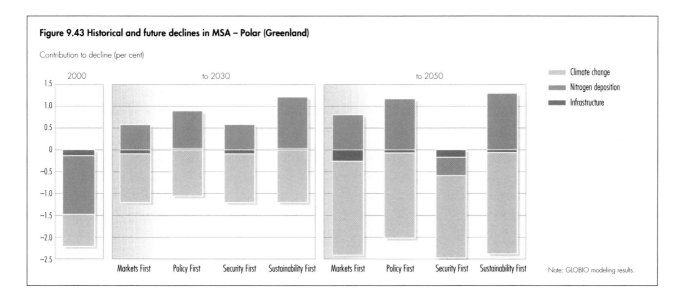

Figure 9.43 Historical and future declines in MSA – Polar (Greenland)

Contribution to decline (per cent)

Legend:
- Climate change
- Nitrogen deposition
- Infrastructure

Note: GLOBIO modeling results.

increasingly includes the Antarctic region in *Markets First*. It also establishes pathways from any other global region to the poles, whether it is hazardous waste or tourists, with profound differences in the scenarios.

Indigenous peoples in the Arctic increasingly face pressures from global climate change and the exploitation of the natural resources, with declining political influence in *Security First*, strong empowerment in *Sustainability First* and surprisingly strong co-management arrangements in *Markets First*. Geopolitical interests increasingly dominate over local and indigenous sovereignty very strongly in *Security First*, but also in *Markets First*. *Sustainability First* promotes decentralized governance systems, and a shift in power towards local communities and indigenous peoples, enabling them to practice adaptive management to sustain their livelihoods as well as human well-being.

The long-term availability of polar resources and ecosystem stability very much depends on the implementation of sustainability principles. The scenarios illustrate how all human activities in the Polar Regions and globally are intertwined, and how only concerted global action can make a difference to the future of the Polar Regions.

RISKS AND OPPORTUNITIES OF THE FUTURE

The *GEO-4* scenarios point to both risks and opportunities in the future. Of particular significance are the risks of crossing thresholds, the potential of reaching turning points in the relationship between people and the environment, and the need to account for interlinkages in pursuing a more sustainable path.

Global change – turning points and thresholds

The hallmarks of global change are discernible in life – the sprawl of cities over the countryside, the manifestation of climate change in warmer winters and increased flood events, and more severe heat waves, and the presence of human-made pollutants in remote regions of the world. While results in this chapter indicate that change will continue, they also show that the rate of change for many key indicators may slow down towards the middle of the century. Changes go on, but the rate of change declines, indicating a potential turning point in human-environment relations. At the same time, the actual level of the changes seen in the scenarios may push us past thresholds in the Earth system, resulting in sudden, abrupt or accelerating changes, which could be potentially irreversible. Examples cited in earlier chapters include the collapse of fisheries, eutrophication and deprivation of oxygen (hypoxia) in aquatic systems, emergence of diseases and pests, introduction and loss of species, large-scale crop failures and climatic changes.

Why do the *GEO-4* scenarios show a slowing down of change, and why do these differ among the scenarios? The answer lies in the trends of the drivers of the scenarios, such as the stabilization of population in *Sustainability First* and the slower growth in total economic activity in *Security First* and *Sustainability First*. Improvements in technology will raise the efficiency of electricity generation, reduce losses in water distribution systems and boost crop yields, albeit at different rates across regions and the scenarios. These and other developments all contribute to slowing the pace of some aspects of global environmental change.

The rate of increase of water withdrawals slows down by the end of the scenario period in all scenarios except *Security First* (see Figure 9.44). Rates of cropland expansion and forest loss steadily decline over the scenario period (see Figures 9.45 and 9.46). Some scenarios also show a slackening in the tempo of species loss, greenhouse gas build-up, and temperature increase (see Figures 9.47–9.49).

Although the tempo of change slows in some cases, the end point of change will not be the same for all scenarios. For example, water withdrawals reach

over 6 000 km² per year under *Market First* but less than 4 000 km² per year under *Policy First* (see Figure 9.44). Also illustrative are the trends of atmospheric CO_2 concentrations and global mean surface temperature. The range in 2050 for CO_2 is from around 475 ppm in *Sustainability First* to over 560 ppm in *Markets First* (see Figure 9.14). For temperature increase, the range is from about 1.7°C above pre-industrial levels in 2050 in *Sustainability First* to about 2.2°C in *Markets First* (see Figure 9.15). The higher figure exceeds the threshold of 2°C (see Chapter 2), beyond which climate change impacts

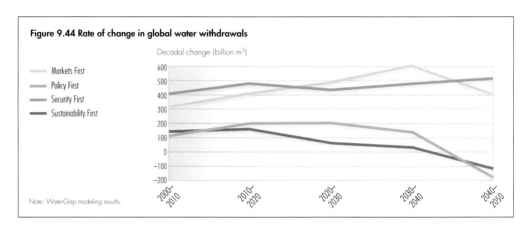

Figure 9.44 Rate of change in global water withdrawals

Note: WaterGap modeling results.

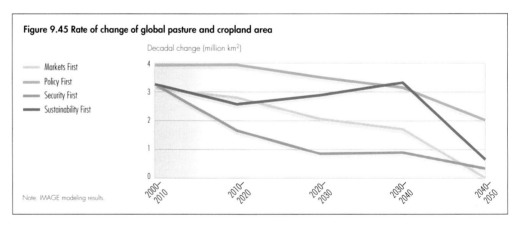

Figure 9.45 Rate of change of global pasture and cropland area

Note: IMAGE modeling results.

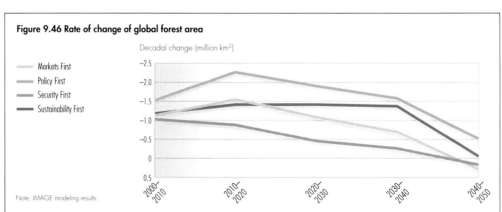

Figure 9.46 Rate of change of global forest area

Note: IMAGE modeling results.

become significantly more severe, and the threat of major irreversible damages becomes more plausible.

Why is this important? A slower rate of change gives hope that the society and nature can more successfully catch up to the pace of change and adjust to it before experiencing many negative consequences. Society has better chance to keep pace with the change by building new infrastructure, natural ecosystems have more time to migrate, conservation policies have a better chance to catch up to the rate of loss of species and society has more time to learn how to adapt.

Conversely, scenarios with a faster pace of change are more likely to come closer to tipping points in the Earth system. What will society reach first: a tempo of change slow enough to adapt to, or levels of change that exceed key thresholds of the Earth system?

Interlinkages

Our Common Future emphasized that "the ability to choose policy paths that are sustainable requires that the ecological dimensions of policy be considered at the same time as the economic, trade, energy, agricultural, industrial and other dimensions – on the same

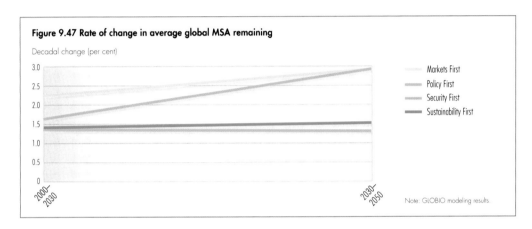

Figure 9.47 Rate of change in average global MSA remaining

Decadal change (per cent)

Markets First
Policy First
Security First
Sustainability First

Note: GLOBIO modeling results.

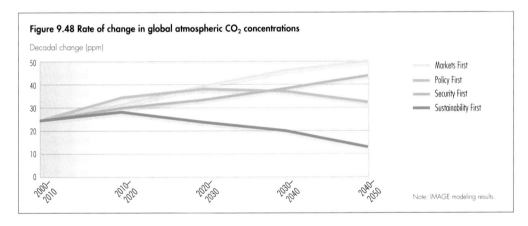

Figure 9.48 Rate of change in global atmospheric CO$_2$ concentrations

Decadal change (ppm)

Markets First
Policy First
Security First
Sustainability First

Note: IMAGE modeling results.

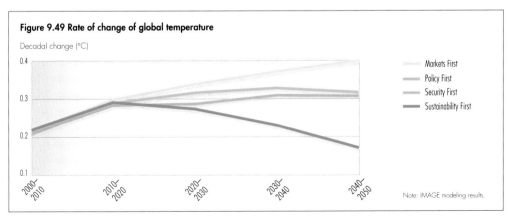

Figure 9.49 Rate of change of global temperature

Decadal change (°C)

Markets First
Policy First
Security First
Sustainability First

Note: IMAGE modeling results.

The scenarios point to both risks and opportunities in the future. There is a need to account for interlinkages in pursuing a more sustainable path.

Credit: Munyaradzi Chenje

agendas and in the same national and international institutions." A recent review suggests that 20 years on, "Our societies and their approaches to challenges remain highly compartmentalized" (WBCSD 2007). Looking forward, the acknowledgement and practice in terms of interlinkages varies significantly across the scenarios. There is a need to address interlinkages among numerous environmental issues, such as air and water pollution, land degradation, climate change, biodiversity loss, and valuing ecosystems goods-and-services. And, there is a need to link environment with development issues, such as extreme poverty and hunger, implementation of the MDGs, and addressing human vulnerability and well-being.

Under *Markets First*, interlinkages are factored in the context of the unfettered functioning of markets. Greater emphasis is placed on economic sectors, with ecosystems goods-and-services considered as primarily production inputs. The implementation of the MEAs largely follows the silos of jurisdictional and administrative boundaries. The economy grows and even more wealth is generated, but human development remains a challenge, as do many environmental issues.

Under *Policy First*, greater effort is made by government to address the complexities of interlinkages, both in the environment on its own as well as in the context of the governance regimes. Climate change is seen as the dominant entry point to address mitigation and adaptation challenges in different areas and over time, rather than symptomatic of the environment-development nexus. While policy-makers give prominence to measures that consider interlinkages, the legal and institutional framework is not adequately reformed to function across national, administrative and special interest boundaries. Competition among regions, countries and institutions still manifests itself, particularly if there are perceived negative socio-economic impacts at national level.

Security First brings new meaning to the Rio Declaration Principle 7 phrase — common but differentiated responsibilities — promoting selective attention to issues, and limiting responsibilities to areas of special interest. For example, where the development of bio-energy accelerates to meet the energy addiction of a few, it does so without considering issues such as agriculture for food security, increased water demand, land-use change and increased use of chemicals. Human and financial resources, as well as governance regimes, are deployed to address the challenges on a selective basis and for the benefit of a few. Some environmental issues are effectively addressed, but do not add up to much when considered in relation to overall environmental degradation. Ultimately, the whole society is put at risk, with greater potential impacts on more vulnerable regions and societies. Development is limited to the minority but only for a limited period as unrest threatens their safe havens.

Under *Sustainability First*, government, civil society, business and industry, the scientific community and other stakeholders work together to address disparate environment and development challenges. The legal and institutional framework is reformed at different levels, bringing coherence across MEAs at the international level and sectoral laws at other levels. MEAs, such as the CBD, CMS, CITES, and Ramsar, achieve greater coherence to ensure not only biodiversity conservation but alleviate the growth of the illegal trade in wildlife and their products. The Basel Convention, and the conventions on prior informed consent on hazardous chemicals (Rotterdam Convention) and on persistent organic pollutants (Stockholm Convention), undertake similar initiatives, and also work closely with the World Trade Organization to address chemicals and waste-related issues. While progress is made in using interlinkages to address the challenges, extensive consultation and drawn-out decision making limit effectiveness in the short-term. The challenge is to maximize the strengths of interlinked approaches and minimize their drawbacks.

CONCLUSIONS

This chapter has presented four scenarios of plausible futures to the year 2050 – *Markets First*, *Policy First*, *Security First* and *Sustainability First* – each exploring how current social, economic and environmental trends may unfold, and what this means for the environment, development and human well-being. The scenarios are fundamentally defined by different policy approaches and societal choices, with their nature and names characterized by the theme that dominates the particular future envisioned, such as what comes *first*. In reality the future, as is the case for the present, will contain elements of each of these scenarios, as well as many others. Still, the scenarios clearly illustrate that the future that will unfold in the long-term will be very dependent on the decisions individuals and society make today. As such, these visions of the future should influence our decisions of today. By providing insights into the challenges and opportunities society will face in the coming half-century, the exploration of these plausible futures can contribute to the discourse about these choices.

The scenarios have been presented at both the global and regional levels, because an understanding of global environmental change and its impacts requires an understanding of what is happening in different regions of the world. What happens in each region is very much influenced by what occurs in other regions and the world as a whole. Still, while there is only one global environment, each region and each person experiences it in their own way. As such, the challenges and opportunities, and even the perspectives on the future differ widely across regions and individuals.

None of the scenarios describes a utopia. Even though some improvements are seen and there is an indication of a slowing of the rate of change in some aspects of global environmental change, some problems remain persistent in all of the scenarios. In particular, climate change and the loss of biodiversity will continue to present significant challenges, and may eventually pose the danger of crossing critical thresholds in the Earth system. Similarly, with respect to human well-being, significant advances are achievable, particularly in *Sustainability First*, but even these will take time, and significant inequities will remain by the end of the scenario horizon.

Furthermore, there are costs and risks in each scenario. These are perhaps most evident in *Security First*, where a narrow definition of security for some is likely to result in increasing vulnerabilities for all. In *Markets First*, both the environment and society move the fastest towards if not beyond tipping points, where sudden, abrupt, accelerating and irreversible changes may occur. This is of particular concern given the uncertainties in the resilience of environmental and social systems. Under *Policy First* and *Sustainability First*, society will achieve a higher material standard of living and greater protection of the environment, but at a significant cost. Indeed, there are particular costs and risks in terms of the actions and approaches taken to address the issues of environment and human well-being. The social and economic costs of these actions may significantly exceed what has been previously assumed, and the lower economic growth seen in the currently well-off regions in *Sustainability First* may not prove to be acceptable. The time required to implement the actions might increase, due to the greater level of bureaucracy foreseen in *Policy First* and the increased level of coordination in *Sustainability First*. Finally, trade-offs may imply that the pursuit of a balanced approach in *Sustainability First* could work against greater progress on any specific target.

Still, to the extent that the scenarios reflect our understanding of the Earth system and environmental governance, they indicate that some approaches are more likely to be effective than others. Specifically, it is important to recognize the trade-offs, synergies and opportunities that exist in addressing the challenges of achieving environmental, development and human well-being goals. This calls for increasing the integration of policies across levels, sectors, and time, strengthening local rights, building capacity among a wide range of groups in society, and improving scientific understanding. The diversity and multiplicity of these trade-offs and opportunities for synergy clearly increase complexity for decision-makers. This is not to imply that this complexity is to be ignored; that would be a misreading of the scenarios as well as the message of *Our Common Future* and the subsequent 20 years. It does, however, point to the need for innovative approaches for exploring the options for action to address the intertwined environmental and developmental challenges the world continues to face. Furthermore, the scenarios point to the need to act quickly. Our common future depends on our actions today, not tomorrow or some time in the future.

TECHNICAL ANNEX

As recognized in the third Global Environment Outlook (*GEO-3*) report and other recent scenario exercises (for example, IPCC 2000, MA 2005, Cosgrove and Rijsberman 2000, and Raskin and others 2002), narratives and modelling complement each other in enriching the overall futures analysis. This annex provides some details about the development of the narratives and the modelling results. However, it is important to note that what is presented here does not fully reflect the effort involved in producing the chapter, and the chapter itself includes only a small portion of the material that has been developed.

Contributors

Hundreds of people and organizations were involved in preparing this chapter, building upon the four scenarios first introduced in *GEO-3*. The following paragraphs highlight the stakeholders and process of developing the *GEO-4* scenarios.

The structure of collaboration followed in the process of developing the chapter provided for an organized means of contribution from a large group of participants and for wider ownership among as many people as possible of the process and its outcomes. The three coordinating lead authors (CLAs) and chapter coordinators oversaw the development of the chapter. Regional team leaders, quantitative modellers, and an expert on facilitating participatory processes comprised the chapter expert group (CEG) and are listed as lead authors (LAs). In addition, primarily for the purpose of providing the regional contributions, a group of about 10 experts per region was chosen by the regional team leaders, in consultation with the regional coordinators of the UNEP Division of Early Warning and Assessment and others. Recognizing the impossibility of the above groups to be truly representative or fully versed in all areas required for the development of the chapter, other regional and modelling experts were also invited to provide a broader range of perspectives and specific expertise. Throughout the process, the team was assisted by Bee Successful (http://www.beesuccessful.com/), a management consultancy with expertise in scenario thinking and participatory methods.

Process

The CLAs and LAs met several times in 2004 and early 2005 to plan the development of the chapter. During the *GEO-4* regional consultations, a strong preference was expressed by participants to retain the basic characteristics of the scenarios, rather than to restart the process. Therefore, the scenarios presented here should be seen as revised and updated versions of those from *GEO-3*, both in terms of the narratives and the quantification (see UNEP/RIVM 2004). Still, they have been influenced by more recent scenario exercises, both those that drew directly from *GEO-3*, e.g. regional studies in Africa (UNEP 2006), and Latin America (UNEP 2004), and those that only marginally considered the scenarios presented in *GEO-3*, most notably the global and sub-global scenarios developed as part of the Millennium Ecosystem Assessment (MA 2005; Lebel and others 2005).

The chapter expert group, along with the seven teams of regional representatives met in Bangkok in September 2005. This was followed by meetings in each of the regions other than North America, in 2006. Further smaller meetings of CEG members were held over the next 18 months to further clarify issues and work out potential inconsistencies between the regional narratives and between the narratives and the quantitative results.

The seven regional teams developed narrative descriptions of each of the four scenarios from the perspective of each region. Taking the drivers and assumptions of the *GEO-3* global scenarios as a starting point, the regional groups worked in parallel to develop rich descriptions of the 'journey' and 'end state' of the four scenarios from a regional perspective. At the same time, each group carefully considered how events or trends in their region might influence, or be influenced by developments in other regions and at the global level. Through a series of iterations, storylines were drafted at both the regional and global levels. In parallel, a suite of advanced state-of-the-art models, described below, was used to develop the quantitative estimates of future environmental change and impacts on human well-being. In order to check the validity and consistency of the scenarios, the narrative teams interacted closely with the global and regional modellers to ensure that the quantitative and qualitative components of the scenarios complemented and reinforced each other. Furthermore, the scenarios were critically reviewed by experts in particular areas, such as energy, many of whom were contributors to other chapters of this report.

A concerted effort was made throughout this process to build regional capacity with respect to scenario development, as well as to make the resulting regional material a central part of the global storylines. In particular, special attention was given to the regional priority issues identified early in the *GEO-4* process and discussed throughout the preceding chapters. These have been tracked through the scenarios presented here.

The Models

Since no single overriding "super model" was available for computing future environmental change and the impacts on human well-being, a suite of advanced global and regional models was assembled for the task. These models have been published in the peer-reviewed scientific literature and have been shown to be useful for linking changes in society with changes in the natural environment. The models were soft-linked with output files from one model being used as inputs to other models. Following standard practice, all of the models are calibrated to historical data up to a common base year, in this case 2000 for most data. Thus, the results presented may show slight deviations across scenarios, as well as from more recent historical data, for the period 2000 to the date of publication of this report, some of which may have been presented in other chapters.

Briefly, the models are as follows:

International Futures (IFs) is a large-scale integrated global modelling system (Hughes and Hillebrand 2006). IFs serves as a thinking tool for the analysis of long-term country-specific, regional, and global futures across multiple and interacting issue areas. The system draws upon standard approaches to modelling specific issue areas whenever possible, extending those as necessary and integrating them across issue areas. For *GEO-4*, IFs provided population trends and the development in GDP and GDP per capita as well as additional information on value added, household consumption, health and education.

IMAGE (Integrated Model to Assess the Global Environment) is a dynamic integrated assessment model for global change developed by the National Institute for Public Health and the Environment (RIVM), The Netherlands (Bouwman and others 2006). IMAGE is used to study a whole range of environmental and global change problems, particularly in the realm of land use change, atmospheric pollution, and climate change. The main objectives of IMAGE are to contribute to scientific understanding and support decision-making by quantifying the relative importance of major processes and interactions in the society-biosphere-climate system. For *GEO-4*, IMAGE provided estimates of energy use, land use, greenhouse gas emissions, and changes in temperature and precipitation.

IMPACT (International Model for Policy Analysis of Agricultural Commodities and Trade) is a representation of a competitive world agricultural market for 32 crop and livestock commodities, including all cereals, soybeans, roots and tubers, meats, milk, eggs, oils, oilcakes and meals, sugar and sweeteners, fruits and vegetables, and fish. It was developed in the early 1990s as a response to concerns about a lack of vision and consensus regarding the actions required to feed the world in the future, reduce poverty, and protect the natural resource base. For *GEO-4*, IMPACT generated projections for crop area, livestock numbers, yield, production, demand for food, feed and other uses, prices, trade and childhood malnutrition.

WaterGAP (Water – Global Assessment and Prognosis) is a global model developed at the Center for Environmental Systems Research of the University of Kassel that computes both water availability and water use on a 0.5° global grid (Alcamo and others 2003a, b; Döll and others 2003). The model aims to provide a basis for an assessment of current water resources and water uses, and for an integrated perspective on the impacts of climate change and socio-economic drivers on the future water sector. For *GEO-4*, WaterGAP provided estimates of water use (for irrigation and in the domestic, manufacturing, and electricity production sectors), water availability, and water stress.

EwE (Ecopath with Ecosim) is an ecological modelling software suite for personal computers of which some components have been under development for nearly two decades. The development is centred at the University of British Columbia's Fishery Centre. The approach is thoroughly documented in the scientific literature, with over 100 ecosystems models developed to date (see www.ecopath.org). EwE uses two main components: Ecopath – a static, mass-balanced snapshot of marine ecosystems, and Ecosim – a time dynamic simulation module for policy exploration that is based on an Ecopath model. For *GEO-4*, EwE provided estimates of catch, profits, and quality of marine fisheries.

The **GLOBIO** model simulates the impact of multiple pressures on biodiversity (Alkemade and others 2006). The model relies on a database of field studies relating magnitude of pressure to magnitude of biodiversity impact. This database includes separate measures of mean species abundance (MSA) and of species richness (MSR) of original species of ecosystems, each in relation to different degrees of pressure. The entries in the database are all derived from peer-reviewed studies, either of change through time in a single plot, or of response in parallel plots undergoing different pressures. An individual study may have reported species richness, mean species abundance, or both. Rows are classified by pressure type, taxon under study, biome and region. For *GEO-4*, GLOBIO provided estimates of changes in mean species abundance for terrestrial ecosystems.

LandSHIFT is an integrated model system that aims at simulating and analysing spatially explicit land use dynamics and their impacts on the environment at global and continental level. The model design is characterized by a highly modular structure that allows the integration of various functional model components. For *GEO-4*, LandSHIFT provided detailed estimates of land use change for Africa.

The CLUE-S (Conversion of Land Use and its Effects) modeling framework, is a tool to downscale projected national land use changes (Verburg and others 2002, Verburg and Veldkamp 2004 and Verburg and others 2004). The framework combines different mechanisms that are important to the land use system in a spatially explicit manner. The model dynamically simulates competition and interactions between land use types and is, therefore, path dependent, resulting in non-linear behaviour that is characteristic for land use systems. For *GEO-4*, CLUE-S provided detailed estimates of land use change for Western and Central Europe.

AIM (the Asia Pacific Integrated Model) is a set of large-scale computer simulation models developed by the National Institute for Environmental Studies in collaboration with Kyoto University and several research institutes in Asia and the Pacific. It assesses policy options for stabilizing global climate and a range of other environmental problems. For *GEO-4*, AIM provided additional estimates of environmental change used in the development of the narratives for Asia and the Pacific.

References

Alcamo, J., Van Vuuren, D., Ringler, C., Alder, J., Bennett, E., Lodge, D., Masui, T., Morita, T., Rosegrant, M., Sala, O., Schulze, K. and Zurek, M. (2005). Chapter 6. Methodology for developing the MA (Millenium Ecosystem Assessment) scenarios. In Carpenter, S., Pingali, P., Bennett, E. and Zurek, M. (eds.) *Ecosystems and Human Well-Being. Volume 2 Scenarios.* Island Press, Washington, DC

Alcamo, J., Döll, P., Henrichs, T., Kaspar, F., Lehner, B., Rösch, T., and Siebert, S. (2003a). Development and testing of the WaterGAP 2 global model of water use and availability. In *Hydrological Sciences* 48 (3):317-337

Alcamo, J., Döll, P., Henrichs, T., Kaspar, F., Lehner, B., Rösch, T. and Siebert, S. (2003b). Global estimation of water withdrawals and availability under current and 'business as usual' conditions. In *Hydrological Sciences* 48 (3):339-348

Alkemade, R., Bakkenes, M., Bobbink, R., Miles, L., Nellemann, C., Simons, H. and Tekelenburg, T. (2006). GLOBIO 3: Framework for the assessment of global terrestrial biodiversity. In Bouwman, A.F., Kram, T. and Klein Goldewijk, K. (eds.) *Integrated Modelling of Global Environmental Change. An Overview of IMAGE 2.4.* Netherlands Environmental Assessment Agency, Bilthoven

Bouwman, A.F., Kram, T. and Klein Goldewijk, K. (2006). *Integrated Modelling of Global Environmental Change: An Overview of Image 2.4.* Netherlands Environmental Assessment Agency, Bilthoven

Butler, C. (2005). Peering into the Fog: Ecological Change, Human Affairs, and the Future. In *EcoHealth* 2:17-21

Butler, C. and Oluoch-Kosura, W. (2005). Human Well-Being across Scenarios. In *Millennium Assessment Ecosystems and Human Well-being. Scenarios: Findings of the Scenarios Working Group.* Island Press, Washington, DC

Cosgrove, W. J. and Rijsberman, F. R. (2000). *World Water Vision: Making water everybody's business.* Earthscan, London

Döll, P., Kaspar, F. and Lehner, B. (2003). A global hydrological model for deriving water availability indicators: model tuning and validation. In *Journal of Hydrology* 270 (1-2):105-134

Draulans, D. and Van Krunkelsven, E. (2002). The impact of war on forest areas in the Democratic Republic of Congo. In *Oryx* 36:35–40

Dudley, J.P., Ginsberg, J.R., Plumptre, A.J., Hart, J.A. and Campos, L.C. (2002). Effects of war and civil strife on wildlife and wildlife habitats. In *Conservation Biology* 16 (2):319–329

Hughes, B. and Hillebrand, E. (2006). *Exploring and Shaping International Futures.* Paradigm Publishers, Boulder, CO

IEA (2006). *World Energy Outlook 2006.* International Energy Agency, Paris

IPCC (2000). *Emission Scenarios.* Cambridge University Press, Cambridge

IPCC (2007a). *Climate Change 2007: Mitigation of Climate: Summary for Policymakers.* Contribution of Working Group III to the Fourth Assessment Report of the Intergovernmental Panel on Climate Change. World Meteorological Organization and United Nations Environment Programme, IPCC Secretariat, Geneva

IPCC (2007b). *Climate Change 2007: The Physical Science Basis: Summary for Policymakers.* Contribution of Working Group I to the Fourth Assessment Report of the Intergovernmental Panel on Climate Change. World Meteorological Organization and United Nations Environment Programme, IPCC Secretariat, Geneva

Lebel. L., Thongbai, P. and Kok, K. (2005). Sub-Global Assessments. In *Millennium Assessment Ecosystems and Human Well-being: Multi-scale Assessments: Findings of the Sub-global Assessments Working Group.* Island Press, Washington, DC

MA (2005). *Ecosystems and Human Well-being: Scenarios: Findings of the Scenarios Working Group of the Millennium Ecosystem Assessment Working Group.* Island Press, Washington, DC

Nelson, G. (2005). Drivers of Change in Ecosystem Condition and Services. In *Millennium Assessment Ecosystems and Human Well-being: Scenarios: Findings of the Scenarios Working Group.* Island Press, Washington, DC

Robinson, J. (2003). Future Subjunctive: Backcasting as Social Learning. In *Futures:* 35, 839-856

Raskin, P., Banuri, T., Gallopin, G., Gutman, P., Hammond, A., Kates, R. and Swart, R. (2002). *Great Transition: The Promise and Lure of the Times Ahead.* Stockholm Environment Institute, Boston, MA

Swart, R. J., Raskin, P. and Robinson, J. (2004). The problem of the future: sustainability science and scenario analysis. In *Global Environmental Change Part A* 14:137-146

UN (2003). *Indicators for Monitoring the Millennium Development Goals: Definitions, Rationale, Concepts and Sources.* United Nations, New York, NY

UNEP (2006). *Africa Environment Outlook 2: Our Environment, Our Wealth.* United Nations Environment Programme, Nairobi

UNEP (2004). *GEO Latin America and the Caribbean Environment Outlook 2003.* United Nations Environment Programme, Nairobi

UNEP and RIVM (2004). *The GEO-3 Scenarios 2002-2032. Quantification and analysis of environmental impacts.* Potting, J. and Bakkes, J. (eds.). UNEP/DEWA/RS.03-4 and RIVM 402001022. United Nations Environment Programme, Nairobi and National Institute for Public Health and the Environment (currently MNP), Bilthoven

UNPD (2007). *World Population Prospects: The 2006 Revision* (in GEO Data Portal). UN Population Division, New York, NY http://www.un.org/esa/population/unpop.htm (last accessed 4 June 2007)

Verburg, P.H., Ritsema-Van Eck, J., De Nijs, T.C.M., Visser, H. and De Jong, K. (2004). A method to analyse neighborhood characteristics of land use patterns. In *Computers, Environment and Urban Systems* 28 (6):667-690

Verburg, P.H., Soepboer, W., Veldkamp, A., Limpiada, R., Espaldon, V. and Sharifah Mastura S.A. (2002). Modeling the Spatial Dynamics of Regional Land Use: the CLUE-S Model. In *Environmental Management* 30 (3):391-405

Verburg, P.H. and Veldkamp, A. (2004). Projecting land use transitions at forest fringes in the Philippines at two spatial scales. In *Landscape Ecology* 19 (1):77-98

WBCSD (2007). *Then & Now: Celebrating the 20th Anniversary of the "Brundtland Report" – 2006 WBCSD Annual Review.* World Business Council for Sustainable Development, Geneva

World Bank (2007). *World Economic Prospects 2007.* The World Bank, Washington, DC

Yohe, G., Adger, W.N., Dowlatabadi, H., Ebi, K., Huq, S., Moran, D., Rothman, D. S., Strzepek, K. and Ziervogel, G. (2005). Recognizing Uncertainties in Evaluating Responses. In *Millennium Ecosystem Assessment (ed.) Ecosystems and Human Well-being: Policy Responses.* Chapter 4. Island Press, Washington

Sustaining Our Common Future

Chapter 10 **From the Periphery to the Core of Decision Making – Options for Action**

While governments are expected to take the lead, other stakeholders are just as important to ensure success in achieving sustainable development. The need couldn't be more urgent and the time couldn't be more opportune, with our enhanced understanding of the challenges we face, to act now to safeguard our own survival and that of future generations.

Chapter **10**

From the Periphery to the Core of Decision Making – Options for Action

Coordinating lead authors: Peter N. King, Marc A. Levy, and George C. Varughese

Lead authors: Asadullah Al-Ajmi, Francisco Brzovic, Guillermo Castro-Herrera, Barbara Clark, Enma Diaz-Lara, Moustapha Kamal Gueye, Klaus Jacob, Said Jalala, Hideyuki Mori, Harald Rensvik, Ola Ullsten, Caleb Wall, and Guang Xia

Contributing authors: Christopher Ambala, Bridget Anderson, Jane Barr, Ivar Baste, Eduardo Brondizio, Munyaradzi Chenje, Marina Chernyak, Paul Clements-Hunt, Irene Dankelman, Sydney Draggan, Patricia Kameri-Mbote, Sylvia Karlsson, Camilo Lagos, Varsha Mehta, Vishal Narain, Halton Peters, Ossama Salem, Valerie Rabesahala, Cristina Rumbaitis del Rio, Mayar Sabet, Jerome Simpson, and David Stanners

Chapter review editors: Steve Bass and Adil Najam

Chapter coordinator: Tessa Goverse

Credit: Munyaradzi Chenje

Main messages

We appear to be living in an era in which the severity of environmental problems is increasing faster than our policy responses. To avoid the threat of catastrophic consequences in the future, we need new policy approaches to change the direction and magnitude of drivers of environmental change and shift environmental policy making to the core of decision making. The main policy conclusions and messages of this chapter are:

Environmental problems can be mapped along a continuum from those where "proven" solutions are available to those where both the understanding of the problem and its solutions are still "emerging." For problems with proven solutions, the cause-and-effect relationships are well known, the scale tends to be local or national, impacts are highly visible and acute, and victims are easily identified. However, the emerging problems (also referred to as "persistent" environmental problems) are rooted in structural causes. Many of the same causes of these environmental problems simultaneously underpin entrenched poverty and over consumption. For these environmental problems, some of the basic science is known about cause-and-effect relationships, but often not enough to predict a point of no return. They often need global or regional responses. Examples include climate change, stratospheric ozone depletion, persistent organic pollutants and heavy metals, tropospheric ozone, acid rain, large-scale deterioration of fisheries, extinction of species, and alien invasive species.

Environmental policy has been successful in solving many environmental issues, especially where marketable technical solutions are available. Such policy success, however, needs to be continually extended, adapted and re-assessed, particularly in parts of the developing world, where many of the environmental problems effectively addressed elsewhere seriously threaten the well-being of billions of people.

The range of policies (the toolbox) for dealing with environmental issues has, in the past 20 years, become more sophisticated and diversified. There are many promising examples showing how this powerful toolbox can be deployed effectively. For instance, many governments have used command-and-control and market-based instruments to achieve environmental goals, community participation techniques to help manage natural resources, and technological advances to implement policy more effectively. Other actors, in the private sector and civil society, have formed innovative voluntary partnerships to contribute to achieving environmental goals.

Success in addressing environmental problems with proven solutions, however, will not solve "the urgent but complex problems bearing on our very survival" that the Brundtland Commission articulated. There is a set of environmental problems for which existing measures and institutional arrangements have systematically demonstrated inadequacies. Achieving significant improvements for a long period on these problems, which emerge from the complex interaction of biological, physical and social systems involving multiple economic sectors and broad segments of society, has been impossible and, for some, the damage may be irreversible.

The search for effective policy responses to these emerging environmental problems has recently focused on options to transform their drivers. Although environmental policy responses have typically focused primarily on reducing pressures, achieving particular environmental states or coping with impacts, policy debates are increasingly concerned with how to address drivers, such as population and economic growth, resource consumption, globalization and social values.

Fortunately, the range of policy options to influence economic drivers is more advanced than at the time of the Brundtland Commission report, *Our Common Future.* These include the use of green taxes, creation of markets for ecosystem services and use of environmental accounting. The analytical foundation for such approaches has been refined, and governments are gaining experience in implementing them, although typically only at relatively small scales.

An organizational focus at all levels on these emerging environmental problems requires the shifting of the environment from the periphery to the core of decision making. The current role that the environment plays in governmental and intergovernmental organizations, and in the private sector could be made more central through structural changes, mainstreaming of environmental concerns into sectoral plans and a more holistic approach to development planning and implementation.

Regular monitoring of policy effectiveness is urgently needed to better understand strengths and weaknesses, and facilitate adaptive management. This infrastructure has not appreciably expanded in the past 20 years, even though policy goals have broadened considerably. Welfare cannot be measured by income only, and aggregate indicators have to take into account the use of natural capital as well. Of particular urgency is an improved scientific understanding of the potential turning points, beyond which reversibility is not assured.

For many problems, the benefits from early and ambitious action outweigh their costs. Both ex-post evaluations of the costs of ignoring warnings as well as the scenarios on the costs of global environmental change show that determined action now is cheaper than waiting for better solutions to emerge. In particular for climate change, our knowledge on the costs of inaction shows a worrying picture even while immediate measures are affordable.

Political decisions need support and legitimacy to be implemented. The knowledge basis for the environmental issues has expanded enormously during the last 20 years. Similarly, the range of options to influence social attitudes, values and knowledge has also expanded. Better environmental education programmes and awareness campaigns, and much more attention to involve various stakeholders will make environmental policies better rooted. An educated and more involved population will be more effective in addressing failures of government and holding institutions to account.

The new environmental policy agenda for the next 20 years and beyond has two tracks:
- **expanding and adapting proven policy approaches to the more conventional environmental problems, especially in lagging countries and regions; and**
- **urgently finding workable solutions for the emerging environmental problems before they reach irreversible turning points.**

Policy-makers now have access to a wide range of innovative approaches to deal with different types of environmental problems. There is an urgent need to make choices that prioritize sustainable development, and to proceed with global, regional, national and local action.

It is imperative for policy-makers to have the tools that help reduce the political risks of making the right decisions for the environment. The political fallout for making a rushed decision that is subsequently proven wrong can be politically damaging, especially if powerful political supporters are adversely affected.

INTRODUCTION

In the two decades since the World Commission on Environment and Development (Brundtland Commission) described a set of "urgent but complex problems bearing on our very survival" (WCED 1987), the global concern over environment and development issues has expanded. However, clear solutions and institutional mechanisms remain poorly defined. The problems identified by the commission have grown more severe, and new problems that were not foreseen have arisen. The main environmental problems described in previous chapters of this report can be categorized along a continuum – from those where "proven" solutions are available, to those where both the understanding of the problem and solutions are still "emerging" (see Figure 10.1).

Problems at the latter end of the continuum share a number of characteristics that make them hard to manage, including complex interactions across global, regional and local scales, long-term dynamics, and multiple stressors and stakeholders (see Chapter 1). Many of these hard to manage problems can be termed "persistent" environmental problems (Jänicke and Volkery 2001). Unfortunately, policy making and institutional reforms remain anchored in the less complex, more manageable environmental challenges of the 1970s, and have not kept pace with the emergence of these persistent environmental problems.

An inventory of environmental policy goals and targets, a review of experience in managing cross-cutting issues, an assessment of the adequacy of multilateral environmental agreements (MEAs), along with the scenario policy analyses in Chapter 9, underpin this review. Evidence shows that there is an urgent need to address the types of environmental problems that may have irreversible consequences, which may make local, regional, or even global environments progressively uninhabitable.

The future policy options point to the need for a two-track approach:

- expanding and adapting proven policy approaches to the more conventional environmental problems, especially in lagging countries and regions; and
- urgently finding workable solutions for the emerging environmental problems before they reach irreversible turning points.

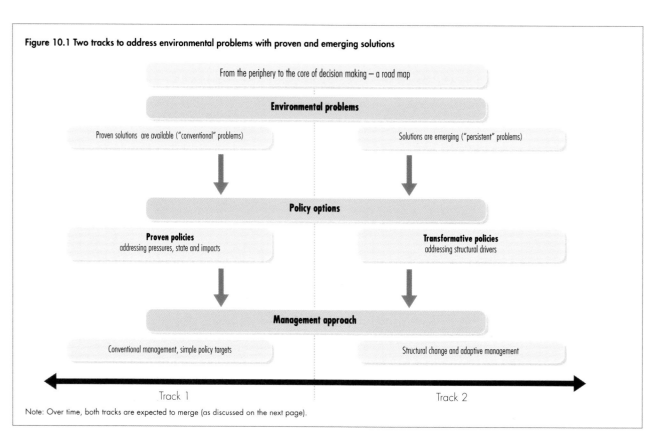

Figure 10.1 Two tracks to address environmental problems with proven and emerging solutions

From the periphery to the core of decision making – a road map

Environmental problems

Proven solutions are available ("conventional" problems) | Solutions are emerging ("persistent" problems)

Policy options

Proven policies addressing pressures, state and impacts | **Transformative policies** addressing structural drivers

Management approach

Conventional management, simple policy targets | Structural change and adaptive management

Track 1 | Track 2

Note: Over time, both tracks are expected to merge (as discussed on the next page).

Over time, both tracks are expected to merge into one, as the environmental policy agenda is progressively moved from the periphery to the core of economic and social development decision making.

For the first track, management and institutional approaches can learn from successful application of environmental policies in other parts of the world. The second track involves dealing with emerging environmental problems, and creating new institutional arrangements based on adaptive management, finding innovative financing mechanisms and improving monitoring, evaluation and social learning. Both tracks, however, need greater focus to address underlying societal and cultural values, increased education, empowerment of citizens and decentralized governance structures.

CURRENT ENVIRONMENTAL POLICY RESPONSES

Management of environmental problems

Environmental problems appear as impacts on nature and human well-being, through the air and atmosphere, in fresh and marine water, and on land. Most aspects of these environmental problems are described in the previous chapters. Eighteen of the key environmental issues discussed in Chapters 2–5 have been organized to illustrate the difficulty of management, and the extent to which the problems can be seen as having reversible or irreversible consequences, making local, regional or even global environments progressively uninhabitable (see Figure 10.2). While it is recognized that other dimensions could be used, GEO-4 has organized the environmental problems in two main clusters along a continuum.

Problems with proven solutions

The cause-and-effect relationships are well known, single sources generally can be identified, the potential victims are often close to those sources and the scale is local or national. Good examples of success stories for solving these environmental problems are available for microbial contamination, harmful local algal blooms, emissions of sulphur, nitrogen oxides, particulate matter, oil spills, local

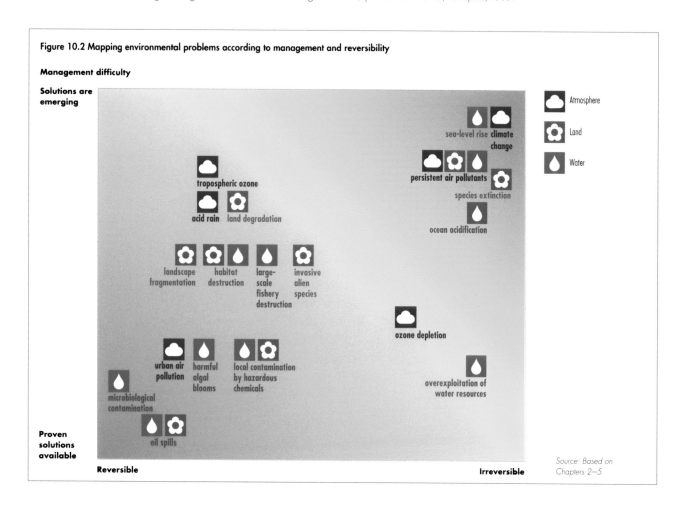

Figure 10.2 Mapping environmental problems according to management and reversibility

Management difficulty

Solutions are emerging

Atmosphere

Land

Water

sea-level rise climate change

tropospheric ozone

persistent air pollutants

species extinction

acid rain land degradation

ocean acidification

landscape fragmentation

habitat destruction

large-scale fishery destruction

invasive alien species

ozone depletion

urban air pollution

harmful algal blooms

local contamination by hazardous chemicals

overexploitation of water resources

microbiological contamination

oil spills

Proven solutions available

Reversible

Irreversible

Source: Based on Chapters 2–5

land degradation, localized habitat destruction, fragmentation of land, and overexploitation of freshwater resources.

Problems with emerging solutions

Some of the basic science about cause-and-effect relationships is known, but often not enough to predict when a turning point or a point of no return will be reached, or exactly how human well-being will be affected. The sources of the problem are quite diffuse and often multisectoral, potential victims are often quite remote from the sources, extremely complex multi-scale ecological processes may be involved, there may be a long time between causes and impacts, and there is a need to implement measures on a very large scale (usually global or regional). Examples include global climate change, stratospheric ozone depletion, persistent organic pollutants and heavy metals, extinction of species, ocean acidification, and introduction of invasive alien species.

The environmental problems at the "emerging solutions" end of the continuum have implications for development, in two fundamental ways:

■ Environmental resources and change create direct opportunities and threats for development (Bass 2006). Natural capital frequently constitutes economically important assets, the management of which has a strong impact on economic growth (Costanza and Daly 1992). Poor countries generally have a higher percentage of their total assets comprised of environmental resources than produced capital (World Bank 2006). Environmental resources frequently affect risk exposures, by mediating or altering natural hazard vulnerability. They frequently play an important role in empowerment of vulnerable social groups, including women; marginalized ethnic, linguistic or regional populations; and the extremely poor. Environmental resources can also play a strong role in shaping the long-term viability of economic development strategies.

■ The diagnosis of the causes of persistent environmental problems shares much in common with similar diagnoses of persistent development problems. In particular, the large gap between proven governance mechanisms and the magnitude and complexity of environmental problems is similarly found in areas where development is lagging.

Therefore, there are strong reasons for coordinating the environment and development agendas. This message is implicit behind the overarching design of major international processes, such as Agenda 21 and the Johannesburg Plan of Implementation, but a major gap remains between both the environment and development agendas (Navarro and others 2005).

The cluster of large-scale, persistent environmental problems has more complex interlinkages, and it is more difficult to get concerted effort at multiple scales to solve the intertwined problems (see Chapter 8). As the Brundtland Commission stated, they are often part of "the downward spiral of linked ecological and economic decline in which many of the poorest nations are trapped" (WCED 1987).

Success stories for solving these kinds of environmental problems are much less common than for the environmental issues identified in the 1970s. In addition, left unattended or uncontrolled, many issues in the first cluster can coalesce and contribute to the persistent problems. For example, expanding local land degradation (see Chapter 3) may result in dust and sandstorms at the regional scale, contributing to atmospheric brown clouds that contribute to global dimming (reduced solar radiation reaching the ground) and impacts on regional monsoons (see Chapter 2).

Elevating environment on the policy agenda

At all points on the continuum, there are significant challenges involved in raising the profile of environmental issues in public policy, but the opportunities are also numerous. Elevating the profile of environmental issues in public policy might involve the following actions.

Raising the profile of the environmental agenda

Although sustainable development has gained general political support, environment remains low on the policy agenda in most day-to-day politics. Poverty reduction, economic growth, security, education and health are clearly the highest priority policy items. Proving that the environment underpins and contributes significantly to all of these high priority issues can raise its political visibility, leading to more political support (Diekmann and Franzen 1999, Carter 2001).

Strengthening integration

Traditionally, environmental policy-makers have not focused on establishing linkages with other important policy agendas, such as poverty reduction, health and security in developing countries, or with economic sectors in developed countries. Phasing out environmentally damaging subsidies may, for instance, release funds for more targeted support for the poor, as well as improve the environment. Integrating environmental policy into other policy areas involves a continuous, adaptive process. End-of-pipe pollution controls in the 1970s led to cleaner production processes in the 1980s and zero-waste factories in the 1990s. Modern environmental policy and legislation needs to follow a similar, stepwise evolutionary path to finding and applying solutions for persistent environmental problems (EEA 2004, EEB 2005).

Setting clear goals and targets, and strengthening monitoring

Political commitments to specific goals and targets are essential to effectively address environmental issues. Developments in this area are often only visible over the medium- to long-term, and tend to escape day-to-day political attention. Therefore, scientific research and monitoring, and information systems need to be maintained at adequate levels, and progress against benchmarks regularly reviewed by an independent body (OECD 2000). The lack of quantifiable targets for Millennium Development Goal (MDG) 7 on environmental sustainability has been one factor in its relatively low profile on the global agenda (UNDP 2005). The need to revisit time-bound targets under MDG 7 would be strategic in strengthening monitoring and accountability.

Reinforcing stakeholder involvement

A participatory approach facilitates collaborative efforts among various stakeholders, engenders a sense of ownership and makes new initiatives more sustainable. An informed population is also more effective in addressing failures of governments, enhancing transparency and holding institutions accountable. Although stakeholder participation often requires additional upfront costs in terms of time and resources it has, particularly at the local level, proven to be a successful instrument and may ultimately result in reduced costs (Eden 1996). However, in many countries and at the international level, the formal right to take part in the decision making process often remains restricted.

Building on small-scale successes

For internationally funded projects and initiatives, the scale of operation is proportional to funds available. Therefore, many environmental initiatives have not been scaled up to the extent where real environmental change may take place (UNESCO 2005a). Once the scale of an environmental problem goes beyond national borders, it is much harder to justify allocating national budgets or bilateral development assistance, creating potential free-rider problems.

Clarifying the role of government

Frequently, environmental ministries are seen as acting more like facilitators than implementers: steering not rowing. Priority could be given to the development

Economic activity is interlinked to land, water, and the atmosphere and environmental policy making must, therefore, involve the integration of all these aspects.

Credit: Ngoma Photos

of more effective policies and policy coherence. Environmental ministries could concentrate more on translating environmental aims and the results of research and monitoring into long-term objectives, priorities, basic legislation and mandatory limits. They should also be charged with reviewing the environmental results for each sector. In turn, sectoral departments need to build the necessary capacity to interpret and internalize environmental priorities into their policies, and take greater responsibility for implementing environmental activities. In some countries, restructuring has already taken place, and environmental units can now be found in the sectoral ministries, although loyalty may remain with sectoral interests (Wilkinson 1997).

Avoiding over-sophisticated legislation
In developed countries, incremental modifications of environmental regulations and lack of involvement of regulatory practitioners in this process make some legislation almost incomprehensible. Room for corruption has been enlarged, and an unnecessary burden has been imposed on industry. When these policy instruments are transferred to developing countries, which often have inadequate capacity to develop innovative, home-grown policies, the excessive level of sophistication makes them impossible to implement. Much clearer and more cost-effective regulations can be set up, drawing, whenever possible, upon capacities of other stakeholders (Cunningham and Grabosky 1998). Ideally, investing in capacity building, and supporting inclusive national legislative development processes will prove more beneficial in the long run.

Tackling hard choices
Many situations exist today where "win-win" solutions are impossible. Objective assessments, backed by freely accessible, high-quality information and public consultation, are needed to weigh trade-offs between potential alternatives. Economic valuation of non-market environmental goods-and-services, and consideration of potential social impacts need to be included in any objective evaluation of alternatives. Political leadership is essential. Delaying decisions may result in needless damage and death (EEA 2001), as well as possible irreversible change for which no trade-offs should be contemplated.

Critical policy gaps and implementation challenges
Successes tempered by policy gaps
The linear, single-source, single medium environmental problems that dominated the agenda at the 1972 Stockholm conference were, for the most part, subjected to increasingly effective management over the following two decades. Environmental ministries were created, national legislation governing air and water quality was implemented, and standards for exposure to toxic chemicals were adopted. Based on the analyses in Chapters 2–8, it can be concluded that nearly all countries now have a set of policy instruments, if not an explicit environmental policy, which provides a platform for improved environmental management (Jordan and others 2003). There is also support for projects and innovative experiments to enhance the capacities of personnel and promote better environmental management in most developing countries.

Considerable effort has been invested in new approaches to environmental policy making (Tews and others 2003). Although there were failures, and many good policies were not implemented, due to institutional constraints, progress has been sustained and significant in a large number of countries. In some urban areas, environmental quality is better today than in the mid-1980s. The main policy gap is in ensuring that policies and organizational arrangements that have worked in some areas are sustained and extended to all (especially developing) countries. While there is an unfinished agenda that affects the well-being of billions of people, the necessary resources and political will to provide the enabling environment are still too often neglected.

Complex problems remain a major policy challenge
By contrast, the complex, multi-source, persistent environmental problems highlighted by the Brundtland Commission, and those that have emerged since have not been effectively managed anywhere (OECD 2001a, Jänicke and Volkery 2001, EEA 2002, Speth 2004). There are no major issues raised in *Our Common Future* for which the foreseeable trends are favourable. Apart from the obvious need to mainstream these problems into national decision making processes, workable policies for dealing with issues that require fundamental transformations in modern societies have yet to emerge.

Despite positive trends observed in some countries, the global environment remains under severe threat, and important ecosystems and environmental functions may be approaching turning points, beyond which the consequences could be disastrous (as shown in earlier chapters of this report). Therefore, there is an urgent need to reinvigorate the environmental dimension of

Box 10.1 Overview of global policy targets

As part of this assessment, policy targets associated with the high-priority global environmental problems analysed in Chapters 2–5 were identified and characterized. Global targets were the primary focus, but sub-global targets that covered large numbers of countries were also analysed.

At the level of *objectives*, or general statements of principle, the global community has articulated clear objectives fairly consistently across all the high-priority problems. However, when it comes to *targets*, or specific, quantifiable, time-bound outcomes, the situation is more uneven. For the most challenging problems, characterized by many of the dimensions of persistence, targets are less common, whereas they are more prevalent among the problems characterized as having proven solutions available. In terms of water, for example, clear targets exist concerning access to piped water and basic sanitation, which are linked to the broader objective of reducing the most pressing aspects of poverty. By contrast, although the objective of integrated watershed management is almost equally widespread, targets concerning how to implement it are rarer. There are clear, widespread targets already embedded in decision making concerning urban air pollution, but this is not the case for indoor air pollution.

The degree to which policy targets are supported by monitoring and evaluation procedures varies considerably. For ozone depletion, for example, there is a robust monitoring programme that measures the atmospheric concentration of ozone-depleting substances, ozone layer thickness, and trends in production, consumption and emissions. By contrast, most of the biodiversity protection targets lack baseline benchmarks and the kind of regular monitoring that would permit tracking of trends.

Most targets aim at improving generic capacities (including adoption of plans, creation of policy frameworks, conducting assessments and setting priorities), or at reducing pressures (lowering emissions, extraction or conversion). It is rarer to find targets that aim at reducing drivers or at achieving specific states. There are some biodiversity targets that target drivers, but none exist in other areas. Regional air pollution in Europe is the best-developed example of a targeting process that focuses on environmental states (in this case, levels of deposition relative to critical loads).

Figure 10.3 Global and regional targets and monitoring programmes

Issue	Targets	Monitoring
Biodiversity loss		
Climate change		
Degradation and loss of forests		
Indoor air pollution		
Integrated Water Resources Management (IWRM)		
Land contamination and pollution		
Land degradation/desertification		
Large-scale marine fisheries		
Long-range air pollution		
POPs		
Stratospheric ozone protection		
Water and sanitation		
Water security		

Targets	Monitoring
■ No targets	■ No regular monitoring
■ Quantitative, time-bound targets; not legally binding	■ Some monitoring takes place, but is less than complete
■ Legally-binding, quantitative, time-bound targets	■ Relevant monitoring taking place globally
Exception: Long-range air pollution assigned yellow; legally-binding targets in Europe only	

Source: Chapters 2–5, review of MEAs at Ecolex 2007, UN 2002a

development, to set realistic goals and targets (see Box 10.1), and to ensure that environmental goals and requirements are integrated into mainstream public policy at global, regional and national levels.

Policy implications of scenarios

The scenarios highlighted in Chapter 9 illustrate the difficulties of responding to persistent environmental problems, and of rapidly changing directions. The environmental implications of the various scenarios illustrate the legacy of past decades and the level of effort required to reverse powerful trends. One of the major policy lessons from the scenarios is that there can be significant delays between changes in human behaviour, including policy choices, and their environmental impacts, specifically:

- much of the environmental change that will occur over the next 50 years has already been set in motion by past and current actions (see also De-Shalit 1995); and
- many of the effects of environmentally relevant policies put into place over the next 50 years will not be apparent until long afterwards. The slow recovery of the ozone "hole" over Antarctica reflects this extended time dimension.

Enormous momentum is built into global economic systems, and many social forces are comfortable with (or profit from) the way the world is today. Combined with the lack of certainty over precisely when ecosystems may pass turning points, it is understandable that shifting trajectories in a deliberative, precautionary manner towards sustainability is so difficult. Nevertheless, the scenarios show:

- the very different outcomes if critical choices are not made in time; and
- the chance to avert global collapse exists if the right choices are made sooner rather than later.

A critical uncertainty in such scenarios is the ability to decouple pollution intensity from economic growth, and to shift towards service industries without lessening economic growth rates (Popper and others 2005).

Implementation challenges

Implementation of good practices needs to be extended to countries that have been unable to keep pace, due to lack of capacity, inadequate finances, neglect or socio-political circumstances. Due to internal or international pressures, most countries have already adopted some policies to address the environmental

Energy use and transport drive industrialization and urbanization. Many countries are now implementing policies to reduce inefficient use of energy, although change can be slow.

Credit: Ngoma Photos

issues with proven solutions. Implementation of these policies, however, remains relatively weak or non-existent in many developing countries. In some cases, it appears as if there is no real intention of implementing the policies, and governments are paying mere lip service to environmental management to pacify lobby groups or donors (Brenton 1994).

In too many countries, environmental policy remains secondary to economic growth. Generally, macro-economic objectives and structural reform have been considered of higher priority than environmental quality. Nowhere has it been possible to integrate economic, ecological and social objectives consistently with a sustainable development model (Swanson and others 2004). Increasing global concerns, such as poverty and security, may even have moved environmental issues further towards the periphery of the political agenda (Stanley Foundation 2004, UN 2005d).

Elevating the agenda to tackle persistent environmental problems impinging on the structural core of societies poses implementation challenges that appear immense. While there are a few examples of countries that have made successful structural changes, worryingly, some countries are even backsliding in implementation of the conventional environmental agenda (Kennedy 2004).

Implementation of environmental policies requiring substantial societal or cultural changes, such as a culture of environmental protection, or structural realignment, will meet with fierce resistance from sectors affected and from some parts of the public. Therefore, governments tend to buy time or defer decisions when such "hard" structural changes in overall policies are required – often until they are inevitable (New Economics Foundation 2006). Hard choices are usually found where the environment and economy intersect or interact, posing structural issues that are difficult to address. The underlying drivers are more entrenched, cross-cutting social and economic problems, with the environment deeply embedded in them.

How important these changes are viewed and how serious governments are about making changes often depend on political ideology and value orientation. To implement such "hard" options, governments have limited opportunities to take a close look at precedents and experience before embarking on them. Often, consideration of social and political costs rather than the lack of funds hinder implementation (Kennedy

2004). For example, removal of agricultural subsidies may have important environmental outcomes, but the political ramifications for making such changes are immense (CEC 2003). Policies designed to yield reduced carbon emissions affect all sectors that use energy. Hence, sectoral agencies and affected stakeholders need to "buy into" environmental policies (NEPP2 1994).

The policies that are easiest to implement are those that do not involve redistribution of wealth or power – often termed "win-win" situations or "soft" options. Many soft options are already being used, such as generating public awareness, setting up organizations, formulating symbolic national legislation and signing weak international conventions. These often create the appearance of action without really tackling the core drivers of the persistent environmental problems.

Although some policy debates are beginning to draw attention to drivers as appropriate focal points for policy intervention (Wiedmann and others 2006, Worldwatch Institute 2004), their representation in global policy fora is in its infancy. In a systematic identification of all global policy targets pertaining to the high priority environmental problems identified in previous chapters, only 2 out of 325 distinct policy targets were aimed at drivers (see Box 10.1). The majority targeted pressures and improvements in coping capacity. The exceptions were targets aimed at promoting sustainable consumption of natural resources in the biodiversity and forest conservation policy areas.

Existing environmental organizations were often not designed to address complex cross-sectoral and transboundary policy implementation. Institutions have been unable to keep up with the fast pace at which economic growth is generating cumulative environmental degradation. As pointed out in the Brundtland Commission report, a holistic approach requires the integration of environmental concerns and measures across all sectors. As persistent environmental problems also affect countries across borders, and become sub-regional, regional or global problems as evident in Chapter 6, coordination and harmonization of implementation approaches raise new organizational challenges.

Improved knowledge management is critical for effective implementation of policies. Although some information regarding these persistent environmental issues is available, it is usually incomplete, and fails to bridge the gap between the technical measures observed and the human impacts that motivate policy-makers. They need clearly and easily understood frameworks, simple metrics and appropriate solutions to act upon. The scientific and academic community communicates the dimensions of such problems to policy-makers, using complex and incomplete measuring tools. While it is relatively easy to provide data on many of the most pressing economic and social outcomes, such as GDP and the Human Development Index, no equivalent concrete measuring tools have been broadly accepted in the environmental domain, although there are several competing options. One review found 23 alternative aggregate environmental indices (OECD 2002a), and several more are under development.

Supporting valuation and measurement initiatives that build up a common platform of understanding of the impact of policies on sustainability, and clearly measure the environmental consequences of economic actions will assist sensible decision making. Consensus on valuation is important, because not all environmental goods-and-services can or should be monetized. Non-monetary valuation indicators that are commonly understood and agreed upon, in conjunction with financial and social indicators, can show the status and trends towards or away from sustainability.

THE FUTURE POLICY FRAMEWORK

A strategic approach

Environmental policy has been successful in solving a wide array of linear, single source, single medium or "conventional" environmental issues, especially where marketable technical solutions have been available, such as chemical replacements for ozone-depleting substances (Hahn and Stavins 1992). However,

persistent environmental problems, such as the rising concentrations of greenhouse gases, the loss of biodiversity, the accumulated contamination of soil and groundwater, and the cumulative effects of dangerous chemicals on human health, are issues where it has been impossible to achieve significant improvements for a long period of time and, for some, the damage may be irreversible (OECD 2001a, Jänicke and Volkery 2001, EEA 2002). Failure to effectively address these persistent problems will undermine or negate all of the impressive achievements in finding solutions to the conventional problems.

Therefore, a two-track strategy is envisaged: adapting and expanding the reach of proven policies, and developing policies to deliver more deeply rooted and structural change at all levels.

Expanding the reach of proven policies

Although a plethora of environmental challenges exist, there are also some effective policies available. Proven successes in environmental policy in other countries can be taken as an encouraging sign in those lagging countries beginning to face up to their own legacy of environmental degradation. Effective policies enhance a particular ecosystem service, and contribute to human well-being without significantly harming other ecosystem services or harming other social groups (UNEP 2006b). Promising responses either do not have a long track record, and thus outcomes are not yet clear, or could become more effective if they were adequately modified. Problematic responses do not meet their goals or harm other ecosystem services or social groups.

Since 1987, the policy landscape has expanded enormously and direct and indirect environmental policies now impinge on virtually all areas of

Table 10.1 Classification of environmental policy instruments				
Command-and-control regulations	**Direct provision by governments**	**Engaging the public and the private sectors**	**Using markets**	**Creating markets**
■ Standards ■ Bans ■ Permits and quotas ■ Zoning ■ Liability ■ Legal redress ■ Flexible regulation	■ Environmental infrastructure ■ Eco-industrial zones or parks ■ National parks, protected areas and recreation facilities ■ Ecosystem rehabilitation	■ Public participation ■ Decentralization ■ Information disclosure ■ Eco-labelling ■ Voluntary agreements ■ Public-private partnerships	■ Removing perverse subsidies ■ Environmental taxes and charges ■ User charges ■ Deposit-refund systems ■ Targeted subsidies ■ Self-monitoring (such as ISO 14000)	■ Property rights ■ Tradeable permits and rights ■ Offset programmes ■ Green procurement ■ Environmental investment funds ■ Seed funds and incentives ■ Payment for ecosystem services

economic activity (Jänicke 2006). One of many categorizations of environmental policies is provided in Table 10.1. The progressive evolution of policies from "command-and-control" to "creating markets" over the past two decades is illustrated in this classification.

The toolbox of policy instruments has been gradually expanded, with much more emphasis on economic instruments, information, communication, and voluntary approaches (Tews and others 2003). These developments are partly related to the fact that the policy focus in the area of pollution control has shifted from the large single polluters (point sources) to more diffuse sources, that can be harder to control (Shortle and others 1998). However, direct regulation (also known as command-and-control) still plays a major role, and is likely to do so in the future (Jaffe and others 2002). Some governments have begun to reform their environmental standards in favour of more ambitious, innovation-friendly systems. For example, the Japanese Top Runner Program on energy efficiency is receiving much attention. In this programme, standards are adapted to the best available technologies, giving a continuous incentive to improve such standards.

Governments will need to continue applying (or threatening to apply) "strong instruments," such as command-and-control regulations, for effective

policy implementation, even if the use of market forces and "soft instruments," such as provision of information, play a more important part than before (Cunningham and Grabosky 1998). An effective toolbox, therefore, has to include a wide variety of instruments, often used in concert, customised to the institutional, social and cultural milieu of the country or region concerned.

The challenge is to find the most efficient policy instrument or mix of instruments for a particular environmental problem in a particular geographic and cultural context. Increasingly, policy-makers are looking at complex models of social, economic and environmental systems to guide policy choices. However, these models themselves are inevitably partial representations of reality. For a number of environmental problems, direct command-and-control regulation will be an effective instrument, and this is therefore widely used today (see Box 10.2). In particular, the instrument is now used far more effectively to specify expected results rather than technical methods. Further, widely agreed technical standards, prescribed by law, may contribute to fair competition in the industry concerned, and also serve as an incentive for gradual technical development and innovation, improving environmental protection. In order to avoid market distortion between competing industries, or globalization-driven pollution havens, internationally agreed standards need to be developed and

To avoid market distortion between competing industries, or globalization-driven pollution havens, internationally agreed standards need to be developed and cautiously applied.

Credit: Ngoma Photos

Box 10.2 Flexible use of policy instruments in Norway

An example of innovative and flexible use of policy instruments involving multiple stakeholders is Norway's regulation on scrapped electrical and electronic products (under the Pollution Control Act and the Product Control Act). An increasing share of the solid waste stream is from the information and communication technology (ICT) sector, with a high content of hazardous materials, such as heavy metals. This waste source is also driving the WEEE (Waste Electric and Electronic Equipment) and RoHS (Restriction of Hazardous Substances) directives of the European Union.

The Norwegian approach involved relevant producers, importers and distributors in a review of the problem from the start, with a scoping study of the volume of such waste and its environmental implications, and a discussion of various means to deal with it. This led to a realization that there was a larger volume of waste than had originally been envisaged, and a proposal from the authorities for new regulations taking effect from 1 July 1999, after wide-ranging public consultation.

Parallel to this regulation, the environmental authorities and the main firms and business associations developed agreements, with fixed dates, commitments and reporting mechanisms, for implementation. These agreements are "voluntary," as firms are free to stay outside or enter into separate agreements (and therefore do not represent a competition issue or a "barrier to entry"), but they are grounded in the regulation, and avoid the "free rider" problem, as well as solving the compliance, control and enforcement issues of concern to business and the authorities.

The agreements involve setting up three waste collection companies by business, for different WEEE waste fractions, and collection of fees to finance the waste collection and treatment systems. The fees are administered by the business partners (collected along with the VAT system, to ensure low administrative costs). Following the introduction of the new policy instruments in 1999, the government in 2005 reported to Parliament that in 2004 "more than 90 per cent" of the total quantity of scrapped electrical and electronic products were collected. Further, the greatest part of the waste collected was recycled, and the hazardous waste components were managed in an environmentally sound manner. This apparently old-fashioned command-and-control instrument has been transformed, in cooperation with the relevant business sectors, and is administered to a large extent through contractual agreements, leaving implementation to the business sector.

Source: Ministry of Environment Norway 2005

---

cautiously applied. While waiting for global action, groups of importers in some markets have already started to set voluntary standards for their own production and supply chains.

A wide range of success factors have been demonstrated as important in best practice policies. Some of the key factors include (Dalal-Clayton and Bass 2002, Volkery and others 2006, Lafferty 2002, OECD 2002b):

■ solid research or science underpinning the policy;

■ high level of political will, usually bipartisan and therefore sustained;

■ multistakeholder involvement, often through formal or informal partnerships;

■ willingness to engage in dialogue with policy opponents;

■ robust systems for mediating conflict;

■ capable, trained staff engaged in implementation;

■ prior systems of monitoring and policy revision agreed, including clauses that mandate periodic revision;

■ legislative backing, combined with an active environmental judiciary;

■ sustainable financing systems, ring-fenced from corruption;

■ evaluation and assessment of policies independent from the rulemaking agent, for example, by advisory committees or public auditors;

■ minimal delays between policy decisions and implementation; and

■ coherence and lack of conflict throughout all government policies.

Finding new, transformative policies

The class of environmental problems still seeking solutions needs innovative policies to address survival or threshold issues. They will challenge existing societal structures, consumption and production patterns, economies, power relationships, and the distribution of wealth (Diamond 2005, Leakey and Lewin 1995, Rees 2003, Speth 2004). There is an urgent need for a fundamental reorientation of public and private policies on environmental issues, and for transformative structural changes (Gelbspan 1997, Lubchenco 1998, Posner 2005, Ehrlich and Ehrlich 2004).

Unfortunately, lack of political will has failed to make environment central to a government's mission (De-Shalit 2000). Modern politics can be characterized as a continuous negotiation among politicians and special interests to get attention for their issues and interests (where the strongest interest often wins). This creates a chaotic situation that can easily focus on short-term, politically expedient gains, rather than long-term sustainable and equitable development (Aidt 1998). As long as politicians and citizens fail to recognize that human well-being depends on a healthy environment, and put issues other than environment among their top priorities, environmental policy-makers can only hope that other policies, such as economic, trade or development policies, will not make the environmental situation worse. Many of the persistent problems are slow to form, initially "invisible," difficult to pin down precisely and inadequately weighted when trade-offs are being considered, failing to get the attention of politicians with short-term horizons (Lehman and Keigwin 1992). However, the political fallout for making a rushed decision that is subsequently proven wrong can be politically damaging, especially if powerful political supporters are adversely affected (UCS 1992, Meadows and others 2004). It is, therefore, imperative for policy-makers to be provided with the tools that help reduce the political risks of making the right decisions for the environment.

For some persistent environmental problems, such as climate change and biodiversity loss, incentives for further environmental degradation are still being promoted, as these are primarily determined by other policy domains and their respective competing objectives (Gelbspan 1997, Wilson 1996, Myers 1997). Despite best intentions, implementation of international environmental agreements by national governments to address such issues is failing, and there are few, if any, sanctions for such failure (Caldwell 1996, Speth 2004).

Environmental policy failures are closely related to the challenge of a more encompassing integration of environmental concerns into other policy sectors (Giddings and others 2002). As environmental issues have become important in all sectors, there is a growing need to converge with economic development policies (see discussions of European efforts at cross-sectoral greening) (Lenschow 2002). However, there is still no robust integrated policy assessment tool (notwithstanding good advances made in Europe) that ensures mainstreaming of environmental issues into all other sectoral policies (Wachter 2005, Steid and Meijers 2004).

In part, environmental problems and mismanagement of natural resources result from not paying the full price for the use of ecosystem services (Pearce 2004). Governments adopt many different objectives that are often competing or even in conflict with each other,

failing to recognize that they all depend on properly functioning ecosystem services. When economic development is given higher priority than environmental protection, policy failure is aggravated by the fact that environmental organizations are often weak, are seen as just another special interest and usually lose out in policy battles. Another complicating factor is the fact that throughout the developing world there is a widespread lack of implementation and enforcement of environmental legislation, due to insufficient administrative capacities (Dutzik 2002).

Ideally, sound science should underpin environmental policy choices. There is little doubt that the knowledge base on the key environmental issues has expanded enormously since 1987, but still too little is known about how close potential turning points are, or how to achieve long-term sustainable development. As noted in *Our Common Future*, "science gives us at least the potential to look deeper into and better understand natural systems" (WCED 1987). The Brundtland Commission observed that scientists were the first to point out the growing risks from the ever-intensifying human activities, and they have continued to play that role in an increasingly coordinated manner.

The Intergovernmental Panel on Climate Change, the Millennium Ecosystem Assessment, Global Environment Outlook, Global Marine Assessment, Global Forest Resources Assessment, Global Biodiversity Assessment, International Assessment of Agricultural Science and Technology for Development (IAASTD), and the Land Degradation Assessment in Drylands (LADA) are indicative of the shared concerns of the global science community, and a willingness to cooperate. These and other assessments have underpinned the MEAs, supported the global summits and conveyed important scientific information to the global community through the media and other means of dissemination. Scientists, statisticians, and people in other disciplines have become increasingly aware of the importance of communicating difficult issues in a form that decision-makers and the public can understand.

However, the almost daily diet of bad news emanating from these studies may have, paradoxically, conditioned the public and decision-makers to always expect predictions of disaster from scientists, despite the evidence that overall human well-being has progressively improved. The unceasing flow of scientific information has itself provided political cover for indecision and delay (Downs 1972, Committee on Risk Assessment of Hazardous Air Pollutants and others 2004). When an isolated piece of good news on the science front, such as bringing back a species from the brink of extinction, is published, it is seized upon as evidence that the scientists are always exaggerating the dangers. The media, in their attempt at balanced reporting, can always find at least one scientist to contradict the general consensus of the majority of scientists, resulting in the common political view that the science is still uncertain, and, therefore, there is no need for precipitous action (Boykoff and Boykoff 2004).

The danger of this balanced, "no action needed yet" approach is that millions of lives might be needlessly lost, human health impaired, or species made extinct. The danger of delayed decisions has been clearly documented in the case of radiation, asbestos, chlorofluorocarbons, and other environmental and human health issues. Despite early warnings from scientists on these issues, it was decades before action was ultimately taken (EEA 2001). Similar delays are being experienced in relation to climate change and biodiversity loss.

The high degree of difficulty in finding innovative policy solutions for these persistent problems can be explained by several factors. The use of natural resources and the release of emissions to the environment are often determined by the logic of industrial production systems and their associated technologies. Hence, sustainable solutions require fundamental changes in industry structure, technologies and input factors for the sectors involved, such as mining, energy, transport, construction and agriculture. The government departments responsible for these sectors see their main duty as providing and securing the environment as a cheap (often free) input for production for their private (or public) sector clients. Such structural problems cannot be solved by environmental policy alone, but, instead, they need coordinated action by different parts of the policy making and implementation process of governments (Jänicke 2006).

International solutions are even more difficult to achieve, however, due to the relatively weak organizational framework and the many veto points that allow interest groups to stop ambitious policies (Caldwell 1996). Even where MEAs have been ratified by national governments, effective implementation is hindered by financial and technical capacity constraints, onerous

Sign of the times; action lags far behind.

Credit: Frans Ijserinkhuijsen

reporting procedures, non-cooperation of non-state actors and attention to other pressing issues (Andresen 2001, Dietz and others 2003).

Effective policy instruments are those that provide long-term signals and incentives on a predictable basis. This is vitally important to the business sector, but also to consumers and households. Publishing long-term plans for how regulations will be tightened is one way of easing changes. To be socially acceptable, redistributive instruments, such as regulatory constraints and environmentally related taxes, and other economic instruments also need to be seen as fair and equitable.

Promising transformative policy options

There are a few promising policy options that demonstrate the power of innovative policies to contribute to the structural changes needed to solve persistent environmental problems. These need to be carefully monitored, and lessons learned disseminated widely and quickly, so that successful policies can be added to the toolbox, always bearing in mind the need for local adaptation and social learning.

Green taxes

A small part of increased tax revenue can be designated for increased energy conservation and energy efficiency measures. Taxing environmental "bads" and subsidizing environmental "goods," while simultaneously achieving income redistribution is typical of the kinds of policies needed to bring the environment to the forefront of political decisions (Andersen and others 2000).

Reduce, Reuse, Recycle (3R) Policy in Japan

The Basic Law for Establishing the Recycling-based Society, enacted in 2000, seeks to lower waste volume (see Table 10.2) To make the law operational, the Fundamental Plan for Establishing a Sound Material-Cycle Society was formulated in 2003 for implementation over 10 years (MOEJ 2005). In addition to calling for greater

recycling, disposal and collecting facilities, the law assigns an extended producer responsibility (EPR) to businesses that produce and sell products. EPR functions through a take-back requirement, deposit refund schemes and the shifting of financial and/or physical responsibility of a product at the post-consumer stage upstream to the producer. A policy on EPR has been introduced for containers, packaging and some household appliances.

The achievements of the policy so far have been encouraging, with an increase in the number of units recovered (post-consumer use) at designated collection sites in 2003 and 2004, of 3 and 10 per cent respectively, compared with 2002 (MOEJ 2005).

The circular economy in China

The circular economy covers production and consumption involving diversified sectors of industry, agriculture and services, as well as the industry of comprehensive recovery and utilization of resources from wastes and scrap (Yuan and others 2006). Production is addressed at three levels in terms of establishment of small-scale cycling, focusing on clean production in enterprises, intermediate-scale cycling in eco-industrial parks, and large-scale cycling in eco-industrial networks in various localities. The circular economy is aimed at the renovation of conventional industrial systems, targeting improvements in resource and energy efficiency and decreasing environmental loads. Steps have also been taken to establish sustainable consumption mechanisms, including the advancement of green procurement by the government.

The government has set the following national targets for 2010 using 2003 indicators as the baseline (China State Council 2005 in UNEP 2006a):

- resource productivity per tonne of energy, iron and other resources increased by 25 per cent;
- energy consumption per unit of GDP decreased by 18 per cent;

Table 10.2 Quantitative targets for Japan's 3R Policy for 2000–2010		
Item	2000 Indicator	2010 Target
Resource productivity	280 000 yen (US$2 500) per tonne	390 000 yen (US$3 500) per tonne (40% improvement)
Target for cyclical use rate	10%	14% (40% improvement)
Target for final disposal amount	56 million tonnes	28 million tonnes (50% reduction)

- average water use efficiency for agricultural irrigation improved by up to 50 per cent;
- reuse rate of industrial solid waste raised above 60 per cent;
- recycle and reuse rate for major renewable resources increased by 65 per cent; and
- final industrial solid waste disposal limited to about 4.5 billion tonnes.

Implementation of the circular economy policy has been fairly recent, involving 13 provinces and 57 cities and counties nationwide. A relatively small number (5 000) of enterprises have passed the assessment for clean production and 32 enterprises have won the title of National Environmentally-friendly Enterprises. China's efforts to decouple economic growth and resource consumption warrant close monitoring over the next few years.

Lead markets for environmental innovations

Environmental innovations are typically developed in "lead markets" (Jacob and others 2005, Jänicke and Jacob 2004, Beise 2001, Meyer-Krahmer 1999).

The emergence of lead markets, such as for the use of wind energy, requires political will, a long-term and integrated strategy and favourable conditions such as for innovation.

Credit: Jim Wark/Still Pictures

These are countries that lead in adopting innovation, and where the penetration of markets is more encompassing than in others. They serve as a model for, and their technologies and related policies are often adopted by other countries. The concept of lead markets has been developed and fruitfully applied for many types of technological innovations, such as the mobile phones that were introduced in Finland, the fax in Japan or the Internet in the United States (Beise 2001). Lead markets for environmental technologies are typically not only stimulated by more pronounced environmental preferences of consumers in that country, but also depend on special promotion measures, or on direct political intervention in the market.

Examples of environmental protection lead markets include the legally enforced introduction of catalytic converters for automobiles in the United States, desulphurization technologies in Japan, Danish support for wind energy, the waste from electrical and electronic equipment directive of the European Union and CFC-free refrigerators in Germany (Jacob and others 2005). Another example is the global distribution of chlorine-free paper. This initially involved political activities by Greenpeace, and support from the USEPA in the United States. There was the introduction of chlorine-free paper whitener in Scandinavian countries, Germany and Austria, and effective political market intervention in Southeast Asian countries (Mol and Sonnenfeld 2000). This shows that political action that stimulates internationally successful innovations is not limited to governments, but that environmental activists can also intervene effectively.

The emergence of lead markets is not a matter of introducing a single policy instrument. Instead, political will, a long-term and integrated strategy, and favourable framework conditions (for example, for innovation) are decisive (Porter and Van der Linde 1995, Jacob and others 2005). Most important is the strong correlation between economic competitiveness and environmental policy performance (Esty and Porter 2000). The development of lead markets requires an innovation-oriented and ambitious environmental policy, integrated in a comprehensive innovation and industrial policy (Meyer-Krahmer 1999). Countries that attain the image of pioneers in environmental policy making are more successful in setting global standards (Porter and van der Linde 1995, Jacob and others 2005).

Innovative solar power has promoted the use of renewable energy.

Credit: Frans Ijserinkhuijsen

Lead markets fulfil a range of functions. From an international perspective, they provide marketable solutions for global environmental problems. Lead markets in high-income countries are able to raise the necessary funds for the development of technologies, which may assist them through teething troubles. By demonstrating both technical and political feasibility, they stimulate other countries and enterprises to adopt their pioneering standards. From a national perspective, ambitious standards or support mechanisms may create a first-mover advantage for domestic industries. Furthermore, ambitious policy measures can attract internationally mobile capital for the development and marketing of environmental innovations. Finally, these economic advantages legitimate the national policy-makers, and an ambitious policy provides them with an attractive, influential role in the global arena.

Transition management in the Netherlands

Against a common failure of environmental policy to effectively transform large technological systems, the concept of transition management has been developed in the Netherlands (Rotmans and others 2001, Kemp and Rotmans 2001, Loorbach 2002, Kemp and Loorbach 2003). The concept focuses on "system innovations," which are defined as fundamental changes of technical, social, regulatory and cultural regimes, which, in their interactions, fulfil specific societal needs, such as transport, food, housing, water and energy. A system change requires co-evolution of technologies, infrastructure, regulations, symbols, knowledge and industrial structure. Historical examples of system innovations are the transition from wind-powered to steam-powered ships, or from wood-based energy to coal-based energy. Such system changes typically require a time frame of 30–40 years (Kemp and Loorbach 2003).

Such a long time frame and the necessary encompassing changes are not manageable by conventional governmental steering. Traditional policy making is segmented in specialized departments, and as is the case for most business actors, is rather short-sighted. Transition management is proposed to provide advanced performance in steering system innovations. However, transition management includes no claim to actually plan transitions, but instead aims to influence the direction and speed of transition processes. The process can be divided into four distinct phases:

- creation of an innovation network (transition arena) for a defined transition problem that includes representatives from government, science, business and NGOs;

- development of integrated visions and images about possible transition paths that span 25–50 years, and, based on these visions, derivation of intermediate objectives;
- the execution of experiments and concerted actions according to the transition agenda (experiments may refer to technologies, regulations or modes of financing); and
- monitoring and evaluation of the process, and implementation of the results of the learning processes.

Successful experiments need to be taken up by the policy process and their diffusion promoted.

Several projects have been underway in the Netherlands since 2001 to experiment with this strategy. Though transition management is not expected to yield immediate results, initiatives in the energy sector indicate that the processes have led to:

- more integration of existing policy options and approaches;
- development of coalitions and networks among stakeholders (from 10 in 2000 to several hundred by the end of 2004);
- more investments (from about US$200 000 in 2000 to US$80 million in 2005) including "relabelled" and additional funds; and
- more attention to the issues with a long-term perspective (Kemp and Loorbach 2003).

Improving consideration of the environment in development decision making

Governments pursue a range of different, sometimes even competing or conflicting objectives. While the division of labour among government departments can be effective and efficient, it is less effective for cross-cutting issues, such as protection of the environment. Even worse, environment is often treated as just one more sector to be balanced against other social objectives, rather than providing the foundation on which all life depends. There has been limited progress in moving environmental considerations from the margins of economic and social decision making, but much more needs to be done.

Environmental policy integration

The need to incorporate environmental concerns into the decision making procedures of non-environmental policies has been a constant challenge for better government. Previously, environmental policy integration (EPI) was the responsibility of environmental agencies alone. However, it proved to be difficult to effectively interfere in the policy domains of other departments. Therefore, a number of countries shifted the responsibility for integrating environmental concerns towards the sectors themselves. This means that government departments that previously were opposed to a comprehensive greening of their policies, such as those responsible for transport, industry, energy and agriculture, must become responsible and accountable for their environmental performance (see Box 10.3).

Such an approach can be seen as "governmental self-regulation." It is up to each department to choose the best means for incorporating environmental objectives in its portfolio of objectives, in a consistent national strategy, and to report on the outcomes. For example, many ministries of industry have established eco-industrial parks or industry clusters with advanced waste treatment systems (UNIDO 2000). To make such a shift in responsibilities work, however, there is a need for high-level commitment by cabinet or parliament, or a clear lead by a designated ministry, and also a need for clear and realistic objectives, indicators and benchmarks, as well as for provisions for

---

Box 10.3 Environment in Tanzania's public expenditure review

Tanzania's National Strategy for Growth and Reduction of Poverty 2005–9 (MKUKUTA) cast aside assumptions employed in earlier strategies about the "priority" status of certain sectors and consequently their protected budgets. It promotes an outcome-based approach, opening the doors to cross-cutting concerns, such as the environment, which had previously been marginal. The key to the door was the Ministry of Finance's public expenditure review (PER) system, which revealed how alternative investments contribute to the planned outcomes:

- environmental investments can support health, agriculture, tourism and industry, and contribute to government revenues;
- there has been significant underpricing and very low revenue collection, especially in fisheries and wildlife;
- some environmentally sensitive "priority" sectors spent nothing on environmental management;
- districts responsible for environmental assets received little of the revenue; and
- fixed government budget formats constrain environmental integration.

The PER case was compelling: the 2006 official environment budget was considerably improved, and the general budget format now requires environmental integration.

Source: Dalal-Clayton and Bass 2006

monitoring contents. The Cardiff Process in the European Union can be seen as one model for this type of EPI (Jacob and Volkery 2004).

Policy appraisal and impact assessment

Tools for incorporating environmental concerns into other sectoral policies include strategic environmental assessments (SEA) (Figure 10.4), regulatory impact assessments (EEA 2004, CEC 2004) and other forms of policy appraisal (see Chapter 8). These instruments aim at identifying possible unwanted side effects and conflicts of interests during the formulation of policies. Typically, plans, programmes and policies are assessed against a number of criteria by the government agency itself. While offering great potential for learning and increased transparency (Stinchcombe and Gibson 2001), the findings are rarely used. The United States and Canada were pioneers in introducing environmental assessments for planned policies in the 1970s. SEA was rediscovered by the European Union in the 1990s. However, SEA application is generally limited to plans, policies and programmes that have a direct impact on the environment (World Bank 2005). Generic policies are usually exempted from the need to conduct an assessment of their environmental impacts, although these could be considerable.

Examples include SEAs of multilateral bank plans and programmes, the United Kingdom's integrated policy appraisal and regulatory impact assessments, the European Union's integrated assessment, and Switzerland's sustainability assessment (Wachter 2005, Steid and Meijers 2004). Recently, there has been a trend towards integrating the requirements to assess impacts, such those as on gender, business, SMEs, environment and the budget, in a single, comprehensive procedure or integrated assessment (IA). Initially, the focus of IA was restricted to minimizing costs for business actors and increasing the efficiency of regulation. This form of regulatory IA did not pay much attention to unintended side effects or non-market effects (Cabinet Office 2005). IA aims at analysing a wide array of generic aspects, such as enhancement of competitiveness, support for small and medium enterprises, consideration of gender aspects or consideration of environmental concerns. Such an integrated perspective aims to reveal conflicts between objectives, or to identify win-win solutions. Denmark, Canada, the Netherlands, Finland, Sweden and the United Kingdom have been main forerunners, although it is significant that Poland is now requiring sustainability impact assessments. This trend reflects a growing insight that side effects, interlinkage effects or non-market effects may have severe implications in other policy areas, and, therefore, need to be taken into account.

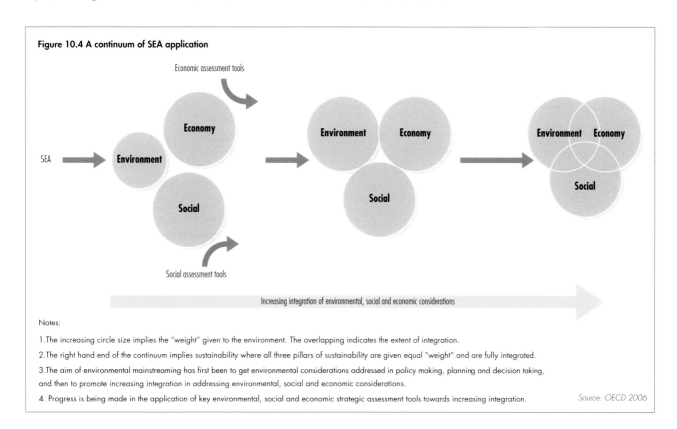

Figure 10.4 A continuum of SEA application

Economic assessment tools

SEA

Environment

Economy

Social

Environment

Economy

Social

Environment

Economy

Social

Social assessment tools

Increasing integration of environmental, social and economic considerations

Notes:

1. The increasing circle size implies the "weight" given to the environment. The overlapping indicates the extent of integration.

2. The right hand end of the continuum implies sustainability where all three pillars of sustainability are given equal "weight" and are fully integrated.

3. The aim of environmental mainstreaming has first been to get environmental considerations addressed in policy making, planning and decision taking, and then to promote increasing integration in addressing environmental, social and economic considerations.

4. Progress is being made in the application of key environmental, social and economic strategic assessment tools towards increasing integration.

Source: OECD 2006

Although IA is a rather generic tool, it has the potential to improve EPI, because it requires ministries or agencies to consider environmental concerns early in the process of policy formulation. Furthermore, these other sectors are required to consult environmental ministries, agencies and relevant stakeholders early in the process. Some initial evaluations of IA schemes, however, demonstrate the possibilities to misuse such approaches to roll back environmental concerns under the rubric of a better regulatory agenda (Wilkinson and others 2004, Environmental Assessment Institute 2006, Jacob and others 2007).

Ultimately, the effectiveness of various forms of environmental assessments will be judged on how they influence policy processes to better manage the environment and enhance human well-being.

Decentralization and delegation

Another innovative approach to integrating environmental concerns in policy making is the inclusion of environmental objectives in controlling systems. New public management gives more discretion to the different units and levels of policy making. In many countries, control by central departments is exerted by adapting controlling mechanisms to delegated governmental units.

There are some generic lessons to be drawn from existing examples of decentralization, and the integration of environmental concerns.

- To close the gap between rhetoric and hard action in sectoral strategies, regular evaluation is necessary. This can be performed by regular reporting to parliament or the cabinet on progress achieved in implementing the plans. In some countries, the national audit office is mandated to audit and report on the environmental and sustainable development performance and financial management of their respective governments. Canada appointed an independent Commissioner of the Environment and Sustainable Development in the Office of the Auditor-General, while New Zealand established an independent Parliamentary Commissioner for the Environment. The evaluation by review through international organizations has proven to be influential in the case of OECD (OECD 2000). The OECD environmental performance reviews also help member states monitor the implementation of their own policies and achievements in meeting their

own targets. Recently, the European Commission initiated a review of the National Strategies for Sustainable Development of its Member States (European Commission 2006).

- The decentralization of responsibility for the environment increases transparency regarding environmental performance and policies of the different governmental sectors.
- The initial momentum for decentralization often comes from a central institution in government, such as the prime minister, the cabinet or the parliament. However, EPI is unlikely to remain prominently on the political agenda of these institutions for long. Therefore, it is necessary for this initial momentum to be used to quickly integrate EPI into regular procedures and institutions of policy making.
- For sustained integration of environmental concerns, it is necessary to couple EPI with the financing mechanisms of government. A number of countries experimented selectively with defining environmental criteria for their spending programmes for infrastructure, and regional and structural development. But, few countries have conducted an in-depth expenditure performance review to reveal spending that is contradictory to environmental objectives (see Box 10.3).

Beyond environmental agencies

Requirements to routinely report on environmental impacts, and appraisal of sectoral policies tend to keep the environment high on the agendas of non-environmental sectors of government. However, for effectiveness these reporting requirements have to be supervised by independent organizations with a strong mandate. In some countries, environment ministries oversee these activities. However, as junior ministries often they cannot prevail over more powerful agencies. In other countries, the responsibility has been shifted to the office of the prime minister. In a few countries (United Kingdom and Germany), national parliaments have set up committees to oversee these activities. Canada and New Zealand mandated the auditors-general to service the parliamentary committees. In some countries, although still underutilized, scientific advisors are assessing environmental policies (and their integration) on a regular basis (Eden 1996) and international policy assessment, comparisons and recommendations have been published in different fields by several research organizations. Environmental ministries do not become obsolete in these approaches, as they have to organize the knowledge base for policy making, provide indicators and data for monitoring and assessment, organize the political process to adopt goals and objectives. Ministries in interested countries could even join forces with the scientific community to utilize experience across the borders, and benchmark environmental performance of different sectors.

It is apparent that environment is moving closer to the core of societal concerns under increasing social pressure upon governments everywhere, and this has already produced a change in the meaning of moving environmental concerns "from the periphery to the core" in decision making. This includes a better understanding of the nature of the existing core of decision making and its drivers, and of the place and role of the environmental issues in it. For too long, the existing core of decision making has been organized around the preservation of a given set of conditions indispensable for the ceaseless accumulation of material wealth. Under that orientation, the environment is necessarily expressed as just another variable of economic policy, implying that nothing more than trade-off decisions are needed. Moving the environment from the periphery to the core of decision making means transforming the core so that eventually the economy and society are reoriented to achieve sustainable environmental quality and human well-being. This reorientation implies major educational, institutional and financial changes.

CONDITIONS FOR SUCCESSFUL IMPLEMENTATION OF A NEW POLICY FRAMEWORK

The drivers-pressures-state-impacts-response (DPSIR) framework is used as a basis for understanding interactions between people and the environment. While the proven problems have often been successfully addressed by targeting a single sector or a single link in the DPSIR chain, persistent problems are more likely to require multisectoral or cross-DPSIR approaches, particularly targeting drivers. The following sections review the types of structural innovations that could form the basis of a more ambitious global policy agenda.

Public awareness, education and learning

Collective learning (Keen and others 2005) and adaptive management (Holling 1978) are management approaches aimed at coming to grips with complexity and uncertainty. Implementers and other stakeholders at different levels are encouraged to collect data and information, and process it in a manner and format that provides feedback and self-learning. Capacity building support is being provided to improve indigenous and/or community-based systems of monitoring, and to relate it to higher levels of information aggregation and decision making. For instance, indigenous knowledge of ecological systems may be included in designing policies, and evaluating the impact of these policies through the use of innovative indicators.

Cotton farmers training centre in Tanzania: feedback from local knowledge will help improve innovation.

Credit: Joerg Boethling/ Still Pictures

Principle 10 of the 1992 Rio Declaration articulates a right to environmental information, decision making and justice. It is often called the "access principle."

While Principle 10 is a very "soft" measure, it has had considerable impact, and has been converted to a "hard" policy in a regional context through the Aarhus Convention, negotiated under the auspices of the UN Economic Commission for Europe (UNECE). Signed in the Danish city of Aarhus in 1998, it became effective in 2001, and by early 2005 had been ratified by 33 countries in Europe and Central Asia. Not only did non-government organizations (NGOs) have an unusually strong influence on the negotiation process, but they have also been given a central role in its operational procedures. Environmental NGOs are represented on the Bureau of the Meeting of the Parties, in follow-up task forces and in the compliance mechanism, which allows the public to submit allegations of non-compliance. Some examples of its provisions are:

- Information has to be made effectively accessible on activities or measures that influence air, water, soil, human health and safety, conditions of life, cultural sites and built structures. For example, each party shall establish a nationwide pollutant release and transfer register (PRTR) on a structured, computerized and publicly accessible database, compiled through standardized reporting.
- Public participation is required in decision making on whether to allow certain types of activities – for example in the energy, mining and waste sectors – and there is an obligation on the decision making body to take due account of such participation, which should also be part of more general decision making on environmental plans and programmes.
- Access to justice is provided for in relation to the review procedures for access to information and public participation, and to challenge breaches in environmental law.

The first report on the status of implementation of the convention indicates that most progress has been made on access to information, a bit less on access to participation and the least on access to justice. This result parallels another study on the implementation of the Rio Principle 10 in nine countries around the world. The convention has the potential to exert influence beyond the UNECE region. It is open to signature by countries outside the region, and the signatories have agreed to promote the application of its principles in international environmental decision making processes and in international organizations related to the environment.

Sources: Petkova and Veit 2000, Petkova and others 2002, UNECE 2005, Wates 2005

For example, the Poorest Areas Civil Society Programme, encompassing 100 of the poorest districts in India, has developed a unique information technology-based monitoring, evaluation and learning (MEAL) system (PACS 2006). With the active participation of more than 440 civil society organizations (CSOs) and 20 000 community-based groups, MEAL synthesizes information from numerous sources, including village profiles and baseline reports, quarterly reports, output tracking, appraisal reports, process reflection, case studies and research documents. The MEAL system has helped to improve programme efficiency, and ensure sharing of knowledge and experiences between participating CSOs and other interested agencies.

Collective learning approaches imply a strong commitment to share information for public awareness and education. It builds public opinion, based on sound and relevant information, leading to participatory decision making, and, ultimately, good governance. Public awareness initiatives may be targeted or broad based. As an example of the latter, the Aarhus Convention establishes rights of the public (individuals and their associations) for access to environmental information, public participation in environmental decision making, and justice (see Box 10.4). Parties to the convention are required to make necessary provisions for public authorities (at the national, regional or local level) to ensure that these rights are effective. South Africa's open information policy is an example of national application of these principles.

Globally, the UN Decade of Education for Sustainable Development is an important initiative to reach out to a broad audience, especially to the younger generation, both in and outside of the school curriculum (UNESCO 2005b). Targeted health and sanitation awareness, coupled with capacity building, empowered poor communities in Kimberly, South Africa, to build sustainable household sanitation (SEI 2004). Similarly the success of the sustainable cities initiative in Curitiba, Brazil was heavily dependent on the awareness building, and involvement of local communities (McKibben 2005).

The environmental performance reviews carried out by international organizations, such as the OECD and the UNECE, and now being prepared by UNECLAC and other UN bodies and organizations at regional level, are important and effective mechanism for strengthening collective learning. Such peer reviews contribute to independent, outside evaluation of the effectiveness, efficiency and equity of environmental policies, with sound, fact-based analysis, and constructive advice and recommendations. They give substance to the goals of accountability, transparency and good governance, and provide a way of exchanging experience and information about best practices and successful policies among countries in a regular and systematic manner (OECD 2000). Peer reviews are very effective in stimulating internal learning, but less effective in conveying learning external to the review area. One way of increasing the learning value is to encourage peer review institutions to do more cross-country comparisons or "benchmarking." This will also

lead to more convergence when it comes to choice
of methodology and terms.

The collective learning approach aligns with the
complex interactions characterizing the ecosystem
approach to environmental management. It recognizes
the need to collect and synthesize information on
ecosystem structure and function, recognize that
different levels in the ecosystem are interrelated and
interdependent, and adopt management strategies that
are ecological, anticipatory and ethical. The concept
of humanity as part of the ecosystem, not separate
from it, is a vital underlying principle of the ecosystem
approach. The health, activities and concerns of local
stakeholders should be viewed as characteristics of
the ecosystem in which they live. It also means that
stakeholders need to be included in decisions that
affect their environment (NRBS 1996).

Monitoring and evaluation

Even where transformative policies are in place and
organizations have been reformed to implement
those policies, it is still necessary to know if the
set goals and targets are being met. Not only
monitoring is needed, but regular assessment and
evaluation in terms of the effectiveness of policies
is important. Statistical departments need to have
their mandates expanded to collect data on policy
implementation. Few countries mandate their national
accounting offices with independent policy evaluation.
International and regional organizations have
developed programmes for policy monitoring and
evaluation, such as the OECD environmental policy
performance reviews (Lehtonen 2005).

Most countries have set up advisory boards, with
experts and stakeholders to provide policy advice

on issues of sustainable development. However,
their mandate and their resources are often limited.
Only a few countries, for example, Austria, France
and Switzerland, have commissioned independent
evaluations of their overall policy performance (Carius
and others 2005, Steurer and Martinuzzi 2005).
While there are some promising steps towards a
systematic and independent policy evaluation beyond
self-reporting, these examples require expansion. Recent
efforts by the European Union, OECD and by UN
agencies to organize evaluation and peer reviews of
national strategies for sustainable development can
bring momentum in the further advancement of such
processes (Dalal-Clayton and Bass 2006, European
Commission 2006). Traditional approaches to
monitoring and evaluation, especially in command-
and-control regimes, have tended to focus on tracking
changes and taking retroactive corrective action. As
a consequence, there has been a resistance from
implementers to report to regulators (Dutzik 2002), and
a tendency to provide only minimal information, often
with emphasis on positive aspects. Even with external
evaluators, who most often spend very short periods
on site, it is difficult to capture the substantive issues.
For persistent environmental problems, indicators need
to be carefully chosen to represent timely change in
underlying drivers.

Organizational reform

Robust organizations are critical for effective
implementation of public policy. In the past two
decades, there has been a diversity of organizational
arrangements. Taking stock is a key component
of evaluation to strengthen effectiveness. Because
environmental problems cut across multiple jurisdictions
and scales, it is necessary to target improvements at
multiple levels.

Global level

The number of organizations, multilateral agreements, agencies, funds and programmes involved in environmental activities has increased significantly since 1972, when UNEP was established by the UN General Assembly (UNGA 1972). The increase has been more evident as a consequence of the follow-up to *Our Common Future* and other international processes. The 1990s was a decade of international conferences, including the Earth Summit in 1992 and global meetings on such issues as gender, population and food. Efforts to enhance system-wide coherence have been a recurrent feature of the governing processes of the evolving United Nations. Chapter 8 contains a diagnosis of global organizational challenges, as well a review of options to improve effectiveness. Reform at the global level is an area of dynamic debate, and crucial to the broader effort to find effective solutions to global environmental problems.

Regional level

At the regional and sub-regional levels, in spite of visible and pressing transboundary environmental issues, there are very few organizational mechanisms that have the capacity to address these complex issues. The European Union is possibly the most advanced, with ambitious agreements and strong enforcement powers of the European Commission. Today, about 80 per cent of the environmental regulations in the member states are rooted in European legislation. The Commission has the right to take action against member states for infringement of European law. There are effective organizational and constitutional means to avoid "a race to the bottom" on environmental standards (CEC 2004).

One example of dealing with a regional issue is acid rain (see Box 10.5). The Convention on Long-range Transboundary Air Pollution, signed in 1979 under the auspices of UNECE, spans from the Russian Federation in the east, to Canada and the United States in the west. A soil protection policy is also being formulated.

The Central America Commission for Environment and Development (CCAD) is headed by ministers who are political leaders in the region, with linkages to other ministers in charge of, for example, agriculture, coastal resource management, urbanization, gender, biodiversity conservation, environmental health, food security, economy, marketing, disaster mitigation, education, tourism, energy and mines, and poverty alleviation. They ensure policy synergies, and harmonize the legal frameworks in the region. There is good experience built up by environmental ministries working together with local government and civil society on interlinkages and cross-cutting issues in the Meso-American Region, which includes Mexico and Central America. Projects include the Meso-American Biological Corridor and the Meso-American Barrier Reef. In Africa, the African Ministerial Conference on the Environment (AMCEN), established in 1985, is a permanent forum where environment ministers meet on a regular basis to discuss environmental topics. ASEAN has no regional environment agency, preferring to work through standing committees. The Commission for Environmental Cooperation (CEC), was created under the North American Agreement on Environmental Cooperation as an environmental "side agreement" to the North American Free Trade Agreement between Mexico, United States and Canada. The CEC's role is to address regional environmental concerns, help prevent potential trade and environmental conflicts, and to promote the effective enforcement of environmental law.

Box 10.5 Acid rain

One of the early defining activities of European environmental regulation was action on the sulphur emissions that contribute to acid rain and damage human health. Removing the worst of acid rain has been a major success story for collaborative European environment policy (see Chapters 2 and 3).

Europe began a programme to address acid emissions after the Stockholm environment conference in 1972. The 1979 UNECE Convention on Long-range Transboundary Air Pollution (CLRTAP) promoted region-wide monitoring and assessment, and created a forum for negotiating regulatory standards. Initial reductions were based on arbitrary reductions from a common baseline. By the late 1980s, Europe had adopted an integrated approach, addressing the problems of acidification, eutrophication and tropospheric ozone. From 1994, regional reduction protocols all addressed these problems through a "critical loads" approach, regulating emissions of sulphur dioxide, nitrogen oxides, ammonia and non-methane volatile organic compounds to improve the protection of the most vulnerable ecosystems. Such an approach was made possible by agreement on a common monitoring system, a political commitment to target critical loads, and decision support tools that enabled negotiators to evaluate alternative regulatory schemes in an integrated manner.

Today, the emission targets set by the European Union are somewhat stricter than those of the CLRTAP. Acid deposition is expected to continue declining, due to the implementation of the NEC Directive and corresponding protocols under the CLRTAP. Based on current projections, EU sulphur dioxide emissions will drop by 51 per cent between 2000 and 2010, when they will be lower than at any time since about 1900.

Sources: EEA 2005, Levy 1995, UNECE 2007

However, such regional organizational arrangements are not available everywhere, or, in some cases where available, they are prevented from functioning effectively by vested interests. East Asia, for example, does not have an organizational mechanism to address transboundary environmental issues, such as acid rain or dust and sandstorms in spite of these problems assuming serious dimensions.

National level

National governments and agencies continue to be the nodal points in negotiating, implementing and enforcing environmental policies. Despite the emergence of non-state actors, and the transfer of some responsibilities to the global, regional, subnational and local levels, governments still control major resources for implementing environmental policies. Most countries have a basic organizational framework for environmental policies, such as environmental ministries, basic laws and agencies to monitor and enforce environmental standards. However, effective implementation at the national level remains a challenge in many countries. Most countries have formulated environmental plans or strategies for sustainable development, with varying degrees of stakeholder participation and scientific rigour (Swanson and others 2004).

A relatively smaller number of countries have made conscious efforts to link their environmental policies with major public budgets. Norway and Canada review their budgets to ascertain the environmental impacts of proposed public spending (OECD 2001b, OECD 2004). The European Union requires an environmental impact assessment for spending on national projects from the structural and regional funds. Despite these examples, the organizational links between the major public budgets and environmental policies remain weak in most countries.

Some countries have established organizations at the national level to facilitate the use of market forces to address environmental problems. As seen in Chapter 2, carbon emissions trading has particularly benefited from these institutional arrangements. While the shift in taxation with a higher burden on energy-intensive industries has encountered stiff resistance from vested interests, ecological tax reforms have stimulated innovation and new employment opportunities.

At the national level, changes in attitudes of governments have been observed, with greater emphasis on stakeholder participation for solving environmental problems. This has been demonstrated by the participation of stakeholders,

Mechanisms to address transboundary environmental issues, such as acid rain or dust and sandstorms, are still not in place, despite these problems assuming serious dimensions.

Credit: sinopictures/viewchina/ Still Pictures

such as the representatives of civil society and the private sector, in joint fora with governments, UN agencies and other international organizations. Some countries have formalized the process of participation. For example, legislation has been passed in Viet Nam and Thailand to include indigenous people in forest management (Enters and others 2000). The Brazilian national system of conservation units recognizes community rights to use and management in a variety of zones, such as conservation areas, extractive reserves and protection forests (Oliveira Costa 2005). Decentralization and the emergence of innovative local governments offer opportunities for social learning and the possibility of scaling up successes (Steid and Meijers 2004, MOEJ 2005).

Emerging organizing principles

Experience over the last few decades from initiatives at the global, regional, national and local levels to address complex environmental and inter-sectoral issues demonstrates some generic principles for public policy formulation and implementation. These include:

- decentralizing power to lower levels of decision making, where it is more timely and meaningful – the subsidiarity principle;
- transferring authority to other stakeholders who have a relative advantage, stake and competence in assuming the responsibility;
- strengthening and reinforcing the normative capacity of agencies operating at a higher level;
- supporting and facilitating the active participation of women, local communities, marginalized and vulnerable groups;
- strengthening the scientific base of monitoring ecosystem health; and
- applying an integrated ecosystem monitoring approach.

Decentralizing power

The principle of subsidiarity states that the higher entity ought not do what the lesser entity can do adequately unless it can do it better. The principle can be used to regulate the exercise of existing competencies, and guide the allocation of competencies. In the context of European integration, both functions can be found. Networks of local authorities, such as the International Council for Local Environmental Initiatives (ICLEI), have also served to shape better practices, for instance in water use and guidelines for green procurement.

Transferring authority to stakeholders

In several countries, a negotiated approach has been tested to engage a wide range of stakeholders in not only planning and consultations, but also in decision making, for example over management of river basins, forests and other natural resources (see Box 10.6). As described in Chapter 4, the negotiated approach, being decentralized and flexible, is effective in making water available at the grassroots level to areas distant from the main water source or delivery system. The negotiated approach empowers local water users, through the creation of formal and informal water management institutions, and the formalization of

---

Box 10.6 The changing role of the state

For many countries, the middle of the 1980s saw the beginning of a transition in the role of the state, its core responsibilities and how it should manage them, with the emergence of various social actors. The changing role of the state led to further political decentralization, economic liberalization and privatization, as well as greater participation of civil society in decision making.

First, the transition translated into devolution of power from the central to the local and provincial governments. About 80 per cent of developing countries are experimenting with some form of decentralization. In virtually all countries, responsibility for local environmental issues, such as air and water pollution, waste management, and land management, belongs to local governments and municipalities. Decentralization reforms range from empowerment of elected local governments with natural resources mandates in Thailand, to the financing of village committees in Cambodia, and emerging co-management arrangements for water and forests in Viet Nam and Laos PDR. While cross-country experience suggests that the impact of decentralization on poverty and the delivery of public services is not straightforward, it is likely to have a positive impact on governance, participation and the efficiency of public service delivery.

Second, on the economic front the erosion of state power translated into large programmes of privatization of state-owned companies, worldwide. The private sector has since become one of the critical actors in facing global challenges such as climate change, and a primary stakeholder in the implementation of flexibility mechanisms allowed by the Kyoto Protocol to the UNFCCC, notably of projects under the Clean Development Mechanism and emissions trading.

Finally, the transition opened the door to civil society and its organizations, especially NGOs, to participate as active stakeholders in political, social, economic and environmental governance. For example, in Porto Alegre, Brazil, budgeting processes now involve consultations with civil society groups. In the United Kingdom, the Women's Budget Group has been invited to review government budget proposals. The Forest Stewardship Council brings together environmental groups, the timber industry, forest workers, indigenous people and community groups in certifying sustainably-harvested timber for export. More than US$7 billion in aid to developing countries now flows through international NGOs, reflecting and supporting a dramatic expansion in the scope and nature of NGO activities. In 2000, there were 37 000 registered international NGOs, one-fifth more than in 1990. More than 2 150 NGOs have consultative status with the UN Economic and Social Council, and 1 550 are associated with the UN Department of Public Information.

Sources: Anheier and others 2001, Dupar and Badenoch 2002, Furtado 2001, Jütting and others 2004, Work undated, World Bank 1997

The role of women in environmental management and sustainable development is vitally important and increasingly recognized. Above, women planting trees in Kenya as part of the Green Belt Movement.

Credit: William Campbell/ Still Pictures

existing knowledge and vision. Simultaneously, it is based on an ecosystem approach and wise use of ecosystems. Scaling up of local initiatives and bringing them to the higher decision making levels is one of the other characteristics of the negotiated approach (Both ENDS and Gomukh 2005).

Strengthening higher-level agencies

Transboundary environmental problems, such as acid rain, haze pollution, desertification, climate change, ozone depletion and loss of migratory species, and the management of shared natural resources pose a unique set of challenges to environmental governance. They highlight the need for decision making processes that go beyond national borders, and illustrate the necessity for creating mechanisms to address these issues at regional and global levels. This process has created new functions for international organizations, as nation states increasingly delegate some of their functions upwards to regional or international organizations to deal with transboundary environmental problems.

Through community legislation, action programmes and 30 years of standard setting, the European Union has established a comprehensive system of environmental protection. This covers issues that range from noise to waste, from conservation of the natural habitat to car exhaust fumes, from chemicals to industrial accidents, and from bathing water to an EU-wide emergency information and help network to deal with environmental disasters, such as oil spills or forest fires. The European Environment Agency (EEA) was set up to help achieve improvement in Europe's environment through the provision of relevant and reliable information to policy-makers and the public. The legislative powers, however, remain with the European Union. Several regional organizations elsewhere have initiated similar although limited, efforts, such as the North American Commission for Environmental Cooperation, the Ministerial Conference on Environment and Development in Asia and the Pacific, and the African Ministerial Conference on the Environment.

Facilitating active participation

Leading up to the 1992 United Nations Conference on Environment and Development, women organized themselves worldwide to have their voices heard in environmental decisions. This resulted in the recognition of women as one of the nine major groups in Agenda 21 for their roles in environmental conservation and sustainable development. In many related processes that followed, such as the meetings of the Commission on Sustainable Development, women fully participate. In these efforts, women often cooperate with other civil society groups, such as indigenous peoples, trade unions and youth, resulting in negotiations that better reflect the interests of local communities, and marginalized and vulnerable groups. As described in Chapter 7, these global processes reflect similar initiatives at regional and national levels.

Strengthening the scientific base of monitoring ecosystem health

Over the last two decades, the tools and techniques for measuring specific environmental parameters have improved considerably. However, the science of understanding ecosystems and profiling ecosystem health at various spatial scales and for different policy domains is still comparatively nascent. The ecological relationships among various environmental parameters are complex. Added to this complexity are the human, social and economic dimensions of ecosystems. It is important to establish meaningful targets and indicators for these dimensions, such as the 2010 biodiversity targets, the Human Development Index and new indicators of ecosystem well-being.

Resilience analysis encourages monitoring systems to detect the proximity of the system to a critical threshold, the amount a system can be disturbed before crossing a threshold, and the ease or difficulty of returning to a previous state once the threshold has been crossed (Walker and others 2004). Measuring these key parameters may be the most cost-effective way of monitoring ecosystem health.

Changes in ecosystem functions have consequences for different sectors of society and for distant generations in terms of human well-being (see Chapter 7). From a policy perspective, it is relevant to track the degree to which these ecosystems can maintain their full capacity to function. The ecosystem health approach serves as a model for diagnosing and monitoring the capacity for maintaining biological and social organization, and the ability to achieve reasonable and sustainable human goals (Nielsen 1999). Yet, ecosystem health is not well monitored in most parts of the world.

Integrated ecosystem monitoring

The climate negotiations over the last decade, as discussed in Chapter 2, have clearly shown the links between a sound scientific basis for policy formulation and the politics of decision making. The science of understanding and profiling ecosystem health and its relationship to persistent environmental problems is invariably going to take some time. In the meantime, a practical approach to integrated ecosystem monitoring that enables policy and decision making is imperative. An integrated monitoring framework will include at least the following steps: identifying ecosystem goals, developing specific management objectives, selecting appropriate and measurable ecosystem indicators, monitoring and assessing the state of the environment, using chosen indicators, and taking appropriate action.

The effectiveness of participatory monitoring and learning is increasingly being recognized. However, this implies that stakeholders at various levels need flexibility to monitor and learn in the method and style with which they are comfortable, and which is most meaningful to them (see Box 10.7). The challenge then becomes how to rationalize and aggregate various kinds of data and information in a way that it is relevant at decision making levels – nationally, regionally or globally. For instance, how will the indigenous practice of monitoring a sacred grove relate to MDG 7 or the Convention on Biological Diversity? At the same time, the

---

Box 10.7 Monitoring implementation of the UNCCD in Niger

Niger, like the other countries that have ratified the UN Convention to Combat Desertification (UNCCD), has committed itself to produce periodic national reports that would take stock of progress made in the framework of the UNCCD implementation. Land degradation processes and dynamics are the subject of regular monitoring in Niger. In the framework of the implementation of the National Plan of Action to combat desertification (PANLCD/GRN), one strategic orientation is to watch and monitor desertification. Among other actions, systematic monitoring of the dynamics of land degradation provides an early warning system to better develop programmes to mitigate the effects of drought and desertification.

The rate of natural resources degradation is assessed especially through field projects and programmes, such as the Desert Margins Programme, which is collecting data on:

- an inventory of endemic, extinct or threatened plant species;
- features of domestic plant and animal biodiversity;
- features of the productive capital (land, vegetation and water), the climate and the socio–economic component at several scales;
- improvement of the understanding of pastoral areas' degradation mechanisms;
- improvement of knowledge regarding wetlands degradation mechanisms; and
- the fight against erosion, and soil fertility management.

Also, in the framework of the Project to Support Training and Assistance in Environment Management (PAFAGE in French) financed by Italy, a National Environmental Information System (SIEN) was set up.

Source: CNEDD 2004

need for capacity building at different levels and technology cooperation needs to be recognized and acted upon.

Defining the frequency of monitoring can also be complex. The life cycles and time spans of environmental and ecosystem changes are much longer than political mandates and generally accepted project or programme time frames. As a consequence, political and programme organizations avoid or delay decision making, since the results may not be visible during their tenure. At the same time, there is also an overload of environmental information contributing to the "noise" in environmental decision making. Ideally, minimal information at different levels has to be available at the right time in a simple format for decision making.

A monitoring protocol that provides flexibility at the lower levels and yet is able to capture information and knowledge for policy and decision making at global, regional and national levels still needs to be developed. At the global level, a comprehensive review of the environment is required about every 3–5 years. This is provided by a range of organizations and processes, including the GEO process. However, a practical approach to integrated ecosystem monitoring and early warning is yet to be incorporated in these initiatives.

Financing the environmental agenda
Financing programmes to address conventional environmental issues, for example pollution control and groundwater depletion, is possible by strict

implementation of "polluter pays" or "user pays" policies. It is also possible through public financing, if the source of the problem is harder to identify or the nature of the environmental good suggests this as the most appropriate approach.

However, financing programmes to eliminate persistent environmental problems is much more complex, since the changes needed involve most of society. There is no single polluter or single pollutant, no single group of identifiable "victims" and often no simple cause-and-effect relationship or dose-response equation (as the problem stems from the "driver" level in the DPSIR framework). Entire sectors, international relationships and the global economy may be involved. While grant funding is limited, capital for investment and loans is currently easily available globally. The limitations are set by higher risks and lower returns on investments in the developing countries that need it most.

There is room for mobilizing financial resources to manage conventional and persistent environmental problems. Agenda 21 (see Chapter 33, Article 13) clearly articulates that financing actions aimed at sustainable development must come from each country's own public and private sectors (UNCED 1992). This has been reaffirmed in several other international instruments, including in the Monterrey Consensus, the final document of the International Conference on Financing for Development (UN 2002b). Several studies have shown that there may be win-win opportunities in phasing out subsidies. For example, an IEA study of eliminating energy subsidies in eight developing countries concluded

Innovative approaches for raising funds for the environmental agenda have been initiated. Above, the Ngorongoro Conservation Area in Tanzania involves the conservation and development of the area's natural resources; the promotion of tourism; and the safeguarding and promotion of the interests of the Maasai people.

Credit: Essling/images.de/ Still Pictures (left); McPHOTO/ Still Pictures (right)

that their annual economic growth would increase by over 0.7 per cent, while CO_2 emissions would go down by nearly 16 per cent (IEA 1999).

Public sector budgets

Countries may have room for increasing the level of government spending on environment (Friends of the Earth 2002). A modest increase would generate significant additional resources provided adequate priority is accorded to environmental issues in national budgets. For example, in Asia and the Pacific, the Asian Development Bank (ADB) has suggested that developing countries allocate at least 1 per cent of GNP to meet their financial requirements for environmentally sound development. At 1 per cent, the region's domestic resource contribution would be about US$26 billion/year (UNESCAP 2001), compared to defence budgets that range up to 6 per cent of GNP (ADB 2001). The European Commission's thematic strategy on air pollution in the EU member states is expected to give a positive return ratio of at least 6:1 (European Commission 2005).

Promising innovative approaches in raising additional funds for a new environmental agenda have also been initiated. Green budgeting, the creation of conservation funds, the introduction

---

Box 10.8 Use of market-based instruments in Europe

The use of environmental taxes and charges has widened since 1996, with more taxes on CO_2, sulphur in fuels, waste disposal and raw materials, and some new product taxes. Only a few tax rates have originally been set on the basis of an assessment of environmental costs as was done for the landfill tax and levy on quarrying of sand, gravel and hard rock in the United Kingdom.

At the regional level, emissions trading has become the instrument highest on the political agenda, with the adoption of the EU Emission Trading Directive, for reducing CO_2 emissions, its incorporation into national laws and the establishment of national emissions allocation plans. The trading system started operating in 2005. There are a number of other trading schemes already in operation, including national emissions trading schemes for CO_2 in Denmark and the United Kingdom, and for NO_x in the Netherlands, certificate trading for green electricity in Belgium and transferable quotas for fisheries management in Estonia, Iceland, Italy and Portugal.

A range of other instruments are either planned or under serious consideration, notably pricing policies for water by 2010 under the EU Water Framework Directive, road charging systems, and the increased use of trading certificates for green electricity. These and other initiatives suggest that the use of market-based instruments is likely to increase in coming years, possibly as part of wider initiatives on environmental tax and subsidies reforms.

Source: Ministry of Environment, Norway 2005

The initially controversial congestion charge introduced in 2003 by the City of London, turned out to be very successful within a year (15 per cent less traffic in the charging zone and 30 per cent reduction in traffic delays).

Credit: Transport for London http://www.cclondon.com/signsandsymbol.shtml

of economic instruments such as user fees and charges, taxation and other forms of payments for the use of ecosystem goods-and-services (see Box 10.8), are among instruments that have been applied sporadically in various countries (ADB 2005, Cunningham and Grabosky 1998). A challenge has been to ensure that revenues collected are reinvested into the resource base, or support other ecosystems (cross-subsidization) rather than being diverted to other non-environmental purposes. Certain instruments, such as carbon taxes, that have a potentially significant impact on industry and national competitiveness, have been less prominent. To date, carbon taxes have only been introduced in about 12 countries worldwide, and their wider adoption has been a very slow process (OECD 2003).

The use of market-based instruments in environmental policy has gained ground substantially in Europe, including countries in Central and Eastern Europe, since the mid-1990s, especially in the areas of taxes, charges and tradeable permits. Comprehensive systems of pollution charges for air and water are being implemented, although the rates tend to be low, because of concerns about people's ability and willingness to pay. Several countries have also introduced resource use and waste taxes. Progress is being made on the wider use of taxes and charges on products, notably for beverage cans and other packaging.

Scandinavian countries and the Netherlands, which started early on environmental tax reform, remain at the forefront of developments. Germany and the United Kingdom have made much progress since the late 1990s. Measures are mainly taken at national or federal level, but increasingly instruments are being applied at lower levels, for example, resource taxes in Flanders and Catalonia and congestion charges in some cities, such as London, and, albeit more modest, Rome and Oslo.

Green taxes and charges

Approaches such as ecological tax reform and "tax shift" have been tried, whereby taxes on energy use and the consumption of other resources are increased while corresponding reductions are made on income tax. When introduced gradually and in ways that are revenue-neutral and easy to administer, such approaches can encourage environmentally-conscious consumption patterns without causing significant negative social distribution effects (Von Weizsäcker and Jesinghaus 1992). Some countries have attempted new ways of raising revenues, including through ecotourism. For example, the Protected Areas Conservation Trust in Belize, in Central America, receives most of its revenue from an airport tax of about US$3.75, paid by all visitors upon departure, together with a 20 per cent commission on cruise ship passenger fees. The British overseas island territory of Turks and Caicos designates 1 per cent of a 9 per cent hotel tax to support the maintenance and protection of the country's protected areas (Emerton and others 2006).

Payment for ecosystem services

Ecosystems such as forests, grasslands and mangroves provide valuable environmental services to society. They include provisioning services that furnish food, water, timber and fibre; regulating services that affect climate, floods, disease, wastes and water quality; cultural services that provide recreational, aesthetic and spiritual benefits; and supporting services, such as soil formation, photosynthesis and nutrient cycling (MA 2003). Biodiversity continues to underpin food security and medicinal goods. Unfortunately, current markets fail to reflect the value of such ecosystems and ecosystem services, creating a "mismatch between market and social prices" (UNEP and LSE 2005, Canadian Boreal Initiative 2005). As a result, ecosystem services are often viewed as free public goods by their beneficiaries. The combined effect results in overexploitation of ecosystems.

A new approach, called "payments for environmental (or ecosystem) services" (PES), attempts to address this problem. PES schemes pay those who engage in meaningful and measurable activities to secure the supply of ecosystem services, while the beneficiaries of the services pay to secure the provision of the services. Many PES schemes have originated in developed countries, particularly in the United States, where it is estimated that the government spends over US$1.7 billion yearly to induce farmers to protect land (USDA 2001). While the conservation goals may be laudable, the trade distorting nature of subsidies should also be considered. In the developing world,

Costa Rica, Brazil, Ecuador and Mexico have pioneered PES schemes to preserve freshwater ecosystems, forests and biodiversity (Kiersch and others 2005). The Wildlife Foundation is securing migration corridors on private land in Kenya through conservation leases at US$1/1 000m²/year (Ferraro and Kiss 2002).

Combined solutions

Three main markets for ecosystem services are emerging:

- watershed management, which may include control of flooding, erosion and sedimentation, protection of water quality, and maintenance of aquatic habitats and dry season flows;
- biodiversity protection, which includes eco-labelled products, ecotourism and payments for conservation of wildlife habitat; and
- carbon sequestration, where international buyers pay for planting new trees or protecting existing forests to absorb carbon, offsetting carbon emissions elsewhere.

Markets for carbon reduction credits are growing rapidly. From US$300 million in 2003 (IFC 2004), they are projected to rise to US$10–40 billion by 2010 (MA 2005). The World Bank alone had nine carbon funds amounting to US$1.7 billion by 2005. A concerted focus on four areas – carbon sequestration, landscape beauty, biodiversity and water – would help to address rural poverty (UNEP and LSE 2005).

While it is widely recognized that market failures need to be corrected, they are not necessarily solved through market solutions alone. A combination of market-based mechanisms and regulatory structures is often needed for markets to work successfully. The cap-and-trade model in the case of carbon emissions is an example of a regulatory framework defining overall emission limits before a market for emission credits could be established (UNEP and LSE 2005).

Financing the bottom of the pyramid

The new approaches to generating additional financial resources, especially through market-based and economic instruments, often have been possible because of an untapped willingness to pay for ecosystem services and environmental quality. For water, studies have shown that the poor often pay more per litre for unsafe, inconvenient and unreliable supplies than the rich pay for safe, publicly-funded piped supplies. Through multiple mechanisms, such as subsidizing bank lending rates, group lending schemes, and combining subsidies with user contributions, there are indications of willingness to pay, even at low level of income, for example in the renewable energy sector (Farhar 1999). Improved support systems for access to credit and markets are needed for the poor to participate.

Managing environmental resources and encouraging conservation efforts through mechanisms that generate employment and revenues in many diverse sectors, such as forest management, biodiversity conservation and investment in sustainable energy projects, have proved effective. Through the Rural Energy Enterprise Development (REED) initiative in Africa, Brazil and China, UNEP, in partnership with the United Nations Foundation and several NGOs, provides early-stage funding and enterprise development services to entrepreneurs who have helped build successful businesses in the supply of clean energy technologies and services to rural and peri-urban areas (UNEP 2006c). Such initiatives have demonstrated that even small-scale financial resources can trigger entrepreneurship and employment generation through environmentally-sound activities. Equally important is contribution to economic diversification and the creation of new markets, especially in slow-growth and poor countries and for local communities, for example women supported by conservation and income generation projects (Jane Goodall Institute 2006). Microfinance and credit for micro-, small- and

---

Box 10.9 Documented returns on environmental investment

Many large economic sectors depend heavily on natural resources and ecosystem services, including agriculture, timber and fisheries. Therefore, investment in protecting environmental assets has the potential to generate tangible economic returns. Pearce (2005) reviewed 400 efforts to quantify such returns. Using conservative assumptions, the following benefit-cost ratios were documented:

- Controlling air pollution: 0.2:1 – 15:1
- Providing clean water and sanitation: 4:1 – 14:1
- Mitigating natural disaster impacts: up to 7:1
- Agroforestry: 1.7:1 – 6.1:1
- Conserving mangrove forests: 1.2:1 – 7.4:1
- Conserving coral reefs: up to 5:1
- Soil conservation: 1.5:1 – 3.3:1
- National parks: 0.6:1 – 8.9:1

Under alternative assumptions, taking into account longer-time frames and broader impacts on poor populations, even higher rates of return were found.

Source: Pearce 2005

medium-sized enterprizes, particularly for those headed by women, have proven to be important means of enhancing access to credit and nurturing small-scale productive activities, especially in rural areas.

Global funding

Several financial mechanisms channelling grant funds have emerged at the international level, including the GEF. Typically, these address problems of global concern (global commons or public goods, such as clean air and biodiversity). There are many areas of environmental stress or degradation, however, where resources can only be mobilized at the domestic or local level. Often a financing scheme can be developed where local resource conservation can pay for itself in the long run, but local communities or domestic financial sources are not in a position to make the initial seed investment (see Box 10.9). In such cases, international loan or grant financing can be prudently utilized for domestic development purposes to "seed the dynamics." In addition to traditional sources of finance, there are many new or revamped mechanisms, such as debt-for-nature swaps, the Clean Development Mechanism, emissions trading, and attempts to create international funds for global public goods such as rainforests and biodiversity.

For many countries, attracting part of the foreign direct investment (FDI) to environmental management is a promising option. Though FDI is largely concentrated in a handful of fast-growing countries, especially in Asia, initiatives by the private sector, including through corporate social responsibility (CSR) and environmental responsibility, have been expanding in many parts of the world. CSR and corporate financing of certain social and environmental activities have been encouraged by global initiatives that have stimulated companies to report not only on their economic activities, but also on their social and environmental performance (GRI 2006 and Box 10.10).

There are some emerging but still controversial proposals, which include proposals for an aviation fuel tax (a long-standing historical omission), and a tax on international currency transactions. Air travel accounts for 3 per cent of global carbon emissions, and it is the fastest growing source of emissions (Global Policy Forum 2006). The IPCC expects air travel to account for 15 per cent of all carbon emissions in 2050 (IPCC 1996, IPCC 1999). In 2000, the European Parliament's Economic and Monetary Affairs Committee confirmed its support for a recommendation to allow the member states to impose a tax on domestic and intra-EU flights (Global Policy Forum 2000).

At the international level, the Initiative against Hunger and Poverty, bringing together Brazil, Chile, France, Germany and Spain has made various proposals on innovative mechanisms of public and private financing, including a proposal for a tax (solidarity levy) on air travel tickets to finance action against hunger and poverty. The initiative received support from 112 countries at the Summit of World

---

Box 10.10 Value at risk revisited

In April 2006, then UN Secretary General Kofi Annan launched the Principles for Responsible Investment (PRI) after ringing the opening bell at the New York Stock Exchange. Six months later, it had 94 institutional investors from 17 countries representing US$5 trillion in investments.

The launch of the principles created the first-ever global network of investors looking at addressing many of the same environmental, social and governance issues as the UN is tasked to address. One of the goals of the PRI community is to work with policy-makers to address issues of long-term importance to both investors and society. Investors representing more than 10 per cent of global capital market value have, therefore, sent the strongest of signals to the marketplace that environment, social and good governance issues count in investment policy making and decision making.

The PRI has evolved because investors have recognized that systemic issues of sustainability are material to long-term investment returns. Since large investors are becoming almost fully diversified, they recognize that the only way they can deliver for their beneficiaries, often pension holders, is to help address systemic issues in the market through shareholder engagement, transparency and better analysis of long-term sustainability risks and opportunities that can affect investments.

But, investors also need help from policy-makers. There are a range of areas where policy-makers could create the necessary environment that would encourage investors to take longer-term views on environmental, social and governance issues. Mandatory disclosure of environmental performance is one such area. Once investors are able to assess the risks involved in various activities, they are able to put pressure on companies to address those risks. But they are unable to do this if they are unaware of what the company is doing. Mandatory disclosure regimes level the playing field, and allow investors to take action when required.

Source: UNEP 2006d

Consumption patterns and global interdependence have contributed to growth in shipping and liberalization of trade.

Credit: Ngoma Photos

Leaders for Action against Hunger and Poverty, held in New York in 2004 (Inter Press Service 2005, UN 2005a) and by 2006, had gained enough momentum to be transformed into an international facility for purchasing medicines. Although many countries have expressed interest, there is a widely shared view that any proposed schemes involving taxes would best be applied nationally but coordinated internationally (UN 2005b).

A tax of about US$6/passenger, with a US$24 surcharge for business class, would generate about US$12 billion a year, about one-fourth of the annual funding shortfall for meeting the Millennium Development Goals (UN 2005c). In 2006, France initiated an additional tax, from US$2.74 for economy class to US$27.40 for business class on national and European flights. On intercontinental flights the tax rises to US$51. The tax is expected to raise about US$266 million a year. In addition to channelling funds for the International Drug Purchase Facility (IDPF-UNITAID), countries may also be interested in joining the initiative to raise funds for environmental purposes (UNITAID 2006).

Tapping international trade

The potential of international trade as a source of finance for sustainable development has been stressed in numerous international fora and instruments (UN 2005b, UN 2002b, WTO 2001). Liberalization of trade in goods-and-services of

interest to developing countries can generate additional financial flows totalling about US$310 billion yearly (UNCTAD 2005). Realizing this potential will depend on success in achieving a rules-based, open, non-discriminatory and equitable multilateral trading system, as well as meaningful trade liberalization that benefits countries at all stages of development.

Estimating the needed resources

Estimates by the World Health Organization (WHO) of the costs and benefits of meeting the MDG targets for water and sanitation total about US$26 billion, with benefit-cost ratios that range from 4 to 14 (Hutton and Haller 2004). Different provisional estimates prepared for the World Bank, though putting the costs at twice the WHO estimates, still result in a benefit-cost ratio of 3.2 to 1, and could save the lives of up to 1 billion children under five years of age from 2015–2020 (Martin-Hurtado 2002). Climate change not accounted for, the sum required over the next 15–20 years to meet the MDG target for ensuring environmental sustainability (MDG 7) is probably between US$60 billion and US$90 billion yearly (Pearce 2005). Comparatively, OECD countries' spending on producer support in agriculture was about US$230 billion in 2000–2002 (Hoekman and others 2002).

For Asia and the Pacific, ADB estimated the annual investment costs required to achieve environmentally sound development based on two scenarios. Under a business-as-usual scenario, the cost would be US$12.9 billion yearly. Under an accelerated progress scenario – one under which developing countries in the region implement the best practices of OECD countries by 2030 – the cost would be US$70.2 billion yearly. A halfway point set between the high and the low estimates would be around US$40 billion yearly (UNESCAP 2001). In addition, repairing the damage done to the land, water, air and living biota was estimated at US$25 billion yearly. Taking into consideration the total financial resources needed and the present level of spending, the financing gap to attain sustainable development in 1997 was about US$30 billion yearly (Rogers and others 1997). Comparatively, military expenditures in the same period (1997) for Central Asia, East Asia and Southeast Asia were estimated at US$120.9 billion (SIPRI 2004).

The cost of inaction

Although there are real costs associated with implementing the measures that will improve the likelihood of successful policy innovation, there are also costs associated with inaction. Both ex-post evaluations of the costs of ignoring warnings as well as scenarios on the costs of global environmental change show that action now is cheaper than waiting for better solutions to emerge. For climate change, for example, our knowledge on the costs of inaction portrays a worrying picture, even while immediate measures are affordable (Stern 2007). Several studies have attempted to measure the effect of the burden of morbidity and mortality due to various environmental causes, in terms of loss of disability adjusted life years (DALYs). Turning DALYs into dollar value produces a global estimate of human capital damage due to environmental causes of over US$2 trillion/year for developing countries alone (Pearce 2005). Using a more conventional income per capita value for developing countries, the total loss of DALYs in the developing world would still be US$200 billion yearly (Pearce 2005). The same studies indicate a significant difference in environmental DALYs in developing relative to developed regions, with the highest cost in developing countries, as a result of greater exposure to environmental damage (Pearce 2005).

Through retrospective analyses of 14 different case studies of the cost of inaction or delayed action to reduce exposure to hazardous agents, the European Environment Agency (EEA 2001) demonstrated that the costs of implementing environmental policy measures are routinely overestimated. As the report indicates, the Netherlands Ministry of Housing and Social Services estimated that the potential benefits of an earlier ban on asbestos in 1965 (compared to the actual ban in 1993) would have saved some 34 000 premature deaths and some US$24 billion in building clean-up and compensation costs. The estimated long-term cost of asbestos to Dutch society was calculated at 56 000 deaths and US$39 billion over the period 1969–2030 (EEA 2001).

All these studies indicate that inaction, delayed action and inappropriate action not only result in higher costs, but unfairly shift the burden of paying for such costs to future generations, in contradiction to the principle of intergenerational equity. Such distributional issues need to be given greater weight in the decision making processes and the estimates of the costs of taking action.

CONCLUSION

Adopting the future policy framework outlined in this report is an opportunity for renewal in the way individuals think about the environment and its impact on their well-being, in the way national decision-makers treat the environmental dimensions of their portfolios, in the way financial resources are mobilized for environmental problems, and in the way the global community organizes itself in the UN system and specialized agencies. Hard to manage, persistent environmental problems will demand complex solutions, and it can be expected that the solutions chosen will, in turn, create new and possibly even more complex problems in their wake. However, the costs of inaction in many of the environmental problems with proven solutions have already become evident. The costs of inaction in dealing with the emerging set of persistent environmental problems are far greater – directly impinging on the future ability of ecosystems to support people.

Therefore, the new environmental policy agenda for the next 20 years and beyond has two tracks:

- expanding and adapting proven policy approaches to the more conventional environmental problems, especially in lagging countries and regions; and
- urgently finding workable solutions for the emerging environmental problems before they reach irreversible turning points.

The latter solutions will generally lie in the "driver" portion of the DPSIR framework used throughout this report. They will strike at the heart of how human societies are structured and relate to nature.

While governments are expected to take the lead, other stakeholders are just as important to ensure success in achieving sustainable development. The need couldn't be more urgent and the time couldn't be more opportune, with our enhanced understanding of the challenges we face, to act now to safeguard our own survival and that of future generations.

References

ADB (2001). *Asian Development Outlook 2001*. Asian Development Bank, Manila

ADB (2005). *Asian Environment Outlook 2005. Making Profits, Protecting Our Planet.* Asian Development Bank, Manila

Aidt, T.S. (1998). Political internalization of economic externalities and environmental policy. In *Journal of Public Economics* 69:1-16

Andersen, M.S., Dengsøe, N. and Pedersen, A.B. (2000). *An Evaluation of the Impacts of Green Taxes in the Nordic Countries.* Centre for Social Research on the Environment, Aarhus University, Aarhus

Andresen, S. (2001). Global Environmental Governance: UN Fragmentation and Co-ordination. In Stokke, O.S. and Thommessen, Ø.B. (eds.) *Yearbook of International Co-operation on Environment and Development.* Earthscan, London

Anheier, H.K., Glasius, M. and Kaldor, M. (eds.) (2001). *Global Civil Society 2001.* Oxford University Press, Oxford

Bass, S. (2006). *Making Poverty Reduction Irreversible: Development Indications of the Millennium Ecosystem Assessment.* International Institute for Environment and Development, London

Beise, M. (2001). *Lead Markets. Country-Specific Success Factors of the Global Diffusion of Innovations.* Physica, Heidelberg

Both ENDS and Gomukh (2005). *River Basin Management; a Negotiated Approach.* Amsterdam and Pune http://www.bothends.org/strategic/RBM-Boek.pdf (last accessed 12 july 2007)

Boykoff, J. and Boykoff, M. (2004). Journalistic Balance as Global Warming Bias: Creating Controversy where Science Finds Consensus. In *Extra!* November/December 2004 http://www.fair.org/index.php?page=1978 (last accessed 12 July 2007)

Brenton, T. (1994). *The Greening of Machiavelli: The Evolution of International Environmental Politics.* Earthscan, London

Cabinet Office (2005). *Regulatory Impact Assessment (RIA) overview.* Department for Business, Enterprise and Regulatory Reform, London http://www.cabinetoffice.gov.uk/regulation/ria/overview/the_ria_process.asp (last accessed 12 July 2007)

Caldwell, L.K. (1996). *International Environmental Policy: From the Twentieth to the Twenty-First Century.* Duke University Press, Durham and London

Canadian Boreal Initiative (2005). *Counting Canada's Natural Capital: Assessing the Real Value of Canada's Boreal Ecosystems.* Canadian Boreal Initiative, Ottawa and Pembina Institute, Drayton Valley

Carius, A., Jacob, K., Jänicke, M. and Hackl, W. (2005). *Evaluation Study on the Implementation of Austria's Sustainable Development Strategy* (in German). Prepared for the Bundesministeriums für Land- und Forstwirtschaft, Umwelt und Wasserwirtschaft http://www.nachhaltigkeit.at/strategie/pdf/Evaluationsbericht_NStrat_Langfassung_06-05-11.pdf (last accessed 12 July 2007)

Carter, N. (2001). *The Politics of the Environment: Ideas, Activism, Policy.* Cambridge University Press, Cambridge

CEC (2003). *Reforming the European Union's Sugar Policy: Summary of Impact Assessment Work.* Commission of the European Communities, Brussels

CEC (2004). *Integrating Environmental Considerations into Other Policy Areas: A Stocktaking of the Cardiff Process.* Commission Working Document. Commission of the European Communities, Brussels

CNEDD (2004). *Troisième rapport national du Niger dans le cadre de la mise en œuvre de la Convention internationale de lutte contre la desertification.* République du Niger, Conseil national de l'environnement pour un développement durable., Niamey

Committee on Risk Assessment of Hazardous Air Pollutants, Board on Environmental Studies and Toxicology, Commission on Life Sciences, National Research Council (2004). Science and Judgment in Risk Assessment. In *National Academy Press*, Washington, DC

Costanza, R. and Daly, H. E. (1992). Natural Capital and Sustainable Development. In *Conservation Biology* 6:37-46

Cunningham, N. and P. Grabosky (1998). *Smart Regulation: Designing Environmental Policy.* Clarendon Press, Oxford

Dalal-Clayton, B. and Bass, S. (2002). *Sustainable Development Strategies – A Resource Book.* Earthscan, London

Dalal-Clayton, B. and Bass, S. (2006). *A review of monitoring mechanisms for national sustainable development strategies.* Environmental Planning Series. International Institute for Environment and Development, London

De-Shalit, A. (1995). *Why Posterity Matters: Environmental Policies and Future Generations.* Routledge, London

De-Shalit, A. (2000). *The Environment: Between Practice and Theory.* Oxford University Press, Oxford

Diamond, J. (2005). *Collapse: How Societies Choose to Fail or Succeed.* Viking, New York, NY

Diekmann, A. and Franzen, A. (1999). The Wealth of Nations and Environmental Concern. In *Environment and Behavior* 31(4):540-549

Dietz, T., Ostrom, E. and Stern, P.C. (2003). The Struggle to Govern the Commons. In *Science* 302 (5652):1907-1912

Downs, A. (1972). Up and Down with Ecology – the "Issue-Attention Cycle." In *Public Interest* 28:38-50 http://www.anthonydowns.com/upanddown.htm (last accessed 25 June 2007)

Dupar, M. and Badenoch, N. (2002). *Environment, livelihoods, and local institutions: Decentralisation in mainland Southeast Asia.* World Resources Institute, Washington, DC

Dutzik, T. (2002). *The State of Environmental Enforcement: The Failure of State Governments to Enforce Environmental Protections and Proposals for Reform.* CoPIRG Foundation, Denver

Ecolex (2007). *Ecolex. A gateway to environmental law.* Operated jointly by FAO, IUCN and UNEP http://ecolex.org/index.php (last accessed 12 July 2007)

Eden, S. (1996). Public participation in environmental policy: Considering scientific, counter-scientific and non-scientific contributions. In *Public Understanding of Science* 5 (3):183-204

EEA (2001). *Late lessons from early warnings: the precautionary principle 1896-2000.* Environmental Issue Report No. 22. European Environment Agency, Copenhagen

EEA (2002). *Environmental Signals 2002 - Benchmarking the millennium.* Environmental Assessment Report No. 9. European Environment Agency, Copenhagen

EEA (2004). *Environmental Policy Integration in Europe: Administrative Culture and Practices.* European Environment Agency, Copenhagen

EEA (2005). *The European Environment: State and Outlook 2005.* European Environment Agency, Copenhagen

EEA (2006). *Using the market for cost-effective environmental policy.* EEA Report 1/2006, European Environment Agency, Copenhagen

EEB (2005). *EU Environmental Policy Handbook: A Critical Analysis of EU Environmental Legislation.* European Environment Bureau, Brussels

Ehrlich, P.R. and Ehrlich, A.H. (2004). *One with Nineveh: Politics, Consumption, and the Human Future.* Island Press, Washington, DC

Emerton, L., Bishop, J. and Thomas, L. (2006). *Sustainable Financing of Protected Areas: A global review of challenges and options.* World Conservation Union (IUCN), Gland http://app.iucn.org/dbtw-wpd/edocs/PAG-013.pdf (last accessed 12 July 2007)

Enters, T., Durst, P.B. and Victor, M. (eds.) (2000). *Decentralization and Devolution of Forest Management in Asia and the Pacific.* RECOFTC Report N.18 and RAP Publication 2000/1. Regional Community Forestry Training Centre for Asia and the Pacific, Bangkok

Environmental Assessment Institute (2006). *Getting Proportions Right – How far should EU Impact Assessments go?* Danish Ministry of Environment, Environmental Assessment Institute, Copenhagen http://imv.net.dynamicweb.dk/Default.aspx?ID=674 (last accessed 12 July 2007)

Esty, D. C. and Porter, M. E. (2000). *Measuring National Environmental Performance and Its Determinants. The Global Competitiveness Report 2000:* 60-75. Harvard University and World Economic Forum. Oxford University Press, New York, NY

European Commission (2005). Cost-Benefit Analysis of the Thematic Strategy of Air Pollution. AEAT/ED48763001/Thematic Strategy. Issue 1. AEA Technology Environment for the European Commission, DG Environment, Brussels http://ec.europa.eu/environment/air/cafe/general/pdf/cba_thematic_strategy_0510.pdf (last accessed 12 July 2007)

European Commission (2006). *A guidebook for peer reviews of national sustainable development strategies.* European Commission, DG Environment, Brussels

Farhar, B.C. (1999). *Willingness to Pay for Electricity from Renewable Resources: A Review of Utility Market Research.* US Department of Energy, National Renewable Energy Laboratory, Golden, CO http://www.nrel.gov/docs/fy99osti/26148.pdf (last accessed 12 July 2007)

Ferraro, P.J. and Kiss, A. (2002). Direct Payments to Conserve Biodiversity. In *Science* 298:29

Friends of the Earth (2002). *Marketing the Earth: The World Bank and Sustainable Development.* Halifax Initiative, Ottawa

Furtado, X. (2001). *Decentralisation and Capacity Development: Understanding the Links and the Implications for Programming.* CIDA Policy Branch No. 4. Occasional Paper Series. Canadian International Development Agency, Ottawa

Gelbspan, R. (1997). *The Heat is On: The High Stakes Battle over the Earth's Threatened Climate.* Addison-Wesley, Reading

Giddings, B., Hopwood, B. and O'Brien, G. (2002). Environment, economy and society: fitting them together into sustainable development. In *Sustainable Development* 10(4):187-196

Global Policy Forum (2000). *European Parliament Supports Move to Tax Aircraft Fuel.* European Report. Global Policy Forum, New York, NY http://www.globalpolicy.org/socecon/glotax/aviation/001213ep.htm (last accessed 12 July 2007)

Global Policy Forum (2006). Aviation Taxes. http://www.globalpolicy.org/socecon/glotax/aviation/index.htm (last accessed 12 July 2007)

GRI (2006). Global Reporting Initiative. http://www.globalreporting.org (last accessed 12 July 2007)

Hahn, R.W. and Stavins, R.N. (1992). Economic Incentives for Environmental Protection: Integrating Theory and Practice. In *The American Economic Review* 82 (9):464-468

Hoekman, B., Ng, F. and Olarreaga, M. (2002). *Reducing Agricultural Tariffs versus Domestic Support: What's More Important for Developing Countries?* World Bank Policy Research Working Paper No. 2918. The World Bank Washington, DC

Holling, C.S. (ed.) (1978). *Adaptive Environmental Assessment and Management.* John Wiley and Sons, New York, NY

Hutton, G. and Haller, L. (2004). *Evaluation of the Costs and Benefits of Water and Sanitation Improvements at the Global Level.* WHO/SDE/WSH/04. 04. World Health Organization, Geneva http://www.who.int/water_sanitation_health/en/wsh0404.pdf (last accessed 12 July 2007)

IEA (1999). *World Energy Outlook – Looking at Energy Subsidies, Getting the Prices Right.* International Energy Agency, Paris

IFC (2004). *2004 Sustainability Report.* International Finance Corporation. Washington, DC

Inter Press Service (2005). *France Begins to Tax Flights for Aid.* November 30. http://www.globalpolicy.org/socecon/glotax/aviation/2005/1130paris.htm (last accessed 12 July 2007)

IPCC (1996). *Technologies, Policies and Measures for Mitigating Climate Change.* IPCC Technical Paper No. 1. Intergovernmental Panel on Climate Change, Geneva

IPCC (1999). *Aviation and the Global Atmosphere.* A Special Report of Working Groups I and III of the Intergovernmental Panel on Climate Change, Cambridge

Jacob, K. and Volkery, A. (2004). Institutions and Instruments for Government Self-Regulation: Environmental Policy Integration in a Cross-Country Perspective. In *Journal of Comparative Policy Analysis* 6(3):291-309

Jacob, K., Beise, M., Blazejczak, J. Edler, D., Haum, R., Jänicke, M., Loew, T., Petschow, U. and Rennings, K. (2005). *Lead Markets of Environmental Innovations.* Physica Verlag, Heidelberg and New York, NY

Jacob, K., Hertin, J. and Volkery, A. (2007). Considering environmental aspects in integrated impact assessment : lessons learned and challenges ahead. In George, C. and Kirkpatrick, C. (eds.) *Impact Assessment and Sustainable Development: European Practice and Experience. Impact Assessment for a New Europe and Beyond.* Edward Elgar, Cheltenham

Jaffe, A.B., Newell, R.G. and Stavins, R.N. (2002). Environmental Policy and Technological Change. In *Environmental and Resource Economics* 22(1-2):41-70

Jane Goodall Institute (2006). http://www.janegoodall.org/ (last accessed 25 June 2007)

Jänicke, M. (2006). Trend Setters in Environmental Policy: The Character and Role of Pioneer Countries. In Jänicke, M. and Jacob, K. (eds.) *Environmental Governance in Global Perspective. New Approaches to Ecological Modernisation.*, Freie Universität Berlin, Berlin

Jänicke, M. and Jacob, K. (2004). Lead Markets for Environmental Innovations: A New Role for the Nation State. In *Global Environmental Politics* 4(1):29-46

Jänicke, M. and Volkery, A. (2001). Persistente Probleme des Umweltschutzes. In *Natur und Kultur* 2 (2):45-59

Jordan, A., Wurzel, R.K.W. and Zito, A.R. (eds.) (2003). *"New" Instruments of Environmental Governance? National Experiences and Prospects.* Frank Cass, London and Portland

Jütting J., Kauffmann, C., Mc Donnell. I., Osterrieder, H., Pinaud, N. and Wegner, L. (2004). *Decentralisation and Poverty in Developing Countries: Exploring the Impact.* OECD Development Centre. Working Paper No. 236. August. Organisation for Economic Co-operation and Development, Paris http://www.oecd.org/dataoecd/40/19/33648213.pdf (last accessed 12 July 2007)

Keen, M., Brown, V.A. and Dyball, R. (eds.) (2005). *Social Learning in Environmental Management: Building a Sustainable Future.* Earthscan, London

Kemp, R. and Loorbach, D. (2003). Governance for Sustainability through Transition Management. Paper for the *EAEPE 2003 Conference,* 7-10 November 2003, Maastricht

Kemp, R. and Rotmans, J. (2001). The Management of the Co-Evolution of Technical, Environmental and Social Systems. Paper for the *International Conference "Towards Environmental Innovation Systems."* 27-29 September 2001, Garmisch Partenkirchen

Kennedy, R.F. Jr. (2004). *Crimes Against Nature.* Harper Collins, New York, NY

Kiersch, B., Hermans, L. and Van Halsema, G. (2005). Payment Schemes for Water-related Environmental Services: A Financial Mechanism for Natural Resources Management - Experiences from Latin America and the Caribbean. Seminar on *Environmental Services and Financing for the Protection and Sustainable Use of Ecosystems,* 10-11 October 2005, Geneva http://www.unece.org/env/water/meetings/payment_ecosystems/Discpapers/FAO.pdf (last accessed 13 July 2007)

Lafferty, W. B. (2002). *Adapting Government Practice to the Goals of Sustainable Development.* Governance for Sustainable Development. Five OECD Case Studies. Organisation for Economic Co-operation and Development, Paris

Leakey, R. and Lewin, R. (1995). *The Sixth Extinction: Patterns of Life and the Future of Humankind.* Doubleday, New York, NY

Lehman, S.J. and Keigwin, L.D. (1992). Sudden changes in North Atlantic circulation during the last deglaciation. In *Nature* 356:757-762

Lehtonen, M. (2005). OECD Environmental Performance Review Programme. In *Evaluation* 11(2):169-188

Lenschow, A. (ed.) (2002). *Environmental Policy Integration: Greening Sectoral Policies in Europe.* James & James/Earthscan, London

Levy, M.A. (1995). International Cooperation to Combat Acid Rain. *Green Globe Yearbook,* Oxford University Press, Oxford

Loorbach, D. (2002). Transition Management - Governance for Sustainability. Paper presented at the *Conference Governance and Sustainability - New challenges for the state, business and civil society, 31 September - 1 October2002,* Berlin

Lubchenco, J. (1998). Entering the Century of the Environment: A New Social Contract for Science. In *Science* 279:491-497

MA (2003). *Ecosystems and human well-being.* Millennium Ecosystem Assessment. Island Press, Washington, DC

MA (2005). *Ecosystems and Human Well-being: Opportunities and Challenges for Business and Industry.* Millennium Ecosystem Assessment/World Resources Institute, Washington, DC

Martin-Hurtado, R. (2002). *Costing the 7th Millennium Development Goal: Ensuring Environmental Sustainability.* Environment Department, The World Bank, Washington, DC

McKibben, B. (2005). Curitiba: A Global Model for Development http://www.commondreams.org (last accessed 25 June 2007)

Meadows, D.H, Randers, J. and Meadows, D.L. (2004). *Limits to Growth: The 30-Year Update.* Chelsea Green Publishing, White River Junction

Meyer-Krahmer, F. (1999). Was bedeutet Globalisierung für Aufgaben und Handlungsspielräume nationaler Innovationspolitiken? In *Innovationspolitik in globalisierten Arenen.* Opladen, Leske und Budrich

Ministry of Environment Norway (2005). Norwegian Parliamentary Bill No: 1 (2005-2006) of 7th October 2005 from the Ministry of Environment (see Chapter 12 on Waste and Recycling)

MOEJ (2005). *Japan's Experience in the Promotion of the 3Rs: For the Establishment of a Sound Material-Cycle Society.* Global Environment Bureau, Ministry of Environment, Tokyo

Mol, A.P.J. and Sonnenfeld, D.A. (eds.) (2000). Special Issue on Ecological Modernisation. In *Environmental Politics* 9(1)

Myers, N. (1997). Mass Extinction and Evolution. In *Science* 24(5338):597-598

Navarro, Y.K., McNeely, J., Melnick. D., Sears, R.R. and Schmidt-Traub, G. (2005) *Environment and Human Well-Being.* Earthscan, London

NEPP2 (1994). *The National Environmental Policy Plan 2 - The Environment: Today's Touchstone.* Ministry of Housing, Spatial Planning and the Environment, Government of the Netherlands, The Hague

New Economics Foundation (2006). *The UK Interdependence Report: How the World Sustains the Nation's Lifestyles and the Price it Pays.* New Economics Foundation, London

Nielsen, N.O. (1999). The Meaning of Health. In *Ecosystem Health 5(2):65-66*

NRBS (1996). *Report of the Northern Rivers Basin Study.* Northern Rivers Basin Study Board, Edmonton

OECD (2000). *Environmental Performance Reviews (First Cycle): Conclusions and Recommendations, 32 countries (1993-2000).* Organisation for Economic Co-operation and Development, Paris

OECD (2001a). *Sustainable Development. Critical Issues.* Organisation for Economic Co-operation and Development, Paris

OECD (2001b). *OECD Environmental Performance Reviews (Second Cycle): Norway.* Organisation for Economic Co-operation and Development, Paris

OECD (2002a). *Aggregated Environmental Indices: Review of Aggregation Methodologies in Use.* Organisation for Economic Co-operation and Development, Paris

OECD (2002b). *Policies to Enhance Sustainable Development. Critical Issues.* Organisation for Economic Co-operation and Development, Paris

OECD (2003). Policies to Reduce Greenhouse Gas Emissions in Industry - Successful Approaches and Lessons Learned: Workshop Report, OECD and IEA Information Paper. In *OECD Papers 4(2), Special Issue on Climate Change. Climate Change Policies: Recent Developments and Long-Term Issues.* COM/ENV/EPOC/IEA/SLT(2003)2:322. Organisation for Economic Co-operation and Development, Paris http://www.oecd.org/dataoecd/8/36/31785351.pdf (last accessed 12 July 2007)

OECD (2004). *OECD Environmental Performance Reviews: Canada.* Organisation for Economic Co-operation and Development, Paris

OECD (2006). *Applying Strategic Environmental Assessment: Good Practice Guidance for Development Co-operation.* Organisation for Economic Co-operation and Development, Paris

Oliveira Costa, J.P. De (2005). Protected Areas Ministro de Estado das relações Exteriores, Brasília http://www.mre.gov.br/cdbrasil/itamaraty/web/ingles/meioamb/arprot/apresent/apresent.htm (last accessed 13 July 2007)

PACS (2006). http://www.empowerpoor.org/ (last accessed 13 July 2007)

Pearce, D.W. (ed.) (2004). *Valuing the Environment in Developing Countries: Case Studies.* Edward Elgar, Cheltenham

Pearce, D.W. (2005). *Investing in Environmental Wealth for Poverty Reduction.* United Nations Development Programme, New York, NY http://www.undp.org/pei/pdfs/InvestingEnvironmentalWealthPovertyReduction.pdf (last accessed 25 June 2007)

Petkova, E. and Veit, P. (2000). *Environmental Accountability Beyond the Nation-State: The Implications of the Aarhus Convention.* World Resources Institute, Washington, DC

Petkova, E., Maurer, C., Henninger, N. and Irwin, F. (2002). *Closing the Gap: Information, Participation and Justice in Decision-making for the Environment.* World Resources Institute, Washington, DC

Popper, S.W., Lempert, R.J. and Bankes, S.C. (2005). Shaping the Future. In *Scientific American* 28 March 2005

Porter, M. E. and Van der Linde, C. (1995). Toward a New Conception of the Environment-Competitiveness Relationship. In *Journal of Economic Perspectives* 9:97-118

Posner, R.A. (2005). *Catastrophe: Risk and Response.* Oxford University Press, Oxford

Rees, M. (2003). *Our Final Hour: A Scientist's Warning: How Terror, Error, and Environmental Disaster Threaten Humankind's Future in this Century-On Earth and Beyond.* Basic Books, New York, NY

Rogers, P., Jalal, K.F., Lohani, B.N., Owens, G.M., Yu, C., Dufournaud, C.M. and Bi, J. (1997). *Measuring Environmental Quality in Asia.* Harvard University and Asian Development Bank, Cambridge

Rotmans, J., Kemp, R. and Van Asselt, M. (2001). More Evolution than Revolution - Transition Management in Public Policy. In *Foresight 3(1):15-31*

SEI (2004). *Ecological Sanitation: Revised and Enlarged Edition.* Stockholm Environment Institute, Stockholm

Shortle, J.S., Abler, D.G. and Horan, R.D. (1998). Research Issues in Nonpoint Pollution Control. In *Environmental and Resource Economics* 11(3-4):571-585

SIPRI (2004). *World and regional military expenditure estimates 1988 – 2006.* Stockholm International Peace Research Institute, Stockholm http://web.sipri.org/contents/milap/milex/mex_wnr_table.html (last accessed 13 July 2007)

Speth, J.G. (2004). *Red Sky at Morning: America and the Crisis of the Global Environment.* Yale University Press, New Haven and London

Stanley Foundation (2004). *Development, Poverty and Security – Issues before the UN's High Level Panel.* http://www.stanleyfoundation.org/publications/report/UNHLP04d.pdf (last accessed 12 July 2007)

Steid, D. and Meijers, E. (2004). Policy integration in practice: some experiences of integrating transport, land-use planning and environmental policies in local government. *Berlin Conference on the Human Dimensions of Global Environmental Change: Greening of Policies – Interlinkages and Policy Integration,* 3-4 December 2004, Berlin

Stern, N. (2007). *The Economics of Climate Change: The Stern Review.* Cambridge University Press, Cambridge

Steurer, R. and Martinuzzi, A. (2005). Towards a New Pattern of Strategy Formation in the Public Sector: First Experiences with National Strategies for Sustainable Development in Europe. In *Environment and Planning C: Government and Policy* 23(3):455-472

Stinchcombe, K. and Gibson, R.B. (2001). Strategic Environmental Assessment as a Means of Pursuing Sustainability: Ten Advantages and Ten Challenges. In *Journal of Environmental Assessment Policy and Management* 3(3):343-372

Swanson, D., Pintér, L., Bregha, F., Volkery, A. and Jacob, K. (2004). *National Strategies for Sustainable Development: Challenges, Approaches and Innovations in Strategic and Coordinated Action.* International Institute for Sustainable Development, Winnipeg, and Deutsche Gesselschaft für Technische Zusammenarbeit (GTZ) GmbH, Eschborn

Tews, K., Busch, P.-O. and Jörgens, H. (2003). The diffusion of new environmental policy instruments. In *European Journal of Political Research* 42(4):569-600

TFL (2004). *Congestion Charging: Update on Scheme Impacts and Operations.* Transport for London, London

UCS (1992). World Scientists' Warning to Humanity (1992). Scientist Statement. Union of Concerned Scientists http://www.ucsusa.org/ucs/about/1992-world-scientists-warning-to-humanity.html (last accessed 13 July 2007)

UN (2002a). *World Summit on Sustainable Development, Johannesburg Plan of Implementation.* United Nations, New York, NY http://www.un.org/esa/sustdev/documents/WSSD_POI_PD/English/POIToc.htm (last accessed 13 July 2007)

UN (2002b). *Report of the International Conference on Financing for Development.* Monterrey, 18-22 March 2002. United Nations VA/CONF.198/II. United Nations, New York, NY

UN (2005a). *Summary by the President of the Economic and Social Council of the Special High-level Meeting of the Council with the Bretton Woods institutions, the World Trade Organization and the United Nations Conference on Trade and Development.* New York, 18 April 2005. A/59/823–E/2005/69. United Nations, New York, NY

UN (2005b). *Summary by the President of the General Assembly of the High-level Dialogue on Financing for Development.* 27-28 June 2005. A/60/219. United Nations, New York, NY

UN (2005c). Address by H. Exc. Mr. Thierry Breton Minister for the Economy, Finance and Industry of France at the *High-level Dialogue on Financing for Development of the General Assembly,* 27 June 2005. United Nations, New York, NY http://www.un.int/france/documents_anglais/050627_ag_financement_developpement_tbreton.htm (last accessed 13 July 2007)

UN (2005d). *In Larger Freedom: Towards Development, Security and Human Rights for All.* United Nations, New York, NY http://www.un.org/largerfreedom/executivesummary.pdf (last accessed 13 July 2007)

UNCED (1992). *Agenda 21 - The United Nations Programme of Action from Rio.* United Nations Conference on Environment and Development, New York, NY

UNCTAD (2005). Statement by Carlos Fortin, Officer-in-Charge of the United Nations Conference on Trade and Development (2004-2005) at the *High-level Dialogue on Financing for Development of the General Assembly,* 27 June 2005. United Nations, New York, NY http://www.unctad.org/Templates/webflyer.asp?docid=6006&intItemID=3551&lang=1 (last accessed 13 July 2007)

UNDP (2002). *Human Development Report 2002: Deepening democracy in a fragmented world.* United Nations Development Programme, New York, NY

UNDP (2005). *Environmental Sustainability in 100 Millennium Development Goals Country Reports.* http://www.unep.org/dec/docs/UNDP_review_of_Environmental_Sustainability.doc (last accessed 13 July 2007)

UNECE (2005). *Synthesis Report on the Status of Implementation of the Convention.* Meeting of the Parties to the Convention on Access to Information, Public Participation in Decision-making and Access to Justice in Environmental Matters, ECE/MP.PP/2005/18, 12 April. United Nations Economic Commission for Europe, Geneva http://www.unece.org/env/pp/reports%20implementation.htm (last accessed 12 July 2007)

UNECE (2007). Convention on Long-range Transboundary Air Pollution (CLRTAP). http://www.unece.org/env/lrtap/ (last accessed 12 July 2007)

UNEP (2006a). *GEO Year Book 2006.* United Nations Environment Programme, Nairobi

UNEP (2006b). *Marine and coastal ecosystems and human well-being: A synthesis report based on the findings of the Millennium Ecosystem Assessment.* United Nations Environment Programme, Nairobi

UNEP (2006c). *Rural Energy Enterprise Development (REED).* United Nations Environment Programme, Paris http://www.uneptie.org/energy/projects/REED/REED_index.htm (last accessed 13 July 2007)

UNEP (2006d). *Principles for Responsible Investment: An investor initiative in partnership with UNEP Finance Initiative and the UN Global Compact* http://www.unpri.org/files/pri.pdf (last accessed 11 July 2007)

UNEP and LSE (2005). Creating Pro-Poor Markets for Ecosystem Services. Concept Note for the *High-Level Brainstorming Workshop "Creating Pro-Poor Markets for Ecosystem Services,"* 10–12 October 2005. United Nations Environment Programme and London School of Economics, London

UNESCAP (2001). *Regional Platform on Sustainable Development for Asia and the Pacific, 3rd Revision.* United Nations Economic and Social Commission for Asia and the Pacific, Phnom Penh

UNESCO (2005a). *"Scaling Up" Good Practices in Girls' Education.* United Nations Educational, Scientific and Cultural Organization, Paris

UNESCO (2005b). *UN Decade of Education for Sustainable Development: Links Between the Global Initiatives in Education.* Technical Paper No. 1. United Nations Educational, Scientific and Cultural Organization, Paris

UNGA (1972). *General Assembly resolution 2997.* United Nations, New York, NY

UNIDO (2000). *Cluster Development and Promotion of Business Development Services (BDS): UNIDO's experience in India.* PSD Technical working papers series Supporting Private industry. United Nations Industrial Development Organization, Vienna http://www.unido.org/en/doc/4809 (last accessed 25 June 2007)

UNITAID (2006). *UNITAID Basic Facts – UNITAID at work.* http://www.unitaid.eu/en/(last accessed 13 June 2007)

USDA (2001). *FI 2001Budget Summary of the.* United States Department of Agriculture http://www.usda.gov/agency/obpa/Budget-Summary/2001/text.htm (last accessed 13 July 2007)

Volkery, A., Swanson, D., Jacob, K., Bregha F. and Pintér L. (2006). Coordination, Challenges and Innovations in 19 National Sustainable Development Strategies. In *World Development* (accepted for publication)

Von Weizsäcker, E.U. and Jesinghaus, J. (1992). *Ecological Tax Reform. A Policy Proposal for Sustainable Development.* Zed Books, London

Wachter, D. (2005). Sustainability Assessment in Switzerland: From Theory to Practice. *EASY-ECHO 2005-2007, First Conference,* Manchester http://www.sustainability.at/easy/?k=conferences&s=manchesterproceedings

Walker, B., Holling, C.S., Carpenter, S.R. and Kinzig, A. (2004). Resilience, adaptability and transformability in social-ecological systems. In *Ecology and Society* 9(2):5

Wates, J. (2005). The Aarhus Convention: a Driving Force for Environmental Democracy. In *JEEPL* (1):2-11.

WCED (1987). *Our Common Future.* World Commission on Environment and Development, Oxford University Press, Oxford

Wiedman, T., Minx, J., Barrett, J. and Wackernagel, M. (2006). Allocating ecological footprints to final consumption categories with input-output analysis. In *Ecological Economics* 56(1):28-48

Wilkinson, D. (1997). Towards sustainability in the European Union? Steps within the European Commission towards integrating the environment into other European Union policy sectors. In *Environmental Politics* 6(1):153-173

Wilkinson, D., Fergusson, M., Bowyer, D., Brown, J., Ladefoged, A., Mokhouse, C. and Zdanowicz, A. (2004). *Sustainable Development in the European Commission's Integrated Impact Assessments for 2003.* Institute for European Environmental Policy, London http://www.ieep.org.uk/publications/pdfs//2004/sustainabledevelopmentineucommission.pdf (last accessed 13 July 2007)

Wilson, E.O. (1996). *In Search of Nature.* Island Press, Washington

Work, R. (undated). *The Role of Participation and Partnership in Decentralised Governance: A Brief Synthesis of Policy Lessons and Recommendations of Nine Country Case Studies on Service Delivery for the Poor.* UNDP-MIT Global Research Programme on Decentralised Governance. United Nations Development Programme, New York, NY

World Bank (1997). *World Development Report 1997: The State in a Changing World.* The World Bank, Washington, DC

World Bank (2005). *Integrating Environmental Considerations into Policy Formulation: Lessons from Policy-based SEA Experience.* The World Bank, Washington, DC

World Bank (2006). *Where is the Wealth of Nations: Measuring Capital for the 21st Century.* The World Bank, Washington, DC

Worldwatch Institute (2004). *State of the World 2004: Consumption by the Numbers.* Worldwatch Institute, Washington, DC

WTO (2001). Ministerial declaration of the *World Trade Organization Ministerial Conference Fourth Session,* 9-14 November 2001, Doha. WT/MIN(01)/DEC/1 http://www.wto.org/english/thewto_e/minist_e/min01_e/mindecl_e.doc (last accessed 13 July 2007)

Yuan, Z., Bi, J. and Moriguichi, Y. (2006). The Circular Economy: A New Development Strategy in China. In *Journal of Industrial Ecology* 10(1-2):4-8

The GEO-4 process

The *GEO-4* assessment has been the most comprehensive since the inception of the GEO processes, following UNEP Governing Council decision of 1995, which requested the preparation of GEO as part of implementing the UNEP overall mandate to keep the world environment under review. Based on a subsequent February 2003 GC decision on the GEO assessment, UNEP has, over the past four years, organized global and regional consultations, first to seek the inputs of policy-makers on the scope and objectives of the assessment, and second, for scientific and policy experts to research and draft the content of the report.

The scope and objectives of the *GEO-4* were defined by two inter-related consultation processes:

■ an intergovernmental consultation, in February 2004, on strengthening the scientific base of UNEP; and

■ an intergovernmental and multistakeholder consultation on *GEO-4*, which was held in February 2005.

The broad based consultative process on strengthening the scientific base of UNEP, which engaged more than 100 governments and 50 partners, identified the following needs:

■ *strengthened interaction between science and policy*, particularly by strengthening the credibility, timeliness, legitimacy and relevance of, and complementarity among environmental assessments so as not to overburden the scientific community;

■ *enhanced focus on scientific interlinkages* between environmental challenges and responses to them, as well as interlinkages between environmental and development challenges as a basis for environmental mainstreaming and development of scenarios and modelling about plausible futures;

■ *improved quantity, quality, interoperability and accessibility of data and information* for most environmental issues, including for early warning related to natural disasters;

■ *enhanced national capacities* in developing countries and countries with economies in transition for data collection and analysis and for environmental monitoring and integrated assessment; and

■ *improved cooperation and synergy* among governments, UN bodies, MEAs and regional environmental, scientific and academic institutions, and networking among national and regional institutions.

The Statement by the Global Intergovernmental and Multi-stakeholder Consultation on the fourth Global Environment Outlook recommended that the *GEO-4* objective, scope and overall outline should provide a global, comprehensive, reliable and scientifically credible, policy-relevant and legitimate up-to-date assessment of and outlook regarding the interaction between environment and society. It stated that the assessment should be in the context of the development of international environmental governance, and its relation to the internationally agreed sustainable development goals and targets since the 1987 report of the World Commission on Environment and Development, analysing, among others the Rio Declaration, Agenda 21, the Millennium Declaration, the Johannesburg Declaration and its Plan of Implementation, and relevant environmental global and regional instruments.

The report also assesses the state-and-trends of the global environment in relation to the drivers and pressures, and the consequences of environmental change for ecosystem services and human well-being as well as on progress and barriers towards meeting commitments under multilateral environmental agreements. Other objectives included:

■ assessing interlinkages between major environmental challenges, and their consequences for policy and technology response options and trade-offs, and assessing opportunities for technology and policy interventions for both mitigating and adapting to environmental change;

■ assessing challenges and opportunities by focusing on certain key cross-cutting issues, and on how environmental degradation can impede progress, with a focus on vulnerable groups, species, ecosystems and locations;

- presenting a global and sub-global outlook, including short-term (up to 2015) and medium-term (up to 2050) scenarios for the major societal pathways, and their consequences for the environment and society; and
- assessing environment for human well-being and prosperity, focusing on the state of knowledge regarding the effectiveness of various approaches to overarching environmental policies.

The *GEO-4* assessment had to answer, in the 10 chapters of the report, more than 30 questions that were identified in the Statement of February 2005.

PARTNERSHIPS

The *GEO-4* assessment combined the widely regarded, bottom-up participatory GEO process with elements from well-proven scientific assessment processes, such as the Millennium Ecosystem Assessment. The GEO assessment has been successful over the past decade due to its strong network of collaborating centres spread across the globe. About 40 centres were involved in the current GEO assessment, and each of them brought different expertise, ranging from thematic to policy analysis. The assessment involved a good regional and gender balance.

Chapter expert groups

The *GEO-4* report has 10 chapters, and for each, an expert working group was established to research, draft, revise and finalize the chapter. Each of the 10 chapter groups included 15-20 individuals: scientists, representatives from GEO collaborating centres, experts nominated by governments, policy practitioners, representatives of UN organizations and GEO Fellows. The experts were nominated, based on their scientific merit and/or policy expertise. UNEP assigned a staff member to each group as chapter coordinator. The expert groups were led by two or three coordinating lead authors, working in close collaboration with the UNEP chapter coordinator. Members of the chapter expert groups were lead authors for the chapters, and specific contributions were made by other specialists (contributing authors).

Chapter review editors

About 20 chapter review editors were identified to review the treatment of comments.

Standing expert groups

The three main standing groups of the assessment are on data, capacity building, and outreach and communications.

GEO Data Working Group

The GEO Data Working Group supported and guided the GEO data component during the production of the GEO assessment. The main focus was on the proper use of indicators, strengthening data capacities in developing regions, filling existing and identifying emerging data gaps, and improving data quality assurance and control. Further development of the GEO data component is closely linked to establishing and strengthening cooperation with new or existing authoritative data providers around the world, focusing on new data and indicators that have become available and are relevant for GEO assessment. A key product is the GEO Data Portal. The Data Portal gives access to a broad collection of harmonized environmental and socio-economic data sets from authoritative sources at global, regional, sub-regional and national levels, and allows data analysis and creation of maps, graphics and tables. Its online database currently holds more than 450 variables. The contents of the Data Portal cover environmental themes, such as climate, forests and freshwater, as well as socio-economic categories, including education, health, economy, population and environmental policies.

Capacity Building Working Group

The Capacity Building Working Group supports, advises and guides GEO capacity building activities. Capacity building has been at the heart of the GEO process since its inception in 1995. Capacity building is achieved through the active participation of developing country experts in *GEO-4*, as well as by hands-on support to governments to produce sub-global reports, supported by:
- development and promotion of the use of integrated assessment tools and methodologies, including use of the GEO Resource Book;
- training and workshops;
- networking and partnerships; and
- GEO Fellowships awarded to students and scholars to work with the GEO process.

Outreach Working Group

The Outreach Working Group, with specialists from the fields of marketing and communication, science, education and technology, has been formed to support and advise UNEP in its outreach activities. Building strong ownership of the report and its findings by involving the media and other key target audiences, and connecting to global networks is the role of Outreach working group.

Government nominees

One of the recommendations of the Intergovernmental and Multi-stakeholder Consultation on the Design and Scope of *GEO-4* (Nairobi, February 2005) was to strengthen the involvement and engagement of expertise present in various countries. In response, UNEP requested governments to nominate experts to participate in *GEO-4*, and 157 nominees, with a wide range of thematic, technical and/or policy expertise, were nominated by 48 governments. Some of the nominees participated in expert working groups.

GEO Fellows

The GEO Fellowship Initiative was established in August 2005, to bring young and qualified professionals, into the *GEO-4* process. The GEO Fellows gained experience from a major environmental assessment process (*GEO-4*), which they can use to contribute to future sub-global or global processes. The Fellows participated in *GEO-4* as contributing authors. Thirty-four fellows, representing 27 countries, were selected from 115 applicants to participate in the *GEO-4* assessment process.

High-level Consultative Group

The High-level Consultative Group on *GEO-4* consists of less than 20 high-level individuals from policy, science, business and civil society backgrounds. The High-level Consultative Group provided guidance on various components of the assessment.

CONSULTATION PROCESS

A key additional feature in the *GEO-4* assessment was the expanded ad hoc intergovernmental and multistakeholder consultative process, which culminated in a final consultation in late September 2007 to review the assessment findings, particularly the Summary for Decision Makers (SDM). The outcome of this consultation was fed into the

UNEP Governing Council/Global Ministerial Environment Forum in 2008. In addition to the two global consultations referred to above, UNEP also organized numerous global and regional meetings to define the environmental issues and also to research and draft the contents of the *GEO-4* assessment. The following are some of the key meetings convened since 2004:

- The *GEO-4* **Planning Meeting** in June 2004 produced outputs on the concept, scope and focus of the report, followed by **Regional meetings** in October 2004 with policy-makers and other stakeholders to consult on the preliminary design and identify key issues for the *GEO-4* report., This culminated in a *GEO-4* **Design Meeting** in November 2004, which developed the draft time schedule and key activities for the period 2005-6.

- The three **Production and Authors' meetings** were convened in 2005 and 2006 to discuss and develop the *GEO-4* chapter outlines and contents, establish the chapter expert groups to research and draft the chapters, review two drafts of the report, and work with chapter review editors in finalizing the report.

- A **Sign-off Meeting of coordinating lead authors** was convened in May 2007 to provide them with a final opportunity to review the full *GEO-4* report before production.

- A meeting of a **Human Well-being Expert Working Group** to discuss and agree on the working definition of human well-being and valuation in the *GEO-4* assessment context.

- A series of more than 20 **chapter production meetings** to prepare, review and revise drafts of the report.

- About 1 000 experts were invited in May 2006 to participate in a comprehensive *GEO-4* **peer review** of the first draft. More than 13 000 comments were received, and were key input to the revision of the different drafts.

- Two **Chapter Review Editors (CREs)** per chapter assessed whether the comments received were adequately addressed by authors in revising the draft.

- **Regional Consultations** were convened in June-July 2006 in all regions to review the regional components of the first draft of *GEO-4*.

- A series of meetings of the *GEO-4* **High-Level Consultative Group** discussed strategic issues related to the assessment, including the sharpening of policy messages and strategic engagement with stakeholders.

- Three meetings of the **Outreach Working Group** were held to develop and implement a communications strategy to publicize the *GEO-4* findings, and engage stakeholders to use such findings in policy processes.
- A series of meetings of the **Capacity Building Working Group** were held to align a training manual on integrated environmental assessment with the new *GEO-4* assessment methodology.

SUMMARY FOR DECISION MAKERS (SDM)

The Summary for Decision Makers (SDM), which is published as a separate document, synthesizes the key scientific findings, gaps and challenges in the form of main messages, which are policy relevant. The SDM highlights the role and contribution of the environment and the services provided by ecosystems to development, including by way of analysing the ecosystems services and human well-being interface, and the complex and dynamic interactions taking place in time and in different spatial dimensions. The SDM contents were considered by governments and other stakeholders at the second Global Intergovernmental and Multi-stakeholder Consultation in September 2007.

Acronyms and Abbreviations

ACSAD	Arabic Centre for the Studies of Arid Zones and Drylands
AEPC	African Environmental Protection Commission
AEPS	Arctic Environmental Protection Strategy
AEWA	African-Eurasian Waterbird Agreement
AIDS	acquired immunodeficiency syndrome
ALGAS	Asia Least Cost Greenhouse Gas Abatement Strategies
AMCEN	African Ministerial Conference on the Environment
AMU	Arab Maghreb Union
ANWR	Arctic National Wildlife Refuge
AoA	Agreement on Agriculture (WTO Uruguay Round)
AOCs	Areas of Concern (Great Lakes, North America)
APELL	Awareness and Preparedness for Emergencies at Local Level
APFM	Associated Programme on Flood Management (WMO and GWP)
ASEAN	Association of Southeast Asian Nations
AU	African Union
BOD	biological oxygen demand
BSE	bovine spongiform encephalopathy
CAB	Centre for Agriculture and Biosciences
CAMP	coastal area management project
CAP	Common Agricultural Policy (European Union)
CARICOM	Caribbean Community
CBC	community-based conservation
CBD	Convention on Biological Diversity
CBO	community-based organization
CCAB-AP	Central American Council for Forests and Protected Areas
CCAMLR	Commission on the Conservation of Antarctic Marine Living Resources
CCFSC	Central Committee for Flood and Storm Control
CEB	UN Chief Executives Board for Coordination
CEC	Commission for Environmental Cooperation (under NAFTA)
CEE	Central and Eastern Europe
CEIT	Countries with Economies in Transition
CEP	Committee for Environmental Protection (Antarctic)
CERES	Coalition for Environmentally Responsible Economics
CFC	chlorofluorocarbon

CGIAR	Consultative Group on International Agricultural Research
CH_4	methane
CIAT	International Centre for Tropical Agriculture
CILSS	Permanent Interstate Committee for Drought Control in the Sahel
CITES	Convention on International Trade in Endangered Species of Wild Fauna and Flora
CLRTAP	Convention on Long-Range Transboundary Air Pollution
CMS	Convention on the Conservation of Migratory Species of Wild Animals
CAN	Comisión Nacional del Agua (Mexico)
CNC	Chinese National Committee
CNG	compressed natural gas
CNROP	National Oceanographic and Fisheries Research Centre (Mauritania)
CO	carbon monoxide
CO_2	carbon dioxide
COP	Conference of the Parties
CPACC	Caribbean Planning for the Adaptation of Global Climate Change
CPF	collaborative partnerships on forests
CRAMRA	Convention on the Regulation of Antarctic Mineral Resource Activities
CRP	Conservation Reserve Program (United States)
CSD	Commission on Sustainable Development
CTBT	Comprehensive Nuclear Test Ban Treaty
CZIMP	Coastal Zone Integrated Management Plan
DALY	disability adjusted life year
DDT	dichlorodiphenyltrichloroethane
DESA	Department of Economic and Social Affairs
DEWA	Division of Early Warning and Assessment (UNEP)
DPSIR	drivers-pressures-state-impact-response
EANET	Acid Deposition Monitoring Network
EBRD	European Bank for Reconstruction and Development
EC	European Community
ECOWAS	Economic Community of West African States
EEZ	exclusive economic zone
EfE	Environment for Europe
EIA	environmental impact assessment

EMEP	Monitoring and Evaluation of the Long-Range Transmission of Air Pollutants in Europe	GMEF	Global Ministerial Environment Forum
EMS	environmental management system	GMO	genetically modified organism
ENSO	El Niño Southern Oscillation	GNI	gross national income
EPC	Emergency Preparedness Canada	GNP	gross national product
EPCRA	Emergency Planning and Community Right-to-Know Act (United States)	GRI	Global Reporting Initiative
EPPR	Emergency Prevention, Preparedness and Response	GRID	Global Resource Information Database
ESA	Endangered Species Act (United States)	GSP	gross state product
ESBM	ecosystem-based management	GWP	Global Water Partnership
ESDP	European Spatial Development Perspective	HCFC	hydrochlorofluorocarbon
ESP	electrostatic precipitator	HDI	Human Development Index
EU	European Union	HELCOM	Helsinki Commission (Baltic Sea)
EVI	environmental vulnerability index	HFC	hydrofluorocarbon
FAD	fish aggregation device	HIPC	heavily indebted poor country
FAO	Food and Agriculture Organization of the United Nations	HIV	human immunodeficiency virus
FDI	foreign direct investment	IABIN	Inter-American Biodiversity Information Network
FDRP	Flood Damage Reduction Program	ICAM	integrated coastal area management
FEMA	Federal Emergency Management Agency (United States)	ICARM	integrated coastal area and river basin management
FEWS	Famine Early Warning System	ICC	International Chamber of Commerce
FEWS NET	Famine Early Warning System Network	ICLEI	International Council for Local Environmental Initiatives
FMCN	Mexican Fund for Nature Conservation	ICM	integrated coastal management
FSC	Forest Stewardship Council	ICRAN	International Coral Reef Action Network
FSU	former Soviet Union	ICRI	International Coral Reef Initiative
FTAA	Free Trade Area for the Americas	ICT	information and communication technology
G7	Group of Seven: Canada, France, Germany, Italy, Japan, United Kingdom, United States	IDNDR	International Decade for Natural Disaster Reduction
G8	Group of Eight: Canada, France, Germany, Italy, Japan, Russian Federation, United Kingdom, United States	IEG	international environmental governance
		IFAD	International Fund for Agricultural Development
		IFF	Intergovernmental Forum on Forests
GATT	General Agreement on Tariffs and Trade	IIASA	International Institute for Applied System Analysis
GAW	Global Atmosphere Watch	IJC	International Joint Commission
GBIF	Global Biodiversity Information Facility	ILBM	integrated lake basin management
GCC	Gulf Cooperation Council	ILEC	International Lake Environment Committee Foundation
GCOS	Global Climate Observing System	ILO	International Labour Organization
GCRMN	Global Coral Reef Monitoring Network	IMF	International Monetary Fund
GDP	gross domestic product	IMO	International Maritime Organization
GEF	Global Environment Facility	INBO	International Network of Basin Organizations
GEMS	Global Environment Monitoring System	INDOEX	Indian Ocean Experiment
GEO	Global Environment Outlook	INEGI	Instituto Nacional de Geografía Estadística e Informática (Mexico)
GISP	Global Invasive Species Programme	IPCC	Intergovernmental Panel on Climate Change
GIWA	Global International Waters Assessment	IPF	Intergovernmental Panel on Forests
GLASOD	Global Assessment of Soil Degradation	IPM	integrated pest management
GLOF	glacial flood outburst lake	IPR	intellectual property rights
GLWQA	Great Lakes Water Quality Agreement	IRBM	integrated river basin management
GM	genetically modified	ISDR	International Strategy for Disaster Reduction
		ISO	International Organization for Standardization
		ITTO	International Tropical Timber Organization

IUCN	World Conservation Union (International Union for the Conservation of Nature and Natural Resources)	NIS	newly independent states	
IWC	International Whaling Commission	NO	nitrogen oxide	
IWRM	integrated water resources management	NO_2	nitrogen dioxide	
IWMI	International Water Management Institute	NO_x	nitrogen oxides	
IYM	International Year of Mountains	N_2O	nitrous oxide	
LADA	Land Degradation Assessment of Drylands	NPK	nitrogen, potassium and phosphorus (fertilizer)	
LCBP	Lake Champlain Basin Program (United States)	NSSD	national strategy for sustainable development	
LIFD	low-income food deficit	O_3	ozone	
LMMA	locally-managed marine area	OAU	Organization for African Unity	
LMO	living modified organism	ODA	official development assistance	
LPG	liquefied petroleum gas	ODS	ozone-depleting substance	
LRT	light rapid transit	OECD	Organisation for Economic Co-operation and Development	
LUCAS	European Land Use/Land Cover Statistical Survey	OCIPEP	Office of Critical Infrastructure and Emergency Preparedness, Canada	
MA	Millennium Ecosystem Assessment			
MAP	Mediterranean Action Plan	OSPAR	Convention for the Protection of the Marine Environment of the North-East Atlantic	
MARPOL	International Convention for the Prevention of Pollution from Ships			
MARS	Major Accident Reporting System	PACD	Plan of Action to Combat Desertification	
MCPFE	Ministerial Conference on the Protection of Forests in Europe	PAME	Protection of the Arctic Marine Environment	
		PCB	polychlorinated biphenyls	
MDGs	Millennium Development Goals	PCP	Permanent Cover Program (Canada)	
MEA	multilateral environmental agreement	PEBLDS	Pan-European Biological and Landscape Diversity Strategy	
MEMAC	Marine Emergency Mutual Aid Centre			
MERCOSUR	Mercado Común del Sur (Southern Common Market)	PEEN	Pan-European Ecological Network	
MPA	marine protected area	PEFC	Pan-European Forest Certification	
MRT	mass rapid transit	PERSGA	Protection of the Environment of the Red Sea and Gulf of Aden	
MSC	Marine Stewardship Council			
NAACO	National Ambient Air Quality Objectives (Canada)	PFRA	Prairie Farm Rehabilitation Administration (Canada)	
NAACS	National Ambient Air Quality Standards (United States)	PICs	Pacific Island Countries	
		PLUARG	Pollution from Land Use Activities Reference Group (Canada, United States)	
NABIN	North American Biodiversity Information Network			
NAFTA	North American Free Trade Agreement	PM	particulate matter. $PM_{2.5}$ has a diameter of 2.5 μm or less. PM_{10} has a diameter of 10 μm or less.	
NARSTO	North American Research Strategy for Tropospheric Ozone			
		POPs	persistent organic pollutants	
NAWMP	North American Waterfowl Management Plan	PRRC	Pasig River Rehabilitation Commission (Philippines)	
NCAR	National Center for Atmospheric Research (United States)	PSR	pressure-state-response	
		RAP	remedial action plan	
NEAP	National Environmental Action Plan	REMPEC	Regional Marine Pollution Emergency Response Centre for the Mediterranean Sea	
NECD	EU Directive on National Emission Ceilings for Certain Atmospheric Pollutants			
		RFMO	regional fish management organization	
NEPA	National Environmental Protection Agency, China	ROPME	Regional Organization for the Protection of the Marine Environment of the sea area surrounded by Bahrain, I.R. Iran, Iraq, Kuwait, Oman, Qatar, Saudi Arabia and the United Arab Emirates	
NEPM	National Environmental Protection Measure, Australia			
NGO	non-governmental organization			
NH_3	ammonia	SACEP	South Asia Cooperative Environment Programme	
NH_x	ammonia and ammonium	SADC	Southern African Development Community	
NSIDC	National Snow and Ice Data Center (United States)	SANAA	National Autonomous Water and Sewage Authority (Honduras)	

SAP	structural adjustment programme
SARA	Species at Risk Act (Canada)
SCOPE	Scientific Committee on Problems of the Environment
SCP	sustainable consumption and production
SEA	strategic environmental assessment
SEI	Stockholm Environment Institute
SIDS	Small Island Developing State or States
SO_2	sulphur dioxide
SO_x	sulphur oxides
SoE	state of the environment
SOPAC	South Pacific Applied Geosciences Commission
SPIRS	Seveso Plants Information Retrieval System
SPM	suspended particulate matter
SPRD	Strategic Planning and Research Department (Singapore)
SST	sea surface temperature
START	system for analysis, research and training
TAI	technology achievement index
TAO	tropical atmospheric ocean
TCA	Treaty for Amazonian Cooperation
TCDD	2,3,7,8-tetrachlorodibenzo-para-dioxin
TEA	Transportation Equity Act
TEK	traditional ecological knowledge
TEN	Trans-European Network
TFAP	Tropical Forestry Action Plan
TOE	tonnes of oil equivalent
TRAFFIC	Trade Records Analysis for Flora and Fauna in International Commerce
TRI	Toxics Release Inventory
TRIPs	trade-related aspects of international property rights
UEBD	Executive Unit for Settlements in Development (Honduras)
UK	United Kingdom
UN	United Nations
UNCCD	United Nations Convention to Combat Desertification
UNCED	United Nations Conference on Environment and Development
UNCHS	United Nations Centre for Human Settlements (now UN-Habitat)
UNCLOS	United Nations Convention on the Law of the Sea
UNCOD	United Nations Conference on Desertification
UNCTAD	United Nations Conference on Trade and Development
UNDAF	United Nations Development Assistance Framework
UNDP	United Nations Development Programme
UNEP	United Nations Environment Programme

UNEP-GPA	United Nations Environment Programme-Programme of Action for the Protection of the Marine Environment from Land-based Activities
UNEP-WCMC	United Nations Environment Programme-World Conservation Monitoring Centre
UNESCO	United Nations Educational, Scientific and Cultural Organization
UNF	United Nations Foundation
UNFCCC	United Nations Framework Convention on Climate Change
UNFF	United Nations Forum on Forests
UNHCR	United Nations High Commission for Refugees
UNICEF	United Nations Children's Fund
UNOCHA	United Nations Office for the Coordination of Humanitarian Affairs
UNSO	United Nations Sudano-Sahelian Office (now UNDP Office to Combat Desertification)
US	United States
USEPA	United States Environmental Protection Agency
USAID	United States Agency for International Development
USFWS	United States Fish and Wildlife Service
USGS	United States Geological Survey
UV	ultraviolet (A and B)
VOC	volatile organic compound
WBCSD	World Business Council for Sustainable Development
WCED	World Commission on Environment and Development
WCD	World Commission on Dams
WCP	World Climate Programme
WCS	World Conservation Strategy
WFD	Water Framework Directive (European Union)
WFP	World Food Programme
WHC	World Heritage Convention
WHO	World Health Organization
WHYCOS	World Hydrological Cycle Observing System
WIPO	World Intellectual Property Organization
WMO	World Meteorological Organization
WRI	World Resources Institute
WSSCC	Water Supply and Sanitation Collaborative Council
WSSD	World Summit on Sustainable Development
WTO	World Trade Organization
WWAP	World Water Assessment Programme
WWC	World Water Council
WWF	World Wide Fund for Nature
ZACPLAN	Zambezi River System Action Plan
ZAMCOM	Zambezi Basin Commission

Contributors

Listed below are individuals and institutions – from governments, collaborating centres, the scientific community and the private sector – who contributed to the *GEO-4* assessment in a variety of ways, and as participants in GEO regional and intergovernmental consultations.

AFRICA:

Anita Abbey, Youth Employment Summit Campaign, M & G Pharmaceuticals Ltd., Ghana

Mamoun Isa Abdelgadir, Ministry of Environment and Physical Development, Sudan

Maisharou Abdou, Ministère de l'Environnement et de la Lutte Contre la Désertification, Niger

Gustave Aboua, Université d'Abobo-Adjamé, Cote d'Ivoire

Melkamu Adisu, Kenya

Vera Akatsa-Bukachi, Kenyatta University, Kenya

Moise Aklé, c/o African Development Bank, Tunisia

Jonathan A. Allotey, Environmental Protection Agency, Ghana

David R. Aniku, Ministry of Environment, Wildlife and Tourism, Botswana

A. K. Armah, University of Ghana, Ghana

Joel Arumadri, National Environment Management Authority, Uganda

Samuel N. Ayonghe, Faculty of Science, University of Buea, Cameroon

Thomas Anatole Bagan, Ministère de l'Environnement, de l'Habitat et de l'Urbanisme, Benin

Philip Olatunde Bankole, Federal Ministry of Environment, Nigeria

Taoufiq Bennouna, Sahara and Sahel Observatory, Tunisia

Jean-Claude Bomba, University of Bangui, Central African Republic

Monday Sussane Businge, Gender and Environmental Law Specialist, Kenya

Adama Diawara, Consulate of Cote d'Ivoire, Kenya

Zephirin Dibi, Permanent Mission of the Republic of Côte d'Ivoire to UNEP, Ethiopia

Ismail Hamdi Mahmoud El-Bagouri, Desert Research Center, Egypt

Moheeb Abd El Sattar Ebrahim, Egyptian Environment Affairs Agency, Egypt

RoseEmma Mamaa Entsua-Mensah, Water Research Institute, Council for Scientific and Industrial Research, Ghana

Sahon Flan, Network for Environment and Sustainable Development in Africa, Côte d'Ivoire

Moustafa M. Fouda, Ministry for State of Environmental Affairs, Egypt

Tanyaradzwa Furusa, ZERO Regional Environment Organisation, Zimbabwe

Cuthbert Z. Gambara, Institute of Mining Research, University of Zimbabwe, Zimbabwe

Donald Gibson, SRK Consulting, South Africa

Elizabeth Gowa, Kenya

Kirilama Gréma, Ministère de l'Environnement et de la Lutte Contre la Désertification, Niger

Caroline Happi, Bureau Régional de l'UICN pour l'Afrique Centrale, Cameroon

Tim Hart, SRK Consulting, South Africa

Ahmed Farghally Hassan, Faculty of Commerce, Cairo University, Egypt

Qongqong Hoohlo, National Environment Management Authority, Lesotho

Pascal Valentin Houenou, Network for Environment and Sustainable Development in Africa, Côte d'Ivoire

Paul Jessen, Ministry of Agriculture, Water and Forestry, Namibia

Wilfred Kadewa (GEO Fellow), University of Malawi, Malawi

Alioune Kane, Universite Cheikh Anta Diop, Senegal

Eucharia U. Kenya, Kenyatta University, Kenya

Darryll Kilian, SRK Consulting, South Africa

Seleman Kisimbo, Division of Environment, The Vice President's Office, United Republic of Tanzania

Michael K. Koech, Kenyatta University, Kenya

Germain Kombo, Ministère de l'Economie Forestière et de l'Environnement, Congo

Mwebihire Kwisenshoni, Permanent Mission of the Republic of Uganda to UNEP, Kenya

Ebenezer Laing, University of Ghana, Ghana

Jones Arthur Lewis (GEO Fellow), Twene Amanfo Secondary School, Ghana

Estherine Lisinge-Fotabong, NEPAD Secretariat, South Africa

Samuel Mabikke, Conservation Trust, Uganda

Lindiwe Mabuza, Council for Scientific and Industrial Research, South Africa

M. Amadou Maiga, Institutionnel de la Gestion des Questions Environnementales, Mali

Nathaniel Makoni, ABS TCM Ltd., Kenya

Peter Manyara (GEO Fellow), Egerton University, Kenya

Deborah Manzolillo Nightingale, Environmental Management Advisors, Kenya

Gerald Makau Masila, British American Tobacco, Kenya

Bora Masumbuko, Network for Environment and Sustainable Development in Africa, Côte d'Ivoire

Simon Mbarire, National Environment Management Authority, Kenya

Likhapha Mbatha, Centre for Applied Legal Studies, University of the Witwatersrand, South Africa

Maria Mbengashe, Biodiversity and Marine International Cooperation, South Africa

John Masalu Phillip Mbogoma, Basel Convention Regional Centre For English-Speaking African Countries, c/o Council for Scientific and Industrial Research, South Africa

Charles Muiruri Mburu, British American Tobacco, Kenya

Salvator Menyimana, Permanent Mission of the Republic of Burundi to UNEP, Kenya

Jean Marie Vianney Minani, Rwanda Environmental Management Authority, Ministry of Land, Environment, Forestry, Water and Mines, Rwanda

Rajendranath Mohabeer, Indian Ocean Commission, Mauritius

Hana Hamadalla Mohammed, High Council of Environment and Natural Resources, Sudan

Crepin Momokama, Agence Internationale pour le Développement de l'Information Environnementale, Gabon

Elizabeth Muller, Council for Scientific and Industrial Research, South Africa

Betty Muragori, IUCN – The World Conservation Union, Kenya

Constansia Musvoto, Institute of Environmental Studies, University of Zimbabwe, Zimbabwe

Shaban Ramadhan Mwinjaka, Division of Environment, The Vice President's Office, United Republic of Tanzania

Omari Iddi Myanza, Ministry of Water, Lake Victoria Environment Management Project, United Republic of Tanzania

Alhassane Savané, Consulate of Cote d'Ivoire, Kenya

Déthié Soumaré Ndiaye, Centre de Suivi Ecologique, Senegal

Jacques-André Ndione, Centre de Suivi Ecologique, Senegal

Martha R. Ngalowera, Division of Environment, The Vice President's Office, United Republic of Tanzania

David Samuel Njiki Njiki, Interim Secretariat of NEPAD Environment Component, Senegal

Musisi Nkambwe, University of Botswana, Botswana

Dumisani Nyoni, Organisation of Rural Associations for Progress, Zimbabwe

Beatrice Nzioka, National Environment Management Authority, Kenya

Tom Okurut, East African Community, United Republic of Tanzania

Ayola Olukanni, Permanent Mission of the Federal Republic of Nigeria to UNEP, Nigeria High Commission, Kenya

Scott E. Omene, Permanent Mission of the Federal Republic of Nigeria to UNEP, Nigeria High Commission, Kenya

Joyce Onyango, National Environment Management Authority, Kenya

Oladele Osibanjo, Basel Convention Regional Coordinating Centre for Africa for Training and Technology Transfer, Federal Ministry of Environment – University of Ibadan, Nigeria

Joseph Qamara, Division of Environment, The Vice President's Office, United Republic of Tanzania

John L. Roberts, Indian Ocean Commission, Mauritius

Vladimir Russo, Ecological Youth of Angola, Angola

Shamseldin M. Salim, Common Market For Eastern and Southern Africa Secretariat, Zambia

Bob Scholes, Council for Scientific and Industrial Research, South Africa

Alinah Segobye, University of Botswana, Botswana

Riziki Silas Shemdoe, Institute of Human Settlements Studies, University College of Lands and Architectural Studies, United Republic of Tanzania

Teresia Katindi Sivi, Institute of Economic Affairs, Kenya

Emelia Sukutu, Environmental Council of Zambia, Zambia

Fanuel Tagwira, Faculty of Agriculture and Natural Resources, Africa University, Zimbabwe

The Department of Environment and Natural Resources of the Ministry of Tourism, Environment and Natural Resources of Zambia, Zambia

Adelaide Tillya, Permanent Mission of United Republic of Tanzania to UNEP, Kenya

Zabeirou Toudjani, Ministère de l'Environnement et de la Lutte Contre la Désertification, Niger

Alamir Sinna Toure, Ministère de l'Environnement et de l'Assainissement, Mali

Evans Mungai Mwangi, University of Nairobi, Kenya

Chantal Will, Council for Scientific and Industrial Research, South Africa

Nico E. Willemse, Ministry of Environment and Tourism, Namibia

Benon Bibbu Yassin, Environmental Affairs Department, Malawi

Ibrahim Zayan, Egypt

Naoual Zoubair, Observatoire National de l'Environnement, Direction des Etudes, de la Planification et de la Prospective, Ministère de l'Aménagement de Territoire, de l'Eau et de l'Environnement, Morocco

Edward H. Zulu, Environmental Council of Zambia, Zambia

ASIA AND THE PACIFIC:

Sanit Aksornkoae, Thailand Environment Institute, Thailand

Mozaharul Alam, Bangladesh Centre for Advanced Studies, Bangladesh

Jayatunga A. Amaraweera, Buddhist and Pali Unversity of Sri Lanka, Sri Lanka

Iswandi Anas, Bogor Agricultural University, Indonesia

Ratnasari Anwar, Ministry of the Environment, Indonesia

Australian Government Department of the Environment and Water Resources, Australia

Lawin Bastian, Ministry of the Environment, Indonesia

Si Soon Beng, Ministry of the Environment and Water Resources, Republic of Singapore

Arvind Anil Boaz, South Asia Cooperative Environment Programme, Sri Lanka

Liana Bratasida, Ministry of the Environment, Indonesia

Chuon Chanrithy, Ministry of Environment, Cambodia

Chaveng Chao, Government and Association Liaison, Bayer Thai Company Limited, Thailand

Weixue Cheng, State Environmental Protection Administration, China

Muhammed Quamrul Islam Chowdhury, Asia-Pacific Forum of Environmental Journalists, Bangladesh

Michael R. Co, Clean Air Initiative for Asian Cities Center, Philippines

Dalilah Dali, Ministry of Natural Resources and Environment, Malaysia

Pham Ngoc Dang, Hanoi University of Civil Engineering, Vietnam

Elenita C. Dano, Third World Network, Philippines

Surakit Darncholvichit, Ministry of Natural Resources and Environment, Thailand

Vikram Dayal, The Energy and Resources Institute, India

Elenda Del Rosario Basug, Department of Environment and Natural Resources, Philippines

Bhujangarao Dharmaji, Ecosystem and Livelihoods Group, IUCN – The World Conservation Union, Sri Lanka

Fiu Mataese Elisara, O le Siosiomaga Society Incorporated, Samoa

Kheirghadam Enayatzamir (GEO Fellow), Soil and Water Science Department, Agriculture Faculty, Tehran University, Islamic Republic of Iran

Neil Ericksen, International Global Change Institute, University of Waikato, New Zealand

Muhammad Eusuf, Bangladesh Centre for Advanced Studies, Bangladesh

Daniel P. Faith, The Australian Museum, Australia

Sota Fukuchi, Ministry of the Environment, Japan

Min Jung Gi, Ministry of the Environment, Republic of Korea

Harka B. Gurung, National Environment Commission, Bhutan

Siti Aini Hanum, Ministry of the Environment, Indonesia

Xiaoxia He, Peking University, c/o State Environmental Protection Administration, China

Saleemul Huq, Bangladesh Centre for Advanced Studies, Bangladesh

Toshiaki Ichinose, National Institute for Environmental Studies, Japan

Saeko Ishihama, Ministry of the Environment, Japan

Zahra Javaherian, Department of the Environment, Islamic Republic of Iran

Suebsthira Jotikasthira, The Industrial Environment Institute, The Federation of Thailand Industries, Thailand

Mahmood A. Khwaja, Sustainable Development Policy Institute, Pakistan

Somkiat Khokiattiwong, Ministry of Natural Resources and Environment, Thailand

Satoshi Kojima, Institute for Global Environmental Strategies, Japan

Santosh Ragavan Kolar (GEO Fellow), The Energy and Resources Institute, India

Pradyumna Kumar Kotta, South Asia Cooperative Environment Programme, Sri Lanka

Margaret Lawton, Landcare Research, New Zealand

Lee Bea Leang, Department of Irrigation and Drainage, Ministry of Natural Resources and Environment, Malaysia

Xinmin Li, State Environmental Protection Administration, China

Ooi Giok Ling, Nanyang Technological University, Republic of Singapore

Qifeng Liu, State Environmental Protection Administration, China

Chou Loke-Ming, National University of Singapore, Republic of Singapore

Shengji Luan, Peking University, c/o State Environmental Protection Administration, China

Ranjith Mahindapala, IUCN Asia Regional Office, Thailand

Sansana Malaiarisson, Thailand Environment Institute, Thailand

Sunil Malla, Asian Institute of Technology, Thailand

Irina Mamieva, Scientific Information Center of Interstate Sustainable Development Commission, Turkmenistan

Melchoir Mataki, The University of the South Pacific, Fiji

Wendy Yap Hwee Min, Association of Southeast Asian Nations Secretariat, Indonesia

Umar Karim Mirza, Pakistan Institute of Engineering and Applied Sciences, Pakistan

Chiaki Mizugaki, Fisheries Agency of Japan, Japan

Hasan Moinuddin, Greater Mekong Subregion Environment Operations Center, Thailand

Kunihiro Moriyasu, Ministry of Land, Infrastructure and Transport, Japan

Hasna J. Moudud, Coastal Area Resource Development and Management Association, Bangladesh

Victor S. Muhandiki, Ritsumeikan University, Japan

Suyanee Nachaiyasit, Ministry of Natural Resources and Environment, Thailand

Rajesh Nair, National Institute for Environmental Studies, Japan

Masahisa Nakamura, Shiga University and International Lake Environment Committee Foundation, Japan

Shuya Nakatsuka, Fisheries Agency of Japan, Japan

Adilbek Nakipov, Ministry of Environment Protection, Republic of Kazakhstan

K. K. Narang, Ministry of Environment and Forests, India

Somrudee Nicro, Thailand Environment Institute, Thailand

Shilpa Nischa, The Energy and Resources Institute, India

Akira Ogihara, Institute for Global Environmental Strategies, Japan

Tomoaki Okuda, Keio University, Japan

Kongsaysy Phommaxay, Science Technology and Environment Agency, Lao People's Democratic Republic

Warasak Phuangcharoen, Ministry of Natural Resources and Environment, Thailand

Chumnarn Pongsri, Mekong River Commission Secretariat, Lao People's Democratic Republic

Bidya Banmali Pradhan, The International Centre for Integrated Mountain Development, GRID-Kathmandu, Nepal

Eric Quincieu, Eco 4 the World, Republic of Singapore

Atiq Rahman, Bangladesh Centre for Advanced Studies, Bangladesh

Danar Dulatovich Raissov, Economic Research Institute, Republic of Kazakhstan

Lakshmi Rao (GEO Fellow), Dorling Kindersley India Pvt. Ltd., India

Karma Rapten, National Environment Commission, Bhutan

Taku Sasaki, Fisheries Agency of Japan, Japan

Ram Manohar Shrestha, Asian Institute of Technology, Thailand

Chiranjeevi L. Shrestha (Vaidya), Freelance Environmentalist, Nepal

Qing Shu, State Environmental Protection Administration, China

Reiko Sodeno, Ministry of the Environment, Japan

Manasa Sovaki, Department of Environment, Ministry of Environment, Fiji

Wijarn Simachaya, Mekong River Commission Secretariat, Lao People's Democratic Republic

Wataru Suzuki, Ministry of the Environment, Japan

Anoop Swarup, Global Knowledge Alliance, Australia

Taeko Takahashi, Institute for Global Environmental Strategies, Japan

Yukari Takamura, National Institute for Environmental Studies, Japan

Pramote Thongkrajaai, Huachiew Chalermprakiet University, Thailand

The Ministry of Forestry, Myanmar

Tawatchai Tingsanchali, Asian Institute of Technology, Thailand

Sujitra Vassanadumrongdee, Thailand Environment Institute, Thailand

Kazuhiro Watanabe, Ministry of the Environment, Japan

Don Wijewardana, International Forestry Consultant, New Zealand

Wipas Wimonsate, Thailand Environment Institute, Thailand

Guang Xia, State Environmental Protection Administration, China

Qinghua Xu, State Environmental Protection Administration, China

Makoto Yamauchi, Fisheries Agency of Japan, Japan

Wang Yi, Chinese Academy of Sciences, China

Ruisheng Yue, State Environmental Protection Administration, China

Ahn Youn-Kwang, Ministry of Environment, Republic of Korea

Jieqing Zhang, State Environmental Protection Administration, China

EUROPE:

Eva Adamová, Department of Environmental Policy and Multilateral Relations, Ministry of the Environment, Czech Republic

Juliane Albjerg, Ministry of Environment, Denmark

Chris Anastasi, British Energy plc, United Kingdom

Georgina Ayre, Department of Environment, Food and Rural Affairs, United Kingdom

Mariam Bakhtadze, Ministry of Environment of Georgia, Georgia

Jan Bakkes, Netherlands Environmental Assessment Agency, The Netherlands

Snorri Baldursson, Icelandic Institute of Natural History, Iceland

Richard Ballaman, Federal Office for the Environment, Switzerland

Anna Ballance, Department for International Development, United Kingdom

C. J. (Kees) Bastmeijer, Faculty of Law, Tilburg University, The Netherlands

Steffen Bauer, German Development Institute, Germany

Rainer Beike, The City Council of Munster, Germany

Stanislav Belikov, All-Russian Research Institute for Nature Protection, Russian Federation

Pascal Bergeret, Ministère de l'agriculture et de la pêche, France

John Michael Bewers, Andorra

Rut Bizková, Ministry of the Environment, Czech Republic

Gunilla Björklund, GeWa Consulting, Sweden

Line Bjørklund Ministry of Environment, Denmark

Antoaneta Boycheva, International Activity Directorate, Ministry of State Policy for Disasters and Accidents, Bulgaria

Anne Burrill, DG Environment, European Commission, Belgium

Elena Cebrian Calvo, European Environment Agency, Denmark

Rada Chalakova, Environmental Strategies and Programmes Department, Ministry of Environment and Water, Bulgaria

Fiona Charlesworth, Department for Environment, Food and Rural Affairs, United Kingdom

Laila Rebecca Chicoine, Bee Successful Limited, United Kingdom

Petru Cocirta, Institute of Ecology and Geography of the Academy of Sciences, Republic of Moldova

Laurence Colinet, Ministère de l'écologie, du développement et de l'aménagement durables, France

Peter Convey, British Antarctic Survey, Natural Environment Research Council, United Kingdom

Wolfgang Cramer, Potsdam Institute for Climate Impact Research, Denmark

Marie Cugny-Seguin, Institut national de l'environnement, France

Angel Danin, National Transport Policy Directorate, Ministry of Transport, Bulgaria

Francois Dejean, European Environment Agency, Denmark

A. J. (Ton) Dietz, Department of Geography, Planning and International Development Studies, University of Amsterdam, Netherlands

Yana Dordina, Russian Assoication of Indigenous Peoples of the North, Russian Federation

Carine Dunand, Federal Office for the Environment, Switzerland

John F. Dunn, DG Environment, European Commission, Belgium

Ida Edwertz, Division for International Affairs, Ministry of the Environment, Sweden

Bob Fairweather, United Kingdom Mission to the United Nations, Switzerland

Malin Falkenmark, Stockholm International Water Institute, Sweden

Jaroslav Fiala, European Environment Agency, Denmark

Richard Fischer, Programme Coordinating Centre of ICP Forests Institute for World Forestry, Germany

Tonje Folkestad, World Wide Fund for Nature, Norway

Karolina Fras, DG Environment, European Commission, Belgium

Atle Fretheim, Ministry of the Environment, Norway

Pierluca Gaglioppa, Nature Reserve Monterano (Rome) – Latium regional Forest Service, Italy

Nadezhda Gaponenko, Analytical Center on Science and Industrial Policy, Russian Academy of Sciences, Russian Federation

Emin Garabaghli, Ministry of Ecology and Natural Resources, Azerbaijan

Anna Rita Gentile, European Environment Agency, Denmark

Amparo Rambla Gil, Ministerio de Medio Ambiente, Spain

Francis Gilbert, The University of Nottingham, United Kingdom

Armelle Giry, Ministère de l'écologie, du développement et de l'aménagement durables, France

Johanna Gloël, University of Tuebingen, Germany

Genady Golubev, Faculty of Geography, Moscow State University, Russian Federation

Elitsa Gotseva, Air Protection Directorate, Ministry of Environment and Water, Bulgaria

Michael Graber, Environmental Consultant, Israel

Alan Grainger, School of Geography, University of Leeds, United Kingdom

Eva-Jane Haden, World Business Council for Sustainable Development, Switzerland

Peter Hadjistoykov, Working Conditions, Management of Crisis and Alternative Service Directorate, Ministry of Labour and Social Policy, Bulgaria

Tomas Hak, Charles University Environmental Centre, Czech Republic

Katrina Hallman, International Secretariat, Swedish Environmental Protection Agency, Sweden

Neil Harris, University of Cambridge, United Kingdom

David Henderson-Howat, Forestry Commission, United Kingdom

Thomas Henrichs, European Environment Agency, Denmark

Rolf Hogan, World Wide Fund for Nature – Convention on Biological Diversity, Switzerland

Ybele Hoogeveen, European Environment Agency, Denmark

Joy A. Kim, Organisation for Economic Cooperation and Development, France

Carlos Solana Ibero, CITES Animals Committee for Europe, Spain

Gytautas Ignatavicius, Environment Protection Agency, Ministry of Environment, Lithuania

Bilyana Ivanova, Ministry of Environment and Water, Bulgaria

Esko Jaakkola, Ministry of Environment, Finland

Andrzej Jagusiewicz, Department of Monitoring, Assessment and Outlooks, Environmental Protection, Poland

Ryszard Janikowski, Institute for Ecology of Industrial Areas, Poland

Dorota Jarosinska, European Environment Agency, Denmark

Karen Jenderedjian, Agency of Bioresources Management, Ministry of Nature Protection, Republic of Armenia

Peder Jensen, European Environment Agency, Denmark

Andre Jol, European Environment Agency, Denmark

Svetlana Jordanova, Energy Efficiency and Environmental Protection Directorate, Ministry of Economy and Energy, Bulgaria

Nazneen Kanji, International Institute for Environment and Development, United Kingdom

Jan Karlsson, European Environment Agency, Denmark

Pawel Kazmierczyk, European Environment Agency, Denmark

Bruno Kestemont, Statistics Belgium, Belgium

Gilles Kleitz, Ministère de l'écologie, du développement et de l'aménagement durables, France

Peter Kristensen, European Environment Agency, Denmark

Alexey Kokorin, World Wildlife Fund – Russian Federation, Russian Federation

Marianne Kroglund, Ministry of the Environment, Norway

Hagen Krohn, University of Tuebingen, Germany

Carmen Lacambra-Segura, Department of Geography, St Edmunds College, University of Cambridge, United Kingdom

Robert Lamb, Federal Office for the Environment, Switzerland

Tor-Björn Larsson, European Environment Agency, Denmark

Patrick Lavelle, Institut de recherche pour le développement, France

Alois Leidwein, Attaché for Agricultural and Environmental Affairs, Permanent Mission of Austria, Switzerland

Øyvind Lone, Ministry of the Environment, Norway

Jacques Loyat, Ministère de l'agriculture et de la pêche, France

Rob Maas, Netherlands Environmental Assessment Agency, The Netherlands

Elena Manvelian, Armenian Women for Health and a Healthy Environment, Republic of Armenia

Pedro Vega Marcote, Facultad de Ciencias de la Educación, Universidad de A Coruña, Spain

Jovanka Maric, Directorate for Environmental Protection, Department for International Cooperation, Ministry of Science and Environmental Protection, Republic of Serbia

Roberto Martin-Hurtado, Environment Directorate, Organisation for Economic Co-operation and Development, France

Miguel Antolin Martinez, Ministerio de Medio Ambiente, Spain

Julian Maslinkov, Climate Change Policy Department, Ministry of Environment and Water, Bulgaria

Jan Mertl, Czech Environmental Information Agency, Czech Republic

Maria Minova, Energy Efficiency and Environmental Protection Directorate, Ministry of Economy and Energy, Bulgaria

Ruben Mnatsakanian, Central European University, Hungary

Richard Moles, Centre for Environmental Research, University of Limerick, Ireland

Miroslav Nikcevic, Directorate for Environmental Protection, Ministry of Science and Environmental Protection, Republic of Serbia

Stefan Norris, World Wildlife Fund International Arctic Programme, Norway

Markus Ohndorff, ETH Zürich Institute for Environmental Decisions, Switzerland

Bernadette O'Regan, Centre for Environmental Research, University of Limerick, Ireland

Olav Orheim, Norwegian Research Council, Norway

Larisa Orlova, Centre for International Projects, Russian Federation

Siddiq Osmani, School of Economics and Politics, University of Ulster, United Kingdom

Paul Pace, Centre for Environmental Education and Research, Malta

Renat Perelet, Institute for Systems Analysis, Russia

Tania Plahay, Department for Environment, Food and Rural Affairs, United Kingdom

Jan Pokorn, Czech Environmental Information Agency, Czech Republic

Franz Xaver Perrez, Federal Office for the Environment, Switzerland

Nicolas Perritaz, Federal Office for the Environment, Switzerland

Hanne K. Petersen, Danish Polar Center, Denmark

Iva Petrova, Energy Market and Restructuring Directorate, Ministry of Economy and Energy, Bulgaria

Marit Victoria Pettersen, Ministry of the Environment, Norway

Attila Rábai, Environmental Informatics Division, Ministry of Environment and Water, Hungary

Hanna Rådberg, Swedish Ecodemics, Sweden

Ortwin Renn, University of Stuttgart, Institute for Social Sciences, Germany

Dominique Richard, Museum National d'Histoire Naturelle, France

Louise Rickard, European Environment Agency, Denmark

Odd Rogne, International Arctic Science Committee, Norway

José Romero, Federal Office for the Environment, Switzerland

Laurence Rouïl, Institut National de l'Environnement Industriel et des Risques, France

Ahmet Saatchi, Marmara University, Turkey

Guillaume Sainteny, Ministère de l'écologie, du développement et de l'aménagement durables, France

Guri Sandborg, Ministry of the Environment, Norway

Sergio Álvarez Sánchez, Ministerio de Medio Ambiente, Spain

Gunnar Sander, European Environment Agency, Denmark

Anna Schin, European Bank for Reconstruction and Development, United Kingdom

Gabriele Schöning, European Environment Agency, Denmark

Astrid Schulz, German Advisory Council on Global Change, WBGU Secretariat, Germany

Stefan Schwarzer, Global Resource Information Database, Geneva, Switzerland

Nino Sharashidze, Ministry for Environmental Protection and Natural Resources of Georgia, Georgia

Sanita Sile, Information Exchange Department, Latvian Environment, Geology and Meteorology Agency, Latvia

Viktor Simoncic, Sivicon, Croatia

Jerome Simpson, The Regional Environmental Center for Central and Eastern Europe, Hungary

Agnieszka Skowronska, Department of Strategic Management and Logistics, Faculty of Regional Economy and Tourism, Academy of Economics in Wroclaw, Poland

Anu Soolep, Estonian Environment Information Centre, Estonia

Danielle Carpenter Sprüngli, World Business Council for Sustainable Development, Switzerland

Rania Spyropoulou, European Environment Agency, Denmark

Lindsay Stringer, School of Environment and Development, University of Manchester, United Kingdom

Larry Stapleton, Department of the Environment, Heritage and Local Government, Environment International, Environmental Protection Agency, Ireland

George Strongylis, DG Environment, European Commission, Belgium

Rob Swart, Netherlands Environmental Assessment Agency, The Netherlands

Elemér Szabo, Ministry of Environment and Water, Hungary

José V. Tarazona, Department of Environment, Spanish National Institute for Agriculture and Food Research and Technology, Spain

Tonnie Tekelenburg, Netherlands Environmental Assessment Agency, The Netherlands

Nevyana Teneva, Water Directorate, Ministry of Environment and Water, Bulgaria

Sideris P. Theocharapoulos, National Agricultural Research Foundation, Greece

Anastasiya Timoshyna, Central European University, Hungary

Ferenc L. Toth, International Atomic Energy Agency, Austria

Camilla Toulmin, International Institute for Environment and Development, United Kingdom

Sébastien Treyer, Ministère de l'écologie, du développement et de l'aménagement durables, France

Milena Tzoleva, Energy Strategies Directorate, Ministry of Economy and Energy, Bulgaria

Edina Vadovics, Central European University, Hungary

Vincent Van den Bergen, Ministry of Housing, Spatial Planning and the Environment, The Netherlands

Kurt van der Herten, DG Environment, European Commission, Belgium

Irina Vangelova, International Activity Directorate, Ministry of State Policy for Disasters and Accidents, Bulgaria

Patrick Van Klaveren, Ministère d'Etat, Monaco

Philip van Notten, International Centre for Integrated Assessment and Sustainable Development, Maastricht University, The Netherlands

Bas van Ruijven, Netherlands Environmental Assessment Agency, The Netherlands

Victoria Rivera Vaquero, Ministerio de Medio Ambiente, Spain

Katya Vasileva, Coordination of Regional Inspectorates of Environment and Water Directorate, Ministry of Environment and Water, Bulgaria

Raimonds Vejonis, Ministry of the Environment of the Republic of Latvia, Latvia

Guus J. M. Velders, Netherlands Environmental Assessment Agency, The Netherlands

Sibylle Vermont, Federal Office for the Environment, Switzerland

Kamil Vilinovic, Environmental Policy and Foreign Affairs Section, Ministry of Environment of the Slovak Republic, Slovakia

Axel Volkery, Environmental Policy Research Unit, Free, University of Berlin, Germany;

Bart Wesselink, Netherlands Environmental Assessment Agency, The Netherlands

Peter D. M. Weesie, University of Groningen, The Netherlands

Wolfgang Weimer-Jehle, University of Stuttgart, Institute for Social Sciences, Germany

Beate Werner, European Environment Agency, Denmark

Mona Mejsen Westergaard, Ministry of Environment, Denmark

Manuel Winograd, European Environment Agency, Denmark

Rebekah Young, World Business Council for Sustainable Development, Switzerland

Dimitry Zamolodchikov, Eco-Accord Center, Russian Federation

Svetlana Zhekova, Mission of Bulgaria to the European Communities, Belgium

Karl-Otto Zentel, Deutsches Komitee Katastrophenvorsorge, Germany

LATIN AMERICA AND THE CARIBBEAN:

Elena Maria Abraham, Instituto Argentino de Investigaciones de las Zonas Áridas, Argentina

Ilan Adler, International Renewable Resources Institute, Mexico

Elaine Gomez Aguilera, Agencia de Medio Ambiente, Ministerio de Ciencia Tecnologia y Medio Ambiente, Cuba

Ollin Ahuehuetl, Mexico

Gisela Alonso, Agencia de Medio Ambiente, Cuba

Germán Andrade, Fundación Humedales, Colombia

Afira Approo, Caribbean Regional Environmental Network, Barbados

Patricia Aquing, Caribbean Environmental Health Institute, Saint Lucia

Carmen Arevalo, Independent Consultant, Colombia

Francisco Arias, Instituto de Investigaciones Marinas y Costeras, Colombia

Dolors Armenteras, Instituto de Investigación de Recursos Biológicos Alexander Von Humboldt, Colombia

Delver Uriel Báez Duarte, Club de Jóvenes Ambientalistas, Nicaragua

Garfield Barnwell, Caribbean Community Secretariat, Guyana

Giselle Beja, Ministerio de Vivienda, Ordenamiento Territorial y Medio Ambiente, Uruguay

Salisha Bellamy, Ministry of Agriculture, Trinidad and Tobago

Jesus Beltran, Centro de Ingenieria y Manejo Ambiental de Bahias y Costas, Cuba

Byron Blake, Independent Consultant, Jamaica

Teresa Borges, Ministerio de Ciencia, Tecnología y Medio Ambiente, Cuba

Rubens Harry Born, Institute for Development, Environment and Peace, Brazil

Eduardo Calvo, Universidad Nacional Mayor de San Marcos, Perú

Mariela C. Cánepa Montalvo, GEO Juvenil Perú, CONAM, Perú

Juan Francisco Castro, Universidad del Pacífico, Perú

Luis Paz Castro, Instituto de Meteorología, Agencia de Medio Ambiente, Ministerio de Ciencia, Tecnología y Medio Ambiente, Cuba

Sonia Catasús, Centro de Estudios Demográficos, Universidad de la Habana, Cuba

Loraine Charles, Bahamas Environment, Science and Technology Commission, Bahamas

Emil Cherrington, Water Center for the Humid Tropics of Latin America and the Caribbean, Panamá

Nancy Chuaca, Consejo Nacional del Ambiente, Perú

Luis Cifuentes, Pontificia Universidad Católica de Chile, Chile

Julio C. Cruz, Mexico

Crispin D'Auvergne, Ministry of Physical Development, Environment and Housing, Saint Lucia

Marly Santos da Silva, Secretaria Executiva, Ministério do Meio Ambiente, Brazil

Guadalupe Menéndez de Flores, Ministerio de Medio Ambiente y Recursos Naturales, El Salvador

Juan Ladrón de Guevara G., Comisión Nacional del Medio Ambiente, Chile

Roberto De La Cruz, Autoridad Nacional del Ambiente, Panamá

Genoveva Clara de Mahieu, Instituto de Medio Ambiente y Ecología, Universidad del Salvador, Argentina

Benita von der Groeben de Oetling, Consejo Nacional de Industriales Ecologistas de México, México

Enma Diaz-Lara, Ministry of Natural Resources and Environment, Guatemala

Jean Max Dimitri Norris, Ministère de l'Environnnement, Haiti

Edgar Ek, Land Information Centre, Ministry of Natural Resources and the Environment, Belize

Daniel Escalona, Ministerio del Ambiente y de los Recursos Naturales Renovables, Venezuela

Argelia Fernández, Agencia de Medio Ambiente, Ministerio de Ciencia Tecnología y Medio Ambiente, Cuba

Margarita Parás Fernández, Centro de Investigación en Geografía y Geomática – Centro GEO, México

Maria E. Fernández, Universidad Nacional Agraria La Molina, Perú

Raúl Figueroa, Instituto Nacional de Estadística Geografia e Informatica, México

Guillaume Fontaine, Facultad Latinoamericana de Ciencias Sociales, Ecuador

Patricia Peralta Gainza, Centro Latino Americano de Ecología Social, Uruguay

Maurício Galinkin, Fundação Centro Brasileiro de Referência e Apoio Cultural, Brazil

Guillermo García, Instituto de Meteorología, Agencia de Medio Ambiente, Ministerio de Ciencia, Tecnología y Medio Ambiente, Cuba

Fernando Gast, Institute Humboldt, Colombia

Héctor Daniel Ginzo, Ministry of Foreign Affairs, Argentina

Deborah Glaser, Island Resources Foundation, United States

Agustín Gómez, Observatorio del Desarrollo, Universidad de Costa Rica, Costa Rica

Alberto Gómez, Centro Uruguayo de Tecnologías Apropiadas, Uruguay

Rosario Gómez, Centro de Investigación de la Universidad del Pacífico, Perú

Claudia A. Gómez Luna, Centro de Educación y Capacitación para el Desarrollo Sustentable, México

Ricardo Grau, Laboratorio de Investigaciones Ecológicas de las Yungas, Universidad Nacional de Tucuman Casilla de Correo, Argentina

Jenny Gruenberger, Liga de Defensa del Medio Ambiente, Bolivia

Eduardo Gudynas, Centro Latino Americano de Ecología Social, Uruguay

Luz Elena Guinand, Secretaría de la Comunidad Andina, Perú

Gonzalo Gutiérrez, Centro Latino Americano de Ecología Social, Uruguay

Alejandro Falcó, Environmental Consultant, Argentina

Sandra Hacon, Fiocruz News Agency, Brazil

Romy Montiel Hernández, Ministerio de Ciencia, Tecnología y Medio Ambiente, Cuba

Laura Hernández-Terrones, Center for Studies on Water, Mexico

Guilherme Pimentel Holtz, Brazilian Institute of The Environment and Renewable Natural Resources, Brazil

Silvio Jablonski, State University of Rio de Janeiro, Brazil

Anita James, Ministry of Agriculture, Forestry and Fisheries, Saint Lucia

Luiz Fernando K. Merico, Instituto Brasileiro do Meio Ambiente e dos Recursos Naturais Renováveis, Brazil

Joanna Noelia Kamiche, Centro de Investigación Universidad del Pacifico, Perú

Elma Kay, University of Belize, Belize

Timothy Killeen, Conservation International, Bolivia

Julian Kenny, National Institute for Space Research, Trinidad and Tobago

Ana Maria Kleymeyer, Office of International Environmental Issues, Secretariat of Environment and Sustainable Development, Argentina

Amoy Lum Kong, The Institute of Marine Affairs, Trinidad and Tobago

David Kullock, Secretaría de Ambiente y Desarrollo Sustentable, Argentina

Iván Lanegra, Consejo Nacional del Ambiente, Perú

Beatriz Leal, Universidad Metropolitana, Venezuela

Kenrick R. Leslie, The Caribbean Community Climate Change Centre, Belize

Juliana León, Mexico

Rafael Lima, Land Information Centre, Ministry of Natural Resources and the Environment, Belize

Juan F. Llanes-Regueiro, Facultad de Economía, Universidad de la Habana, Cuba

Fernando Antonio Lyrio Silva, Ministerio de Medio Ambiente, Brazil

Manuel Madriz, Association of Caribbean States, Trinidad and Tobago

Vicente Paeile Marambio, Comisión Nacional del Medio Ambiente, Chile

Laneydi Martínez, Centro de Investigaciones de la Economía Mundial, Cuba

Osvaldo Martínez, Centro de Investigaciones de la Economía Mundial, Cuba

Arturo Flores Martinez, Secretaría de Medio Ambiente y Recursos Naturales, México

Juan Mario Martínez, Agencia de Medio Ambiente, Cuba

Julio Torres Martínez, Observatorio Cubano de Ciencia y Tecnología, Academia de Ciencias de Cuba, Cuba

Rosina Methol, Centro Latino Americano de Ecología Social, Uruguay

Napoleao Miranda, ISER Parceria 21, Universidad Federal Fluminese, Brazil

Elizabeth Mohammed, Fisheries Division, Ministry of Agriculture, Trinidad and Tobago

Maria da Piedade Morais, Department of Regional and Urban Studies, Institute of Applied Economic Research, Brazil

Amílcar Morales, Centro de Investigación en Geografía y Geomática – Centro GEO, México

Cristóbal Díaz Morejón, Ministerio de Ciencia, Tecnología y Medio Ambiente, Cuba

Evandro Mateus Moretto, Ministerio de Medio Ambiente, Brazil

Scott Agustín Muller, Conservación y Desarrollo Sostenible en Acción, Panamá

Javier Palacios Neri, Secretaría de Medio Ambiente y Recursos Naturales, México

Jorge Madeira Nogueira, Universidade de Brasília, Brazil

Kenneth Ochoa, Organización Juvenil Ambiental, Universidad El Bosque, Colombia

Luis Oliveros, Organizción del Tratado de Cooperación Amazónica, Brazil

Carlos Sandoval Olvera, Consejo Nacional de Industriales Ecologistas de México, México

Hazel Oxenford, Centre for Resource Management and Environmental Studies, University of the West Indies, Barbados

Raúl Estrada Oyuela, Ministry of Foreign Affairs, International Trade and Worship, Argentina

Elena Palacios, Fundación Ecológica Universal, Argentina

Margarita Paras, Centro de Investigación en Geografía y Geomática, Ing. J. L. Tamayo, México

Martín Pardo, Centro Latino Americano de Ecología Social, Uruguay

Wendel Parham, Caribbean Agricultural Research Development Institute, Trinidad and Tobago

Araceli Parra, Consejo Nacional de Industriales Ecologistas de México, México

Lino Rubén Pérez, Agencia de Información Nacional, Cuba

Joel Bernardo Pérez Fernández, Water Center for the Humid Tropics of Latin America and the Caribbean, Panamá

Alejandro Mohar Ponce, Centro de Investigación en Geografía y Geomática – Centro GEO, México

Carlos Costa Posada, Instituto de Hidrología, Meteorología y Estudios Ambientales Instituto de Colombia, Colombia

Armando José Coelho Quixada Pereira, Instituto Brasileiro do Meio Ambiente e dos Recursos Naturais Renováveis, Brazil

Lorena Aguilar Revelo, IUCN – The World Conservation Union, Costa Rica

Sonia Reyes-Packe, Dirección de Servicios Externos, Facultad de Arquitectura, Diseño y Estudios Urbanos, Pontificia Universidad Católica de Chile, Chile

Evelia Rivera-Arriaga, Centro de Ecología, Pesquerías y Oceanografía del Golfo de México, Universidad Autónoma de Campeche, México

César Edgardo Rodríguez Ortega, Secretaría de Medio Ambiente y Recursos Naturales, México

Mario Rojas, Oficial de Cooperación y Relaciones Internacionales, Ministerio del Ambiente y Energía, Costa Rica

Marisabel Romaggi, Escuela de Ingeniería Ambiental, Facultad de Ecología y Recursos Naturales, Universidad Andrés Bello, Chile

Emilio Lebre-La Rovere, Federal University of Rio de Janeiro, Brazil

Francisco José Ruiz, Organización del Tratado de Cooperación Amazónica, Brazil

Tricia Sabessar, The Cropper Foundation, Trinidad and Tobago

Dalia Maria Salabarria Fernandez, Centre for Environmental Information Management and Education, Ministry of Science Technology and Environment, Cuba

José Somoza, Instituto Nacional de Investigaciones Económicas, Cuba

Juan Carlos Sanchez, Universidad Metropolitana, Venezuela

Orlando Rey Santos, Ministerio de Ciencia, Tecnología y Medio Ambiente, Cuba

Muriel Saragoussi, Ministerio de Medio Ambiente, Brazil

Amrikha Singh, Ministry of Housing, Lands and the Environment, Barbados

Avelino G. Suárez, Instituto de Ecología y Sistemática, Agencia de Medio Ambiente, Ministerio de Ciencia, Tecnología y Medio Ambiente, Cuba

José Roberto Solórzano, University of Denver, El Salvador

Felipe Omar Tapia, Centro de Investigación en Geografía y Geomática "Ing. Jorge L. Tamayo" A.C., México

Rodrigo Tarté, International Center for Sustainable Development at the City of Knowledge, Panamá

Adrian Ricardo Trotman, Caribbean Institute for Meteorology and Hydrology, Barbados

Miyuki Alcázar V., Mexico

Virginia Vásquez, Coastal Zone Management Authority and Institute, Belize

Raúl Garrido Vázquez, Ministry of Science, Technology and Environment, Cuba

Gerardo Bocco Verdinelli, Investigación de Ordenamiento Ecológico y Conservación de los Ecosistemas, Instituto Nacional de Ecología, México

Carolina Villalba, Centro Latino Americano de Ecología Social, Uruguay

Paola Visca, Centro Latino Americano de Ecología Social, Uruguay

Leslie Walling, Mainstreaming Adaptation to Climate Change Project, The Caribbean Community, Belize

Marcos Ximenes, Instituto de Pesquisa Ambiental da Amazonia, Brazil

Gustavo Adolfo Yamada, Universidad de Pacífico, Perú

Bolívar Zambrano, Autoridad Nacional del Ambiente, Panamá

Anna Zuchetti, Grupo GEA "Emprendemos el Cambio", Perú

NORTH AMERICA:

Sherburne Abbott, American Association for the Advancement of Science, United States

Arun George Abraham, Department of Political Science, University of Pennsylvania

John T. Ackerman, Department of International Security and Military Studies, Air Command and Staff College, United States

Patrick Adams, Statistics Canada, Environmental Accounts and Statistics Division, Environment Canada, Canada

Kwaku Agyei, Natural Resources Canada, Canada

Marie-Annick Amyot, Natural Resources Canada, Canada

John C. Anderson, Environment Canada, Canada

Robert Arnot, Environment Canada, Canada

Ghassem R. Asrar, National Aeronautics and Space Administration, United States

Richard Ballhorn, Department of Foreign Affairs and International Trade, Canada

Bill Bertera, Water Environment Federation, United States

Greg Block, Northwestern School of Law and Clark College, United States

Erik Bluemel, Georgetown University Law Center, United States

Wayne Bond, National Indicators and Reporting Office, Environment Canada, Canada

Denis Bourque, Canadian Space Agency, Canada

Birgit Braune, Environment Canada, Canada

William Brennan, National Oceanic and Atmospheric Agency, United States

Morley Brownstein, Environmental Health Centre, Health Canada, Canada

Angle Bruce, Environment Canada, Canada

Elizabeth Bush, Environment Canada, Canada

John Calder, National Oceanic and Atmospheric Administration, United States

Richard J. Calnan, United States Geological Survey, United States

Celina Campbell, Natural Resources Canada, Canada

Hilda Candanedo, Autoridad Nacional del Ambiente, Panamá

F. Stuart Chapin, III, University of Alaska Fairbanks, United States

Audrey R. Chapman, American Association for the Advancement of Science, United States

Julie Charbonneau, Strategic Information Integration, Environment Canada, Canada

Franklin G. Cardy, Canada

John Carey, Environment Canada, Canada

Chantal Line Carpentier, Commission for Environmental Cooperation of North America, Canada

Amy Cassara, World Resources Institute, United States

Gilbert Castellanos, Office of International Environmental Policy, United States Environmental Protection Agency, United States

Bob Chen, Center for International Earth Science Information Network, United States

Eileen Claussen, Pew Center of Global Climate Change and Strategies for the Global Environment, United States

Steve Cobham, Environment Canada, Canada

Nancy Colleton, Institute for Global Environmental Strategies, United States

Paul K. Conkin, Vanderbilt University, United States

Richard Connor, Unisféra, Canada

Luke Copland, University of Ottawa, Canada

Sylvie Côté, Environment Canada, Canada

Carmelle J. Cote, Environmental Systems Research Institute, Inc., United States

Philippe Crabbé, Institute for the Environment, University of Ottawa, Canada

Rob Cross, Environment Canada, Canada

Howard J. Diamond, National Oceanic and Atmospheric Agency, United States

Martin Dieu, United States Environmental Protection Agency, United States

Chuck Dull, United States Forest Service, United States

Alex de Sherbinin, Center for International Earth Science Information Network, United States

Joanne Egan, Environment Canada, Canada

Roger Ehrhardt, Canadian International Development Agency, Canada

Mark Erneste, United States Geological Survey, United States

Victoria Evans, Office of Air Quality Planning and Standards, United States Environmental Protection Agency, United States

Terry Fenge, Terry Fenge Consulting, Canada

Eugene A. Fosnight, United States Geological Survey, United States

Amy A. Fraenkel, United States Senate Committee on Commerce, Science and Transportation, United States

Bernard Funston, Sustainable Development Working Group of the Arctic Council Secretariat, Canada

Tim Gabor, Mount Sinai Hospital, Canada

Brigitte Gagne, Canadian Environmental Network, Canada

Wei Gao, Colorado State University, United States

David K. Garman, United States Department of Energy, United States

David Gauthier, Canadian Plains Research Center, Canada

Sylvie M. Gauthier, Natural Resources Canada, Canada

Aubry Gerald, Canadian Environmental Assessment Agency, Canada

Mike Gill, Circumpolar Biodiversity Monitoring Program, Canada

Michael H. Glantz, Center for Capacity Building, University Corporation for Atmospheric Research, United States

Jerome C. Glenn, American Council for the United Nations University, United States

Victoria Gofman, Aleut International Association, United States

Jean-François Gobeil, Environment Canada, Canada

Bernard D. Goldstein, Graduate School of Public Health, University of Pittsburgh, United States

Peter Graham, Natural Resources Canada, Canada

Don Greer, Canadian Water Resources Association, Ontario Ministry of Natural Resources, Canada

Charles G. Groat, United States Geological Survey, United States

Charles Gurney, United States Department of State, United States

Leonie Haimson, Class Size Matters Campaign, United States

Veena Halliwell, Transport Canada, Canada

David Hallman, World Council of Churches' Climate Change Programme, United Church of Canada, Canada

Nancy Hamzawi, Environment Canada, Canada

Chris Hanlon, Environment Canada, Canada

Kelley Hansen, United States Department of State, United States

Selwin Hart, Permanent Mission of Barbados to the United Nations, United States

Tracy Hart, The World Bank, United States

Alan Hecht, United States Environmental Protection Agency, United Sates

Ole Hendrickson, Environment Canada, Canada

Kerri Henry, Environment Canada, Canada

John Herity, IUCN – The World Conservation Union, Canada

Hans Herrmann, Commission for Environmental Cooperation of North America, Canada

Janet Hohn, United States Fish and Wildlife Service, United States

Annette Teresa Huber-lee, Boston Office, Stockholm Environment Institute, United States

Nathaniel Hultman, School of Foreign Service, Georgetown University, United States

Henry P. Huntington, Huntington Consulting, United States

Gary Ironside, Environment Canada, Canada

Irwin Itzkovitch, Earth Science Sector, Natural Resources Canada, Canada

Kirsten Jaglo, United States Department of State, United States

Robin James, Strategic Engagement, Climate Change International, Environment Canada, Canada

Lawrence Jaworski, Water Environment Federation, United States

David J. Jhirad, World Resources Institute, United States

Matt Jones, Climate Change International, Environment Canada, Canada

Glenn P. Juday, University of Alaska, United States

Shashi Kant, Faculty of Forestry, University of Toronto, Canada

John Karau, Fisheries and Oceans, Canada

Terry J. Keating, Office of Air and Radiation, United States Environmental Protection Agency, United States

Norine Kennedy, United States Council on International Business, United States

John Kineman, National Oceanic and Atmospheric Agency, United States

Ken Korporal, Canadian GEO Secretariat, Environment Canada, Canada

Sarah Kyle, Sustainable Development Strategy, Sustainable Policy, Environment Canada, Canada

Nicole Ladouceur, Environment Canada, Canada

Tom Laughlin, National Oceanic and Atmospheric Agency, United States

Conrad C. Lautenbacher, Jr., National Oceanic and Atmospheric Agency, United States

Philippe Le Prestre, Institut Hydro-Québec en Environnement, Development Société, Canada

Song Li, Global Environmental Facility Secretariat, United States

Kathryn Lindsay, Environmental Reporting Branch, Knowledge Integration Directorate, Environment Canada, Canada

Steve Lonergan, University of Victoria, Canada

Thomas E. Lovejoy, The John Heinz III Center for Science, Economics and Environment, United States

Sarah Lukie, McKenna Long, United States

H. Gyde Lund, Forest Information Services, United States

Ron Lyen, Natural Resources Canada, Canada

Daniel Magraw, Center for International Environmental Law, United States

Mark Mallory, Canadian Wildlife Service, Environment Canada, Canada

Tim Marta, Agriculture and Agri-Food Canada, Canada

Margaret McCauley, Office of Environmental and Scientific Affairs, United States Department of State, United States

Elizabeth McLanahan, National Oceanic and Atmospheric Administration, United States

Claudia A. McMurray, Bureau of Oceans and International Environmental and Scientific Affairs, United States Department of State, United States

John Robert McNeill, School of Foreign Service, Georgetown University, United States

Terence McRae, Knowledge Integration Strategies, Strategic Information Integration, Environment Canada, Canada

John Melack, Bren School of Environmental Science and Management, University of California, United States

Jerry Melillo, Ecosystem Center, Marine Biological Laboratory, United States

Roberta B. Miller, Center for International Earth Science Information Network, United States

Rebecca Milo, Meteorological Service of Canada, Environment Canada, Canada

Adrian Mohareb, Natural Resources Canada, Canada

Jim Moseley, United States Department of Agriculture, United States

Melissa Dawn Newhook, Natural Resources Canada, Canada

Kate Newman, World Wide Fund for Nature, United States

Dennis O'Farrell, National Indicators and Reporting Office, Environment Canada, Canada

Dean Stinson O'Gorman, Environment Canada, Canada

Maureen O'Neil, International Development Research Centre, Canada

Katia Opalka, Commission for Environmental Cooperation of North America, Canada

Gordon H. Orians, Department of Biology, University of Washington, United States

Yuga Juma Onziga, Environmental Centre for New Canadians, Canada

László Pintér, International Institute for Sustainable Development, Canada

Robert Prescott-Allen, Canada

Gary Pringle, Foreign Affairs Canada, Canada

David Renne, National Renewable Energy Laboratory, United States

Christina Paradiso, Environment Canada, Canada

Anjali Pathmanathan, Center for International Environmental Law, United States

Corey Peabody, Industry Canada, Canada

Kenneth Peel, Council on Environmental Quality, United States

Luc Pelletier, Environment Canada, Canada

Sajjadur Syed Rahman, Canadian International Development Agency, Canada

David J. Rapport, The School of Rural Planning Development, University of Guelph, Canada

Paul Raskin, Boston Office, Stockholm Environment Institute, United States

John Reed, Secretariat of the Working Group on Environmental Auditing of the International Organization of the Supreme Audit Institutions, Office of the Auditor General of Canada, Canada

Carmen Revenga, Global Priorities Group, The Nature Conservancy, United States

Sandra Ribey, Natural Resources Canada, Canada

Douglas Richardson, Association of American Geographers, United States

Brian Roberts, Indian and Northern Affairs, Canada

Keith Robinson, Agriculture Canada, Canada

David Runnalls, International Institute for Sustainable Development, Canada

Paul Salah, Economic and Social Research Institute, Canada

Peter D. Saundry, National Council for Science and the Environment, United States

Mark Schaefer, NatureServe, United States

Karl F. Schmidt, Johnson and Johnson, United States

Jackie Scott, Natural Resources Canada, Canada

Nancy Seymour, Consumer and Commercial Products, Environmental Stewardship Branch, Environment Canada, Canada

Hua Shi, Global Resource Information Database, Sioux Falls, United States

Emmy Simmons, United States Agency for International Development, United States

Andrea Dalledone Siqueira, Indiana University, United States

Risa Smith, Environment Canada, Canada

Sharon Smith, Natural Resources Canada, Canada

William Sonntag, United States Environmental Protection Agency, United States

Janet Stephenson, Natural Resources Canada, Canada

David Suzuki, David Suzuki Foundation, Canada

Darren Swanson, International Institute for Sustainable Development, Canada

Hongmao Tang, AMEC Earth and Environmental, Canada

Fraser Taylor, International Steering Committee for Global Mapping, Carleton University, Canada

Ian D. Thompson, Natural Resources Canada, Canada

Jeffrey Thornton, International Environmental Management Services Limited, United States

John R. Townshend, University of Maryland, United States

Woody Turner, National Aeronautics and Space Administration, United States

Mathis Wackernagel, Global Footprint Network, United States

Lawrence A. White, Algonquin College, Canada

Loise Vallieres, Canadian International Development Agency, Canada

Richard Verbisky, Environment Canada, Canada

Charles Weiss, School of Foreign Service, Georgetown University, United States

Doug Wright, Commission for Environmental Cooperation of North America, Canada

Ruth Waldick, Environment Canada, Canada

John D. Waugh, IUCN – The World Conservation Union, United States

WEST ASIA:

Directorate General of Environment, Ministry of Environment, Lebanon

Iman Abdulrahim, Conference Services Centre, Syrian Arab Republic

Ziad Hamzah Abu-Ghararah, Meteorology and Environment Protection, Saudi Arabia

Emad Adly, Arab Network for Environment and Development, Egypt

Mohammed Bin Sulaiman Al-Abry, Ministry of Regional Municipalities, Environment and Water Resources, Sultanate of Oman

Suzan Mohammed Al-Ajjawi, Public Commission for the Protection of Marine Resources, Environment and Wildlife, Bahrain

Fahmi Al-Ali, Gulf Cooperation Council, Saudi Arabia

Badria Al-Awadhi, Arab Regional Center for Environmental Law, Kuwait

Abdul Rahman Al-Awadi, Regional Organization for the Protection of the Marine Environment, Kuwait

Hanan S. Haider Alawi, Public Commission for the Protection of Marine Resources, Environment and Wildlife, Bahrain

Ziyad Al-Alawneh, Ministry of Environment, Jordan

Eman Al-Banna, Environment Friends Society, Bahrain

Ahmed Mohammed Al-Hamadeh, The Emirates Centre for Strategic Studies and Research, United Arab Emirates

Ali Jassim M. Al-Hesabi, Public Commission for the Protection of Marine Resources, Environment and Wildlife, Bahrain

Jaber E. Al-Jabri, Environmental Research and Wildlife Development Agency, United Arab Emirates

Mohammed Al-Jawdar, Environment Agency – Abu Dhabi, United Arab Emirates

Nada Al-Khalili, Al-Reem Environmental Consultation and Ecotourism, Bahrain

Ahlam Al-Marzouqi, Environment Agency, United Arab Emirates

Hamad Eisa Al-Matroushi, Federal Environmental Agency, United Arab Emirates

Ahmed Al-Mohammad, General Commission Environmental Affairs, Syrian Arab Republic

Khawla Al-Muhannadi, Environment Friends Society, Bahrain

Abdullah Al-Ali Al-Nuaim, Arab Urban Development Institute, Saudi Arabia

Safia Saad Al-Rumaihi, Bahrain Radio and TV Cooperation, Bahrain

Ahmed Al-Salloum, Arab Urban Development Institute, Saudi Arabia

Abdulkader Mohammed Al-Sari, Natural Resources and Environment Research Institute, King Abdulaziz City for Science and Technology, Saudi Arabia

Abdulrahman Hassan Hashem Al-Shehari, Department of GIS, Environment Protection Authority, Yemen

Mohanned S. Al-Sheriadeh, University of Bahrain, Bahrain

Mahmoud Al-Sibai, The Arab Centre for the Studies of Arid Zones and Drylands, Syrian Arab Republic

Ibrahim N. Al-Zu'bi, Emirates Diving Association, United Arab Emirates

Feras Asfour, Ministry of Local Administration and Environment, Syrian Arab Republic

Sarah Ben Arfa, PHE Gulf, Bahrain

Abdulla Saleh Babaqi, Sana'a University, Yemen

Yousif H. Edan, Arabian Gulf University, Bahrain

Alia El-Husseini, Lebanon IUCN National Committee, Lebanon

Karim El-Jisr, ECODIT LIBAN, c/o Ministry of Environment, Lebanon, Lebanon

Reem Aref Fayyad, Department of Guidance and Assessment, Ministry of Environment, Lebanon

Ibrahim Abdel Gelil, Arabian Gulf University, Bahrain

Bashar A. Hamdoon, Arab Science and Technology Foundation, United Arab Emirates

Waleed Hamza, Emirates Environmental Group, United Arab Emirates

Meena Kadhimi, Bahrain Women Society, Bahrain

Maher Suleiman Khaleel, Arab Forests and Range Institute, Syrian Arab Republic

Fadia Kiwan, Institute of Political Sciences, Saint Joseph University, Lebanon

Lamya Faisal Mohamed, Environmental Management Program, Arabian Gulf University, Bahrain

Abdullah Abdulkader Naseer, Arab NGO Network for Environment and Development, Saudi Arabia

Najib Saab, Al-Bia Wal Tanmia Environment and Development, Lebanon

Mohammed Y. Saidam, Environment Monitoring and Research Central Unit, Royal Scientific Society, Jordan

Taysir Toman, Environment Quality Authority, Palestine National Authority, Occupied Palestinian Territories

Shahira Hassan Ahmed Wahbi, Department of Environment, Housing and Sustainable Development, Council of Arab Ministers Responsible for the Environment, Egypt

Batir M. Wardam, Ministry of Environment, Jordan

Abdel Nasser H. Zaied, Arabian Gulf University, Bahrain

INTERGOVERNMENTAL AND MULTI-STAKEHOLDER CONSULTATION:

Yousef Abu-Safieh, Environmental Quality Authority, Palestine National Authority, Occupied Palestinian Territories

Jeanne Josette Acacha Akoha, Ministère de l'Environnement de l'Habitat et de l'Urbanisme, Benin

Meshgan Mohamed Al Awar, Zayed International Prize for the Environment, United Arab Emirates

Salem Al-Dhaheri, Federal Environmental Agency, United Arab Emirates

Hussein Alawi Al-Gunied, Ministry of Water and Environment for Environmental Affairs, Yemen

Mohammed Bin Saif Sulaiyam Al-Kalbani, Ministry of Regional Municipalities, Environment and Water Resources, Sultanate of Oman

Cholpon Alibakieva, Ministry of Ecology and Emergency Situations, Republic of Kyrgyzstan

Zahwa Mohammed Al-Kuwari, Public Commission for the Protection of Marine Resources, Environment & Wildlife, Bahrain

Said Al-Numairy, Federal Environmental Agency, United Arab Emirates

Khawlah Mohammed Al-Obaidan, Environment Public Authority, Kuwait

Muthanna A. Wahab Wahab Al-Omar, Deputy Minister for Technical Affairs, Republic of Iraq

Mario Andino, Ministry of Environment, Ecuador

Gonzalo Javier Asencio Angulo, National Environmental Commission, Chile

Mahaman Laminou Attaou, Ministère de l'Hydraulique, de l'Environnement de la Lutte Contre la Désertification, Niger

Rajen Awotar, Environment Liaison Centre International, Mauritius

Christoph Bail, Delegation of the European Commission – Kenya and Somalia, Kenya

Mogos Woldeyohannes Bairu, Department of Environment, Ministry of Lands, Water and Environment, Eritrea

Maria Caridad Balaguer Labrada, Ministry of Foreign Affairs, Cuba

Abbas Naji Balasem, Ministry of Environment, Republic of Iraq

Kurbangeldi Balliyev, Unit of Scientific, Technological Problems and International Co-operation Science-Information Centre, Ministry of Nature Protection of Turkmenistan, Turkmenistan

W. M. S. Bandara, Sri Lanka High Commission, Kenya

Stephen Bates, Department of the Environment and Heritage, Australia

Theo A. M. Beckers, Telos Research Center for Sustainable Development, Tilburg University, The Netherlands

Dzaba-Boungou Benjamin, Ministère de l'Economie Forestière et de l'Environnement, Congo

Nalini Bhat, Ministry of Environment and Forests, India

Peter Koefoed Bjørnsen, National Environmental Research Institute, Ministry of the Environment, Denmark

Adriana Maria Bonilla, Latin American Faculty of Social Sciences, Costa Rica

Valerie Brachya, Ministry of the Environment, Israel

Liana Bratasida Ministry of the Environment, Indonesia

Andrea Brusco, Environmental and Sustainable Development Promotion, Ministry de Salud y Ambiente, Argentina

Cesar Buitrago, Instituto de Hidrologia, Meteorologia y Estudios Ambientales, Instituto de Colombia, Colombia

Robin Carter, Department for Environment, Food and Rural Affairs, Ministry of State for Environment and Agri-Environment, United Kingdom

Sergio Castellari, Ministry for the Environment and Territory, Italy

Enid Chaverri-Tapia, Ministry of Environment and Energy of Costa Rica, Costa Rica

Chris Reid Cocklin, Monash Environment Institute, Australia

Victor Manuel do Sacramento Bonfi, Ministério dos Recursos Naturais e Meio Ambiente, Saõ Tomé and Principe

Stela Bucatari Drucioc, Ministry of Ecology and Natural Resources, Republic of Moldova

Ould Bahneine El Hadrami, Islamic Republic of Mauritania

James Emmons Coleman, Environmental Protection Agency, Liberia

Loraine Cox, The Bahamas Environment, Science and Technology Commission, Ministry of Health and Environment, The Bahamas

Rodolfo Roa, Ministerio del Ambiente y de los Recursos Naturales, Venezuela

Raouf Hashem Dabbas, Ministry of Environment, Jordan

Oludayo O. Dada, Department of Pollution Control and Environmental Health, Federal Ministry of Environment, Federal Secretariat, Nigeria

Allan Dauchi, Ministry of Tourism, Environment and Natural Resources, Zambia

Adama Diawara, Consul Honoraire de la Republique de Côte D'Ivoire au Kenya, Kenya

Didier Dogley, Ministry of Environment and Natural Resources, Seychelles

Sébastien Dusabeyezu, Rwanda Environmental Management Authority, Ministry of Lands, Environment, Forestry, Water and Mines, Rwanda

Fatma Salah El Din El Mallah, Department of Environment and Sustainable Development, League of Arab States, Egypt

Davaa Erdenebulgan, Ministry of Nature and Environment, Mongolia

Indhira Euamonchat, Office of Natural Resources and Environmental Policy and Planning, Ministry of Natural Resources and Environment, Thailand

Jan Willem Erisman, Energy Research Centre of the Netherlands, The Netherlands

Caroline Eugene, Sustainable Development and Environment Unit, Ministry of Physical Development, Environment and Housing, Saint Lucia

Fariq Farzaliyev, Ministry of Ecology and Natural Resources, Azerbaijan

Qasim Hersi Farah, Ministry of Environment and Disaster Management, Somalia

Liban Sheikh Mahmoud Farah, Federal Environmental Agency, United Arab Emirates

Veronique Plocq Fichelet, Scientific Committee on Problems of the Environment, France

Seif Eddine Fliss, The Embassy of Tunisia in Addis Ababa, Ethiopia

Cheikh Fofana, Secrétariat Intérimaire du Volet Environnement du NEPAD, Senegal

Cornel Glorea Gabrian, Ministry of Environment and Water Management, Romania

Jorge Mario García Fernández, Centro de Información, Gestión y Educación y Educatión Ambiental, Ministerio de Ciencia, Tecología y Medio Ambiente, Cuba

Sameer Jameel Ghazi, Meteorology and Environment, Saudi Arabia

Tran Hong Ha, Vietnam Environment Protection Agency, Ministry of Natural Resources and Environment, Viet Nam

Nadhir Hamada, Ministry of Environmental and Sustainable Development, Tunisia

Mohamed Salem Hamouda, Environment General Authority, Libyan Arab Jamahiriya

Hempel Gotthilf, Berater des Präsidenten des Senats für den Wissenschaftsstandort Bremen, Germany

Keri Herman, National Environment Service, Cook Islands

Paul Hofseth, Ministry of the Environment, Norway

Rustam Ibragimov, State of Committee for Nature Protection, Republic of Uzbekistan

Khan M. Ibrahim Hossain, Ministry of Environment and Forests, Government of the People's Republic of Bangladesh

Moheeb A. El Sattar Ibrahim, Egyptian Environmental Affairs Agency, Egypt

Lorna Inniss, Ministry of Housing, Lands and the Environment, Barbados

Nikola Ru Inski, Faculty of Mechanical Engineering and Naval Architecture, University of Zagreb, Croatia

Adélaïde Itoua, Attaché Forêts, Faune et Environnement, Congo

Said Jalala, Environment Quality Authority, Occupied Palestinian Territory

Christopher Joseph, Ministry of Health, Social Security, Environment and Eccelesiastic Relations, Grenada

Volney Zanardi Júnior, Ministry of the Environment, Brazil

Etienne Kayengeyenge, Department de l'Environnement et du Tourism L'Aménagement du Territoire et de l'Environnement, Environnement et du Tourism, Burundi

Keobang A. Keola, Cabinet of Science, Technology and Environment Agency, Lao People's Democratic Republic

Mootaz Ahmadein Khalil, Ministry of Foreign Affairs, Egypt

Bernard Yao Koffi, Ministère de l'Environnement, Cote d'Ivoire

Tomyeba Komi, Ministry of Environment and Natural Resources, Togo

Margarita Korkhmazyan, Department of International Cooperation, Ministry of Nature Protection, Republic of Armenia

Pradyumna Kumar Kotta, South Asia Cooperative Environment Programme, Sri Lanka

Izabela Elzbieta Kurdusiewicz, Ministry of the Environment, Poland

Daniel Lago, Maoni Network, Kenya

Aminath Latheefa, Ministry of Environment and Construction, Maldives

Stephen Law, Environment Liaison Centre International, South Africa

P. M. Leelaratne, Ministry of Environment and Natural Resources, Sri Lanka

Rithirak Long, Ministry of Environment, Cambodia

Sharon Lindo, Ministry of Natural Resources and the Environment, Belize

Fernando Lugris, Permanent Mission of Uruguay to the United Nations, Switzerland

Rejoice Mabudhafasi, Department of Environmental Affairs and Tourism, Republic of South Africa

Oualbadet Magomna, Chad

Sylla Mamadouba, Ministrère de l'Environnement, Republic of Guinea

Seraphin Mamyle-Dane, Ministry in Charge of Environment, Central African Republic

Blessing Manale, Department of Environmental Affairs and Tourism, Republic of South Africa

Alena Marková, Department of Strategies, Ministry of the Environment of the Czech Republic, Czech Republic

Chrispen Maseva, Department of Natural Resources, Zimbabwe

Maurice B. Masumbuko, Ministère de l'Environnement, Democratic Republic of Congo

Lyborn Matsila, Department of Environmental Affairs and Tourism, Republic of South Africa

Mary Fosi Mbantenkhu, Ministry of Environment and Forestry, Cameroon

Dave A. McIntosh, Environmental Management Authority, Trinidad and Tobago

Lamed Mendoza, Intergubernmental y de Múltiples Autoridad Nacional del Ambiente Cooperation Tecnica Internacional, Panamá

Raymond D. Mendoza, Department of Environment and Natural Resources, Philippines

José Santos Mendoza Arteaga, Ministerio del Ambiente Y Los Recursos Naturales, Nicaragua

Samuel Kitamirike Mikenga, World Wildlife fund International, Kenya

Rita Mishaan, Ministry of Environment and Natural Resources, Guatemala

Bedrich Moldan, Charles University Environmental Centre, Czech Republic

Santaram Mooloo, Ministry of Environment and National Development Unit, Mauritius

Majid Shafiepour Motlagh, Department of Environment, Environmental Research Centre, Islamic Republic of Iran

John Mugabe, African Commission on Science and Technology, South Africa

Telly Eugene Muramira, National Environment Management Authority, Uganda

Dali Najeh, Ministry of Environment and Sustainable Development, Tunisia

Timur Nazarov, Department of Ecological Monitoring and Standards, State Committee for Environmental Protection and Forestry, Tajikistan

Przemyslaw Niesiolowski, Permanent Mission of the Poland to UNEP

Faraja Gideon Ngerageza, The Vice President's Office, United Republic of Tanzania

Raharimaniraka Lydie Norohanta, Ministry of Environment, Water and Forests, Madagascar

Kenneth Ochoa, Organization Juvenil Ambiental, Colombia

Herine A. Ochola, Environment Liaison Centre International, Kenya

Rodrigue Abourou Otogo, Directeur des Etudes du Contentieux et du Droit de l'Envrironnement, Gabon

Monique Ndongo Ouli, Ministry of Environment and Nature Protection, Cameroon

Pedro Luis Pedroso, Permanent Mission of Cuba, Cuba

Detelina Peicheva, Ministry of Environment and Water, Bulgaria

Reinaldo Garcia Perera, Embassy of Cuba, Kenya

Carlos Humberto Pineda, Secretaria de Recursos Naturales y Ambiente, Honduras

Peter Prokosch, GRID Arendal, Norway

Navin P. Rajagobal, Ministry of the Environment and Water Resources, Republic of Singapore

Victor Rezepov, Centre for International Projects, Russian Federation

Cyril Ritchie, Environment Liaison Centre International, Switzerland

Rosalud Jing Rosa, Environment Liaison Centre International, Italy

Thomas Rosswall, International Council for Science, France

Uiloi F. Samani, Ministry of the Environment, Tonga

Mariano Castro Sánchez-Moreno, Consejo Nacional del Ambiente, Perú

Kaj Harald Sanders, Ministry of Housing, Spatial Planning and the Environment, Netherlands

Carlos Santos, Ministry of Urban Affairs and Environment, Angola

Momodou B. Sarr, National Environment Agency, Gambia

Alhassane Savane, Consulate of Cote d'Ivoire

Gerald Musoke Sawula, National Environment Management Authority, Uganda

Tan Nguan Sen, Public Utilities Board, Republic of Singapore

Manuel Leão Silva de Carvalho, Ministry of the Environment, Agriculture and Fisheries, Republic of Cape de Verde

Mohamed Adel Smaoui, Permanent Mission of Tunisia to UNEP, Federal Democratic Republic of Ethiopia

Kerry Smith, Department of the Environment and Heritage, Australia

Miroslav Spasojevic, International Cooperation and European Integration, Directorate for Environment Protection, Ministry for Science and Environment Protection of Serbia, Serbia and Montenegro

Katri Tuulikki Suomi, Ministry of the Environment, Finland

Hamid Tarofi, Embassy of Iran, Kenya

Tshering Tashi, National Environment Commission Secretariat, Bhutan

Tukabu Teroroko, Ministry of Environment, Lands and Agriculture Development, Kiribati

Testaye Woldeyes, Environmental Protection Authority, Ethiopia

Nicholas Thomas, Environmental Systems Research Institute, United States

Alain Edouard Traore, Secrétaire Permanent du Conseil National pour l'Enviornnement et le Développement Durable, Burkina Faso

Lourenço António Vaz, General Directorate of Environment, Guinea-Bissau

Sani Dawaki Usman, Department of Planning, Research and Statistics, Federal Ministry of Environment, Federal Secretariat, Nigeria

Geneviéve Verbrugge, Direction Générale de l'Administration, Service des affaires internationals, Ministère de l'Ecologie et du Développement Durable, France

Jameson Dukuza Vilakati, Swaziland Environment Authority, Ministry of Tourism, Environment and Communications, Swaziland

Eric Vindimian, Ministère de l'écologie, du développement et de l'aménagement durables, France

Aboubaker Douale Waiss, Ministère de L'Habitat, de l'Urbanisme, de l'Environnement et de l'Amenagement du Territoire, Republic de Djibouti

Shahira Hassan Ahmed Wahbi, Division of Resources and Investment, League of Arab States, Egypt

Elisabeth Wickstrom, Swedish Environmental Protection Agency, Sweden

Alf Willis, Department of Environmental Affairs and Tourism, Republic of South Africa

Théophile Worou, Ministère de l'Environnement de l'Habitat et de l'Urbanisme, Benin

Carlos Lopes Ximenes, Ministry of Development and Environment, Timor (East)

Huang Yi, Peking University, China

B. Zaimov, Ministry of Foreign Affairs of Blugaria, Bulgaria

Daniel Ziegerer, Federal Office for the Environment, Switzerland

UNITED NATIONS ENVIRONMENT PROGRAMME

Peter Acquah
Martin Adriaanse*
Awatif Ahmed Alif
Siren Al-Majali
Abdul Elah Al-Wadaee
Ahmad Basel Al-Yousfi
Lars Rosendal Appelquist
Charles Arden-Clarke
Andreas Arlt [Secretariat for the Basel Convention]
Edgar Arredondo
Maria Eugenia Arreola
Franck Attere
Esther Berube
Luis Betanzos
An Bollen
Matthew Broughton
Alberto T. Calcagno
John Carstensen
Paul Clements-Hunt
Twinkle Chopra
Luisa Colasimone [Coordinating Unit for the Mediterranean Action Plan]
Ludgarde Coppens
Emily Corcoran
Julia Crause
Tamara Curll [Secretariat for the Vienna Convention for the Protection of the Ozone Layer and for the Montreal Protocol]
James S. Curlin
Mogens Dyhr-Nielsen [United Nations Environment Programme Collaborating Centre on Water and Environment]
Ayman Taha El-Talouny
Kamala Ernest
Ngina Fernandez
Silvia Ferratini
Hilary French
Betty Gachao
Louise Gallagher
Ahmad Ghosn
Marco Gonzalez [Secretariat for the Vienna Convention for the Protection of the Ozone Layer and for the Montreal Protocol]
Matthew Gubb
Julien Haarman
Abdul-Majeid Haddad
Batyr Hadjiyev
Stefan Hain
Lauren E. Haney
Peter Herkenrath
Ivonne Higuero
Arab Hoballah [Coordinating Unit for the Mediterranean Action Plan]
Robert Höft [Secretariat of the Convention on Biological Diversity]
Teresa Hurtado
Melanie Hutchinson
Yuwaree In-na
Niels Henrik Ipsen [Collaborating Centre on Water and Environment]
Mylvakanam Iyngararasn
David Jensen
Bob Kagumaho Kakuyo
Charuwan Kalyangkura [Regional Coordinating Unit for the East Asian Seas Action Plan]
Valerie Kapos
Aida Karazhanova
Nonglak Kasemsant
Elizabeth Khaka
Johnson U. Kitheka

Arnold Kreilhuber
Nipa Laithong
Christian Lambrechts
Bernadete Lange
Achira Leophairatana
Fredrick Lerionka
Kaj Madsen
Ken Maguire
Elizabeth Masibo
Robyn Matravers
Emilie Mazzacurati
Desta Mebratu
Mushtaq Ahmed Memon
Danapakorn Mirahong
Ting Aung Moe
Erika Monnati
Cristina Montenegro
David Morgan [Secretariat of the Convention on International Trade in Endangered Species]
Andrew Morton
Elizabeth Maruma Mrema
Onesmus Mutava
Fatou Ndoye
Hiroshi Noshimiya
Werner Obermeyer
Akpezi Ogbuigwe
David Ombisi
Joanna Pajkowska
Janos Pasztor
Hassan Partow
Pravina Patel
Cecilia Pineda
Mahesh Pradhan
Daniel Puig
Mark Radka
Anisur Rahman
Purna Rajbhandari
Richard Robarts
Adelaida Bonomin Roman
Hiba Sadaka
Bayasgalan Sanduijav
Vincente Santiago-Fandino
Pinya Sarasas
Rajendra M. Shende
Fulai Sheng
Otto Simonett
Subrato Sinha
Angele Luh Sy
Gulmira Tolibaeva
Dechen Tsering
Rie Tsutsumi
Aniseh Vadiee
Sonia Valdivia
Maliza Van Eeden
Hanneke Van Lavieren
Anja Von Moltke
Monika G. Wehrle-MacDevette
Willem Wijnstekers [Secretariat of the Convention on International Trade in Endangered Species]
Matthew Woods
Grant Wroe-Street
Saule Yessimova

OTHER UNITED NATIONS BODIES

Mohamed J. Abdulrazzak, United Nations Educational, Scientific and Cultural Organization
Mohammed Ahmed Al-Aawah, United Nations Educational, Scientific and Cultural Organization
Mohammed H. Al-Sharif, United Nations Development Programme
Jörn Birkmann, United Nations University Institute for Environment and Human Security
Sandra Bos, United Nations Human Settlements Programme
Caros Corvalan, World Health Organization
Phillip Dobie, United Nations Development Programme
Glenn Dolcemascolo, United Nations International Strategy for Disaster Reduction
Henrik Oksfeldt Enevoldsen, Intergovernmental Oceanographic Commission of United Nations Educational, Scientific and Cultural Organization;
Nejib Friji, United Nations Information Centre
Sonia Gonzalez, United Nations Development Programme
Robert Hamwey, United Nations Conference on Trade and Development
Maharufa Hossain, United Nations Human Settlements Programme
Masakazu Ichimura, United Nations Economic and Social Commission for Asia and the Pacific
Rokho Kim, World Health Organization European Centre for Environment and Health, Germany
Melinda L. Kimble, United Nations Foundation
Anne Klen, United Nations Human Settlements Programme
Iris Knabe, United Nations Human Settlements Programme
Mikhail. G. Kokine, United Nations Economic Commission for Europe
Ousmane Laye, United Nations Economic Commission for Africa
Sarah Lowder, United Nations Economic and Social Commission for Asia and the Pacific
Silvia Llosa, United Nations International Strategy for Disaster Reduction
Festus Luboyera [United Nations Framework Convention on Climate Change Secretariat]
Ole Lyse, United Nations Human Settlements Programme
Leslie Malone, World Meteorological Organization
Mariana Mansur, United Nations Development Programme
Anthony Mitchell, United Nations Economic Commission for Latin America and the Caribbean
S. Njoroge, World Meteorological Organization Subregional Office for Eastern and Southern Africa
Joseph Opio-Odongo, United Nations Development Programme, Regional Service Centre for Eastern and Southern Africa Drylands Development Centre
Nohoalani Hitomi Rankine, United Nations Economic and Social Commission for Asia and the Pacific
Xin Ren, Seretariat of the United Nations Framework Convention on Climate Change
Ulrika Richardson, United Nations Development Programme
Tarek Sadek, United Nations Economic and Social Commission for Western Asia
Trevor Sankey, United Nations Educational, Scientific and Cultural Organization
Halldor Thorgeirsson, United Nations Framework Convention on Climate Change
Rasna Warah, United Nations Human Settlements Programme
Ulrich Wieland, United Nations Statistics Division

*since moved or retired

Glossary

This glossary is compiled from citations in different chapters, and draws from glossaries and other resources available on the websites of the following organizations, networks and projects:

American Meteorological Society, Center for Transportation Excellence (United States), Charles Darwin University (Australia), Consultative Group on International Agricultural Research, Convention on Wetlands of International Importance especially as Waterfowl Habitat, Europe's Information Society, European Environment Agency, European Nuclear Society, Food and Agriculture Organization of the United Nations, Foundation for Research, Science and Technology (New Zealand), Global Footprint Network, GreenFacts Glossary, Intergovernmental Panel on Climate Change, International Centre for Research in Agroforestry, International Comparison Programme, International Research Institute for Climate and Society at Columbia University (United States), International Strategy for Disaster Reduction, Lyme Disease Foundation (United States), Millennium Ecosystem Assessment, Illinois Clean Coal Institute (United States), National Safety Council (United States), Natsource (United States), The Organisation for Economic Co-operation and Development, Professional Development for Livelihoods (United Kingdom), SafariX eTextbooks Online, Redefining Progress (United States), The Edwards Aquifer Website (United States), TheFreeDictionary.com, The World Bank, UN Convention to Combat Desertification in Countries Experiencing Serious Drought and/or Desertification, Particularly in Africa, UN Development Programme, UN Framework Convention on Climate Change, UN Industrial Development Organization, UN Statistics Division, US Department of Agriculture, US Department of the Interior, US Department of Transportation, US Energy Information Administration, US Environmental Protection Agency, US Geological Survey, Water Quality Association (United States), Wikipedia and World Health Organization.

Term	Definition
Abundance	The number of individuals or related measure of quantity (such as biomass) in a population, community or spatial unit.
Acid deposition	Any form of deposition on water, land and other surfaces that increases their acidity by contamination with acid pollutants, such as sulphur oxides, sulphates, nitrogen oxides and nitrates, or ammonium compounds. The deposition can be either dry (as in the adsorption of acid pollutants to particles) or wet (as in acid precipitation).
Acidification	Change in environment's natural chemical balance caused by an increase in the concentration of acidic elements.
Acidity	A measure of how acid a solution may be. A solution with a pH of less than 7.0 is considered acidic.
Adaptation	Adjustment in natural or human systems to a new or changing environment, including anticipatory and reactive adaptation, private and public adaptation, and autonomous and planned adaptation.
Adaptive capacity	The potential or ability of a system, region or community to adapt to the effects or impacts of a particular set of changes. Enhancement of adaptive capacity represents a practical means of coping with changes and uncertainties, reducing vulnerabilities and promoting sustainable development.
Aerosols	A collection of airborne solid or liquid particles, with a typical size between 0.01 and 10 µm, that reside in the atmosphere for at least several hours. Aerosols may be of either natural or anthropogenic origin.
Afforestation	Establishment of forest plantations on land that is not classified as forest.
Algal beds	Reef top surface feature dominated by algae cover, usually brown algae (such as Sargassum or Turbinaria).
Alien species (also non-native, non-indigenous, foreign, exotic)	Species introduced outside its normal distribution.
Aquaculture	The farming of aquatic organisms in inland and coastal areas, involving intervention in the rearing process to enhance production and the individual or corporate ownership of the stock being cultivated.
Aquatic ecosystem	Basic ecological unit composed of living and non-living elements interacting in an aqueous milieu.
Aquifer	An underground geological formation or group of formations, containing usable amounts of groundwater that can supply wells and springs.

Term	Definition
Arable land	Land under temporary crops (double-cropped areas are counted only once), temporary meadows for mowing or pasture, land under market and kitchen gardens, and land temporarily fallow (less than five years). The abandoned land resulting from shifting cultivation is not included in this category.
Archetype of vulnerability	A specific, representative pattern of the interactions between environmental change and human well-being.
Aridity index	The long-term mean of the ratio of mean annual precipitation to mean annual potential evapotranspiration in a given area.
Benthic organism	The biota living on or very near the bottom of the sea, river or lake.
Bioaccumulation	The increase in concentration of a chemical in organisms that reside in contaminated environments. Also used to describe the progressive increase in the amount of a chemical in an organism resulting from rates of absorption of a substance in excess of its metabolism and excretion.
Biocapacity	The capacity of ecosystems to produce useful biological materials and to absorb waste materials generated by humans, using current management schemes and extraction technologies. The biocapacity of an area is calculated by multiplying the actual physical area by the yield factor and the appropriate equivalence factor. Biocapacity is usually expressed in units of global hectares.
Biochemical oxygen demand (BOD)	The amount of dissolved oxygen, in milligrams per litre, necessary for the decomposition of organic matter by micro-organisms, such as bacteria. Measurement of BOD is used to determine the level of organic pollution of a stream or lake. The greater the BOD, the greater the degree of water pollution.
Biodiversity (a contraction of biological diversity)	The variety of life on Earth, including diversity at the genetic level, among species and among ecosystems and habitats. It includes diversity in abundance, distribution and in behaviour. Biodiversity also incorporates human cultural diversity, which can both be affected by the same drivers as biodiversity, and itself has impacts on the diversity of genes, other species and ecosystems.
Biofuel	Fuel produced from dry organic matter or combustible oils from plants, such as alcohol from fermented sugar, black liquor from the paper manufacturing process, wood and soybean oil.
Biogas	Gas, rich in methane, which is produced by the fermentation of animal dung, human sewage or crop residues in an airtight container.
Biomass	Organic material, both above ground and below ground, and both living and dead, such as trees, crops, grasses, tree litter and roots.
Biome	The largest unit of ecosystem classification that is convenient to recognize below the global level. Terrestrial biomes are typically based on dominant vegetation structure (such as forest and grassland). Ecosystems within a biome function in a broadly similar way, although they may have very different species composition. For example, all forests share certain properties regarding nutrient cycling, disturbance and biomass that are different from the properties of grasslands.
Biotechnology (modern)	The application of *in vitro* nucleic acid techniques, including recombinant deoxyribonucleic acid (DNA) and direct injection of nucleic acid into cells or organelles, or fusion of cells beyond the taxonomic family, that overcome natural physiological, reproductive or recombination barriers and that are not techniques used in traditional breeding and selection.
Bleaching (of coral reefs)	A phenomenon occurring when corals under stress expel their mutualistic microscopic algae, called zooxanthellae. This results in a severe decrease or even total loss of photosynthetic pigments. Since most reef building corals have white calcium carbonate skeletons, the latter show through the corals' tissue and the coral reef appears bleached.
Blue water	Surface water and groundwater that is available for irrigation, urban and industrial use and environmental flows.
Bus rapid transit (BRT)	A passenger traffic system that builds on the quality of rail transit and the flexibility of buses. The BRT combines intelligent transportation systems technology, priority for transit, cleaner and quieter vehicles, rapid and convenient fare collection, and integration with land use policy.
Canopy cover (also called crown closure or crown cover)	The percentage of the ground covered by a vertical projection of the outermost perimeter of the natural spread of the foliage of plants. Cannot exceed 100 per cent.
Cap and trade (system)	A regulatory or management system that sets a target level for emissions or natural resource use, and, after distributing shares in that quota, lets trading in those permits determine their price.
Capital	Resource that can be mobilized in the pursuit of an individual's goals. Thus, we can think of natural capital (natural resources such as land and water), physical capital (technology and artefacts), social capital (social relationships, networks and ties), financial capital (money in a bank, loans and credit), human capital (education and skills).
Carbon market	A set of institutions, regulations, project registration systems and trading entities that has emerged from the Kyoto Protocol. The protocol sets limits on total emissions by the world's major economies, as a prescribed number of "emission units." The protocol also allows countries that have emissions units to spare – emissions permitted but not "used" – to sell this excess capacity to countries that are over their targets. This is called the "carbon market," because carbon dioxide is the most widely-produced greenhouse gas, and because emissions of other greenhouse gases will be recorded and counted in terms of their "carbon dioxide equivalents."
Carbon sequestration	The process of increasing the carbon content of a reservoir other than the atmosphere.
Catchment (area)	The area of land bounded by watersheds draining into a river, basin or reservoir. *See also Drainage basin.*
Clean technology (also environmentally sound technology)	Manufacturing process or product technology that reduces pollution or waste, energy use or material use in comparison to the technology that it replaces. In clean as opposed to "end-of-pipe" technology, the environmental equipment is integrated into the production process.

Term	Definition
Climate change	Any change in climate over time, whether due to natural variability or as a result of human activity. (The UN Framework Convention on Climate Change defines climate change as "a change of climate which is attributed directly or indirectly to human activity that alters the composition of the global atmosphere and which is in addition to natural climate variability observed over comparable time periods.")
Climate variability	Variations in the mean state and other statistics (such as standard deviations and the occurrence of extremes) of the climate on all temporal and spatial scales beyond that of individual weather events. Variability may be due to natural internal processes in the climate system (internal variability), or to variations in natural or anthropogenic external forcing (external variability).
Coal washing	Removal of pyritic sulphur from coal through traditional coal pre-separation procedures of float/sink separation. Also, cleaning the coal with substances that enhance combustion efficiency and reduce potential pollutants.
Conservation tillage	Breaking the soil surface without turning over the soil.
Conventional environmental problems	Environmental problems for which the cause-and-effect relationships are well known, single sources generally can be identified, the potential victims are often close to those sources and the scale is local or national. There are good examples of solutions to "conventional" problems such as microbial contamination, harmful local algal blooms, emissions of sulphur and nitrogen oxides, and particulate matter, oil spills, local land degradation, localized habitat destruction, fragmentation of land, and overexploitation of freshwater resources. *See also Persistent environmental problems and environmental problems.*
Coping capacity	The degree to which adjustments in practices, processes or structures can moderate or offset the potential for damage, or take advantage of opportunities.
Cost-benefit analysis	A technique designed to determine the feasibility of a project or plan by quantifying its costs and benefits.
Cross-cutting issue	An issue that cannot be adequately understood or explained without reference to the interactions of several dimensions that are usually treated separately for policy purposes. For example, in some environmental problems economic, social, cultural and political dimensions interact with one another to define the ways and means through which society interacts with nature, and the consequences of these interactions for both.
Cultural services	The non-material benefits people obtain from ecosystems, including spiritual enrichment, cognitive development, recreation and aesthetic experience.
Dead zone	A part of a water body so low in oxygen that normal life cannot survive. The low oxygen conditions usually result from eutrophication caused by fertilizer run-off from land.
Deforestation	Conversion of forested land to non-forest areas.
Desertification	This is land degradation in arid, semi-arid and dry sub-humid areas resulting from various factors, including climatic variations and human activities. It involves crossing thresholds beyond which the underpinning ecosystem cannot restore itself, but requires ever-greater external resources for recovery.
Disability Adjusted Life Years (DALYs)	A health gap measure that extends the concept of potential years of life lost due to premature death to include equivalent years of healthy life lost in states of less than full health, broadly termed disability. One DALY represents the loss of one year of equivalent full health.
Disaster risk management	The systematic process of using administrative decisions, organization, operational skills and capacities to implement policies, strategies and coping capacities of the society and communities to lessen the impacts of natural hazards and related environmental and technological disasters.
Disaster risk reduction	The conceptual framework of elements considered with the possibilities to minimize vulnerabilities and disaster risks throughout a society, to avoid (prevention) or to limit (mitigation and preparedness) the adverse impacts of hazards, within the broad context of sustainable development.
Drainage basin (also called watershed, river basin or catchment)	Land area where precipitation runs off into streams, rivers, lakes and reservoirs. It is a land feature that can be identified by tracing a line along the highest elevations between two areas on a map, often a ridge.
Drylands	Areas characterized by lack of water, which constrain two major, interlinked ecosystem services: primary production and nutrient cycling. Four dryland sub-types are widely recognized: dry sub-humid, semi-arid, arid and hyper-arid, showing an increasing level of aridity or moisture deficit. Formally, this definition includes all land where the aridity index value is less than 0.65. *See also Aridity index.*
Early warning	The provision of timely and effective information, through identified institutions, that allows individuals exposed to a hazard to take action to avoid or reduce their risk and prepare an effective response.
E-business (electronic business)	Both e-commerce (buying and selling online) and the restructuring of business processes to make the best use of digital technologies.
Eco-labelling	A voluntary method of certification of environmental quality (of a product) and/or environmental performance of a process based on life cycle considerations and agreed sets of criteria and standards.
Ecological footprint	An index of the area of productive land and aquatic ecosystems required to produce the resources used and to assimilate the wastes produced by a defined population at a specified material standard of living, wherever on Earth that land may be located.
Ecological security	A condition of ecological safety that ensures access to a sustainable flow of provisioning, regulating and cultural services needed by local communities to meet their basic capabilities.

Term	Definition
Ecosystem	A dynamic complex of plant, animal and micro-organism communities and their non-living environment, interacting as a functional unit.
Ecosystem approach	A strategy for the integrated management of land, water and living resources that promotes conservation and sustainable use in an equitable way. An ecosystem approach is based on the application of appropriate scientific methods, focused on levels of biological organization, which encompass the essential structure, processes, functions and interactions among organisms and their environment. It recognizes that humans, with their cultural diversity, are an integral component of many ecosystems.
Ecosystem assessment	A social process through which the findings of science concerning the causes of ecosystem change, their consequences for human well-being, and management and policy options are used to advise decision-makers. *See also environmental assessment and Strategic environmental assessment.*
Ecosystem function	An intrinsic ecosystem characteristic related to the set of conditions and processes whereby an ecosystem maintains its integrity (such as primary productivity, food chain and biogeochemical cycles). Ecosystem functions include such processes as decomposition, production, nutrient cycling, and fluxes of nutrients and energy.
Ecosystem health	The degree to which ecological factors and their interactions are reasonably complete and functioning for continued resilience, productivity and renewal of the ecosystem.
Ecosystem management	An approach to maintaining or restoring the composition, structure, function and delivery of services of natural and modified ecosystems for the goal of achieving sustainability. It is based on an adaptive, collaboratively developed vision of desired future conditions that integrates ecological, socio-economic, and institutional perspectives, applied within a geographic framework, and defined primarily by natural ecological boundaries.
Ecosystem resilience	The level of disturbance that an ecosystem can undergo without crossing a threshold into a different structure or with different outputs. Resilience depends on ecological dynamics as well as human organizational and institutional capacity to understand, manage and respond to these dynamics.
Ecosystem services	The benefits people obtain from ecosystems. These include provisioning services, such as food and water, regulating services, such as flood and disease control, cultural services, such as spiritual, recreational and cultural benefits, and supporting services, such as nutrient cycling, that maintain the conditions for life on Earth. Sometimes called ecosystem goods-and-services.
Effluent	In issues of water quality, refers to liquid waste (treated or untreated) discharged to the environment from sources such as industrial process and sewage treatment plants.
El Niño (also El Niño-Southern Oscillation (ENSO))	In its original sense, it is a warm water current that periodically flows along the coast of Ecuador and Peru, disrupting the local fishery. This oceanic event is associated with a fluctuation of the intertropical surface pressure pattern and circulation in the Indian and Pacific oceans, called the Southern Oscillation. This coupled atmosphere-ocean phenomenon is collectively known as El Niño-Southern Oscillation, or ENSO. During an El Niño event, the prevailing trade winds weaken and the equatorial countercurrent strengthens, causing warm surface waters in the Indonesian area to flow eastward to overlie the cold waters of the Peru current off South America. This event has great impact on the wind, sea surface temperature and precipitation patterns in the tropical Pacific. It has climatic effects throughout the Pacific region and in many other parts of the world. The opposite of an *El Niño* event is called *La Niña*.
Emission inventory	Details the amounts and types of pollutants released into the environment.
Endangered species	A species is endangered when the best available evidence indicates that it meets any of the criteria A to E specified for the endangered category of the IUCN Red List, and is therefore considered to be facing a very high risk of extinction in the wild.
Endemic species	Species native to, and restricted to, a particular geographical region.
Endemism	The fraction of species that is endemic relative to the total number of species found in a specific area.
End-of-pipe technology	Technology to capture or to transform emissions after they have formed without changing the production process. This includes scrubbers on smokestacks, catalytic converters on automobile tailpipes and wastewater treatment.
Energy intensity	Ratio of energy consumption and economic or physical output. At the national level, energy intensity is the ratio of total domestic primary energy consumption or final energy consumption to gross domestic product or physical output. Lower energy intensity shows greater efficiency in energy use.
Energy efficiency	Using less energy to achieve the same output or goal.
Environmental assessment (EA)	An environmental assessment is the entire process of undertaking a critical and objective evaluation and analysis of information designed to support decision making. It applies the judgment of experts to existing knowledge to provide scientifically credible answers to policy relevant questions, quantifying where possible the level of confidence. It reduces complexity but adds value by summarizing, synthesizing and building scenarios, and identifies consensus by sorting out what is known and widely accepted from what is not known or not agreed. It sensitizes the scientific community to policy needs and the policy community to the scientific basis for action.
Environmental health	Those aspects of human health and disease that are determined by factors in the environment. It also refers to the theory and practice of assessing and controlling factors in the environment that can potentially affect health. Environmental health includes both the direct pathological effects of chemicals, radiation and some biological agents, and the effects (often indirect) on health and well-being of the broad physical, psychological, social and aesthetic environment. This includes housing, urban development, land use and transport.
Environmental impact assessment (EIA)	An environmental impact assessment (EIA) is an analytical process or procedure that systematically examines the possible environmental consequences of the implementation of a given activity (project). The aim is to ensure that the environmental implications of decisions related to a given activity are taken into account before the decisions are made.

Term	Definition
Environmental policy	A policy initiative aimed at addressing environmental problems and challenges.
Environmental problems	Environmental problems are human and/or natural influences on ecosystems that lead to a constraint, cutback or even a cessation of their functioning. They may be broadly categorized into environmental problems with proven solutions, and problems with emerging solutions. *See also conventional environmental problems and persistent environmental problems.*
Equity	Fairness of rights, distribution and access. Depending on context, this can refer to resources, services or power.
Estuary	Area at the mouth of a river where it broadens into the sea, and where fresh and seawater intermingle to produce brackish water. The estuarine environment is very rich in wildlife, particularly aquatic, but it is very vulnerable to damage as a result of human activities.
Eutrophication	The degradation of water quality due to enrichment by nutrients, primarily nitrogen and phosphorus, which results in excessive plant (principally algae) growth and decay. Eutrophication of a lake normally contributes to its slow evolution into a bog or marsh and ultimately to dry land. Eutrophication may be accelerated by human activities that speed up the ageing process.
Evapotranspiration	Combined loss of water by evaporation from the soil or surface water, and transpiration from plants and animals.
E-waste (electronic waste)	A generic term encompassing various forms of electrical and electronic equipment that has ceased to be of value and is disposed of. A practical definition of e-waste is "any electrically powered appliance that fails to satisfy the current owner for its originally intended purpose."
External cost	A cost that is not included in the market price of the goods-and-services being produced. In other words, a cost not borne by those who create it, such as the cost of cleaning up contamination caused by discharge of pollution into the environment.
Fine particle	Particulate matter suspended in the atmosphere less than 2.5 μm in size ($PM_{2.5}$).
Flue gas desulphurization	A technology that employs a sorbent, usually lime or limestone, to remove sulphur dioxide from the gases produced by burning fossil fuels. Flue gas desulphurization is current state-of-the art technology for major SO_2 emitters, like power plants.
Forest	Land spanning more than 0.5 hectares with trees higher than 5 metres and a canopy cover of more than 10 per cent, or trees able to reach these thresholds in situ. It does not include land that is predominantly under agricultural or urban land use.
Forest degradation	Changes within the forest that negatively affect the structure or function of the stand or site, and thereby lower the capacity to supply products and/or services.
Forest management	The processes of planning and implementing practices for the stewardship and use of forests and other wooded land aimed at achieving specific environmental, economic, social and/or cultural objectives.
Forest plantation	Forest stands established by planting and/or seeding in the process of afforestation or reforestation. They are either of introduced species (all planted stands), or intensively managed stands of indigenous species, which meet all the following criteria: one or two species at plantation, even age class and regular spacing. "Planted forest" is another term used for plantation.
Fossil fuel	Coal, natural gas and petroleum products (such as oil) formed from the decayed bodies of animals and plants that died millions of years ago.
Freedom	The range of options a person has in deciding the kind of life to lead.
Fuel cell	A device that converts the energy of a chemical reaction directly into electrical energy. It produces electricity from external supplies of fuel (such as hydrogen on the anode side) and oxidant (such as oxygen on the cathode side). These react in the presence of an electrolyte. A fuel cell can operate virtually continuously as long as the necessary flows are maintained. Fuel cells differ from batteries in that they consume reactant, which must be replenished, while batteries store electrical energy chemically in a closed system. One great advantage of fuel cells is that they generate electricity with very little pollution — much of the hydrogen and oxygen used in generating electricity ultimately combine to form water. Fuel cells are being developed as power sources for motor vehicles, as well as stationary power sources.
Fuel switching	One of the simplest approaches to the control of acid gas emissions, involving the replacement of high-sulphur fuels with low-sulphur alternatives. The most common form of fuel switching is the replacement of high-sulphur coal with a low-sulphur coal. Coal may also be replaced entirely by oil or natural gas.
Genetic diversity	The variety of genes within a particular species, variety or breed.
Geographic information system	A computerized system organizing data sets through a geographical referencing of all data included in its collections.
Global (international) environmental governance	The assemblage of laws and institutions that regulate society-nature interactions and shape environmental outcomes.
Global warming	Changes in the surface air temperature, referred to as the global temperature, brought about by the enhanced greenhouse effect, which is induced by emission of greenhouse gases into the air.
Globalization	The increasing integration of economies and societies around the world, particularly through trade and financial flows, and the transfer of culture and technology.
Governance	The manner in which society exercises control over resources. It denotes the mechanisms through which control over resources is defined and access is regulated. For example, there is governance through the state, the market, or through civil society groups and local organizations. Governance is exercised through institutions: laws, property rights systems and forms of social organization.
Green procurement	Taking environmental aspects into consideration in public and institutional procurement.

Term	Definition
Green tax	Tax with a potentially positive environmental impact. It includes energy taxes, transport taxes, and taxes on pollution and resources. They are also called environmental taxes. Green taxes are meant to reduce environmental burden by increasing prices, and by shifting the basis of taxation from labour and capital to energy and natural resources.
Green water	That fraction of rainfall that is stored in the soil and is available for the growth of plants.
Greenhouse effect	Greenhouse gases possess high emissivity at specific infrared wavelengths. Atmospheric infrared radiation is emitted to all sides by those greenhouse gases, including downward to the Earth's surface. Thus greenhouse gases add more heat within the surface-troposphere system, leading to an increase of the temperature. Atmospheric radiation is strongly coupled to the temperature of the level at which it is emitted. In the troposphere the temperature generally decreases with height. Effectively, infrared radiation emitted to space originates from an altitude with a temperature of, on average, -19°C, in balance with the net incoming solar radiation, whereas the Earth's surface is kept at a much higher temperature of, on average, +14°C. An increase in the concentration of greenhouse gases leads to an increased infrared opacity of the atmosphere, and therefore to an effective radiation into space from a higher altitude at a lower temperature. This causes a radiative forcing, an imbalance that can only be compensated for by an increase of the temperature of the surface-troposphere system. This is the enhanced greenhouse effect.
Greenhouse gases (GHGs)	Gaseous constituents of the atmosphere, both natural and anthropogenic, that absorb and emit radiation at specific wavelengths within the spectrum of infrared radiation emitted by the Earth's surface, the atmosphere and clouds. This property causes the greenhouse effect. Water vapour (H_2O), carbon dioxide (CO_2), nitrous oxide (N_2O), methane (CH_4) and ozone (O_3) are the primary greenhouse gases in the Earth's atmosphere. There are human-made greenhouse gases in the atmosphere, such as the halocarbons and other chlorine and bromine containing substances. Beside CO_2, N_2O and CH_4, the Kyoto Protocol deals with sulphur hexafluoride (SF_6), hydrofluorocarbons (HFCs) and perfluorocarbons (PFCs).
Grey water	Wastewater other than sewage, such as sink drainage or washing machine discharge.
Groundwater	Water that flows or seeps downward and saturates soil or rock, supplying springs and wells. The upper surface of the saturate zone is called the water table.
Habitat	(1) The place or type of site where an organism or population naturally occurs. (2) Terrestrial or aquatic areas distinguished by geographic, abiotic and biotic features, whether entirely natural or semi-natural.
Hazard	A potentially damaging physical event, phenomenon or human activity that may cause the loss of life or injury, property damage, social and economic disruption or environmental degradation.
Hazardous waste	By-products of society that can pose a substantial or potential hazard to human health or the environment when improperly managed. Substances classified as hazardous wastes possess at least one of four characteristics: ignitability, corrosivity, reactivity or toxicity, or appear on special lists.
Heavy metals	A group name for metals and semimetals (metalloids), such as arsenic, cadmium, chromium, copper, lead, mercury, nickel and zinc, that have been associated with contamination and potential toxicity.
High seas	The oceans outside of national jurisdictions, lying beyond each nation's exclusive economic zone or other territorial waters.
Human health	A state of complete physical, mental and social well-being, and not merely the absence of disease or infirmity.
Human well-being	The extent to which individuals have the ability to live the kinds of lives they have reason to value; the opportunities people have to achieve their aspirations. Basic components of human well-being include: security, material needs, health and social relations (see Box 1.2 in Chapter 1).
Hydrological cycle	Succession of stages undergone by water in its passage from the atmosphere to the earth and its return to the atmosphere. The stages include evaporation from land, sea or inland water, condensation to form clouds, precipitation, accumulation in the soil or in water bodies, and re-evaporation.
Income poverty	A measure of deprivation of well-being focusing solely on per capita or household income.
Inorganic contaminants	Mineral-based compounds, such as metals, nitrates and asbestos, that naturally occur in some parts of the environment, but can also enter the environment as a result of human activities.
Integrated coastal zone management (ICZM)	Approaches that integrate economic, social, and ecological perspectives for the management of coastal resources and areas.
Institutions	Regularized patterns of interaction by which society organizes itself: the rules, practices and conventions that structure human interaction. The term is wide and encompassing, and could be taken to include law, social relationships, property rights and tenurial systems, norms, beliefs, customs and codes of conduct as much as multilateral environmental agreements, international conventions and financing mechanisms. Institutions could be formal (explicit, written, often having the sanction of the state) or informal (unwritten, implied, tacit, mutually agreed and accepted). Formal institutions include law, international environmental agreements, bylaws and memoranda of understanding. Informal institutions include unwritten rules, codes of conduct and value systems. The term institutions should be distinguished from organizations.
Integrated ecosystem monitoring	The intermittent (regular or irregular) surveillance to ascertain the extent of compliance with a predetermined standard or the degree of deviation from an expected norm.

Term	Definition
Integrated water resources management (IWRM)	A process which promotes the coordinated development and management of water, land and related resources, in order to maximize the resultant economic and social welfare in an equitable manner without compromising the sustainability of vital ecosystems (see Box 4.10 in Chapter 4).
Interlinkages	The cause-effect chains that cross the boundaries of current environmental and environment-development challenges.
Intrinsic value	The value of someone or something in and for itself, irrespective of its utility for people.
Invasive alien species	An alien species whose establishment and spread modifies ecosystems, habitats or species.
Kyoto Protocol	A protocol to the 1992 UN Framework Convention on Climate Change (UNFCCC) adopted at the Third Session of the Conference of the Parties to the UNFCCC in 1997 in Kyoto, Japan. It contains legally binding commitments, in addition to those included in the UNFCCC. Countries included in Annex B of the protocol (most OECD countries and countries with economies in transition) agreed to control their national anthropogenic emissions of greenhouse gases (CO_2, CH_4, N_2O, HFCs, PFCs and SF_6) so that the total emissions from these countries would be at least 5 per cent below 1990 levels in the commitment period, 2008 to 2012. The protocol expires in 2012.
Kuznets curve (environmental)	An inverted U relationship between per capita income and some environmental pollution indicators. This relationship suggests that environmental pollution increases in the early stages of growth, until socio-economic needs are met, but it eventually decreases as income exceeds a certain level and funds can be allocated for reducing and preventing pollution. In practice, the relation holds for a few air and water pollutants with local effects, but there is scant evidence that the same is true for other indicators of environmental degradation, such as anthropogenic greenhouse gas emissions.
La Niña	A cooling of the ocean surface off the western coast of South America, occurring periodically every 4-12 years and affecting Pacific and other weather patterns.
Land cover	The physical coverage of land, usually expressed in terms of vegetation cover or lack of it. Influenced by but not synonymous with land use.
Land degradation	The loss of biological or economic productivity and complexity in croplands, pastures and woodlands. It is due mainly to climate variability and unsustainable human activity.
Land use	The human use of land for a certain purpose. Influenced by, but not synonymous with, land cover.
Leachate drainage	A solution containing contaminants picked up through the leaching of soil.
Lead markets for environmental innovations	Countries that are earlier in the introduction of environmental innovation and with more widespread diffusion of the innovations. If these countries serve as an example or model for other countries and their innovations are distributed elsewhere as well, these countries are lead markets.
Legitimacy	Measure of the political acceptability or perceived fairness. State law has its legitimacy in the state; local law and practices work on a system of social sanction, in that they derive their legitimacy from a system of social organization and relationships.
Lifetime (in the atmosphere)	The approximate amount of time it would take for the anthropogenic increment to an atmospheric pollutant concentration to return to its natural level (assuming emissions cease) as a result of either being converted to another chemical compound or being taken out of the atmosphere via a sink. Average lifetimes can vary from about a week (sulphate aerosols) to more than a century (CFCs, carbon dioxide). A specific lifetime cannot be given for carbon dioxide, because it is continuously cycled between the atmosphere, oceans and land biosphere, and its net removal from the atmosphere involves a range of processes with different timescales.
Lyme disease	A multi-system bacterial infection caused by the spirochete Borrelia burgdoferi. These spirochetes are maintained in nature in the bodies of wild animals, and transmitted from one animal to another through the bite of an infected tick. People and pets are incidental hosts to ticks.
Mainstreaming	Mainstreaming the environment into development policy making means that environmental considerations are considered in the design of policies for development.
Marine protected area (MPA)	A geographically defined marine area that is designated or regulated and managed to achieve specific conservation objectives.
Mega-cities	Urban areas with more than 10 million inhabitants.
Mitigation	Structural and non-structural measures undertaken to limit the adverse impact of natural hazards, environmental degradation and technological hazards.
Monitoring (environmental)	Continuous or regular standardized measurement and observation of the environment (air, water, soil, land use, biota).
Multilateral environmental agreements (MEAs)	Treaties, conventions, protocols and contracts among several states to jointly agree on activities regarding specified environmental problems.
Natural capital	Natural assets in their role of providing natural resource inputs and environmental services for economic production. Natural capital includes land, minerals and fossil fuels, solar energy, water, living organisms, and the services provided by the interactions of all these elements in ecological systems.
Nitrogen deposition	The input of reactive nitrogen, mainly derived from nitrogen oxides and ammonia emissions, from the atmosphere into the biosphere.

Term	Definition
Non-point source of pollution	A pollution source that is diffused (so without a single point of origin or not introduced into a receiving stream from a specific outlet). Common non-point sources are agriculture, forestry, city streets, mining, construction, dams, channels, land disposal and landfills, and saltwater intrusion.
Non-wood forest product (NWFP)	A product of biological origin other than wood derived from forests, other wooded land and trees outside forests. Examples include, food, fodder, medicine, rubber and handicrafts.
Normalized difference vegetation index (NDVI)	Also referred to as the greenness index. It is a non-linear transformation of the red and near infra-red bands of reflected light measured by Earth-observation satellites, calculated as the difference between the red and near infra-red bands divided by the sum. Since the near infra-red waveband is strongly absorbed by chlorophyll, NDVI is related to the percentage of vegetation cover and green biomass.
Nutrient loading	Quantity of nutrients entering an ecosystem in a given period of time.
Nutrient pollution	Contamination of water resources by excessive inputs of nutrients.
Nutrients	The approximately 20 chemical elements known to be essential for the growth of living organisms, including nitrogen, sulphur, phosphorous and carbon.
Oil sands	A complex mixture of sand, water and clay trapping very heavy oil, known as bitumen.
Organizations	Bodies of individuals with a specified common objective. Organizations could be political organizations (political parties, governments and ministries), economic organizations (federations of industry), social organizations (NGOs and self-help groups) or religious organizations (church and religious trusts). The term organizations should be distinguished from institutions.
Overexploitation	The excessive use of raw materials without considering the long-term ecological impacts of such use.
Ozone hole	A sharp seasonal decrease in stratospheric ozone concentration that occurs over the Antarctic, generally between August and November. First detected in the late 1970s, the ozone hole continues to appear every year.
Ozone layer	Very dilute atmospheric concentration of ozone found at an altitude of 10-50 kilometres above the earth's surface.
Ozone-depletion potential	A relative index indicating the extent to which a chemical may cause ozone depletion. The reference level of 1 is the potential of CFC-11 and CFC-12 to cause ozone depletion.
Ozone-depleting substance (ODS)	Any substance with an ozone depletion potential greater than 0 that can deplete the stratospheric ozone layer.
Parklands	Integrated tree-crop-livestock systems that are common throughout the Sahel.
Participatory approach	Securing an adequate and equal opportunity for people to place questions on the agenda and to express their preferences about the final outcome during decision making to all group members. Participation can occur directly or through legitimate representatives. Participation may range from consultation to the obligation of achieving a consensus.
Pastoralism, pastoral system	The use of domestic animals as a primary means for obtaining resources from habitats.
Pathogen	A disease causing micro-organism, bacterium or virus.
Payment for environmental services	Appropriate mechanisms for matching the demand for environmental services with the incentives of land users whose actions modify the supply of those environmental services.
Peatlands	Wetlands where the soil is highly organic, because it is formed mostly from incompletely decomposed plants.
Pelagic ecosystem	That relating to, or living or occurring in the open sea.
Percolation	Flow of a liquid through an unsaturated porous medium.
Perennial stream	A stream that flows from source to mouth throughout the year.
Permafrost	Soil, silt and rock located in perpetually cold areas, and that remains frozen year-round.
Persistent environmental problems	Some of the basic science about cause-and-effect relationships is known, but often not enough to predict when a turning point or a point of no return will be reached, or exactly how human well-being will be affected. The sources of the problem are quite diffuse and often multisectoral, potential victims are often quite remote from the sources, extremely complex multi-scale ecological processes may be involved, there may be a long time between causes and impacts, and there is a need to implement measures on a very large scale (usually global or regional). Examples include global climate change, stratospheric ozone depletion, persistent organic pollutants and heavy metals, extinction of species, ocean acidification, and introduction of alien species. See also Conventional environmental problems and Environmental problems.
Persistent organic pollutants (POPs)	Chemicals that remain intact in the environment for long periods, become widely distributed geographically, accumulate in the fatty tissue of living organisms and are toxic to people and wildlife. POPs circulate globally and can cause damage wherever they travel.
Photochemical reaction	A chemical reaction brought about by the light energy of the sun. The reaction of nitrogen oxides with hydrocarbons in the presence of sunlight to form ozone is an example of a photochemical reaction.
Phytoplankton	Microscopically small plants that float or swim weakly in fresh or saltwater bodies.

Term	Definition
Pluralism (legal or institutional)	Coexistence of more than one legal or institutional system with regard to the same set of activities. For instance, state law may coexist with customary law and practices, social relationships and local systems of property rights and tenurial systems. Legal or institutional pluralism provides an analytical framework, for instance, for the analysis of the interface of formal and informal institutions.
Point source of pollution	The term covers stationary sources, such as sewage treatment plants, power plants and other industrial establishments, and other, single identifiable sources of pollution, such as pipes, ditches, ships, ore pits and smokestacks.
Policy	Any form of intervention or societal response. This includes not only statements of intent, such as a water policy or forest policy, but also other forms of intervention, such as the use of economic instruments, market creation, subsidies, institutional reform, legal reform, decentralization and institutional development. Policy can be seen as a tool for the exercise of governance. When such an intervention is enforced by the state, it is called public policy.
Policy space	An area of policy formulation and/or implementation. For example, health, education, environment and transportation can all be seen as policy spaces.
Pollutant	Any substance that causes harm to the environment when it mixes with soil, water or air.
Pollution	The presence of minerals, chemicals or physical properties at levels that exceed the values deemed to define a boundary between "good or acceptable" and "poor or unacceptable" quality, which is a function of the specific pollutant.
Poverty	The pronounced deprivation of well-being.
Precautionary approach	The management concept stating that in cases "where there are threats of serious or irreversible damage, lack of full scientific certainly shall not be used as a reason for postponing cost-effective measures to prevent environmental degradation."
Precision agriculture	Farming practices that adapt to local variability of soil and terrain within every management unit, rather than ignoring variability. This term is also used to describe automated techniques employed for such practices.
Prediction	The act of attempting to produce a description of the expected future, or the description itself, such as "it will be 30 degrees tomorrow, so we will go to the beach."
Primary energy	Energy embodied in natural resources (such as coal, crude oil, sunlight or uranium) that has not undergone any anthropogenic conversion or transformation.
Primary pollutant	Air pollutant emitted directly from a source.
Projection	The act of attempting to produce a description of the future subject to assumptions about certain preconditions, or the description itself, such as "assuming it is 30 degrees tomorrow, we will go to the beach."
Provisioning services	The products obtained from ecosystems, including, for example, genetic resources, food and fibre, and freshwater.
Purchasing power parity (PPP)	The number of currency units required to purchase the amount of goods and services equivalent to what can be bought with one unit of the currency of the base country, for example, the US dollar.
Rangeland	An area where the main land use is related to the support of grazing or browsing mammals, such as cattle, sheep, goats, camels or antelope.
Reforestation	Planting of forests on lands that have previously contained forest, but have since been converted to some other use.
Regulating services	The benefits obtained from the regulation of ecosystems processes, including, for example, the regulation of climate, water and some human diseases.
Renewable energy source	An energy source that does not rely on finite stocks of fuels. The most widely known renewable source is hydropower; other renewable sources are biomass, solar, tidal, wave and wind energy.
Resilience	The capacity of a system, community or society potentially exposed to hazards to adapt by resisting or changing in order to reach and maintain an acceptable level of functioning and structure.
Resistance	The capacity of a system to withstand the impacts of drivers without displacement from its present state.
Riparian	Related to, living or located on the bank of a natural watercourse, usually a river, but sometimes a lake, tidewater or enclosed sea.
ROPME Sea Area	The sea area surrounded by the eight Member States of ROPME (Regional Organization for the Protection of the Marine Environment): Bahrain, I.R. Iran, Iraq, Kuwait, Oman, Qatar, Saudi Arabia and the United Arab Emirates.
Rules and norms	A part of the umbrella concept of institutions. While the distinction is a bit thin, rules could be considered to be directions for behaviour that can both be explicit or implicit. Norms are an accepted standard or a way of behaving or doing things that most people agree with.
Run-off	A portion of rainfall, melted snow or irrigation water that flows across the ground's surface and is eventually returned to streams. Run-off can pick up pollutants from air or land and carry them to receiving waters.
Sahel	A loosely defined strip of transitional vegetation that separates the Sahara desert from the tropical savannahs to the south. The region is used for farming and grazing, and because of the difficult environmental conditions that exist at the border of the desert, the region is very sensitive to human-induced land cover change. It includes parts of Senegal, the Gambia, Mauritania, Mali, Niger, Nigeria, Burkina Faso, Cameroon and Chad.
Salinization	The buildup of salts in soils.

Term	Definition
Savannah	A tropical or subtropical region of grassland and other drought-resistant (xerophilous) vegetation. This type of growth occurs in regions that have a long dry season (usually "winter dry"), but a heavy rainy season, and continuously high temperatures.
Scale	The spatial, temporal (quantitative or analytical) dimension used to measure and study any phenomena. Specific points on a scale can thus be considered levels (such as local, regional, national and international).
Scenario	A description of how the future may unfold based on "if-then" propositions, typically consisting of a representation of an initial situation, a description of the key drivers and changes that lead to a particular future state. For example, "given that we are on holiday at the coast, if it is 30 degrees tomorrow, we will go to the beach".
Seagrass bed	Benthic community, usually on shallow, sandy or muddy bottoms of sea dominated by grass-like marine plants.
Secondary energy	Form of energy generated by conversion of primary energies, such as electricity from gas, nuclear energy, coal or oil, fuel oil and gasoline from mineral oil, or coke and coke oven gas from coal.
Secondary forest	Forest regenerated largely through natural processes after significant human or natural disturbance of the original forest vegetation.
Secondary pollutant	Not directly emitted as such, but forms when other pollutants (primary pollutants) react in the atmosphere.
Security	Relates to personal and environmental security. It includes access to natural and other resources, and freedom from violence, crime and war, as well as security from natural and human-caused disasters.
Sediment	Solid material that originates mostly from disintegrated rocks and is transported by, suspended in or deposited from water.
Sediment load	The amount of non-dissolved matter that passes through a given river cross section per unit time.
Sedimentation	Strictly, the act or process of depositing sediment from suspension in water. Broadly, all the processes whereby particles of rock material are accumulated to form sedimentary deposits. Sedimentation, as commonly used, involves not only aqueous but also glacial, aeolian and organic agents.
Shared waters	Water resources shared by two or more governmental jurisdictions.
Siltation	The deposition of finely divided soil and rock particles on the bottom of stream and riverbeds and reservoirs.
Smog	Classically a combination of smoke and fog in which products of combustion, such as hydrocarbons, particulate matter and oxides of sulphur and nitrogen, occur in concentrations that are harmful to human beings and other organisms. More commonly, it occurs as photochemical smog, produced when sunlight acts on nitrogen oxides and hydrocarbons to produce tropospheric ozone.
Soft law	Non-legally binding instruments, such as guidelines, standards, criteria, codes of practice, resolutions, and principles or declarations established to implement national or international laws.
Soil acidification	A naturally occurring process in humid climates that has long been the subject of research, whose findings suggest that acid precipitation affects the productivity of terrestrial plants. The process is summarized as follows: as soil becomes more acidic basic cations (such as Ca^{2+}, Mg^{2+}) in the soil are exchanged by hydrogen ions or solubilized metals. The basic cations, now in solution, can be leached through the soil. As time progresses, the soil becomes less fertile and more acidic. Resultant decreases in soil pH cause reduced, less active populations of soil micro-organisms, which, in turn, slows decomposition of plant residues and cycling of essential plant nutrients.
Species	An interbreeding group of organisms that is reproductively isolated from all other organisms, although there are many partial exceptions to this rule in particular taxa. Operationally, the term species is a generally agreed fundamental taxonomic unit, based on morphological or genetic similarity that once described and accepted is associated with a unique scientific name.
Species diversity	Biodiversity at the species level, often combining aspects of species richness, their relative abundance and their dissimilarity.
Species richness/ abundance	The number of species within a given sample, community or area.
Strategic environmental assessment (SEA)	SEA is undertaken for plans, programmes and policies. It helps decision makers reach a better understanding of how environmental, social and economic considerations fit together. SEA has been described as a range of "analytical and participatory approaches that aim to integrate environmental considerations into policies, plans and programmes and evaluate the interlinkages with economic and social considerations."
Subsidiarity, principle of	The notion of devolving decision making authority to the lowest appropriate level.
Subspecies	A population that is distinct from, and partially reproductively isolated from other populations of a species, but which has not yet diverged sufficiently to make interbreeding impossible.
Supporting services	Ecosystem services that are necessary for the production of all other ecosystem services. Some examples include biomass production, production of atmospheric oxygen, soil formation and retention, nutrient cycling, water cycling, and provisioning of habitat.
Surface water	All water naturally open to the atmosphere, including rivers, lakes, reservoirs, streams, impoundments, seas and estuaries. The term also covers springs, wells or other collectors of water that are directly influenced by surface waters.
Susceptible drylands	Susceptible drylands refer to arid, semi-arid and dry sub-humid areas. Hyper-arid areas (the true deserts, with an aridity index of less than 0.05) are not considered to be susceptible to desertification because of their very low biological activity and limited opportunities for human activity. See also Drylands and Aridity index.
Sustainability	A characteristic or state whereby the needs of the present and local population can be met without compromising the ability of future generations or populations in other locations to meet their needs.

Term	Definition
Sustainable development	Development that meets the needs of the present generation without compromising the ability of future generations to meet their own needs.
Tailings	Residue of raw materials or waste separated out during the processing of crops, mineral ores or oil sands.
Taxon (pl. taxa)	The named classification unit to which individuals or sets of species are assigned. Higher taxa are those above the species level. For example, the common mouse, *Mus musculus*, belongs to the Genus *Mus*, the Family *Muridae*, and the Class *Mammalia*.
Taxonomy	A system of nested categories (taxa) reflecting evolutionary relationships or morphological similarities.
Technology	Physical artefacts or the bodies of knowledge of which they are an expression. Examples are water extraction structures, such as tube wells, renewable energy technologies and traditional knowledge. Technology and institutions are related. Any technology has a set of practices, rules and regulations surrounding its use, access, distribution and management.
Technology barrier	An identified gap in available technology that needs to be filled (for which capability has to be created) in order for proposed product, process or service developments to take place.
Technology transfer	A broad set of processes covering the flows of know-how, experience and equipment among different stakeholders.
Thermohaline circulation (THC)	Large-scale density-driven circulation in the ocean, caused by differences in temperature and salinity. In the North Atlantic, the thermohaline circulation consists of warm surface water flowing northward and cold deep water flowing southward, resulting in a net poleward transport of heat. The surface water sinks in highly restricted sinking regions located in high latitudes. Also referred to as the (global) ocean conveyer belt or the meridional overturning circulation (MOC).
Threshold	A point or level at which new properties emerge in an ecological, economic or other system, invalidating predictions based on mathematical relationships that apply at lower levels.
Tipping point	The tipping point is the critical point in an evolving situation that leads to a new and irreversible development.
Total maximum daily load	The amount of pollution that a water body can receive and still maintain water quality standards and beneficial uses.
Traditional or local ecological knowledge	A cumulative body of knowledge, know-how, practices or representations maintained or developed by peoples with extended histories of interaction with the natural environment.
Traditional use (of natural resources)	Exploitation of natural resources by indigenous users, or non-indigenous residents using traditional methods. Local use refers to exploitation by local residents.
Trophic level	Successive stages of nourishment as represented by the links of the food chain. According to a grossly simplified scheme the primary producers (phytoplankton) constitute the first trophic level, herbivorous zooplankton the second trophic level and carnivorous organisms the third trophic level.
Urban sprawl	The decentralization of the urban core through the unlimited outward extension of dispersed development beyond the urban fringe, where low density residential and commercial development exacerbates fragmentation of powers over land use.
Urban systems	Built environments with a high human population density. Operationally defined as human settlements with a minimum population density commonly in the range of 400–1 000 persons per square kilometre, minimum size of typically between 1 000 and 5 000 people, and maximum (non-)agricultural employment usually in the range of 50–75 per cent.
Urbanization	An increase in the proportion of the population living in urban areas.
Voluntary agreement	An agreement between government and business, or a unilateral private sector commitment that is acknowledged by the government, aimed at achieving environmental objectives or improving environmental performance.
Vulnerability	An intrinsic feature of people at risk. It is a function of exposure, sensitivity to impacts of the specific unit exposed (such as a watershed, island, household, village, city or country), and the ability or inability to cope or adapt. It is multi-dimensional, multidisciplinary, multisectoral and dynamic. The exposure is to hazards such as drought, conflict or extreme price fluctuations, and also to underlying socio-economic, institutional and environmental conditions.
Wastewater treatment	Any of the mechanical, biological or chemical processes used to modify the quality of wastewater in order to reduce pollution levels.
Water quality	The chemical, physical and biological characteristics of water, usually in respect to its suitability for a particular purpose.
Water scarcity	Occurs when annual water supplies drop below 1 000 m³ per person, or when more than 40 per cent of available water is used.
Water stress	Occurs when low water supplies limit food production and economic development, and affect human health. An area is experiencing water stress when annual water supplies drop below 1 700 m³ per person.
Water table	The top of the water surface in the saturated part of an aquifer.
West Nile virus	The mosquito-borne virus that causes West Nile fever. One of the flaviviruses, a family of viruses also responsible for dengue, yellow fever and tick-borne encephalitis.
Wetland	Area of marsh, fen, peatland, bog or water, whether natural or artificial, permanent or temporary, with water that is static or flowing, fresh, brackish or salt, including areas of marine water to a depth at low tide that does not exceed 6 metres.
Woodland	Wooded land, which is not classified as forest, spanning more than 0.5 hectares, with trees higher than 5 metres and a canopy cover of 5-10 per cent, or trees able to reach these thresholds *in situ*, or with a combined cover of shrubs, bushes and trees above 10 per cent. It does not include areas used predominantly for agricultural or urban purposes.

Index